Flora of North America

Flora of North America

North of Mexico

Edited by FLORA OF NORTH AMERICA EDITORIAL COMMITTEE

VOLUME 2

Pteridophytes and Gymnosperms

PSILOTOPHYTA *(Whisk Ferns)*

LYCOPODIOPHYTA *(Club-mosses)*

EQUISETOPHYTA *(Horsetails)*

POLYPODIOPHYTA *(Ferns)*

CYCADOPHYTA *(Cycads)*

GINKGOPHYTA *(Ginkgos)*

CONIFEROPHYTA *(Conifers)*

GNETOPHYTA *(Gnetophytes)*

NEW YORK OXFORD · OXFORD UNIVERSITY PRESS · 1993

Oxford University Press

Oxford   New York   Toronto
Delhi   Bombay   Calcutta   Madras   Karachi
Kuala Lumpur   Singapore   Hong Kong   Tokyo
Nairobi   Dar es Salaam   Cape Town
Melbourne   Auckland   Madrid

and associated companies in
Berlin   Ibadan

Published by Oxford University Press, Inc.,
198 Madison Avenue, New York, New York 10016-4314

Oxford is a registered trademark of Oxford University Press

Library of Congress Cataloging-in-Publication Data
(Revised for volume 2)
Flora of North America north of Mexico
edited by Flora of North America Editorial Committee.
Includes bibliographical references and indexes.
Contents: v. 1. Introduction—v. 2. Pteridophytes and gymnosperms.
ISBN 0-19-508242-7 (v. 2)
1. Botany—North America.
2. Botany—United States.
3. Botany—Canada.
I. Flora of North America Editorial Committee.
QK110.F55   1993   581.97   92-30459

9 8 7 6 5 4 3

Printed in the United States of America
on acid-free paper

Contents

For their partial funding
of the Flora of North America Project,
we gratefully acknowledge and thank:

National Science Foundation
The Pew Charitable Trusts
The Caleb C. and Julia W. Dula Foundation
The Surdna Foundation
The Robert and Lucile Packard Foundation
National Fish and Wildlife Foundation
ARCO Foundation
The William and Flora Hewlett Foundation
Edward Chase Garvey Memorial Foundation

MEMBER INSTITUTIONS

Flora of North America Association

Arnold Arboretum
Jamaica Plain, Massachusetts

Biosystematics Research Institute
Canada Agriculture
Ottawa, Ontario

Canadian Museum of Nature
Ottawa, Ontario

Carnegie Museum of Natural
    History
Pittsburgh, Pennsylvania

East Central University
Ada, Oklahoma

Field Museum of Natural History
Chicago, Illinois

Fish and Wildlife Service
United States Dept. of the Interior
Washington, D.C.

Harvard University Herbaria
Cambridge, Massachusetts

Hunt Institute for Botanical
    Documentation
Carnegie Mellon University
Pittsburgh, Pennsylvania

Jacksonville State University
Jacksonville, Alabama

Jardin Botanique de Montréal
Montréal, Quebec

Kansas State University
Manhattan, Kansas

Missouri Botanical Garden
St. Louis, Missouri

New Mexico State University
Las Cruces, New Mexico

New York State Museum
Albany, New York

Northern Kentucky University
Highland Heights, Kentucky

Royal Ontario Museum
Toronto, Ontario

Southern Illinois University
Carbondale, Illinois

The New York Botanical Garden
Bronx, New York

The University of British Columbia
Vancouver, British Columbia

The University of Texas at Austin
Austin, Texas

Université de Montréal
Montreal, Quebec

University of Alaska
Fairbanks, Alaska

University of Alberta
Edmonton, Alberta

University of California
Berkeley, California

University of California
Davis, California

University of Idaho
Moscow, Idaho

University of Illinois
Urbana-Champaign, Illinois

University of Iowa
Iowa City, Iowa

University of Kansas
Lawrence, Kansas

University of Michigan
Ann Arbor, Michigan

University of Oklahoma
Norman, Oklahoma

University of Ottawa
Ottawa, Ontario

University of Southwestern
    Louisiana
Lafayette, Louisiana

University of Tennessee
Knoxville, Tennessee

University of Western Ontario
London, Ontario

University of Wyoming
Laramie, Wyoming

Utah State University
Logan, Utah

Special Thanks

Many people have worked countless hours to bring volumes 1 and 2 of *Flora of North America North of Mexico* to publication simultaneously. I particularly thank some of the key people involved in this gargantuan undertaking.

Alan R. Smith and Warren H. Wagner Jr. were Special Pteridophyte Editors. The gymnosperm treatments were edited by John W. Thieret. James E. Eckenwalder was Special Gymnosperm Editor.

Preparation of manuscripts for publication has been an immense task. All technical editing and most of the word processing has been done by Helen Jeude. Bruce Ford, Denis Kearns, Bruce Parfitt, and Alan Whittemore helped check many aspects of the manuscripts.

John Kartesz provided some distributional information for the gymnosperms. Robbin Moran and George Yatskievych gave help and advice on all aspects of preparation of the pteridophyte treatments. Ihsan Al-Shehbaz was a constant source of technical and scientific information and guidance.

Preparation of distribution maps was overseen by Judith Unger, and she did final editing of maps. Illustrations for the taxonomic treatments in volume 2 were drawn by Laurie Lange and Bobbi Angell; John Myers assisted with preparation of the titles; Bruce Parfitt coordinated the illustration work.

All taxonomic treatments for volume 2 were reviewed by Larry E. Morse, Peggy Olwell, and Michael O'Neal to determine which taxa should be indicated as "of conservation concern."

Editorial Committee members who voluntarily reviewed every manuscript in its final stage of preparation for publication—and who became known as the "final five"—were Theodore M. Barkley, John G. Packer, Richard W. Spellenberg, John L. Strother, and R. David Whetstone.

I am personally grateful to project staff, editorial committee members, authors, and reviewers for the tremendous amount of work they have done and for their enthusiasm for and support of Flora of North America.

Nancy R. Morin
Convening Editor

Acknowledgments

Flora of North America North of Mexico is the product of creative thinking and hard work on the part of many people. This project owes its start to those who attended an organizing meeting on 30 April and 1 May 1982 at the Missouri Botanical Garden: G. Argus, D. Bates, W. Burger, S. Hatch, N. Holmgren, T. Jacobsen, M. Johnston, R. Kiger, J. Massey, J. McNeill, R. Mecklenburg, N. Morin, J. Phipps, D. Porter, J. Reveal, S. Shetler, F. Utech, and G. Webster.

The final form of this project has been designed, in large part, by the Editorial Committee. This committee has changed its makeup slightly over the years, but overall it has been a remarkably stable group. By 1986 the Editorial Committee consisted of G. Argus, T. Barkley, D. Boufford, L. Brouillet, D. Henderson, R. Kiger, B. MacBryde, J. McNeill, N. Morin, J. Packer, J. Phipps, L. Shultz, R. Spellenberg, J. Strother, J. Thieret, G. Webster, and D. Whetstone. Since then, Henderson and MacBryde have resigned from the committee and have been replaced by R. Hartman and J. Fay. Later J. Estes, M. Johnston, D. Murray, G. Straley, and R. Thompson were added to the committee as permanent members, and J. Eckenwalder, A. Smith, and W. H. Wagner joined as advisory editors for volumes 1 and 2. In 1991, bryophytes were added to the project, and the following bryologists were added to the Editorial Committee: W. Buck, M. Crosby, J. Engel, M. Hicks, D. Horton, N. Miller, B. Murray, W. Reese, R. Stotler, B. Thiers, and D. Vitt.

The Flora of North America project is coordinated from the Organizational Center, located at the Missouri Botanical Garden. Support by the Missouri Botanical Garden for this Center has been instrumental in the success of the project, and we are grateful to Peter H. Raven, Marshall R. Crosby, Enrique Forero, and W. D. Stevens for committing institutional resources and for contributing their own time and energy to Flora of North America. The institutions at which editors work provide more than half of the overall cost of the project through in-kind support. We are grateful to all of the foundations and individuals who have financially supported Flora of North America. In particular, The Pew Charitable Trusts and the National Science Foundation have supported the project from its very early, developmental phases. We are particularly grateful to program directors at the National Science Foundation for their help and ideas.

Coordination of the activities of the project's many editors and hundreds of authors and reviewers has, in itself, been an enormous task. Project coordinators, who have also been coeditors of the Flora of North America Newsletter, have been, in succession, Carol L. Blaney, Kay Tomlinson, and Judith Unger. Lois Ganss, Eloise Halliday, Mary Lawrence, Nan Winkelmeyer, and Eleanor Zeller have provided secretarial support. Editors at other institutions have been assisted by Donna M. Connelly, Darlene O'Neill, Carolyn Parker, Kimberley Perez, Robert Rhode, and Frederick H. Utech. Undergraduate interns, some of whom were funded by the Research Experiences for Undergraduates program of the National Science Foundation, have been a tremendous help in all aspects of the project. They have been: Carol Davit, Sandra S. Howell, Jennifer R. Milburn, Lisa Nodolf, Stacy A. Oglesbee, Susan Reiter, Viviana M. Tharp, and Larry A. Turner.

The Flora of North America taxonomic database is an integral part of Missouri Botanical Garden's database system. This system was designed by Marshall R. Crosby and Robert E. Magill, and the programs were written by Magill. The computer staff at the Garden, M. Christine McMahon, Deborah Kama, and John Satterfield, have been helpful with many aspects of Flora of North America. The bibliographic database, held at the Hunt Institute for Botanical Documentation, was designed by Robert W. Kiger. Development of the Flora of North America database was initially overseen by G. E. Gibbs Russell and is now being designed and managed by Deborah Kama. Character lists have been compiled by A. Whittemore and J. Bruhl. Data entry has been done by Kimberly Lindsey and more recently by Kathleen Janus and Judith McMurtry. Gibbs Russell, with Flora of North America undergraduate intern Larry A. Turner, designed and analyzed a questionnaire on floristic databases, the report of which has been an important reference for determining the status and potential compatibilty of floristic databases.

The following institutions signed a Memorandum of Cooperation as members of the Flora of North America Association: Arnold Arboretum, Jamaica Plain; Biosystematics Research Institute, Canada Agriculture; Carnegie Museum of Natural History; Field Museum of Natural History; Harvard University Herbaria, Cambridge; Hunt Institute for Botanical Documentation; Jacksonville State University, Alabama; Kansas State University, Manhattan; Missouri Botanical Garden, St. Louis; Montreal Botanical Garden; National Museum of Natural Sciences, Ottawa; New Mexico State University, Las Cruces; Northern Kentucky University; Office of Scientific Authority, U.S. Fish and Wildlife Service; The New York Botanical Garden, Bronx; Université de Montréal; University of Alaska, Fairbanks; University of Alberta, Edmonton; University of California, Berkeley; University of California, Davis; University of Idaho, Moscow; University of Illinois at Urbana-Champaign; University of Kansas, Lawrence; University of Ottawa; University of Western Ontario, London, Ontario; and University of Wyoming, Laramie.

Members of the Flora of North America Project Advisory Panel were: F. A. Almeda Jr., California Academy of Sciences; D. M. Bates, Liberty Hyde Bailey Hortorium; T. P. Bennett, Academy of Natural Sciences; A. Bouchard, Jardin Botanique de Montréal; W. C. Burger, Field Museum of Natural History; T. Duncan, University of California, Berkeley; C. G. Gruchy, National Museum of Natural Sciences, Canada; V. L. Harms, University of Saskatchewan; J. Hickman, University of California, Berkeley; A. G. Jones, University of Illinois; R. W. Kiger, Hunt Institute for Botanical Documentation; M. M. Littler, National Museum of Natural History; J. Massey, University of North Carolina, Chapel Hill; G. A. Mulligan, Vascular Plant Herbarium, Ottawa; G. B. Ownbey, University of Minnesota; J. G. Packer, University of Alberta; D. H. Pfister, Harvard University Herbaria; G. T. Prance, New York Botanical Garden; R. F. Scagel, University of British Columbia; R. L. Shaffer, University of Michigan; R. F. Thorne, Rancho Santa Ana Botanic Garden; B. L. Turner, University of Texas.

Members of the Flora of North America Project Database Consulting Group were: Guy Baillargeon, Biosystematics Research Centre, Ottawa; Chris Beecher, NAPRALERT, Chicago; Warren Brigham, Illinois Natural History Survey; Christian Burks, Genbank, Los Alamos; Theodore J. Crovello, California State University, Los Angeles; Thomas Duncan, University of California, Berkeley; Janet Gomon, National Museum of Natural History, Washington, D.C.; Ronald L. Hartman, Rocky Mountain Herbarium; Maureen Kelley, BIOSIS, Philadelphia; Robert W. Kiger, Hunt Institute for Botanical Documentation; Kenneth M. King, EDUCOM, Washington, D.C.; Robert Magill, Missouri Botanical Garden; Jim Ostell, National Library of Medicine, Bethesda; David Raber, Case Ware Inc., Costa Mesa; Beryl Simpson, University of Texas; Frederick Springsteel, University of Missouri, Columbia; Kerry S. Walter, Center for Plant Conservation, Jamaica Plain.

We are deeply grateful to all these people and institutions for their hard work and continuing support on behalf of the Flora of North America project.

N. R. M.

We thank the following reviewers for their help with volume 2.

Ray Angelo

D. S. Barrington

James Bartel

Donald M. Britton

Ralph E. Brooks

George Buddell

Adolf Ceska

Carl F. Chuey

David deLaubenfels

Theodore R. Dudley

James R. Griffin

Vernon L. Harms

Noel H. Holmgren

Walter S. Judd

Ronald Lanner

Elbert L. Little

Clifton E. Nauman

B. Ernie Nelson

William H. Parker

James H. Peck

Robert A. Price

John Silba

W. Carl Taylor

R. Dale Thomas

Alice F. Tryon

David H. Wagner

Thomas L. Wendt

Michael D. Windham

George Yatskievych

Contributors

Robert P. Adams
Department of Biology
Baylor University
Waco, Texas

Edward R. Alverson
The Nature Conservancy
Eugene Public Works Department
Engineering Division
Eugene, Oregon

Elisabeth G. Andrews
Department of Botany
University of Kansas
Lawrence, Kansas

Tim A. Atkinson
Carolina Biological Supply Company
Burlington, North Carolina

Joseph M. Beitel, deceased
New York Botanical Garden
Bronx, New York

Dale M. Benham
Department of Biology
Nebraska Wesleyan University
Lincoln, Nebraska

Donald M. Britton
Department of Molecular Biology
  and Genetics
University of Guelph
Guelph, Ontario

Daniel F. Brunton
Ottawa, Ontario

Kenton L. Chambers
Botany Department
Oregon State University
Corvallis, Oregon

Raymond Cranfill
Graham and James
San Francisco, California

James E. Eckenwalder
Department of Botany
University of Toronto
Toronto, Ontario

A. Murray Evans
Department of Botany
University of Tennessee
Knoxville, Tennessee

Donald R. Farrar
Department of Botany
Iowa State University
Ames, Iowa

Christopher H. Haufler
Department of Botany
University of Kansas
Lawrence, Kansas

R. L. Hauke
Botany Department
University of Rhode Island
Kingston, Rhode Island

R. James Hickey
Department of Botany
Miami University
Oxford, Ohio

Matthew H. Hils
Department of Biology
Hiram College
Hiram, Ohio

Richard S. Hunt
Forestry Canada
Pacific Forest Centre
Victoria, British Columbia

Carol A. Jacobs
Jacobs & Associates
Ames, Iowa

David M. Johnson
Department of Botany-Microbiology
Ohio Wesleyan University
Delaware, Ohio

Masahiro Kato
Botanical Gardens
Faculty of Science
University of Tokyo
Tokyo, Japan

Robert Kral
Department of General Biology
Vanderbilt University
Nashville, Tennessee

Karl U. Kramer
Botanischer Garten und Institut für
  Systematische Botanik
Universität Zürich
Zürich, Switzerland

Garrie P. Landry
Department of Biology
University of Southwest Louisiana
Lafayette, Lousiana

Frank A. Lang
Department of Biology
South Oregon State College
Ashland, Oregon

Barney Lipscomb
*Botanical Research
Institute of Texas, Inc.
Fort Worth, Texas*

Robert M. Lloyd
*Department of Botany
Ohio University
Athens, Ohio*

Neil T. Luebke
*Milwaukee Public Museum
Milwaukee, Wisconsin*

Thomas A. Lumpkin
*Department of Agronomy and Soils
Washington State University
Pullman, Washington*

David C. Michener
*University of Michigan
Botanical Gardens
Ann Arbor, Michigan*

John T. Mickel
*New York Botanical Garden
Bronx, New York*

James D. Montgomery
*Ecology III, Inc.
Berwick, Pennsylvania*

Robbin C. Moran
*Missouri Botanical Garden
St. Louis, Missouri*

Clifton E. Nauman
Knoxville, Tennessee

Cathy A. Paris
*Department of Botany
University of Vermont
Burlington, Vermont*

William H. Parker
*School of Forestry
Lakehead University
Thunder Bay, Ontario*

James H. Peck
*Department of Biology
University of Arkansas
Little Rock, Arkansas*

Kathleen M. Pryer
*Botany Department
Duke University
Durham, North Carolina*

Eric W. Rabe
*Department of Botany
University of Kansas
Lawrence, Kansas*

Alan R. Smith
*University Herbarium
University of California
Berkeley, California*

Dennis W. Stevenson
*The New York Botanical Garden
Bronx, New York*

Ronald J. Taylor
*Biology Department
Western Washington University
Bellingham, Washington*

W. Carl Taylor
*Botany Department
Milwaukee Public Museum
Milwaukee, Wisconsin*

John W. Thieret
*Department of Biological Sciences
Northern Kentucky University
Highland Heights, Kentucky*

Iván A. Valdespino
*The New York Botanical Garden
Bronx, New York*

David H. Wagner
*Biology Department
University of Oregon
Eugene, Oregon*

Florence S. Wagner
*Department of Biology
University of Michigan
Ann Arbor, Michigan*

Warren H. Wagner Jr.
*Department of Biology
University of Michigan
Ann Arbor, Michigan*

Frank D. Watson
*Department of Biology
St. Andrews College
Laurinburg, North Carolina*

Charles R. Werth
*Department of Biological Sciences
Texas Tech University
Lubbock, Texas*

R. David Whetstone
*Department of Biology
Jacksonville State University
Jacksonville, Alabama*

S. A. Whitmore
*Department of Biological Sciences
University of California
Santa Barbara, California*

Michael D. Windham
*Utah Museum of Natural History
University of Utah
Salt Lake City, Utah*

Richard P. Wunderlin
*Department of Biology
University of South Florida
Tampa, Florida*

George Yatskievych
*Natural History Division
Missouri Department of Conservation
St. Louis, Missouri*

Flora of North America

Introduction

Nancy R. Morin

Scope of the Work

Flora of North America North of Mexico is a synoptic floristic account of the plants of North America north of Mexico: the continental United States of America (including the Florida Keys and Aleutian Islands), Canada, Greenland (Kalâtdlit-Nunât), and St. Pierre and Miquelon. The flora is intended to serve both as a means of identifying plants within the region and as a systematic conspectus of the North American flora. Taxa and geographical areas in need of further study also are identified in the flora.

Contents · General

The published flora includes identification keys, summaries of habitats and geographic ranges, synonymies, descriptions, chromosome numbers, phenological information, and other biological observations. Each volume will contain a bibliography and an index to the taxa included in the volume. A comprehensive, consolidated bibliography and comprehensive index will be published in the last volume. The treatments, written and reviewed by experts from throughout the systematic botanical community, are based on original observations of herbarium specimens and, whenever possible, on living plants. These observations are supplemented by critical reviews of the literature.

Taxonomic Sequence of the Volumes:

Volume 1   Introductory Essays
Volume 2   Pteridophytes and Gymnosperms
Volume 3   Magnoliidae and Hamamelidae
Volume 4   Caryophyllidae

Basic Concepts

Our goal has been and continues to be to make the flora as clear, concise, and informative as practicable so that it can be an important resource for both botanists and nonbotanists. To this end, we are attempting to be consistent in style and content from the first volume to the last. Readers may assume that a term has the same meaning each time it appears and that, within groups, descriptions may be compared directly with one another. Any departures from consistent usage will be explicitly noted in the treatments (see also References).

Treatments are intended to reflect current knowledge of taxa throughout their ranges worldwide, and classifications are therefore based on all available evidence. Where there are notable differences of opinion about the classification of a group, appropriate references are mentioned in the discussion of the group.

Documentation and arguments supporting significantly revised classifications will have been published separately in botanical journals before publication of the pertinent volume of the flora. Similarly, all new names, names for new taxa, and new combinations will have been published prior to their use in the flora. No nomenclatural innovations will be published intentionally in the flora. Journals and series in which papers relevant to *Flora of North America* have been or may be published include the *American Fern Journal, Brittonia, Canadian Journal of Botany, North American Flora, Novon, Systematic Botany, Systematic Botany Monographs,* and *Taxon,* among others.

Contents of Treatments

Treatments are intended to be succinct and diagnostic but adequately descriptive. Characters and character states used in the keys are repeated in the descriptions. Descriptions of related taxa are directly comparable.

Taxa treated in full include native species, native species thought to be recently extinct, hybrids that are well established (or frequent), and waifs or cultivated plants that are found frequently outside cultivation. Taxa mentioned only in discussions include waifs or naturalized plants now known only from isolated old records and some nonnative, economically important or extensively cultivated plants, particularly when they are relatives of native species. Excluded names and taxa are listed at ends of the appropriate sections, e.g., species at the end of genus, genera at the end of family.

With few exceptions, taxa are presented in taxonomic sequence. If an author is unable to produce a classification, the taxa are arranged alphabetically, and the reasons are given in the discussion.

Treatments of hybrids follow that of one of the putative parents. Hybrid complexes are treated at the ends of their genera, after the descriptions of species.

We have attempted to keep terminology as simple as accuracy permits. Common English equivalents have been used in place of Latin or Latinized terms or other specialized terminology whenever the correct meaning could be conveyed in approximately the same space, e.g., "pitted" rather than "foveolate," but "striate" rather than "with fine longitudinal lines." Specialized terms that are used are defined in the generic or family descriptions and, in some cases, are illustrated.

References

The primary and authoritative reference for descriptive morphological terms for *Flora of North America North of Mexico* will be a forthcoming revision of *A Guide for Contributors to Flora North America, Part II: An Outline and Glossary of Terms for Morphological and Habitat Description* (D. M. Porter et al., 1973), which is now (1993) under final review by Kiger and Porter. Until this glossary becomes available, editors and authors are obliged to use other references. For the most part they have used definitions found in the glossaries by G. H. M. Lawrence (1951) *Taxonomy of Vascular Plants,* for general terms; D. B. Lellinger (1985), *A Field Manual of the Ferns & Fern-allies of the United States & Canada,* for terms peculiar to ferns and fern allies; or R. E. Magill (1990), Glossarium polyglottum bryologiae, a multilingual glossary for bryology, for bryophytes.

Authoritative general reference works used for style are *Chicago Manual of Style,* ed. 13 (University of Chicago Press 1982); *Webster's New Geographical Dictionary* (Merriam-Webster 1988); and *The Random House Dictionary of the English Language,* ed. 2, unabridged (S. B. Flexner and L. C. Hauck 1987). *B-P-H/S. Botanico-Periodicum-Huntianum/Supplementum* (G. D. R. Bridson and E. R. Smith 1991) has been used for abbreviations of titles of serials; and *Taxonomic Literature,* ed. 2 [*TL-2*] (F. A. Stafleu and R. S. Cowan 1976—1988) and the supplement by F. A. Stafleu and E. A. Mennega (1992) have been used for abbreviations of titles of books.

Graphic Elements

Distribution maps for taxonomic treatments in volume 2 have been prepared by the Washington University Medical School Computer Graphics unit from draft maps supplied by authors. The base map is a Lambert equal-area projection. Maps in the introductory chapters have been drawn by this unit using the same base map as that used in the treatments.

Each genus, and approximately one out of three species, is illustrated. The illustrations may be of typical or of unusual species, or they may show diagnostic traits or complex structures. Most illustrations have been drawn from herbarium specimens selected by the authors. In some cases living material or photographs have been used. Data on specimens that were used and parts that were illustrated have been recorded. This information, together with the archivally preserved original drawings, is deposited in the Missouri Botanical Garden Library and is available for scholarly study.

Specific Information in Treatments

Keys

Keys are included for all ranks if two or more taxa are treated. For dioecious species, keys are designed for use with either staminate or pistillate plants. Keys are also designed to facilitate identification of taxa that flower before leaves appear. More than one key may be given, and for some groups tabular comparisons may be presented in addition to keys.

Nomenclatural Information

Basionyms, with author and literature citation, are given when pertinent. Other synonyms in common use are listed in alphabetical order, without literature citations.

Vernacular names in common usage are given in the appropriate language. In general, such names have not been created for use in the flora. Those preferred by governmental or conservation agencies usually are listed first.

The last names of authors of taxonomic names have been spelled out. The conventions of *Authors of Plant Names* by R. K. Brummitt and C. E. Powell (1992) have been used as a guide for including first initials to discriminate individuals who share surnames.

If only one infraspecific taxon within a species occurs in the flora area, nomenclatural information (literature citation, basionym with literature citation, relevant synonyms) is given for the species, as is information on the number of infraspecific taxa in the species and their distribution worldwide, if known. A description and detailed distributional information are given only for the infraspecific taxon.

Descriptions

Character states common to all taxa are treated in the description of the taxon at the next higher rank. For example, if corolla color is yellow for all species treated within a genus, that character state is given in the generic description. Characters used in keys are repeated in the descriptions. Characteristics are given as they occur in plants from the flora area. Notable characteristics that occur in plants from outside the flora area are given in square brackets or are included in a brief discussion at the end of the description. In families with one genus and one or more species, the family description is given as usual, the genus description is condensed, and the species description is as usual.

In reading descriptions of vascular plants, the reader may assume, unless otherwise noted, that: the plant is green, photosynthetic, and reproductively mature; a woody plant is perennial; stems are erect; roots are fibrous; leaves are simple and petiolate. Because measurements and elevations are almost always approximate, modifiers such as "about," "circa," or "±" are usually omitted.

Arrangements of elements within descriptions of taxa are from base to apex, proximal to distal, abaxial to adaxial. General features such as persistence, habit, nutrition, and sexuality are given first. For a particular structure or organ system, description of parts follows the order: presence, number, position/insertion, arrangement, orientation, connation, coherence, adnation, adherence. Features of a whole organ follow the order: color, architecture, shape, dimensions (length, width, thickness, mass), texture, surface, vesture. Unless otherwise noted, dimensions are length × width. If only one dimension is given, it is length or height. All measurements are given in metric units. Measurements usually are based on dried specimens but these should not differ significantly from the measurements actually found in fresh or living material.

Chromosome numbers generally are given only if published, documented counts are available from North American material or from an adjacent region. Among pteridophytes, however, little if any variation in chromosome number is found within the geographical range of a species, and therefore counts have been included even when made from plants collected outside the flora. No new counts are published in the flora. Chromosome counts from non-sporophyte tissue have been converted to the 2n form. A literature reference for each reported chromosome number is available in the Flora of North America database (see below).

Flowering time and often fruiting time or time of sporulation are given by season, sometimes qualified by early, mid, or late. Elevation generally is rounded to the nearest 100 m. Mean sea level is shown as 0 m, with the understanding that this is approximate. Elevation often is omitted from herbarium specimen labels, particularly for collections made where the topography is not remarkable, and therefore elevation is sometimes not known for a given taxon.

The term "introduced" is defined broadly to refer to plants released deliberately or accidentally into the flora and that now exist as wild plants in areas in which they were not recorded as native in the past. The nature of introduced populations is discussed as far as understood.

If a taxon is globally rare or if its continued existence is threatened in some way, the words "of conservation interest" appear before the statements of elevation and geographic range. Taxa thought to have become extinct during the period of permanent European settlement, i.e., the past 500 years, are included in the flora. Reviews of treatments of such taxa by various conservation agencies are coordinated by the Center for Plant Conservation at the Missouri Botanical Garden and by the Nature Conservancy.

Range maps are given for each species or infraspecific taxon. We have assumed that details such as "northeastern Florida" are apparent on the map; consequently, directional qualifiers are not given in the list of territories, provinces, and states. Authors are expected to have seen at least one specimen documenting each state record and have been urged to examine as many specimens as possible from throughout the range of each taxon. Additional information about distribution may be given in the discussion.

Distributions are stated in the following order: Greenland; St. Pierre and Miquelon; Canada (provinces and territories in alphabetic order); United States (states in alphabetic order); Mexico (11 northern states may be listed specifically, in alphabetical order); West Indies; Bermuda; Central America (Guatemala, Belize, Honduras, El Salvador, Nicaragua, Costa Rica, Panama); South America; Eurasia, or Europe; Asia (including Indonesia); Africa; Pacific Islands; Australia; Arctic; Antarctic.

Discussion

The discussion section includes information on economic uses, weediness, special patterns of endemism, and any other notable aspects of the taxon. Toxicity, if known, also is mentioned in the discussion, and pertinent literature is cited.

Selected References

Major references used in preparation of a treatment or containing critical information about a taxon are cited after the discussion. These, and other works that are referred to briefly in the discussion or elsewhere, are included in the bibliography at the end of the volume and in the consolidated bibliography in the last volume of the flora.

The Database

In addition to appearing in the published flora, the information compiled by the project is being entered into a computerized database. This database will allow easy access to, and sorting and comparison of, large amounts of floristic information. The computerized storage of these data will also make it possible to manipulate floristic information in novel ways. Detailed morphological descriptions, geographical ranges, and other information not published in the flora itself will be stored in this form. The Missouri Botanical Garden maintains this taxonomic database as a permanent resource, and the Hunt Institute similarly maintains a corresponding bibliographic database. Data from both of these databases are available to interested users through printed copy, magnetic tapes, and other media.

PTERIDOPHYTES

KEY TO PTERIDOPHYTE FAMILIES

Alan R. Smith

Characters used to circumscribe fern families often relate to cryptic features, e.g., presence or absence of annulus and its orientation, length and diameter (number of cells) of the sporangial stalk, nature of the spores (whether all one kind or dissimilar on a given plant), spore shape, color, ornamentation, and number per sporangium, stem and petiole cross-section anatomy, and adaxial grooving (or lack thereof) of the rachis and costae. In addition, chromosome number, a trait almost never assessable by those doing identifications, is often considered in circumscription of families. As much as possible, these definitive characters have been avoided in the key below, which makes the key artificial in part.

In the flora area, species in Hymenophyllaceae, Grammitidaceae, and Vittariaceae sometimes exist primarily or entirely as independently reproducing populations of gametophytes. See species descriptions in those families.

1. Leaves grasslike, blade not expanded; spore-bearing structures embedded in leaf bases or on short stalks 1–2 mm at leaf bases, at or below ground level; plants of aquatic, semi-aquatic, or vernally wet habitats.
   2. Leaves tightly clustered on compact 2- or 3-lobed corms, rarely on short-creeping stems, not circinate; sporangia embedded singly in swollen leaf bases. . . . . 4. Isoëtaceae *(Isoëtes)*, p. 64
   2. Leaves borne on filiform, short-creeping stems, circinate; sporangia numerous in globose, hairy sporocarps attached by stalk 1–2 mm at bases of leaves. . . . .
      . . . . 23. Marsileaceae *(Pilularia)*, p. 331
1. Leaves scalelike, needlelike, or with expanded blades, if grasslike then sporangia at tips of leaves; spore-bearing structures usually borne aboveground (except *Marsilea*), at or near leaf bases, or on blade margins or surfaces, occasionally on modified erect stalks or in hardened sporocarps; plants variously aquatic, terrestrial, or epiphytic.
   3. Plants bearing inconspicuous scalelike or needlelike leaves (microphylls) less than 2 cm, or leaves longer and rushlike, all leaves with single unbranched vein; sporangia (synangia in *Psilotum*) occurring singly in leaf axils or aggregated in conelike structures, or pendent from peltate sporophylls in terminal cones.

4. Stems (and branches, if any) jointed, usually fluted, hollow, often rough from silica deposited in cells; leaves borne in whorls at each node, fused at base to form sheath but with free tips that may be caducous; sporangia aggregated into terminal strobili with polygonal segments. 5. Equisetaceae *(Equisetum)*, p. 76

4. Stems and branches, if any, not obviously jointed, not fluted or hollow, lacking silica; leaves spirally or oppositely arranged; sporangia variously arranged.

    5. Leaves ca. 1–2 mm, scalelike, generally 5–30 cm apart, borne on nearly naked, repeatedly dichotomous stems; sporangia fused into 3-lobed clusters (synangia). . . . . 1. Psilotaceae *(Psilotum)*, p. 16

    5. Leaves generally greater than 2 mm, closely placed and usually less than 5 mm apart, often less than 2 mm apart; sporangia separate, not fused into 3-lobed structures.

        6. Sporangia borne singly in leaf axils, the leaves unmodified or modified and aggregated in cylindrical strobili mostly 3–25 mm wide at branch tips; spores of 1 size, less than 50 μm in diam. 2. Lycopodiaceae, p. 18

        6. Sporangia commonly borne in flattened or 4-sided strobili 1–2.5(–3.5) mm wide at branch tips (except *Selaginella selaginoides* with cylindrical cones 4–6 mm wide); spores of 2 sizes, large megaspores greater than 300 μm diam., borne singly or in groups to 4, and minute microspores in uncountable masses. . . . . 3. Selaginellaceae *(Selaginella)*, p. 38

3. Plants commonly bearing expanded leaves (megaphylls) with branched vasculature in blades, or leaves reduced to nonlaminate vascular tissue with terminal sporangium-bearing lobes in some Schizaeaceae, or leaves rounded to ovate, crowded, and floating in *Azolla* and *Salvinia*; sporangia borne in clusters (sori) of various shapes, hardened sporocarps, or covering blade surface.

    7. Plants floating in water or sometimes rooted in mud at edges of ponds or streams, or in wet meadows and vernally inundated areas; plants less than 30(–50) cm.

        8. Sporangia borne on erect, nonlaminate stalks often equaling or exceeding sterile blades, fertile stalks arising at base of or below sterile blades; spores all 1 kind. . . . . 6. Ophioglossaceae, p. 85

        8. Sporangia borne in hardened sporocarps or on blade tissue, never on erect non-laminate stalks; spores all 1 kind or of 2 kinds, 1 distinctly larger than the other.

            9. Blades more than 10 cm, pinnately lobed or divided, dimorphic, fertile blades taller and with narrower segments; petioles commonly inflated; spores all 1 kind. . . . . 12. Parkeriaceae *(Ceratopteris)*, p. 119

            9. Blades less than 10 cm, not pinnately lobed or divided, monomorphic; petioles filiform or absent; spores of 2 kinds borne in specialized sporocarps.

                10. Photosynthetic blades 4-parted in cloverlike fashion on filiform petioles; leaves borne on short- to long-creeping stems usually rooted in mud. . . . . 23. Marsileaceae *(Marsilea)*, p. 331

                10. Photosynthetic blades undivided, not cloverlike, entire, round to oval or ovate; leaves borne on usually floating stems.

                    11. Blades mostly 5–15 mm, round to oval, with conspicuous hairs adaxially. . . . . 24. Salviniaceae *(Salvinia)*, p. 336

                    11. Blades mostly less than 1 mm, ovate, papillate but lacking hairs adaxially. . . . . 25. Azollaceae *(Azolla)*, p. 338

    7. Plants terrestrial, on rock, or epiphytic, if rooted in mud then leaves erect and more than 30 cm.

        12. Sporangia fused laterally into 2-rowed, long-stalked linear units (synangia) and opening by double row of pores or slits; sterile portions of blades entire (except *Cheiroglossa*). . . . . 6. Ophioglossaceae, p. 85

        12. Sporangia discrete, not fused into synangia; sterile portions of blades often divided.

13. Leaves with rachis twining, high-climbing, pinna midribs dichotomously forked or pseudodichotomously forked (i.e., with dormant hairy bud at fork); sporangia borne on lateral lobes of pinnules, each sporangium subtended by indusiumlike flap. . . . . 10. Lygodiaceae *(Lygodium)*, p. 114
13. Leaves not twining, occasionally scandent, leaf branching various (stems hemiepiphytic and climbing in *Maxonia*); sporangia never both borne singly *and* subtended by indusiumlike flap.
   14. Blades pseudodichotomously branched (i.e., at many forks with hairy dormant bud between branches), scrambling or trailing; sori exindusiate, sporangia commonly 4–8(–15) per sorus; spores more than 200 per sporangium. . . . . 8. Gleicheniaceae *(Dicranopteris)*, p. 110
   14. Blades not pseudodichotomously branched (usually pinnate, palmate, or undivided), lacking dormant, hairy buds, rarely creeping (*Hypolepis, Pteridium*); sori exindusiate or indusiate, sporangia usually more than 15 per sorus; spore number various.
      15. Sporangial capsules (excluding stalks) greater than 0.4 mm, opening by action of subterminal annulus or by thin-walled and poorly developed lateral patch, or annulus seemingly lacking; sporangia usually with several hundreds to thousands of spores; sporangia sessile or with very short stalk with 4 or more cells in cross section; sporangia usually borne on nonlaminate (not green) tissue or on naked or specialized stalks.
         16. Leaves without expanded blades, grasslike; fertile blades terminated by pinnately branched or palmately arranged lobes (sporophores) that bear sporangia; annulus subapical. . . 9. Schizaeaceae, p. 112
         16. Leaves with expanded sterile blades, not grasslike; fertile blades not as above, if sporangia borne on specialized stalks or lobes then these not terminal; annulus subapical, lateral, or absent.
            17. Sporangial capsules pear-shaped, ca. 0.5 mm, borne on 2 nonlaminate, branched, spikelike stalks (highly modified pair of basal pinnae), or leaves wholly dimorphic and less than 10 cm; annulus subapical and girdling sporangium. . . . . 11. Anemiaceae *(Anemia)*, p. 117
            17. Sporangial capsules globose, 0.5–1 mm diam., borne either on single nonlaminate, branched stalk attached below or at base of blade, or on specialized areas at middle or tip of leaf with otherwise "normal" blade, or leaves dimorphic and more than 50 cm; annulus not apical and not girdling sporangium.
               18. Fertile portions of leaves inserted near base of sterile portion, long-stalked; sterile blades usually less than 30 cm; spores transparent, thousands per sporangium. . . . . 6. Ophioglossaceae *(Botrychium)*, p. 85
               18. Fertile portions of leaves apical or in middle of sterile blades, or fertile and sterile leaves dimorphic; sterile blades, or those with sterile portions, usually greater than 30 cm; spores green, hundreds per sporangium. . . . . 7. Osmundaceae *(Osmunda)*, p. 107
      15. Sporangial capsules (excluding stalks) less than 0.4 mm, opening by action of vertical or slightly oblique thick-walled annulus, usually containing 16–64 spores; sporangia short- to long-stalked with 1–3 rows of stalk cells in cross section, rarely sessile; sporangia usually borne on blade tissue, rarely on nonlaminate tissue.

19. Blades 1-cell thick between veins, translucent, lacking stomates; sporangia borne within tubular, conic, or 2-valved marginal involucres (indusia); leaves minute to small, 0.5–20 (–40) cm; scales absent at stem apex, hairs sometimes present. . . . . 15. Hymenophyllaceae, p. 190

19. Blades 3–many cells thick between veins, usually opaque, stomates present; sporangia borne otherwise; leaves small to large, usually greater than 20 cm; scales (rarely hairs) present at stem apices.

   20. Plants hemiepiphytic, with stems rooted in ground and then climbing trees. . . . . 20. Dryopteridaceae *(Maxonia)*, p. 246

   20. Plants terrestrial, on rock, or strictly epiphytic.

      21. Blades linear, 1–3 mm wide, entire; sori linear in 2 submarginal grooves, 1 on each side of midrib, without indusium; stem scales strongly clathrate (lattice-like, with dark lateral walls and thin, translucent surface walls); pendent epiphytes. . . . . 14. Vittariaceae *(Vittaria)*, p. 187

      21. Blades usually pinnatifid or more divided, if simple then more than 5 mm wide; sori generally round or oblong; stem scales not clathrate (except *Asplenium*); terrestrial, if epiphytic, not pendent.. . . . . See couplet 22

22. Sori elongate, in 1 row on each side of and immediately adjacent to costae (and costules in more divided blades), extending nearly entire length of pinnae or end-to-end (chainlike); indusia introrse, with openings facing costae (or costules) and away from margins . . . . 18. Blechnaceae, p. 223

22. Sori elongate to round, usually many per pinna, if elongate and parallel to costae then not immediately adjacent to them; indusia (when present) opening variously, commonly extrorse.

   23. Stems and petiole bases bearing hairs 1 cell wide, lacking scales 2 or more cells wide; sori marginal or nearly so; indusia cuplike, attached proximally and sometimes along sides, or formed by revolute blade margins. . . . . 16. Dennstaedtiaceae, p. 198

   23. Stems and petiole bases bearing scales several cells wide; sori marginal or well back from margin, or sporangia covering blade surfaces; indusia linear, reniform, or peltate, rarely cuplike, or absent, sometimes replaced by false indusium formed by reflexed, recurved, or revolute blade margins.

      24. Sori elongate along veins, never marginal, usually with linear indusium.

         25. Scales on stems and petiole bases clathrate; sori generally along 1 side of vein only; sporangial stalks 1 cell thick; petioles with 2 back-to-back C-shaped vascular bundles in cross section, these fused into an X-shape distally. . . . . 19. Aspleniaceae *(Asplenium)*, p. 228

         25. Scales on stems and petiole bases not clathrate; sori often along both sides of veins or curved around end of vein; sporangial stalks 2–3 cells thick; petiole vasculature various in cross section, if bundles 2 then these crescent-shaped and fused into a U-shape distally. . . . . 20. Dryopteridaceae (part), p. 246

      24. Sori variously shaped, often round, sometimes marginal, or sporangia covering surfaces, if elongate along veins then without indusium.

         26. Blades pinnatifid, less than 5 cm × 5 mm; sporangial stalks 1 cell thick; spores trilete, green; [rare, known in flora area from single population of gametophytes and sterile sporophytes in North Carolina]. . . . . 21. Grammitidaceae *(Grammitis)*, p. 309

         26. Blades simple, pinnatifid, 1-pinnate, or more divided, commonly more than 5 cm × 5 mm; sporangial stalks 2–3 cells thick; spores trilete or monolete, rarely green.

27. Blades simple, pinnatifid, or pinnatisect nearly to rachis, rarely 1-pinnate, lobes (pinnae) entire or nearly so; leaves borne on short phyllopodia and cleanly abscising at this junction with age; sori without indusium; spores bilateral, colorless or yellowish, often transparent. . . . 22. Polypodiaceae, p. 312

27. Blades palmate, 1-pinnate, or commonly more divided, if pinnatisect, at least some of the lobes themselves pinnatifid; leaves not borne on phyllopodia, petioles continuous, not cleanly abscising with age; sori with or without indusium; spores bilateral to tetrahedral-globose, variously colored, usually opaque.

    28. Sori marginal or submarginal, sometimes in continuous bands and/or covered by revolute or recurved margins of segments; spores globose-tetrahedral, trilete. . . . . 13. Pteridaceae (part), p. 122

    28. Sori medial on veins or covering surfaces of fertile segments, if submarginal, then not elongate or covered by revolute or recurved margins; spores various.

        29. Sporangia completely covering surfaces of fertile segments.

            30. Stems short, petiole bases contiguous; blades variously dissected, if 1-pinnate then greater than 1 m; petioles not green-winged. . . . . 13. Pteridaceae (part), p. 122

            30. Stems long-creeping, petiole bases well separated; blades 1-pinnate, less than 0.5 m; petioles green-winged. . . . . 20. Dryopteridaceae (Lomariopsis), p. 246

        29. Sporangia along veins or aggregated into discrete, round or oblong sori.

            31. Costae, veins, and sometimes tissue between veins with hairs sparse to dense, needlelike or sometimes stellate, transparent, never obscuring abaxial leaf surface; blades 1-pinnate, 2-pinnatifid, or 1-pinnate-pinnatifid (2–3-pinnate in Macrothelypteris); petioles in cross section with 2 crescent-shaped vascular bundles at base, these united into a U-shape distally; spores bilateral, monolete. . 17. Thelypteridaceae, p. 206

            31. Costae, veins, and tissue between veins lacking needlelike, transparent hairs, or (in Bommeria) hairs so dense as to obscure abaxial leaf surface, sometimes with brownish hairlike scales (proscales) or glandular trichomes; blades variously divided; petiole vasculature various, with 1–many vascular bundles; spores various.

                32. Sori without indusium, sporangia extending along veins; leaves monomorphic; blades pentagonal or (in Pityrogramma) lanceolate; blades abaxially whitish to yellowish with powder or exudate (except Bommeria and 2 spp. of Argyrochosma); petioles with 1 (2 in Pityrogramma) vascular bundle; spores trilete. . . . . 13. Pteridaceae (part), p. 122

                32. Sori with indusium, or if without indusium then round (indusia concealed by recurved blade margins in Onoclea and Matteuccia, which have dimorphic, fertile and sterile, leaves); blades generally lanceolate or deltate, lacking powdery covering; petioles with 2–many vascular bundles; spores monolete. . . . . 20. Dryopteridaceae, p. 246

1. PSILOTACEAE Kanitz

• Whisk-fern Family

John W. Thieret

Plants perennial, terrestrial or epiphytic, with corallike, rhizoid-bearing, branched, subterranean axes. **Roots** absent. **Aerial shoots** simple or dichotomously branched; appendages leaflike or bractlike, alternate to subopposite, veinless or 1-veined, less than 1 cm. **Synangia** globose, of 2–3 fused, homosporous eusporangia, solitary in axils of shoot appendages, dehiscing loculicidally. **Spores** many, reniform, not green. **Gametophytes** subterranean, mycotrophic, fleshy, elongate, and branched.

Genera 2, species 4–8 (1 genus, 1 species in the flora): worldwide in tropical regions.

1. PSILOTUM Swartz, J. Bot. (Schrader) 1800(2): 8, 109. 1801 · Whisk-fern [Greek *psilos*, naked, referring to the plant's leafless aerial shoots]

Plants terrestrial, sometimes epiphytic. **Aerial shoots** often clumped, simple proximally, dichotomously branched distally, 3(–several)-ridged. **Appendages** minute, bractlike, borne distally on ridges of aerial shoots, sterile appendages subulate, those subtending synangia 2-lobed. **Synangia** ± globose, obscurely 3-lobed.

Species 2 or 3 (1 in the flora): tropical and warm temperate regions.

SELECTED REFERENCES Cooper-Driver, G. 1977. Chemical evidence for separating the Psilotaceae from the Filicales. Science 198: 1260–1262. White, R. A., D. W. Bierhorst, P. G. Gensel, D. R. Kaplan, and W. H. Wagner Jr. 1977. Taxonomic and morphological relationships of the Psilotaceae: A symposium. Brittonia 29: 1–68.

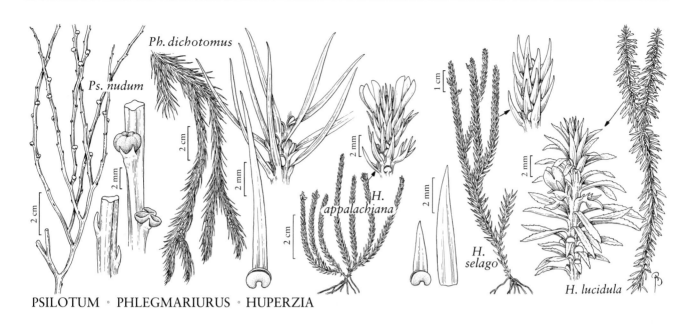

PSILOTUM · PHLEGMARIURUS · HUPERZIA

1. Psilotum nudum (Linnaeus) Palisot de Beauvois, Prodr. Aethéogam., 106, 112. 1805

Lycopodium nudum Linnaeus, Sp. Pl. 2: 1100. 1753

Aerial shoots to 50 cm, 4 mm diam. at base. **Appendages** 0.7–2.5 mm (sterile), 1–1.5 mm (fertile). **Synangia** yellowish to greenish yellow, 2–3 mm wide.

Low to mesic woods, thickets, swamps, hammocks, rocky slopes; 0–50 m (1100 m in Ariz.); Ala., Ariz., Fla., Ga., La., Miss., S.C., Tex.; Mexico; West Indies in the Antilles; Central America; South America; tropical Asia; tropical Africa.

Reports of *Psilotum nudum* in New Mexico and in west Texas remain unverified. A wide range of chromosome numbers on material outside the flora has been reported for this species: n = ca. 46–56, 104, 210. Most counts, however, are based on n = 52 (reviewed in P. J. Brownsey and J. D. Lovis 1987). No count is available from North American material.

Psilotum nudum occurs as a minor weed in greenhouses. Many horticultural forms—including one without appendages and with terminal synangia (A. S. Rouffa 1971)—are grown, especially in Japan.

2. LYCOPODIACEAE Mirbel

· Club-moss Family

Warren H. Wagner Jr.

Joseph M. Beitel

Plants terrestrial, on rock, or epiphytic. **Roots** emerging near origin, or growing through cortex and emergent some distance from origin. **Horizontal stems** present or absent, mainly protostelic, in some species becoming actino- or plectostelic, on substrate surface or subterranean, or forming stolons. **Upright shoots** simple or branched, usually conspicuously leafy at least at base; abscising gemmae formed by reduced lateral shoots. **Lateral shoots** present or absent, simple or branched, branching pattern dichotomous and sometimes pseudomonopodial; leaves uniform or dimorphic or trimorphic. **Upright and lateral shoots** round or flat in cross section; leaves on subterranean parts flat, appressed, nonphotosynthetic, and scalelike; leaves on aerial parts appressed, ascending, or spreading, with 1 central unbranched vein, needlelike to lanceolate to ovate, remote to dense and imbricate, with or without basal and/or mucilage canals. **Strobili** sessile or stalked, upright, nodding, or pendent. **Sporangia** solitary, adaxial near leaf base or axillary; subtending leaves (sporophylls) unmodified and photosynthetic to much modified, nonphotosynthetic, reduced, and aggregated in strobili; sporangia reniform to globose, thick-walled with hundreds of spores, outer walls variously modified. **Spores** all 1 kind, trilete, thick-walled, surfaces pitted to small-grooved, rugulate, or reticulate. **Gametophytes** subterranean and nonphotosynthetic or surficial and photosynthetic.

Genera 10–15, species 350–400 (7 genera, 27 species in the flora): worldwide.

The Lycopodiaceae are an extremely diverse, ancient family. The family may contain even more than the estimated 400 species because the tropical members and the very large genus *Phlegmariurus* are still poorly known. The relationships among genera of Lycopodiaceae are not well understood because large evolutionary gaps exist among most genera. Some of the genera, notably *Diphasiastrum, Huperzia,* and *Lycopodiella,* exhibit extensive interspecific hybridization, which has caused much taxonomic confusion in the past. Differences in expressions of many of the generic characters are subtle, and some of the characters are microscopic.

SELECTED REFERENCES Øllgaard, B. 1987. A revised classification of the Lycopodiaceae s. lat. Opera Bot. 92: 153–178. Øllgaard, B. 1989. Index of the Lycopodiaceae. Biol. Skr. 34: 1–135. Øllgaard, B. 1990. Lycopodiaceae. In: K. Kubitzki et al., eds. 1990+. The Families and Genera of Vascular Plants. 1+ vol. Berlin etc. Vol. 1, pp. 31–39. Wagner, F. S. 1992. Cytological problems in *Lycopodium* sens. lat. Ann. Missouri Bot. Gard. 79: 718–729. Wagner, W. H. Jr. and J. M. Beitel. 1992. Generic classification of modern North American Lycopodiaceae. Ann. Missouri Bot. Gard. 79: 676–686.

1. Horizontal stems absent; upright parts of shoots clustered; roots traveling in stem cortex some distance before emerging; sporangia borne in axils of unmodified leaves; spores pitted to small-grooved.
    2. Leafy gemmae and gemmiferous branchlets absent; spore sides at equator convex with acute to blunt angles; mainly tropical, epiphytic. . . . . 1. *Phlegmariurus*, p. 19
    2. Leafy gemmae and gemmiferous branchlets present; spore sides at equator concave with truncate angles; mainly temperate and subarctic, terrestrial or on rock. . . . . 2. *Huperzia*, p. 20
1. Horizontal stems present; upright shoot systems alternating along rhizome; roots emerging from where they originate; sporangia borne in axils of highly modified, reduced sporophylls aggregated into upright or nodding or pendent strobili; spores reticulate or rugulate.
    3. Strobili borne on distinct peduncles or sessile; peduncles, if present, bearing remote, reduced leaves; spores reticulate; gametophytes subterranean, nonphotosynthetic; mainly dry uplands.
        4. Ultimate shoots (including leaves) 5–12 mm diam.; rounded (flattened in *Lycopodium obscurum*); leaves 6-ranked or more, not imbricate; peduncles, if present, falsely appearing to have 1 main branch (pseudomonopodial) and alternate; gametophytes disc-shaped. . . . . 3. *Lycopodium*, p. 25
        4. Ultimate shoots (including leaves) 2–6 mm diam., quadrate to flattened (except in *D. sitchense*, which is round-branched); leaves mostly 4–5-ranked, mostly imbricate (except in *D. sitchense*); peduncles, if present, dichotomously branched; gametophytes carrot-shaped. . . . . 4. *Diphasiastrum*, p. 28
    3. Strobili erect on leafy peduncles (or nonleafy peduncles in *Pseudolycopodiella*) or nodding or pendent on lateral shoots; peduncles, if present, bearing closely spaced, unreduced leaves; spores rugulate; gametophytes on substrate surface, photosynthetic; mainly wetlands.
        5. Upright shoots many branched; leaves linear to needlelike; strobili nodding or pendent at lateral shoot tips. . . . . 5. *Palhinhaea*, p. 33
        5. Upright shoots not branched; leaves linear-lanceolate to lanceolate; strobili erect on upright shoots.
            6. Peduncles nearly bare, with few, scattered, scalelike leaves; sporophylls shorter than leaves of peduncles; horizontal stems with lateral leaves larger than median leaves and lying flat on substrate. . . . . 6. *Pseudolycopodiella*, p. 34
            6. Peduncles leafy with crowded unmodified leaves; sporophylls equaling or longer than leaves of peduncles; horizontal stems with leaves monomorphic, supine or arching. . . . . 7. *Lycopodiella*, p. 34

1. PHLEGMARIURUS Holub, Preslia 36: 17, 21. 1964 · Hanging fir-moss [based on epithet of *Lycopodium phlegmaria;* Greek *phlegma*, flame, and *oura*, tail; in reference to the tasslelike fertile portions of the plant]

Plants epiphytic or terrestrial, pendent [erect]. **Roots** produced in distal parts of shoots, migrating downward in cortex to emerge in substrate. **Horizontal stems** absent [present]. **Shoots** clustered, round in cross section, dichotomously branching, 1.5–5 mm diam. **Leaves** not in distinct ranks [ranked], imbricate or not, monomorphic (or sporophylls sometimes differentiated), appressed, ascending to spreading, linear to subulate, margins mostly entire [denticulate]. **Gemmiferous branchlets and gemmae** absent. **Sporangia** reniform, borne individually in axils of undifferentiated or slightly differentiated sporophylls [differentiated sporophylls in strobili]. **Spores** pitted to small-grooved, sides at equator convex and angles acute. **Gameto-**

phytes nonphotosynthetic, mycorrhizal, mainly growing buried in humus on trees, branched, paraphysate with uniseriate hairs; ring meristem absent. $x = 68$.

Species over 300 (1 in the flora): in tropical areas worldwide.

As construed here, *Phlegmariurus* is very large and diverse. Many groups have been tentatively suggested in this genus (B. Øllgaard 1987). Some of these ultimately may be recognized as subgenera or even genera.

1. **Phlegmariurus dichotomus** (Jacquin) W. H. Wagner & Beitel, Novon 3: 305. 1993 · Hanging fir-moss

Lycopodium dichotomum Jacquin, Enum. Stirp. Vindob., 314. 1762; *Huperzia dichotoma* (Jacquin) Trevisan

Shoots pendent, 1–3-forked, clustered at base, 10–30 cm, long-lived. **Roots** densely covered with fine grayish uniseriate hairs, to 1 mm. **Leaves** spreading to ascending, 15–20 × 0.5–12 mm, very gradually narrowed from base to long-attenuate apex; margins entire. **Fertile leaves** reduced, 7–12 × 0.2–0.5 mm. **Sporangia** 2–3 times width of subtending leaf.

Epiphytic on trunks and branches of trees in tropical and subtropical regions; Fla.; Mexico; West Indies; Central America; South America.

In the flora, *Phlegmariurus dichotomus* is known only from Big Cypress Swamp, Florida.

2. **HUPERZIA** Bernhardi, J. Bot. (Schrader) 1800(2): 126. 1801 · Gemma fir-moss [for Johann Peter Huperz (d. 1816), a German fern horticulturist]

Plants terrestrial or on rock, erect to decumbent. **Roots** produced in apical portions of shoot, migrating downward in cortex to emerge at soil level. **Horizontal stems** absent. **Shoots** determinate (entire plant dying after several years of spore production) or indeterminate (entire plant not dying after several years), clustered to decumbent, round in cross section, equally dichotomously branched, 2–16 mm diam. including leaves. **Leaves** not in distinct ranks [ranked], not imbricate [imbricate], appressed, ascending to spreading, triangular, lanceolate to oblanceolate, monomorphic or varying in size according to seasonal growth patterns; juvenile (basal or proximal) leaves mostly larger than mature (terminal or distal) leaves, margins irregularly dentate to entire, ± irregularly roughened by papillae formed by marginal cells. **Gemmiferous branchlets and attached gemmae** formed among leaves in same phyllotactic spiral, gemmae articulate and abscising at maturity, deltoid, 2.5–6 × 3–6 mm, with 4 leaves flattened into 1 plane, 2 large lateral leaves, and 1 abaxial, 1 adaxial leaf. **Sporangia** reniform, borne individually at adaxial base of unmodified or reduced leaf, fertile leaves in zones or scattered along shoot. **Spores** pitted to shallowly grooved, sides at equator concave, angles truncate. **Gametophytes** nonphotosynthetic, mycorrhizal, subterranean, unbranched, linear to elliptic in outline, paraphyses numerous, uniseriate; ring meristem absent. $x = 67, 68$.

Species 10–15 (7 in the flora): temperate, alpine, and arctic regions, and tropical Asian mountains.

We distinguish the temperate and arctic *Huperzia* as a distinct genus because of its many differences from the tropical epiphytes, in particular the remarkably complex and specialized shoots, gemmiferous branchlets and gemmae, and the unbranched gametophytes.

In the area of central to northwestern Canada, numerous scattered collections of plants of this genus were formerly identified as *Huperzia selago*. Although many of them may be that species, a strong possibility exists that other, rather similar species may be represented. A

Huperzia

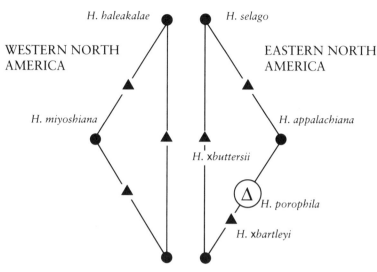

Relationships in *Huperzia*. Solid circles represent parental taxa; circled triangle represents an allopolyploid; and solid triangles represent hybrids reproducing by gemmae.

careful study of materials from Saskatchewan, northern Alberta, northern Manitoba, and Mackenzie and Keewatin districts of the Northwest Territories, and perhaps a wider area, is needed. Collectors are urged to take samples from as many localities as possible. The general area is shown in the map for *H. selago* (cf. W. J. Cody and D. M. Britton 1989).

Hybrids between species of North American *Huperzia* are extremely common. They are recognizable as hybrids because, in this genus (as opposed to *Diphasiastrum* and *Lycopodiella*), they usually have abortive spores that vary greatly in size and shape. The abortion of spores can be recognized readily under a high-powered dissecting microscope or a low-powered compound microscope. In all cases, the hybrids are intermediate in characters: if one parent has numerous teeth on the leaf margins and the other has entire margins, the hybrid will have few, inconspicuous teeth; if one parent has oblanceolate leaves, the other subulate leaves, the hybrid will have narrowly lanceolate leaves. In our descriptions we have not included the shape and dimensions of the gemmae of the different species, but they do differ in size, overall outlines, and gemmae leaf shapes; the hybrids are intermediate in these characters.

In the flora, *Huperzia* includes at least one sexual allopolyploid and six sterile hybrid taxa that reproduce by gemmae. Because of the subtle differences among the basic species and their propensity to form vegetatively reproducing hybrids, most *Huperzia* species and hybrids (except *H. lucidula*) have been confused with *H. selago*, populations of which are confined to northeastern North America. Many of the sterile hybrids often occur at considerable distances from the parents, suggesting that wind may disperse the gemmae. The aerodynamics involved would be an interesting subject for study. Spores of two species coming together at the same site by long-distance dispersal may also explain some disjunctions.

SELECTED REFERENCES Stevenson, D. W. 1976. Observations on phyllotaxis, stelar morphology, the shoot apex and gemmae of *Lycopodium lucidulum* Michaux (Lycopodiaceae). Bot. J. Linn. Soc. 72: 81–100. Ulrike, R. 1987. Growth patterns of gemmlings of *Lycopodium lucidulum*. Amer. Fern J. 77: 50–57. Waterway, M. J. 1986. A reevaluation of *Lycopodium porophilum* and its relationship to *L. lucidulum* (Lycopodiaceae). Syst. Bot. 11: 263–276.

1. Leaves narrowly obovate, teeth 1–8, irregular; stomates abaxial; spores (23–)24–26(–29) μm. . . . . 3. *Huperzia lucidula*
1. Leaves lanceolate or oblanceolate, entire or with 1–3 low teeth; stomates on both surfaces; spores 25–41 μm.
  2. Largest leaves oblanceolate; shoots 12–20 cm; mainly terrestrial in shaded forests along streams. . . . . 5. *Huperzia occidentalis*
  2. Largest leaves lanceolate or widest at base, or sides nearly parallel; shoots mainly 8–15 cm (except *H. miyoshiana* to 25 cm); mainly of rocky cliffs and talus habitats, or acidic bogs, ditches, meadows, and marshes.
    3. Largest leaves lanceolate with sides nearly parallel much of length; stomates 1–25 per 1/2 leaf on adaxial surface. 6. *Huperzia porophila*
    3. Largest leaves lanceolate to ovate or nearly triangular and widest at base or sides nearly parallel much of length; stomates more than 30 per 1/2 leaf on adaxial surface.
      4. Shoots with weak annual constrictions; gemmiferous branchlets and gemmae formed in 1 pseudowhorl at end of annual growth. . . . . 7. *Huperzia selago*
      4. Shoots without annual constrictions; gemmiferous branchlets and gemmae produced in 1–3 pseudowhorls at end of annual growth or throughout mature shoots.
        5. Mature distal leaves mostly 2–3.5 mm, somewhat triangular. . . . . . 1. *Huperzia appalachiana*
        5. Mature distal leaves mostly 3.5–5.5 mm, lanceolate or ovate.
          6. Mature shoots 12–18(–25) cm; gemmae in 2–3 pseudowhorls at end of annual growth; gemma lateral leaves 1.2–1.8 mm wide; mostly mossy boulders and marshes in conifer forests. . . . . 4. *Huperzia miyoshiana*
          6. Mature shoots 8–11 cm; gemmiferous branchlets and gemmae produced throughout mature portion of shoot; gemma lateral leaves 1.5–2 mm wide; alpine and subalpine mossy meadows. . . . . 2. *Huperzia haleakalae*

1. Huperzia appalachiana Beitel & Mickel, Amer. Fern J. 82: 45. 1992 · Mountain fir-moss, lycopode des Appalaches

Shoots erect, determinate, 6–10 cm, clustered to rarely shortly decumbent, decumbent portion to 1 cm; leaves in mature distal portion markedly smaller than leaves in juvenile proximal portion; annual constrictions absent; juvenile growth erect. **Leaves** ascending to spreading (juvenile portion) or ascending to appressed (mature portion), green to yellow green, not lustrous; leaves in juvenile portion narrowly triangular, 4–6 mm, widest at base; leaves in mature portion narrowly triangular, 2–3.5 mm; margins entire or with occasional papillae; stomates present on both surfaces, numerous (35–60 per 1/2 leaf) on adaxial surface. **Gemmiferous branchlets** produced throughout mature portion; gemmae 3–4 × 2.5–3.5 mm; lateral leaves 0.5–1 mm wide, narrowly acute. **Spores** 29–35 μm.

On damp, acidic, igneous rocks in alpine zone or exposed cliffs and talus slopes elsewhere; 800–2300 m, lower (600–1200 m) along coast of Atlantic Ocean and Lake Superior; Greenland; St. Pierre and Miquelon; Nfld., N.S., Ont., Que.; Ga., Maine, Mass., Mich., Minn., N.H., N.Y., N.C., S.C., Tenn., Vt., Va.; possibly Europe.

2. Huperzia haleakalae (Brackenridge) Holub, Folia Geobot. Phytotax. 20: 73. 1985 · Alpine fir-moss

Lycopodium haleakalae Brackenridge in Wilkes, U.S. Expl. Exped. 16: 321. 1854; *L. selago* Linnaeus var. *haleakalae* (Brackenridge) Warburg; *Urostachys haleakalae* (Brackenridge) Nessel

Shoots erect, indeterminate, 8–11 cm, short- to long-decumbent, 3–8 cm; leaves in mature portion smaller than in juvenile portion; annual constrictions absent; juvenile growth erect. **Leaves** spreading-ascending to appressed, yellow-green to yellow-brown, greener at stem tip (top 1 cm), lustrous (as if covered with clear yellow varnish); leaves in juvenile portion lanceolate, 4.5–6(–7) mm, apex acute; leaves in mature portion ovate, 3–4 mm, apex acute; margins entire, with scattered papillae; stomates present on both surfaces, numerous (38–84 per 1/2 leaf) on adaxial surface. **Gemmiferous branchlets** produced throughout mature portion of shoot; gemmae 3–4 × 3.5 mm, lateral leaves 1.5–2 mm wide, broadly acute to obtuse. **Spores** 31–41 μm.

Terrestrial in exposed, moist meadows and mossy heaths in alpine and subalpine zones; (160–)1300–1700(–3600) m; Alta., B.C., Yukon; Alaska, Colo., Mont., Wash., Wyo.; Asia in Siberia; Pacific Islands in Hawaii.

Huperzia haleakalae has been found in Hawaii in Haleakala Crater, on Maui, at 2300 meters, but it is extremely rare or possibly extinct there.

3. **Huperzia lucidula** (Michaux) Trevisan, Atti Soc. Ital. Sci. Nat. 17: 248. 1875 · Shining fir-moss, lycopode brillant

Lycopodium lucidulum Michaux, Fl. Bor.-Amer. 2: 284. 1803; *Urostachys lucidulus* (Michaux) Nessel

Shoots erect, indeterminate, 14–20(–100) cm, becoming long-decumbent, with long, trailing, senescent portion turning brown; juvenile and mature portions strictions due to formation of winter bud; juvenile growth erect. **Leaves** spreading to reflexed, dark green, lustrous; largest leaves narrowly obovate, leaves broadest at or above middle, 7–11 mm, margins papillate, teeth 1–8, irregular, large; smallest leaves (at annual constrictions) narrowly lanceolate, 3–6 mm; stomates on abaxial surface only. **Gemmiferous branchlets** produced in 1 pseudowhorl at end of each annual growth cycle; gemmae 4–6 × 3–6 mm, lateral leaves 1.5–2.5 mm wide, broadly obtuse with distinct mucro. **Spores** 23–29 μm. $2n = 134$.

Terrestrial in shaded conifer forests and mixed hardwoods, rarely on rock on shady mossy acidic sandstone; 0–1800 m; St. Pierre and Miquelon; Man., N.B., Nfld., N.S., Ont., P.E.I., Que.; Ala., Ark., Conn., Del., Ga., Ill., Ind., Iowa, Ky., Maine, Md., Mass., Mich., Minn., Mo., N.H., N.J., N.Y., N.C., Ohio, Pa., R.I., S.C., Tenn., Vt., Va., W.Va., Wis.

Huperzia ×*bartleyi* (Cusick) Kartesz & Gandhi, a sterile hybrid between *H. lucidula* and *H. porophila*, occurs throughout the range of *H. porophila* and is discussed under that species. *Huperzia* ×*buttersii* (Abbe) Kartesz & Gandhi is a hybrid between *H. lucidula* and *H. selago*.

4. **Huperzia miyoshiana** (Makino) Ching, Acta Bot. Yunnan. 3(3): 303, 304. 1981 · Pacific fir-moss, lycopode de Miyoshi

Lycopodium miyoshianum Makino, Bot. Mag. (Tokyo) 12: 36. 1898; *L. selago* Linnaeus subsp. *miyoshianum* (Makino) Calder & Roy L. Taylor; *L. selago* Linnaeus var. *miyoshianum* (Makino) Makino; *L. tenuifolium* Herter; *Urostachys miyoshiana* (Makino) Nessel

Shoots erect, determinate, 12–18(–25) cm, clustered to long-decumbent, to 8 cm; leaves in mature portion smaller than leaves in juvenile portion; annual constrictions absent; juvenile growth sharply down-curled forming ± 1/2 circle, leaves at tip appressed to form pointed apex. **Leaves** in juvenile (proximal) portion spreading-reflexed (shade) to appressed-ascending (sun), in mature (distal) portion spreading-ascending (shade) to appressed-ascending (sun), light green to yellow, lustrous; leaves in juvenile portion narrowly lanceolate, parallel-sided, 4.5–7 mm; leaves in mature portion triangular, widest at base, 3.5–5.5 mm; margins entire; stomates present on both surfaces, numerous, 35–80 per 1/2 leaf on adaxial surface. **Gemmiferous branchlets** produced in 2–3 pseudowhorls at end of annual growth; gemmae 3.5–5 × 3–4 mm, lateral leaves 1.25–1.75 mm wide, acute with acuminate tip. **Spores** 25–34 μm.

On rock or terrestrial on moss-covered boulders in talus slopes, cliffs, near waterfalls, marshes in conifer forest; 0–1600(–1800) m; B.C., Nfld.; Alaska, Wash.; Asia in Japan, Korea, Siberia.

5. **Huperzia occidentalis** (Clute) Kartesz & Gandhi, Phytologia 70: 201. 1991

Lycopodium lucidulum Michaux forma *occidentale* Clute, Fern Bull. 11: 13. 1903; *L. lucidulum* Michaux var. *occidentale* (Clute) L. R. Wilson

Shoots erect, indeterminate, 12–20 cm, becoming long-decumbent, 4–20 cm; leaves in mature portion slightly smaller than in juvenile portion; distinct annual constrictions present; juvenile growth erect. **Leaves** reflexed (juvenile portion) or spreading to reflexed (mature portion), light green to whitish green, lustrous; largest leaves oblanceolate, broadest at 1/2–3/4 length, 6–10 mm; smallest leaves narrowly triangular, broadest at base, 4–7 mm; margins with small papillae; stomates present on both surfaces, numerous (36–80 per 1/2 leaf) on adaxial surface. **Gemmiferous branchlets** produced in 1 pseudowhorl at end of annual growth; gemmae 4–4.5 × 3.5–4 mm; lateral leaves broadly obtuse, widest above middle, 1.25–1.5 mm wide. **Spores** 30–38 μm.

Terrestrial in shaded conifer forests and swamps, often along streams and in marshes; 10–1000(–2000) m; Alta., B.C., Yukon; Alaska, Idaho, Mont., Oreg., Wash.

Huperzia occidentalis is similar to the eastern *H. lucidula* and occupies similar habitats.

6. Huperzia porophila (F. E. Lloyd & L. Underwood) Holub, Folia Geobot. Phytotax. 20: 76. 1985

Lycopodium porophilum F. E. Lloyd & L. Underwood, Bull. Torrey Bot. Club 27(4): 150. 1900; *L. lucidulum* Michaux var. *porophilum* (F. E. Lloyd & L. Underwood) Clute; *L. selago* var. *porophilum* (F. E. Lloyd & L. Underwood) Clute; *Urostachys lucidulus* (Michaux) Nessel var. *porophilus* (F. E. Lloyd & L. Underwood) Nessel

Shoots erect, determinate, occasionally indeterminate, 12–15 cm, clustered to short-decumbent, leaves of mature portion slightly smaller than leaves of juvenile portion; annual constrictions distinct to indistinct; juvenile growth erect. **Leaves** reflexed at base, ascending at stem apex (forming cluster) and spreading for most of stem length, sparse, yellow-green (juvenile portion) to yellow-green to green (mature portion), lustrous; largest leaves lanceolate with roughly parallel sides, 5–8 mm; smallest leaves triangular, widest at base, 3–6 mm; margins almost entire with low papillae or a few large teeth; stomates on both surfaces, few (1–25 per 1/2 leaf) on adaxial surface. **Gemmiferous branchlets** produced in 1–3 pseudowhorls at end of annual growth; gemmae 4–5 × 3–4 mm; lateral leaves 1–1.5 mm wide, acute, widest above middle. **Spores** 29–38 μm.

On damp, shaded, acidic sandstone, rarely on shale or exposed sandstone; 50–1200 m; Ala., Ill., Ind., Iowa, Ky., Minn., Mo., Ohio, Pa., Tenn., Va., W.Va., Wis.

Huperzia porophila occurs in several disjunct populations, most notably in the Driftless Area of Iowa, Minnesota, and Wisconsin. A sterile hybrid between *H.*

porophila and *H. lucidula*, *H.* × *bartleyi*, occurs throughout the range of *H. porophila*. It reproduces vegetatively by gemmae and is common in habitats intermediate between those of the two parents. In Ohio the hybrid may occupy sites to which *H. porophila* would otherwise be restricted, thus reducing the sites available for colonization by *H. porophila* (A. W. Cusick 1987).

7. Huperzia selago (Linnaeus) Bernhardi ex Schrank & Martius, Hort. Reg. Monac., 3. 1829 · Northern fir-moss, lycopode sélagine

Lycopodium selago Linnaeus, Sp. Pl. 2: 1102. 1753; *Plananthus selago* (Linnaeus) Palisot de Beauvois; *Urostachys selago* (Linnaeus) Herter

Shoots erect, indeterminate, 8–12 cm, becoming short-decumbent; leaves of mature portion slightly smaller than leaves of juvenile portion (more pronounced in sun form); indistinct annual constrictions present (more pronounced in shade form); juvenile growth erect. **Leaves** spreading-ascending (shade) to appressed-ascending (sun) in mature portion, more reflexed in juvenile portion, green (shade) to yellow-green (sun), lustrous; largest leaves triangular, widest at base, 4–7.5 mm; smallest leaves lanceolate, 3.5–5 mm; margins almost entire, papillate; stomates on both surfaces, numerous on adaxial surface, 30–90 per 1/2 leaf. **Gemmiferous branchlets** produced in 1 pseudowhorl at end of annual growth; gemmae 4–5 × 3–4.5 mm, lateral leaves 1.5–2 mm wide, broadly acute. **Spores** 29–37 μm. $2n = 268$.

Terrestrial in sandy borrow pits, ditches, lakeshore swales, and conifer swamps, rarely on acidic, igneous rock or calcareous coast cliffs; 0–700 m, rarely to 1600 m; St. Pierre and Miquelon; Alta., B.C., Man., N.B., Nfld., N.W.T., N.S., Ont., P.E.I., Que., Sask., Yukon; Conn., Maine, Mass., Mich., Minn., N.H., N.Y., Ohio, R.I., Vt., Wis.; Europe; Asia.

Plants from Greenland formerly identified as *Huperzia selago* are *H. appalachiana*.

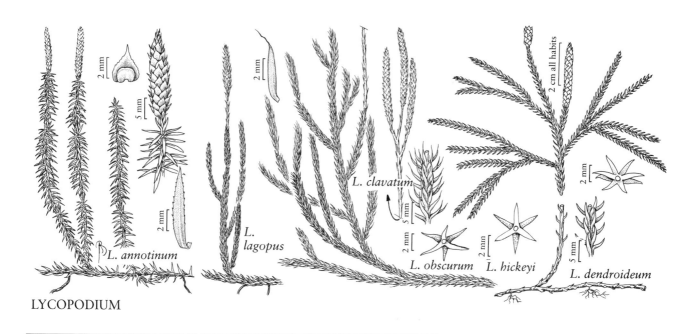

LYCOPODIUM

3. LYCOPODIUM Linnaeus, Sp. Pl. 2: 1100. 1753; Gen. Pl. ed. 5, 486. 1754 · Club-moss

[Greek *lykos,* wolf, and *pous, podes,* foot; in reference to the resemblance of the branch tips to a wolf's paw]

Plants mainly trailing on ground. **Roots** emerging from point of origin on underside of main stems. **Horizontal stems** on substrate surface or subterranean, long-creeping. **Upright shoots** scattered along horizontal stem, 5–16 mm diam., round or flat in cross section, unbranched or with 1–4 lateral branchlets. **Leaves** not imbricate, linear to linear-lanceolate; leaves on horizontal stems scattered, appressed, membranous; leaves on lateral branchlets mostly 6-ranked or more, monomorphic with few exceptions, appressed, ascending to spreading, margins entire to dentate. **Gemmiferous branchlets and gemmae** absent. **Strobili** single and sessile or multiple and pedunculate, apex blunt to acute; peduncle, when present, conspicuously leafy; sporophylls extremely reduced, much shorter than peduncle or stem leaves. **Sporangia** reniform. **Spores** reticulate, sides at equator convex, angles acute. **Gametophytes** nonphotosynthetic, mycorrhizal, subterranean, flat and irregularly button-shaped, with ring meristem around circumference. $x = 34$.

Species 15–25 (6 in the flora): mainly temperate and subarctic.

In striking contrast to *Diphasiastrum, Huperzia,* and *Lycopodiella,* interspecific hybridization is practically unknown in *Lycopodium.* Many of the species now recognized in *Lycopodium* have been segregated from *Lycopodium clavatum, L. annotinum,* and *L. jussiaei* Desvaux ex Poiret. The three groups given in the key below should probably be treated as subgenera.

SELECTED REFERENCES Hickey, R. J. 1977. The *Lycopodium obscurum* complex in North America. Amer. Fern J. 67: 45–49. Wagner, W. H. Jr., J. M. Beitel, and R. C. Moran. 1989. *Lycopodium hickeyi*: A new species of North American clubmoss. Amer. Fern J. 79: 119–121.

1. Strobili pedunculate; upright shoots with 2–5 branches, not treelike; leaves with hair tips
1–4 mm (these may fall off early, but remain at shoot apices) (*L. clavatum* group).
    2. Strobili mostly solitary on peduncle, if paired then nearly lacking pedicels; leaves 3–5
mm, ascending to appressed; branches 2–3(–4), mostly upright. . . . . 2. *Lycopodium lagopus*
    2. Strobili 2–5, borne on loosely alternate pedicels, 0.5–0.8 cm; leaves 4–6 mm, spread-
ing to somewhat ascending; branches 3–6, mostly oblique or spreading. . . . . 3. *Lycopodium clavatum*
1. Strobili sessile; upright shoots either unbranched or much branched to produce treelike
habit; leaves lacking hair tips.
    3. Strobili single at top of upright shoot; shoot unbranched or branched 1–2 times; hori-
zontal stems on substrate surface (*L. annotinum* group). . . . . 1. *Lycopodium annotinum*
    3. Strobili 1–7 at top of many-branched, upright, treelike shoot; horizontal stems subter-
ranean (*L. dendroideum* group).
      4. Lateral shoots flat in cross section, leaves unequal in size, lateral leaves spreading
and twisted, adaxial surfaces facing upward, proximal leaves much reduced; leaves
on main axis dark green, tightly appressed. . . . . 6. *Lycopodium obscurum*
      4. Lateral shoots round in cross section, leaves equal in size, none twisted, adaxial leaf
surfaces all facing stem, proximal leaves not reduced; leaves on main axis light or
dark green, spreading or appressed.
        5. Leaf ranks 1 on upperside of lateral branch, 2 on each side, and 1 on underside;
leaves of main axis below branches dark green, tightly appressed, soft to touch.
. . . . 5. *Lycopodium hickeyi*
        5. Leaf ranks 2 on top of lateral branch, 1 on each side, and 2 on underside; leaves
of main axis below branches pale green, spreading, prickly to touch. . . . . . .
. . . . 4. *Lycopodium dendroideum*

Lycopodium annotinum group

1. Lycopodium annotinum Linnaeus, Sp. Pl. 2: 1103.
1753 · Bristly club-moss, lycopode innovant

Horizontal stems on substrate surface. **Upright shoots** clustered, mainly unbranched or sparsely branching mainly at base, 1.2–1.6 cm diam.; annual bud constrictions abrupt and conspicuous. **Lateral branchlets** few and like upright shoots but annual bud constrictions absent. **Leaves** spreading to reflexed, dark green, linear-lanceolate, (2.5–)5–8 × 0.6–1.2 mm; margins closely and shallowly dentate mainly in distal 1/2; apex sharply pointed, lacking hair tip. **Strobili** solitary, sessile on shoots, 15–30 × 3.5–4.5 mm. **Sporophylls** (1.5–)3.5 × 0.7(–2) mm, abruptly narrowed to pointed tip. $2n = 68$.

Swampy or moist coniferous forests, mountain forests, and exposed grassy or rocky sites; 0–1850 m;

Greenland; St. Pierre and Miquelon; Alta., B.C., Man., N.B., Nfld., N.W.T., N.S., Ont., P.E.I., Que., Sask., Yukon; Alaska, Ariz., Colo., Conn., Idaho, Ky., Maine, Md., Mass., Mich., Minn., Mont., N.H., N.J., N.Mex., N.Y., N.C., Ohio, Oreg., Pa., R.I., Tenn., Utah, Vt., Va., Wash., Wis., Wyo.

This widespread and common club-moss has been divided into various forms or varieties, some of which have been treated as species. Present evidence supports the hypothesis that these are environmentally induced forms, the most distinctive of which has been called *Lycopodium annotinum* var. *alpestre* C. Hartman, with leaves only 2.5–6 mm, very leathery, entire-margined, and appressed. Plants intermediate between this and *L. annotinum* var. *annotinum* is a form that has been called var. *pungens* (Bachelot de la Pylaie) Desvaux, an invalid name. Both are found in cold, bleak, northern or high elevation habitats. The species should be studied in detail to determine whether it contains any groups that should be recognized taxonomically.

Lycopodium clavatum group

2. Lycopodium lagopus (Laestadius ex C. Hartman) G. Zinserling ex Kuzeneva-Prochorova, Fl. Murmansk. Obl. 1: 80. 1953 · One-cone club-moss, lycopode patte-de-lapin

Lycopodium clavatum Linnaeus var. *lagopus* Laestadius ex C. Hartman, Handb. Skand. Fl. ed. 7, 313. 1858; *L. clavatum* var. *brevispicatum* Peck; *L. clavatum* var. *integerrimum* Spring; *L. clavatum* var. *megastachyon* Fernald & Bissel; *L. clavatum* var. *monostachyon* Hooker & Greville

Horizontal stems on substrate surface. **Upright shoots** clustered, shoots 0.5–0.8 cm diam., dominant main shoot branches 2–3(–4), mostly in lower 1/2. **Lateral branchlets** few and like upright shoots; annual bud constrictions abrupt and conspicuous, shoots 0.5–0.8 cm wide, branches mostly erect. **Leaves** ascending to appressed, medium green, 3–5 × 0.4–0.7 mm; margins entire; apex with narrow hair tip 1–3 mm. **Peduncles** 3.5–12.5 cm, with remote pseudowhorls of appressed leaves, unbranched. **Strobili** solitary (if double, usually nearly sessile), 20–55 × 3–5 mm. **Sporophylls** 1.5–2.5 mm, apex rather gradually reduced to hair tip. $2n = 68$.

More or less exposed, grassy fields and openings in second-growth woods; 50–1500 m; Greenland; Alta., B.C., Man., N.B., Nfld., N.W.T., N.S., Ont., P.E.I., Que., Sask., Yukon; Alaska, Maine, Mich., Minn., N.H., N.Y., Vt., Wis.; Eurasia.

Lycopodium lagopus is generally more northern than its sister species, *L. clavatum* (W. J. Cody and D. M. Britton 1989). Where they come together, however, they can grow side by side (even in southern Michigan) and maintain their distinctions.

3. Lycopodium clavatum Linnaeus, Sp. Pl. 2: 1101. 1753 · Common club-moss, lycopode claviforme

Lycopodium clavatum var. *subremotum* Victorin

Horizontal stems on substrate surface. **Upright shoots** clustered, 0.6–1.2 cm diam., dominant main shoot with 3–6 branches mostly in lower 1/2. **Lateral branchlets** few and like upright shoots; annual bud constrictions abrupt, branchlets mostly spreading. **Leaves** spreading, often somewhat ascending in distal 1/3 of branches, medium green, linear, 4–6 × 0.4–0.8 mm; margins entire; apex with narrow hair tip 2.5–4 mm. **Peduncles** 3.5–12.5 cm, with remote pseudowhorls of appressed leaves, loosely branched into 2–5 alternate stalks, 0.5–0.8 cm. **Strobili** 2–5 on alternate stalks (if double, usually with stalks 5–8 mm), 15–25 × 3–6 mm. **Sporophylls** 1.5–2.5 mm, apex abruptly reduced to hair tip. $2n = 68$.

Fields and woods; 100–1800 m; St. Pierre and Miquelon; B.C., Man., N.B., Nfld., N.S., Ont., P.E.I., Que., Sask.; Alaska, Calif., Conn., Ga., Idaho, Ill., Ind., Ky., Maine, Md., Mass., Mich., Minn., Mont., N.H., N.J., N.Y., N.C., Ohio, Oreg., Pa., R.I., Tenn., Vt., Va., Wash., W.Va., Wis.; Mexico; West Indies; Central America; South America; Europe; Asia; Africa; Pacific Islands.

Plants found in eastern North America have been called *Lycopodium clavatum* var. *clavatum;* those in the western part of the range, which have been called *L. clavatum* var. *integrifolium* Goldie, are distinguished by early shedding of the characteristic hairs on the leaf tips.

Lycopodium dendroideum group

4. Lycopodium dendroideum Michaux, Fl. Bor.-Amer. 2: 282. 1803 · Prickly tree club-moss, lycopode dendroïde

Lycopodium obscurum Linnaeus var. *dendroideum* (Michaux) D. C. Eaton

Horizontal stems subterranean. **Upright shoots** treelike, many branched, branchlets numerous and strongly differentiated; annual bud constrictions absent; leaves spreading, pale green below lateral branchlets, prickly needlelike, 3.5–4 × 0.9–1 mm. **Lateral branchlets** round in cross section, 5–8 mm diam.; annual bud constrictions inconspicuous; leaves spreading to ascending, pale green, in 6 ranks, 2 upperside, 2 lateral, and 2 underside, equal in size, linear, 2.4–5.5 × 0.5–1.2 mm; margins entire; apex acuminate, lacking hair tip. **Strobili** sessile, 1–7 on tip of upright shoot, 12–55 mm. **Sporophylls** 3.5 × 3.5 mm, apex short, acute, abruptly narrowing. $2n = 68$.

Dry woodlands and second-growth shrubby areas; 50–1800 m; St. Pierre and Miquelon; Alta., B.C., Man., N.B., Nfld., N.W.T., N.S., Ont., P.E.I., Que., Sask., Yukon; Alaska, Conn., Idaho, Iowa, Maine, Mass., Mich., Minn., Mo., Mont., N.H., N.Y., Pa., S.Dak., Vt., Va., Wash., W.Va., Wis., Wyo.; Asia.

5. Lycopodium hickeyi W. H. Wagner, Beitel, & R. C. Moran, Amer. Fern J. 79: 119–121. 1989 · Hickey's tree club-moss, lycopode de Hickey

Lycopodium obscurum Linnaeus var. *isophyllum* Hickey

Horizontal stems subterranean. **Upright shoots** treelike, many branched, branchlets numerous and strongly differentiated; annual bud constrictions absent; leaves on main axis below lateral branchlets tightly appressed, dark green, needlelike, 3.5–4.5 × 0.5–0.6 mm, soft. **Lateral branchlets** round in cross section, 4–7 mm diam.; annual bud constrictions inconspicuous; leaves ascending, in 6 ranks, 1 on upperside, 4 lateral, and 1 on underside, equal in size, linear, widest in middle; margins entire; apex acuminate, lacking hair tip. **Strobili** sessile, 1–7 per upright shoot, 15–65 mm. **Sporophylls** 3–3.5 × 2–2.5 mm, apex long, gradually narrowing to tip. $2n = 68$.

Mainly in hardwood forests and second-growth, shrubby habitats; 0–1600 m; N.B., Nfld., N.S., Ont., P.E.I., Que., Sask.; Conn., Ind., Ky., Maine, Md., Mass., Mich., Minn., N.H., N.J., N.Y., N.C., Ohio, Pa., R.I., Tenn., Vt., Va., W.Va., Wis.

The range of *Lycopodium hickeyi* overlaps with that of *L. obscurum* and extends considerably north and west of that species. Although the arrangement of the leaf ranks is similar to that of *L. obscurum*, the leaf dimorphy and the ascending orientation and absence of twisting of the leaves are diagnostic. Where ranges of two or three species overlap, individual species retain their identities, indicating that their critical differences have a genetic basis.

6. Lycopodium obscurum Linnaeus, Sp. Pl. 2: 1102. 1753 · Flat-branched tree club-moss, lycopode obscur

Horizontal stems subterranean. **Upright shoots** treelike, many branched, branchlets numerous and strongly differentiated; annual bud constrictions absent; leaves on main stem below lateral branchlets tightly appressed, needlelike, 3.5–4.5 × 0.6–0.7 mm, soft. **Lateral branchlets** flat in cross section, 6–9 mm diam.; annual bud constrictions inconspicuous. **Leaves** in 6 ranks, 1 on upperside, 4 lateral, ascending-spreading, 2.5–5.5 × 0.5–1.2 mm, and 1 on underside, much smaller than others, ascending; margins entire; apex lacking hair tip. **Strobili** 1–6 at tip of shoot, sessile, 12–60 mm. **Sporophylls** 3 × 2 mm, apex long, gradually narrowing. $2n = 68$.

Rich hardwood forests and successional shrubby areas; 0–1600 m; N.B., N.S., Ont., Que.; Ala., Conn., Del., Ga., Ky., Maine, Md., Mass., Mich., Ohio, Pa., N.H., N.J., N.Y., N.C., R.I., S.C., Tenn., Vt., Va., W.Va., Wis.

4. DIPHASIASTRUM Holub, Preslia 47: 104. 1975 · False *Diphasium* [*Diphasium,* a generic name, and *-astrum,* incomplete resemblance]

Plants terrestrial, mainly trailing on ground with erect shoots. **Roots** emerging immediately on underside of main stems. **Horizontal stems** on substrate surface to subterranean, long-creeping. **Upright shoots** quadrate to flattened (except in *D. sitchense*), 2–6 mm diam., usually with 2–5 lateral branchlets on main erect stem; lateral branchlets leafy, ± flat in cross section. **Leaves** on horizontal stems somewhat distant, appressed, linear to lanceolate, thin, scalelike; leaves on ultimate branchlets appressed to divergent, linear-lanceolate to nearly filiform, usually almost scalelike and mostly imbricate, in 4 ranks, leaves of lateral ranks larger, more spreading than those of upperside and underside ranks (except in *Diphasiastrum sitchense* with 5 ranks of uniform nonimbricate leaves). **Gemmiferous branchlets and gemmae** absent. **Strobili** solitary and sessile or multiple and stalked, apex blunt, acute, or with sterile apical projection; peduncle conspicuously leafy; sporophylls shorter than peduncle leaves. **Sporangia** reniform. **Spores** reticulate, sides at equator convex, angles acute. **Gametophytes** nonphotosynthetic, mycorrhizal, carrot-shaped, paraphyses absent; ring meristem present. $x = 23$.

Species 15–20 (11 taxa in the flora, including 5 species and 6 fertile hybrids): mainly north temperate and subarctic.

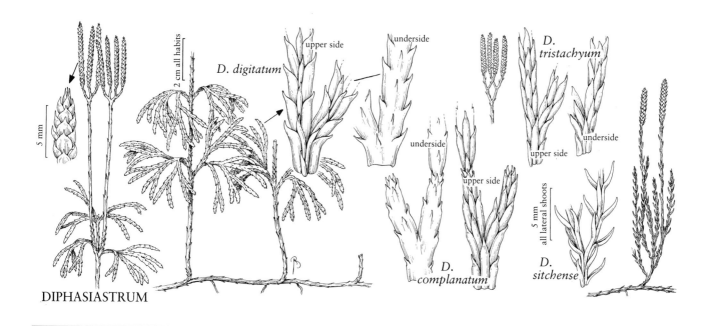

DIPHASIASTRUM

Diphasiastrum is remarkable in North America for its ability to form numerous homoploid, apparently fertile interspecific hybrids, some of which are frequent to common and must be reckoned with in floristic studies. Field and laboratory studies of these hybrids and their parents are needed for determination of the population dynamics of *Diphasiastrum* and to answer questions as to why the parental species retain their genetic identities.

SELECTED REFERENCES Beitel, J. M. 1979. The clubmosses *Lycopodium sitchense* and *L. sabinaefolium* in the upper Great Lakes area. Michigan Bot. 18: 3–13. Holub, J. 1975. *Diphasiastrum*, a new genus in Lycopodiaceae. Preslia 14: 97–100. Wilce, J. H. 1965. Section *Complanata* of the genus *Lycopodium*. Beih. Nova Hedwigia 19: i–ix, 1–233, plate 40.

1. Upright shoots to 18 cm, mostly less than 12 cm including base; strobili mostly sessile.
    2. Ultimate branchlets approximately square in cross section; leaves 4-ranked, strongly overlapping. . . . . 1. *Diphasiastrum alpinum*
    2. Ultimate branchlets round in cross section; leaves 5-ranked, not overlapping. . . . . .
        . . . . 4. *Diphasiastrum sitchense*
1. Upright shoots, 8–50 cm, mostly more than 12 cm, including base; strobili stalked.
    3. Ultimate branchlets cordlike, nearly square in cross section, usually bluish glaucous; underside leaves approximately equal in size to lateral and upperside leaves. . . . . . .
        . . . . 5. *Diphasiastrum tristachyum*
    3. Ultimate branchlets narrowly bladelike, flat in cross section, usually green; underside leaves much smaller than lateral and upperside leaves.
        4. Branchlets irregular, with conspicuous annual bud constrictions; peduncles, if present, regularly forked; strobili mostly 15–25 mm, lacking sterile tips. . . . .
            . . . . 2. *Diphasiastrum complanatum*
        4. Branchlets very regularly fan-shaped, lacking conspicuous annual bud constrictions; peduncles mostly branching abruptly at base to produce false whorl of strobili; strobili mostly 20–35 mm, many with sterile tips. . . . . 3. *Diphasiastrum digitatum*

1. **Diphasiastrum alpinum** (Linnaeus) Holub, Preslia 47: 107. 1975 · Alpine club-moss, lycopode alpin

Lycopodium alpinum Linnaeus, Sp. Pl. 2: 1104. 1753

Horizontal stems mainly shallowly buried, 0.5–3 mm wide, sometimes emerging, 1.1–2.2 mm wide; leaves appressed, spatulate to lanceolate, 1.5–3.8 × 0.5–1.4 mm, apices truncate. **Upright shoots** 6–14 cm, clustered, fasciculate, branching successively 3–5 times; leaves on upright main stem ascending, deltate-ovate, 3.5–4 × 0.8 mm, apices needlelike. **Branchlets** square in cross section, 1.8–4 mm wide, annual bud constrictions abrupt and conspicuous; underside often glaucous, concave; upperside green, dull to faintly shiny, convex. **Leaves** on branchlets 4-ranked, overlapping; upperside leaves appressed, lanceolate, 3–5.8 mm, free portion of blades 1.7–2.9 × 0.1–1.1 mm; lateral leaves strongly divergent, 3.3–6.5 × 1.8–2.4 mm, margins revolute; underside leaves well developed, perpendicular to stem, 1.3–3.3 × 0.6–1.3 mm, unique in genus in having base contracted, blade flaring, and margins becoming parallel. **Peduncles** absent. **Strobili** solitary, 5–30 × 2–4 mm, sterile tips absent. **Sporophylls** deltate to nearly cordate, 2.2–3.5 × 1.6–3 mm, apices gradually tapering. $2n = 46$.

Dry conifer or mixed forests, grassy mountain slopes; Greenland; Alta., B.C., Nfld., N.W.T., Que., Yukon; Alaska, Colo., Idaho, Mont., Wash.; Europe; Asia in Japan.

The branchlet leaves of *Diphasiastrum alpinum* are unique in the genus, and the trowel-shaped underside leaves with their flared and rolled blades and contracted bases are particularly unusual. The leaves of the other North American species are much simpler in shape and contour.

2. **Diphasiastrum complanatum** (Linnaeus) Holub, Preslia 47: 108. 1975 · Northern running-pine, lycopode aplati

Lycopodium complanatum Linnaeus, Sp. Pl. 2: 1104. 1753; *L. complanatum* var. *canadense* Victorin

Horizontal stems on substrate surface or shallowly buried in litter, 1.1–2.2 mm wide; leaves appressed, linear to narrowly lanceolate, 1.4–4 × 0.5–1.2 mm, apices acute. **Upright shoots** 8–44 cm, branching irregularly successively to 5 times; leaves on upright main stem appressed with decurrent base, narrowly lanceolate, 1.2–3.2 × 0.5–1.1 mm, apex acute to acuminate. **Branchlets** flat in cross section, narrowly bladelike, 1.8–4 mm wide, annual bud constrictions abrupt and conspicuous; upperside green, faintly shiny, flat; underside dull, pale, flat. **Leaves** on branchlets 4-ranked; upperside leaves appressed, linear-lanceolate, free portion of blades 0.7–2 × 0.5–1.2 mm; lateral leaves appressed, 2.6–7.3 × 0.8–2.1 mm; underside leaves weakly developed, appressed, narrowly deltate, 0.7–1.5 × 0.4–0.9 mm. **Peduncles** 1–2 on each upright shoot, 0.5–8.5 × 0.4–0.9 cm; leaves spirally arranged to nearly whorled, linear-lanceolate, 1.4–4.1 × 0.4–1 mm, apex acute to blunt. **Stalks** forked at uniform distances. **Strobili** 1–2(–4), 8.3–32 × 2–3 mm, apex blunt, sterile tip absent. **Sporophylls** broadly deltate to nearly cordate, 2–3 × 2–2.4 mm, apex abruptly tapering. $2n = 46$.

Dry open coniferous or mixed forest alpine slopes, 0–2000 m; Greenland; St. Pierre and Miquelon; Alta., B.C., Man., N.B., Nfld., N.W.T., N.S., Ont., P.E.I., Que., Sask., Yukon; Alaska, Idaho, Maine, Mich., Minn., Mont., N.H., N.Y., Vt., Wash., Wis., Wyo.; circumboreal.

Diphasiastrum complanatum forms a hybrid with *D. digitatum* that is seemingly uncommon and has never received a binomial designation. It is probably far more common than collections indicate, however. Superficially, the hybrid resembles both parents and is often confused with them. Collections are known from Ontario, Quebec, Connecticut, Maine, Michigan, Minnesota, New Hampshire, Vermont, and Wisconsin.

3. **Diphasiastrum digitatum** (Dillenius ex A. Braun) Holub, Preslia 47: 108. 1975 · Southern running-pine, lycopode en éventail

Lycopodium digitatum Dillenius ex A. Braun, Amer. J. Sci. Arts, ser. 2, 6: 81. 1848; *L. complanatum* Linnaeus var. *flabelliforme* Fernald; *L. flabelliforme* (Fernald) Blanchard

Horizontal stems on substrate surface, 1.3–2.7 mm wide; leaves appressed to ascending, linear to narrowly lanceolate, 1.8–4.5 × 0.6–1.2 mm, apex acute, scarious, often lost. **Upright shoots** 15–50 cm, branching regularly successively to 3 times; leaves appressed with decurrent base, subulate, 1.8–3.5 × 0.6–1 mm, apex acute. **Branchlets** flat in cross section, narrowly bladelike, 2.8–3.9 mm wide, annual bud constrictions very rare; underside dull, pale, flat; upperside green, flat, shiny. **Leaves** of branchlets 4-ranked; upperside leaves appressed, linear-lanceolate, free portion of blade 0.7–1.5 × 0.5–0.9 mm; lateral leaves appressed to spreading (spreading especially in juvenile stages), 3.1–5.5 × 1–2 mm; underside leaves very weakly developed, spreading, narrowly deltate, 0.3–1 × 0.3–0.7 mm, apex pointed. **Peduncles** mostly 2, 4.4–12.5 × 0.1–1.3 cm; leaves usually somewhat whorled, linear-lanceolate to nearly filiform, 2–3.3 × 0.5–0.9 mm, apex blunt to acute. **Stalks** mostly pseudowhorled, 2-forked, forks basal. **Strobili** 2–4 per upright shoot, 14–40 × 2–3 mm exclusive of elongate sterile tip; sterile tips to 11 mm (occurring on ca. 50% of specimens), apex blunt to acute if sterile tip is absent. **Sporophylls** deltate, 1.7–2.6 × 1.8–2.8 mm, apex abruptly tapering. $2n = 46$.

Coniferous and hardwood forests and second growth, shrubby or open fields; 0–1500 m; St. Pierre and Miquelon; N.B., Nfld., N.S., Ont., P.E.I., Que.; Ala., Ark., Conn., Del., D.C., Ga., Ill., Ind., Iowa, Ky., Maine, Md., Mass., Mich., Minn., Mo., N.H., N.J., N.Y., N.C., Ohio, Pa., R.I., S.C., Tenn., Va., W.Va., Wis.

An endemic in eastern North America, *Diphasiastrum digitatum* is the most abundant species of *Diphasiastrum* on the continent, much used for decoration as wreaths. It was long confused with the circumboreal *D. complanatum*.

4. Diphasiastrum sitchense (Ruprecht) Holub, Preslia 47: 108. 1975 · Sitka club-moss, lycopode de Sitka

Lycopodium sitchense Ruprecht, Beitr. Pflanzenk. Russ. Reiches 3: 30. 1845

Horizontal stems on substrate surface or shallowly buried, 1–2.7 mm wide; leaves appressed, broadly lanceolate, 1.8–3.2 × 0.5–1 mm, apex blunt. **Upright shoots** clustered and branching mostly at base, 5.5–17.5 cm; leaves appressed, broadly lanceolate, 1.8–3.2 × 0.5–1 mm, apex acuminate. **Branchlets** dark green, somewhat shiny, round in cross section, 1.7–2.5 mm wide, annual bud constrictions inconspicuous. **Leaves** on branchlets monomorphic, 5-ranked, not overlapping, appressed to spreading-ascending, incurved, free portion of blades 3.4–5.6 × 0.4–0.9 mm, widest at middle, apex sharply pointed. **Peduncles** absent or rarely 1 cm. **Stalks** absent. **Strobili** solitary on upright shoots, 4.5–38 × 3–5 mm, gradually narrowing to rounded tip. **Sporophylls** deltate, 1.8–3.6 × 1.7–2.8 mm; apex rounded. $2n = 46$.

Alpine meadows, open rocky barrens, conifer woods; 200–2000 m; Greenland; St. Pierre and Miquelon; Alta., B.C., Man., N.B., Nfld., N.S., Ont., P.E.I., Que., Sask., Yukon; Alaska, Idaho, Maine, Mont., N.H., N.Y., Oreg., Vt., Wash.; Asia in Kamchatka, Japan.

The mature shoots in *Diphasiastrum sitchense* resemble the juvenile phases of the other species. The

unique, round, 5-ranked leaves may represent an early developmental state.

The hybrid *Diphasiastrum alpinum* × *sitchense* is very rare. It is known from Greenland, British Columbia, Newfoundland, Montana, Oregon, and Washington. Specimens of *D. sitchense* from Greenland, Newfoundland, and Washington cited by J. H. Wilce (1965) are actually this hybrid.

5. Diphasiastrum tristachyum (Pursh) Holub, Preslia 47: 108. 1975 · Blue ground-cedar, lycopode à trois épis

Lycopodium tristachyum Pursh, Fl. Amer. Sept. 2: 653. 1814

Horizontal stems deeply (5–12 cm) buried, 1.5–3.2 mm wide; leaves spatulate to somewhat obovate, 1.8–3.5 × 1.1–1.5, apex faintly erose to irregularly lobed. **Upright shoots** clustered, branching near base, 17–36 cm; leaves monomorphic, appressed, subulate, 1.9–3.4 × 0.6–1 mm, apex acute. **Branchlets** square with rounded angles in cross section, 1–2.2 mm wide, annual bud constrictions abrupt and conspicuous; upperside convex, bluish to whitish green. **Leaves** on branchlets 4-ranked, upperside leaves appressed, needlelike, free portion of blade 1–1.7 × 0.5–0.9 mm; lateral leaves appressed, 3.4–7.2 × 1.1–2 mm; underside leaves appressed, somewhat decurrent, 1–2 × 0.4–0.7 mm. **Peduncles** (1–)3, 4–15 × 0.4–1 mm; leaves remote, scattered, decurrent, free tips ascending, subulate, 2–3 × 0.2–0.25 mm. **Stalks** mostly formed by successive forking of peduncle, branches uniformly spaced. **Strobili** (2–)3–4(–7), 10–28 × 2–3 mm, apex round-tipped, sterile tips absent. **Sporophylls** deltate, 2.2–3.5 × 1.6–3 mm, apex gradually tapering. $2n = 46$.

Sterile, acidic soils in open conifer and oak forests; 50–1800 m; Man., N.B., Nfld., N.S., Ont., P.E.I., Que.; Ala., Conn., Del., D.C., Ga., Ind., Ky., Maine, Md., Mass., Mich., Minn., Mo., N.H., N.J., N.Y., N.C., Ohio, Pa., R.I., S.C., Tenn., Vt., Va., W.Va., Wis.; Europe; Asia in w China.

The distinctive *Diphasiastrum tristachyum* has narrow, rounded branches and dull, bluish white color. It is a parent in more hybrid combinations than any other North American *Diphasiastrum*.

The reticulogram shows the known pattern of interspecific hybridization in *Diphasiastrum*. The hybrids are discussed in detail by J. H. Wilce (1965), and their cytology is summarized by F. S. Wagner (1992). The best known North American hybrids are the four involving *D. tristachyum*. All of the hybrids have apparently normal meiosis and spores.

Diphasiastrum ×*zeilleri* (Rouy) Holub (= *D. com-*

Diphasiastrum

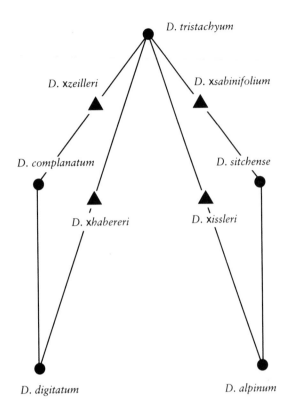

Relationships in *Diphasiastrum*. Solid circles represent parental taxa, and solid triangles represent hybrids presumed to be fertile.

planatum × *tristachyum*) is a frequent plant in areas of distributional overlap between the parents, especially in north central and western Minnesota jackpine forests.

Diphasiastrum × *habereri* (House) Holub (= *D. digitatum* × *tristachyum*) has been overlooked and confused with both parents in zones of overlap. It is found occasionally to frequently in habitats like those of the parents, not necessarily growing close to them.

Diphasiastrum × *issleri* (Rouy) Holub (= *D. alpinum* × *tristachyum*) is a rare hybrid in North America, reported only from Maine, but much more widespread in Europe.

Diphasiastrum × *sabinifolium* (Willdenow) Holub (= *D. sitchense* × *tristachyum*) is widespread and frequent in eastern Canada. This hybrid is commonly confused with *D. sitchense*. It is highly variable, and some individuals approach one or the other parent in morphology (W. J. Cody and D. M. Britton 1989). In the flora, the populations are mainly disjunct from the main range of *D. sitchense*, including those in Ontario, Quebec, Maine, Michigan, New Hampshire, New York, Pennsylvania, and Vermont.

Two other North American nothospecies are *Diphasiastrum complanatum* × *digitatum* and *D. alpinum* × *sitchense*.

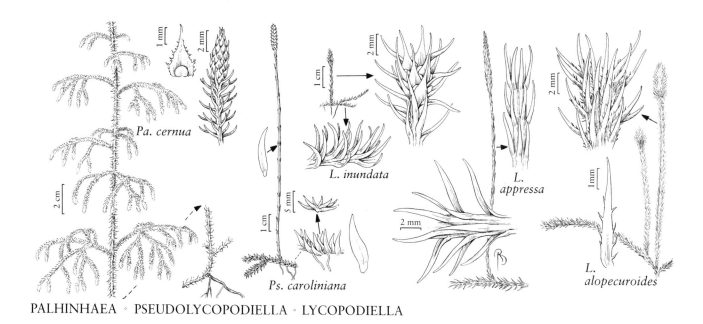

Pa. cernua

L. inundata

L. appressa

Ps. caroliniana

L. alopecuroides

PALHINHAEA · PSEUDOLYCOPODIELLA · LYCOPODIELLA

5. PALHINHAEA Vasconcellos & Franco, Bol. Soc. Brot., ser. 2, 41: 24. 1967 · Nodding club-moss [for R. T. Palhinha (1871–1950), a Portuguese botanist]

Plants on substrate surface, resembling small, many-branched tree. **Roots** emerging immediately on underside of stems. **Horizontal stems** branching, stolonlike, rooting where they touch ground. **Upright shoots** usually 1 to each arch of horizontal stems, to 3.5 mm diam., unequally, dichotomously branched with well-differentiated lateral branchlets much branched. **Leaves** not in distinct ranks, not imbricate, monomorphic, generally ascending, linear-needlelike, margin entire. **Gemmiferous branchlets and gemmae** absent. **Strobili** solitary, sessile, nodding to pendent, tip blunt; sporophylls smaller than vegetative leaves. **Sporangia** nearly globose. **Spore** surface rugulate; sides convex at equator. **Gametophytes** photosynthetic, on substrate surface, pincushion-shaped; ring meristem absent.

Species 10–15 (1 in the flora): mainly pantropical and subtropical.

Palhinhaea is a very common genus in Lycopodiaceae worldwide.

1. Palhinhaea cernua (Linnaeus) Vasconcellos & Franco, Bol. Soc. Brot., ser. 2, 41: 25. 1967 · Nodding club-moss

Lycopodium cernuum Linnaeus, Sp. Pl. 2: 1103. 1753
Roots clustered at soil contacts of horizontal stem, 5–30 × 0.4–0.6 cm. **Horizontal stems** with leaves remote, linear-needlelike, recurved at base, upcurved at apex, 2.9–3.1 × 0.1–0.15 mm. **Upright shoots** to 45 × 3.5 mm, gradually diminishing and branched successively 3 times, 3–12 cm from base to form complex treelike habit.

Lateral branchlets spreading-ascending, drooping at tips, 0.2–0.4 mm wide, leaves needlelike, recurved basally, upcurved apically, 2–2.5 × 0.1–0.2 mm, leaves and stems sometimes hairy near strobili. **Strobili** nodding at 60–80° to subtending vegetative branch, 0.4–0.8 × 0.15–0.2 cm. **Sporophylls** trowel-shaped, 1.5–2 × 0.7–0.8 mm, margins fringed, teeth to 0.2 mm, mostly branched.

Wet depressions and ditches in pinelands, road banks; 0–100 m; Ala., Fla., Ga., La., Miss., S.C.

Palhinhaea cernua is a showy plant. This is probably the world's most abundant club-moss. The species overwinters as buried stem tips, the rest of the plant dying.

6. PSEUDOLYCOPODIELLA Holub, Folia Geobot. Phytotax. 18: 441. 1983

· [Greek *pseudo-*, false, and Latin *-ella*, diminutive, meaning false little *Lycopodium*, because of its resemblance to *Lycopodiella*]

Plants creeping on wet substrates. **Roots** emerging immediately on underside of stems. **Horizontal stems** on substrate surface, short-creeping. **Upright shoots** not branched, forming sparsely leafy peduncles scattered along horizontal stem, nearly naked, 9–11 mm diam. **Leaves** of horizontal stems not in distinct ranks, not imbricate, dimorphic, margins entire; lateral leaves narrowly linear, nearly subulate, median leaves 1/2–2/3 shorter than lateral leaves, ascending. **Gemmiferous branchlets and gemmae** absent. **Strobili** solitary, not conspicuously differentiated from peduncle, tip blunt; peduncle nearly naked with scattered minute leaves, 0.9–3 mm diam.; sporophylls much shorter than peduncle leaves. **Sporangia** reniform. **Spores** rugulate, sides at equator convex, angles acute. **Gametophytes** photosynthetic, on substrate surface, tuber-shaped and lobed; ring meristem absent. $x = 35$.

Species 12 (1 in the flora): widespread.

SELECTED REFERENCE Bruce, J. G. 1976. Comparative studies of *Lycopodium carolinianum*. Amer. Fern J. 66: 125–137.

1. Pseudolycopodiella caroliniana (Linnaeus) Holub, Folia Geobot. Phytotax. 18: 442. 1983 · Slender bog club-moss

Lycopodium carolinianum Linnaeus, Sp. Pl. 2: 1104. 1753

Horizontal stems flat in cross section, evergreen, tightly anchored by roots, year's growth 3–12 × 0.9–1.1 cm; leaves 3.5–6 × 1.2–2 mm, facing upward, spreading and appearing 2-ranked, broadly adnate, slightly subulate, median leaves 3.5–4 × 0.3–0.6 mm, nearly erect. **Peduncle** 1–(2–3), 5–30 cm, leaves 2–3 × 0.3–0.6 mm, in remote appressed pseudowhorls. **Strobilus** 1.5–9 × 0.3–0.5 mm. $2n = 70, 140$.

Wet meadows, ditches, pinewoods, and bogs, often with sphagnum moss; 0–50 m; Ala., Del., Fla., Ga., Ky., La., Md., Mass., Miss., N.J., N.Y., N.C., Pa., S.C., Tex., Va.; Mexico; Central America; South America; Asia; Africa.

7. LYCOPODIELLA Holub, Preslia 36: 20, 22. 1964 · Bog club-moss [*Lycopodium*, a genus name, and *-ella*, diminutive]

Plants creeping on wet ground. **Roots** emerging immediately on underside of stems. **Horizontal stems** on substrate surface, supine or arching. **Upright shoots** forming very leafy peduncles scattered along horizontal stems, 2–9 mm diam., unbranched. **Gemmiferous branchlets and gemmae** absent. **Strobili** solitary, fully differentiated from peduncle or peduncle not differentiated, tip blunt to ± acute; peduncle leafy, leaves not in distinct ranks, not imbricate, usually monomorphic, linear-lanceolate, margins commonly with a few teeth; sporophylls generally longer than peduncle leaves. **Sporangia** nearly globose. **Spores** rugulate, sides at equator convex, angles acute. **Gametophytes** photosynthetic, on substrate surface, pincushion-shaped; ring meristem absent. $x = 78$.

Species 8–10 (6 in the flora): north temperate region and tropical America.

This concept of *Lycopodiella* excludes the segregate genera *Pseudolycopodiella* (including *Lycopodium carolinianum*) and *Palhinhaea* (including *Lycopodium cernuum*). It has been treated as *Lepidotis* Palisot de Beauvois ex Mirbel, but this is a later name for *Lycopodium*. Species

Lycopodiella

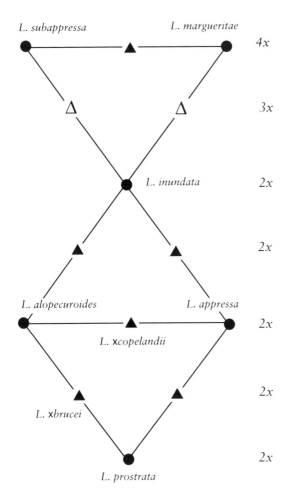

L. subappressa   L. margueritae   4x

3x

L. inundata   2x

2x

L. alopecuroides   L. appressa   2x

L. ×copelandii

2x

L. ×brucei

Relationships in *Lycopodiella*. Solid circles represent parental taxa; solid triangles represent hybrids presumed to be fertile; and open triangles represent sterile hybrids.

2x

L. prostrata

of *Lycopodiella* hybridize readily (see reticulogram). Hybrids between species of the same ploidy level are apparently fertile with normal meiosis and spores, but those between different ploidy levels are sterile (J. G. Bruce 1975).

SELECTED REFERENCES Bruce, J. G. 1975. Systematics and Morphology of Subgenus *Lepidotus* of the Genus *Lycopodium* (Lycopodiaceae). Ph.D. thesis. University of Michigan. Bruce, J. G., W. H. Wagner Jr., and J. M. Beitel. 1991. Two new species of bog clubmoss, *Lycopodiella* (Lycopodiaceae) from southwestern Michigan. Michigan Bot. 30: 3–10. Snyder, L. H. Jr. and J. G. Bruce. 1986. Field Guide to the Ferns and Other Pteridophytes of Georgia. Athens, Ga.

1. Fertile shoots mostly 3.5–6 cm; sporophylls spreading; mainly north of 45° N latitude and high in mountains southward. . . . . 1. *Lycopodiella inundata*
1. Fertile shoots 4–45 cm but mostly 8–35 cm; sporophylls spreading or appressed; mainly south of 45° N latitude at low elevations.
   2. Strobili 0–2 mm thicker than upright shoot; horizontal stem leaves with marginal teeth absent or sparse.
      3. Upright shoots 15–40 × 0.3–0.4 cm; horizontal stem leaves with scattered marginal teeth. . . . . 2. *Lycopodiella appressa*

3. Upright shoots 9–15 × 0.4–0.7 cm; horizontal stem leaves usually without marginal teeth. . . . . 3. *Lycopodiella subappressa*
2. Strobili 3–6 mm thicker than upright shoot; horizontal stem leaves commonly with marginal teeth.

4. Horizontal stems (excluding leaves) 2–4 mm diam., strongly arching above substrate, largest leaves 0.5–0.7 mm wide. . . . . 4. *Lycopodiella alopecuroides*
4. Horizontal stems (excluding leaves) 1.3–2.2 mm diam., prostrate, largest leaves 0.8–1.8 mm wide.

5. Upright shoots 18–35 cm; horizontal stems 1.3–1.6 mm diam.; strobili 15–20 mm wide, with incurved, spreading leaves; sporophyll margins with 1–5 teeth. . . . . 5. *Lycopodiella prostrata*
5. Upright shoots 13–17 cm; horizontal stems 1.8–2.2 mm diam.; strobili 4–9 mm wide with incurved, ascending leaves; sporophyll margins lacking obvious teeth. . . . . 6. *Lycopodiella margueritae*

1. Lycopodiella inundata (Linnaeus) Holub, Preslia 36: 21. 1964 · Northern bog club-moss, lycopode inondé

Lycopodium inundatum Linnaeus, Sp. Pl. 2: 1102. 1753
Horizontal stems flat on ground, 3–12 × 0.5–0.9 cm; stems (excluding leaves) slender, 0.5–0.9 mm diam.; leaves monomorphic, spreading, upcurved, 5–6 × 0.5–0.7 mm, margins entire. **Peduncles** 1(–2) per plant, 3.5–6 × 0.4–0.7 cm; strobilus length 1/2–1/3 total height; leaves spreading, 5–6 × 0.5–0.8 mm, margins rarely toothed. **Strobili** 10–20 × 2.5–5.5 mm. **Sporophylls** spreading to spreading-ascending, 4.5–5 × 0.5–0.9 mm, margins rarely toothed. 2*n* = 156.

Bogs, lakeshores, marshes, lichens, borrow pits; 0–2000 m; St. Pierre and Miquelon; Alta., B.C., N.B., Nfld., N.S., Ont., P.E.I., Sask., Que.; Alaska, Conn., Idaho, Ill., Ind., Ky., Maine, Md., Mass., Mich., Minn., Mont., N.H., N.J., N.Y., Ohio, Pa., R.I., Vt., Va., Wash., W.Va., Wis.; Eurasia.

2. Lycopodiella appressa (Chapman) Cranfill, Amer. Fern J. 71: 97. 1981 · Appressed bog club-moss

Lycopodium inundatum Linnaeus var. *appressum* Chapman, Bot. Gaz. 3: 20. 1878; *L. appressum* (Chapman) F. E. Lloyd & L. Underwood; *L. inundatum* var. *bigelovii* Tuckerman
Horizontal stems flat on ground, 15–45 × 0.4–0.6 cm; stems (excluding leaves) thick, 1.5–2 mm wide; leaves monomorphic, appressed, 5–7 × 0.8–1

mm, marginal teeth 0–7 per side. **Peduncles** 1–7 per plant, 13–40 × 0.3–0.4 cm; strobilus 1/3–1/6 total length; leaves appressed, 3–5.3 × 0.3–0.5 mm, marginal teeth 0–3 per side. **Strobili** 25–60 × 3–4 mm. **Sporophylls** appressed, incurved, 3.5–5 × 0.3 mm, marginal teeth absent. 2*n* = 156.

Bogs, lakeshores, marshes, ditches, borrow pits; 0–100 m; N.B., Nfld., N.S., P.E.I.; Ala., Ark., Conn., Del., Fla., Ga., Ill., Kans., Ky., La., Maine, Md., Mass., Miss., Mo., N.H., N.J., N.Y., N.C., Okla., Pa., R.I., S.C., Tenn., Tex., Va.; West Indies in Cuba.

3. Lycopodiella subappressa J. G. Bruce, W. H. Wagner, & Beitel, Michigan Bot. 30: 4. 1991 · Northern appressed club-moss

Horizontal stems flat on ground, 4–17 × 0.3–0.8 cm, stems (excluding leaves) thick, 1–1.5 mm diam.; leaves monomorphic, vertically ascending, further ascending when dry, 4–6 × 0.8–1 mm, marginal teeth absent. **Upright shoots** 1, 9–13 × 0.4–0.5 cm; strobilus 1/5–1/3(–1/2) total length; leaves strongly appressed, 3.5–6 × 0.3–0.8 mm, marginal teeth absent. **Strobili** 20–40 × 4–8 mm. **Sporophylls** appressed, 3–4 × 0.2–0.5 mm, marginal teeth absent. 2*n* = 312.

Wet, acidic ditches and borrow pits; of conservation concern; 0–200 m; Mich.

The distinctive *Lycopodiella subappressa* and *L. margueritae* are sexual polyploids. The full geographic distribution of *L. subappressa* is still unknown.

4. Lycopodiella alopecuroides (Linnaeus) Cranfill, Amer. Fern J. 71: 97. 1981 · Foxtail bog club-moss

Lycopodium alopecuroides Linnaeus, Sp. Pl. 2: 1102. 1753

Horizontal stems strongly arching, 5–40 × 0.8–1.1 cm, rooting at tip; stems (excluding leaves) 2–4 mm diam.; leaves monomorphic, spreading to ascending, 5–7 × 0.5–0.7 mm, marginal teeth 1–7 per side. **Peduncles** 1–3 per plant, 6–30 × 0.2–0.3 cm; strobilus 1/3–1/7 total length; leaves spreading to ascending, 6–7 × 0.3–0.5 mm, marginal teeth 1–10 per side. **Strobili** 20–60 × 12–20 mm. **Sporophylls** wide-spreading, 6–7 × 0.5–0.9 mm, marginal teeth 1–5 per side in proximal 1/2. $2n = 156$.

Bogs, marshes, ditches, borrow pits; 0–600 m; Ala., Ark., Conn., Del., Fla., Ga., Ky., La., Md., Mass., Miss., N.J., N.Y., N.C., Pa., R.I., S.C., Tex., Va.; West Indies in Cuba.

5. Lycopodiella prostrata (R. M. Harper) Cranfill, Amer. Fern J. 71: 97. 1981 · Prostrate bog club-moss

Lycopodium prostratum R. M. Harper, Bull. Torrey Bot. Club 33: 229. 1906; *L. inundatum* Linnaeus var. *pinnatum* Chapman

Horizontal stems flat on ground, 10–45 × 1.2–1.9 cm; stems (excluding leaves) 1–1.5 mm diam.; leaves dimorphic, upperside leaves smaller, 4–5 × 0.4–0.6 mm, ascending; lateral leaves horizontal, perpendicular to stem to slightly reflexed, 7–8 × 0.7–1.8 mm, marginal teeth 1–10 per side, mostly in proximal 1/2, with many basal teeth. **Peduncles** 1(–2) per plant, 18–35 × 0.5–0.9 cm; strobilus 1/4–1/10 total length; leaves appressed to ascending, 5–8 × 0.3–0.6 mm, marginal teeth 1–4 per side. **Strobili** 50–80 × 15–19 mm. **Sporophylls** spreading to somewhat ascending (sometimes reflexed at maturity), 7–9 × 0.3–0.5 cm, marginal teeth 1–5 per side. $2n = 156$.

Roadside ditches, wet pine barrens; 0–50 m; Ala., Ark., Fla., Ga., La., Miss., N.C., S.C., Tex.

6. Lycopodiella margueritae J. G. Bruce, W. H. Wagner, & Beitel, Michigan Bot. 30: 9. 1991 · Northern prostrate club-moss

Horizontal stems flat on ground, 10–18 × 1–1.6 cm, stems (excluding leaves) thick, 1.8–2.2 mm diam.; leaves monomorphic, spreading, and nearly perpendicular to stem, 6–13 × 0.8–1.2 mm; marginal teeth 3–4 per side, mainly on proximal 1/2. **Upright shoots** 1(–2) per plant, 13–17 × 0.3–0.7 cm; strobilus 1/2–1/3 total length; leaves initially divergent, then incurved, almost appressed, 5–6 × 0.4–0.8 mm, marginal teeth 0–2 per side. **Strobili** 5–8 × 0.4–0.9 cm. **Sporophylls** appressed, incurved, 4–6 × 0.4–0.5 mm, marginal teeth absent. $2n = 312$.

Wet, acidic ditches and borrow pits; of conservation concern; Mich.

Lycopodiella margueritae forms apparently fertile hybrids with the other tetraploid species, *L. subappressa*. Its hybrids with the diploid *L. inundata* are sterile, however. *Lycopodiella margueritae* can be distinguished from *L. appressa* and *L. subappressa* by the rather thick and large strobili, 1/3–1/2 the total length of upright shoots, and the more spreading leaves of the strobili and upright shoots.

3. SELAGINELLACEAE Willkomm

• Spike-moss family

Iván A. Valdespino

Plants herbaceous, annual or perennial, sometimes remaining green over winter. **Stems** leafy, branching dichotomously, regularly or irregularly forked or branched, protostelic (sometimes with many protosteles or meristeles), siphonostelic, or actino-plectostelic. **Rhizophores** (modified leafless shoots producing roots) present or absent, geotropic, borne on stems at branch forks, throughout, or confined to base of stems. **Leaves** on 1 plant dimorphic or monomorphic, small, with adaxial ligule near base, single-veined [rarely veins forked]. **Strobili** (clusters of overlapping sporophylls) sometimes ill-defined, terminal [lateral], cylindric, quadrangular, or flattened. **Sporophylls** (fertile leaves) monomorphic or adjacently different, slightly or highly differentiated from vegetative (sterile) leaves. **Sporangia** short-stalked, solitary in axil of sporophylls, opening by distal slits. **Spores** of 2 types (plants heterosporous), megaspores (1–2–)4, large, microspores numerous (hundreds), minute.

Genera 1, over 700 species (38 species in the flora): worldwide, primarily in tropical and subtropical regions.

Selaginellaceae traditionally include only one genus of living plants, *Selaginella* (A. C. Jermy 1990b; R. M. Tryon and A. F. Tryon 1982). Some authors (O. Kuntze 1891–1898, vol. 2, pp. 824–827; W. Rothmaler 1944), however, have segregated other genera based on generic concepts established by A. Palisot de Beauvois (1805, pp. 95–114), who recognized four genera. A. F. Spring (1850) combined the four genera into the broadly defined genus *Selaginella*. Spring's generic delimitation has resulted in misinterpretations that created many nomenclatural problems and partly led to the continued recognition of only one genus. Nevertheless, species in *Selaginella* fall into at least three well-defined groups, all present in North America, that may be recognized as genera based on anatomy, embryology, morphology and arrangement of the leaves and sporophylls, and morphology and symmetry of the strobilus. North American Selaginellaceae, which represent only a small portion of the family, are treated here in *Selaginella*, pending a full revision of the family worldwide.

Species in the fossil genus *Selaginellites* Zeller, which dates to the Carboniferous period, presumably are congeneric with *Selaginella*. Among the fern allies, Selaginellaceae are related

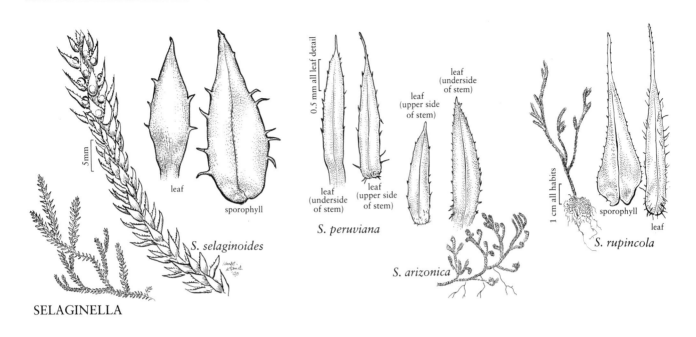

only distantly to the other lycopod families, Lycopodiaceae and Isoëtaceae (R. M. Tryon and A. F. Tryon 1982).

SELECTED REFERENCES Baker, J. G. 1887. Handbook of the Fern-allies: A Synopsis of the Genera and Species of the Natural Orders Equisetaceae, Lycopodiaceae, Selaginellaceae, Rhizocarpeae. London. Jermy, A. C. 1990b. Selaginellaceae. In: K. Kubitzki et al., eds. 1990+. The Families and Genera of Vascular Plants. Berlin etc. Vol. 1, pp. 39–45. Tryon, R. M. and A. F. Tryon. 1982. Ferns and Allied Plants, with Special Reference to Tropical America. New York.

1. SELAGINELLA Palisot de Beauvois, Prodr. Aethéogam., 101. 1805, name conserved · Spike-moss, sélaginelle [from *Selago,* an ancient name for *Lycopodium,* a genus resembling *Selaginella,* and Latin *-ella,* diminutive suffix]

Plants terrestrial, on rock, or rarely hemiepiphytic (initially terrestrial, becoming epiphytic) or epiphytic (in *S. oregana*). **Stems** prostrate, creeping, decumbent, cespitose, climbing, or fully erect, articulate or not, slightly to greatly branched. **Rhizophores** usually present, stout or filiform. **Roots** branching several times dichotomously from rhizophore tips. **Leaves** on aerial stems dimorphic or monomorphic; if monomorphic, then linear to narrowly lanceolate, highly overlapping, spirally arranged; if leaves on aerial stem dimorphic, then round or oblong to lanceolate, arranged in 4 ranks, 2 ranks of larger spreading lateral leaves and 2 ranks of smaller, appressed, and ascending median leaves, often with axillary leaf at base of each branching dichotomy. **Megasporangia** lobed to ovoid; microsporangia reniform to ovoid. **Megaspores** tetrahedral, ovoid, or globose, variously sculptured, (127–)200–1360 μm diam.; microspores tetrahedral, variously sculptured, 20–75 μm diam. $x = 7, 8, 9, 10, 11, 12.$

Species probably more than 700 (38 in the flora): worldwide, mainly tropical and subtropical regions.

The generic and infrageneric classification of *Selaginella* is controversial, and more than one genus may be recognized (see R. E. G. Pichi-Sermolli 1971 for information on generic synonyms). A. C. Jermy (1986, 1990b) proposed a subgeneric classification similar to that of J. G. Baker (1883, 1887). Despite some reservations, I consider Jermy's system useful for our purpose; therefore it is followed here. Three of the five subgenera proposed by Jermy occur in the flora area: subg. *Selaginella*, subg. *Tetragonostachys*, and subg. *Stachygynandrum*. One of the species in the flora, *Selaginella eatonii* (see discussion), may eventually prove to be best classified within a fourth, subg. *Heterostachys* Baker.

Some characteristics used in the keys and descriptions are best observed in fresh specimens or by soaking a sample of a dried specimen in water, using material at branch forks or buds. This is particularly true for members of subg. *Tetragonostachys*. Use a minimum of 20× (40–60× better) magnification and take measurements of both young and old leaves. Measurements of leaf length include the bristle and the most basal portion.

Selaginella subg. *Tetragonostachys* has a tendency for stem and leaves close to the substrate surface to be morphologically different from those on the side away from the substrate. In this case, the leaves on the side of the axis away from the surface are called *upperside leaves*, and those on the side toward the surface are called *underside leaves*. Otherwise, the leaves are designated only as *leaves*. In the subg. *Stachygynandrum*, however, which has complete structural differentiation between stem sides, the upper leaves are called *median leaves*, and the lower ones are called *lateral leaves*.

SELECTED REFERENCES Alston, A. H. G. 1955. The heterophyllous *Selaginellae* of continental North America. Bull. Brit. Mus. (Nat. Hist.), Bot. 1(8): 219–274. Baker, J. G. 1883. A synopsis of the genus *Selaginella*, pt. 1. J. Bot. 21: 1–5. Horner, H. T. Jr. and H. J. Arnott. 1963. Sporangial arrangement in North American species of *Selaginella*. Bot. Gaz. 124: 371–383. Jermy, A. C. 1986. Subgeneric names in *Selaginella*. Fern Gaz. 13: 117–118. Koller, A. L. and S. E. Scheckler. 1986. Variation in microsporangia and microspore dispersal in *Selaginella*. Amer. J. Bot. 73: 1274–1288. Tryon, A. F. 1949. Spores of the genus *Selaginella* in North America, north of Mexico. Ann. Missouri Bot. Gard. 36: 413–431.

Key to the subgenera of *Selaginella*

1. Leaves on aerial stems dimorphic, arranged in 4 ranks (2 median, 2 lateral), axillary leaves present at branching points; rhizophores present. . . . . 1c. *Selaginella* subg. *Stachygynandrum*, p. 57
1. Leaves on aerial stems monomorphic, not in distinct ranks, axillary leaves absent at branching points; rhizophores present or absent.
    2. Strobili cylindric; sporophylls spreading; leaves thin, soft, margins short-spiny; stomates throughout abaxial surface of leaf; rhizophores absent. . . . . 1a. *Selaginella* subg. *Selaginella* (*S. selaginoides*), p. 40
    2. Strobili quadrangular; sporophylls usually appressed; leaves thick or fleshy (seldom thin), firm, margins dentate, serrate, or ciliate (never spiny); stomates in abaxial groove; rhizophores present. . . . . 1b. *Selaginella* subg. *Tetragonostachys*, p. 41

1a. SELAGINELLA Palisot de Beauvois subg. SELAGINELLA

Stems radially symmetric, not articulate; creeping stems with closely spaced, dichotomous forks; upright stems unbranched; vessel elements absent. **Rhizophores** absent. **Leaves** loosely spirally arranged, monomorphic, lanceolate to lanceolate-triangular, thin, soft; margins short-spiny; stomates scattered throughout abaxial surface; axillary leaves absent. **Strobili** solitary, cylindric. **Sporophylls** loosely spirally arranged, larger and slightly different shape from vegetative leaves, most sporophylls fertile, megasporophylls and microsporophylls same size, auricles absent.

Species 2 (1 in the flora): circumboreal and Hawaii.

SELECTED REFERENCES Karrfalt, E. E. 1981. The comparative and developmental morphology of the root system of *Selaginella selaginoides* (L.) Link. Amer. J. Bot. 68: 224–253. Page, C. N. 1989. Compression and slingshot megaspore ejection in *Selaginella selaginoides*—A new phenomenon in pteridophytes. Fern Gaz. 13: 267–275.

1. Selaginella selaginoides (Linnaeus) Palisot de Beauvois ex Martius & Schrank, Hort. Reg. Monac. 1: 182. 1829 · Northern spike-moss, prickly mountain-moss, sélaginelle fausse-sélagine

Lycopodium selaginoides Linnaeus, Sp. Pl. 2: 1101. 1753

Plants on rock or terrestrial, forming loose to dense mats. **Stems** not readily fragmenting, tips not upturned; creeping stems filiform, indeterminate, branching dichotomously; upright stems stout, unbranched (3–10 cm aboveground), terminating in simple strobili. **Leaves** green, lanceolate, 3–4.5 × 0.75–1.2 mm (smaller on horizontal stems, 1/3 less than those on upright stems); abaxial groove absent; base decurrent, forming saclike structure with stem; margins with soft spiny projections, 0.1–0.2 mm; apex acuminate to subulate. **Strobili** (1–)2–3(–5) cm; sporophylls lanceolate-triangular, 4.5–6 × 1.15–1.5 mm, lacking abaxial ridges. $2n = 18$.

Wet places, among mossy stream banks, lakeshores, bogs, and wet talus slopes, in neutral to alkaline soil; 600–2900(–3800) m; Greenland; St. Pierre and Miquelon; Alta., B.C., Man., N.B., Nfld., N.W.T., N.S., Ont., P.E.I., Que., Sask., Yukon; Alaska, Colo., Idaho, Maine, Mich., Minn., Mont., Nev., N.Y., Wis., Wyo.; Eurasia; nw Africa in the Canary Islands.

Selaginella selaginoides is reported to have strobili with basal megasporangia and apical microsporangia (H. T. Horner Jr. and H. J. Arnott 1963). Some individuals, however, have megasporangia at the tip of the strobili. *Selaginella selaginoides* is generally thought to be a primitive member of the genus (F. O. Bower 1908; T. L. Phillips and G. A. Leisman 1966; R. M. Tryon 1955), but certain of its characteristics may be derived. It is unique in having an active megaspore dispersal mechanism, termed "compression and slingshot megaspore ejection" (C. N. Page 1989), and it has a peculiar root position and development (E. E. Karrfalt 1981) probably found elsewhere only in the closely related species *S. deflexa* Brackenridge of Hawaii. Both features may be derived rather than primitive.

1b. SELAGINELLA Palisot de Beauvois subg. **TETRAGONOSTACHYS** Jermy, Fern Gaz. 13: 118. 1986

Stems radially symmetric or upperside and underside structurally different, not articulate, prostrate, creeping, or erect, few to many branched; vessel elements present. **Rhizophores** borne on upperside of stems, throughout stem length or confined to base of stem. **Leaves** monomorphic, tightly appressed and spirally arranged, or upperside and underside leaves slightly differentiated; all leaves linear or linear-lanceolate, thick or fleshy (seldom thin); margins dentate, serrate, or ciliate; abaxial groove with stomates, these arranged along vein; axillary leaves absent. **Strobili** quadrangular. **Sporophylls** differentiated from vegetative leaves, most sporophylls fertile; megasporophylls and microsporophylls same size, in 4 alternating ranks, appressed, base usually with 2 diverging flaps or auricles, auricles protecting sporangia below.

Species ca. 50 (26 species and 1 hybrid in the flora): widespread, Mexico, West Indies, Central America, South America, Asia, Africa including Madagascar.

This treatment of subg. *Tetragonostachys* generally follows that of R. M. Tryon (1955). Some problems, however, remain to be resolved, particularly for *Selaginella arenicola* and *S. densa*, in which taxa have been recognized at the infraspecific (R. M. Tryon 1955) and specific (L. H. Snyder Jr. and J. G. Bruce 1986; J. W. Thieret 1980; J. M. Beitel and W. R. Buck, pers. comm.) levels. Based on examination of a wide range of specimens in the *S. arenicola* complex, I recognize two species, one of which contains two subspecies. In the *S. densa* complex, I recognize three well-defined species: *S. densa, S. scopulorum,* and *S. standleyi.* The phylogenetic relationships among the different series proposed by Tryon need further study.

Within the series *Arenicolae* R. M. Tryon, a tendency toward structural differentiation occurs between the stem's upperside and its underside (e.g., *S. rupincola*). This feature may link series *Arenicolae* to species such as *S. hansenii* and *S. wrightii,* which I place in the series *Eremophilae* R. M. Tryon and which may represent an early intermediate stage toward full stem differentiation. According to Tryon, this "transitional stage" is primitive within the series *Eremophilae,* where it is found in *S. peruviana.*

The occurrence of hybrids within *Selaginella* subg. *Tetragonostachys* is best shown in *S. × neomexicana.* Hybridization may be a more common phenomenon, however, than previously acknowledged.

SELECTED REFERENCES　Clausen, R. T. 1946. *Selaginella* subgenus *Euselaginella,* in the southeastern United States. Amer. Fern J. 36: 65–82. Tryon, R. M. 1955. *Selaginella rupestris* and its allies. Ann. Missouri Bot. Gard. 42: 1–99, plates 1–6. Tryon, R. M. 1971. The process of evolutionary migration in species of *Selaginella.* Brittonia 23: 89–100. Van Eseltine, G. P. 1918. The allies of *Selaginella rupestris* in the southeastern United States. Contr. U.S. Natl. Herb. 20(5): 159–172.

Key to the species of *Selaginella* subg. *Tetragonostachys*

1. Stems prostrate, undersides and uppersides differentiated; leaves conspicuously to slightly dimorphic; rhizophores throughout stem length.
   2. Underside and upperside leaves abruptly adnate to stem; leaves slightly dimorphic.
      3. Apex of leaves with white or whitish bristle 0.5–1.4 mm; marginal cilia white to whitish, strongly ascending; leaves green, usually with red spots or wholly reddish wine-colored; sporophylls ovate-deltate, short-attenuate toward apex. . . . . . 6. *Selaginella hansenii*
      3. Apex of leaves with yellowish bristle 0.2–0.5 mm or absent; marginal cilia transparent, spreading; leaves green, never reddish; sporophylls lanceolate, long-attenuate toward apex. . . . . 5. *Selaginella wrightii*
   2. Underside leaves decurrent, upperside leaves abruptly adnate to stem; leaves strongly to moderately dimorphic.
      4. Leaves with tortuous (twisted) bristle at tip, becoming acute to mucronate; upperside leaves lanceolate; plants forming dense mats. . . . . 2. *Selaginella eremophila*
      4. Leaves acute to bristled, bristle straight; upperside leaves linear-lanceolate; plants forming rather loose mats.
         5. Underside leaves lanceolate, widest at middle; leaf apex acute or with short and flattened bristle 0.1–0.3 mm (mostly at branch tips or buds); sporophylls acute to acuminate. . . . . 3. *Selaginella arizonica*
         5. Underside leaves narrowly linear-lanceolate, widest at base; leaf apex with long, round bristle 0.3–0.8 mm; sporophylls bristle-tipped. . . . . 4. *Selaginella peruviana*
1. Stems pendent, erect, ascending, or rarely prostrate, radially symmetric or undersides and uppersides slightly differentiated (if so, leaves decurrent as in *S. densa* complex); leaves not to rarely dimorphic; rhizophores throughout stem length or restricted to stem base.
   6. Plants epiphytic, seldom terrestrial; aerial stems long-pendent, usually forming festoon-like or overlapping mats; leaves loosely appressed; strobili 1–6 cm. . . . . 20. *Selaginella oregana*
   6. Plants on rock or terrestrial, never epiphytic; aerial stems erect, ascending, long- to short-creeping, decumbent, radially symmetric or slightly differentiated, forming long- or short-spreading mats, cushionlike mats, or cespitose mats; leaves usually tightly appressed; strobili 0.2–4.5(–9) cm.
      7. Aerial stems erect or ascending, sometimes decumbent or creeping; rhizome or rhizomatous stem present; budlike arrested branches usually present on rhizome or lowermost aerial stem.
         8. Base of leaf abruptly adnate.
            9. Leaf margins short-ciliate throughout, cilia 0.02–0.08 mm; leaf base cordate to almost peltate. . . . . 7. *Selaginella bigelovii*
            9. Leaf margins long-ciliate at least at base, cilia 0.06–0.2 mm; leaf base rounded.
               10. Leaf bristle 0.65–1.85 mm; marginal cilia long and spreading throughout, 0.1–0.2 mm; sporophylls strongly tapering toward apex. . . . 8. *Selaginella rupincola*

10. Leaf bristle 0.3–0.46 mm; marginal cilia long and spreading at base, short
     to dentiform and ascending toward apex, 0.06–0.17 mm; sporophylls
     not strongly tapering. . . . . 9. *Selaginella* ×*neomexicana*
8. Base of leaf decurrent or long-decurrent.
   11. Scalelike leaves on rhizome loosely appressed; stem leaves without hairs along
        abaxial groove.
       12. Scalelike leaves on rhizome incurved; leaf apex bristle-tipped; sporo-
            phyll bristle-tipped. . . . . 10. *Selaginella weatherbiana*
       12. Scalelike leaves on rhizome straight; leaf apex acute or obtuse; sporo-
            phyll apex acute to obtuse. . . . . 11. *Selaginella viridissima*
   11. Scalelike leaves on rhizome tightly appressed or rhizome absent; stem leaves
        with hairs along abaxial groove.
       13. Leaf bristle tortuous (twisted); leaf base glabrous; abaxial groove and
            ridges on leaf not prominent, often obscure; strobili 0.4–0.6 cm. . . .
            . . . . 12. *Selaginella tortipila*
       13. Leaf bristle straight, never twisted; leaf base pubescent; abaxial groove
            and ridges on leaf prominent; strobili (0.5–)1–3(–3.5) cm.
           14. Underground (rhizomatous) stem leaves scalelike; rhizophores mostly
                subterranean; sporophyll base glabrous; leaf and sporophyll apices
                glabrous. . . . . 13. *Selaginella arenicola*
           14. Underground (rhizomatous) stem leaves not scalelike; rhizophores
                mostly aerial; sporophyll base pubescent; leaf and sporophyll apices
                often puberulent. . . . . 14. *Selaginella acanthonota*
7. Aerial stems creeping or decumbent, never erect, radially symmetric or lower stem
   and upper stem slightly differentiated; rhizome or rhizomatous stem absent; budlike
   arrested branches absent.
   15. Leaves on main stem adnate to stem (distinct from stem in color), bases rounded
        or seldom slightly decurrent and cuneate (on underside leaves or in plants from
        wet places).
       16. Leaf apex abruptly short- to long-bristled, bristle puberulent or sometimes
            entire, (0.16–)0.2–0.46(–0.9) mm; leaves in whorls of 4; strobili often paired,
            1–4.5(–9) cm. 21. *Selaginella wallacei*
       16. Leaf apex blunt or acute to acuminate or seldom short-bristled, bristle if
            present entire, 0.03–0.45 mm; leaves in whorls of 3; strobili usually soli-
            tary, 0.2–3 cm.
           17. Leaves tightly appressed, apex keeled, mucro or bristle if present 0.03–
                0.45 mm; stems radially symmetric; strobili (0.6–)1–3 cm. . . . 17. *Selaginella mutica*
           17. Leaves loosely appressed, apex plane, not bearing bristle; stems slightly
                structurally differentiated; strobili 0.2–0.4 cm. . . . . 18. *Selaginella cinerascens*
   15. Leaves on main stem decurrent (not distinct from stem in color), bases cuneate
        or oblique (seldom adnate and rounded on upperside).
       18. Main stems with upperside and underside slightly differentiated; upperside
            and underside leaves unequal in size, bases decurrent and oblique (*S. densa*
            complex).
           19. Leaf apex bearing conspicuously puberulent bristle, (1–)1.25–1.9 mm;
                leaf margins usually long-ciliate, cilia 0.07–0.17(–0.2) mm; sporophyll
                margins entirely ciliate. . . . . 26. *Selaginella densa*
           19. Leaf apex bearing slightly puberulent or entire bristle, 0.4–1.25 mm;
                leaf margins relatively short-ciliate, cilia 0.02–0.07(–0.15) mm; spo-
                rophyll margins short-ciliate or denticulate in parts.
               20. Sporophylls deltate-ovate; apex keeled, strongly truncate in pro-
                    file; bristle usually yellowish; margins short-ciliate to denticulate
                    on distal 3/4. . . . . 28. *Selaginella standleyi*

20. Sporophylls ovate-lanceolate to lanceolate or seldom ovate; apex attenuate or slightly keeled, not truncate in profile; bristle usually whitish transparent, seldom yellowish (in old leaves); margins short-ciliate to denticulate on proximal 1/2, lacking cilia toward apex. . . . . 27. *Selaginella scopulorum*

18. Main stems radially symmetric; leaves equal in size, if stem slightly differentiated then leaf bases decurrent and cuneate.

   21. Leaves on main stems in alternate pseudowhorls of 5 or 6.

      22. Leaves on main stem in alternate pseudowhorls of 5; leaf base decurrent and cuneate on upperside; leaf apex truncate in profile; sporophyll apex truncate in profile. . . . . 16. *Selaginella sibirica*

      22. Leaves on main stem in alternate pseudowhorls of 6; leaf base sometimes adnate and rounded on upperside; leaf apex attenuate in profile; sporophyll apex not truncate in profile. . . . . . 15. *Selaginella rupestris*

   21. Leaves on main stem in alternate pseudowhorls of 4.

      23. Lateral branches spreading, not ascending; stems forming festoon-like mats or rarely compact mats; dry stems not readily fragmenting; strobili sometimes paired. . . . . 19. *Selaginella underwoodii*

      23. Lateral branches usually strongly ascending; stems forming compact cushionlike, usually rounded mats or less often loose mats; dry stems readily fragmenting or not; strobili solitary.

         24. Leaf apex bearing bristle 0.5–1.4 mm; leaf base pubescent; sporophyll base often pubescent. . . . . 23. *Selaginella asprella*

         24. Leaf apex mucronate, blunt or acute, bearing bristle or mucro 0–0.6 mm; leaf base glabrous, seldom pubescent; sporophyll base always glabrous.

            25. Dry stems not readily fragmenting; lateral branches 1–3-forked; leaf apex strongly keeled. . . . . 22. *Selaginella watsonii*

            25. Dry stems readily fragmenting; lateral branches 1-forked; leaf apex keeled, slightly attenuate or obtuse.

               26. Leaf apex bearing puberulent bristle 0.2–0.6 mm; leaves not in well-defined alternate pseudowhorls .. . . 24. *Selaginella leucobryoides*

               26. Leaf apex blunt, acute or only ending in very short entire bristle or mucro 0–0.4 mm; leaves in defined alternate pseudowhorls. . . . . 25. *Selaginella utahensis*

2. Selaginella eremophila Maxon, Smithsonian Misc. Collect. 72: 3–5. 1920 · Desert spike-moss

Plants on rock or terrestrial, forming dense mats. **Stems** not readily fragmenting, prostrate, upperside and underside structurally different, irregularly forked; branches determinate, tips upturned. **Rhizophores** borne on upperside of stems, throughout stem length, 0.2 mm diam. **Leaves** conspicuously dimorphic, in 8 ranks, tightly appressed, ascending, green; abaxial ridges present; apex with deciduous, twisted, transparent bristle ± 0.3 mm, becoming acute to slightly mucronate in oldest branches. **Underside leaves** lanceolate to lanceolate-elliptic (on central ranks) or falcate (on marginal ranks), 2–2.7 × 0.5–0.7 mm; base decurrent, glabrous; margins ciliate, cilia transparent to opaque, spreading, 0.04–0.1 mm. **Upperside leaves** lanceolate, 1.3–1.4 × 0.3–0.4 mm; base abruptly adnate, pubescent, hairs often running along groove; margins ciliate, cilia transparent to opaque, spreading, ca. 0.1 mm. **Strobili** solitary, 3–8 mm; sporophylls ovate-deltate, abaxial ridges not prominent, base glabrous, margins ciliate, apex acute to mucronate.

Rocky and sandy slopes, in open rock or crevices or in soil; 130–1000 m; Ariz., Calif.; Mexico in Baja California.

Selaginella eremophila is most closely related to the Mexican *S. parishii* L. Underwood and *S. landii* Greenman & Pfeiffer. In *S. eremophila* and the following two species, *S. arizonica* and *S. peruviana*, the leaves are arranged in 8 conspicuous ranks: 3 underside (2 marginal, 1 central), 2 lateral, and 3 upperside (2 marginal, 1 central).

3. Selaginella arizonica Maxon, Smithsonian Misc. Collect. 72: 5–6. 1920 · Arizona spike-moss

Plants on rock or terrestrial, forming rather loose mats. **Stems** not readily fragmenting, prostrate, upperside and underside structurally different, irregularly forked, branches determinate, tips upturned in extremely dry conditions. **Rhizophores** borne on upperside of stem throughout, 0.25–0.3 mm diam. **Leaves** conspicuously dimorphic, in 8 ranks, tightly appressed to ascending, green; abaxial ridges present; apex with transparent to opaque, flattened bristle 0.1–0.3 mm, sometimes becoming acute (by breaking off of bristle). **Underside leaves** lanceolate, 2–2.5 × 0.5–0.6 mm; base decurrent, glabrous; margins ciliate, cilia transparent to opaque, spreading or ascending, 0.06–0.13 mm. **Upperside leaves** linear-lanceolate to slightly falcate (on marginal ranks), 1.9–2.25 × 0.4–0.55 mm; base abruptly adnate, pubescent or glabrous; margins ciliate, cilia transparent to opaque, spreading, 0.06–0.15 mm. **Strobili** solitary, 5–10 mm; sporophylls ovate-deltate, abaxial ridges not prominent, base glabrous, margins short-ciliate to denticulate, apex acute.

In rock crevices or on gravel, on sandstone, igneous, or rarely limestone substrates; 600–2000 m; Ariz., Tex.; Mexico in Baja California, Sonora.

Selaginella arizonica can be further distinguished from the similar *S. peruviana* by its broad, thin underside leaves. In *S. peruviana* the underside leaves are narrow and fleshy.

4. Selaginella peruviana (J. Milde) Hieronymus, Hedwigia 39: 307. 1900 · Peruvian spike-moss

Selaginella rupestris (Linnaeus) Spring forma *peruviana* J. Milde, Fil. Eur., 263. 1867; *S. sheldonii* Maxon

Plants on rock or terrestrial, forming loose mats. **Stems** not readily fragmenting, prostrate, upperside and underside structurally different, irregularly forked, branches determinate, tips upturned. **Rhizophores** borne on upperside of stems, throughout stem length, 0.23–0.33 mm diam. **Leaves** dimorphic, arranged in 8 ranks, tightly appressed, ascending, green; abaxial ridges present; apex with persistent, whitish, terete bristle 0.3–0.8 mm. **Underside leaves** narrowly linear-lanceolate (on central ranks) to falcate (on marginal ranks), 2.5–4 × 0.4–0.6 mm; base decurrent (oblique on marginal ranks), pubescent (sometimes glabrous); margins ciliate, cilia transparent to opaque, spreading at base, ascending toward apex, 0.1–0.15 mm. **Upperside leaves** linear-lanceolate (on central ranks) to falcate (on marginal ranks), 2.3–2.75 × 0.5–0.55 mm; base abruptly adnate, pubescent; margins ciliate, cilia transparent to opaque, ascending or spreading, 0.08–0.16 mm. **Strobili** solitary, 0.5–2 cm; sporophylls ovate-deltate to ovate, abaxial ridges not prominent, base glabrous, margins short-ciliate, apex bristled.

Rocky slopes, rock crevices, ledges of sandstone or igneous cliffs, less often on sandy or clay soil; 1300–2300 m; N.Mex., Okla., Tex.; Mexico; South America.

R. M. Tryon (1955) reported an elevation range of 600–3000 m for *Selaginella peruviana* in the United States.

5. Selaginella wrightii Hieronymus, Hedwigia 39: 298. 1900 · Wright's spike-moss

Plants on rock, forming loose to dense mats. **Stems** not readily fragmenting, prostrate, upperside and underside structurally different, irregularly forked, branches determinate, tips upturned. **Rhizophores** borne on upperside of stems, throughout stem length, 0.25–0.37 mm diam. **Leaves** dimorphic, arranged in 8 ranks, tightly appressed, ascending, green; abaxial ridges absent; apex with yellowish bristle 0.2–0.5 mm, becoming denticulate (by breaking off of bristle); bristle usually more persistent in underside leaves. **Underside leaves** narrowly linear-lanceolate to falcate (on marginal ranks), 3.5–4.5(–5) × 0.55–0.7 mm; base abruptly adnate or slightly decurrent, usually pubescent, sometimes glabrous; margins ciliate, cilia transparent, spreading, 0.12–0.26 mm. **Upperside leaves** linear-lanceolate, 3.3–3.85 × 0.6–0.75 mm; base abruptly adnate, pubescent; margins ciliate, cilia transparent, spreading, 0.12–0.26 mm. **Strobili** solitary, (0.7–)1.5–2(–2.6) cm; sporophylls lanceolate, abaxial ridges not prominent, base glabrous, margins ciliate, apex strongly tapering, bristle obscure.

On exposed or shaded limestone cliffs; of conservation concern; 800–2300 m; N.Mex., Tex.; Mexico.

Of other species in the flora, *Selaginella wrightii* seems to be allied to *S. hansenii*. The structural differentiation of the stem, adjacently different leaves, and upturned branch tips align the two species to the series *Eremophilae*. *Selaginella wrightii* is a calciphile, according to R. M. Tryon (1955).

6. Selaginella hansenii Hieronymus, Hedwigia 39: 301. 1900 · Hansen's spike-moss

Plants terrestrial, forming loose to clustered mats. **Stems** not readily fragmenting, prostrate, upperside and underside structurally different, irregularly forked, branches determinate, tips upturned. **Rhizophores** borne on upperside of stems, throughout stem length, 0.25–0.45 mm diam. **Leaves** with underside leaves slightly longer and narrower than upperside leaves, otherwise monomorphic, not clearly ranked, tightly appressed, ascending, green or green with red spots, or reddish, linear-lanceolate (underside) to linear-triangular (upperside), (2–)3–4.5 × 0.5–0.6 mm; abaxial ridges present; base abruptly adnate, pubescent (sometimes glabrous); margins ciliate, cilia white to white opaque, strongly appressed and ascending, 0.03–0.1 mm; apex with bristle white to white opaque, 0.5–1.4 mm (those on underside leaves sometimes 1/4–1/2 longer than those on upperside leaves). **Strobili** solitary, 5–7 mm; sporophylls ovate-deltate to ovate-triangular, abaxial ridges not prominent, base glabrous, margins short-ciliate, apex bristled.

Cliffs and rocky slopes or on igneous rock; 330–1350 m; Calif.

Leaf dimorphism in *Selaginella hansenii* is only slightly and inconsistently expressed; the upperside leaves tend to be more lanceolate, short, and slightly thick, whereas the underside leaves tend to be more linear, longer, and thinner, but in some specimens the leaves are monomorphic. Red leaves are rare within *Selaginella* subg. *Tetragonostachys*, otherwise found in the flora only occasionally in *S. rupestris*. Such leaves are more common in *S. steyermarkii* Alston from southern Mexico and Guatemala and *S. sartorii* Hieronymus from Mexico.

7. Selaginella bigelovii L. Underwood, Bull. Torrey Bot. Club 25: 130. 1898 · Bigelow's spike-moss, bushy spike-moss

Plants on rock or terrestrial, forming clumps. **Stems** radially symmetric, underground (rhizomatous) and aerial, not readily fragmenting, irregularly forked; rhizomatous and aerial stems often with 1 branch arrested, budlike, tips straight; aerial stems erect or occasionally ascending. **Rhizophores** borne on upperside of stems, restricted to rhizomes and lower 1/3 of aerial stems, 0.3–0.4 mm diam. **Leaves** dimorphic, not clearly ranked. **Rhizomatous stem leaves** persistent, tightly appressed, scalelike. **Aerial stem leaves** appressed, ascending, green, linear-lanceolate to narrowly lanceolate, 2.2–3.8 × 0.29–0.4 (–0.75) mm; abaxial ridges present; base abruptly adnate, cordate to almost peltate, pubescent or sometimes glabrous; margins short-ciliate at base, denticulate toward apex, cilia white to transparent or greenish, spreading at base, ascending toward apex, 0.02–0.08 mm; apex keeled, bristled; bristle puberulent, rough, transparent to whitish, 0.23–0.75 mm. **Strobili** solitary, (0.4–)1–1.5 cm; sporophylls ovate-lanceolate to lanceolate, abaxial ridges not prominent, base glabrous, margins short-ciliate to denticulate, apex bristled. $2n = 18$.

Exposed rock crevices, cliffs, boulders, sandstone or igneous rock, serpentine, or gravelly soil; 0–2000 m; Calif.; Mexico in Baja California.

Selaginella bigelovii is a member of the series *Arenicolae* (R. M. Tryon 1955) and is closely related to *S. rupincola* (see discussion). It may be confused with *S. ×neomexicana*. *Selaginella bigelovii*, however, always has well-developed megasporangia with most of the megaspores and microspores well formed, whereas *S. ×neomexicana* is a presumed sterile hybrid that does not form megaspores, seldom forms microspores, and usually has most sporangia misshapen (R. M. Tryon 1955). Moreover, *S. ×neomexicana* has not been reported from either California or Baja California.

8. Selaginella rupincola L. Underwood, Bull. Torrey Bot. Club 25: 129. 1898

Plants on rock or terrestrial, forming loose clumps. **Stems** radially symmetric, underground (rhizomatous) and aerial, not readily fragmenting, irregularly forked; both rhizomatous and aerial stems often with 1 branch arrested, budlike, tips straight; rhizomatous stems hard to distinguish on wholly creeping plants; aerial stems erect or ascending, sometimes decumbent to slightly creeping, budlike arrested branches restricted mostly near stem base. **Rhizophores** borne on upperside of stems, restricted to lower stems or throughout stem length, 0.3–0.5 mm diam. **Leaves** dimorphic, not clearly ranked. **Rhizomatous stem leaves** persistent or deciduous, tightly appressed, scalelike. **Aerial stem leaves** appressed, ascending, green, linear-lanceolate, 3–4.7 × 0.45–0.65 mm; abaxial ridges present; base abruptly adnate, rounded, pubescent; margins long-ciliate, cilia white to whitish, spreading, 0.1–0.2 mm; apex not keeled to slightly keeled; bristle white to whitish or yellowish to greenish near base, puberulent, 0.65–1.85 mm (1/3–

1/2 length of leaves). **Strobili** solitary, 0.5–2.5(–3.5) cm; sporophylls lanceolate, strongly tapering toward tip, abaxial ridges prominent, base glabrous, margins short-ciliate, apex long-bristled.

Exposed ledges and rock, steep slopes, rock crevices or gravelly soil; 1000–2000 m; Ariz., N.Mex., Tex.; Mexico.

Selaginella rupincola is allied to *S. bigelovii*. It is one of the presumed parents of *S. ×neomexicana* (see discussion). In addition to characteristics given, it can be separated from *S. bigelovii* in having hairs often running along the ridges of the abaxial groove, whereas *S. bigelovii* has nonhairy ridges on the abaxial groove.

9. Selaginella ×neomexicana Maxon, Smithsonian Misc. Collect. 72: 2. 1920

Plants on rock, forming clumps. **Stems** radially symmetric, underground (rhizomatous) and aerial, not readily fragmenting, irregularly forked; rhizomatous and aerial stems often with 1 branch arrested, budlike, tips straight; rhizomatous stems sometimes difficult to distinguish, without obvious living budlike branches; aerial stems erect to ascending, budlike branches mostly restricted to stem base (more conspicuous in ascending stems). **Rhizophores** borne on upperside of stems, restricted to lower 1/2 on erect stems or throughout stem length on ascending stems, 0.2–0.3 mm diam. **Leaves** dimorphic, not clearly ranked. **Rhizomatous stem leaves** deciduous or persistent on base of emergent aerial stem, abruptly adnate, pubescent. **Aerial stem leaves** appressed, ascending, green, linear-lanceolate, 1.9–2.7 × 0.36–0.46 mm; abaxial ridges present; base abruptly adnate, rounded, pubescent; margins long-ciliate, cilia white, whitish to transparent or opaque, long and spreading at base, short to dentiform and ascending toward apex, 0.06–0.17 mm; apex keeled; bristle whitish to white, greenish to yellowish opaque, slightly puberulent, 0.3–0.46 mm. **Strobili** solitary, (0.5–)1–3 cm; sporophylls ovate-lanceolate to lanceolate, abaxial ridges prominent, base glabrous, margins denticulate, apex keeled, short-bristled.

On canyon rock; 1400–1700(–2000) m; Ariz., N.Mex., Tex.

Selaginella ×neomexicana is treated here as a hybrid, following R. M. Tryon (1955). Plants of this hybrid lack megaspores and megasporangia and have misshapen microsporangia. Several hypotheses for its origin have been advanced. It is clearly allied to *S. rupincola*, with which it shares white, long, spreading, marginal cilia on the leaves, hairs sometimes running along the ridges of the abaxial groove of the leaf, obscure rhizomatous underground stems, and buds mostly restricted to the base of aerial stems. Tryon (1955) suggested that the two presumed parents were *S. rupincola* and *S. mutica*, because *S. ×neomexicana* has been found growing with *S. mutica* (usually var. *limitanea*). The usually strongly keeled apex in *S. ×neomexicana* is a feature of *S. mutica*, and the range of *S. ×neomexicana* is within the range of the two presumed parents. *Selaginella underwoodii* might conceivably be the second parent instead; its range overlaps the ranges of the putative hybrid and *S. rupincola*. It is possible that *S. ×neomexicana* may represent an asexual race of *S. rupincola*. More detailed studies are necessary to determine the reproductive biology and cytology of this presumed hybrid and to assess its relationships.

10. Selaginella weatherbiana R. M. Tryon, Amer. Fern J. 40: 69. 1950 · Weatherby's spike-moss

Plants on rock, forming clumps. **Stems** radially symmetric, underground (rhizomatous) and aerial, not readily fragmenting, irregularly forked; rhizomatous and aerial stems often with 1 branch arrested, budlike, tips straight; aerial stems erect, less often ascending, cespitose, stout, branches not conspicuously arrested, budlike branches mostly near base. **Rhizophores** borne on upperside of stems, mostly restricted to rhizomatous stems or to lower 1/2 of aerial stems, 0.16–0.26(–3) mm diam. **Leaves** dimorphic, not clearly ranked. **Rhizomatous stem leaves** persistent, loosely appressed, ascending, often incurved, scalelike. **Aerial stem leaves** tightly appressed, ascending, green, linear-lanceolate to narrowly lanceolate, 1.7–2.4 × 0.36–0.43 mm; abaxial ridges prominent; base cuneate and decurrent on main stem or rounded and abruptly adnate on apical branch portions, glabrous or pubescent; margins short-ciliate at base, cilia transparent, spreading, denticulate, and ascending toward apex, 0.03–0.06 mm; apex keeled; bristle transparent to opaque or yellowish to brownish (on old leaves), puberulent to smooth, 0.3–0.6(–0.7) mm. **Strobili** solitary, (0.7–)1–3 cm; sporophylls narrowly ovate-lanceolate to lanceolate, abaxial ridges prominent, base glabrous, margins denticulate to short-ciliate, apex keeled, bristled. $2n = 18$.

Exposed or shaded granitic rock outcrops, ledges, cliffs, or in rock crevices; of conservation concern; 1600–3000 m; Colo., N.Mex.

One of the most striking features of *Selaginella weatherbiana* is that at branch forks the larger branch continues to grow as a vegetative shoot, and the smaller one usually forms a strobilus. Therefore, the strobili appear to be lateral rather than terminal. *Selaginella weatherbiana* grows in close association with *S. underwoodii* (R. M. Tryon 1955). The two species (as well as *S. mutica* var. *mutica*) are very often mixed on herbarium specimens.

11. Selaginella viridissima Weatherby, J. Arnold Arbor. 24: 326. 1943 · Slender spike-moss

Selaginella coryi Weatherby

Plants on rock, forming clumps or mounds. **Stems** radially symmetric, underground (rhizomatous) and aerial, not readily fragmenting, irregularly forked; rhizomatous and aerial stems often with 1 branch arrested, budlike, tips straight; aerial stems mainly erect, seldom ascending, with budlike arrested branches throughout stem length. **Rhizophores** borne on upperside of stems, restricted to rhizomatous stems and lower 1/4 of aerial stems, 0.16–0.3 mm diam. **Leaves** dimorphic, not clearly ranked. **Rhizomatous stem leaves** loosely appressed, straight, scalelike. **Aerial stem leaves** appressed, ascending, green, linear-lanceolate to narrowly lanceolate, 1.8–2.1 × 0.49–0.56 mm; abaxial ridges prominent; base cuneate and decurrent to slightly rounded and adnate, glabrous; margins denticulate to very short-ciliate, cilia transparent, spreading to ascending toward apex, 0.02–0.04 mm; apex acute or seldom blunt. **Strobili** solitary, 0.5–1.2(–2.5) cm; sporophylls deltate-ovate to ovate-lanceolate, abaxial ridges prominent, base glabrous, margins denticulate, apex acute to obtuse.

Shaded cliffs, slopes, rock crevices, and igneous rock; of conservation concern; 1650–2300 m; Tex.; Mexico in Coahuila.

In Texas *Selaginella viridissima* is known only from the Chisos Mountains.

12. Selaginella tortipila A. Braun, Ann. Sci. Nat., Bot., sér. 5, 3: 271. 1865 · Kinky-hair spike-moss

Plants on rock or terrestrial, forming compact clumps or mounds. **Stems** radially symmetric, underground (rhizomatous) and aerial, not readily fragmenting, irregularly forked; rhizomatous and aerial stems often with 1 branch arrested, budlike, tips straight; aerial stems erect or ascending to decumbent, budlike branches throughout. **Rhizophores** borne on upperside of stems, restricted to rhizomatous stems or to lowermost base of aerial stems, 0.2–0.3 mm diam. **Leaves** dimorphic, in alternate pseudowhorls of 5. **Rhizomatous stem leaves** strongly appressed, overlapping, scalelike. **Aerial stem leaves** tightly appressed, ascending, green, narrowly lanceolate to linear-lanceolate, (2.5–)3–4.5 × 0.4–0.7 mm; abaxial ridges inconspicuous or more visible from apex to middle of leaf; base cuneate, decurrent, glabrous; margins short-ciliate to denticulate or entire, cilia transparent, spreading, 0.02–0.06(–0.08) mm; apex keeled (more so in dry leaves); bristle transparent or yellowish to brownish near base, puberulent, twisted, persistent or falling off early, 1.2–1.7 mm (1/3–1/2 length of leaves). **Strobili** solitary, 4–6 mm; sporophylls ovate-lanceolate to lanceolate, abaxial ridges obvious, base glabrous, margins denticulate, apex keeled, long-bristled, bristle twisted.

Exposed granite slopes and rock, rock crevices or soil, less often in shaded sites; 600–1500 m; Ga., N.C., S.C., Tenn.

Selaginella tortipila, a very distinct species, is probably without close relatives in the flora but may be distantly related to *S. rupestris*. The two irregularly forked branches are particularly unusual: the larger one forms the strobilus while the smaller becomes arrested and forms either a budlike branch or grows and divides again to form a vegetative shoot.

13. Selaginella arenicola L. Underwood, Bull. Torrey Bot. Club 25: 541. 1898 · Sand spike-moss

Plants terrestrial or on rock, forming clumps. **Stems** radially symmetric, underground (rhizomatous) and aerial, not readily fragmenting, irregularly forked; rhizomatous and aerial stems often with 1 branch arrested, budlike, tips straight; rhizomatous stems mostly ascending; aerial stems erect or ascending. **Rhizophores** borne on upperside of stems, restricted to rhizomatous stems, 0.2–0.33 mm diam. **Leaves** dimorphic, in pseudowhorls of 4. **Rhizomatous stem leaves** persistent, appressed, scalelike. **Aerial stem leaves** tightly or somewhat loosely appressed, ascending, green, narrowly triangular-lanceolate or narrowly lanceolate, 2–3 × 0.4–0.5 mm; abaxial ridges present; base cuneate, strongly decurrent, pubescent or glabrescent; hairs restricted to base; margins short-ciliate, cilia transparent, scattered, spreading at base, dentiform and ascending toward apex, 0.02–0.07 mm; apex plane, attenuate; bristle white to whitish, straight, coarsely puberulent, 0.25–0.85(–0.9) mm. **Strobili** solitary, (0.5–)1–3(–3.5) cm; sporophylls ovate-lanceolate to lanceolate, often abruptly tapering toward apex, abaxial ridges not prominent, base glabrous, rarely with few hairs, margins ciliate, apex often recurved, bristled.

Subspecies 2: only in the flora.

Selaginella arenicola and related species have been considered as forming a species complex. This interpretation has been the center of much taxonomic controversy (R. M. Tryon 1955; G. P. Van Eseltine 1918). Tryon recognized one species in the complex, *S. arenicola*, with three subspecies: subsp. *arenicola*, subsp. *riddellii*, and subsp. *acanthonata*. Other authors (e.g., R. T. Clausen 1946) treated the subspecies as species. I recognize two well-defined species within this complex, *S. arenicola* and *S. acanthonota*, which are readily distinguishable by the characteristics given in the key. Some specimens reported by R. M. Tryon (1955) as intermediate between *S. arenicola* and *S. acanthonota* appear to be hybrids between *S. acanthonota* and *S. rupestris*. In particular, more detailed studies are needed to assess whether populations from Georgia are hybrids or variants of *S. acanthonota* or of *S. rupestris*. Future studies are also needed to determine relationships and proper taxonomic rank of *Selaginella arenicola* subsp. *arenicola* and subsp. *riddellii*, which are provisionally recognized here.

1. Leaves mostly tightly appressed; base conspicuously pubescent; strobili distinctly larger in diameter than subtending stem; sporophyll apex often recurved. . . . . 13a. *Selaginella arenicola* subsp. *arenicola*
1. Leaves usually loosely appressed; base very often glabrescent; strobili not distinctly larger in diameter than subtending stem; sporophyll apex usually straight. . . . . 13b. *Selaginella arenicola* subsp. *riddellii*

13a. Selaginella arenicola L. Underwood subsp. arenicola

Leaves (aerial stems) usually tightly appressed; base conspicuously pubescent. **Strobili** sometimes with apical vegetative growth, distinctly larger in diameter than subtending stem. **Sporophylls** abruptly tapering toward apex; apex often recurved.

Mostly on dry, exposed sand dunes, white sand, or sandy soil; 0 m; Fla., Ga.

Selaginella arenicola subsp. *arenicola* usually has more slender (1 mm diam.) stems than subsp. *riddelli* (stems more than 1 mm diam.). In Georgia, many forms intermediate between the two subspecies have been reported.

13b. Selaginella arenicola L. Underwood subsp. riddellii

(Van Eseltine) R. M. Tryon, Ann. Missouri Bot. Gard. 42: 24. 1955 · Riddell's spike-moss

Selaginella riddellii Van Eseltine, Contr. U.S. Natl. Herb. 20: 162. 1918

Leaves (aerial stems) usually loosely appressed; base often glabrescent. **Strobili** sometimes with apical vegetative growth, not distinctly larger than subtending stem. **Sporophylls** usually not abruptly tapering toward apex; apex straight.

Mostly on granite outcrops, granite boulders, or gravelly or sandy soil; 0–200 m; Ala., Ark., Ga., La., Okla., Tex.

The specimens of *Selaginella arenicola* subsp. *riddellii* from Texas are more constant in their morphologic characteristics and thus more easily separated from subsp. *arenicola*.

14. Selaginella acanthonota L. Underwood, Torreya 2:

172. 1902 · Spiny spike-moss, sandy spike-moss

Selaginella arenicola L. Underwood subsp. *acanthonota* (L. Underwood) R. M. Tryon; *S. floridana* Maxon; *S. funiformis* Van Eseltine; *S. humifusa* Van Eseltine; *S. rupestris* (Linnaeus) Spring var. *acanthonota* (L. Underwood) Clute

Plants terrestrial, less often on rock, forming close clumps. **Stems** radially symmetric, underground (rhizomatous) and aerial, not readily fragmenting, irregularly forked; rhizomatous and aerial stems often with 1 branch arrested, budlike, tips straight; rhizomatous stems with budlike branches, these sometimes inconspicuous; aerial stems erect or ascending, lateral branches conspicuously determinate. **Rhizophores** borne on upperside of stems, restricted to rhizomatous stems or lowermost base of aerial stems (seldom on distal 2/3, if so, short), mostly aerial, 0.25–0.43 mm diam. **Leaves** monomorphic, in pseudowhorls of 4 or 5, tightly appressed, ascending, green, narrowly triangular-lanceolate or narrowly lanceolate, 2–3.25 × 0.4–0.6 (–0.7) mm; abaxial ridges present; base rounded to cuneate, slightly decurrent to adnate, pubescent; margins ciliate, cilia transparent, spreading at base, dentiform, ascending toward apex, 0.02–0.1 mm; apex plane, attenuate or seldom slightly keeled; bristle white or whitish to transparent, sometimes with brownish to reddish band at base marking breaking point (in old leaves), straight, puberulent, (0.35–)0.5–1.4 mm. **Strobili** solitary, (0.5–)1–3(–3.5) cm; sporophylls ovate-lanceolate to lanceolate, abaxial ridges not prominent, base pubescent, margins ciliate, apex bristled.

Pine barrens, sand pine-oak scrubs, dry sandy hill or dunes, open white sandy soil, white sand, or sandstone rock; 0 m; Fla., Ga., N.C., S.C.

Selaginella acanthonota is a member of the *S. arenicola* complex, a taxonomically difficult group. Specimens of *S. acanthonota* from the northern part of its range (e.g., North Carolina, South Carolina, and Georgia) tend to have rather prostrate underground (rhizomatous) stems, with ascending to erect, short aerial stems. Those from Florida have rather ascending underground (rhizomatous) stems and more slender aerial stems. *Selaginella acanthonota*, in addition to features given in the description, is characterized by having hairs running lengthwise along or at least to the proximal half of the ridges bordering the abaxial groove of the leaves and sporophylls, and, usually, puberulent leaves and sporophyll apices. The hairs on the ridges sometimes break off easily or are somewhat enclosed within the abaxial groove (when the ridges close as a response to dryness), but they can be seen under a microscope. More systematic studies are needed within *S. acanthonota* and the entire *S. arenicola* complex.

15. Selaginella rupestris (Linnaeus) Spring, Flora 21: 182. 1838 · Rock spike-moss, dwarf spike-moss, sélaginelle des rochers, sélaginelle rupestre

Lycopodium rupestris Linnaeus, Sp. Pl. 2: 1101. 1753

Plants on rock or terrestrial, forming long or spreading mats or rarely cushionlike mats. **Stems** radially symmetric, long to moderately short-creeping to decumbent, not readily fragmenting, irregularly forked, without budlike arrested branches, tips straight; main stem indeterminate, lateral branches conspicuously or inconspicuously determinate, sometimes ascending, 1–3-forked. **Rhizophores** borne on upperside of stems, throughout stem length, 0.25–0.45 mm diam. **Leaves** monomorphic, in alternate pseudowhorls of 6 (on main stem) to 4 (on lateral branches), tightly appressed, ascending, green, occasionally reddish, linear or linear-lanceolate, 2.5–4(–4.5) × 0.45–0.6 mm; abaxial ridges well defined; base cuneate and decurrent on underside to rounded and adnate on upperside, pubescent or glabrous; margins long-ciliate, cilia transparent, spreading, (0.05–) 0.07–0.17 mm; apex slightly keeled, mostly attenuate; bristle white, whitish, or transparent, puberulent, 0.45–1(–1.5) mm. **Strobili** solitary, 0.5–3.5 cm; sporophylls deltate-ovate to ovate-lanceolate, strongly tapering or not toward apex, abaxial ridges well defined, base glabrous, margins ciliate to slightly dentate, apex slightly keeled, not truncate in profile, long-bristled. $2n = 18$.

Dry ledges, sea cliffs, limestone, open fire-barrens, sandstone, granite outcrops, exposed rock, rock crevices, sandy or gravelly soil or grassy meadows; 0–1900 m; Greenland; Alta., Man., N.B., N.S., Ont., Que., Sask.; Ala., Ark., Conn., Del., Ga., Ill., Ind., Iowa., Ky., Kans., Maine, Md., Mass., Mich., Minn., Miss., Mo., Nebr., N.H., N.J., N.Y., N.C., Ohio., Okla., Pa., R.I., S.C., S.Dak., Tenn., Vt., Va., W.Va., Wis., Wyo.

Selaginella rupestris has the widest range of any selaginella in the flora. It is variable in many characteristics, e.g., the hairiness of the margins (which sometimes are not hairy), leaf base pubescence, and shape of sporophylls. The variation in sporangial distribution pattern in the strobili and the number of megaspores per megasporangium are important for understanding reproduction and relationships in this species. Very often the strobili are wholly megasporangiate, with only 1–2 megaspores per megasporangium, suggesting asexual reproduction. In other cases, both types of sporangia are present in a strobilus, suggesting sexual reproduction. R. M. Tryon (1971) correlated sporangial and spore distribution patterns in *S. rupestris* with distributional ranges and concluded that there are four races. Race A has 4 megaspores per megasporangium, has microsporangia, is sexual, and is distributed from southeastern Pennsylvania south to Georgia and Alabama. Race B has 1–2(–4) megaspores per megasporangium, has microsporangia, has an unknown type of reproduction, and has the same range as Race A, but it extends into New York. Race C has 1–2 megaspores per megasporangium, has microsporangia, is probably asexual, and is distributed throughout the species range. Race D has 1–2 megaspores per megasporangium, lacks microsporangia, is therefore asexual, and is found throughout the species range except where Race A occurs. These patterns suggest the presence of more than one species within *S. rupestris* in the broad sense. More studies, especially cytologic, are needed, as well as fieldwork, in order to understand these relationships and the variability in *S. rupestris*. Among the species in the flora, *S. rupestris* seems to be most closely related to *S. sibirica* (see discussion) and may also be allied to the *S. arenicola* complex (see discussion).

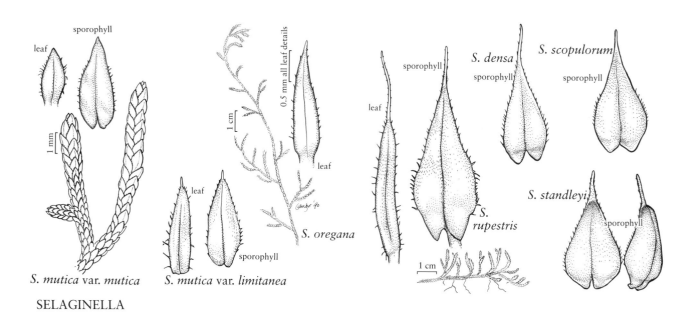

S. mutica var. mutica

S. mutica var. limitanea

SELAGINELLA

S. oregana

S. densa

S. scopulorum

S. rupestris

S. standleyi

16. Selaginella sibirica (J. Milde) Hieronymus, Hedwigia 39: 290. 1900 · Siberian spike-moss

Selaginella rupestris (Linnaeus) Spring forma *sibirica* J. Milde, Fil. Eur., 262. 1867

Plants on rock or terrestrial, forming discrete long-spreading mats or seldom cushionlike mats. **Stems** radially symmetric, creeping or decumbent, not readily fragmenting, irregularly forked, without budlike arrested branches, tips straight; main stem indeterminate, lateral branches conspicuously or inconspicuously determinate, often strongly ascending, 1–3-forked. **Rhizophores** borne on upperside of stems, throughout stem length, 0.2–0.37 mm diam. **Leaves** monomorphic, in alternate pseudowhorls of 5, tightly appressed, ascending, green, linear-lanceolate to narrowly lanceolate, 2–3.5 × 0.35–0.5 mm (smaller on lateral branches); abaxial ridges prominent; base cuneate and decurrent to rounded and adnate on young lateral branches or buds, glabrous or sometimes pubescent; margins long-ciliate, cilia transparent, spreading to ascending, 0.07–0.17 mm; apex keeled, truncate in profile, obtuse to attenuate; bristle white to whitish or transparent, puberulent, 0.45–0.8 mm. **Strobili** solitary, 0.5–2.5 cm; sporophylls deltate-ovate to ovate-lanceolate, abaxial ridges well defined, base glabrous, margins ciliate, apex truncate in profile, bristled. $2n = 18$.

Dry, alpine, rocky slopes, rock crevices, granite rock, limestone boulders, sandstone, bare open grassy tundra; 130–2400 m; N.W.T., Yukon; Alaska; Asia in Japan and the former Soviet republics.

Selaginella sibirica is most closely allied to *S. rupestris*. In addition to differences noted in the descriptions, it can be distinguished from *S. rupestris* by the numerous marginal cilia on the leaves and by the transparent sporophyll margins; *S. rupestris* has a variable number (usually few) of marginal cilia and nontransparent sporophyll margins.

17. Selaginella mutica D. C. Eaton ex L. Underwood, Bull. Torrey Bot. Club 25: 128. 1898

Plants on rock or terrestrial, forming loose mats. **Stems** radially symmetric, long- to short-creeping, not readily fragmenting, ± regularly forked, without budlike arrested branches, tips straight; main stem indeterminate, lateral branches determinate, 1–2-forked. **Rhizophores** borne on upperside of stems, throughout stem length, 0.13–0.23 mm diam. **Leaves** monomorphic, in ± alternate pseudowhorls of 3, tightly appressed, ascending, green, lanceolate to linear-lanceolate or lanceolate-elliptic, 1–2 × 0.45–0.6 mm; abaxial ridges well defined; base rounded and adnate, sometimes slightly decurrent, pubescent or glabrous; margins ciliate to denticulate, cilia transparent, spreading or ascending, 0.03–0.17 mm; apex keeled, obtuse or slightly attenuate, nearly truncate in profile, blunt to short-bristled;

bristle transparent to greenish transparent or whitish, smooth, 0.06–0.45 mm. **Strobili** solitary, (0.6–)1–3 cm; sporophylls ovate-lanceolate, ovate-elliptic, or deltate-ovate, abaxial ridges well defined, base glabrous, margins ciliate to denticulate, apex strongly to slightly keeled, short-bristled to blunt. $2n = 18$.

Varieties 2: only in the flora.

Selaginella mutica, *S. underwoodii* (R. M. Tryon 1955; C. A. Weatherby 1943), and *S. wallacei* all have similar patterns of variation. Study is needed to assess to what degree such variability is caused by environmental or genetic factors. Within *S. mutica*, two rather distinct, morphologic extremes are recognized here as varieties. Many specimens having leaves with spreading, long, marginal cilia and a short, broken, apical bristle have been considered intermediate between the two varieties, but they belong in *S. mutica* var. *mutica*.

Selaginella mutica may be one of the parent species of the putative hybrid species *S. ×neomexicana* (see discussion). *Selaginella mutica* is often found growing in the same habitat with *S. underwoodii*, *S. ×neomexicana*, and *S. weatherbiana*. According to R. M. Tryon (1955), where the two grow together, *S. mutica* mats gradually entirely replace mats of *S. underwoodii* over time. *Selaginella mutica* is sometimes confused with *S. viridissima*.

1. Margins of sporophylls usually long-ciliate, seldom denticulate, cilia spreading; apex of leaves with or without bristle, 0.03–0.06 mm; leaf margins long-ciliate, rarely denticulate, cilia spreading. . . . . . 17a. *Selaginella mutica* var. *mutica*
1. Margins of sporophylls mostly very short-ciliate to denticulate, cilia and teeth ascending; apex of leaves with bristle 0.2–0.45 mm; leaf margins short-ciliate to denticulate, cilia and teeth ascending. . . . . 17b. *Selaginella mutica* var. *limitanea*

17a. Selaginella mutica D. C. Eaton ex L. Underwood var. **mutica**

Leaf apex with or without bristle, 0.03–0.06 mm; margins long-ciliate, rarely denticulate, cilia spreading. **Sporophyll** margins usually long-ciliate, seldom denticulate, cilia spreading.

Exposed or sheltered rocky bluffs, cliffs, ledges, on igneous, sandstone, or limestone; 1400–4300 m; Ariz., Colo., N.Mex., Utah, Tex., Wyo.

17b. Selaginella mutica D. C. Eaton ex L. Underwood var. **limitanea** Weatherby, J. Arnold Arbor. 25: 414. 1944

Leaf apex bristles 0.02–0.45 mm; margins short-ciliate (rarely long-ciliate) to denticulate, cilia and teeth ascending. **Sporophyll** margins mostly with very short cilia to denticulate, cilia and teeth ascending.

Sheltered cliffs, rocky hillsides, igneous rock; 1300–2400 m; Ariz., N.Mex., Tex.

18. Selaginella cinerascens A. A. Eaton, Fern Bull. 7: 33. 1899 · Gray spike-moss

Plants terrestrial, forming loose to compact mats. **Stems** creeping, not readily fragmenting, upperside and underside structurally slightly different, irregularly forked, without budlike arrested branches, tips straight; main stem indeterminate, lateral branches determinate, ascending, 1–2-forked. **Rhizophores** borne on upperside of stems, throughout stem length, (0.17–)0.2–0.3 mm diam. **Leaves** monomorphic, not in defined pseudowhorls, loosely appressed, ascending, green, linear-lanceolate, (1–)2.5–3 × (0.25–)0.4–0.6 mm (leaves in secondary and tertiary branches smaller); abaxial ridges inconspicuous; base rounded and adnate or cuneate and slightly decurrent, glabrous, seldom pubescent; margins short-ciliate, cilia transparent, scattered, ascending, 0.02–0.75 mm; apex plane, blunt, acute to slightly acuminate (not distinctly bristled). **Strobili** solitary, 2–4 mm; sporophylls deltate-ovate to lanceolate-ovate, abaxial ridges not prominent, base glabrous, margins short-ciliate, apex not keeled, acute.

Dry open places of clay soil, clayey-sandy soil, or in shade under shrubs and trees; 0–200 m; Calif.; Mexico in Baja California.

The light brown and grayish mats, short lateral branches, narrow stem, and short strobili distinguish *Selaginella cinerascens* from all other species in the flora, in which it has no close relatives. R. M. Tryon (1955) related *S. cinerascens* to *S. arsenei* Weatherby from Mexico. *Selaginella cinerascens* also closely resembles *S. nivea* Alston from Madagascar. In California *S. cinerascens* is known only from San Diego County.

19. **Selaginella underwoodii** Hieronymus in Engler & Prantl, Nat. Pflanzenfam. 1: 714. 1901 · Underwood's spike-moss

Selaginella fendleri (L. Underwood) Hieronymus 1900, not Baker 1883

Plants on rock, forming loose festoonlike mats or rarely compact mats. **Stems** radially symmetric, long-creeping, short-creeping, or pendent, not readily fragmenting, irregularly forked, tips straight; main stem indeterminate, lateral branches determinate, spreading, 1–2-forked. **Rhizophores** borne on upperside of stems, throughout stem length, 0.15–0.27(–0.3) mm diam. **Leaves** monomorphic, in alternate pseudowhorls of 4 (on main stem and older lateral branches) or 3 (on young lateral branches and secondary branches), loosely appressed, ascending, green, linear to linear-lanceolate or narrowly triangular-lanceolate, (2–)2.5–3.4 × 0.45–0.5(–0.7) mm; abaxial ridges prominent; base mostly cuneate and decurrent, rarely rounded and adnate (on young branches), pubescent or glabrous; margins entire to denticulate or very short-ciliate, cilia transparent, scattered, mostly ascending, dentiform toward apex, 0.02–0.07 mm; apex keeled, slightly attenuate, short- to long-bristled; bristle transparent greenish to greenish-yellowish, rarely white, smooth, seldom slightly puberulent, sometimes breaking off, 0.25–0.7(–1) mm. **Strobili** sometimes paired, 0.5–3.5 cm; sporophylls lanceolate to ovate-lanceolate, abaxial ridges prominent, base glabrous, with prominent auricles (no other species has such prominent auricles), margins entire or very short-ciliate to denticulate, apex keeled, short- to long-bristled.

Moist or shaded cliffs, rocky slopes, rock crevices, granitic outcrops, hanging over granite cliffs, sandstone or limestone ledges; (800–)1500–3000(–4000) m; Ariz., Colo., N.Mex., Okla., Tex., Utah, Wyo.; Mexico in Chihuahua and Nuevo León.

R. M. Tryon (1971) reported that the bristle on *Selaginella underwoodii* leaves is longer (to 1.44 mm) in the southern part of the range and shorter (to 0.43 mm) northward and in central Arizona. *Selaginella underwoodii* seems to be closely related to *S. oregana*, perhaps sharing a common ancestor.

20. **Selaginella oregana** D. C. Eaton in S. Watson, Bot. California 2: 350. 1880 · Oregon spike-moss

Plants usually epiphytic, less often terrestrial, forming festoonlike mats. **Stems** radially symmetric, long-pendent, not readily fragmenting, irregularly forked, without budlike arrested branches, tips straight; main stem indeterminate, lateral branches determinate, ascending, 1-forked. **Rhizophores** borne on upperside of stems, restricted to base of pendent stems, or borne throughout on terrestrial stems, 0.13–0.2 mm diam. **Leaves** monomorphic, in alternate pseudowhorls of 4 (on main stem) and 3 (on lateral branches and secondary branches), loosely appressed, ascending, green, narrowly triangular-lanceolate to linear-lanceolate, 2–3.35 × 0.4–0.6 mm; abaxial ridges prominent, often flanked by two bands of cells (several rows wide) with whitish papillae (only in *S. oregana*, better seen on dry leaves); base cuneate and strongly decurrent on main stems and lateral branches or rounded and slightly decurrent to adnate on secondary branches, glabrous (seldom pubescent); margins entire or with very short cilia or denticulate, cilia few, transparent, scattered, ascending to slightly spreading, dentiform toward apex, 0.02–0.04 mm; apex slightly keeled, long-attenuate, short-bristled; bristle (hard to distinguish from apex) transparent or greenish transparent to yellowish or brownish (in old leaves), smooth, sometimes breaking off, (0.07–)0.17–0.4 mm. **Strobili** often paired, 1–6 cm; sporophylls lanceolate to narrowly ovate-lanceolate, abaxial ridges prominent, base glabrous, margins short-ciliate to denticulate (at middle), entire toward both base and apex, apex keeled to plane, short-bristled or merely long-attenuate.

Pendent on trunks and branches of mossy trees (*Acer macrophyllum* Pursh, *Populus trichocarpa* Torrey & A. Gray ex Hooker, and *Alnus rubra* Bongard) or on deep-shaded and moist rocky banks; of conservation concern; 0–200 m; B.C.; Calif., Oreg., Wash.

Selaginella oregana, one of the most distinct species in the flora, is easily distinguished by its usually long, epiphytic-pendent stems, slightly loose strobili, and curled branches (in dry specimens). In the flora, *S. oregana* is most closely related to *S. underwoodii*. It is sometimes confused with *S. wallacei* (see discussion), and it shares some characteristics with the Mexican species, *S. extensa* L. Underwood. In *S. oregana*, very often where a branch fork occurs, one of the branches is arrested (R. M. Tryon 1955). The strobili of *S. oregana* are among the longest in the flora, and they often show several novel features. Very often the apex of a strobilus undergoes a period of vegetative growth, thus becoming a vegetative shoot, and after an interval the apex reverts to the fertile condition, forming a strobilus again. In other cases, the strobilus forks, giving rise to two new strobili.

21. Selaginella wallacei Hieronymus, Hedwigia 39: 297. 1900 · Wallace's spike-moss

Plants on rock or terrestrial, forming loose or compact mats. **Stems** radially symmetric, creeping or decumbent, not readily fragmenting, irregularly forked, without budlike arrested branches, tips straight; main stem long, indeterminate, lateral branches determinate, ascending, 1–2-forked. **Rhizophores** borne on upperside of stems, throughout stem length, 0.23–0.36(–0.4) mm diam. **Leaves** monomorphic, in ± alternate pseudowhorls of 4, tightly or loosely appressed, ascending, green, linear-lanceolate, (1.5–)1.8–3.5 × 0.39–0.66 mm; abaxial ridges well defined; base rounded and adnate or cuneate and slightly decurrent on fleshy, loosely appressed stem leaves (from wet places), pubescent, seldom glabrous; margins short-ciliate to denticulate, cilia transparent, spreading at base, dentiform, and ascending toward apex, 0.03–0.06(–0.1) mm; apex keeled and obtuse, sometimes attenuate or plane and attenuate, abruptly short- to long-bristled; bristle transparent to whitish, puberulent, sometimes breaking off, (0.16–)0.2–0.46(–0.9) mm. **Strobili** often paired, 1–4.5(–9) cm; sporophylls deltate-ovate (mostly on exposed and compact mats) or lanceolate-ovate (on loose, spreading mats from wet places), abaxial ridges well defined, base glabrous, margins short-ciliate to denticulate, apex keeled, abruptly short-bristled, seldom tapering into bristle.

On dry, exposed cliffs, rocky slopes, rocky knolls, or sandy-gravelly soil or on moist, shaded, rocky banks or in meadows; 0–2000 m; Alta., B.C.; Calif., Idaho, Mont., Oreg., Wash.

Selaginella wallacei is extremely variable depending on its habitat (R. M. Tryon 1955). Plants in dry, exposed conditions have short stems, form compact mats with tightly appressed leaves adnate to the stem, and have a rather keeled, abruptly bristled apex. Plants from moist habitats have long stems, form rather moderately long-creeping mats, and have less appressed, decurrent, fleshy leaves, with a more plane-attenuate apex that gradually tapers into a bristle. Plants from exposed, dry conditions sometimes are confused with *S. scopulorum*, but they have a keeled apex with well-defined ridges on the abaxial groove whereas in *S. scopulorum* the leaf apex is ± plane and attenuate, and the ridges on the abaxial groove are not prominent. Plants from moist

habitats somewhat resemble plants of *S. underwoodii*.

R. M. Tryon (1955) found strobili 9 cm long in *Selaginella wallacei*, the longest strobili known within subg. *Tetragonostachys* and comparable only to those of *S. oregana*.

22. Selaginella watsonii L. Underwood, Bull. Torrey Bot. Club 25: 127. 1898 · Alpine spike-moss, Watson's spike-moss

Plants on rock or terrestrial, forming long or compact cushionlike mats. **Stems** radially symmetric, decumbent to long-creeping, not readily fragmenting, irregularly forked, without budlike arrested branches, tips straight; main stem conspicuously determinate, lateral branches conspicuously or inconspicuously determinate, strongly ascending, 1–3-forked. **Rhizophores** borne on upperside of stems, throughout stem length, 0.35–0.55 mm diam. **Leaves** monomorphic, in alternate pseudowhorls of 4, tightly or loosely appressed, ascending, green, linear-lanceolate, (2.5–)3–4 × 0.5–0.7 mm; abaxial ridges prominent; base cuneate, decurrent, glabrous or sometimes pubescent; margins entire or short-ciliate, cilia transparent, scattered, spreading, 0.05–0.1 mm; apex strongly keeled, obtuse, abruptly bristled; bristle whitish to transparent, smooth, 0.25–0.5 mm. **Strobili** solitary, 0.5–3(–3.5) cm; sporophylls lanceolate to ovate-lanceolate, abaxial ridges well defined, base glabrous, margins entire, rarely dentate, apex strongly keeled to truncate in profile, short-bristled.

On exposed or shaded cliffs, rocky slopes, rock crevices, granite boulders, quartzite rock, gravelly or sandy soil, alpine meadows, or swampy grounds; 1800–4300 m; Ariz., Calif., Idaho, Mont., Nev., Oreg., Utah, Wyo.

R. M. Tryon (1955) suggested that *Selaginella watsonii* is a possible ancestor of (or shares a common ancestor with) *S. leucobryoides*, *S. asprella*, and *S. utahensis* (see discussion).

23. Selaginella asprella Maxon, Smithsonian Misc. Collect. 72: 6. 1920 · Bluish spike-moss

Plants on rock or terrestrial, forming cushionlike or loose mats. **Stems** decumbent to short-creeping, dry stem readily fragmenting, irregularly forking, without budlike arrested branches, tips straight; main stem upperside and underside structurally slightly different, inconspicuously indeterminate, lateral branches radially symmetric, determinate or not, often strongly ascending on cushionlike mats, 1–2-forked. **Rhizophores** borne on upperside of stems, throughout stem length, 0.2–0.4 mm diam. **Leaves** monomorphic, in alternate pseudowhorls of 4, tightly appressed, ascending, green, narrowly triangular-lanceolate to linear-lanceolate, (2–)2.5–4 × 0.45–0.7 (–0.8) mm (smaller on young buds); abaxial ridges present; base cuneate and decurrent or sometimes rounded and adnate on young buds, pubescent (hairs often covering 1/4 of leaf length abaxially); margins ciliate, cilia transparent to whitish, spreading, 0.7–0.15 mm; apex keeled, attenuate or obtuse, bristled; bristle white or transparent, puberulent, 0.5–1.4 mm. **Strobili** solitary, 0.4–1.5(–2) cm; sporophylls lanceolate and strongly tapering to apex or deltate-ovate to ovate-lanceolate, abaxial ridges moderately defined, base pubescent or glabrous, margins short-ciliate to dentate, apex keeled or plane, bristled.

Limestone ridges, dry rocky slopes, igneous rock, exposed cliffs or gravelly soil; of conservation concern; 900–2700 m; Calif.; Mexico in Baja California.

Selaginella asprella may be confused with *S. leucobryoides* particularly because of its readily fragmenting stems.

24. Selaginella leucobryoides Maxon, Smithsonian Misc. Collect. 75: 8. 1920 · Mojave spike-moss

Plants on rock, forming rounded cushionlike mats. **Stems** decumbent to short-creeping, dry stems readily fragmenting; irregularly forked, without budlike arrested branches, tips straight; main stem upper side and underside structurally slightly different, inconspicuously indeterminate, lateral branches radially symmetric, determinate, strongly ascending, 1-forked. **Rhizophores** borne on upperside of stems, throughout stem length, 0.2–0.35 mm diam. **Leaves** monomorphic, ± in alternate pseudowhorls of 4, tightly appressed, ascending, green, linear-oblong to linear-lanceolate, sometimes falcate on lateral rows (on main stem), 2–4.5 × 0.5–0.65 mm (usually smaller on young ascending branches); abaxial ridges present; base cuneate and decurrent (rounded and adnate on young branches), glabrous; margins short-ciliate, cilia transparent, scattered, spreading at base to ascending and dentiform toward apex, 0.07–0.15 mm; apex slightly attenuate and bristled or obtuse and abruptly bristled; bristle whitish or transparent, puberulent, 0.2–0.6 mm. **Strobili** solitary, 0.4–1.5 cm; sporophylls deltate-ovate or lanceolate, abaxial ridges moderately defined, base glabrous, margins short-ciliate to denticulate, apex acuminate with very short bristle.

In rock crevices or on exposed rock; 800–2800 m; Ariz., Calif., Nev.

Selaginella leucobryoides has very tightly intertwined stems that readily fragment, a characteristic shared with *S. utahensis* and *S. asprella*. *Selaginella leucobryoides* is very closely related to, and difficult to separate from, *S. utahensis* (see discussion).

25. Selaginella utahensis Flowers, Amer. Fern J. 39: 83. 1949 · Utah spike-moss

Plants on rock or terrestrial, forming cushionlike mats. **Stems** decumbent to short-creeping, dry stems readily fragmenting, irregularly forked, without budlike arrested branches, tips straight; main stem upperside and underside structurally slightly different, inconspicuously indeterminate, lateral branches radially symmetric, determinate, strongly ascending, 1-forked. **Rhizophores** borne on upperside of stems, throughout stem length, 0.2–0.33 mm diam. **Leaves** monomorphic, in alternate pseudowhorls of 4, tightly appressed, ascending, green, linear-oblong to linear-lanceolate, seldom lanceolate-elliptic, sometimes falcate on lateral ranks (on main stem), 2–4.25 × 0.45–0.75(–1) mm (smaller on young ascending branches); abaxial ridges present; base cuneate and decurrent, rarely rounded and adnate, glabrous, seldom slightly pubescent; margins short-ciliate or denticulate to entire, cilia few, transparent, scattered, ascending to spreading, 0.02–0.1 mm; apex keeled, attenuate or obtuse, blunt or acute or ending in a very short bristle or mucro; bristle or mucro transparent to opaque, yellowish or whitish, smooth, 0–0.4 mm. **Strobili** solitary, 0.5–2 cm; sporophylls lanceolate to ovate-lanceolate, abaxial ridges moderately defined, base glabrous, margins short-ciliate to denticulate, apex short-bristled.

Dry sandstone crevices, sandy soil or clay soil; of conservation concern; 1300–2300 m; Nev., Utah.

Selaginella utahensis is very closely related to, and can be easily confused with, *S. leucobryoides*. The leaf apex of *S. utahensis* is sometimes blunt, smooth, and rather opaque, or with a very short bristle or mucro, and its leaves are in defined alternate pseudowhorls of four. In contrast, *S. leucobryoides* has obvious whitish, puberulent (rough) bristles, and the leaves are not in well-defined alternate pseudowhorls. The two species overlap in range and expressions of morphologic characters. They are treated here as separate species until additional studies can be carried out to determine whether or not they represent ecological variations of the same species or distinct species.

26. Selaginella densa Rydberg, Mem. New York Bot. Gard. 1: 7. 1900 · Prairie club-moss, Rocky Mountains spike-moss

Selaginella rupestris (Linnaeus) Spring var. *densa* (Rydberg) Clute

Plants terrestrial or on rock, forming cushionlike or loose mats. **Stems** decumbent or creeping, not readily fragmenting, irregularly forked, without budlike arrested branches, tips straight; main stem upperside and underside structurally slightly different, conspicuously indeterminate, lateral branches radially symmetric, conspicuously or inconspicuously determinate, strongly ascending, 2–3-forked. **Rhizophores** borne on upperside of stems, throughout stem length, 0.2–0.35 mm diam. **Leaves** essentially monomorphic, in poorly defined pseudowhorls of 5 or 6, tightly appressed, ascending, green, linear to linear-lanceolate, (2.7–)3–5 × 0.4–0.7 mm (upperside leaves smaller than underside ones, also smaller on ascending buds); abaxial ridges present; base long-decurrent, oblique, and glabrous on underside leaves, slightly decurrent, oblique, and sometimes pubescent on upperside leaves; margins long-ciliate, cilia transparent, mostly ascending or spreading on proximal 1/2, ascending on distal 1/2, 0.07–0.17(–0.2) mm; apex slightly keeled to plane, rather obtuse, abruptly long-bristled; bristle white or transparent, puberulent, (1–)1.25–1.9 mm. **Strobili** solitary, (0.5–)1–3(–4) cm; sporophylls ovate-lanceolate, seldom ovate, abaxial ridges well defined, base glabrous, margins ciliate entire length or dentate near tip, apex usually long-bristled.

Prairies, alpine meadows, dry rocky slopes, rock crevices, sandstone, quartzite or granite rock, and dry gravelly, clayey or sandy soil; 1100–4000 m; Alta., B.C., Man., Ont., Sask.; Ariz., Colo., Idaho, Mont., Nebr., N.Mex., N.Dak., S.Dak., Utah, Wyo.

Selaginella densa has been treated as including three varieties: var. *densa*, var. *scopulorum* (Maxon) R. M. Tryon, and var. *standleyi* (Maxon) R. M. Tryon (R. M. Tryon 1955), which are recognized here at the species level. Intermediates between *S. densa* and the other two species of the group may represent ecological variations of the species, hybrids between species within the complex, or hybrids with other closely related species, such as *S. watsonii*. This group is in need of detailed systematic studies. Megasporangia with only two well-developed megaspores have been observed, which may indicate apogamy and the presence of different races as found in *S. rupestris*.

27. Selaginella scopulorum Maxon, Amer. Fern J. 11: 36. 1921

Selaginella densa Rydberg var. *scopulorum* (Maxon) R. M. Tryon; *S. engelmannii* Hieronymus var. *scopulorum* (Maxon) C. F. Reed

Plants terrestrial or on rock, forming cushionlike or rather loose mats. **Stems** decumbent or creeping, not readily fragmenting, irregularly forked, without budlike arrested branches, tips straight; main stem upperside and underside structurally slightly different, conspicuously or inconspicuously indeterminate, lateral branches radially symmetric, conspicuously determinate, strongly ascending, 1–2-forked. **Rhizophores** borne on upperside of stems, throughout stem length, 0.25–0.45 mm diam. **Leaves** monomorphic, in poorly defined pseudowhorls of 4 or 6, tightly appressed, ascending, green, linear-lanceolate to linear, in lateral ranks sometimes falcate, 2.5–4(–4.3) × 0.5–0.75 mm (upperside leaves smaller than underside, smaller also on ascending buds); abaxial ridges present; base (on main stem) decurrent, oblique, and glabrous on underside leaves, slightly decurrent to adnate, oblique, and glabrous or rarely puberulent on upperside leaves; margins usually short-ciliate, cilia transparent, spreading or ascending at base, denticulate and ascending on distal 2/3, 0.02–0.07(–0.15) mm; apex plane or sometimes slightly keeled, obtuse to attenuate, abruptly bristled; bristle whitish, transparent to opaque, with few teeth or smooth, 0.5–1.1 mm. **Strobili** solitary, (0.5–)1–3(–4.5) cm; sporophylls ovate-lanceolate, lanceolate, or seldom ovate, usually tapering toward apex, abaxial ridges well defined, base glabrous, margins proximally short-ciliate to denticulate, lacking cilia apically, apex usually attenuate or slightly keeled, short-bristled.

Rocky alpine tundra, subalpine meadows, dry cliffs, rocky slopes, rock crevices, granitic outcrops and ledges, sandstone outcrops, in soil pockets among rocks, or sandy or granitic soil; 700–3700 m; Alta., B.C.; Ariz., Calif., Colo., Idaho, Mont., N.Mex., Oreg., Utah, Wash., Wyo.

Selaginella scopulorum is a member of the *S. densa* complex, in which there is a clear need for more systematic studies. Some specimens of *S. scopulorum* from Montana, Wyoming, and Colorado have more conspicuous whitish bristles than those elsewhere and are difficult to distinguish from *S. densa*.

28. Selaginella standleyi Maxon, Smithsonian Misc. Collect. 72: 9. 1920 · Standley's spike-moss

Selaginella densa Rydberg var. *standleyi* (Maxon) R. M. Tryon

Plants terrestrial or on rock, forming cushionlike or rather short, loose mats. **Stems** decumbent to short-creeping, not readily fragmenting, irregularly forked, without budlike arrested branches, tips straight; main stem upperside and underside structurally slightly different, inconspicuously indeterminate, lateral branches radially symmetric, determinate, strongly ascending, 1-forked. **Rhizophores** borne on upperside of stems throughout stem length, 0.2–0.35 mm diam. **Leaves** monomorphic, in poorly defined pseudowhorls of 5 or 6, tightly appressed, ascending, green, linear, linear-oblong or linear-lanceolate, (2.5–)3–4.5 mm (smaller on upperside leaves and in ascending buds); abaxial ridges present; base decurrent, oblique, glabrous or rarely pubescent; margins short-ciliate to denticulate, cilia transparent, scattered, spreading to ascending, 0.05–0.07 (–0.1) mm; apex keeled, obtuse, rather abruptly bristled; bristle usually yellowish or transparent to opaque, slightly puberulent or smooth, (0.4–)0.7–1.25 mm. **Strobili** solitary, 0.5–1(–2.3) cm; sporophylls deltate-ovate, rarely ovate-lanceolate, abaxial ridges well defined, base glabrous, margins short-ciliate to denticulate on distal 3/4, apex keeled, strongly truncate in profile, abruptly bristled.

Rock crevices, granitic outcrops, gravelly soil, bare soil, or alpine meadows; 2000–3700 m; Alta., B.C.; Alaska, Colo., Mont., Wyo.

R. M. Tryon (1955) reported an elevation range of 1500–4660 m for *Selaginella standleyi.* I have not seen specimens from these lower and higher elevations.

Selaginella standleyi is a member of the *S. densa* complex. It has sometimes been confused with *S. watsonii* and *S. sibirica;* it is, however, rather easy to distinguish by leaf and strobilus characters.

1c. SELAGINELLA Palisot de Beauvois subg. **STACHYGYNANDRUM** (Palisot de Beauvois) Baker, J. Bot. 21: 3. 1883

Stachygynandrum Palisot de Beauvois ex Mirbel in J. Lamarck & C. de Mirbel, Hist. Nat. Veg. 3: 477. 1803

Stems creeping, cespitose, erect, climbing, or vinelike, upperside and underside structurally different, usually highly branched, articulate or not; vessel elements absent. **Rhizophores** borne on underside or upperside, or in axils of branch forks, always present. **Leaves** of different shapes and sizes, in 4 ranks (2 median, 2 lateral), linear-lanceolate to ovate, thin and papery to thick and stiff, margins entire, dentate, or ciliate; stomates arranged throughout abaxial surface or sometimes also on adaxial surface; axillary leaves present. **Strobili** usually quadrangular, sometimes flattened. **Sporophylls** highly differentiated from the vegetative leaves, most sporophylls fertile (strobili with only 1 basal megasporangium in *S. kraussiana*), seldom with megasporophylls larger than microsporophylls, in 4 alternating ranks, with or without prominent auricles.

Species probably more than 650 (11 in the flora): tropical and subtropical regions.

Selaginella subg. *Stachygynandrum* is treated here in the broad sense, including *Selaginella eatonii* and species belonging to the *Selaginella apoda* complex.

Previous to J. G. Baker (1883, 1887), and more recently to A. C. Jermy (1986, 1990b), all species with dimorphic vegetative leaves were placed in *Selaginella* subg. *Heterophyllum* (Spring) Hieronymus. These species are now referred to each of the currently recognized subgenera (e.g., subg. *Stachygynandrum*, subg. *Heterostachys*), according to Jermy's system. *Selaginella* subg. *Stachygynandrum* has priority over subg. *Heterophyllum*.

SELECTED REFERENCES Buck, W. R. 1977. A new species of *Selaginella* in the *Selaginella apoda* complex. Canad. J. Bot. 55: 366–371. Buck, W. R. 1978. The taxonomic status of *Selaginella eatonii.* Amer. Fern J. 68: 33–36. Buck, W. R. and T. W. Lucansky. 1976. An anatomical and morphological comparison of *Selaginella apoda* and *Selaginella ludoviciana.* Bull. Torrey Bot. Club

103: 9–16. Somers, P. 1982. A unique type of microsporangium in *Selaginella* series *Articulatae*. Amer. Fern J. 72: 88–92. Somers, P. and W. R. Buck. 1975. *Selaginella ludoviciana, S. apoda* and their hybrids in the southeastern United States. Amer. Fern J. 65: 76–82. Webster, T. R. 1990. *Selaginella apoda* × *ludoviciana*, a synthesized hybrid spikemoss. Amer. J. Bot. 77(6,suppl.): 108.

Key to the species of *Selaginella* subg. *Stachygynandrum*

1. Stems articulate or swollen; rhizophores on upperside of stems; strobili with only 1 basal megasporangium; megasporophylls larger than microsporophylls. . . . . .35. *Selaginella kraussiana*
1. Stems not articulate; rhizophores on underside of stems, axillary, or seldom on upperside (*S. lepidophylla*); strobili with many megasporangia, megasporangia basal or throughout length of strobilus; megasporophylls and microsporophylls similar (except in *S. apoda* complex).
  2. Stems tufted, curling inward when dry, rosette-forming; leaves thick and stiff; growing in seasonally dry and exposed areas.
    3. Median leaves ovate-deltate to deltate, apex acuminate; margins with wide transparent portion, ciliate at base and dentate or ciliate toward apex; base nearly cordate. . . . . 29. *Selaginella lepidophylla*
    3. Median leaves lanceolate (slightly falcate), apex long-bristled, bristle 1/3–1/2 length of leaf; margins with narrow transparent portion, entire or with scattered teeth; base rounded to truncate. . . . . 30. *Selaginella pilifera*
  2. Stems prostrate, creeping or erect, never rosette-forming; leaves delicate, papery; growing in mild humid areas, very often in shady places.
    4. Plants with large aerial, erect, vinelike (high climbing) stems, or bushy.
      5. Stems mostly erect, covered with stiff hairs; median leaves long-acuminate to bristled; base of axillary leaves truncate. . . . . 34. *Selaginella braunii*
      5. Stems vinelike, climbing, or shrublike, without hairs; median leaves acute, tip rounded; base of axillary leaves with 2 auricles. . . . . 32. *Selaginella willdenowii*
    4. Plants with rather short-branched, prostrate or creeping stems.
      6. Margins of lateral leaves entire or basally ciliate; stems long-creeping, lateral branches 2–3-forked.
        7. Median leaves cuspidate with 2 ciliate auricles at base, outer auricle larger and more ciliate than inner auricle; lateral leaves ciliate at base; margins green. . . . . 31. *Selaginella douglasii*
        7. Median leaves acuminate, without auricles; lateral leaves not ciliate; margins transparent. . . . . 33. *Selaginella uncinata*
      6. Margins of lateral leaves dentate to serrate; stem short-creeping or long-creeping (as in *S. ludoviciana*), lateral branches 1–2-forked.
        8. Median leaves linear or narrowly lanceolate, 0.7–1.2 mm; apex long-bristled; plants minute, forming small mats less than 6 cm diam. . . . . 36. *Selaginella eatonii*
        8. Median leaves ovate to ovate-lanceolate, (1–)1.25–1.5(–1.8) mm; apex acute to acuminate; plants not minute, forming mats larger than 6 cm diam.
          9. Leaf margins with 3–5 rows of transparent cells; stomates of lateral leaves confined to midrib region on adaxial surface. . . . . 37. *Selaginella ludoviciana*
          9. Leaf margins undifferentiated in color or with 1(–2) rows of slightly paler cells; stomates of lateral leaves scattered over adaxial surface.
            10. Median leaves with long-attenuate, veined apices. . . . . 39. *Selaginella eclipes*
            10. Median leaves with acute apices or, if attenuate, usually keeled and without midrib extension. . . . . 38. *Selaginella apoda*

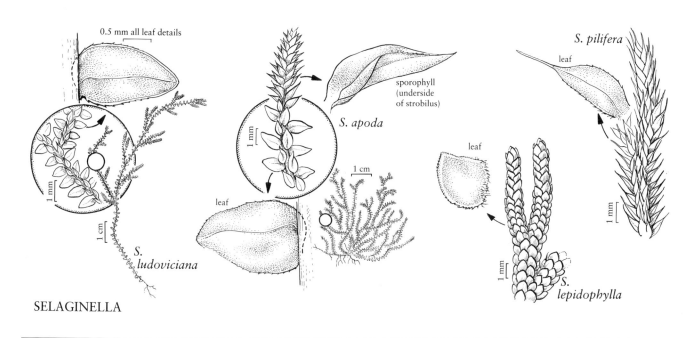

0.5 mm all leaf details

S. pilifera

leaf

sporophyll
(underside
of strobilus)

S. apoda

leaf

leaf

1 mm

1 cm

S. ludoviciana

S. lepidophylla

SELAGINELLA

29. Selaginella lepidophylla (Hooker & Greville) Spring in Martius et al., Fl. Bras. 1(2): 126. 1840 · Resurrection plant

Lycopodium lepidophyllum Hooker & Greville, Icon. Filic. 2(9): 162. 1830

Plants terrestrial or on rock, forming rosettes. **Main (central) stem** spirally compact, branched, branches 2–3-forked, prostrate, flat when moist, curling inward when dry (ball-like), not artic-ulate, weakly puberulent. **Rhizophores** borne on upper-side of stems, restricted to basal part of rosette, 0.3–0.5 mm diam. **Leaves** thick and stiff. **Lateral leaves** yellow to reddish on abaxial surface, green on adaxial surface, overlapping, ascending, deltate to deltate-ovate, 2–2.2 × (1–)1.7–1.8 mm; base nearly cordate, pubescent; margins transparent, ciliate toward base, dentate to ciliate toward apex; apex rounded. **Median leaves** broadly ovate, 1.5–1.7 × 1.4–1.5 mm; base nearly cordate to truncate, pubescent; margins transparent, ciliate toward base, dentate to ciliate toward apex; apex abruptly acuminate (short-cuspidate) to obtuse. **Strobili** solitary, 3–12 mm; sporophylls monomorphic, deltate-ovate, slightly keeled, keel not dentate, base pubescent, margins transparent, short-ciliate at base, denticulate toward apex, apex acuminate to acute.

Dry places on rocky soil or on limestone talus; of conservation concern; 900–2000 m; N.Mex., Tex.; Mexico.

Selaginella lepidophylla is sold as a commonly grown house plant and is cultivated in greenhouses. When dry, lateral branches of desiccated plants curl inward; upon rehydration, they uncurl and resume normal growth, even after years of being dry. Among the species in the flora, it is allied to *S. pilifera*.

30. Selaginella pilifera A. Braun, Index Seminum (Berlin), App., 20. 1857 · Resurrection plant

Selaginella pilifera A. Braun var. *pringlei* (Baker) C. V. Morton

Plants terrestrial or on rock, forming rosettes. **Main (central) stem** spirally compact, branched, branches 2–3-forked, prostrate, flat when moist, almost flat when dry, not articulate, glabrous. **Rhizophores** borne on underside of stems, restricted to base of rosette, 4–5 mm diam. **Leaves** thick and stiff. **Lateral leaves** overlapping, as-cending, green, elliptic to elliptic-ovate, (2–)3–3.5 × 0.8–1 mm; base cordate, with 2 ciliate lobes or auri-cles; margins transparent, acroscopic margins short-ciliate at base, dentate toward apex; basiscopic margins entire to scattered dentate; apex bristle 1/3–1/2 length of leaf blade. **Median leaves** peltate, oblique-lanceolate, 2–3 × 0.7–1 mm; base rounded to truncate, pubes-cent; margins green to slightly transparent, inner margins dentate, outer margins entire or slightly dentate; apex bristle 1/3 length of leaf blade. **Strobili** solitary, 5–10 mm; sporophylls monomorphic, lanceolate-ovate, slightly keeled, keel not dentate, base pubescent, margins transparent to greenish, short-ciliate to denticulate, apex long-bristled. $2n = 20$.

Dry rocky soil, rock crevices, limestone rock, and cliff faces; of conservation concern; 1500–2500 m; N. Mex., Tex.; n Mexico.

Selaginella lepidophylla Mettenius is a misapplied name.

The long-bristled leaf apex of *Selaginella pilifera* is unique among New World xerophytic members of subg. *Stachygynandrum* series *Circinatae* Spring. The closest relative of *S. pilifera* is *S. gypsophila* A. R. Smith & T. Reeves, which is from Nuevo León, Mexico, and differs by having obtuse leaf apices. Further studies are needed to determine whether *S. gypsophila* represents a well-differentiated species or an environmental variant of *S. pilifera*.

31. **Selaginella douglasii** (Hooker & Greville) Spring, Bull. Acad. Roy. Sci. Bruxelles 10: 138. 1843 · Douglas's spike-moss

Lycopodium douglasii Hooker & Greville, Bot. Misc. 2: 396. 1831 **Plants** on rock or terrestrial, forming loose mats. **Stems** long-creeping, branched, branches 2–3-forked, flat, not articulate, glabrous. **Rhizophores** borne on underside of stems throughout stem length or restricted to proximal ± 2/3 of main stem or axillary throughout stem, 0.2–0.4 mm diam. **Leaves** delicate and papery. **Lateral leaves** spreading or slightly ascending, distant, shiny green becoming shiny brown, with orange or red spot or entirely reddish, ovate to ovate-oblong or oblong, 1.5–3.2 × (1–)1.5–2.2 mm; base auriculate, basiscopic auricle conspicuous, acroscopic auricle inconspicuous or base ± rounded; margins green, ciliate toward auricles, otherwise entire; apex rounded to obtuse or truncate. **Median leaves** ovate-oblong, (1.8–)2–2.2 × 1–1.3 mm; base auriculate, outer auricle larger than inner one; margins green, ciliate at auricles, otherwise entire; apex abruptly cuspidate to bristled. **Strobili** paired, 0.6–1.1 cm; sporophylls monomorphic, ovate-lanceolate, keeled, keel not dentate, base glabrous, margins green, entire or with a few scattered, short cilia, apex acute to acuminate.

Rocky slopes, mossy rock, rock crevices, in partial shade, often along river banks; 100–800 m; Idaho, Oreg., Wash.

Selaginella douglasii, with no close relatives in the flora, is easy to identify by its shiny green leaves when young, turning shiny light brown when old, with an orange to red spot at the base, or totally reddish. Its closest relative is the Mexican *S. delicatissima* Linden ex A. Braun.

32. **Selaginella willdenowii** (Desvaux ex Poiret) Baker, Gard. Chron., 783, 950. 1867 · Vine spike-moss

Lycopodium willdenowii Desvaux ex Poiret in Lamarck et al., Encycl., Suppl. 3: 552. 1814 **Plants** terrestrial, vinelike or shrublike. **Stems** high-climbing, many times branched, branches 4–5-forked, flat, not articulate, glabrous. **Rhizophores** borne on upperside or underside of stems throughout stem length, 2–3 mm diam. **Leaves** delicate, papery. **Lateral leaves** distant, iridescent, blue-green, ovate to oblong, (2.5–)3–4 × (1–)1.5–2 mm (leaves on tertiary stems ± 1/3 smaller); basiscopic base rounded, acroscopic base with whitish, long, downward-curving auricle; margins transparent (whitish and shiny when dry), entire; apex rounded or obtuse. **Median leaves** falcate-lanceolate or oblique-ovate, 2.4–2.7 × 0.9–1.3 mm; base auriculate, outer auricle larger than inner; margins transparent, entire; apex obtuse. **Strobili** solitary, 0.5–2 cm; sporophylls monomorphic, cordate to ovate-deltate, base glabrous, margins green, entire, apex slightly cuspidate. $2n = 20$.

Hammocks; 0–50 m; introduced; Fla.; West Indies; Central America; Asia, native to Burma, Malaysia, Indonesia, and the Philippines.

Selaginella willdenowii is cultivated principally as a garden plant; it escapes and becomes naturalized in southern Florida. It is now widely distributed and naturalized in many regions in tropical and subtropical America. Its bushy to vinelike habit and blue-green, iridescent leaves are unusual. The iridescence is apparently caused by the effect of thin film interference filters in the leaf epidermis (D. W. Lee 1977). Lee pointed out that the convex epidermal cells in this species may focus light into a single, distal, large chloroplast, possibly adaptations for the improvement of photosynthetic efficiency at the forest floor level.

Selaginella willdenowii is related to *S. uncinata* (Desvaux ex Poiret) Spring and to *S. plana* (Desvaux ex Poiret) Hieronymus, which has been reported in Florida (O. Lakela and R. W. Long 1976) but apparently has not become naturalized. *Selaginella plana* is an erect plant; the secondary branches have obovate-oblong axillary leaves with the apices acute to slightly acuminate, lateral leaves with rounded apices, and median leaves obtuse to rounded. The sporophylls are ovate-lanceolate, with serrate to short-ciliate and very distinctive, white transparent margins.

33. Selaginella uncinata (Desvaux ex Poiret) Spring, Bull. Acad. Roy. Sci. Bruxelles 10: 141. 1843 · Blue spikemoss, peacock spike-moss

Lycopodium uncinatum Desvaux ex Poiret in Lamarck et al., Encycl., Suppl. 3: 558. 1814

Plants terrestrial, forming diffuse mats. **Stems** long-creeping, branched, branches 3-forked, flat, not articulate, glabrous. **Rhizophores** axillary, mostly at stem base or apex, 0.3–0.4 mm diam.

Leaves delicate, papery. **Lateral leaves** distant, iridescent, green to blue-green, ovate-oblong, 3–4.2 × 1.5–2.5 mm; basiscopic base with small auricle, acroscopic base overlapping stem; margins conspicuously transparent, entire; apex acute to obtuse. **Median leaves** ovate-lanceolate, 2.2–3.5 × 1.2–1.8 mm; base with outer auricle; margins transparent, entire; apex acuminate. **Strobili** solitary, 0.5–1.5 cm; sporophylls monomorphic, lanceolate to narrowly ovate-lanceolate, strongly tapering toward apex, keeled, keel not dentate, base glabrous, margins transparent, entire, apex long-acuminate. $2n = 18$.

Hammocks in shade near streams; 0–50 m; introduced; Fla., Ga., La.; Asia in China.

Selaginella uncinata is widely cultivated outdoors along the Gulf Coast of the United States and in greenhouses and nurseries. It is a native of southern China and is closely allied to *S. delicatula* (Desvaux ex Poiret) Alston, also in part from China.

34. Selaginella braunii Baker, Gard. Chron., 1120. 1867

Braun's spike-moss

Plants terrestrial, tree-shaped. **Stems** erect, highly branched, branches 4-forked, flat, not articulate, hispid. **Rhizophores** not seen. **Leaves** rugose. **Lateral leaves** distant, green, ovate-oblong, 2–2.5 × 0.7–1 mm; base truncate, basiscopically forming very short wing; margins green to slightly transparent, usually revolute, crenate; apex obtuse to rounded. **Median leaves** lanceolate-oblong to asymmetric, 2–3 × 1 mm; base peltate, rounded; margins green to slightly transparent, usually revolute, crenate; apex long-acuminate. **Strobili** solitary, 2–8 mm; sporophylls monomorphic, cordate to deltate-ovate, base glabrous, margins green, slightly crenate, apex cuspidate to acuminate. $2n = 20$.

Habitat unknown; introduced; 0–50 m; Ala., Ga., La., N.C.; native of China.

Selaginella braunii is cultivated in greenhouses, nurseries, and gardens mainly in Florida, Georgia, Louisiana, and Texas. It is related to *S. ostenfeldii* Hieronymus of Thailand, Burma, and Indochina, and to *S. mairei* H. Leveillé, also from Burma and from Indochina.

35. Selaginella kraussiana (Kunze) A. Braun, Index Seminum (Berlin), App., 22. 1860 · Mat spike-moss

Lycopodium kraussianum Kunze, Linnaea 18: 114. 1844

Plants terrestrial, forming diffuse mats. **Stems** long-creeping, branched, branches 3-forked, flat, articulate, glabrous. **Rhizophores** borne on upper side of stems throughout stem length, 1–3 mm diam. **Leaves** delicate, papery.

Lateral leaves nearly perpendicular to stem, well spaced, green, lanceolate, 2.5–3.6 × 0.8–1.2 mm; base rounded; margins slightly transparent to green, dentate; apex acute. **Median leaves** lanceolate to linear-lanceolate, 2–2.7 × 0.6–0.8 mm; base with small outer auricle; margins slightly transparent to green, dentate; apex acuminate. **Strobili** solitary, 0.2–2.5 cm, with only 1 megasporangium, megasporangium basal; sporophylls keeled, dentate, strongly tapering toward apex, base glabrous, margins denticulate, apex acuminate; megasporophylls larger than microsporophylls, in groups of 4, 2 like vegetative leaves, 2 like sporophylls, of the latter 1 large, lanceolate-elliptic, 1 smaller, falcate-lanceolate; microsporophylls lanceolate to narrowly ovate-lanceolate. $2n = 20$.

Moist areas, riverbanks, lake margins, lawns; 0–50 m; introduced; Ala., Ga.

Selaginella kraussiana has escaped from cultivation and is naturalized in central Georgia, and probably farther south and west. It has been reported as far north as coastal central California and northern Virginia (D. B. Lellinger 1985), but I have not seen specimens from these areas. *Selaginella kraussiana* is frequently cultivated and has several cultivars. It is widely used in morphologic and anatomic research and for teaching purposes. This species belongs to the series *Articulatae* Spring, a very distinct group of heterophyllous selaginellas with rhizophores on the upperside of the stem, special microsporangium type and dehiscence (P. Somers 1982), basal megasporangia, the largest megaspores in the genus, mostly spiny microspores, and usually more than one meristele. These and other characteristics suggest that series *Articulatae* probably deserves subgeneric ranking.

36. Selaginella eatonii Hieronymus ex Small, Ferns Trop. Florida, 67. 1918 · Eaton's spike-moss

Diplostachyum eatonii (Hieronymus ex Small) Small

Plants terrestrial, forming tiny (1–4 cm), dense clumps. **Stems** short-creeping, unbranched or few-forked, flat, not articulate, glabrous. **Rhizophores** axillary, 0.02–0.04 mm diam. **Leaves** delicate, papery. **Lateral leaves** spreading, well spaced or crowded toward stem tip, green, ovate to ovate-oblong, 1–1.5 × 0.5–0.9 mm; base rounded; margins transparent, serrate; apex acute. **Median leaves** lanceolate, 0.8–1.2 × 0.3–0.35 mm; base oblique; margins transparent, serrate; apex bristled; bristle to 1/3 length of leaf. **Strobili** solitary, 2–3 mm; sporophylls ovate-lanceolate, strongly keeled toward tip, keel dentate, base glabrous, margins serrate, apex long-acuminate.

Hammocks and sink holes in limestone soil; of conservation concern; 0 m; Fla.; West Indies in the Bahamas.

Selaginella eatonii is a minute species, easy to distinguish by its long-bristled leaf apex, iridescent leaf surface, and somewhat transparent sporophylls on the underside of the stem. It does not have any close relatives among the species in the flora. *Selaginella eatonii* may best be placed within subg. *Heterostachys*, with which it shares flattened strobili, keeled sporophylls, and a partial laminar flap on the sporophylls. *Selaginella eatonii* does not have strongly dimorphic sporophylls or a very well-defined laminar flap, and therefore I prefer to treat it here with the other heterophyllous species of subg. *Stachygynandrum* until a detailed study of subg. *Heterostachys* can be made.

37. Selaginella ludoviciana (A. Braun) A. Braun, Ann. Sci. Nat., Bot., sér 4, 13: 58. 1860 · Gulf spike-moss, Louisiana spike-moss

Lycopodium ludovicianum A. Braun, Index Seminum (Berlin), App., 12. 1858; *Diplostachyum ludovicianum* (A. Braun) Small

Plants terrestrial, forming diffuse mats. **Stems** long-creeping, usually ascending, sparsely branched, branches mostly simple or 1-forked, flat, not articulate, glabrous. **Rhizophores** axillary, 0.1–0.2 mm diam. **Leaves** delicate, papery. **Lateral leaves** well spaced, green, ovate to ovate-lanceolate, 1.6–2.65 × 0.98–1.64 mm; base slightly cordate; margins transparent, serrate; apex acute to slightly obtuse, conspicuously ending in teeth. **Median leaves** spaced, ovate-lanceolate to narrowly ovate-lanceolate (on basal stems), 1.3–2 × 0.4–0.8 mm; base oblique on inner side, rounded and prominent on outer side; margins transparent, serrate; apex long-acuminate to bristled. **Strobili** solitary or paired, lax, flattened, 0.4–0.7(–1.5) cm; sporophylls very strongly keeled, keel dentate, base slightly cordate to rounded, margins transparent, sparsely serrate, apex acuminate; megasporophylls larger and wider than microsporophylls, usually on underside of strobili.

Swamps, stream banks, ditch banks, or moist ravines of calcareous ledges; 0–50 m; Ala., Fla., Ga., La., Miss.

Among the species in the flora, *Selaginella ludoviciana* is most closely related to *S. apoda* (see discussion) and has often been included in *S. apoda*. W. R. Buck and T. W. Lucansky (1976) concluded that two species should be recognized based on anatomic and morphologic data. A close examination of distribution of sporangia in the strobili in many specimens reveals that sporangial arrangement may be more variable than reported by H. T. Horner Jr. and H. J. Arnott (1963). All species in the *S. apoda* complex (see *S. apoda* and *S. eclipes* for discussion on the complex) have sporophylls with fused blade tissue.

38. Selaginella apoda (Linnaeus) Spring in Martius et al., Fl. Bras. 1(2): 119. 1840 (as *Selaginella apus*) · Meadow spike-moss, sélaginelle apode

Lycopodium apodum Linnaeus, Sp. Pl. 2: 1105. 1753; *Diplostachyum apodum* (Linnaeus) Palisot de Beauvois

Plants terrestrial, forming loose or clustered mats. **Stems** prostrate to short-creeping, sparsely branched, branches mostly simple or 1-forked, flat, not articulate, glabrous. **Rhizophores** axillary, throughout stem length or restricted to proximal 1/3 of stem, 0.05–0.1 mm diam. **Leaves** delicate, papery. **Lateral leaves** distant, green, ovate to ovate-lanceolate, 1.35–2.25 × 0.75–1.35 mm; base slightly cordate; margins green or with 1 row of transparent cells, serrate; apex acute, ending in teeth. **Median leaves** ovate-lanceolate, 1–1.6 × 0.45–0.7 mm; base oblique on inner side, rounded and prominent on outer side; margins green or with row of transparent cells, serrate; apex straight, acuminate to long-acuminate. **Strobili** paired or solitary, lax, flattened, 1–2 cm; sporophylls ovate to ovate-deltate, strongly keeled, keel dentate, base slightly cordate to rounded, margins with scattered teeth, apex acuminate; megasporophylls larger and wider than microsporophylls, usually on underside of strobili.

Swamps, meadows, marshes, pastures, damp lawns, open woods, and stream banks, in basic to acidic soil; 0–100 m; Ala., Ark., Conn., Del., Fla., Ga., Ill., Ind., Ky., La., Maine, Md., Mass., Miss., Mo., N.H., N.J., N.Y., N.C., Ohio, Okla., Pa., R.I., S.C., Tenn., Tex., Vt., Va., W.Va.; Mexico in Chihuahua, s to Chiapas.

Selaginella apoda is the central component of a taxonomically difficult species complex of eastern North America. It is closely related to *S. eclipes* (see discussion) and *S. ludoviciana*. Naturally occurring and experimental hybrids between *S. apoda* and *S. ludoviciana* have been reported (P. Somers and W. R. Buck 1975; T. R. Webster 1990). Also, some evidence indicates that hybrids may occur between *S. apoda* and *S. eclipes*. More studies are needed in this complex.

The species in the *S. apoda* complex may be best classified under subg. *Homostachys* of J. G. Baker (1883, 1887), with which they share flattened strobili and larger sporophylls (megasporophylls) that are usually in the same plane as the vegetative lateral leaves. They are, however, treated here with the other heterophyllous species of subg. *Stachygynandrum* until a reassessment of the classification of the genus *Selaginella* can be made.

39. **Selaginella eclipes** W. R. Buck, Canad. J. Bot. 55: 366. 1977 · Buck's meadow spike-moss, hidden spike-moss

Plants terrestrial, forming loose to dense mats. **Stems** short-creeping, branched, branches 1–2-forked, flat, not articulate, glabrous. **Rhizophores** throughout stem length, 0.06–0.1 mm diam. **Leaves** papery, delicate. **Lateral leaves** nearly perpendicular to stem, green, ovate to ovate-elliptic, 1–2 × 0.5–1.3 mm; base rounded to slightly subcordate; margins slightly transparent, serrate; apex acute. **Median leaves** ovate to ovate-lanceolate or lanceolate, 1–1.8 × 0.4–0.8 mm; base rounded to oblique; margins green, serrate; apex abruptly tapered, long-acuminate to bristled, frequently transparent, midrib extending into apex. **Strobili** solitary or paired, lax, flattened, 1–4 cm; sporophylls ovate to ovate-deltate, strongly keeled, keel dentate, base glabrous, slightly cordate to rounded, margins serrate, apex acuminate; megasporophylls larger and wider than microsporophylls, usually on underside of strobili.

Moist to wet, calcareous habitats, swamps, meadows, pastures, open woods, or rarely on rock; 0–100 m; Ont., Que.; Ark., Ill., Ind., Iowa, Mich., Mo., N.Y., Okla., Wis.

Selaginella eclipes, a member of the *S. apoda* complex, may prove to be better treated as a subspecies of *S. apoda* (W. R. Buck 1977). It is recognized here at the specific level to highlight the problems within this species complex. Further research is needed to elucidate the relationships among the species of the complex.

4. ISOËTACEAE Reichenbach

• Quillwort Family

W. Carl Taylor

Neil T. Luebke

Donald M. Britton

R. James Hickey

Daniel F. Brunton

Plants tufted, grasslike, heterosporous (megaspores and microspores not alike), perennial; evergreen aquatics to ephemeral terrestrials. **Rootstock** brown, cormlike, lobed. **Roots** arising along central groove separating each rootstock lobe, simple or dichotomously branched, containing eccentric vascular strand and surrounding lacuna. **Leaves** linear, simple, spirally or distichously arranged, dilated toward base, tapering to apex, containing 4 transversely septate longitudinal lacunae, a central collateral vascular strand, and frequently several peripheral fibrous bundles; ligule inserted above sporangium. **Megasporophylls and microsporophylls** usually borne in alternating cycles; hardened scales and phyllopodia occasionally surrounding leaves. **Sporangia** solitary, adaxial, embedded in basal cavity of leaf, velum (thin flap extending downward over sporangium) partly to completely covering adaxial surface of sporangium; megasporangium with several to hundreds of megaspores; microsporangium with thousands of microspores. **Megagametophytes** white, endosporic, exposed when megaspore opens along proximal ridges; archegonia 1 to several, indicated by quartets of brownish neck cells. **Microgametophytes** 9-celled, endosporic, antheridium releasing 4 multitailed spermatozoids.

Genus 1, species ca. 150 (24 species in the flora): nearly worldwide.

SELECTED REFERENCES Jermy, A. C. 1990. Isoëtaceae. In: K. Kubitzki et al., eds. 1990+. The Families and Genera of Vascular Plants. 1+ vol. Berlin etc. Vol. 1, pp. 26–31. Pfeiffer, N. E. 1922. Monograph of the Isoëtaceae. Ann. Missouri Bot. Gard. 9: 79–233. Reed, C. F. 1953. Index Isoëtales. Bol. Soc. Brot., ser. 2a, 27: 5–72. Tryon, R. M. and A. F. Tryon. 1982. Ferns and Allied Plants, with Special Reference to Tropical America. New York, Heidelberg, and Berlin.

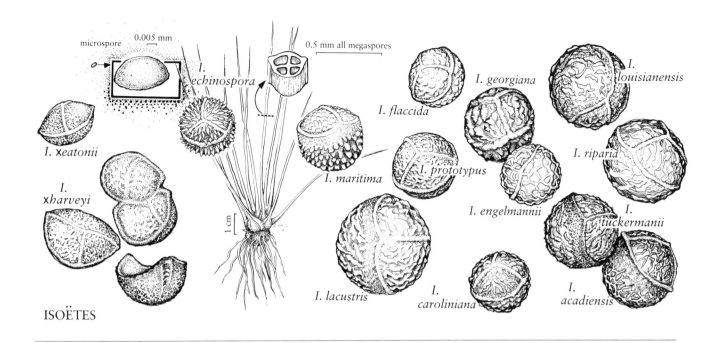

I. echinospora

microspore 0.005 mm

0.5 mm all megaspores

I. xeatonii

I. xharveyi

I. maritima

I. flaccida

I. georgiana

I. louisianensis

I. prototypus

I. riparia

I. engelmannii

I. tuckermanii

ISOËTES

I. lacustris

I. caroliniana

I. acadiensis

1. ISOËTES Linnaeus, Sp. Pl. 2: 1100. 1753; Gen. Pl. ed. 5, 486. 1754 · Quillwort

[Greek *isos,* equal, and *etos,* year, referring to evergreen habit of some species]

Rootstock 2(–3)-lobed, nearly globose to horizontally spindle-shaped and proliferous, corky. **Leaves** several to many, erect to spreading, straight to recurved, 1–100 cm; ligule deltate to cordiform, 1–6 mm, membranous. **Sporangia** ovoid to ellipsoid or oblong, 3–15 mm, walls unpigmented or brown-streaked to completely brown, traversed internally by trabeculae (internal partitions). **Megaspores** white, gray, or black, globose, mostly 300–700 µm diam., trilete, each with equatorial ridge and 3 converging proximal ridges, smooth or with spines, tubercles, or ridges. **Microspores** grayish or brownish in mass, reniform, mostly 20–50 µm, monolete, smooth or textured with spines, tubercles, or ridges. $x = 11$.

Species ca. 150 (24 in the flora): nearly worldwide.

Geography, habitat, megaspore texture, spore size, and velum provide features that will distinguish *Isoëtes* taxa. The northeastern, southeastern, and western regions of North America contain endemic species. Within these geographic regions, habitats are important for identification. Some species are submerged or emergent aquatics growing in permanent lakes, ponds, streams, estuaries, or bogs, or on persistently wet soil. Others are temporary aquatics that pass into dormancy as the pools and streams that they inhabit dry. A few are seasonal terrestrials, actively growing in spring when the soil is saturated.

Texture and size of mature, dry megaspores are usually required for identification. A 10× hand lens will adequately resolve megaspore textures of some species, but magnification of 30× or more is required for others. A compound microscope fitted with an ocular micrometer is necessary to determine spore size. Twenty spores should be measured to determine their average size.

Normally, megaspores are globose and marked with four bold ridges: an equatorial ridge encircling the spore and three radial ridges converging at the proximal pole of the spore.

Between these ridges, megaspore textures may be echinate, cristate, reticulate, rugulate, tuberculate, or nearly smooth. A zone around the spore along the distal side of the equatorial ridge is called the girdle. The girdle is obscure when it is textured like the rest of the spore or it is distinguishable when it is textured differently.

Plants collected early in the growing season possess small, fragile, yellowish white, immature megaspores that have an underdeveloped smooth or mealy surface. Such specimens are difficult to identify and have been a source of taxonomic confusion. Mature spores may be found in decaying leaf bases or in soil around the rootstock throughout the year, but plants collected soon after their growing season are easiest to identify because spores are well developed and still within sporangia.

The velum, a thin flap of tissue completely or partially covering the adaxial wall of the sporangium, also has diagnostic value. Seven taxa in this flora have a velum covering the entire sporangium wall (see illustration of *Isoëtes nuttallii* leaf base). The others here have a velum covering less than three-fourths of the sporangium wall (see illustration of *I. howellii* leaf base). Although the velum is usually membranous, it is reasonably durable and its coverage can be determined by using forceps to lift it off the sporangium wall.

Interspecific hybrids, which are frequent, have confused clear distinctions between species. These hybrids, recognized by often flattened, malformed spores of variable size, shape, and texture, may be expected where two or more species occur together.

Species of *Isoëtes* appear to have evolved in two ways, either by ecological isolation and genetic divergence as taxa adapted to terrestrial or aquatic habitats or through interspecific hybridization and chromosome doubling as divergent species migrated, possibly via waterfowl, into the same aquatic habitats. Recognition of primary diploid species, interspecific hybrids, and allopolyploids reduces confusion in the identification of *Isoëtes* species.

SELECTED REFERENCES   Boom, B. M. 1982. Synopsis of *Isoëtes* in the southeastern United States. Castanea 47: 38–59. Cody, W. J. and D. M. Britton. 1989. Ferns and Fern Allies of Canada. Ottawa. Eaton, A. A. 1900. The genus *Isoëtes* in New England. Fernwort Pap. 2: 1–16. Engelmann, G. 1882. The genus *Isoëtes* in North America. Trans. Acad. Sci. St. Louis 4: 358–390. Hickey, R. J. 1986. *Isoëtes* megaspore surface morphology: Nomenclature, variation, and systematic importance. Amer. Fern J. 76: 1–16. Hickey, R. J., W. C. Taylor, and N. T. Luebke. 1989. The species concept in Pteridophyta with special reference to *Isoëtes*. Amer. Fern J. 79: 78–89. Kott, L. S. and D. M. Britton. 1983. Spore morphology and taxonomy of *Isoëtes* in northeastern North America. Canad. J. Bot. 61: 3140–3163. Reed, C. F. 1965. *Isoëtes* in southeastern United States. Phytologia 12: 369–400. Soper, J. H. and S. Rao. 1958. *Isoëtes* in eastern Canada. Amer. Fern J. 48: 97–102. Taylor, T. M. C. 1970. Pacific Northwest Ferns and Their Allies. Toronto. Taylor, W. C. and R. J. Hickey. 1992. Habitat, evolution, and speciation of *Isoëtes*. Ann. Missouri Bot. Gard. 79: 613–622. Taylor, W. C., N. T. Luebke, and M. B. Smith. 1985. Speciation and hybridization in North American quillworts. Proc. Roy. Soc. Edinburgh, B 86: 259–263.

1. Megaspores with echinate texture.
   2. Megaspores echinate with thin, sharp spines; girdle obscure; microspores averaging less than 30 μm; circumboreal. . . . . 1. *Isoëtes echinospora*
   2. Megaspores echinate to short-cristate with stout, blunt spines and crests; girdle densely echinate; microspores averaging more than 30 μm; Alaska, British Columbia, Washington. . . . . 2. *Isoëtes maritima*
1. Megaspores with cristate, reticulate, rugulate, tuberculate, or smooth textures.
   3. Plants e of Rocky Mountains.
      4. Plants submerged or emergent aquatics, of permanent lakes, ponds, streams, estuaries, and bogs, or of persistently wet soil.
         5. Megaspores averaging more than 600 μm diam.; plants submerged aquatics. . . 3. *Isoëtes lacustris*
         5. Megaspores averaging less than 600 μm diam.; plants submerged or emergent aquatics.
            6. Velum covering more than 1/2 of sporangium.

7. Velum covering less than 3/4 of sporangium.

    8. Sporangium wall brown-streaked; megaspores cristate to reticulate with broad jagged ridges; plants of Georgia. . . . . 4. *Isoëtes georgiana*

    8. Sporangium wall unpigmented; megaspores cristate to reticulate with broken lamellate ridges; plants of North Carolina, Tennessee, Virginia, and West Virginia. . . . . 5. *Isoëtes caroliniana*

7. Velum covering entire sporangium.

    9. Leaves dark green, rigid; plants of New Brunswick and Nova Scotia, and Maine. . . . . 6. *Isoëtes prototypus*

    9. Leaves bright green, pliant; plants of Florida and Georgia. . . . . . . 7. *Isoëtes flaccida*

6. Velum covering less than 1/2 of sporangium.

    10. Megaspores averaging less than 500 μm diam.; reticulate with unbroken lamellate ridges. . . . . 8. *Isoëtes engelmannii*

    10. Megaspores averaging more than 500 μm diam.; reticulate, rugulate, or cristate with isolated or broken ridges.

        11. Megaspores with densely papillate or smooth girdle; leaves olive green to reddish brown.

            12. Megaspores with densely papillate girdle, reticulate or cristate with ridges having irregular and roughened crests. . . . . . 9. *Isoëtes tuckermanii*

            12. Megaspores with smooth girdle, rugulate to reticulate with ridges having rounded and smooth crests. . . . . 10. *Isoëtes acadiensis*

        11. Megaspores with obscure girdle; leaves bright green.

            13. Megaspores cristate, with isolated and branching lamellate ridges; plants of ne North America and e seaboard. . . . . 11. *Isoëtes riparia*

            13. Megaspores cristate to reticulate, with branched and interconnected ridges; plants of s United States.

                14. Megaspores with thick ridges; plants of Louisiana. . . . . 12. *Isoëtes louisianensis*

                14. Megaspores with thin ridges or lamellae; plants of Georgia. . . . . 13. *Isoëtes boomii*

4. Plants terrestrial or becoming so, of seasonally saturated soil, temporary pools, and streams.

    15. Velum covering entire sporangium; megaspores gray, brown, or black; plants of temporary pools on granite outcrops.

        16. Megaspores obscurely rugulate; plants of Texas. . . . . 14. *Isoëtes lithophila*

        16. Megaspores tuberculate; plants of Georgia.

            17. Leaves distichously arranged; rootstock horizontally elongated, mat-forming. . . . . 15. *Isoëtes tegetiformans*

            17. Leaves spirally arranged; rootstock nearly globose, not mat-forming. . . . . 16. *Isoëtes melanospora*

15. Velum covering less than 3/4 of sporangium; megaspores white; plants of varied habitats.

    18. Megaspores averaging more than 450 μm diam.; plants of calcareous soil. . . . . 17. *Isoëtes butleri*

    18. Megaspores averaging less than 450 μm diam.; plants of noncalcareous soil.

        19. Megaspores obscurely rugulate; leaves pale to lustrous black toward base; plants of e, c United States. . . . . 18. *Isoëtes melanopoda*

        19. Megaspores boldly tuberculate to rugulate; leaves pale to dark brown toward base; plants of se United States. . . . . 19. *Isoëtes virginica*

3. Plants of Rocky Mountains and further west.

  20. Plants submerged or emergent aquatics, of persistent lakes and pools.

    21. Megaspores averaging less than 500 μm diam.; leaves abruptly tapering to fine tip. . . . . 20. *Isoëtes bolanderi*

21. Megaspores averaging more than 500 μm diam.; leaves gradually tapering to
    tip. . . . . 21. *Isoëtes occidentalis*

20. Plants terrestrial or becoming so, of seasonally saturated soil, temporary streams,
    vernal pools.

    22. Velum covering less than 3/4 of sporangium; sporangium wall brown-streaked
    to completely brown. 22. *Isoëtes howellii*

    22. Velum covering entire sporangium; sporangium wall unpigmented.

        23. Plants of seasonally saturated soil, temporary streams; leaves generally
       more than 8 cm, rigid, almost brittle; megaspores averaging more than
       350 μm diam. . . . . 23. *Isoëtes nuttallii*

        23. Plants of vernal pools; leaves generally less than 8 cm, pliant; megaspores
       averaging less than 350 μm diam. . . . . 24. *Isoëtes orcuttii*

1. Isoëtes echinospora Durieu, Bull. Soc. Bot. France 8:
164. 1861· Spiny-spored quillwort, isoète à spores
épineuses

Isoëtes braunii Durieu; *I. echino-*
spora subsp. *muricata* (Durieu)
A. Löve & D. Löve; *I. echino-*
spora var. *boottii* Engelmann; *I.*
echinospora var. *braunii* (Durieu)
Engelmann; *I. echinospora* var.
muricata (Durieu) Engelmann; *I.*
muricata Durieu; *Isoëtes setacea*
Lamarck subsp. *muricata*
(Durieu) Holub

Plants usually aquatic, occasionally emergent. **Root-**
stock nearly globose, 2-lobed. **Leaves** ± deciduous,
bright green to reddish green, pale toward base, spirally arranged, to 25(–40) cm, pliant, gradually tapering toward tip. **Velum** covering less than 1/2 of sporangium. **Sporangium wall** ± brown-streaked. **Megaspores**
white, 400–550 μm diam., echinate with thin, sharp
spines; girdle obscure. **Microspores** gray to light brown
in mass, 20–30 μm, smooth to spinulose. $2n = 22$.

Spores mature in late summer. Emergent or in shal-
low, cool, oligotrophic water of slightly acidic lakes,
ponds, and streams; Greenland; Alta., B.C., Man., N.B.,
Nfld., N.W.T., N.S., Ont., P.E.I., Que., Sask., Yukon;
Alaska, Calif., Colo., Conn., Idaho, Maine, Mass.,
Mich., Minn., Mont., N.H., N.J., N.Y., Ohio, Oreg.,
Pa., R.I., Utah, Vt., Wash., Wis.; circumboreal.

North American plants of *Isoëtes echinospora*, which
bear stomata, have been called *I. muricata* or *I. echi-*
nospora var. *braunii* to distinguish them from European plants of *I. echinospora*, which do not have stomata.

Isoëtes echinospora is a distinct species but has considerable variation, especially in size, color, and form
of leaves. It is the most commonly encountered quillwort in oligotrophic, noncalcareous lakes and ponds of
northeastern North America.

Isoëtes echinospora hybridizes with *I. bolanderi*; *I.*
engelmannii [= *I.* × *eatonii* Dodge (later synonym =

I. × *gravesii* A. A. Eaton)]; *I. lacustris* [= *I.* × *hickeyi*
Taylor & Luebke]; *I. maritima*; *I. riparia* [= *I.* × *dodgei*
A. A. Eaton]; and *I. tuckermanii.*

2. Isoëtes maritima L. Underwood, Bot. Gaz. 13: 94.
1888 · Maritime quillwort

Isoëtes beringensis Komarov;
I. echinospora Durieu subsp.
maritima (L. Underwood)
A. Löve; *I. echinospora* var.
maritima (L. Underwood) A. A.
Eaton; *I. macounii* A. A. Eaton
Plants usually aquatic, occasionally emergent. **Rootstock** nearly
globose, 2-lobed. **Leaves** ±
deciduous, bright green, pale toward base, spirally
arranged, to 12 cm, pliant, gradually tapering toward
tip. **Velum** covering less than 1/2 of sporangium.
Sporangium wall ± brown-streaked. **Megaspores** white,
380–600 μm diam., echinate to short-cristate with stout,
blunt spines and crests; girdle densely echinate.
Microspores gray to brown in mass, 30–40 μm,
papillose. $2n = 44$.

Spores mature late summer. Lakes and streams; B.C.;
Alaska, Wash.

Isoëtes maritima hybridizes with *I. echinospora* and
I. occidentalis [= *I.* × *truncata* (A. A. Eaton) Clute].

3. Isoëtes lacustris Linnaeus, Sp. Pl. 2: 1100. 1753
· Lake quillwort, isoète lacustre

Isoëtes hieroglyphica A. A.
Eaton; *I. macrospora* Durieu
Plants aquatic, submerged.
Rootstock nearly globose, 2-
lobed. **Leaves** evergreen, dark
green to reddish green, pale
brown toward base, spirally
arranged, to 25 cm, rigid,
abruptly tapering to tip. **Velum**
covering less than 1/2 of sporangium. **Sporangium wall**
± brown-streaked. **Megaspores** white, 550–750 μm

diam., cristate to reticulate with branching to anastomosing ridges; girdle densely papillate or rarely smooth. **Microspores** gray in mass, 33–45 μm, papillose. $2n = 110$.

Spores mature late summer. Cool, oligotrophic, slightly acidic lakes and streams; Greenland; Man., N.B., Nfld., N.S., Ont., P.E.I., Que., Sask.; Conn., Maine, Mass., Mich., Minn., N.H., N.Y., R.I., Tenn., Vt., Va., Wis.; n,c Europe.

Populations of *Isoëtes lacustris* in Tennessee and Virginia are disjunct.

North American plants of *Isoëtes lacustris* have been segregated as *I. macrospora*; both taxa are decaploids ($2n = 110$) and have similar leaf and spore morphology. They cannot reliably be distinguished from each other except on the basis of geography.

Isoëtes lacustris is a totally submerged aquatic. Plants have been found at depths of more than 3 m. Plants with rugulate megaspores bearing smooth, rounded ridges and a smooth girdle have been called *I. hieroglyphica* A. A. Eaton [*I. macrospora* f. *hieroglyphica* (A. A. Eaton) N. E. Pfeiffer].

Isoëtes lacustris hybridizes with *I. echinospora* [= *I. ×hickeyi* W. C. Taylor & Luebke]; *I. engelmannii*; *I. riparia* [= *I. ×jeffreyi* D. M. Britton & Brunton]; and with *I. tuckermanii* [= *I. ×harveyi* A. A. Eaton].

4. Isoëtes georgiana Luebke, Amer. Fern J. 82: 24. 1992 · Georgia quillwort

Plants aquatic, emergent. **Rootstock** nearly globose, 2-lobed. **Leaves** deciduous, olive green, pale toward base, spirally arranged, to 40 cm, pliant, gradually tapering toward tip. **Velum** covering less than 3/4 of sporangium. **Sporangium wall** ± brown-streaked. **Megaspores** white, 450–650 μm diam., cristate to reticulate with broad, jagged ridges; girdle obscure. **Microspores** light brown in mass, 23–33 μm, papillose. $2n = 66$.

Spores mature late summer. Creeks; of conservation concern; Ga.

Isoëtes georgiana is known only from Worth County, Georgia. An early collection from this locality was reported by B. M. Boom (1982) to be *I. engelmannii* × *piedmontana*, but the broad velum and bold texture of the megaspores suggest a possible involvement of *I. flaccida*.

5. Isoëtes caroliniana (A. A. Eaton) Luebke, Amer. Fern J. 82: 26. 1992 · Carolina quillwort

Isoëtes engelmannii A. Braun var. *caroliniana* A. A. Eaton, Fern Bull. 8: 60. 1900

Plants aquatic, emergent. **Rootstock** nearly globose, 2-lobed. **Leaves** evergreen, bright green, pale toward base, spirally arranged, to 60 cm, pliant, gradually tapering to tip. **Velum** covering less than 3/4 of sporangium. **Sporangium wall** unpigmented. **Megaspores** white, 400–550 μm diam., cristate to reticulate with broken lamellate ridges; girdle obscure. **Microspores** light brown in mass, 24–34 μm, spinulose. $2n = 22$.

Spores mature summer. Emergent or in shallow water of upland lakes and bogs; N.C., Tenn., Va., W.Va.

6. Isoëtes prototypus D. M. Britton, Canad. J. Bot. 69: 278. 1991 · Prototype quillwort

Plants aquatic, submerged. **Rootstock** nearly globose, 2-lobed. **Leaves** evergreen, dark green, pale reddish brown toward base, spirally arranged, to 12 cm, rigid, gradually tapering to tip. **Velum** covering entire sporangium. **Sporangium wall** unpigmented, entirely enclosed in translucent saccate membrane. **Megaspores** white, 425–575 μm diam., obscurely rugulate with molded and wavy ridges; girdle obscure. **Microspores** light brown in mass, 24–32 μm, spinulose. $2n = 22$.

Spores mature in summer. Deep water of cold, oligotrophic, acidic lakes; of conservation concern; N.B., N.S.; Maine.

7. Isoëtes flaccida A. Braun, Flora 29: 178. 1846 · Florida quillwort

Isoëtes chapmanii (Engelmann) Small

Plants aquatic, emergent. **Rootstock** nearly globose, 2–3-lobed. **Leaves** evergreen, bright green, pale toward base, spirally arranged, to 60 cm, pliant, gradually tapering to tip. **Velum** covering entire sporangium. **Sporangium wall** unpigmented. **Megaspores** white, 250–500 μm diam., tuberculate to rugulate; girdle smooth. **Microspores** light brown in mass, 25–33 μm, papillose. $2n = 22$.

Spores mature in summer. Emergent or in shallow

water of lakes, ponds, streams, ditches, and marshes; Fla., Ga.

Isoëtes flaccida includes individuals with a wide range of megaspore sizes and textures. Although megaspores are fairly uniform among individuals within a population, considerable variation occurs between populations. A local Florida population with megaspores that are nearly smooth on the proximal faces and low-tuberculate on the distal hemisphere has been described as *I. flaccida* var. *chapmanii* Engelmann. *Isoëtes flaccida* var. *alata* (Small) N. E. Pfeiffer has included plants in which the megaspores are densely tuberculate to rugulate on the proximal faces and have bold, anastomosing ridges on the distal hemispheres.

Isoëtes flaccida hybridizes with *I. engelmannii*.

8. Isoëtes engelmannii A. Braun, Flora 29: 178. 1846 · Engelmann's quillwort

Isoëtes valida (Engelmann) Clute
Plants aquatic, emergent. **Rootstock** nearly globose, 2-lobed. **Leaves** evergreen, bright green, pale toward base, spirally arranged, to 60(–90) cm, pliant, gradually tapering to tip. **Velum** covering less than 1/4 of sporangium. **Sporangium wall** usually unpigmented, occasionally ± brown-streaked. **Megaspores** white, 400–560 μm diam., reticulate, with unbroken lamellate ridges; girdle obscure. **Microspores** gray in mass, 20–30 μm, smooth to papillose. $2n = 22$.

Spores mature in summer. Emergent or in shallow water of lakes, ponds, streams, and ditches; Ont.; Ala., Ark., Conn., Del., Fla., Ga., Ill., Ind., Ky., Md., Mass., Mich., Mo., N.H., N.J., N.Y., N.C., Ohio, Pa., R.I., S.C., Tenn., Vt., Va., W.Va.

Isoëtes engelmannii is the most widely distributed quillwort of eastern North America. Plants with larger megaspores, ranging from 480–560 μm, have been called *I. engelmannii* var. *georgiana* Engelmann. This variety may represent a tetraploid cytotype. A tetraploid population of *I. engelmannii* ($2n = 44$) from northern Florida has larger megaspores characteristic of var. *georgiana*.

Isoëtes engelmannii hybridizes with *I. echinospora* [= *I. ×eatonii* Dodge (later synonym = *I. ×gravesii* A. A. Eaton)]; *I. flaccida*; *I. tuckermanii* [= *I. ×foveolata* A. A. Eaton ex Dodge]; *I. lacustris*; and *I. riparia* [= *I. ×brittonii* Brunton & W. C. Taylor].

9. Isoëtes tuckermanii A. Braun ex Engelmann in A. Gray, Manual ed. 5, 676. 1867 · Tuckerman's quillwort, isoète de Tuckerman

Plants aquatic, occasionally emergent. **Rootstock** nearly globose 2(–3)-lobed. **Leaves** evergreen, olive green to reddish brown, pale toward base, spirally arranged, to 20 cm, pliant to rigid, gradually tapering to tip. **Velum** covering less than 1/2 of sporangium. **Sporangium wall** brown-streaked. **Megaspores** white, 450–650 μm diam., reticulate with ridges having irregular and roughened crests; girdle densely papillate. **Microspores** gray in mass, 25–35 μm, spinulose. $2n = 44$.

Spores mature in late summer. Slightly acid lakes, ponds, and streams; N.B., Nfld., N.S., Ont., Que.; Conn., Maine, Mass., N.H., N.J., N.Y., R.I., Vt.

Isoëtes tuckermanii may be an allotetraploid like *I. riparia*, but it appears to have had different parental species. Variation may also have resulted from multiple allotetraploid origins. Typical plants of *I. tuckermanii* have thin, soft leaves. More northern plants, referred to as *I. tuckermanii* var. *borealis* A. A. Eaton, have thick, stiff leaves.

Isoëtes tuckermanii hybridizes with *I. echinospora*; *I. engelmannii* [= *I. ×foveolata* A. A. Eaton ex Dodge]; and *I. lacustris* [= *I. ×harveyi* A. A. Eaton].

10. Isoëtes acadiensis Kott, Canad. J. Bot. 59: 2592. 1981 · Acadian quillwort

Plants aquatic, submerged. **Rootstock** nearly globose 2(–3)-lobed. **Leaves** evergreen, olive green to reddish brown, pale toward base, spirally arranged, to 20 cm, pliant, gradually tapering to tip. **Velum** covering less than 1/2 of sporangium. **Sporangium wall** brown-streaked. **Megaspores** white, 400–650 μm diam., rugulate to reticulate with ridges having rounded or smooth crests; girdle smooth. **Microspores** gray in mass, 25–35 μm, spinulose. $2n = 44$.

Spores mature in late summer. Emergent or in shallow water of slightly acid lakes, ponds, and streams; N.B., Nfld., N.S.; Mass., Maine, N.H., N.Y.

Isoëtes acadiensis can be distinguished from *I. tuckermanii* by its megaspore texture. Where these two taxa grow together, plants bearing megaspores with intermediate patterns occur. In Atlantic Canada, *I. acadiensis* typically occupies shallower water than *I. tuckermanii*.

Isoëtes acadiensis hybridizes with *I. lacustris*.

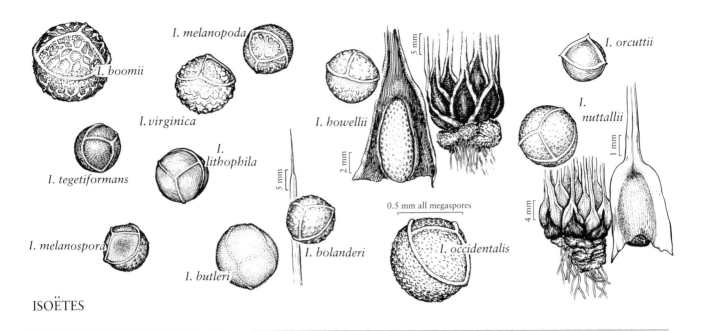

ISOËTES

11. Isoëtes riparia Engelmann ex A. Braun, Flora 29: 178. 1846 · Shore quillwort, isoète riparien

Isoëtes canadensis (Engelmann) A. A. Eaton; *I. saccharata* Engelmann

Plants usually aquatic, occasionally emergent. **Rootstock** nearly globose, 2-lobed. **Leaves** evergreen, bright green, pale toward base, spirally arranged, often twisted, to 35(−50) cm, pliant, gradually tapering to tip. **Velum** covering less than 1/2 of sporangium. **Sporangium wall** brown-streaked to completely brown. **Megaspores** white, 450–650 μm diam., cristate with isolated and branching, lamellate ridges; girdle obscure. **Microspores** gray to light brown in mass, 30–40 μm, tuberculate. $2n = 44$.

Spores mature late summer. Margins of lakes, ponds, and streams; tidal shores and estuaries; calcareous to slightly acidic substrates in fresh, usually oligotrophic, water; Ont., Que.; Conn., Del., Fla., Maine, Md., Mass., N.H., N.J., N.Y., N.C., Pa., R.I., S.C., Vt., Va., W.Va.

Megaspore morphology, electrophoretic profiles of leaf enzymes, and chromosome number provide evidence that *Isoëtes riparia* is an allotetraploid, formed by chromosome doubling in *I.* ×*eatonii*, the primary diploid hybrid between *I. echinospora* and *I. engelmannii*. The considerable variation in leaf and spore morphology possibly results from multiple allotetraploid origins.

Isoëtes riparia hybridizes with *I. echinospora* [= *I.*
×*dodgei* A. A. Eaton]; *I. engelmannii* [= *I.* ×*brittonii* Brunton & W. C. Taylor]; *I. lacustris* [= *I.* ×*jeffreyi* D. M. Britton & Brunton]; and *I. tuckermanii*.

12. Isoëtes louisianensis Thieret, Sida 5: 129. 1973 · Louisiana quillwort

Plants aquatic, emergent. **Rootstock** nearly globose, 2-lobed. **Leaves** evergreen, bright green, pale toward base, spirally arranged, to 40 cm, pliant, gradually tapering to tip. **Velum** covering less than 1/2 of sporangium. **Sporangium wall** brown-streaked. **Megaspores** white, 500–625 μm diam., cristate with thick ridges; girdle obscure. **Microspores** brown in mass, 25–35 μm, spinulose. $2n = 44$.

Spores mature winter–early spring. Creeks; of conservation concern; La.

Isoëtes louisianensis is known from drainage streams in St. Tammany and Washington parishes in southeastern Louisiana. Based on the reticulate texture of the megaspore, it is possible that *I. louisianensis* represents an allotetraploid with *I. engelmannii* as one of its parents. *Isoëtes louisianensis* may represent *I. engelmannii* × *melanopoda* (B. M. Boom 1982). Spores are uniform in size and texture and readily germinate in culture.

13. Isoëtes boomii Luebke, Amer. Fern J. 82: 23. 1992 · Boom's quillwort

Plants aquatic, emergent. **Rootstock** nearly globose, 2-lobed. **Leaves** deciduous, bright green, pale toward base, spirally arranged, to 45 cm, pliant, gradually tapering toward tip. **Velum** covering less than 1/2 of sporangium. **Sporangium wall** ± brown-streaked. **Megaspores** white, 460–610 μm diam., cristate to reticulate with thin ridges; girdle obscure. **Microspores** light gray in mass, 25–30 μm, papillose. $2n = 66$.

Spores mature in late summer. Flowing water in low woods; of conservation concern; Ga.

Isoëtes boomii is known only from low, wet woods in Laurens County, Georgia. Early collections bearing only microspores were identified as *I. flaccida*. These were reported as *I. flaccida* × *piedmontana* and as having a velum coverage of 70–90% (B. M. Boom 1982), but recent collections have a velum coverage of less than 50%. The megaspore texture of broken lamellae and a narrow velum coverage do not indicate an *I. flaccida* influence.

14. Isoëtes lithophila N. Pfeiffer, Ann. Missouri Bot. Gard. 9: 135. 1922 · Rock quillwort

Plants becoming terrestrial. **Rootstock** nearly globose, 2-lobed. **Leaves** deciduous, bright green, pale toward base, spirally arranged, to 12(–20) cm, pliant, gradually tapering to tip. **Velum** covering entire sporangium. **Sporangium wall** unpigmented. **Megaspores** light gray to gray-brown, 290–360 μm diam., obscurely rugulate with low ridges; girdle obscure. **Microspores** brown in mass, 30–33 μm, tuberculate to spinulose. $2n = 22$.

Spores mature in winter or spring. Shallow depressions on granite outcrops; of conservation concern; Tex.

Isoëtes lithophila is known from granite outcrops in Burnett, Mason, and Llano counties in south central Texas.

15. Isoëtes tegetiformans Rury, Amer. Fern J. 68: 100. 1978 · Mat-forming Merlin's-grass

Plants becoming terrestrial. **Rootstock** horizontally elongate, mat-forming, often proliferous. **Leaves** deciduous to nearly evergreen, bright green, pale toward base, distichously arranged, to 4 cm, pliant, gradually tapering to tip. **Velum** covering entire sporangium. **Sporangium wall** unpigmented. **Megaspores** dark gray, 275–370 μm diam., tuberculate with low and distinct tubercules; girdle obscure. **Microspores** brown in mass, 26–33 μm, spinulose. $2n = 22$.

Spores mature late winter and spring. Shallow pools on granite outcrops; of conservation concern; Ga.

Isoëtes tegetiformans is a unique mat-forming quillwort distinguished by its distichous leaf arrangement, horizontally elongate rootstock, and nondichotomous roots. (See discussion under *I. melanospora*.)

16. Isoëtes melanospora Engelmann, Trans. Acad. Sci. St. Louis 3: 395. 1877 · Black-spored quillwort

Plants becoming terrestrial. **Rootstock** nearly globose, 2-lobed. **Leaves** deciduous, bright green, pale toward base, spirally arranged, to 7 cm, pliant, gradually tapering to tip. **Velum** covering entire sporangium. **Sporangium wall** unpigmented. **Megaspores** gray to black, 350–480 μm diam., tuberculate with distinct to confluent tubercles; girdle obscure. **Microspores** brown in mass, 26–31 μm, smooth to papillose. $2n = 22$.

Spores mature in late spring. Shallow depressions on granite outcrops; of conservation concern; Ga.

A population of plants in which the individuals have a velum incompletely covering the sporangium, a brown-pigmented sporangium wall, and dark brown megaspores occurs in Lancaster County, South Carolina. These plants appear to be transitional between *Isoëtes melanospora* and *I. virginica*.

Plants of *Isoëtes melanospora* and *I. tegetiformans* typically grow actively after fall and winter precipitation. New leaves can appear a few days after the soil becomes saturated by rain, but normally they shrivel and disappear during summer when the soil dries.

17. Isoëtes butleri Engelmann, Bot. Gaz. 3: 1. 1878
· Butler's quillwort

Plants terrestrial. **Rootstock** nearly globose, 2-lobed. **Leaves** deciduous, dull green to gray-green or yellow green, pale toward base, spirally arranged, to 15(–30) cm, pliant, tapering gradually to tip. **Velum** covering less than 1/4 of sporangium. **Sporangium wall** ± brown-streaked. **Megaspores** white, (360–)480–650 μm diam., obscurely tuberculate; girdle obscure. **Microspores** light brown in mass, 27–37 μm, papillose. $2n = 22$.

Spores mature in late spring. Calcareous soil; limestone cedar glades, barrens; Ala., Ark., Ga., Ill., Kans., Ky., Mo., Okla., Tenn., Tex.

Populations of *Isoëtes butleri* in Illinois and Texas are disjunct.

Isoëtes butleri is locally abundant in open areas on alkaline soils saturated by water from early spring rains. The leaves yellow, wither, and disappear by late spring.

18. Isoëtes melanopoda Gay & Durieu, Bull. Soc. Bot. France 11: 102. 1864 · Black-footed quillwort

Plants terrestrial or becoming so. **Rootstock** nearly globose, 2-lobed. **Leaves** deciduous, bright green, pale to lustrous black toward base, spirally arranged, to 40 cm, pliant, tapering gradually to tip. **Velum** covering less than 3/4 of sporangium. **Sporangium wall** brown-streaked. **Megaspores** white, 280–440 μm diam., obscurely rugulate with low ridges, rarely tuberculate or reticulate; girdle obscure. **Microspores** gray in mass, 20–30 μm, spinulose. $2n = 22$.

Spores mature in late spring. Noncalcareous soil; meadows, fields, ditches, soil pockets on rock outcrops; Ark., Ga., Ill., Ind., Iowa, Kans., Ky., La., Minn., Miss., Mo., Nebr., N.J., N.C., Okla., S.C., S.Dak., Tenn., Tex., Utah, Va.

Populations of *Isoëtes melanopoda* in New Jersey, North Carolina, South Carolina, Utah, and Virginia are disjunct.

Plants of *Isoëtes melanopoda* with black leaf bases are typical. Plants without black leaf bases have been called *I. melanopoda* f. *pallida* (Engelmann) Fernald. Variation also occurs in megaspore morphology. A collection from Dallas County, Texas, has lustrous black leaf bases and boldly rugulate spores with ridges that occasionally anastomose. In many respects, *I. melanopoda* is similar to *I. howellii*.

19. Isoëtes virginica N. E. Pfeiffer, Claytonia 3: 29. 1937 · Virginia quillwort

Isoëtes virginica var. *piedmontana* N. E. Pfeiffer; *I. piedmontana* (N. E. Pfeiffer) C. F. Reed
Plants becoming terrestrial. **Rootstock** nearly globose, 2-lobed. **Leaves** deciduous, dull green, pale to dark brown toward base, spirally arranged, less than 30 cm, pliant, gradually tapering to tip. **Velum** covering less than 1/3 of sporangium. **Sporangium wall** brown-streaked to completely brown. **Megaspores** white, 400–480 μm diam., boldly tuberculate to rugulate with cristate, occasionally branched to anastomosing ridges; girdle obscure. **Microspores** light brown in mass, 27–33 μm, spinulose. $2n = 22, 44$.

Spores mature late in spring–early summer. Mud flats and depressions on and around granite outcrops; Ala., Ga., S.C., Va.

Short plants with 15 to 50 leaves and a tendency toward lower and more open megaspore ornamentation have been called *Isoëtes virginica* var. *piedmontana* Pfeiffer.

20. Isoëtes bolanderi Engelmann, Amer. Naturalist 8: 214. 1874 · Bolander's quillwort

Isoëtes bolanderi var. *parryi* Engelmann; *I. californica* Engelmann; *I. pygmaea* Engelmann
Plants aquatic, occasionally emergent. **Rootstock** nearly globose, 2-lobed. **Leaves** deciduous, bright green, pale brown toward base, spirally arranged, to 20 cm, pliant, abruptly tapering to fine tip. **Velum** covering less than 1/2 of sporangium. **Sporangium wall** ± brown-streaked. **Megaspores** white, 300–500 μm diam., rugulate to tuberculate; girdle obscure. **Microspores** brown in mass, 20–30 μm, spinulose. $2n = 22$.

Spores mature late summer. Alpine or subalpine lakes and ponds; Alta., B.C.; Ariz., Calif., Colo., Idaho, Mont., Nev., N.Mex., Oreg., Utah, Wash., Wyo.

Small plants with leaves less than 2.5 cm have been called *Isoëtes bolanderi* var. *pygmaea* (Engelmann) Clute.

Isoëtes bolanderi hybridizes with *I. echinospora* and *I. occidentalis*.

21. Isoëtes occidentalis L. F. Henderson, Bull. Torrey Bot. Club 27: 358. 1900 · Western quillwort

Isoëtes flettii (A. A. Eaton) N. E. Pfeiffer; *I. lacustris* Linnaeus var. *paupercula* Engelmann; *I. paupercula* (Engelmann) A. A. Eaton; *I. piperi* A. A. Eaton

Plants aquatic. **Rootstock** nearly globose, 2-lobed. **Leaves** evergreen, dark green, pale toward base, spirally arranged, to 20 cm, rigid, gradually tapering to tip. **Velum** covering less than 1/2 of sporangium. **Sporangium wall** unpigmented. **Megaspores** 500–700 μm diam., cristate, tuberculate, rugulate, or echinate with ridges, tubercles, or spines; girdle smooth. **Microspores** brown in mass, 35–45 μm, papillose to spinulose. $2n = 66$.

Spores mature in late summer. Lakes; B.C; Alaska, Calif., Colo., Idaho, Mont., Oreg., Utah, Wash., Wyo.

Megaspores of *Isoëtes occidentalis* are variable in wall pattern. Populations exist with rugulate or tuberculate megaspores and other populations with cristate to echinate megaspores. Plants with thin-walled megaspores that crack easily have been called *I. paupercula*. Populations in which megaspores have short ridges and tubercles in a band along the equator have been called *I. flettii*. Populations with broad-based tubercles on the megaspores have been called *I. piperi*. The variation in megaspore pattern may indicate multiple allopolyploid origins for *I. occidentalis*.

The general aspect of *Isoëtes occidentalis* and its tough, dark green leaves suggested to early workers an affinity with *I. lacustris*.

Isoëtes occidentalis hybridizes with *I. bolanderi*, *I. echinospora*, and *I. maritima* [= *I.* × *truncata* (A. A. Eaton) Clute].

22. Isoëtes howellii Engelmann, Trans. Acad. Sci. St. Louis 4: 385. 1882 · Howell's quillwort

Isoëtes melanopoda Gay & Durieu var. *californica* A. A. Eaton; *I. minima* A. A. Eaton; *I. nuda* Engelmann; *I. underwoodii* L. F. Henderson

Plants becoming terrestrial. **Rootstock** nearly globose, 2-lobed. **Leaves** deciduous, bright green, pale to brown or lustrous black toward base, spirally arranged, to 25 cm, pliant, gradually tapering to tip. **Velum** covering less than 1/2 of sporangium. **Sporangium wall** brown-streaked to completely brown. **Megaspores** white, 300–500 μm diam.,

obscurely tuberculate to rugulate; girdle smooth. **Microspores** brown in mass, 25–35 μm, spinulose. $2n = 22$.

Spores mature in late spring and summer. Wet depressions, vernal pools, and lake margins; B.C.; Calif., Idaho, Mont., Oreg., Wash.

In many respects, *Isoëtes howellii* appears similar to *I. melanopoda*. Small plants with leaves less than 10 cm and megaspores less than 420 μm diam. have been called *I. howellii* var. *minima* (A. A. Eaton) N. E. Pfeiffer.

23. Isoëtes nuttallii A. Braun, Amer. Naturalist 8: 215. 1874 · Nuttall's quillwort

Isoëtes opaca Nuttall; *I. suksdorfii* Baker

Plants terrestrial or becoming so. **Rootstock** nearly globose, 2(–3)-lobed. **Leaves** deciduous, dull green to gray-green or yellow-green, pale toward base, spirally arranged, to 20 cm, ± rigid, gradually tapering to tip, often surrounded at base by several black scales to ca. 5 mm, spirally arranged, rigid, almost brittle. **Velum** covering entire sporangium. **Sporangium wall** unpigmented. **Megaspores** white, 360–600 μm diam., lustrous, ± smooth to tuberculate; girdle obscure. **Microspores** brown in mass, 20–30 μm, papillose. $2n = 22$.

Spores mature in late spring–early summer. Seasonally wet soil, temporary streams; mostly 0–1500 m; B.C; Calif., Oreg., Wash.

24. Isoëtes orcuttii A. A. Eaton, Fern Bull. 8: 13. 1900 · Orcutt's quillwort

Isoëtes nuttallii A. Braun var. *orcuttii* (A. A. Eaton) Clute

Plants becoming terrestrial. **Rootstock** nearly globose 2(–3)-lobed. **Leaves** deciduous, bright green, pale toward base, less than 8 cm, gradually tapering to tip, often surrounded at base by several black scales to ca. 5 mm, spirally arranged, pliant, gradually tapering to tip. **Velum** covering entire sporangium. **Sporangium wall** unpigmented. **Megaspores** white to gray, lustrous, 200–380 μm diam., ± smooth to obscurely tuberculate; girdle obscure. **Microspores** brown in mass, 20–30 μm, smooth to papillose. $2n = 22$.

Spores mature in spring. Vernal pools; mostly 0–1500 m; Calif.; Mexico in Baja California.

Isoëtes orcuttii, a vernal pool endemic, may be difficult to distinguish from *I. nuttallii*, which occurs in a wider range of habitats. Plants of *I. orcuttii* are generally smaller than those of *I. nuttallii*, which has longer, thicker, less flexible leaves and larger megaspores.

5. EQUISETACEAE Michaux ex DeCandolle • Horsetail Family

Richard L. Hauke

Plants with jointed stems, with distinct nodes. **Leaves** small, whorled, fused into sheaths; tips remaining free, toothlike. **Sporangia** borne on peltate sporophylls aggregated in cones 0.3–10 cm. **Spores** green (except white in hybrids), all 1 kind. **Gametophytes** green, terrestrial, unisexual; male gametophytes smaller than female.

Genus 1, species 15 (11 species in the flora): nearly worldwide.

1. EQUISETUM Linnaeus, Sp. Pl. 2: 1061. 1753; Gen. Pl. ed. 5, 484. 1754 · Horsetail, scouring rush, prêle [Latin *equis*, horse, and *seta*, bristle, referring to the coarse black roots of *E. fluviatile*]

Plants perennial, rhizomatous. **Aerial stems** annual or perennial. **Stems** with hollow center and series of small carinal (under the ridges) and larger vallecular (under the valleys) canals. **Leaves** in whorls, fused part of length into sheaths. **Stem ridges** traversing length of internode and continuing into sheaths, terminating in sheath teeth. **Branches** when present borne at nodes, erupting through base of subtending sheath. **Cones** terminal on green stems or, in some species, terminating special, reproductive, brown stems, composed of whorls of peltate sporophylls; cone apices rounded or sharply pointed; sporangia 5–10 per sporophyll, pendent, attached to inner surface of sporophylls, elongate, dehiscing longitudinally. $x = 108$.

Species 15 (11 in the flora): nearly worldwide.

Equisetum occurs in moist places such as riverbanks, lakeshores, roadsides, ditches, seepage areas, meadows, marshes, and wet woodlands. Aerial stems of *Equisetum* vary considerably in habit and appearance, even on individual plants, because of environmentally induced modifications affecting height and branching. Many taxonomically trivial varieties and forms have been named. For an extended discussion of this, see R. L. Hauke (1966). Four widespread, named hybrids are treated in the key and fully described below.

Equisetum

Subgenus *Hippochaete*

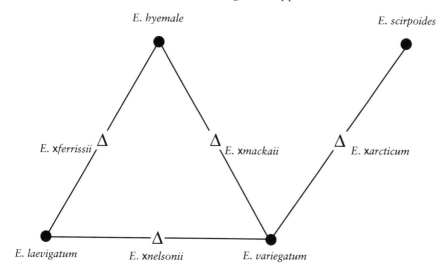

Subgenus *Equisetum*

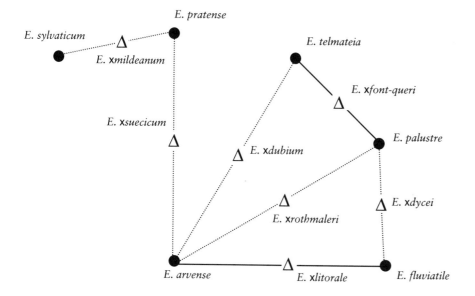

Relationships in *Equisetum*. Solid circles represent parental taxa; triangles represent sterile hybrids mostly reproducing vegetatively; and broken lines represent hybridization known in Europe but not yet known in North America.

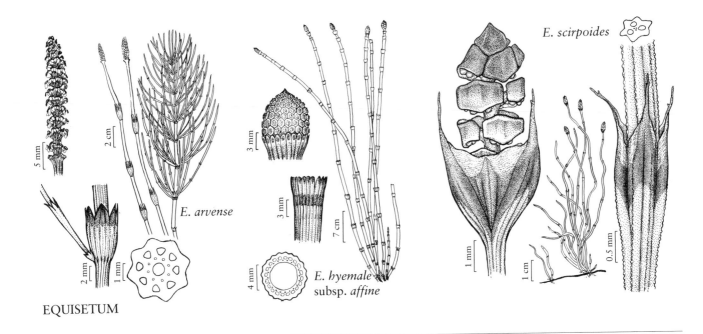

E. arvense

E. hyemale subsp. *affine*

E. scirpoides

EQUISETUM

In species descriptions and in the key, length and width are given for the leaf sheath, excluding the free teeth. If the length and width of flattened sheaths are approximately equal and the sides are straight, the sheath is more or less square in face view, i.e., about as long as broad; if the length is greater than the width and the sides are straight, the sheath is more or less elongate in face view, i.e., longer than broad; if the length is greater than the width and the sides are slightly convex, the sheath is elliptic in face view. Stomates are usually visible at 20× magnification.

Reticulation in *Equisetum* is summarized in the reticulograms, which show the known and expected hybrids in North America. Most of those in *Equisetum* subg. *Equisetum* are still unknown in North America, but they should be sought, especially north of 45° N latitude. According to W. J. Cody and D. M. Britton (1989), *E.* ×*font-queri* occurs rarely in British Columbia and materials possibly representing *E.* ×*arcticum* Rothmaler have been taken in the Richardson Mountain region of Mackenzie. R. L. Hauke (1978) cited collections of *E.* ×*font-queri* from British Columbia and California.

SELECTED REFERENCES   Hauke, R. L. 1966. A systematic study of *Equisetum arvense*. Nova Hedwigia 13: 81–109. Hauke, R. L. 1979. *Equisetum ramosissimum* in North America. Amer. Fern J. 69: 1–5.

1. Aerial stems persisting only 1 year or less, usually with regular whorls of branches; stomates on surface, scattered or in bands; cone apex rounded. . . . . .1a. *Equisetum* subg. *Equisetum*
    2. Aerial stems unbranched.
       3. Aerial stems green.
          4. Sheaths square in face view; teeth more than 11 per sheath, often black throughout or with narrow white margins, 2–3 mm. . . . . 1. *Equisetum fluviatile*
          4. Sheaths elongate in face view; teeth fewer than 11 per sheath, with prominent white margins and dark centers, 2–5 mm. . . . . 2. *Equisetum palustre*
       3. Aerial stems not green.
          5. Sheath teeth reddish, papery, coherent into 3–4 large groups. . . . . . . 6. *Equisetum sylvaticum*

5. Sheath teeth black or brown, firm, separate or coherent in more than 4 small
    groups.
    6. Aerial stems with stomates, persistent, becoming green and branched. . .   7. *Equisetum pratense*
    6. Aerial stems lacking stomates, dying back after spores are shed.
        7. Sheath teeth more than 14. 3. *Equisetum telmateia*
        7. Sheath teeth fewer than 14. . . . . 4. *Equisetum arvense*
2. Aerial stems branched with regular whorls of branches.
    8. First internode of each branch shorter than subtending stem sheath; branch valleys
    rounded.
        9. Branches solid, branch ridges furrowed. . . . . 3. *Equisetum telmateia*
        9. Branches hollow, branch ridges rounded.
            10. Sheaths square in face view; teeth more than 11 per sheath, dark, occasion-
        ally with narrow white margins, 2–3 mm. . . . . 1. *Equisetum fluviatile*
            10. Sheaths elongate in face view; teeth fewer than 11 per sheath, with promi-
        nent white margins and dark centers, 2–5 mm. . . . . 2. *Equisetum palustre*
    8. First internode of each branch equal to or longer than subtending stem sheath; branch
    valleys channeled.
        11. Aerial stem sheath teeth reddish, papery, coherent in 3–4 large groups; stem
        branches also branched. . . . . 6. *Equisetum sylvaticum*
        11. Aerial stem sheath teeth dark, firm, separate or coherent in more than 4 small
        groups; stem branches unbranched.
            12. Branch sheath teeth deltate; branches spreading. . . . . 7. *Equisetum pratense*
            12. Branch sheath teeth attenuate; branches ascending.
                13. Lowest whorls of branches with 1st internode longer than sheath; spores
            green, spheric. . . . . 4. *Equisetum arvense*
                13. Lowest whorls of branches with 1st internode nearly equal to sheath;
            spores white, misshapen. . . . . 5. *Equisetum* × *litorale*
1. Aerial stems persisting more than a year (except some *Equisetum laevigatum*), usually
    unbranched; stomates sunken, in single lines; cone apex pointed (except some *E. laevigatum*).
    . . . . 1b. *Equisetum* subg. *Hippochaete*
14. Aerial stems regularly branched; stomatal lines occasionally doubled. . . . 8. *Equisetum ramosissimum*
14. Aerial stems unbranched or with scattered branches; stomatal lines always single.
    15. Cone apex rounded; aerial stems annual. . . . . 9. *Equisetum laevigatum*
    15. Cone apex pointed; aerial stems perennial (at least persisting over winter in Cal-
    ifornia populations of *E. laevigatum*).
        16. Spores white, misshapen.
            17. Sheaths green; teeth persistent. . . . . 14. *Equisetum* × *nelsonii*
            17. Sheaths dark-girdled; teeth persistent or shed.
                18. Teeth 14 or fewer per sheath, persistent. . . . . 13. *Equisetum* × *mackaii*
                18. Teeth more than 14 per sheath, usually shed. . . . . 11. *Equisetum* × *ferrissii*
        16. Spores green, spheric.
            19. Sheaths dark-girdled at most nodes of stem; teeth 14 or more per sheath,
        usually shed; articulation line visible. . . . . 10. *Equisetum hyemale*
            19. Sheaths green or obscurely girdled at nodes near base of stem; teeth 32
        or fewer per sheath, usually persistent, shed in some species; articulation
        line lacking.
                20. Teeth 3–32 per sheath; stem ridges same number as teeth; aerial
            stems erect and straight.
                    21. Sheath teeth usually shed; cone apex rounded to apiculate with
                blunt tip; stem ridges flattened or ± convex. . . . . . . 9. *Equisetum laevigatum*
                    21. Sheath teeth usually persistent throughout; cone apex sharply
                apiculate; stem ridges minutely grooved. . . . . 12. *Equisetum variegatum*
              20. Teeth 3 per sheath; stem ridges 6; aerial stems inclined and tortuous.
                . . . . 15. *Equisetum scirpoides*

1a. EQUISETUM Linnaeus subg. EQUISETUM · Horsetail

Aerial stems persisting for 1 year or less, with stomates on surfaces scattered or in bands on each side of valleys. **Branches** in regular whorls, except for unbranched forms of *Equisetum fluviatile* and *E. palustre* and brown fertile stems of dimorphic species. **Cones** rounded at apex.

Species 8 (6 in the flora): North America, s Central America, w South America, Europe, Asia.

SELECTED REFERENCE   Hauke, R. L. 1978. A taxonomic monograph of *Equisetum* subgenus *Equisetum*. Nova Hedwigia 30: 385–455.

1. **Equisetum fluviatile** Linnaeus, Sp. Pl. 2: 1062. 1753 · River horsetail, pipes, prêle fluviatile

Equisetum limosum Linnaeus

Aerial stems monomorphic, green, unbranched or branched, 35–115 cm; hollow center large, to 9/10 stem diam.; vallecular canals absent. **Sheaths** squarish in face view, ca. 4–10 × 4–10 mm; teeth black, occasionally with narrow white border, 12–24, narrow, 2–3 mm. **Branches** when present only from midstem nodes, spreading, hollow, ridges 4–6, valleys rounded; 1st internode of each branch shorter than subtending stem sheath; sheath teeth narrow. 2*n* = 216.

Cones maturing in summer. Standing in water, in ponds, ditches, marshes, swales; 0–1500 m; St. Pierre and Miquelon; Alta., B.C., Man., N.B., Nfld., N.W.T., N.S., Ont., P.E.I., Que., Sask., Yukon; Alaska including the Aleutian Islands, Conn., Del., D.C., Idaho, Ill., Ind., Iowa, Maine, Md., Mass., Mich., Minn., Mont., Nebr., N.H., N.J., N.Y., N.Dak., Ohio, Oreg., Pa., R.I., S.Dak., Vt., Va., Wash., W.Va., Wis., Wyo.; Eurasia s to n Italy, China, Korea, Japan.

2. **Equisetum palustre** Linnaeus, Sp. Pl. 2: 1061. 1753 · Marsh horsetail, prêle des marais

Equisetum palustre var. *americanum* Victorin

Aerial stems monomorphic, green, branched or unbranched, 20–80 cm; hollow center small, to 1/3 stem diam.; vallecular canals nearly as large. **Sheaths** elongate, 4–9 × 2–5 mm; teeth dark, 5–10, narrow, 2–5 mm, margins white, scarious. **Branches** when present only from midstem nodes, spreading, hollow; ridges 4–6; valleys rounded; 1st internode of each branch shorter than subtending stem sheath; sheath teeth narrow. 2*n* = 216.

Cones maturing in summer. Marshes and swamps; 0–1500 m; Alta., B.C., Man., N.B., Nfld., N.W.T., N.S., Ont., P.E.I., Que., Sask., Yukon; Alaska, Calif., Idaho, Maine, Mich., Minn., Mont., N.H., N.Y., N.Dak., Oreg., Pa., Vt., Wash., Wis.; Eurasia s to Himalayas, n China, Korea, Japan.

The name *Equisetum palustre* var. *americanum* has been used for specimens from the flora that have longer teeth than those from Eurasia.

3. **Equisetum telmateia** Ehrhart, Hannover. Mag. 21: 287. 1783 · Giant horsetail

Subspecies 2 (1 in the flora): North America, Europe, n Africa, w Asia.

3a. **Equisetum telmateia** Ehrhart subsp. **braunii** (J. Milde) Hauke, Nova Hedwigia 30: 434. 1978

Equisetum braunii J. Milde, Verh. K. K. Zool.-Bot. Ges. Wien 12: 515. 1862; *E. telmateia* var. *braunii* (J. Milde) J. Milde

Aerial stems dimorphic; vegetative stems green, branched, 30–100(–200) cm; hollow center 2/3–3/4 stem diam. **Sheaths** squarish in face view, 7–18 × 5–13 mm; teeth green proximally and dark distally, 14–30(–36), 3–12 mm. **Branches** in regular whorls, ascending to spreading, solid; ridges 4–5, furrowed; valleys rounded; 1st internode of each branch shorter than subtending stem sheath; sheath teeth attenuate. **Fertile stems** brown, lacking stomates, unbranched, shorter (17–45 cm) than vegetative stems, with larger (15–40 mm) sheaths, fleshy, ephemeral.

Cones maturing in early spring. Coastal marshes, stream banks, ditches, and other wet places; 0–1000 m; B.C.; Alaska, Calif., Oreg., Wash.

Equisetum telmateia subsp. *telmateia*, from Europe, northern Africa, and the Middle East, differs primarily in having main aerial stem with white internodes, lacking both green tissue and stomates but having green, whorled branches.

4. Equisetum arvense Linnaeus, Sp. Pl. 2: 1061. 1753
· Field horsetail, common horsetail, prêle des champs

Aerial stems dimorphic; vegetative stems green, branched, 2–60(–100) cm; hollow center 1/3–2/3 stem diam. **Sheaths** squarish in face view, 2–5(–10) × 2–5 (–9) mm; teeth dark, 4–14, narrow, 1–3.5 mm, often cohering in pairs. **Branches** in regular whorls, ascending, solid; ridges 3–4; valleys channeled; 1st internode of each branch longer than subtending stem sheath; sheath teeth attenuate. **Fertile stems** brown, lacking stomates, unbranched, shorter than vegetative stems, with larger sheaths, fleshy, ephemeral. $2n = $ ca. 216.

Cones maturing in early spring. Roadsides, riverbanks, fields, marshes, pastures, tundra; 0–3200 m; Greenland; St. Pierre and Miquelon; Alta., B.C., Man., N.B., Nfld., N.W.T., N.S., Ont., P.E.I., Que., Sask., Yukon; all states except Fla., La., Miss., S.C.; Eurasia s to Himalayas, c China, Korea, Japan.

Among the many infraspecific taxa that have been named in this species, *Equisetum arvense* var. *boreale* Bongard has been most generally accepted and has been applied to plants with tall, erect stems with 3-ridged branches. Because both 3-ridged and 4-ridged branches may occur on a single stem, the variety *boreale* is not recognized here as distinct (R. L. Hauke 1966).

5. Equisetum ×litorale Kühlewein ex Ruprecht, Beitr. Pflanzenk. Russ. Reiches 4: 91. 1845, as a species
· Prêle littorale

Aerial stems monomorphic, green branched or occasionally unbranched, 2–70 cm; hollow center 2/3–4/5 stem diam. **Sheaths** somewhat elongate, 3.5–8 × 2.5–6 mm; teeth dark, 7–14, narrow, 1–3 mm. **Branches** mostly from midstem nodes, ascending to spreading, solid; ridges 4–5; valleys channeled; proximal whorls with 1st internode of each branch equal to subtending stem sheath, distal whorls with 1st internode of each branch longer than stem sheath; sheath teeth attenuate.

Cones maturing in early summer, but misshapen spores not shed. Ditches, stream banks, wet meadows; 0–1000 m; Alta., B.C., Man., N.B., Nfld., N.W.T., N.S., Ont., P.E.I., Que., Sask., Yukon; Alaska, Conn., Del., Idaho, Ill., Ind., Iowa, Maine, Md., Mass., Mich., Minn., Mont., Nebr., N.H., N.J., N.Y., N.Dak., Ohio, Oreg., Pa., R.I., S.Dak., Vt., Va., Wash., W.Va., Wis., Wyo.

Equisetum ×litorale is a hybrid between *E. arvense* and *E. fluviatile*. It should be expected where the parents coexist. This hybrid has been mistaken for *Equisetum palustre*; the solid branches with long first internodes and channeled valleys distinguish it from that species.

6. Equisetum sylvaticum Linnaeus, Sp. Pl. 2: 1061. 1753
· Wood horsetail, prêle des bois

Aerial stems dimorphic; vegetative stems brownish to green, branched, 25–70 cm; hollow center 1/6–1/3 stem diam. **Sheaths** squarish in face view, 3–6 × 2.5–6 mm; teeth reddish, 8–18, papery, 3–10 mm, coherent in 3–4 large groups. **Branches** in regular whorls, delicate, arching, branched, solid; ridges 3–4; valleys channeled; 1st internode of each branch longer than subtending stem sheath; sheath teeth attenuate. **Fertile stems** brown, with stomates, initially unbranched, persisting and becoming branched and green after spore discharge. $2n = 216$.

Cones maturing in late spring. Moist forests; 0–2800 m; Greenland; St. Pierre and Miquelon; Alta., B.C., Man., N.B., Nfld., N.W.T., N.S., Ont., P.E.I., Que., Sask., Yukon; Alaska, Conn., Del., Idaho, Iowa, Maine, Md., Mass., Mich., Minn., N.H., N.J., N.Y., N.Dak., Ohio, Pa., R.I., S.Dak., Vt., Va., Wash., W.Va., Wis.; Europe; n Asia to ne China, Japan in Hokkaido.

7. Equisetum pratense Ehrhart, Hannover. Mag. 22: 138. 1784 · Meadow horsetail, prêle des prés

Aerial stems dimorphic; vegetative stems green, branched, 16–50 cm; hollow center 1/6–1/3 stem diam. **Sheaths** somewhat elongate, 3–5 × 2–4 mm; teeth 8–18, narrow, 1.5–4 mm, centers dark and margins white. **Branches** in regular whorls, horizontal to drooping, solid; ridges 3; valleys channeled; 1st internode of each branch equal to or longer than subtending stem sheath; sheath teeth deltate. **Fertile stems** brown, with stomates, initially unbranched, persisting and becoming branched and green after spore discharge.

Cones maturing in late spring. Meadows, wet woodlands; 0–2000 m; Alta., B.C., Man., N.B., Nfld., N.W.T., N.S., Ont., P.E.I., Que., Sask., Yukon; Alaska, Conn., Ill., Iowa, Maine, Mass., Mich., Minn., N.H., N.Y., N.Dak., Vt., Wis.; n Eurasia to ne China, Japan in Hokkaido.

1b. EQUISETUM Linnaeus subg. HIPPOCHAETE (J. Milde) Baker, Handb. Fern-allies, 3. 1887 · Scouring rush

Hippochaete J. Milde, Bot. Zeitung (Berlin) 23: 297. 1865

Aerial stems persisting more than a year (except *Equisetum laevigatum*), with sunken stomates in single lines on each side of stem valleys; branches generally lacking or few (except *E. ramosissimum*). **Cones** sharply pointed at apex (except *E. laevigatum*).

Species 7 (5 in the flora): nearly worldwide.

The only species of this subgenus in the flora that is regularly branched, bearing several branches per midstem node, is *Equisetum ramosissimum*. The others, though normally unbranched, with age or injury may develop one or a few branches, usually from the proximal or most distal nodes.

SELECTED REFERENCE   Hauke, R. L. 1963. A taxonomic monograph of *Equisetum* subgenus *Hippochaete*. Beih. Nova Hedwigia 8: 1–123.

8. Equisetum ramosissimum Desfontaines, Fl. Atlant. 2: 398. 1799

Subspecies 2 (1 in the flora): North America, Europe, Asia, Africa, Pacific Islands.

8a. Equisetum ramosissimum Desfontaines subsp. ramosissimum

Aerial stems persisting more than a year, regularly branched, 32–250 cm; lines of stomates occasionally doubled; ridges 10–16. **Sheaths** greatly elongate, 8–12 × 3–6 mm; proximal sheaths brown with darker girdle, distal sheaths green; teeth 10–16, not articulate but often thin and drying. **Cones** pointed at their apex; spores green, spheric.

Cones maturing in summer. Moist sandy or clay areas; 0–100 m; introduced; Fla., La., N.C.; s,c Europe; Asia; Africa.

Equisetum ramosissimum subsp. *ramosissimum* apparently was introduced from Europe with ballast (R. L. Hauke 1979). *Equisetum ramosissimum* subsp. *debile* (Roxburgh) Hauke occurs in southeast Asia and the southern Pacific Islands. Where the ranges of the two subspecies overlap, fertile, morphologically intermediate individuals are found (R. L. Hauke 1963).

9. Equisetum laevigatum A. Braun, Amer. J. Sci. Arts 46: 87. 1844 · Smooth scouring rush

Equisetum funstonii A. A. Eaton, *E. kansanum* J. H. Schaffner

Aerial stems lasting less than a year, occasionally overwintering in the southwestern United States, usually unbranched, 20–150 cm; lines of stomates single; ridges 10–32. **Sheaths** green, elongate, 7–15 × 3–9 mm; teeth 10–32, articulate and usually shed early, leaving dark rim on sheath. **Cone** apex rounded to apiculate with blunt tip; spores green, spheric. $2n = 216$.

Cones maturing in spring–early summer. Moist prairies, riverbanks, roadsides; 1530–3500 m; Alta., B.C., Man., Ont., Que., Sask.; Ariz., Ark., Calif., Colo., Idaho, Ill., Ind., Iowa, Kans., Mich., Minn., Mo., Mont., Nebr., Nev., N.Mex., N.Dak., Ohio, Okla., Oreg., S.Dak., Tex., Utah, Wash., Wis., Wyo.; n Mexico including Baja California.

Schaffner named this species *Equisetum kansanum* because he applied the name *E. laevigatum* to what we now know is the hybrid *E.* ×*ferrissii*. The coarser-stemmed, occasionally persistent forms in the southwestern United States have been called *Equisetum funstonii*.

10. Equisetum hyemale Linnaeus, Sp. Pl. 2: 1062. 1753 · Common scouring rush, prêle d'hiver

Subspecies 2 (1 in the flora): North America, Mexico, Central America in Guatemala, Europe, Asia.

10a. Equisetum hyemale Linnaeus subsp. **affine** (Engelmann) Calder & Roy L. Taylor, Canad. J. Bot. 43: 1387. 1965

Equisetum robustum A. Braun var. *affine* Engelmann, Amer. J. Sci. Arts 46: 88. 1844; *E. hyemale* var. *affine* (Engelmann) A. A. Eaton; *E. hyemale* var. *californicum* J. Milde; *E. hyemale* var. *pseudohyemale* (Farwell) C. V. Morton; *E. hyemale* var. *robustum* (A. Braun) A. A. Eaton; *E. prealtum* Rafinesque

Aerial stems persisting more than a year, unbranched, 18–220 cm; lines of stomates single; ridges 14–50. **Sheaths** when mature dark-girdled, brown to gray above girdle, squarish in face view, 4.5–17 × 3.5–18 mm; teeth 14–50, articulate and promptly shed or persistent. **Cone** apex pointed; spores green, spheric. $2n = 216$.

Cones maturing in summer, old stems sometimes developing branches with cones in spring. Moist roadsides, riverbanks, lakeshores, woodlands; 0–3000 m; Alta., B.C., Man., N.B., Nfld., N.W.T., N.S., Ont., P.E.I., Que., Sask., Yukon; Alaska including the Aleutian Islands, all other states; Mexico; Central America in Guatemala.

In southern and central to western regions plants tend to be taller and have more persistent teeth (*Equisetum robustum, E. prealtum*); in the Far West they often have bituberculate ridges (*E. hyemale* var. *californicum*). *Equisetum hyemale* subsp. *hyemale* is found in Europe and Asia to northwestern China in Xinjiang.

11. Equisetum × ferrissii Clute, Fern Bull. 12: 22. 1904, as a species

Equisetum hyemale Linnaeus var. *elatum* (Engelmann) C. V. Morton; *E. hyemale* var. *intermedium* A. A. Eaton

Aerial stems having basal part persisting over winter, unbranched, 20–180 cm; lines of stomates single; ridges 14–32. **Sheaths** elongate, 7–17 × 3–12 mm, becoming dark-girdled with age; teeth 14–32, articulate and promptly shed or persistent. **Cone** apex pointed; spores white, misshapen.

Cones maturing in late spring–early summer but spores not shed. Moist lakeshores, riverbanks, roadsides, prairies; 0–2500 m; Alta., B.C., Ont., Que., Sask.; Ariz., Ark., Calif., Colo., Conn., Del., D.C., Idaho, Ill., Ind., Iowa, Kans., Maine, Md., Mass., Mich., Minn., Mo., Mont., Nebr., Nev., N.H., N.J., N.Mex., N.Y.,

N.C., N.Dak., Ohio, Okla., Oreg., Pa., R.I., S.Dak., Tex., Utah, Vt., Va., Wash., W.Va., Wis., Wyo.; n Mexico including Baja California.

The hybrid between *Equisetum hyemale* and *E. laevigatum, E. ×ferrissii*, was mistaken for *E. laevigatum* by Schaffner and some subsequent authors. Although sterile, it exists outside the range of *E. laevigatum*, and apparently it is dispersed vegetatively (R. L. Hauke 1963). Perhaps it has persisted in some areas from a time when the parents were both there. *Equisetum ×ferrissii* has been reported from Maine, New Hampshire, Rhode Island, South Dakota, and Vermont, but I have not seen specimens from those states.

12. Equisetum variegatum Schleicher ex F. Weber & D. Mohr, Bot. Taschenbuch, 60, 447. 1807 · Variegated scouring rush, prêle panachée

Aerial stems persisting more than a year, unbranched, 6–55 cm; lines of stomates single; ridges 3–12. **Sheaths** green with black apical band, spreading, 1–6 × 1–5 mm; teeth 3–14, not articulate. **Cone** apex pointed; spores green, spheric. $2n = 216$.

Subspecies 2 (2 in the flora): North America, Europe, Asia.

1. Sheath teeth erect, with prominent white margins. . . 12a. *Equisetum variegatum* subsp. *variegatum*
1. Sheath teeth incurved, with obscure margins or all black. . . . . 12b. *Equisetum variegatum* subsp. *alaskanum*

12a. Equisetum variegatum Schleicher ex F. Weber & D. Mohr subsp. **variegatum**

Stems 6–48 cm; teeth 3–12, erect, centers dark, margins white, prominent.

Cones maturing in late summer, or cones overwintering and shedding spores in spring. Lakeshores, riverbanks, ditches, wet woods, tundra; 0–3500 m; Greenland; St. Pierre and Miquelon; Alta., B.C., Man., N.B., Nfld., N.W.T., Ont., P.E.I., Que., Sask., Yukon; Alaska including the Aleutian Islands, Colo., Conn., Idaho, Ill., Ind., Maine, Mass., Mich., Minn., Mont., N.H., N.Y., Oreg., S.Dak., Utah, Vt., Wash., Wis., Wyo.; Europe; n Asia to Kamchatka.

12b. Equisetum variegatum Schleicher ex F. Weber & D. Mohr subsp. **alaskanum** (A. A. Eaton) Hultén, Acta Univ. Lund. 37(1): 59. 1941

Equisetum variegatum var. *alaskanum* A. A. Eaton in Merriam, Harriman Alaska Exped. 5: 390. 1904

Stems 26–55 cm; teeth all black or with obscure white margins, 8–14, incurved.

Cones maturing in late summer, or cones overwintering and shedding spores in spring. Lakeshores, riverbanks, ditches, wet woods, tundra; 0–1500 m; B.C., Yukon; Alaska including the Aleutian Islands, Wash.

13. Equisetum ×mackaii (Newman) Brichan, Phytologist 1: 369. 1843 (Nov. 1842)

Equisetum hyemale Linnaeus var. *mackaii* Newman, Phytologist 1: 305. 1843 (Sept. 1842); *E. hyemale* subsp. *trachyodon* A. Braun; *E. ×trachyodon* (A. Braun) Koch; *E. variegatum* Schleicher ex F. Weber & D. Mohr var. *jesupii* A. A. Eaton

Aerial stems persisting more than a year, unbranched, 20–86 cm; lines of stomates single; ridges 7–16. **Sheaths** black or black-girdled and white distally, appressed, elongate, 3.5–8 × 2–5.5 mm; teeth 7–16, centers dark, margins white, prominent, not articulate, tip usually brown, long, filiform. **Cone** apex pointed; spores white, misshapen.

Cones maturing in late summer, but misshapen spores not shed. Lakeshores, riverbanks, marshes; 0–500 m; Greenland; B.C., N.B., Nfld., Ont., Que., Sask.; Conn., Ill., Ind., Maine, Mich., Minn., Mont., N.H., N.J., N.Y., Ohio, Oreg., Vt., Wis.; n Europe.

The hybrid between *Equisetum hyemale* and *E. variegatum*, *E. ×mackaii*, is often mistaken for small forms of *E. hyemale*. I have not seen specimens for the reports from Connecticut and New Hampshire.

14. Equisetum ×nelsonii (A. A. Eaton) J. H. Schaffner, Amer. Fern J. 16: 46. 1926, as a species

Equisetum variegatum Schleicher ex F. Weber & D. Mohr var. *nelsonii* A. A. Eaton, Fern Bull. 12: 41. 1904

Aerial stems persisting less than a year or only the proximal part overwintering, unbranched, 20–60 cm; lines of stomates single; ridges 6–14. **Sheaths** green, elongate, 3.5–7.5 × 2–4 mm; teeth 6–14, centers brown and margins white, prominent, tip usually brown, long, filiform. **Cone** apex pointed; spores white, misshapen.

Cones maturing in early summer, but spores not shed. Lakeshores, riverbanks; Ont., Que.; Ill., Ind., Mich., Minn., Mont., N.Y., Oreg., Wis., Wyo.

Equisetum ×nelsonii, the hybrid between *E. laevigatum* and *E. variegatum*, is often mistaken for small forms of *E. ×ferrissii*.

15. Equisetum scirpoides Michaux, Fl. Bor.-Amer. 2: 281. 1803 · Dwarf scouring rush, prêle faux-scirpe

Aerial stems persisting more than a year, unbranched, tortuous, 2.5–28 cm; lines of stomates single; ridges 6. **Sheaths** green proximally, black distally, elliptic in face view, 1–2.5 × 0.75–1.5 mm; teeth 3, dark with white margins, not articulate to sheath. **Cone** apex pointed; spores green, spheric. $2n = 216$.

Cones maturing in summer, or cones overwintering and shedding spores in spring. Wet woods, peat bogs, tundra; 0–1000 m; Greenland; St. Pierre and Miquelon; Alta., B.C., Man., N.B., Nfld., N.W.T., N.S., Ont., P.E.I., Que., Sask., Yukon; Alaska, Idaho, Ill., Iowa, Maine, Mass., Mich., Minn., Mont., N.H., N.Y., S.Dak., Vt., Wash., Wis.; n Eurasia.

6. OPHIOGLOSSACEAE C. Agardh
· Adder's-tongue Family

Warren H. Wagner Jr.

Florence S. Wagner

Plants perennials, terrestrial or epiphytic. **Roots** lacking root hairs, unbranched or with a few narrow lateral branches, in 1 species dichotomously branched. **Stems** simple, unbranched, upright, with eustelic vascular tissue. **Leaf bases** dilated, clasping, forming sheath, open or fused, surrounding successive leaf primordia; primordia glabrous or with long, uniseriate hairs. **Leaves** 1(–2) per stem, with common stalk divided into sterile, laminate, photosynthetic portion (trophophore) and fertile, spore-bearing portion (sporophore). **Trophophore blades** compound to simple, rarely absent, veins anastomosing or free, pinnate, or arranged like ribs of fan. **Indument** absent or of widely scattered, long, uniseriate hairs, especially on petioles and rachises. **Sporophores** pinnately branched or simple. **Sporangia** exposed or embedded, 0.5–1.5 mm diam., thick-walled, with thousands of spores. **Spores** all 1 kind, trilete, thick-walled, surface rugate, tuberculate, baculate (with projecting rods usually higher than wide), sometimes joined in delicate network, mostly with ± warty surface. **Gametophytes** not green, usually fleshy, round or linear, subterranean, mycorrhizal.

Genera 5, species ca. 70–80 (3 genera, 38 species in the flora): nearly worldwide.

Ophioglossaceae comprise two clearly defined subfamilies, Botrychioideae and Ophioglossoideae, which are sometimes recognized as distinct families. Ophioglossaceae may be only distantly related to the ferns and more closely related to Marattiales and certain seedplants, especially Cycadales, in such characteristics as stelar type, cork cambium, dilated leaf bases, conduplicate vernation, intercalary leaf growth, collateral leaf traces, circular-bordered pits, eusporangia, massive gametophytes, sunken archegonia, and presence in some species of endoscopic embryos.

1. Blades mostly pinnately divided or lobed; veins free; margins entire to dentate to lacerate; sporangial clusters pinnately branched, sporangia sessile or terminating short stalks. . . 1. *Botrychium*, p. 86
1. Blades undivided or palmately lobed; veins anastomosing; margins entire; sporangial clusters with sporangia embedded in compact linear spike.

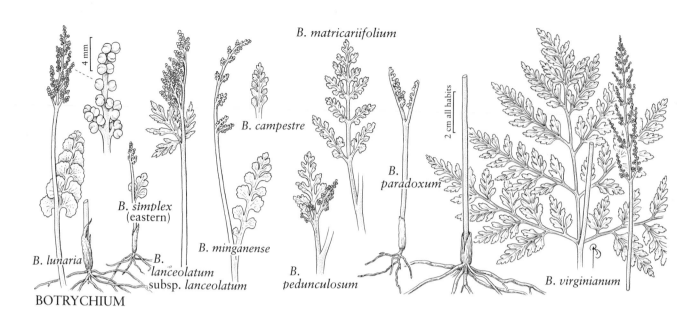

B. matricariifolium

B. campestre

B. simplex (eastern)

B. minganense

B. lunaria

B. lanceolatum subsp. *lanceolatum*

B. pedunculosum

B. paradoxum

B. virginianum

BOTRYCHIUM

2. Trophophore blades simple, unlobed, to 4.5 cm wide; main areoles mostly less than 6 mm wide; sporophore 1 per leaf at ground level or above ground level, or absent; plants terrestrial. . . . . 2. *Ophioglossum*, p. 102
2. Trophophore blades palmately lobed, to 30 cm wide; main areoles large, mostly more than 30 mm; sporophores several to many per leaf at base of blade; plants epiphytic. . . . . 3. *Cheiroglossa*, p. 106

1. BOTRYCHIUM   Swartz, J. Bot. (Schrader) 1800(2): 8, 110. 1801 · Grapefern, moonwort, botryche [Latin *botry,* bunch (of grapes), and *-oides,* like; in reference to the sporangial clusters]

Plants terrestrial. **Roots** occasionally branching laterally, yellowish to black, 0.5–2 mm diam., smooth or with corky ridges, not proliferous. **Stems** upright, forming caudex to 5 mm thick; gemmae absent or minute, spheric. **Trophophores** ascending to perpendicular to stem, sessile or stalked; blades linear, oblong, or deltate, simple to 5-pinnate, 4–25 × 1–35 cm. **Pinnae** (reduced to segments in many species) spreading to ascending, fan-shaped to lanceolate to linear; margins entire to dentate to lacerate, apex rounded or acute; veins free, arranged like ribs of fan or pinnate. **Sporophores** normally 1 per leaf, 1–3-pinnate, long-stalked, borne at ground level to high on common stalk. **Sporangial clusters** with sporangia sessile to short-stalked, almost completely exposed, borne in 2 rows on pinnate (except in very small plants) sporophore branches. **Gametophytes** broadly ovate, unbranched, 1–3 × 1–10 mm. $x = 44,$ 45, 92.

Species 50–60 (30 in the flora): nearly worldwide.

The greatest diversity in *Botrychium* is at high latitudes and high elevations, mostly in disturbed meadows and woods. Extensive field and laboratory research has revealed unexpected diversity in North America, especially in subg. *Botrychium*. For accurate identification, a substantial number of carefully spread and pressed leaves are usually needed because of the large amount of variation found in most species. Taking many samples will have little effect on the population as long as the underground shoots and roots are left intact. Approximately a dozen sterile hybrid combinations have been encountered but they are very infrequent.

The range maps south of Canada reflect mostly local occurrences at high elevations (1000–3700 m) in the mountains. The ranges for many of the species are probably more extensive and continuous than indicated by our present knowledge.

SELECTED REFERENCES Clausen, R. T. 1938. A monograph of the Ophioglossaceae. Mem. Torrey Bot. Club. 19(2): 1–177. Wagner, F. S. 1993. Chromosomes of North American grapeferns and moonworts (Ophioglossaceae: *Botrychium*). Contr. Univ. Michigan Herb. 19: 83–92. Wagner, W. H. Jr. and F. S. Wagner. 1983. Genus communities as a systematic tool in the study of New World *Botrychium* (Ophioglossaceae). Taxon 32: 51–63. Wagner, W. H. Jr. and F. S. Wagner. 1990. Moonworts *(Botrychium* subg. *Botrychium)* of the upper Great Lakes region. Contr. Univ. Michigan Herb. 17: 313–325.

1. Leaf blades deltate, mostly 5–25 cm, commonly sterile, sporophores absent or misshapen; plants mostly over 12 cm; leaf sheaths open or closed.
   2. Trophophore blade thin, herbaceous; leaf sheaths open; sporophores, when present, arising from base of trophophore blade high on common stalk; leaves absent during winter (subg. *Osmundopteris*). . . . . 1. *Botrychium virginianum*
   2. Trophophore blades herbaceous or thick-papery to leathery; leaf sheaths closed; sporophores, when present, arising near ground from basal portion of common stalk; leaves present during winter (subg. *Sceptridium*).
      3. Trophophores prostrate, blades commonly 2 per plant; roots yellowish, to 30 per plant; sporophore stalks and midrib broadly flattened, fleshy; leaves dying in early spring, new leaves appearing in late fall (sect. *Hiemobotrychium*)
      . . . . 2. *Botrychium lunarioides*
      3. Trophophores erect or ascending, blades commonly 1 per plant; roots blackish, to 15 per plant; sporophore stalks and midrib only slightly flattened, not fleshy; leaves appearing in late spring and lasting until following spring (subg. *Sceptridium* sect. *Sceptridium*).
         4. Basal pinnae mostly long-stalked and remotely alternate; pinnule venation nearly like ribs of fan but with short midrib; blades dull gray-green. . . . . . . 5. *Botrychium jenmanii*
         4. Basal pinnae short-stalked and mostly subopposite; pinnule venation pinnate, with strong midrib; blades bluish green or green to dark green.
            5. Terminal pinnules larger than lateral pinnules; pinnae undivided except in proximal 1/2–3/4.
               6. Trophophore blades usually 2–3-pinnate, terminal pinnae elongate and nearly parallel-sided; leaves mostly remaining green during winter. . . .
               . . . . 3. *Botrychium biternatum*
               6. Trophophore blades usually 2–4-pinnate, pinnules trowel-shaped or ovate (rarely linear), apex rounded to acute; leaves green or bronze during winter.
                  7. Pinnules obliquely ovate, margins finely denticulate to crenulate, apex rounded to acute; trophophore blades green in winter. . . . . . 7. *Botrychium oneidense*
                  7. Pinnules obliquely trowel-shaped or linear, margins denticulate to lacerate or coarsely cut, apex acute; trophophore blades bronze in winter if exposed. 4. *Botrychium dissectum*
            5. Terminal pinnules similar to or only slightly larger than lateral pinnules; pinnae divided to tip.
               8. Segments of blades rounded, nearly entire, plane; texture leathery; n North America. . . . . 6. *Botrychium multifidum*
               8. Segments of blades angular, ± dentate, somewhat channeled and concave abaxially; texture semiherbaceous; Great Lakes and St. Lawrence Seaway region. . . . . .8. *Botrychium rugulosum*

1. Leaf blades mainly oblong to linear, mostly 2–4 cm, all fertile, sporophores always present; plants to 15 cm, mostly less than 10 cm; leaf sheaths closed (subg. *Botrychium*).
    9. Trophophores linear to linear-oblong, simple to lobed, lobes rounded to square and angular, stalks usually 1/3–2/3 length of trophophore; plants in deep shade under shrubs and trees.
        10. Segments rounded; plants herbaceous. . . . . 29. *Botrychium simplex*
        10. Segments angular; plants succulent.
            11. Surfaces shiny yellow-green (alive); blade apex undissected or with 2–3 lobes, cutting fairly regular; trophophore stalks very succulent; Great Lakes region. . . . . 21. *Botrychium mormo*
            11. Surfaces dull gray-green (alive); blade apex dissected with 3–5 lobes or projections, cutting somewhat irregular; trophophore stalks somewhat succulent; far w North American mountains. . . . . 20. *Botrychium montanum*
    9. Trophophores linear to deltate (narrowly oblong in *Botrychium minganese*), pinnate, rarely simple, lobes, if present, of various shapes, stalk usually less than 1/4 length of trophophore; plants usually in exposed sites.
        12. Distance between 1st and 2d pinna pairs greater than that between 2d and 3d pairs; segments asymmetric, enlarged on acroscopic side.
            13. Apex of blade undivided or coarsely divided; pinnae ovate–fan-shaped, margins shallowly sinuate; small leaves frequently simple or nearly so; large mature blades subternate to ternate; sporophores 1-pinnate; nearly circumboreal. . . . . 29. *Botrychium simplex*
            13. Apex of blade always finely divided; pinnae fan-shaped to narrowly spatulate, margins crenate to dentate or jagged; small leaves always deeply lobed or pinnate; large mature blades oblong-linear; sporophores 2–3-pinnate; w Minnesota prairie. . . . . 14. *Botrychium gallicomontanum*
        12. Distance between 1st and 2d pinna pairs same or slightly more than between 2d and 3d pairs; segments asymmetric to symmetric.
            14. Trophophores present; basal pinnae or segments with venation like ribs of fan, midrib absent; basal pinnae fan-shaped to spatulate.
                15. Trophophore blades ovate to deltate.
                    16. Sporophores 3–5 times length of trophophores, arising at or just above leaf sheath; blades bright green, pinnae remote or approximate, fan-shaped, papery; widespread, w North America. . . . . 29. *Botrychium simplex*
                    16. Sporophores 1–1.5 times length of trophophores, arising high on common stalk; blades dull whitish green, pinnae overlapping, cuneate, leathery; sw Oregon. . . . . 28. *Botrychium pumicola*
                15. Trophophore blades oblong to oblong-lanceolate.
                  17. Basal pinnae broadly fan-shaped.
                    18. Plants herbaceous; trophophores on most plants less than 4 × 1.5 cm; pinnae 2–5 pairs, well separated; margins commonly crenate to dentate; sporophores 1.3–3 times length of trophophore; damp sites; w North America. . . . . .12. *Botrychium crenulatum*
                    18. Plants fleshy; trophophores on most plants more than 5 × 2 cm; pinnae 4–9 pairs, approximate to overlapping; margins usually entire to undulate, rarely dentate; sporophores 0.8–2 times length of trophophore; dry sites; widespread. . . . . .17. *Botrychium lunaria*
                17. Basal pinnae narrowly fan-shaped, or cuneate to lanceolate or linear.
                  19. Pinnae strongly ascending; margins conspicuously dentate-lacerate. . . . . 10. *Botrychium ascendens*
                  19. Pinnae spreading or only moderately ascending; outer margins entire to crenate or rarely dentate.

20. Trophophores ± folded longitudinally when alive, usually to 4 × 1 cm; pinnae to 5 pairs, most proximal pinnae 2-lobed.
    21. Blades very fleshy; sporophores usually less than 1.5 times length of trophophores; pinnae mostly linear; basal pinna lobes usually ± equal; appearing in spring. . . 11. *Botrychium campestre*
    21. Blades herbaceous; sporophores usually 1.5–4 times length of trophophores; pinnae asymmetrically fan-shaped; basal pinna lobes unequal; appearing in late spring. . . . . 22. *Botrychium pallidum*
20. Trophophores flat or folded only at base when alive, usually to 10 × 2.5 cm; pinnae to 10 pairs, basal pinnae unlobed or if lobed, not usually 2-cleft.
    22. Blades narrowly oblong, firm to herbaceous; pinnae nearly spheric to fan-shaped; margins shallowly crenate; proximal sporophore branches 1-pinnate. . . 19. *Botrychium minganense*
    22. Blades narrowly deltate, leathery; pinnae spatulate to linear-spatulate; margins entire to very coarsely and ir-regularly dentate; most proximal sporophore branches usually 2-pinnate. . . . . 30. *Botrychium spathulatum*
14. Trophophores present or replaced by sporophore; if present, basal pinnae or segment venation pinnate, midrib present, oblanceolate to linear to lanceolate to ovate.
    23. Trophophore replaced by sporophore, yielding 2 sporophores. . . . . . .
    . . . . 23. *Botrychium paradoxum*
    23. Trophophore present, fully distinct from sporophore.
        24. Trophophore blades deltate; sporophores divided proximally into several equally long branches. . . . . 16. *Botrychium lanceolatum*
        24. Trophophore blades ovate to oblong (nearly deltate in *Botrychium hesperium*, deltate-oblong in *B. pinnatum*); sporophores with single midrib or 1 dominant midrib and 2 smaller ribs.
        25. Trophophore stalk long, equal to length of trophophore rachis; blade mostly ovate-oblong to deltate-oblong; basal pinnae ovate-rhombic; nw North America. . . . . 24. *Botrychium pedunculosum*
        25. Trophophore stalk short to nearly absent, to 1/4 length of tro-phophore rachis; blade mainly oblong-lanceolate to triangular; most basal pinnae elongate, oblanceolate to oblong to linear or linear-lanceolate.
            26. Large trophophore blades nearly deltate, basal pinna pair elongate; pinnae distal to basal pair approximate to overlap-ping; segments and lobes rounded at apex. . . . . 15. *Botrychium hesperium*
            26. Large trophophore blades mostly oblong-deltate to ovate-oblong; basal pinna pair not elongate; pinnae distal to basal pair remote to approximate; segments and lobes truncate, rounded, or acute at apex.
                27. Sporophores long, 1–3 times length of trophophore; blades dull, blue to green.
                    28. Pinnae ovate to lanceolate, blunt, shallowly to deeply lobed. . . . . 18. *Botrychium matricariifolium*
                    28. Pinnae oblanceolate to linear-lanceolate, acuminate, entire to shallowly lobed. . . . . 9. *Botrychium acuminatum*
                27. Sporophores short, only 1–2 times length of tropho-phore; blades shiny, bright green.

29. Pinnae acute, oblanceolate to narrowly spatulate, rarely more than 2-lobed; pinnae well separated. . . . . 13. *Botrychium echo*
29. Pinnae mostly with rounded apex, ovate to broadly spatulate, to 6-lobed; pinnae approximate to overlapping.
      30. Pinnae of mature trophophores nearly as wide as long, with slightly pointed tips, costa rudimentary, veins otherwise ± like ribs of fan; basal pinnae with only shallow, narrow sinuses and 1–3 lobes. . . . . 26. *Botrychium boreale*
      30. Pinnae of mature trophophores considerably longer than wide, mostly with blunt tips, veins mainly pinnate; basal pinnae with deep, ± wide sinuses and 3–8 lobes.
            31. Pinnae ascending, usually somewhat overlapping; trophophore blades leathery, somewhat shiny; Lake Superior region . . . . 27. *Botrychium pseudopinnatum*
            31. Pinnae ± ascending to ± horizontal, usually approximate to somewhat remote; trophophore blades papery, shiny; nw North America. . . . . 25. *Botrychium pinnatum*

1a. BOTRYCHIUM Swartz subg. OSMUNDOPTERIS (J. Milde) R. T. Clausen, Mem. Torrey Bot. Club 19(2): 93. 1938

Botrychium Swartz sect. *Osmundopteris* J. Milde, Verh. Zool.- Bot. Ges. Wien 19(2): 96. 1869

Roots 15 or fewer, yellow to brown, 0.5–2 mm diam. 1 cm from base. **Plants** over 12 cm. **Common stalk** lacking idioblasts (elongate, spindle-shaped cells in vascular strand). **Trophophore** sessile and erect, arising from middle to distal portion of common stalk well above ground level; blade usually 1 per plant, seasonal, absent during winter, deltate, 3–4-pinnate, 5–25 cm wide when mature, thin, herbaceous. **Leaf primordia** densely hairy, hairs 1.5–2 mm. **Leaf sheath** open. **Pinna** divisions reduced in size to tip. **First pinnule** on basal pinnae usually borne acroscopically. **Sporophores** long-stalked, arising from middle to distal portion of common stalk well above ground level, commonly absent due to abortion in early development or seasonal, stalks only slightly flattened, not fleshy, 0.5–0.8 mm wide. $x = 92$.

Species 2–3 (1 in the flora): worldwide.

1. Botrychium virginianum (Linnaeus) Swartz, J. Bot. (Schrader) 1800(2): 111. 1801 · Rattlesnake fern, common grapefern, botryche de Virginie

Osmunda virginiana Linnaeus, Sp. Pl. 2: 1064. 1753

Trophophore sessile; blade pale green, 3–4-pinnate, to 25 × 33 cm, thin, herbaceous. **Pinnae** to 12 pairs, usually approximate to overlapping, slightly ascending, distance between 1st and 2d pinnae not or slightly more than between 2d and 3d pairs, lanceolate, divided to tip. **Pinnules** lanceolate and deeply lobed, lobes linear, serrate, apex pointed, venation pinnate, midrib present. **Sporophores** 2-pinnate, 0.5–1.5(–2) times length of trophophore. $2n = 184$.

Leaves seasonal, appearing in early spring and dying in late summer. Common to abundant, especially in shaded forests and shrubby second growth, rare or absent in arid regions; 0–1500 m; Alta., B.C., Man., N.B., Nfld., N.W.T., N.S., Ont., P.E.I., Que., Sask., Yukon; all states except Calif.; Mexico; Central America; South America in Brazil, Colombia, Ecuador, Peru; Eurasia.

Botrychium virginianum is the most widespread *Botrychium* in North America.

1b. Botrychium Swartz subg. Sceptridium (Lyon) R. T. Clausen sect. Hiemobotrychium W. H. Wagner, Novon 2: 267. 1992

Roots 20–30, yellow to brown, 0.6–1 mm diam. 1 cm from base. **Plants** over 6 cm. **Common stalk** with tracheidal idioblasts. **Trophophore** short-stalked and prostrate, arising from basal portion of common stalk near ground level; blades usually 2 per plant, appearing in late fall and dying in early spring, deltate, 2–3-pinnate, mostly 5–8 cm wide when mature, herbaceous. **Leaf primordia** with scattered hairs, hairs 0.6–1 mm. **Leaf sheath** closed. **Pinna** divisions reduced to tip. **First pinnule** on basal pinnae usually borne basiscopically. **Sporophores** long-stalked, arising from basal portion of common stalk near ground level, commonly absent without primordia; stalks and rachis broadly flattened, fleshy, to 1.4 mm wide. $x = 45$.

Species 1: endemic to se United States.

Tracheidal idioblasts in *Botrychium* sect. *Hiemobotrychium* are enormous, elongate, spindle-shaped cells with helical or annular thickenings and are scattered along the vascular strand. They are unique among ferns.

2. Botrychium lunarioides (Michaux) Swartz, Syn. Fil., 172. 1806 · Winter grapefern, prostrate grapefern

Botrypus lunarioides Michaux, Fl. Bor.-Amer. 2: 274. 1803

Trophophore stalk 0.1–1 cm; blades usually pale green, plane, 2–3-pinnate, to 8 × 12 cm, often much smaller, fleshy. **Pinnae** to 5 pairs, usually well separated, horizontal, distance between 1st and 2d pinnae not or slightly more than between 2d and 3d pairs, divided to tip. **Pinnules** fan-shaped, margins denticulate, apex rounded, venation like ribs of fan, midrib absent. **Sporophores** 2-pinnate, 1–2 times length of trophophores. $2n = 90$.

Leaves appearing in late fall and dying in early spring. In open grassy places in prairies, cemeteries, and weedy roadsides; 0–250 m; Ala., Ark., Fla., Ga., La., Miss., N.C., S.C., Tex.

The stalk and proximal part of rachis of *Botrychium lunarioides* contains huge tracheidal idioblasts with annular thickenings, visible in cleared leaves (H. J. Arnott 1960). Another peculiarity of this species is the tendency for the sporophores to remain curled in late fall and early winter and to become erect in February. *Botrychium lunarioides* is often associated with *Schizachyrium scoparius* Michaux and *Ophioglossum crotalophoroides* Walter. The name *B. biternatum* was misapplied by L. Underwood to *B. lunarioides*.

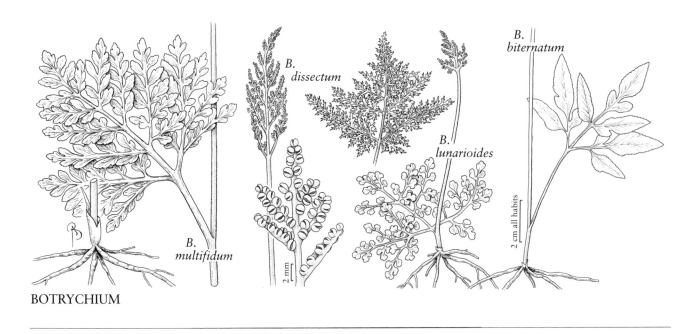

B. dissectum

B. biternatum

B. lunarioides

B. multifidum

2 mm

2 cm all habits

BOTRYCHIUM

1c. BOTRYCHIUM Swartz subg. SCEPTRIDIUM (Lyon) R. T. Clausen sect. SCEPTRIDIUM Mem. Torrey Bot. Club 19(2): 24.1938

Sceptridium Lyon, Bot. Gaz. 40: 457. 1905

Roots usually 10 or less, blackish, 1–4 mm diam. 1 cm from base. **Plants** over 12 cm. **Common stalk** lacking idioblasts. **Trophophore** erect, long-stalked, arising from basal portion of common stalk near ground level; blade usually 1 per plant, appearing in spring, dying the following spring, present during winter, deltate, 2–4-pinnate, mostly 5–25 cm wide when mature, herbaceous or leathery. **Leaf primordia** densely hairy, hairs 1.5–2 mm. **Leaf sheath** closed. **Pinna** divisions reduced in size to tip (except *B. biternatum, B. dissectum,* and *B. oneidense* with undivided tips larger than lateral pinnae). **First pinnule** on basal pinnae usually borne basiscopically. **Sporophore** long-stalked, arising from basal portion of common stalk near ground level, commonly absent or represented by hairy rudiment due to abortion, or falling off, stalks and rachis only slightly flattened, not fleshy, 0.5–0.8 mm wide. $x = 45$.

Species 13 (6 in the flora): worldwide.

3. Botrychium biternatum (Savigny) L. Underwood, Bot. Gaz. 22: 407. 1896 · Sparse-lobed grapefern, southern grapefern

Osmunda biternata Savigny in Lamarck et al., Encycl. 4: 650. 1797; *Botrychium dissectum* Sprengel var. *tenuifolium* (L. Underwood) Farwell; *B. tenuifolium* L. Underwood

Trophophore stalk 4–20 cm, 2–2.5 times length of trophophore rachis; blade green to dark green, plane, 2–3-pinnate, to 18 × 28 cm, herbaceous. **Pinnae** to 7 pairs, usually remote, horizontal, distance be-tween 1st and 2d pinnae not or slightly more than be-tween 2d and 3d pairs, undivided except in proximal 2/3–1/2. **Pinnules** elongate, obliquely lanceolate to nar-rowly lanceolate, margins nearly parallel and finely denticulate, apex short-acuminate, venation pinnate. **Sporophores** 1–2-pinnate, 2–3 times length of tropho-phore. $2n = 90$.

Leaves green over winter, sporophores seasonal, new leaves appearing in late spring–early summer. Frequent in low woods and brushy fields, 0–600 m; Ala., Ark., Del., Fla., Ga., Ill., Ind., Ky., La., Md., Miss., Mo., N.J., N.C., Ohio, Okla., Pa., S.C., Tenn., Tex., Va., W.Va.

Botrychium biternatum often grows with *B. dissec-*

tum and *B. jenmanii*. The name *B. biternatum* was misapplied by L. Underwood to *B. lunarioides* (W. H. Wagner Jr. 1961).

4. Botrychium dissectum Sprengel, Anleit. Kenntn. Gew. 3: 172. 1804 · Dissected grapefern, botryche découpé

Botrychium obliquum Muhlenberg in Willdenow

Trophophore stalk 3–15 cm, 1.5–2.5 times length of trophophore rachis; blade shiny green, often bronze in winter, plane to convex, 3–4-pinnate, to 20 × 30 cm, leathery. **Pinnae** to 10 pairs, approximate to remote, slightly ascending, distance between 1st and 2d pinnae not or slightly more than between 2d and 3d pairs, undivided except in proximal 2/3–3/4. **Pinnules** usually obliquely angular–trowel-shaped to widely trowel-shaped to obliquely round-lanceolate to ovate and pointed, margins denticulate to lacerate to coarsely cut halfway or wholly into linear-divergent segments in some populations, venation pinnate. **Sporophores** 2–3-pinnate, 1.5–2.5 times length of trophophore. $2n = 90$.

Leaves green over winter, new leaves appearing in late spring. In variety of habitats, open grassy areas to deep forest; 0–1500 m; N.B., N.S., Ont., Que.; Ala., Ark., Conn., Del., D.C., Fla., Ga., Ill., Ind., Iowa, Kans., Ky., La., Maine, Md., Mass., Mich., Miss., Mo., N.H., N.J., N.Y., N.C., Ohio, Pa., R.I., S.C., Tenn., Tex., Vt., Va., W.Va., Wis.; West Indies in the Antilles.

Botrychium dissectum is highly variable, even within the same population. In Florida and along the Gulf Coast, the extremely lacerate form is absent, and the blade segments are usually strongly angular, trowel-shaped, and dentate. In eastern Kentucky and central Tennessee in forested valleys, on shale and limestone soils, plants have narrowly linear, somewhat blunt-tipped segments with a more or less whitish gray central line above the veins. This variant, which grows with *B. dissectum,* may deserve recognition as a distinct species.

5. Botrychium jenmanii L. Underwood, Fern Bull. 8: 59. 1900 · Alabama grapefern

Botrychium alabamense Maxon

Trophophore stalk 2–15 cm, 0.8–1.2 times length of trophophore rachis; blade somewhat dull gray-green, plane, 3-pinnate, to 18 × 26 cm, herbaceous. **Pinnae** to 5 pairs, well separated, slightly descending to ascending, distance between 1st and 2d pinnae not or slightly more than between 2d and 3d pairs;

basal pinnae remotely alternate and long-stalked (basal pinnae in all other botrychiums are opposite to subopposite and short-stalked), divided to tip. **Pinnules** ovate to fan-shaped, margins uniformly denticulate, apex rounded, venation like ribs of fan with short midrib. **Sporophores** 2-pinnate, 1.2–2.5 times length of trophophore. $2n = 180$.

Leaves green over winter, arising at variable times during last half of summer, meiosis as late as September. Woods and grassy places; 50–500 m; Ala., Fla., Ga., La., Miss., N.C., S.C., Tenn., Va.; West Indies in Greater Antilles.

Botrychium jenmanii occurs in a variety of habitats. In hardwoods and especially pine woods, it is associated with *B. biternatum*; in open grassy places and lawns it is found with *B. lunarioides*. In many ways, *B. jenmanii* is intermediate between *B. biternatum* and *B. lunarioides*, and it is possibly their allopolyploid derivative (W. H. Wagner Jr. 1968). It is the only tetraploid among New World members of subg. *Sceptidrium*.

6. Botrychium multifidum (S. G. Gmelin) Ruprecht, Bemerk. Botrychium, 40. 1859 · Leather grapefern, botryche à feuille couchée

Osmunda multifida S. G. Gmelin, Novi Comment. Acad. Sci. Imp. Petrop. 12: 517. 1768; *Botrychium californicum* L. Underwood; *B. coulteri* L. Underwood; *B. silaifolium* C. Presl

Trophophore stalk 2–15 cm, 0.3–1.2 times length of trophophore rachis; blade shiny green, plane, ternate, 2–3-pinnate, to 25 × 35 cm, leathery. **Pinnae** to 10 pairs, approximate to remote, horizontal to ascending, distance between 1st and 2d pinnae not or slightly more than between 2d and 3d pairs, divided to tip. **Pinnules** obliquely ovate, rounded, margins usually ± entire to shallowly crenulate, sometimes inconspicuously and shallowly denticulate, apex rounded, venation pinnate. **Sporophores** 2–3-pinnate, 1.2 times length of trophophore. $2n = 90$.

Leaves green over winter, appearing in spring. Widespread mainly in fields; 0–3000 m; Greenland; Alta., B.C., Man., N.B., Nfld., N.W.T., N.S., Ont., Que., Sask.; Alaska, Ariz., Calif., Colo., Conn., Idaho, Ill., Ind., Iowa, Maine, Mass., Mich., Minn., Mont., Nev., N.H., N.J., N.Y., N.Dak., Ohio, Oreg., Pa., R.I., S.Dak., Utah, Vt., Va., Wash., W.Va., Wis., Wyo.; Europe; nw Asia.

Botrychium multifidum is rather similar to *B. robustum* (Ruprecht) L. Underwood of Japan, eastern China, and the former Soviet republics. Specimens identified as *B. robustum* have been collected on Unalaska Island.

7. **Botrychium oneidense** (Gilbert) House, Amer. Midl. Naturalist 7: 126. 1905 · Blunt-lobed grapefern, botryche du lac Onéida

Botrychium ternatum (Thunberg) Swartz var. *oneidense* Gilbert, Fern Bull. 9: 27. 1901; *B. dissectum* Sprengel var. *oneidense* (Gilbert) Farwell; *B. multifidum* (S. G. Gmelin) Ruprecht var. *oneidense* (Gilbert) Farwell

Trophophore stalk 2–15 cm, 1.5–2.5 times length of blade rachis; blade dull bluish green, ± plane, 2–3-pinnate, to 15 × 20 cm, ± leathery. **Pinnae** to 5 pairs, usually remote, horizontal to ascending, distance between 1st and 2d pinnae not or slightly more than between 2d and 3d pairs, undivided except in proximal 2/3–3/4. **Pinnules** obliquely ovate, margins finely crenulate to denticulate, apex rounded to acute, venation pinnate. **Sporophores** 2–3-pinnate, 1.5–2.5 times length of trophophore. $2n = 90$.

Leaves green over winter, sporophores seasonal, new leaves appearing in spring. In moist, shady, acidic woods and swamps; 0–1200 m; N.B., Ont., Que.; Conn., Del., D.C., Ind., Ky., Maine, Md., Mass., Mich., Minn., N.H., N.J., N.Y., N.C., Ohio, Pa., R.I., Vt., Va., W.Va., Wis.

Botrychium oneidense commonly occurs with *B. dissectum* and *B. multifidum*. Young individuals of both may resemble *B. oneidense* (W. H. Wagner Jr. 1961b).

8. **Botrychium rugulosum** W. H. Wagner, Contr. Univ. Michigan Herb. 15: 315. 1982 · St. Lawrence grapefern, botryche du St. Laurent

Botrychium multifidum (S. G. Gmelin) Ruprecht forma *dentatum* R. M. Tryon

Trophophore stalk 2 to 15 cm, 1–2.5 times length of trophophore rachis; blade green, finely rugulose and convex distally, 2–4-pinnate, to 15 × 26 cm, somewhat herbaceous. **Pinnae** to 9 pairs, usually approximate, horizontal to ascending, distance between 1st and 2d pinnae not or slightly more than between 2d and 3d pairs, divided to tip. **Pinnules** obliquely and angularly trowel-shaped to spatulate, margins usually denticulate, apex acute, venation pinnate. **Sporophores** 2-pinnate, 1–2 times length of trophophore. $2n = 90$.

Leaves green over winter, appearing in midspring. In open fields and secondary forests over wide range in vicinity of St. Lawrence Seaway; 200–1000 m; Ont., Que.; Mich., Minn., N.Y., Vt., Wis.

The name "rugulosum" refers to the tendency of the segments to become more or less wrinkled and convex. *Botrychium rugulosum* occurs with *B. dissectum*, *B. multifidum*, and rarely *B. oneidense*. It is often found in small stands of only 5–10 individuals, but some populations number over 100.

1d. BOTRYCHIUM Swartz subg. BOTRYCHIUM

Roots usually 10 or fewer, yellow or brown, 0.5–1.5 mm diam. 1 cm from base. **Plants** less than 15 cm. **Common stalk** lacking idioblasts. **Trophophore** short-stalked or nearly sessile (long-stalked in forms of *B. simplex* and *B. pedunculosum*) arising from the middle or high on common stalk (low in some individuals of *B. montanum*, *B. mormo*, and *B. simplex*), blade usually 1 per plant, appearing in spring and dying in summer or fall, absent during winter, mostly linear to oblong to oblong deltate (deltate in *B. lanceolatum*), lobed to 1–2 (–3)-pinnate, mostly less than 2.5 cm wide when mature, herbaceous to leathery. **Leaf primordia** glabrous. **Leaf sheath** closed. **Pinna** lobes and segments, when present, asymmetric, either borne basiscopically or acroscopically. **Sporophores** long- or short-stalked, arising from middle to distal portion of common stalk, well above ground level (low on common stalk, near ground level in some forms of *B. simplex*), always present, stalks and rachis only slightly flattened, not fleshy, 0.5–2 mm wide. $x = 45$.

Species ca. 25 (21 in the flora): worldwide.

Trophophore blades of most species in *Botrychium* subg. *Botrychium* are divided into lobes or segments and are not truly pinnate except in those species with fan-shaped segments. Very rarely, blades are more than 1-pinnate (western form of *B. simplex*).

9. **Botrychium acuminatum** W. H. Wagner, Contr. Univ. Michigan Herb. 17: 321. 1990 · Pointed moonwort

Trophophore stalk 0–20 mm, 0.1–0.5 times length of trophophore rachis; blade dull, glaucescent, ovate-oblong, 1-pinnate, to 6 × 5 cm, firm. **Pinnae** to 6 pairs, spreading to ascending, mostly separated by pinna width, distance between 1st and 2d pinnae not or slightly more than between 2d and 3d pairs, basal pinna pair approximately equal in size and cutting to next adjacent pair, narrowly oblanceolate to linear-lanceolate, simple to irregularly divided to tip, margins commonly ± entire and with 1–several shallow and irregular lobes, apex acuminate, venation pinnate. **Sporophores** 1–2-pinnate, 1.4–2 times length of trophophore. $2n = 180$.

Leaves appearing in midspring, dying in fall. Lake Superior region in sand dunes in shade, old fields, grassy railroad sidings, and roadside ditches; of conservation concern; 200–500 m; Ont.; Mich.

Botrychium acuminatum grows with various other moonworts, particularly its nearest relatives, *B. hesperium* and *B. matricariifolium*.

10. **Botrychium ascendens** W. H. Wagner, Amer. Fern J. 76: 36, figs. 1, 2. 1986 · Upswept moonwort

Trophophore stalk 3–10 mm, 1/6 length of trophophore rachis; blade yellow-green, oblong to oblong-lanceolate, 1-pinnate, to 6 × 1.5 cm, thin but firm. **Pinnae** to 5 pairs, strongly ascending, well separated, distance between 1st and 2d pinnae not or slightly more than between 2d and 3d pairs, basal pinna pair approximately equal in size and cutting to adjacent pair, obliquely narrowly cuneate, undivided to tip, margins sharply denticulate and often shallowly incised, apex rounded, venation like ribs of fan, midrib absent. **Sporophores** 2-pinnate at base of sporangial cluster, 1.3–2 times length of trophophore. $2n = 180$.

Leaves appearing in late spring to midsummer. In grassy fields, widely scattered; 0–2500 m; B.C., Ont., Yukon; Alaska, Calif., Mont., Nev., Oreg., Wyo.

Botrychium ascendens is a distinctive little moonwort that grows with *B. crenulatum*, *B. lunaria*, and *B. minganense*. This species and *B. pedunculosum* are the only grapeferns that often have extra sporangia on the proximal pinnae.

11. **Botrychium campestre** W. H. Wagner & Farrar, Amer. Fern J. 76: 39, figs. 2, 4, 5. 1986 · Prairie moonwort, botryche champêtre

Trophophore stalk usually absent but sometimes broadly tapered to 10 mm in forms with coalesced proximal pinnae; blade glaucescent, oblong, longitudinally folded when alive, 1-pinnate, to 4 × 1.3 cm, very fleshy. **Pinnae** to 5(–9) pairs, spreading, usually remote, separated 1–3 times pinna width, in some populations irregularly and extensively fused with considerable webbing along rachis, distance between 1st and 2d pinnae not or slightly more than between 2d and 3d pairs, basal pinna pair approximately equal in size and cutting to the adjacent pair, mostly linear to linear-spatulate, undivided to tip, margins crenulate to dentate, usually notched or cleft into 2 or several segments, apex rounded to acute, venation like ribs of fan, midrib absent. **Sporophores** 1 (–2, rarely)-pinnate, 1–1.5 times length of trophophore. $2n = 90$.

Leaves appearing in early spring and dying in late spring and early summer, long before those of associated moonworts. Extremely inconspicuous in prairies, dunes, grassy railroad sidings, and fields over limestone; of conservation concern; 50–1200 m; Alta., Ont., Sask.; Colo., Iowa, Mich., Minn., Mont., Nebr., N.Y., N.Dak., S.Dak., Wis., Wyo.

Botrychium campestre is one of four moonwort species that commonly produce dense clusters of minute, spheric gemmae at the root bases. Peculiar forms of *B. campestre* with coalescent pinnae are found on dunes in the vicinity of Lake Michigan.

12. **Botrychium crenulatum** W. H. Wagner, Amer. Fern J. 71: 21. 1981 · Dainty moonwort

Trophophore stalk 0.5–7 mm; blade yellow-green, oblong, 1-pinnate, to 6 × 2 cm, thin, herbaceous. **Pinnae** to 5 pairs, spreading, well separated, distance between 1st and 2d pinnae not or slightly more than between 2d and 3d pairs, basal pinna pair approximately equal in size and cutting to adjacent pair, broadly fan-shaped, undivided to tip, margins mainly crenulate to dentate, proximal pinnae with 1 or more shallow incisions, apex rounded, apical lobe linear to linear-cuneate, well separated from adjacent lobes, venation like ribs of fan, midrib absent. **Sporophores** 1–2-pinnate, 1.3–3 times length of trophophore. $2n = 90$.

Leaves appearing in mid to late spring, dying in late summer; in extremely dry years of shorter duration or not appearing at all. Local in marshy and springy areas; 1200–2500 m; Ariz., Calif., Idaho, Mont., Oreg., Nev., Utah, Wash., Wyo.

Botrychium crenulatum is commonly associated with *B. simplex* in California. In the Wallowa Mountains of Oregon it occurs with *B. ascendens, B. lunaria,* and *B. minganense.*

13. Botrychium echo W. H. Wagner, Amer. Fern J. 73: 57. 1983 · Echo moonwort

Trophophore stalk 0–4 mm; blade shiny green, broadly oblong to oblong-deltate, 1–2-pinnate, to 4 × 3 cm, firm. Pinnae to 4 pairs, spreading or only moderately ascending, well separated, distance between 1st and 2d pinnae not or slightly more than between 2d and 3d pairs, basal pinna pair approximately equal in size and cutting to adjacent pair, oblanceolate to linear-spatulate, ± parallel-sided, divided to tip, shallowly lobed or rarely 2-cleft, basal pinna cleft into single basiscopic projection and large acroscopic projection, margins entire, apex acute, venation pinnate. Sporophores 1–2 pinnate, 1–2 times length of trophophore rachis. $2n = 180$.

Leaves appearing in June, dying in September. Grassy mountain slopes, snow fields, road ditches, and sand dunes; of conservation concern; 2500–3700 m; Ariz., Colo., Utah.

Botrychium echo is one of four moonwort species that commonly produce clusters of minute, spheric gemmae at the root bases. This species tends to have a reddish brown stripe along the common stalk from the base of the trophophore stalk.

14. Botrychium gallicomontanum Farrar & Johnson-Groh, Amer. Fern J. 81: 1, figs. 1, 2, 4. 1991 · Frenchman's Bluff moonwort

Trophophore stalk 1–8 mm; blade yellow-green, ovate to oblong-linear, 1-pinnate, to 3 × 0.9 cm, firm, glaucescent. Pinnae to 6 pairs, strongly ascending, well separated, distance between the 1st and 2d pinnae considerably greater than between 2d and 3d pairs, basal pinna pair approximately equal in size and cutting to adjacent pair, fan-shaped to narrowly spatulate, often asymmetric, with distal portion longer than and arching over proximal portion, undivided to tip, rarely 2-cleft, margins entire to irregularly cleft, apex rounded, venation like ribs of

fan, midrib absent. Sporophores 2–3-pinnate, 1.5–3 times length of trophophore.

Leaves appearing in midspring, dying in summer. Prairies; of conservation concern; 300 m; Minn.

Botrychium gallicomontanum is known only from one locality in western Minnesota, where it grows with both *B. campestre* and *B. simplex.* It is intermediate between them in the spacing, shape, and stalk length of the pinnae. This is one of four moonwort species that commonly produce dense clusters of minute, spheric gemmae at the root bases. This moonwort is probably an allopolyploid of the two associated species.

15. Botrychium hesperium (Maxon & R. T. Clausen) W. H. Wagner & Lellinger, Amer. Fern J. 71: 92. 1981 · Western moonwort

Botrychium matricariifolium (Döll) W. D. J. Koch subsp. *hesperium* Maxon & R. T. Clausen, Mem. Torrey Bot. Club 19(2): 88, fig. 15. 1938

Trophophore stalk 0–3(–10) mm, to 1/4 length of trophophore rachis; blade ± gray-green, dull, oblong-linear to deltate, 1–2-pinnate, to 6 × 5 cm, firm. Pinnae to 6 pairs, ascending, usually approximate or overlapping except in shade forms, distance between 1st and 2d pinnae not or slightly more than between 2d and 3d pairs, basal pinna pair commonly much larger and more divided than adjacent pair, lobed to tip, basal pair oblong to oblong-lanceolate with lobed margins, remainder broadly spatulate with entire margins or 1 or more shallow lobes, apex rounded, venation pinnate. Sporophores 1–3 pinnate, 2–3 times length of trophophore. $2n = 180$.

Leaves appearing in midspring, dying in early fall. Grassy mountain slopes, snow fields, road ditches with willows, and sand dunes; 200–2800 m; Alta., B.C., Ont., Sask.; Ariz., Colo., Idaho, Mich., Mont., Utah, Wyo.

In the Rocky Mountains *Botrychium hesperium* grows often with *B. echo,* and in the Lake Superior region, with *B. acuminatum* and *B. matricariifolium.*

16. Botrychium lanceolatum (S. G. Gmelin) Angström, Bot. Not. 1854: 68. 1854 · Triangle moonwort, botryche élancé

Osmunda lanceolata S. G. Gmelin, Novi Comment. Acad. Sci. Imp. Petrop. 12: 516. 1768

Trophophore stalk 0–1 mm; blade dull to shiny green to dark green, deltate, 1–2-pinnate, to 6 × 7 cm. Pinnae to 5 pairs, ascending, approximate, distance between 1st and 2d pinnae not or slightly more than between 2d and 3d pairs, linear to broadly lanceolate, entire to divided to tip, margins with distinct lobes or segments, apex acute to rounded, venation pinnate.

Sporophores 1–3-pinnate, 1–2.5 times length of trophophore, divided into several equally long branches (all other botrychiums have a single stalk or 1 dominant and 2 smaller). $2n = 90$.

Subspecies 2 (2 in the flora): North America, Eurasia.

1. Trophophore blade green to pale yellow-green, broad, coarse, succulent; middle and terminal segments usually more than 2 mm wide exclusive of lobes. . . . .
. . . . 16a. *Botrychium lanceolatum* subsp. *lanceolatum*
1. Trophophore blade dark green, slender, delicate but firm; middle segments usually less than 2 mm wide exclusive of lobes. . . . .
16b. *Botrychium lanceolatum* subsp. *angustisegmentum*

16a. Botrychium lanceolatum (S. G. Gmelin) Angström subsp. **lanceolatum**

Plants stout. **Leaf base**, including sheath, to 1.2 cm diam., often swollen 1–6 times width of common stalk (dried). **Common stalk** green, trophophore blade medium green to yellow-green, somewhat shiny. **Segments** wide, ultimate lobes broadly ovate-oblong, 2–3 mm wide, blade tissue leathery-succulent. **Pinnae** and lobes somewhat blunt. **Sporangia** 0.9–1.3 mm diam., exposed and not immersed, mostly approximate or slightly separated and covering sporangiophore midrib.

Leaves usually drying up in midsummer together with other associated species. Mainly open fields; 0–3700 m; Greenland; Alta., B.C., Nfld., Que., Sask., Yukon; Alaska, Ariz., Colo., Idaho, Mont., Nev., N.Mex., Oreg., Utah, Wash., Wyo.; Eurasia.

16b. Botrychium lanceolatum (S. G. Gmelin) Angström subsp. **angustisegmentum** (Pease & A. H. Moore) R. T. Clausen, Bull. Torrey Bot. Club 64: 280. 1937 · Narrow triangle moonwort

Botrychium lanceolatum var. *angustisegmentum* Pease & A. H. Moore, Rhodora 8: 229. 1906; *B. angustisegmentum* (Pease & A. H. Moore) Fernald

Plants slender. **Leaf base**, including sheath, to 0.7 cm diam., usually swollen only 1–3 times width of common stalk (dried). **Common stalk** dark brownish green, trophophore blade dark green, very shiny. **Segments** narrow, ultimate lobes narrowly linear-oblong, 1–2 mm wide, blade tissue thin but firm. **Pinnae** and lobes mostly sharply pointed. **Sporangia** 0.7–1 mm diam., slightly immersed, mostly

± separated and exposing sporangiophore midrib.

Leaves appearing in late spring or early summer, releasing spores later than most associated species, and dying as late as October. Mainly in shaded woods; 0–1200 m; St. Pierre and Miquelon; Alta., B.C., N.B., Nfld., N.S., Ont., Que., Sask., Yukon; Conn., Ky., Maine, Mass., Mich., Minn., Mont., N.H., N.J., N.Y., Ohio, Pa., R.I., Tenn., Vt., Va., W.Va., Wis.

17. Botrychium lunaria (Linnaeus) Swartz, J. Bot. (Schrader) 1800(2): 110. 1801 · Common moonwort, botryche lunaire

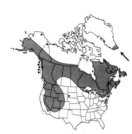

Osmunda lunaria Linnaeus, Sp. Pl. 2: 1064. 1753; *Botrychium onondagense* L. Underwood

Trophophore stalk 0–1 mm; blade dark green, oblong, 1-pinnate, to 10 × 4 cm, thick, fleshy. **Pinnae** to 9 pairs, spreading, mostly overlapping except in shaded forest forms, distance between 1st and 2d pinnae not or slightly more than between 2d and 3d pairs, basal pinna pair approximately equal in size and cutting to adjacent pair, broadly fan-shaped, undivided to tip, margins mainly entire or undulate, rarely dentate, apical lobe usually cuneate to spatulate, notched, approximate to adjacent lobes, apex rounded, venation like ribs of fan, midribs absent. **Sporophores** 1–2-pinnate, 0.8–2 times length of trophophore. $2n = 90$.

Leaves appearing in spring, dying in latter half of summer. Open fields, occasionally forests in southern occurrences; 0–3700 m; Greenland; St. Pierre and Miquelon; Alta., B.C., Man., N.B., Nfld., N.W.T., N.S., Ont., P.E.I., Que., Sask., Yukon; Alaska, Ariz., Calif., Colo., Idaho, Maine, Mass., Mich., Minn., Mont., Nev., N.H., N.Mex., N.Y., Oreg., Pa., S.Dak., Utah, Vt., Wash., Wis., Wyo.; s South America; Eurasia; Pacific Islands in New Zealand; Australia.

Botrychium lunaria grows with many other species of *Botrychium*, occasionally hybridizing with them. This species, geographically the most widespread of the moonworts, has notably uniform morphology.

18. **Botrychium matricariifolium** (Döll) A. Braun ex W. D. J. Koch, Syn. Deut. Schweiz. Fl. ed. 2, 7: 1009. 1847 · Daisy-leaf moonwort, botryche à feuille de matricaire

Botrychium lunaria (Linnaeus) Swartz var. *matricariifolium* Döll, Rhein. Fl., 24. 1843

Trophophore stalk 0–5 mm, to 1/6 length of trophophore rachis; blade dull, ± glaucescent green, oblong, 1–3-pinnate, to 10 × 9 cm, firm. **Pinnae** to 7 pairs, somewhat ascending, approximate to slightly remote, distance between 1st and 2d pinnae not or slightly more than between 2d and 3d pairs, basal pinna pair approximately equal in size and cutting to adjacent pairs, spatulate-ovate to narrowly ovate, divided to tip, ultimate segments squarish to linear, rounded to square to acute at apex, margins entire to lobed to fully dissected, apex rounded to acute, venation pinnate. **Sporophores** 1–3-pinnate, 1.3–2.4 times length of trophophore. $2n = 180$.

Leaves appearing in spring, dying in late summer. Old fields, secondary growth woods; 0–1200 m; St. Pierre and Miquelon; N.B., Nfld., N.S., Ont., Que.; Conn., Del., D.C., Ill., Ind., Iowa, Ky., Maine, Md., Mass., Mich., Minn., N.H., N.J., N.Y., N.C., Ohio, Pa., R.I., Tenn., Vt., Va., W.Va., Wis.; Europe.

19. **Botrychium minganense** Victorin, Proc. & Trans. Roy. Soc. Canada, ser. 3, 21: 331. 1927 · Mingan moonwort, botryche de Mingan

Botrychium lunaria (Linnaeus) Swartz var. *minganense* (Victorin) Dole

Trophophore stalk 0–2 cm, 0 to 1/5 length of trophophore rachis; blade dull green, oblong to linear, 1-pinnate, to 10 × 2.5 cm, firm to herbaceous. **Pinnae** to 10 pairs, horizontal to slightly spreading, approximate to remote, distance between 1st and 2d pinnae not or slightly more than between 2d and 3d pairs, basal pinna pair approximately equal in size and cutting to adjacent pair, occasionally basal pinnae and/or some distal pinnae elongate, lobed to tip, nearly circular, fan-shaped or ovate, sides somewhat concave, margins nearly entire, shallowly crenate, occasionally pinnately lobed or divided, apex rounded, venation like ribs of fan with short midrib. **Sporophores** 1-pinnate, 2-pinnate in very large, robust plants, 1.5–2.5 times length of trophophore. $2n = 180$.

Leaves appearing in spring through summer. Widely scattered; 0–3700 m; St. Pierre and Miquelon; Alta.,

B.C., Man., N.B., Nfld., N.W.T., N.S., Ont., P.E.I., Que., Sask., Yukon; Alaska, Ariz., Calif., Colo., Idaho, Maine, Mich., Minn., Mont., Nev., N.H., N.Y., N.Dak., Oreg., Utah, Vt., Wash., Wis., Wyo.

Specimens of *Botrychium minganense* have sometimes been misidentified as *B. dusenii* (H. Christ) Alston, a South American species.

20. **Botrychium montanum** W. H. Wagner, Amer. Fern J. 71: 29. 1981 · Western goblin

Trophophore stalk 0.3–2 cm, 0.2–0.5 times length of rachis; blade dull, glaucous, gray-green, mostly linear, lobed to 1-pinnate, to 6 × 0.7 cm, somewhat succulent. **Pinnae** or lobes to 6 pairs, ascending, mostly widely separated, distance between 1st and 2d pinnae not or slightly more than between 2d and 3d pairs, extremely variable in outline, linear to cuneate, undivided to tip, margins entire to coarsely dentate, distal pinnae or blade tip cut into 3–5 lobes, apex angular, venation like ribs of fan, midrib absent. **Sporophores** 1-pinnate, 1.5–4.5 times length of sporophore. $2n = 90$.

Leaves appearing in late spring to late summer. Dark coniferous forests, usually near swamps and streams; 1000–2000 m; B.C.; Calif., Mont., Oreg., Wash.

Botrychium montanum may come to be recognized as a subspecies of the eastern *B. mormo*, from which it differs in lacking an attached gametophyte and in having a more dissected trophophore apex, glaucous blades, shorter trophophore stalk, earlier seasonal development, and full opening of sporangia when mature. *Botrychium montanum* and *B. mormo* differ from *B. simplex* in being more robust and in having truncate, adnate, lateral lobes on the trophophore blade and dentate to deeply cleft trophophore apex.

21. **Botrychium mormo** W. H. Wagner, Amer. Fern J. 71: 26. 1981 · Little goblin

Trophophore stalk 0.2–2.5 cm, usually 0.3–0.6 times length of trophophore rachis; blade yellow-green to green, shiny, linear to linear-spatulate, lobed (rarely 1-pinnate), to 5 × 1 cm, very succulent. **Pinnae** or lobes to 3 pairs, ascending, mostly widely separated, usually fused together in distal half of trophophore, distance between 1st and 2d pinnae not or slightly more than between 2d and 3d pairs, extremely variable in outline, linear to fan-shaped, undivided to tip, margins entire to coarsely dentate, proximal pinnae or blade tip not dentate or with 2–3 shallow, broad teeth, apex angular, venation like ribs of fan, midrib absent. **Sporophores** 1-pinnate, 0.2–3 cm, 0.2–3.5 times length of trophophore. $2n = 90$.

Leaves appearing in late spring to fall. Extremely sporadic, in rich northern basswood, beech, sugar maple forest; of conservation concern; 300–600 m; Mich., Minn., Wis.

The highly seasonal appearance of *Botrychium mormo* is more like a fungus carpophore than a moonwort. In wet years both mature and juvenile plants are fairly easy to find in known localities; in dry years they do not appear aboveground. *Botrychium mormo* may be eaten and dispersed by animals. The succulent nature of the whole plant may make it attractive to herbivores, and because the sporangia do not open, passage through animal digestive tracts may be required to facilitate the release of the spores. Another unusual feature of *B. mormo* is the tendency for gametophytes to persist on mature sporophytes.

22. Botrychium pallidum W. H. Wagner, Amer. Fern J. 80: 74. 1990 · Pale moonwort, botryche pâle

Trophophore stalk 2–8 mm, 0–1/5 length of trophophore rachis; blade glaucous, pale green to whitish, oblong, ± longitudinally folded when alive, 1-pinnate, to 4 × 1 cm, herbaceous. **Pinnae** to 5 pairs, ascending, approximate, distance between 1st and 2d pinnae not or slightly more than between 2d and 3d pairs, basal pinna pair approximately equal in size and cutting to adjacent pair, fan-shaped, strongly asymmetric, lobed to divided to tip, margins entire to irregularly crenate-dentate, largest pinnae often split into 2 unequal lobes, apex rounded, venation like ribs of fan, midrib absent. **Sporophores** 1–2-pinnate, 1.5–4 times length of trophophore. $2n = 90$.

Leaves appearing in late spring and early summer. Sporadic, mainly in open fields but also in shaded places; of conservation concern; 0–2600 m; Man., Ont., Que., Sask.; Colo., Maine, Mich.

A usually tiny plant, *Botrychium pallidum* is separable from dwarfed and narrow sun forms of *B. minganense* by the peculiar, often folded pinnae and pale green to whitish color. It has been found growing with *B. campestre, B. echo, B. hesperium, B. lunaria, B. matricariifolium, B. minganense,* and *B. spathulatum.* Its small size may cause it to be overlooked. This is one of four moonwort species that commonly produce dense clusters of minute, spheric gemmae at the root bases.

23. Botrychium paradoxum W. H. Wagner, Amer. Fern J. 71: 24. 1981 · Paradox moonwort

Trophophores converted entirely to second fertile segment, stalk 1/2 length of fertile segment. **Sporophores** double, 2 per leaf, 1-pinnate, 0.5–4 cm. $2n = 180$.

Sporophores in June to August. Difficult to detect, plants usually hidden under other vegetation, in snowfields, secondary growth pastures; of conservation concern; 1500–3000 m; Alta., B.C., Sask.; Mont., Utah.

The leaf structure of *Botrychium paradoxum* is uniform and unique. Very rare teratological individuals of other moonwort species may have trophophores partially or wholly transformed into sporophores.

Botrychium ×*watertonense* W. H. Wagner, known only from one locality in western Alberta, is the sterile hybrid of *B. hesperium* and *B. paradoxum.* It can be identified by its trophophore pinnae; all are bordered with sporangia. It may reproduce by some unknown mechanism, such as unreduced spores (W. H. Wagner Jr., F. S. Wagner, et al. 1984).

24. Botrychium pedunculosum W. H. Wagner, Amer. Fern J. 76: 43, figs. 2, 7. 1986 · Stalked moonwort

Trophophore stalk 8–26 mm, to 1.1 times length of trophophore rachis; blade dull green, ovate-oblong to deltate-oblong, 1-pinnate, to 4.5 × 2 cm, leathery. **Pinnae** to 5 pairs, somewhat ascending, approximate to well separated, distance between 1st and 2d pinnae not or slightly more than between 2d and 3d pairs, basal pinna pair approximately equal in size and cutting to adjacent pair, ovate-rhombic to spatulate, lobed to tip, margin entire to irregularly lobed, apex rounded to acute, venation pinnate. **Sporophores** 1–3-pinnate, 2–4 times length of trophophore. $2n = 180$.

Leaves appearing in late spring, dying in early fall. Brushy secondary-growth habitats along streams and roadsides; of conservation concern; 300–1000 m; Alta., B.C., Sask.; Oreg.

The common stalk on this species tends to be reddish brown. The presence of extra sporangia on the proximal pinnae is known only in *Botrychium pedunculosum* and *B. ascendens.*

Botrychium pedunculosum grows with other moonworts, *B. lanceolatum, B. lunaria, B. minganense,* and *B. pinnatum.* It has not been found in association with the rather similar and much more common *B. hesperium.*

25. Botrychium pinnatum H. St. John, Amer. Fern J. 19: 11. 1929 · Northwestern moonwort

Botrychium boreale J. Milde subsp. *obtusilobum* (Ruprecht) R. T. Clausen

Trophophore stalk 0–2 mm, 0 to 0.1 times length of trophophore rachis; blade bright shiny green, oblong-deltate, 1–2-pinnate, to 8 × 5 cm, papery. **Pinnae** to 7 pairs, only slightly ascending, approximate to overlapping, distance between 1st and 2d pinnae not or slightly more than between 2d and 3d pairs, basal pinna pair approximately equal in size and cutting to adjacent pair, obliquely ovate to lanceolate-oblong, to spatulate, deeply and regularly lobed or pinnulate, lobed to tip, margins entire to very shallowly crenate, apex truncate to somewhat acute, venation pinnate. **Sporophores** 2-pinnate, 1–2 times length of trophophore. $2n = 180$.

Leaves appearing in June to August. Grassy slopes, streambanks, woods; 0–2500 m; Alta., B.C., N.W.T., Yukon; Alaska, Calif., Colo., Idaho, Mont., Nev., Oreg., Utah, Wash., Wyo.

Botrychium pinnatum is most commonly associated with *B. lanceolatum* and *B. lunaria*. Specimens of *B. pinnatum* have been misidentified as *Botrychium boreale*.

26. Botrychium boreale J. Milde, Bot. Zeit. 15(51): 880. 1857 · Northern moonwort

Trophophore stalk sessile or nearly so; blade shiny green, ovate-deltate, 1–2-pinnate, to 6 cm, fleshy. **Pinnae** to 6 pairs, ascending, mostly overlapping, distance between 1st and 2d pinnae only slightly greater than between 2d and 3d pairs, basal pinna pair usually considerably larger than adjacent pair, obliquely rhomboidal to oblanceolate-spatulate, mostly shallowly lobed to rarely pinnate, margins entire to very narrowly shallowly crenate, apex pointed, venation pinnate only at bases of proximal pinnae, otherwise ± like ribs of fan. **Sporophores** 1–2-pinnate, 1–1.5 times length of trophophore.

Leaves appearing in July and August. Dry meadows, south-facing slopes; 200–600 m; Greenland.

This well-marked northern Eurasian species is best known in Scandinavia, where it occurs most commonly with *Botrychium lunaria*, with which it occasionally hybridizes.

27. Botrychium pseudopinnatum W. H. Wagner, Contr. Univ. Michigan Herb. 17: 322. 1990 · False northwestern moonwort

Trophophore stalk 0–3 mm, 0 to 0.2 times length of trophophore rachis; blade dark green, somewhat shiny, oblong, 1–2-pinnate, to 4.5 × 2.5 cm, leathery. **Pinnae** to 6 pairs, ascending, approximate to overlapping, distance between 1st and 2d pinnae not or slightly more than between 2d and 3d pairs, basal pinna pair approximately equal in size and cutting to adjacent pair, obliquely ovate to lanceolate-oblong to spatulate, deeply and regularly lobed or pinnulate, lobed to tip, margins entire to very shallowly crenate, apex truncate, venation pinnate. **Sporophores** 2-pinnate, 1–2 times length of trophophore. $2n = 270$.

Leaves appearing in late spring to early fall. Sandy soil; of conservation concern; 300–500 m; Ont.

Botrychium pseudopinnatum is the only known hexaploid in *Botrychium* subg. *Botrychium*. It differs from *B. pinnatum* in smaller size, in narrower trophophore with relatively shorter and more oblique and ascending pinnae, in longer trophophore stalk, and in blade color, texture, and luster.

28. Botrychium pumicola Coville in L. Underwood, Native Ferns ed. 6, 69. 1900 · Pumice moonwort

Trophophore stalk 0–10 mm, 0.1–0.5 times length of trophophore rachis; blade dull, glaucous, whitish green, deltate, 2-pinnate, 4 × 6 cm, thickly leathery. **Pinnae** to 6 pairs, overlapping, strongly ascending, distance between 1st and 2d pinnae not or slightly more than between 2d and 3d pairs, asymmetrically cuneate, basal pinna pair often divided into 2 unequal parts, lobed to tip, margins entire, sinuate to shallowly crenate, apex rounded to truncate, venation pinnate. **Sporophores** 1–3-pinnate, 1–1.5 times length of trophophore. $2n = 90$.

Leaves appearing in summer. Pumice scree; of conservation concern; 1900–2500 m; Oreg.

Botrychium pumicola is a famous narrow endemic known from only a few colonies on fully exposed pumice scree on the sides of and in the general vicinity of Crater Lake, Klamath and Deschutes counties, Oregon. This plant has a very congested appearance with an extremely compact sporangial cluster and overlapping pinnae. Like most other members of subg. *Botrychium*, the trophophore is located high on the common stalk, but the common stalk is subterranean, giving the impression that the leaf originates near ground level. *Botrychium pumicola* has been found growing with *B. lanceolatum* and *B. simplex*.

29. Botrychium simplex E. Hitchcock, Amer. J. Sci. Arts 6: 103, plate 8. 1823 · Least moonwort, botryche simple

Botrychium simplex E. Hitchcock var. *tenebrosum* (A. A. Eaton) R. T. Clausen; *B. tenebrosum* A. A. Eaton

Trophophore stalk 0–3 cm, 0–1.5 times length of trophophore rachis; blade dull to bright green to whitish green, linear to ovate-oblong to oblong to fully triangular with pinnae arranged ternately, simple to 2 (–3)-pinnate, to 7 × 0.2 cm, fleshy to thin, papery or herbaceous. **Pinnae** or well-developed lobes to 7 pairs, spreading to ascending, approximate to widely separated, distance between 1st and 2d pinnae frequently greater than between 2d and 3d pairs, basal pinna pair commonly much larger and more complex than adjacent pair, cuneate to fan-shaped, strongly asymmetric, undivided to divided to tip, basiscopic margins ± perpendicular to rachis, acroscopic margins strongly ascending, basal pinnae often divided into 2 unequal parts, margins usually entire or shallowly sinuate, apex rounded, undivided and boat-shaped to strongly divided and plane, venation pinnate or like ribs of fan, with midrib. **Sporophores** mainly 1-pinnate, 1–8 times length of trophophores. $2n = 90$.

Leaves appearing midspring to early fall. Dry fields, marshes, bogs, swamps, roadside ditches; 0–2200 m; Greenland; Alta., B.C., N.B., Nfld., N.S., Ont., Que., Sask.; Calif., Colo., Conn., Del., D.C., Idaho, Ill., Ind., Iowa, Maine, Md., Mass., Mich., Minn., Mont., Nev., N.H., N.J., N.Y., N.C., Ohio, Oreg., Pa., R.I., Utah, Vt., Va., Wash., W.Va., Wis., Wyo.; Europe.

The many environmental forms and juvenile stages of *Botrychium simplex* have resulted in the naming of numerous, mostly taxonomically worthless, infraspecific taxa. The western montane populations in the flora from Colorado to north Saskatchewan and westard are evidently distinctive, however, and may warrant subspecies or species status.

Mature, full-sized plants of these can be distinguished as follows:

Eastern *Botrychium simplex:* Sporophore 1–4 times length of trophophores, arising from well-developed common stalk from below middle to near top, well above leaf sheath; trophophore nonternate or if subternate, lateral pinnae smaller than central pinnae and simple to merely lobed (rarely pinnate); pinnae usually adnate to rachis, rounded and ovate to spatulate, segment sides at angles mostly less than 90°; trophophore tip undivided; texture papery to herbaceous; common in upland fields.

Western *Botrychium simplex:* Sporophore 3–8 times length of trophophore, mostly arising directly from top of leaf sheath, common stalk much reduced to absent; trophophore ternate with 3 equal segments (rarely nonternate, then resembling single segment of ternate blade); pinnae usually strongly contracted at base to stalked, angular to fan-shaped, segment sides at angles mostly more than 90°, like those of *B. lunaria*; trophophore tip divided, usually in 3 parts including narrow central lobe; texture thin, herbaceous; habitats mainly along marshy margins and in meadows.

The eastern, typical *Botrychium simplex* has a common woodland and swamp shade form (*B. tenebrosum* A. A. Eaton) that appears to be a persistent juvenile. It is small and extremely slender, the trophophore simple, rudimentary, and attached near the top of an exaggerated common stalk. Many intermediates between this and more typical forms exist, however, and the variation appears to be the result of different growing conditions. The persistent western juvenile counterpart differs in the generally lower attachment of the trophophore (not necessarily on the top of the sheath), greater length of the trophophore, and more herbaceous texture.

30. Botrychium spathulatum W. H. Wagner, Amer. Fern J. 80: 77. 1990 · Spatulate moonwort, botryche spatulé

Trophophore stalk 0–1 mm; blade shiny yellow-green, narrowly deltate, flat, 1-pinnate, to 8 × 2.5 cm, thick, leathery. **Pinnae** to 8 pairs, ascending, remote, distance between 1st and 2d pinnae not or slightly more than between 2d and 3d pairs, basal pinna pair approximately equal in size and cutting to adjacent pair, mostly narrowly spatulate to linear-spatulate and rounded or ± 2-cleft, lobed to unlobed to tip, margins mainly entire or occasionally irregularly and shallowly incised, apex rounded-notched, venation like ribs of fan, midrib absent. **Sporophores** 1–2-pinnate, 1.2–2 times length of trophophore. $2n = 180$.

Leaves appearing late spring through summer. Sand dunes, old fields, and grassy railroad sidings; 0–2000 m; Alta., B.C., N.B., N.W.T., Ont., P.E.I., Que., Yukon; Alaska, Mich., Mont.

Botrychium spathulatum has long been confused with the more common *B. minganense*, with which it often grows in the Lake Superior region. The leaves appear later in *B. spathulatum* than in *B. minganense*.

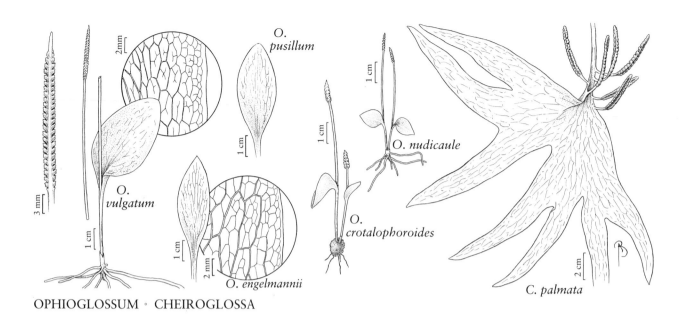

OPHIOGLOSSUM · CHEIROGLOSSA

2. OPHIOGLOSSUM Linnaeus, Sp. Pl. 2: 1062. 1753; Gen. Pl. ed. 5, 484. 1754

· Adder's-tongue [Greek *ophis,* snake, and *glossa,* tongue, in reference to the sporophore tip]

Plants terrestrial. **Roots** unbranched, whitish yellow to black, 0.1–1.5 mm diam., smooth, commonly proliferous and forming clones. **Stems** upright, forming caudex, to 1.6 cm thick (cormlike in *O. crotalophoroides*); gemmae absent. **Trophophores** erect to prostrate, blades nearly sessile or stalked, lanceolate to cordate, simple, 0.4–100 mm × 0.3–45 mm; margins entire; apex rounded, acute, or apiculate; veins anastomosing, main areoles to 15 × 4 mm, but mostly less than 5 × 3 mm. **Sporophores** 1 per leaf, simple, stalked, borne from ground level to well above ground at base of trophophore but commonly absent, leaf made up of only trophophyll. **Sporangial clusters** with sporangia in 2 rows, deeply sunken in simple, linear or oblong, fleshy sporophore tip, tip usually ± apiculate. **Gametophytes** brown to white, narrowly linear, unbranched, 2–20 × 1–3 mm diam. $x = 30$.

Species 25–30 (7 in the flora): nearly worldwide, mainly tropical and subtropical.

Ophioglossum occurs mostly in disturbed, open, grassy habitats. It is often overlooked because of superficial resemblance to seedlings of monocotyledonous plants. The intensive and careful field studies of R. D. Thomas (W. H. Wagner Jr., C. M. Allen, and G. P. Landry 1984) have greatly extended our knowledge of adder's-tongue distributions in North America. The chromosome numbers of *Ophioglossum* are the highest known in all vascular plants; numbers as high as $2n = 1200+$ have been reported (A. Löve et al. 1977).

1. Stems spheric, cormlike, fleshy, 3–12 mm diam.; leaves arising from deep cavity in top of stem; trophophore blades usually lying flat on ground, deltate to cordate. . . . . 2. *Ophioglossum crotalophoroides*
1. Stems upright, 1–5 mm diam., seldom more; leaves arising at top of stem; trophophore blades erect to spreading, mostly ovate to lanceolate.

2. Trophophore veins in larger leaves forming small areoles within larger areoles.
   3. Blades commonly folded when alive, ovate to ovate-lanceolate, to 10 × 4.5 cm; roots 0.5–1.5 mm diam.; sporophores 1.3–2.5 times length of trophophore; dried blades uniformly green without pale central band. . . . . 3. *Ophioglossum engelmannii*
   3. Blades plane when alive, ovate to lanceolate, to 4.5 × 1.7 cm; roots 0.2–0.8 mm diam.; sporophores 2–6 times length of trophophore; dried blades commonly with pale central band. . . . . 4. *Ophioglossum nudicaule* (large extreme)
2. Trophophore veins forming only branching or nonbranching, free, included veinlets within larger areoles.
   4. Blades rounded at apex, apiculum absent, ovate-lanceolate, oblanceolate, ovate, to trowel-shaped, to 10 × 3.5 cm; adult leaves usually 1 per stem, appearing in a single flush once per year.
      5. Blades dark green, somewhat shiny (alive), ovate to trowel-shaped, firm, base abruptly tapering; damp forest habitats; se United States, Mexican highlands. . . . . 7. *Ophioglossum vulgatum*
      5. Blades pale green, dull (alive), obovate to oblanceolate or ovate, herbaceous, base gradually tapering; open pastures, meadows, marshes, ditches; n North America. . . . . 6. *Ophioglossum pusillum*
   4. Blades acute at apex, with ± developed apiculum or apiculum absent, ovate-oblong, ovate-deltate, or ovate-lanceolate, to 4.5 × 1 cm (6 × 3 mm occasionally in *O. petiolatum*); adult leaves commonly 2–3 per stem, appearing in 1 or more flushes per year, depending on rains.
      6. Roots dark brown, usually fewer than 8 per shoot, major roots generally straight, 0.8–1.3 mm diam.; blade apiculum mostly absent; venation usually coarse. . . . . 5. *Ophioglossum petiolatum*
      6. Roots yellowish to pale brown, usually more than 12 per shoot, 0.2–1 mm diam.; blade apiculum absent or present; venation usually fine, intricate.
         7. Roots 0.5–1 mm diam.; sporophores 1–2.5 times length of trophophore; trophophores 0.4–4.3 × 0.3–1 cm, commonly folded when alive; blades thick, herbaceous; California. . . . . 1. *Ophioglossum californicum*
         7. Roots 0.2–0.8 mm diam.; sporophores 2–6 times length of trophophore; trophophores as small as 0.4 × 0.3 cm, usually plane when alive; blades thin, herbaceous; se United States. . . . . 4. *Ophioglossum nudicaule* (small extreme)

1. Ophioglossum californicum Prantl, Ber. Deutsch. Bot. Ges. 1: 355. 1883 · California adder's-tongue

Ophioglossum lusitanicum Linnaeus subsp. *californicum* (Prantl) R. T. Clausen

Roots to 16 per plant, pale brown, 0.5–1 mm diam., producing proliferations. **Stem** upright, to 1.6 cm, 5 mm diam., commonly 2 leaves per stem. **Trophophore stalk** 0–1.8 cm, to 2.5 times length of trophophore blade. **Trophophore blade** erect to spreading, commonly ± folded when alive, green, dull, without pale central band when dried, to 4.3 × 1 cm (rarely 0.4 × 0.3 mm), herbaceous, thick, gradually tapering to base, apex attenuate; venation complex-reticulate, with numerous parallel narrow areoles, each with 1–several included veinlets. **Sporophores** arising near ground level, 1–2.5 times length of trophophore; sporangial clusters 8–15 × 1–3 mm, with 8–15 pairs of sporangia, apiculum 0.3–1 mm.

Leaves appearing in late winter and early spring; apparently absent during dry years. Open grassy fields and prairies; 50–300 m; Calif.; Mexico.

Ophioglossum californicum differs from the Old World species *O. lusitanicum* in that *O. lusitanicum* has a narrowly linear to linear-oblanceolate trophophore that is 1/4 to 1/2 as wide as long; *O. lusitanicum* also has a much simpler venation and usually lacks an apiculum.

2. Ophioglossum crotalophoroides Walter, Fl. Carol., 256. 1788 · Bulbous adder's-tongue

Roots to 20 per plant, blackish, usually extremely narrow, often almost hairlike, less than 0.1 mm diam., proliferations not reported. **Stems** spheric, 3–12 mm diam., succulent, cormlike with perforation at apex, apical meristem located at bottom of cavity through which leaves emerge at top, leaves 2 per stem. **Trophophore stalk** to 0.6 cm, 0.1–0.2 times as long as trophophore blade. **Trophophore blade** lying nearly flat on ground, not folded longitudinally, pale green throughout, deltate to cordate, to 3 × 2 cm, contracted abruptly at truncate to cordate base, apex with apiculum. **Venation** coarsely reticulate with included veinlets. **Sporophores** arising at ground level, 1–5 times as long as trophophore; sporangial clusters usually short, less than 1 cm, 2–3 mm wide, with 3–8 pairs of sporangia, apiculum to 1.5 mm.

Leaves appearing mainly in late winter and early spring, sometimes also appearing later in season after heavy rains. Second-growth fields, vacant lots, roadside ditches, and lawns; 0–100 m; Ala., Ark., Fla., Ga., La., Miss., Mo., N.C., S.C., Tex.; widespread in tropical highlands; Mexico; West Indies; Central America; South America.

Ophioglossum crotalophoroides is very remarkable morphologically for its highly modified stem and threadlike nonproliferous roots. The gametophyte is disclike (M. R. Mesler 1973). It is especially common in lawns and cemeteries in the southeastern United States.

3. Ophioglossum engelmannii Prantl, Ber. Deutsch. Bot. Ges. 1: 351. 1883 · Limestone adder's-tongue

Roots to 25 per plant, tan to brown, 0.5–1.5 mm diam., straight, producing proliferations. **Stem** upright, to 1.5 cm, 4 mm diam., leaves 1–2 per stem. **Trophophore stalk** to 0.1 cm, 0.01 times length of blade. **Trophophore blade** erect to spreading, commonly ± folded when alive, uniformly pale green throughout when dried, dull, ovate to ovate-lanceolate, 10 × 4.5 cm, firm, herbaceous, base narrowed abruptly, apex with apiculum to 0.8 mm; venation complex-reticulate, veinlets forming numerous, very tiny, secondary areoles within the major areoles. **Sporophores** arising at ground level, 1.3–2.5 times as long as trophophore; sporangial clusters 2–4 × 0.13–0.31 cm, pairs of sporangia 20–40, apiculum 0–1.3 mm.

Leaves appearing early–late spring, often with second flush later in season following summer rains. Mostly in soil over limestone in open fields, pastures, and cedar glades; 50–1000 m; Ala., Ariz., Ark., Fla., Ga., Ill., Ind., Kans., Ky., La., Miss., Mo., Ohio, Okla., N.Mex., N.C., Tenn., Tex., Va., W.Va.; Mexico; Central America.

4. Ophioglossum nudicaule Linnaeus f., Suppl. Pl., 443. 1781 · Slender adder's-tongue

Ophioglossum dendroneuron E. P. St. John; *O. ellipticum* Hooker & Greville; *O. mononeuron* E. P. St. John

Roots yellowish to pale brown, to 15 per plant, 0.2–0.8 mm diam., proliferous at wide intervals. **Stem** upright, 0.2–1.2 cm, 1–5 mm diam., commonly 2–3 leaves per stem. **Trophophore stalk** to 0.8 cm, 0.1–0.2 times length of trophophore blade. **Trophophore blade** spreading, usually plane when alive, green, dull, largest leaves drying with pale central band, ovate to lanceolate, thin, blades less than 0.4 × 0.3 cm in many colonies but blades large, to 4.5 × 1.7 cm in other colonies, herbaceous, base gradually tapered, apex with short apiculum; venation finely complex-reticulate, areoles with only included veinlets in smaller blades but with numerous secondary areoles in largest blades. **Sporophores** arising at or near ground level, 2–6 times as long as trophophore; sporangial clusters 0.5–1.5 cm, 1.5 mm or less wide, mostly with 5–12 pairs of sporangia, apiculum 0.5–1 mm.

Leaves appearing in latter half of winter and early spring, sometimes with second flush in same year after heavy rains. Second-growth fields, vacant lots, roadside ditches, and lawns; 0–90 m; Ala., Ark., Fla., Ga., La., Miss., N.C., Okla., S.C., Tex.; Mexico; West Indies; Central America; South America; Asia; Africa; Pacific Islands.

Ophioglossum nudicaule is much less common than *O. crotalophoroides*; they often occur together and are found in the same or similar habitats. The gametophytes of *O. nudicaule* are typical for the genus (M. R. Mesler et al. 1975). A given colony may be made up of small, medium, or large plants (W. H. Wagner Jr., C. M. Allen, and G. P. Landry 1984).

5. Ophioglossum petiolatum Hooker, Exot. Fl. 1: 56. 1823 · Stalked adder's-tongue

Roots dark brown, to 8 per plant, 0.8–1.3 mm diam., producing proliferations. **Stem** upright, 0.3–1 cm, 1.5–2.5 mm diam., 2–3 leaves per stem. **Trophophore stalk** 0–3 mm, 0–0.1 times length of blade. **Trophophore blade** erect to spreading, usually plane or nearly so when alive, gray-green, dull, ovate to trowel-shaped, to 6 × 3 cm, fleshy, cuneate to truncate to nearly cordate at base, contracted gradually to acute apex, apiculum mostly absent; venation coarse, reticulate, areoles large with few free or anastomosing included veinlets. **Sporophores** arising at ground level, 0.8–7 times length of trophophore; sporangial clusters to 4 × 0.35 cm, with up to 30 pairs of sporangia, apiculum 0.3–1.2 mm.

Leaves appearing during wet periods. Plants sometimes weedy in lawns, ditches, and around buildings; 0–90 m; introduced; Ala., Ark., Fla., Ga., La., Miss., Mo., N.C., Okla., S.C., Tex., Va.; West Indies; Mexico; n South America; Asia; Pacific Islands.

Ophioglossum petiolatum grows readily in pots, making it suitable for botany instruction. Earliest records in North America date from 1900 to 1930, suggesting that it is probably introduced.

6. Ophioglossum pusillum Rafinesque, Précis Découv. Somiol., 46. 1814 · Northern adder's-tongue, herbe-sans-couture

Ophioglossum vulgatum Linnaeus var. *pseudopodum* (S. F. Blake) Farwell

Roots yellow to tan, to 15 per plant, 0.3–1 mm diam., producing proliferations. **Stem** upright, to 2 cm, 3 mm diam., 1 leaf per stem. **Trophophore stalk** expanding gradually into blade. **Trophophore blade** erect or spreading, usually plane when alive, pale green, dull, mostly oblanceolate to obovate to ovate, widest point in middle, to 10 × 3.5 cm, soft, herbaceous, base tapering gradually, apex rounded; venation complex-reticulate, with included free veinlets in areoles. **Sporophores** arising at ground level, 2.5–4.5 times length of trophophore; sporangial clusters 20–45 × 1–4 mm, with 10–40 pairs of sporangia, apiculum 1–2 mm. $2n = 960$.

Leaves appearing midspring. Frequent and widespread, open fens, marsh edges, pastures, and grassy shores and roadside ditches, north of the southern boundary of Wisconsin glaciation; 100–2000 m; B.C., N.B., N.S., Ont., Que.; Alaska, Calif., Conn., Idaho, Ill., Ind., Iowa, Ky., Maine, Md., Mass., Mich., Minn., Mont., Nebr., N.H., N.J., N.Y., N.Dak., Ohio, Oreg., Pa., R.I., S.Dak., Vt., Va., Wash., W.Va., Wis.

Ophioglossum pusillum is inconspicuous and may be much more common than collections indicate. It differs from *O. vulgatum* in having an ephemeral, membranous basal sheath.

7. Ophioglossum vulgatum Linnaeus, Sp. Pl. 2: 1062. 1753 · Southern adder's-tongue, herbe sans couture

Ophioglossum pycnostichum (Fernald) A. Löve & D. Löve; *O. vulgatum* var. *pycnostichum* Fernald

Roots to 20 per plant, 0.3–0.9 mm diam., producing proliferations. **Stem** upright, to 1 cm, 3 mm diam., leaves 1 per stem. **Trophophore stalk** formed abruptly at base, to 5 mm, sometimes more, 0.05 times length of trophophore blade. **Trophophore blade** erect to spreading, usually plane when alive, dark green, somewhat shiny, mostly ovate to ovate–trowel-shaped, widest in proximal half, to 10 × 4 cm, firm, herbaceous, base tapering abruptly, apex rounded; venation complex-reticulate with included free veinlets in areoles. **Sporophores** arising at ground level, stalk 2–4 times length of trophophore; sporangial clusters 20–40 × 1–4 mm, with 10–35 pairs of sporangia, apiculum 1–1.5 mm. $2n =$ ca. 1320.

Leaves appearing spring–early summer. Shaded secondary woods, rich wooded slopes, forested bottomlands, and floodplain woods, south of Wisconsin glaciation; 0–800 m; Ala., Ariz., Ark., Del., Fla., Ga., Ill., Ind., Ky., La., Md., Mich., Miss., Mo., N.J., N.C., Ohio, Okla., Pa., S.C., Tenn., Tex., Va., W.Va.; Mexico; Eurasia.

In addition to characteristics given in the key, *Ophioglossum vulgatum* differs from *O. pusillum* in having an unusually persistent leathery basal leaf sheath (B. W. McAlpin 1971; W. H. Wagner Jr. 1971b) rather than an ephemeral one and in having smaller spores (mostly 35–45 μm in *O. vulgatum* compared with 50–60 μm in *O. pusillum*). The chromosome number of *O. vulgatum* in India and Europe has been reported as $2n = 480$, and that may be the number of most North America populations, which are small spored. In the Appalachians, however, a distinctive large-spored form has a chromosome number of $2n =$ ca. 1320.

3. CHEIROGLOSSA C. Presl, Suppl. Tent. Pterid., 56. 1845 · Hand fern, dwarf staghorn [Greek *cheir,* hand, and *glossa,* tongue, in reference to the palmately lobed trophophores and the linear sporophores]

Plants epiphytic. **Roots** dichotomously divided at wide intervals, pale yellowish to brown, 1–1.5 mm diam., smooth, proliferous. **Stems** pendent, to 1.5 mm thick but appearing thicker because of persistent leaf bases, gemmae absent. **Trophophores** arching or pendent, blades stalked, palmately lobed, broadly fan-shaped, to 45 × 30 cm; margins entire; apex rounded to acute; veins anastomosing, major areoles very large, to 35 × 8 mm. **Sporophores** numerous, to 10, arising on each side of base of trophophore blade and apex of trophophore stalk. **Sporangial clusters** with sporangia in 2 rows embedded in compact, linear spike, apiculum absent. **Gametophytes** brown to white, cylindric, repeatedly branched, branches 2–20, 1–4 mm thick, and the whole ± stellate structure sometimes reaching 20 mm.

Species 1 (1 in the flora): very scattered and local in tropics, North America, Mexico, West Indies, Central America, South America, Asia, Africa.

Cheiroglossa, which is sometimes treated as a subgenus of *Ophioglossum,* is widely divergent from *Ophioglossum* in many characteristics, including the epiphytic habit, dichotomous roots, hairy stem apex, pendent and very large trophophore, palmately lobed blade, extremely large areoles, sporophores multiple and arising from sides of blade base, and the much branched gametophytes.

1. **Cheiroglossa palmata** (Linnaeus) C. Presl, Suppl. Tent. Pterid., 57. 1845 · Hand fern

Ophioglossum palmatum Linnaeus, Sp. Pl. 2: 1063. 1753
Roots tan, to 15 per plant, 1–1.5 mm diam., proliferating in humus or underneath palm leaf bases. **Stems** 0.5–1.5 cm, covered near apex with dense tuft of white to light brown, multicellular hairs to 7 mm. **Trophophore blades** green, shiny, to 45 × 30 cm, proximal margins diverging 90°–150° from stalk, with up to 7 rounded to mostly linear, acute lobes; venation complex-reticulate, very coarse, veinlets included in (usually elongate) very large major areoles to 35 × 8 mm, veinlets free or sometimes forming individual secondary areoles; blades firm, herbaceous. **Trophophore stalk** well defined, 1–2.5 times as long as trophophore blade. **Sporophores** single and central only on youngest or smallest leaves, normally to 10 per leaf, arising closely spaced from both sides of base of trophophore blade and top of stalk, sporangial clusters 1–7 cm × 2.5–3 mm, apiculum absent.

Evergreen. Among leaf bases on palmetto trunks [*Serenoa repens* (Bartram) Small], mainly in hammocks and swamps; 0–50 m; Fla.; Mexico; West Indies; Central America; South America; disjunct in Asia in s Vietnam; Africa in Madagascar, Seychelles, and Réunion.

This very remarkable plant is becoming extremely rare in Florida.

7. OSMUNDACEAE Berchtold & J. Presl
• Royal Fern Family

R. David Whetstone

T. A. Atkinson

Plants terrestrial, herbaceous, frequently in clumps. **Stems** creeping, beset with old petiole bases and black fibrous roots; scales absent; older stems seldom persisting. **Leaves** monomorphic or dimorphic. **Blades** 1–2-pinnate (2-pinnatifid); rachises grooved. **Pinnae** monomorphic or dimorphic. **Indument** of reddish to light brown hairs. **Veins** dichotomous, running to margins. **Sori** absent; sporangia born on slightly modified fertile segments of blades also possessing fully expanded pinnae, or sporangia covering blades lacking green expanded pinnae, clustered in marginal zones, indusia lacking. **Spores** green, all alike. **Gametophytes** green, aboveground, obcordate to elongate.

Genera 3, species 16–36 (1 genus with 3 species and 1 hybrid in the flora): nearly worldwide, temperate and tropical regions.

Osmundaceae are considered intermediate in several respects between eusporangiate and leptosporangiate ferns. In the absence of sori, simultaneous maturation of spores, and development of sporangia from several initial cells, they are much like eusporangiate ferns. Their large prothalli with projecting antheridia are similar to those of leptosporangiate ferns.

SELECTED REFERENCES Bobrov, A. E. 1967. The family Osmundaceae (R. Br.) Kaulf. Its taxonomy and geography. Bot. Zhurn. (Moscow & Leningrad) 52: 1600–1610. Hewitson, W. 1962. Comparative morphology of the Osmundaceae. Ann. Missouri Bot. Gard. 49: 57–93.

1. OSMUNDA Linnaeus, Sp. Pl. 2: 1063. 1753; Gen. Pl. ed. 5, 484. 1754 [Saxon, *Osmunder,* name for Thor, god of war]

Plants terrestrial. **Stems** creeping; tips often somewhat erect. **Leaves** dimorphic; fertile leaves erect, often notably smaller than sterile leaves in length and width. **Blades** 1–2-pinnate; pinnae monomorphic to dimorphic, pinnatifid or pinnate.

Species 10 (3 in the flora): nearly worldwide, tropical and temperate regions.

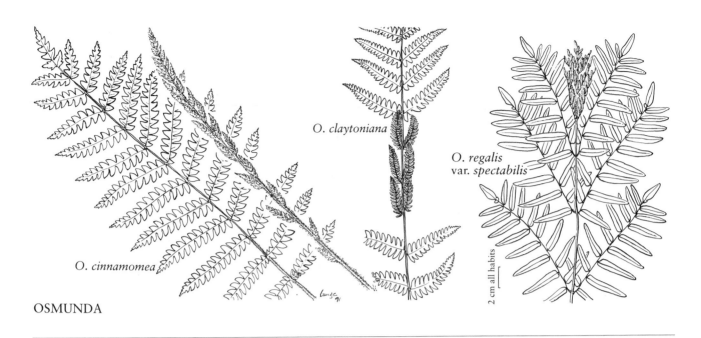

O. claytoniana

O. regalis var. *spectabilis*

O. cinnamomea

2 cm all habits

OSMUNDA

1. Fertile leaves with pinnae monomorphic, all spore-bearing; tuft of hairs persistent on abaxial surface of pinnae near base. . . . . 1. *Osmunda cinnamomea*
1. Fertile leaves with pinnae dimorphic, some spore-bearing, some not; tuft of hairs absent on abaxial surface of pinnae near base.
    2. Sterile leaves 2-pinnate with pinnules sessile. . . . . 4. *Osmunda* ×*ruggii*
    2. Sterile leaves pinnate-pinnatifid or 2-pinnate with pinnules stalked.
        3. Fertile pinnae apical, sterile leaves 2-pinnate. . . . . 3a. *Osmunda regalis* var. *spectabilis*
        3. Fertile pinnae medial, sterile leaves pinnate-pinnatifid. . . . . 2. *Osmunda claytoniana*

1. Osmunda cinnamomea Linnaeus, Sp. Pl. 2: 1066. 1753 · Cinnamon fern, osmonde canelle

Osmunda cinnamomea var. *glandulosa* Waters

Leaves pinnate-pinnatifid; petioles slightly shorter than blades, not winged, with light brown hairs when young, glabrate with age. **Sterile leaves** ovate to lanceolate, ca. 0.3–1.5 m; pinnae broadly oblong with persistent tuft of hairs on abaxial surface at base; ultimate segments with base obtuse, margins entire, apex usually mucronate. **Fertile leaves** with no expanded pinnae, green, becoming brownish, shorter and narrower than sterile leaves, withering after sporulation. **Sporangia** brown. $2n = 44$.

Sporulation spring–early summer (late summer, early winter in Florida). Moist areas, acidic soils, frequently in vernal seeps; 0–2300 m; St. Pierre and Miquelon; N.B., Nfld., N.S., Ont., P.E.I., Que.; Ala., Ark., Conn., Del., Fla., Ga., Ill., Ind., Iowa, Ky., La., Maine, Md., Mass., Mich., Minn., Miss., Mo., N.H., N.J., N.Y., N.C., Ohio, Okla., Pa., R.I., S.C., Tenn., Tex., Vt., Va., W.Va., Wis.; Mexico; West Indies; Central America; South America; Asia.

Many forms of *Osmunda cinnamomea* have been described from within the flora area. It is widely cultivated as an ornamental.

2. Osmunda claytoniana Linnaeus, Sp. Pl. 2: 1066. 1753 · Interrupted fern, osmonde de Clayton

Leaves pinnate-pinnatifid; petioles ca. 1/3 length of blades, winged, with light brown hairs, becoming glabrate. **Sterile leaves** elliptic to oblong, ca. 0.5–1 m; pinnae broadly oblong, lacking persistent tuft of hairs at base; ultimate segments with base truncate, margins entire, apex rounded. **Fertile leaves** with greatly reduced, sporangia-bearing medial pinnae that wither early, giving appearance of no middle pinnae (hence the vernacular name, interrupted fern). **Sporangia** greenish, turning dark brown. $2n = 44$.

Sporulation early spring–midsummer; 0–2300 m; St. Pierre and Miquelon; Man., N.B., Nfld., N.S., Ont., P.E.I., Que.; Conn., Ill., Ind., Iowa, Ky., Maine, Md., Mass., Mich., Minn., Mo., N.H., N.J., N.Y., N.C., Ohio, Pa., R.I., Tenn., Vt., Va., W.Va., Wis.; Asia

Osmunda claytoniana is sparingly cultivated as an ornamental.

3. Osmunda regalis Linnaeus, Sp. Pl. 2: 1065. 1753

· Royal fern, osmonde royale

Varieties 4–5 (1 in the flora): North America, West Indies, Bermuda, Central America, South America, Europe, Asia, Africa.

Osmunda regalis var. *regalis* is widely distributed in Europe and Asia and is widely cultivated as an ornamental.

3a. Osmunda regalis Linnaeus var. spectabilis

(Willdenow) A. Gray, Manual ed. 2, 600. 1856

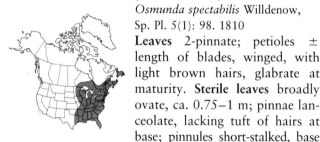

Osmunda spectabilis Willdenow, Sp. Pl. 5(1): 98. 1810

Leaves 2-pinnate; petioles ± length of blades, winged, with light brown hairs, glabrate at maturity. **Sterile leaves** broadly ovate, ca. 0.75–1 m; pinnae lanceolate, lacking tuft of hairs at base; pinnules short-stalked, base oblique to somewhat truncate, margins subentire to remotely dentate, apex acute to rounded. **Fertile leaves** with greatly reduced sporangia-bearing pinnae at apex. **Sporangia** greenish, turning red, then rusty brown. $2n = 44$.

Sporulation early spring–midsummer; 0–2300 m; St. Pierre and Miquelon; N.B., Nfld., N.S., Ont., P.E.I., Que.; Ala., Ark., Conn., Del., Fla., Ga., Ill., Ind., Iowa, Ky., La., Maine, Md., Mass., Mich., Minn., Miss., Mo., N.H., N.J., N.Y., N.C., Ohio, Okla., Pa., R.I., S.C., Tenn., Tex., Vt., Va., W.Va., Wis.

The chloroplasts within the spores give the young sporangia their green color. As the spores mature and are shed, the sporangia change color to a distinctive rusty brown.

4. Osmunda ×ruggii R. M. Tryon, Amer. Fern J. 30: 65. 1940

Leaves 2-pinnate; petioles ca. 1/3–1/2 length of blades, not winged, with light brown hairs, becoming glabrate. **Sterile leaves** ovate to ovate-lanceolate, 0.50–0.75 m; pinnae lanceolate, lacking persistent tuft of hairs at base; pinnules sessile, base narrowly adnate to costae, margins nearly entire, apex obtuse to acute. **Fertile leaves** with greatly reduced, sporangia-bearing pinnae at apex, midleaf, or both, or comprising most of leaf, notably smaller than sterile leaves. (Fertile leaves not known from natural populations). $2n = 44$.

Sporulation late spring–early summer; 650 m; Conn., Va.

Osmunda ×ruggii is a sterile, natural hybrid between *O. claytoniana* and *O. regalis* and one of only two interspecific hybrids known to occur in Osmundaceae. This nothospecies is known from two natural populations despite the widespread sympatry of parental stocks. The first reported population in Fairfield County, Connecticut, has not been relocated and is most likely extirpated. A single natural population is known to exist in Craig County, Virginia; it is estimated that its age could be greater than 1100 years.

8. GLEICHENIACEAE C. Presl
• Forking Fern Family

Clifton E. Nauman

Plants coarse, terrestrial. **Stems** long-creeping, forked, stele protostelic (a solid rod of vascular tissue with phloem surrounding xylem), covered with scales or hairs. **Leaves** monomorphic, large, scrambling or trailing, 1–many times forked. **Petiole** not articulate to stem, with pair of opposite pinnae and arrested bud at apex, or rachis continuing and producing 2 or more pairs of opposite pinnae. **Pinnae** 1–several times forked, with arrested bud at each fork; indument of simple, branched, or stellate hairs [scales]. **Veins** free, 1–4-forked. **Sori** round, indusia absent. **Sporangia** 2–many on slightly elevated receptacle, sessile to subsessile; annulus complete, transverse, medial, longitudinally dehiscent; spores 120–800 per sporangium. **Spores** all alike, whitish to yellowish, bilateral to globose, monolete or trilete, generally smooth without elaborately ornamented surface. **Gametophyte** borne aboveground, green, obcordate to elongate.

Genera 4, species ca. 140 (1 genus, 1 species in the flora): nearly worldwide in tropical to subtropical regions.

SELECTED REFERENCES Holttum, R. E. 1957. Morphology, growth habit, and classification in the family Gleicheniaceae. Phytomorphology 7: 168–184. Maxon, W. R. 1909. Gleicheniaceae. In: N. L. Britton et al., eds. 1905–1972. North American Flora. New York. Vol. 16, pp. 53–63. Underwood, L. M. 1907. American ferns—VIII. A preliminary review of the North American Gleicheniaceae. Bull. Torrey Bot. Club 34: 243–262.

1. DICRANOPTERIS Bernhardi, Neues J. Bot. 1(2): 38. 1805 · Forking ferns [Greek *dikranos,* twice-forked, and *pteris,* fern, derived from *pteron,* feather, in reference to the leaf architecture]

Stems long-creeping; hairs many celled, rigid to lax. **Leaves** usually separated several centimeters, of apparently indeterminate growth; pinnae opposite, in 1–several pairs, each pinna 1–many times forked, equal, each fork bearing arrested apex (bud) covered with tuft of hairs (all other axes glabrous) and pair of stipulelike appendages. **Penultimate segments** pectinate,

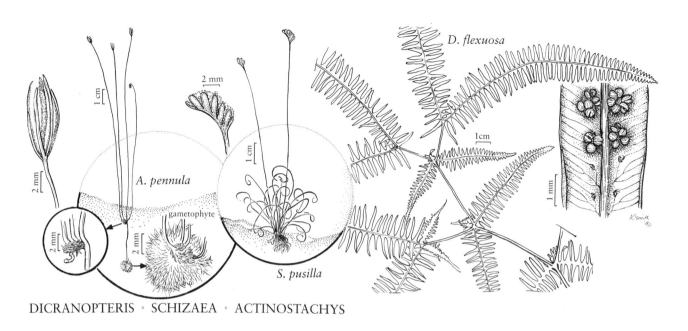

DICRANOPTERIS · SCHIZAEA · ACTINOSTACHYS

in ± equal pairs, usually ascending. **Veins** 2–4-forked. **Sori** with 6–15 or more sporangia, not paraphysate. **Spores** trilete.

Species 8–12 (1 in the flora): nearly worldwide in tropical and subtropical regions.

1. **Dicranopteris flexuosa** (Schrader) L. Underwood, Bull. Torrey Bot. Club 34: 254. 1907 · Forked fern, net fern

Mertensia flexuosa Schrader, Gött. Gel. Anz. 1824: 863. 1824; *Gleichenia flexuosa* (Schrader) Mettenius

Stems 2–5 mm diam.; hairs reddish brown to chestnut brown, falling off early. **Leaves** to more than 1 m, axes straw-colored, glabrous except at petiole base. **Penultimate segments** sessile, divergent to ascending, in ± equal pairs, deeply pinnatisect, lanceolate to oblanceolate, to 30 × 6 cm wide, leathery, glabrous, whitish waxy or glaucous abaxially. **Ultimate segments** linear, slightly dilated proximally, margins strongly revolute, apex retuse. **Sori** nearer midrib than margin; sporangia usually 6–12 per sorus.

Terrestrial and climbing on open slopes in drainage ditches; widespread, tropical; 0 m; Ala., Fla.; Mexico; Central America; South America.

Plants in Florida tend to be depauperate when compared to tropical populations; individual leaves are smaller and plants seldom form dense thickets. *Dicranopteris flexuosa* appears to be a natural element in the flora (J. R. Burkhalter 1985; R. Moyroud and C. E. Nauman 1989). Plants may not persist very long, however, as evidenced by Alabama and some Florida populations that are no longer extant.

9. SCHIZAEACEAE Kaulfuss

• Curly-grass Family

Warren H. Wagner Jr.

Plants terrestrial or epiphytic. **Roots** numerous. **Stems** mainly erect, covered with many stiff hairs 1–3 cells long, with simple siphonostele (hollow vascular cylinder). **Leaves** tufted, monomorphic or dimorphic. **Petioles** much longer than blades, blades reduced to tiny apical fistlike or radiating groups of rudimentary fertile pinnae ("digits"). **Petioles** sometimes repeatedly dichotomous, in some species webbed between branches to form fan-shaped false blades. **Sporangia** arranged in 1–4 ranks on abaxial surface of digits with revolute margins. **Annulus** subapical, composed of 1–(2–3) layer(s) of thickened cells. **Gametophytes** subterranean and not green or borne aboveground and green, tuberlike or flattened, cordate or filamentous. **Spores** bilateral, monolete.

Genera 2–3, species 30 (2 genera, 2 species in the flora): nearly worldwide, mainly in tropical areas.

SELECTED REFERENCES Bierhorst, D. W. 1971. Morphology of Vascular Plants. New York. Nauman, C. E. 1987. Schizaeaceae in Florida. Sida 12: 69–74.

1. Fertile blades pinnate, segments arising along rachis on distal 3–10 mm; sterile leaves curling. . . . . 1. *Schizaea*, p. 112
1. Fertile blades falsely digitate, segments appearing to arise at single point; sterile leaves absent. . . . . 2. *Actinostachys*, p. 113

1. SCHIZAEA Smith, Mém. Acad. Roy. Sci. (Turin) 5: 419, plate 9, fig. 9. 1793, name

conserved · Curly-grass fern [Greek *schizein*, split, i.e., split into narrow lobes]

Plants terrestrial. **Roots** blackish, very narrow, tangled, practically glabrous. **Stems** upright to reclining; hairs few celled, uniseriate. **Leaves** all fertile or dimorphic with some sterile, lacking blades, mainly unbranched, long-petioled. **Fertile blades** folded and fistlike or pectinate, pinnate, segments (digits) arising along rachis on distal 3–10 mm; sterile leaves straight or often curling. **Sporangia** in 2 rows. **Gametophytes** growing mixed with mosses, green (photosynthetic), delicately filamentous, lacking hairs. $x = 77, 94, 103$.

Species 10 (1 in the flora): mainly tropical.

1. Schizaea pusilla Pursh, Fl. Amer. Sept. 2: 657. 1814 · Curly-grass fern

Plants minute, rarely more than 10 cm. **Stems** 0.3–3 cm, covered by leaves or leaf bases. **Leaves** dimorphic. **Sterile leaves** ± coiled and spreading, very numerous, somewhat flattened, 1–5 cm × 0.2–0.3 mm. **Fertile leaves** upright, 1–12 cm; petioles long, straight, filiform; fertile blades apical, short, folded, 2–8 × 1–3 mm, segments 3–8 pairs, with multicellular hairs along margins. **Gametophyte** an algalike, branching filament. $2n = 206$.

Very local in bogs, wet, sandy depressions, crevices of ledges along shores; St. Pierre and Miquelon; N.B., Nfld., N.S.; Del., N.J., N.Y.

One of the most famous plants of the New Jersey Pine Barrens, this peculiar little plant has attracted much interest (W. H. Wagner Jr. 1963). It is abundant in southern Newfoundland. Plants from eastern Canada are shorter and denser than those from New Jersey. The same or a closely related species has been reported from Peru (R. G. Stolze 1987).

2. ACTINOSTACHYS Wallich, Numer. List., 1. 1829 · Ray spiked fern [Greek *aktis*, ray, and *stachys*, spike, referring to the arrangement of the fertile segments

Plants terrestrial. **Roots** dark, fibrous, covered with dark, stiff hairs, 2–3 mm. **Stems** upright; hairs uniseriate. **Leaves** all fertile (even youngest), unbranched, long-petioled. **Blades** falsely digitate, reduced to 2–many erect to spreading terminal rays; rays appearing to be whorled but actually borne on very short rachis. **Sporangia** in 2–4 rows. **Gametophytes** subterranean, not green, tuberlike, brown-hairy. $x = 134, 140$.

Species 20 (1 in the flora): nearly worldwide in tropical regions.

SELECTED REFERENCE Wagner, W. H. Jr. and V. Quevedo. 1985. Polymorphism in *Actinostachys pennula* (Swartz) Hooker and the taxonomic status of *A. germanii* (Fée) Prantl. [Abstract.] Amer. J. Bot. 72: 927–928.

1. Actinostachys pennula (Swartz) Hooker, Gen. Fil., plate 111A. 1842 · Ray spiked fern

Schizaea pennula Swartz, Syn. Fil., 150, 379. 1806; *Actinostachys germanii* Fée; *Schizaea germanii* (Fée) Prantl

Plants extremely variable in size, 5–50 cm, those in leaf mold (only form known in flora) usually small, those in open sunny areas much larger. **Leaves** 1–60; rays 1–12, 4–40 mm; petioles 0.5–2 cm. $2n = 168$ (Trinidad).

On and around rotten stumps and decomposing litter in damp forests and open baylands; Fla.; West Indies; Central America; South America.

Small individuals of *Actinostachys pennula* have sometimes been called *A. germanii*, but plants conforming to *A. germanii* are merely at the end of a morphological series; they are probably juvenile. Each juvenile has a large bulbous gametophyte remaining attached (a so-called "tuber") that is sometimes mistaken for the stem. Not only is the persistent gametophyte mistaken for a stem, but in the large forms of typical *A. pennula* the stem is not a solitary structure. Instead it resembles a compact, intergrown bush, made up of numerous tiny, narrow stems, usually less than 1 mm diam., that proliferate from old leaf bases—a type of cauline organization apparently unknown in any other living fern but possibly found in the fossil genus *Tempskya*. Measurements for plants in the North American flora are all in the lower parts of the range for the species. *Actinostachys pennula* is among North America's most unusual ferns, a highly treasured species, sought by field botanists who must, however, respect its rarity.

10. LYGODIACEAE C. Presl
• Climbing Ferns

Clifton E. Nauman

Plants terrestrial. **Stems** subterranean, protostelic; indument of dark, dense hairs. **Leaves** vinelike, of indeterminate growth. **Pinnae** reduced to short stalks, each bearing a pair of opposite pinnules, usually with an often dormant apical bud. **Sporangia** in 2 rows, 1 on each side of midvein of contracted, oblong, marginal lobes of ultimate segments, covered by hoodlike flap of tissue serving as indusium. **Spores** tetrahedral-globose, trilete, rarely monolete. **Gametophytes** terrestrial, cordate, glabrous.

Genus 1, species ca. 40 (3 species in the flora): tropical regions nearly worldwide and temperate regions of North America and Asia, s Africa, Pacific Islands in New Zealand.

1. LYGODIUM Swartz, J. Bot. (Schrader) 1800(2): 7, 106. 1801, name conserved
• Climbing ferns [Greek *lygodes,* flexible, in reference to the twining rachis]

Plants terrestrial. **Stems** branched, slender. **Leaves** often more than several meters, 2-pinnate or more divided, climbing by means of twining rachis; fertile pinnae borne toward apex of fertile leaves. **Blades** of short, alternate primary pinnae. **Pinnules** ± entire to palmately or pinnately lobed; fertile and sterile pinnae similar or fertile pinnae greatly contracted.

Species ca. 40 (3 in the flora): nearly worldwide, mostly tropical regions, a few species in temperate regions, North America, Asia in Japan, s Africa, Pacific Islands in New Zealand.

SELECTED REFERENCES Beckner, J. 1968. *Lygodium microphyllum,* another fern escaped in Florida. Amer. Fern J. 58: 93–94. Nauman, C. E. 1987. Schizaeaceae in Florida. Sida 12: 69–74. Nauman, C. E. and D. F. Austin. 1978. Spread of the exotic fern *Lygodium microphyllum* in Florida. Amer. Fern J. 68: 65–66.

1. Pinnules palmately lobed, sterile tissue nearly absent on fertile lobes; petioles borne 1–4
   cm apart. . . . . 1. *Lygodium palmatum*
1. Pinnules 1-pinnate to palmately lobed, sterile tissue present on fertile lobes; petioles borne
   less than 1 cm apart.

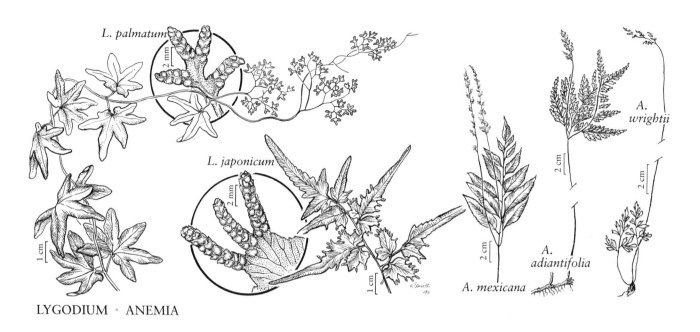

LYGODIUM · ANEMIA

2. Pinnules usually 1-pinnate, primary divisions mostly undivided, usually without basal lobes; lobes absent or rounded, auriculate, not directed toward leaf apex; ultimate segments articulate to petiolules, leaving wiry stalks when detached; blade tissue glabrous abaxially. . . . . 2. *Lygodium microphyllum*
2. Pinnules usually more than 1-pinnate, primary divisions pinnately to palmately lobed or divided, or if undivided with distinct basal lobes; lobes directed toward leaf apex; ultimate segments not articulate to petiolules, not leaving wiry stalks when detached; blade tissue sparsely to moderately pubescent abaxially. . . . . 3. *Lygodium japonicum*

1. **Lygodium palmatum** (Bernhardi) Swartz, Syn. Fil., 154. 1806 · Hartford fern, American climbing fern, creeping fern, Windsor fern

Gisopteris palmata Bernhardi, J. Bot. (Schrader) 1800(2): 129, plate 2(1). 1801

Stems long-creeping. **Leaves** to ca. 3 m. **Petioles** borne 1–4 cm apart, 9–15 cm. **Sterile pinnae** on 1–2 cm stalks, very broadly ovate, deeply and palmately 3–7-lobed, 1–4 × 2–6 cm; ultimate lobes triangular-elongate to oblong; lobe apex acute to blunt or rounded; segments not articulate to petiolules, not leaving wiry stalks when detached; blade tissue pubescent abaxially with transparent hairs. **Fertile pinnae** stalked to ca. 1.5 cm, then ± irregularly 3–5-forked or lobed, ultimate divisions palmately and sometimes irregularly lobed, smaller than sterile pinnules, 2–4 × 2–8 cm, otherwise similar; ultimate segments narrowly triangular to linear-triangular.

Terrestrial in woods, thickets, and bog margins in humus-rich, acid soils; Ala., Conn., Del., Ga., Ky., Maine, Md., Mass., Miss., N.H., N.J., N.Y., N.C., Ohio, Pa., R.I., S.C., Tenn., Vt., Va., W.Va.

Lygodium palmatum is generally local and rare except for the Cumberland Plateau of Kentucky and Tennessee where it is locally abundant in poorly drained, acidic soils, especially after disturbance (J. M. Shaver 1954; R. Cranfill 1980). Other authors have reported this species from Florida (O. Lakela and R. W. Long 1976), but I have not seen specimens (C. E. Nauman 1987). This species is not tolerant of shading.

2. **Lygodium microphyllum** (Cavanilles) R. Brown, Prodr., 162. 1810 · Small-leaved climbing fern

Ugena microphylla Cavanilles, Icon. 6: 76, plate 595. 1801

Stems creeping. **Leaves** to ca. 10 m. **Petioles** borne 2–5 mm apart, 7–25 cm. **Sterile pinnae** on 0.5–1.5 cm stalks, oblong, 1-pinnate, 5–12 × 3–6 cm; ultimate segments triangular-lanceolate to oblong-lanceolate, truncate to shallowly cordate or somewhat auriculate proximally, usually not lobed, but if lobed, lobes rounded at apex and not directed toward leaf apex; segment apex rounded-acute to obtuse; segments articulate to petiolules, leaving wiry stalks when detached; blade tissue glabrous abaxially. **Fertile pinnae** on 0.5–1 cm stalks, oblong, 1-pinnate, 3–14 × 2.5–6 cm; ultimate segments ovate to lanceolate-oblong, fringed with fertile lobes, otherwise similar to sterile segments.

Terrestrial on riverbanks, swamps (especially cypress swamps), cabbage palm hammocks, and other wet, disturbed sites; introduced; Fla.; Asia.

Lygodium microphyllum is native to southeastern Asia and recently naturalized. The species may be very abundant locally and may climb to a height of 9 meters in trees. Sometimes it forms thick mats covering considerable areas at ground level (J. Beckner 1968; C. E. Nauman and D. F. Austin 1978).

3. **Lygodium japonicum** (Thunberg ex Murray) Swartz, J. Bot. (Schrader) 1800(2): 106. 1801 · Japanese climbing fern

Ophioglossum japonicum Thunberg ex Murray, Syst. Veg. ed. 14, 926. May–June 1784

Stems creeping. **Leaves** to ca. 3 (–30) m. **Petioles** borne 2–7 mm apart, 10–35 cm. **Sterile pinnae** on 1.5–3.5 cm stalks, triangular to lanceolate, 2–3-pinnate, 6–15 × 5.5–15 cm; ultimate segments lanceolate, lobed or divided proximally; lobes usually acute at tip and directed toward apex; segment apices long-attenuate to acute; segments not articulate to petiolules, not leaving wiry stalks when detached; blade tissue pubescent abaxially with short, curved hairs. **Fertile pinnae** on 1–2 cm stalks, lanceolate-triangular, 2–3-pinnate, 5–18 × 4–14 cm; ultimate segments ovate to lanceolate, fringed with fertile lobes, otherwise similar to sterile segments.

Terrestrial in wet woods, marshes, roadside ditches, riverbanks, and other wet, disturbed sites in circumneutral soil; introduced; Ala., Ark., Fla., Ga., La., Miss., N.C., S.C., Tex.; Asia in China, Japan.

Lygodium japonicum is native to eastern Asia. It is commonly naturalized or escaped from cultivation. It has been reported as weedy in southern Alabama and Florida where its dense canopy can eliminate underlying vegetation.

11. ANEMIACEAE Link

John T. Mickel

Plants terrestrial or on rock. **Stems** compact or short-creeping, horizontal, solenostelic (having phloem on both sides of xylem) or dictyostelic (having complex nets of xylem), clothed with orange to reddish brown hairs. **Leaves** erect [rarely forming a flat rosette], partially to entirely dimorphic. **Sporangia** in 2 rows on ultimate segments of fertile pinnae, sessile, oblong; annulus apical. **Spores** tetrahedral-globose, with parallel or rarely anastomosing ridges. **Gametophytes** terrestrial, green, cordate with unequal lobes.

Genera 2, species 119 (1 genus, 3 species in the flora): widespread in tropical and subtropical regions.

1. ANEMIA Swartz, Syn. Fil., 6, 155. 1806, name conserved [Greek *aneimon*, without clothing, referring to the absence of blade protection for the sporangia]

Stems short-creeping, horizontal, clothed with dark hairs. **Leaves** partially dimorphic with sporangia restricted to erect, dissected, most proximal pair of pinnae arising from petiole just below sterile part of blade or leaves fully dimorphic and blade tissue lacking on fertile leaves. **Blade** 1–3-pinnate, papery to leathery. **Veins** free [anastomosing]. $x = 38$.

Species 117 (3 in the flora): tropical and subtropical regions, North America, Mexico, West Indies, Central America, South America, 1 in Asia in s India, 10 in Africa.

Anemias are most abundant in Brazil (ca. 70 spp.) and have a secondary center of diversity in Mexico (20 spp.). They are limited in the flora to peninsular Florida and the Edwards Plateau, Texas. All 3 species belong to the calciphilic subgenus *Anemiorrhiza*.

SELECTED REFERENCES Mickel, J. T. 1962. Monographic study of the fern genus *Anemia* subgenus *Coptophyllum*. Iowa State Coll. J. Sci. 36: 349–482. Mickel, J. T. 1981. The fern genus *Anemia* (Schizaeaceae) subgenus *Anemiorrhiza*. Brittonia 33: 413–429. Mickel, J. T. 1982. The genus *Anemia* (Schizaeaceae) in Mexico. Brittonia 34: 388–413. Walker, T. G. 1962. The *Anemia adiantifolia* complex in Jamaica. New Phytol. 61: 291–298.

1. Blades 1-pinnate; Texas. . . . . 2. *Anemia mexicana*
1. Blades 2–3-pinnate; Florida.
    2. Leaves 2-pinnate, pinnae of leaf all fertile or all sterile; sporangia on all pinnae of fertile
       leaves, blade tissue lacking; sterile leaves 4–10 cm. . . . . 3. *Anemia wrightii*
    2. Leaves 3-pinnate, often with dimorphic pinnae; sporangia limited to proximal pair of
       pinnae; sterile leaves (excluding erect fertile pinnae) 17–60 cm. . . . . 1. *Anemia adiantifolia*

1. **Anemia adiantifolia** (Linnaeus) Swartz, Syn. Fil., 157.
   1806 · Pine fern

Osmunda adiantifolia Linnaeus,
Sp. Pl. 2: 1065. 1753
Stems ca. 2 mm diam. **Leaves**
partially dimorphic (sporangia
limited to proximal pair of pinnae), 17–85 × 7–35 cm, sterile
leaves (excluding erect fertile
pinnae) 17–60 cm. **Petiole** straw-
colored to chestnut brown, 1/2–
2/3 length of leaf, 1–1.4 mm wide, hirsute to glabrous.
Blade deltate, 3-pinnate, leathery. **Pinnae** 10–18 pairs,
alternate to subopposite, segments oblanceolate, base
cuneate, margins minutely denticulate, apex obtuse, pilose with stiff white hairs. **Fertile pinnae** usually taller
than sterile blades. **Spores** with ridges ± parallel, distant. $2n = 76, 114, 152$.

   Terrestrial on open to lightly shaded, rocky slopes
and in hammocks and pine woods, often on limestone;
0–30 m; Fla.; Mexico; West Indies in the Antilles,
Trinidad; Central America; South America to Ecuador
and Brazil.

2. **Anemia mexicana** Klotzsch, Linnaea 18: 526. 1844

Stems ca. 2 mm diam. **Leaves**
partially dimorphic (sporangia
limited to proximal pair of pinnae), (15–)22–45 × 8–16 cm.
Petiole straw-colored, 1/2–2/3
length of leaf, ca. 1 mm wide,
glabrate. **Blade** deltate-lanceolate, 1-pinnate, somewhat leathery. **Pinnae** 4–7 pairs, mostly alternate, lanceolate, base truncate, margins minutely
serrulate, proximal margins often slightly excavate, apex
acuminate, hirsute with minute white hairs to glabrous
on abaxial surface, glabrous adaxially. **Fertile pinnae**
usually taller than sterile blades. **Spores** with ridges
parallel and closely placed. $2n = 76$.

   Lightly shaded limestone outcrops of the Edwards
Plateau; 400–500 m; Tex.; n Mexico.

3. **Anemia wrightii** Baker in Hooker & Baker, Syn. Fil. 10:
   435. 1868

Stems ca. 1 mm diam. **Leaves** entirely dimorphic. **Sterile leaves** 4–
10 × 1.4–2.5 cm. **Petiole** green
to straw-colored, 1/2–3/4 length
of leaf, ca. 0.3 mm wide, glabrous. **Blade** deltate, 2-pinnate,
papery to somewhat leathery.
Pinnae 2–4 pairs, alternate to
subopposite, segments oblanceo-
late, base cuneate, margins minutely denticulate, apex
obtuse, sparsely pilose with stiff white hairs. **Fertile leaves**
6–25 cm, much taller than sterile leaves. **Petiole** 3/4–
9/10 length of leaf. **Pinnae** compact, short-petiolulate,
1–6 mm. **Spores** with ridges anastomosing.

   Rare around exposed to lightly shaded solution holes
and limestone sinks; 0 m; Fla.; West Indies in Bahamas, Cuba.

   Anemia wrightii has been misidentified in Florida as
A. cicutaria Kunze ex Sprengel (of Mexico in Yucatan;
West Indies in the Bahamas, Cuba; and Central America in Guatemala), but that species has 3-pinnate sterile
blades, abundant blade hairs, and fertile leaves that are
more open with 4–5 pairs of long-petiolulate pinnae.

12. PARKERIACEAE Hooker

• Water Fern Family

Robert M. Lloyd

Plants short-lived, rooted or floating. **Stems** erect, bearing a few thin scales. **Leaves** dimorphic, erect to spreading, with adventitious bud initials or small plantlets along margins; proximal leaves sterile, herbaceous, venation reticulate, without included veinlets; distal leaves fertile, larger than sterile leaves; segments revolute, covering sporangia. **Sporangia** abaxial, thin-walled, with mixed development, bearing 0–71 indurate annulus cells. **Spores** yellow, 16 or 32 per sporangium, 70–150 μm diam. **Gametophytes** terrestrial, green, asymmetrically cordate.

Genus 1, species 3–4 (3 species in the flora): tropical, subtropical, and warm temperate regions worldwide.

SELECTED REFERENCES Benedict, R. C. 1909. The genus *Ceratopteris*: A preliminary revision. Bull. Torrey Bot. Club 36: 463–476. Lloyd, R. M. 1974. Systematics of the genus *Ceratopteris* Brongn. (Parkeriaceae) II. Taxonomy. Brittonia 26(2): 139–160.

1. CERATOPTERIS Brongniart, Bull. Sci. Soc. Philom. Paris, sér. 3, 8: 186. 1821

· Antler ferns, water ferns, floating ferns [Greek *cerato*, horned, and *pteris*, fern, referring to the antlerlike fertile leaf]

Parkeria Hooker, Exot. Fl. 2: plate 147. 1825

Plants 2–120 cm. **Stems** 1–10 mm diam., sparsely scaly. **Leaves** erect. **Proximal leaves** broad, simple and lobed to 4-pinnate, glabrous; distal leaves finely dissected with linear ultimate segments. **Sporangia** nearly sessile, usually densely packed, globose. **Spores** tetrahedral.

Species 3–4 (3 in the flora): tropical, subtropical, and warm temperate regions worldwide.

SELECTED REFERENCES DeVol, C. E. 1957. The geographical distribution of *Ceratopteris pteridoides*. Amer. Fern J. 47: 67–72. Hickok, L. G. 1977. Cytological relationships between three diploid species of the fern genus *Ceratopteris* Brongn. Canad. J. Bot. 55: 1660–1667. Hickok, L. G. and E. J. Klekowski Jr. 1974. Inchoate speciation in *Ceratopteris*: An analysis of the synthesized hybrid C. *richardii* × C. *pteridoides*. Evolution 28: 439–446.

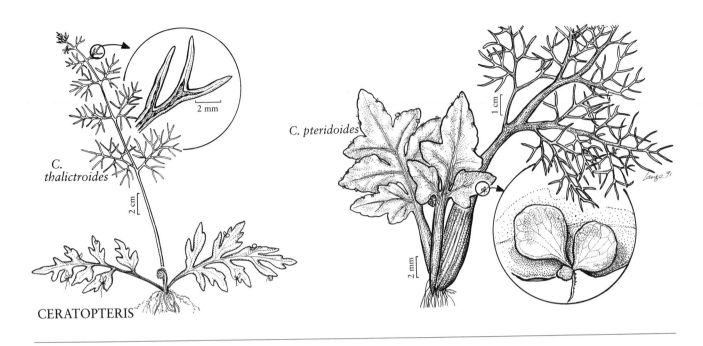

1. Sterile leaves simple, palmately or pinnately lobed, sometimes 2–4-pinnate, deltate to cordate to ovate; proximal pinnae or veins of lobes usually opposite; petioles frequently inflated. Fertile leaves usually deltate to cordate to reniform; sporangia lacking annulus or annulus well developed, indurate cells 0–10(–40). . . . . 2. *Ceratopteris pteridoides*
1. Sterile leaves (1–)2–3-pinnate, lanceolate, lance-ovate, ovate to deltate; proximal pinnae usually alternate; petioles usually not inflated. Fertile leaves lanceolate, sometimes ovate, deltate, or cordate; sporangia with well-developed annulus, indurate cells 13–71.
   2. Sporangia with 32 spores. . . . . 1. *Ceratopteris thalictroides*
   2. Sporangia with 16 spores. . . . . 3. *Ceratopteris richardii*

1. Ceratopteris thalictroides (Linnaeus) Brongniart, Bull. Sci. Soc. Philom. Paris, sér. 3. 8: 186. 1821

Acrostichum thalictroides Linnaeus, Sp. Pl. 2: 1070. 1753

Plants usually rooted in soil. **Sterile leaves** lanceolate to lance-ovate to ovate to deltate or cordate. **Petiole of sterile leaf** 1–31 cm, not inflated. **Blade of sterile leaf** 1–3-pinnate, 2–41 × 2–20 cm; segments lobed or incised, elliptic to lanceolate to ovate or deltate, to 12.5 cm; proximal pinnae ± alternate. **Fertile leaves** lanceolate to ovate to deltate or cordate, 2–117 × 2–48 cm. **Petiole of fertile leaf** 1–46 cm. **Blade of fertile leaf** 3–4-pinnate proximally, 2-pinnate distally; terminal segments linear. **Sporangia** usually crowded between segment midvein and revolute margin, with 13–71 indurate annulus cells. **Spores** 32 per sporangium, 96–124 μm diam. $2n = 154, 156$.

Aquatic to semiaquatic in swamps, bogs, canals, ponds, lakes, ditches, marshes; 0–200 m; Calif., Fla., La., Tex.; worldwide in tropical areas except Africa.

Ceratopteris thalictroides is common in Florida but rare elsewhere. It is tetraploid (*n* = 77, 78), the two cytotypes reproductively isolated. It can be distinguished from the diploid *C. richardii* on the basis of spore number per sporangium. The single population in southern California may have been a recent introduction and apparently has not persisted. Several populations are of hybrid origin, with reduced spore viability and irregular meiotic pairing. These include some in southern Florida and Texas.

2. **Ceratopteris pteridoides** (Hooker) Hieronymus, Bot. Jahrb. Syst. 34: 561. 1905

Parkeria pteridoides Hooker, Exot. Fl. 2: plate 147. 1825; *Ceratopteris lockhartii* (Hooker & Greville) Kunze

Plants floating or rooted. **Sterile leaves** deltate to cordate to ovate. **Petiole of sterile leaf** 1–19 cm, usually inflated, in some near base, but in most inflated nearer blades. **Blade of sterile leaf** 2–4-pinnate, 5–33 × 4–29 cm, simple and palmately 3-lobed (ternate), or pinnately 5-lobed or pinnate near base; proximal pinnae or veins of lobes usually opposite. **Fertile leaves** deltate to cordate to reniform, 9–50 × 8–36(–50) cm. **Petiole of fertile leaf** 4–25 cm. **Blade of fertile leaf** 1–4-pinnate; terminal segments narrow, linear. **Sporangia** usually crowded between segment midvein and revolute margin, with 0–10(–40) indurate annulus cells. **Spores** 32 per sporangium, 70–100 μm diam. 2*n* = 78.

Aquatic to semiaquatic; in swamps, bogs, canals, ponds, lakes, ditches, marshes; 0–25 m; Fla., La.; West Indies; Central America; South America; se Asia in Vietnam.

Ceratopteris pteridoides is usually easily recognized by its sterile leaf morphology, which varies considerably with habitat. Leaves intermediate between sterile and fertile are fairly common, with various degrees of laminar development of the fertile segments. Some fertile leaves have quite broad segments with rows of sporangia along the margins only. *Ceratopteris pteridoides* is sexual and diploid and is incompletely reproductively isolated from the diploid *C. richardii*. Hybrids synthesized by L. G. Hickok (1977) result in 40% viable spores.

3. **Ceratopteris richardii** Brongniart, Dict. Class. Hist. Nat. 3: 351. 1823

Ceratopteris deltoidea Benedict

Plants floating or rooted. **Sterile leaves** lanceolate to deltate to ovate. **Petiole of sterile leaf** 1–11 cm, not inflated; small leaves lobed to pinnate, segments or pinnae with entire to somewhat incised margins, larger leaves 2-pinnate-pinnatifid with deeply incised pinnae. **Blade of sterile leaf** 3–16 × 2.5–17 cm, pinnae deltate to ovate; proximal pinnae usually alternate. **Fertile leaves** lanceolate to deltate to ovate, to 19 × 12 cm. **Petiole of fertile leaf** 6–9 cm. **Blade of fertile leaf** 2–3-pinnate; terminal segments narrow. **Sporangia** scattered to densely crowded between midvein and revolute margin, with 20–40 or more indurate annulus cells. **Spores** 16 per sporangium, 107–150 μm diam. 2*n* = 78.

Aquatic to semiaquatic; lakes and ponds; 0 m; La.; West Indies; Central America in Guatemala; South America; Africa.

According to L. G. Hickok (1977), *Ceratopteris richardii* is diploid. Morphologically, specimens from the United States are difficult to distinguish from tetraploid *C. thalictroides*. The primary characteristic distinguishing *C. richardii* is its 16-spored sporangia. Herbarium specimens with 16-spored sporangia from the West Indies and Latin America have variable morphology ranging from that of *C. pteridoides* to that of *C. thalictroides* (R. M. Lloyd 1974, fig. 6). Some specimens have both 16- and 32-spored sporangia. This suggests multiple origins for *C. richardii*. Because reproductive isolation is incomplete among the diploid taxa, and highly fertile F$_2$ segregates of various morphologic types occur, further work is needed to determine the nature, origin, and distinctness of *C. richardii*.

13. PTERIDACEAE Reichenbach

• Maidenhair Fern Family

Michael D. Windham

Plants perennial [annual], on rock or terrestrial, of small (rarely large) stature. **Stems** compact to creeping, branched or unbranched, dictyostelic, bearing hairs and/or scales. **Leaves** monomorphic to dimorphic, circinate or noncircinate in bud. **Petioles** usually with persistent scales proximally, lacking spines; vascular bundles 1–several, roundish or crescent-shaped in cross section. **Blades** 1–6-pinnate, without laminar buds. **Indument** on petioles, rachises, costae, and blades, rarely absent or commonly of hairs, glands, and/or scales, occasionally of white or yellow farina. **Veins** pinnate or parallel in ultimate segments of blades, simple or forked, free or infrequently anastomosing in complex patterns. **Sori** borne abaxially on veins, often confluent with age and forming a continuous submarginal band, or sporangia densely covering abaxial surface (acrostichoid); receptacle not or only slightly elevated. **Indusia** (when present) formed by reflexed, recurved, or revolute leaf margin (false indusium). **Sporangia** stalk of 2–3 rows of cells; annulus vertical, interrupted by stalk; spores 64 or 32 (rarely 16) per sporangium. **Spores** all 1 kind, brown, black, or gray (rarely yellow), globose to globose-tetrahedral or trigonal, occasionally with prominent equatorial ridge, trilete, variously ornamented (usually cristate or rugose). **Gametophytes** green, aboveground, obcordate to reniform, sometimes asymmetric, usually glabrous (glandular-farinose in *Notholaena*); archegonia and antheridia borne on abaxial surface, antheridia 3-celled.

Genera ca. 40, species ca. 1000 (13 genera, 90 species in the flora): worldwide.

Considerable disagreement exists concerning the circumscription and proper name of this family. The taxa comprising the Pteridaceae in this treatment were assigned to the Sinopteridaceae and Pteridaceae by D. B. Lellinger (1985) and were included in five families by R. E. G. Pichi-Sermolli (1977). The broad concept followed here is similar (except for the exclusion of *Ceratopteris*) to that espoused by R. M. Tryon and A. F. Tryon (1982), who applied the name Pteridaceae to the group. Until very recently, the newer name Adiantaceae was more commonly used.

As represented in North America, Pteridaceae comprise three major evolutionary lines (the

122

adiantoids, the pteroids, and the cheilanthoids). Characteristics holding the family together include abaxial (usually submarginal) sori that lack indusia or are protected by a reflexed or revolute leaf margin, spores that are usually globose-tetrahedral and trilete, and chromosome base numbers of 30 or 29 (rarely 27). The xeric-adapted members of the family (particularly the cheilanthoids) have undergone extensive parallel and convergent evolution, and they have frustrated attempts to produce a natural generic classification based on macromorphologic characteristics alone. Although some workers have aggregated species into a few large genera (e.g., J. T. Mickel 1979b), most tend to recognize smaller segregate genera based on a combination of morphologic, chromosomal, and biochemical data. The latter approach seems to provide a more useful, evolutionarily informative classification and is the one adopted here. *Aspidotis* and *Notholaena* are maintained here as distinct from *Cheilanthes,* and three recently described genera (*Argyrochosma, Astrolepis,* and *Pentagramma*) have been incorporated into the treatment. The reasons for these changes in generic circumscription are discussed under the individual genera.

SELECTED REFERENCES Lellinger, D. B. 1985. A Field Manual of the Ferns & Fern-allies of the United States & Canada. Washington. Mickel, J. T. 1979b. The fern genus *Cheilanthes* in the continental United States. Phytologia 41: 431–437. Pichi-Sermolli, R. E. G. 1977. Tentamen pteridophytorum genera in taxonomicum ordinem redigendi. Webbia 31: 313–512. Tryon, R. M. and A. F. Tryon. 1982. Ferns and Allied Plants, with Special Reference to Tropical America. New York, Heidelberg, and Berlin.

1. Sporangia borne directly on reflexed marginal lobes of ultimate segments, lobes separate and distinct; veins of ultimate blade segments prominent, dichotomously branched, essentially parallel distally. . . . . 1. *Adiantum,* p. 125
1. Sporangia borne on abaxial leaf surface or, if seemingly attached to marginal lobes of ultimate segments, lobes confluent and poorly defined; veins of ultimate blade segments obscure or, if prominent, pinnately branched and more divergent distally.
  2. Sporangia covering entire abaxial surface on fertile pinnae; veins strongly anastomosing throughout, forming several rows of areoles between costa and margin; mature leaves usually more than 1 m. . . . . 2. *Acrostichum,* p. 130
  2. Sporangia confined to marginal sori or scattered along veins but not covering entire abaxial surface of pinnae; veins free or rarely anastomosing, not forming several rows of areoles between costa and margin; mature leaves less than 1 m (occasionally longer in *Pteris* and *Pityrogramma*).
    3. Petioles longitudinally ridged and 2–3-grooved, containing 2 or more distinct vascular bundles (usually 1 in *Pteris*); spores with prominent equatorial ridge (lacking in *Pityrogramma trifoliata*); plants of disturbed or mesic habitats, mostly on southeastern coastal plain.
      4. Petioles green, straw-colored, or light brown distally; blades glabrous or sparsely pubescent, lacking white or yellow farina on abaxial surface; sporangia submarginal, often covered by reflexed leaf margin (false indusium). . . . . 3. *Pteris,* p. 132
      4. Petioles black or dark brown throughout; blades with white or yellow farina on abaxial surface (this sometimes lost in heat-treated specimens); sporangia following veins for most of length, not covered by reflexed leaf margin (false indusium). . . . . 4. *Pityrogramma,* p. 135
    3. Petioles rounded, flattened, or with single longitudinal groove adaxially, containing single vascular bundle (2 in *Astrolepis,* which has leaves covered with fringed or stellate scales); spores lacking prominent equatorial ridge; plants of rocky, mostly xeric habitats in continental interior, rarely found on southeastern coastal plain.
      5. Leaves strongly dimorphic, fertile leaves obviously longer than the sterile, with narrow, elongate, usually revolute ultimate segments; petioles green to straw-colored distally, essentially glabrous; mature spores usually yellow. . . . 5. *Cryptogramma,* p. 137

5. Leaves monomorphic to weakly dimorphic, either all alike and fertile or with a few sterile leaves poorly differentiated from the fertile; petioles brown to black and glabrous or pubescent, or if lighter, then sparsely to densely pubescent; mature spores brown to black, rarely yellowish.

  6. Blades 1-pinnate to pinnate-pinnatifid throughout, abaxial surface densely covered with fringed or stellate scales; petioles containing 2 vascular bundles . . . . 6. *Astrolepis*, p. 140

  6. Blades 2–5-pinnate proximally or, if less divided, then abaxial surfaces glabrous or pubescent; petioles containing single vascular bundle.

    7. Blades with white or yellow farina on abaxial surface (concealed beneath fringed or stellate scales in *Notholaena aschenborniana*); stem scales either uniformly dark brown to black or bicolored with dark central stripe; ultimate segments sessile or subsessile, usually adnate to midrib.

      8. Sporangia confined to modified vein tips located near margin of ultimate segments; spores black to dark brown, appearing ± globose; gametophytes glandular-farinose. . . . . 7. *Notholaena*, p. 143

      8. Sporangia following veins for most of length, sometimes nearly covering abaxial surface of ultimate segments; spores tan to brown, distinctly trigonal; gametophytes lacking farina-producing glands. . . . . 8. *Pentagramma*, p. 149

    7. Blades without white or yellow farina on abaxial surface; stem scales yellow or brown to black; ultimate segments sessile or stalked or, if farinose (in *Argyrochosma*), then stem scales uniformly tan or brown and ultimate segments distinctly stalked.

      9. Blades conspicuously pubescent and/or scaly, hairs or scales often concealing abaxial surface, farina lacking.

        10. Sporangia following veins for most of length; blades deeply pinnate-pinnatifid and pentagonal. . . . . 9. *Bommeria*, p. 151

        10. Sporangia submarginal; blades 2–4-pinnate or, if pinnate-pinnatifid, then linear to lanceolate. . . . . 10. *Cheilanthes*, p. 152

      9. Blades glabrous or sparsely hirsute-pubescent or with whitish farina on abaxial surface, not conspicuously pubescent or scaly.

        11. Ultimate segments of blades linear to lanceolate; blades glabrous, lustrous on adaxial surface and often striate; false indusia broad, scarious, appearing inframarginal (obscurely so in *A. californica*), strongly differentiated from segment margin. . . . . 11. *Aspidotis*, p. 170

        11. Ultimate segments narrowly elliptic to round (occasionally linear); blades glabrous, farinose or sparsely pubescent, usually dull on adaxial surface, not striate; false indusia (when present) usually greenish or whitish, narrow, formed by revolute segment margin and poorly differentiated from it.

          12. Stem scales strongly bicolored, or if concolored, then largest ultimate segments more than 4 mm wide. . . . . 13. *Pellaea*, p. 175

          12. Stem scales concolored to weakly bicolored; ultimate segments usually less than 4 mm wide.

            13. Blades with whitish farina on abaxial surface or if glabrous, then ultimate segments somewhat cordate at base and attached to lustrous, dark-colored stalks. . . 12. *Argyrochosma*, p. 171

            13. Blades sparsely pubescent or glabrous, lacking whitish farina on abaxial surface; ultimate segments mostly rounded to truncate or cuneate at base, sessile or attached to dull greenish stalks. 10. *Cheilanthes*, p. 152

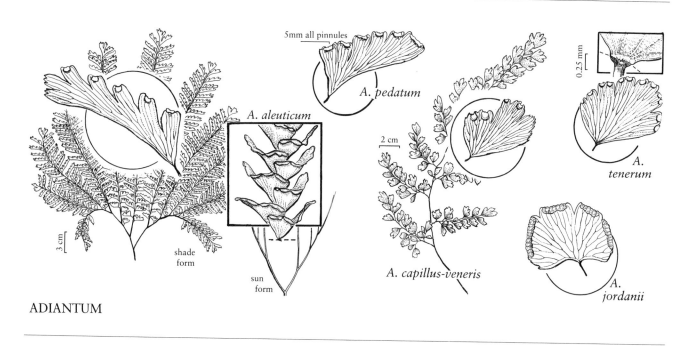

A. aleuticum

5mm all pinnules

A. pedatum

shade form

sun form

2 cm

A. capillus-veneris

0.25 mm

A. tenerum

A. jordanii

ADIANTUM

1. ADIANTUM Linnaeus, Sp. Pl. 2: 1094. 1753; Gen. Pl. ed 5, 485. 1754 · Maidenhair fern [Greek *adiantos,* unwetted, for the glabrous leaves, which shed raindrops]

Cathy A. Paris

Plants terrestrial or on rock. **Stems** short- to long-creeping or suberect, branched; scales deep tawny yellow to dark reddish brown [black], concolored or bicolored, linear-lanceolate to lanceolate, margins entire, erose-ciliate, or minutely dentate. **Leaves** monomorphic to somewhat dimorphic, densely clustered to closely spaced [distant], 15–110 cm. **Petiole** chestnut brown to dark purple or blackish, with single groove adaxially, glabrous, hispid, or strigose, with 1 or 2 vascular bundles. **Blade** lanceolate, ovate, trowel-shaped, or fan-shaped, 1–4(–9)-pinnate proximally, membranaceous to papery, both surfaces commonly glabrous (2 species with scattered hairs), adaxially dull or shiny, not striate; rachis straight or flexuous. **Ultimate segments** subsessile to short-stalked (stalks terminating in cupulelike swelling at base of pinna in *A. tenerum*), round, fan-shaped, rhombic, or oblong, 3–29 mm wide; base truncate to cuneate, free from costa; stalk dark, often lustrous; fertile segments with marginal lobes recurved to form false indusia. **Veins** of ultimate segments conspicuous, free, ± dichotomously forking near base and well above segment base [anastomosing in a few tropical species], parallel distally. **False indusia** light gray-green or brown to dark brown, narrow, 0.6–1 mm wide, marginal, concealing sporangia until sporangia dehisce. **Sporangia** submarginal, borne along or sometimes also between veins on abaxial surface of false indusium, paraphyses and glands absent. **Spores** yellow or yellowish brown, tetrahedral-globose, trilete, rugulate to rugose or tuberculate, equatorial ridge absent. $x = 29, 30$.

Species ca. 150–200 (9 in the flora): nearly worldwide except at latitudes greater than 60°.

Most diverse in Andean South America, *Adiantum* is primarily a tropical genus; of the nine species occurring in the flora, *A. melanoleucum, A. tenerum,* and *A. tricholepis* are strictly subtropical. *Adiantum hispidulum* occurs only as an escape from cultivation. The genus is absent from dry areas in the interior of the continent.

Adiantum is a very clearly circumscribed genus of ferns, the character state "sporangia borne on abaxial surface of false indusium" being both necessary and sufficient to define it. Within this large and widespread genus, however, species relationships are mostly unknown. An evolutionary classification of the group is indeed much needed (R. M. Tryon and A. F. Tryon 1982).

SELECTED REFERENCES Fernald, M. L. 1950b. *Adiantum capillus-veneris* in the United States. Rhodora 52: 201–208. Paris, C. A. 1991. *Adiantum viridimontanum*, a new maidenhair fern in eastern North America. Rhodora 93: 105–122. Paris, C. A. and M. D. Windham. 1988. A biosystematic investigation of the *Adiantum pedatum* complex in eastern North America. Syst. Bot. 13: 240–255. Wagner, W. H. Jr. 1956. A natural hybrid, × *Adiantum tracyi* C. C. Hall. Madroño 13: 195–205.

1. Segments at middle of penultimate divisions of blades ± fan-shaped, rhombic, transversely oblong, or nearly round, about as long as broad.
   2. Dark color of stalks extending into base of ultimate segments. . . . . 1. *Adiantum capillus-veneris*
   2. Dark color of stalks ending ± abruptly at base of ultimate segments.
      3. Segment stalks terminating in small, cupulelike swelling at base of ultimate segments.
      . . . . .2. *Adiantum tenerum*
      3. Segment stalks not terminating in small, cupulelike swelling at base of ultimate segments.
         4. Ultimate segments glabrous. . . . . 3. *Adiantum jordanii*
         4. Ultimate segments hirsute. 4. *Adiantum tricholepis*
1. Segments at middle of penultimate divisions of blades ± oblong or long-triangular, at least 2 times as long as broad (rarely, reniform).
   5. Rachises hispid or strigose; blades pinnate (occasionally pseudopedate in *Adiantum hispidulum*).
      6. Ultimate segments with scattered multicelled hairs; rachises hispid; false indusia ± round. . . . . 5. *Adiantum hispidulum*
      6. Ultimate segments glabrous; rachises strigose; false indusia crescent-shaped. . . . .
      . . . . .6. *Adiantum melanoleucum*
   5. Rachises glabrous; blades pseudopedate.
      7. Segments at middle of penultimate divisions of blades ± oblong; leaves lax-arching, blades fan-shaped.
         8. Segments at middle of penultimate divisions of blades generally less than 3.2 times as long as broad, apices with rounded, crenulate or crenate-denticulate lobes, lobes separated by shallow sinuses 0.1–2(–3.7) mm, segment stalks ca. 0.6–0.9 mm. . . . . 7. *Adiantum pedatum*
         8. Segments at middle of penultimate divisions usually more than 3.2 times as long as broad, apices with sharply denticulate, angular lobes, lobes separated by deep sinuses 0.6–4 mm, segment stalks to 0.6 mm. . . . . 8. *Adiantum aleuticum*
      7. Segments at middle of penultimate divisions ± long-triangular or reniform; leaves arching to stiffly erect, blades fan-shaped to funnel-shaped.
         9. Central ultimate segments on stalks less than 0.9 mm; false indusia mostly less than 3.5 mm. . . . . 8. *Adiantum aleuticum*
         9. Central ultimate segments on stalks generally greater than 0.9 mm; false indusia mostly exceeding 3.5 mm. . . . . 9. *Adiantum viridimontanum*

1. Adiantum capillus-veneris Linnaeus, Sp. Pl. 2: 1096.
1753 · Venus's-hair fern, southern maidenhair

Adiantum capillus-veneris var.
modestum (L. Underwood)
Fernald; *A. capillus-veneris* var.
protrusum Fernald; *A. capillus-
veneris* var. *rimicola* (Slosson)
Fernald

Stems short-creeping; scales golden brown to medium brown, concolored, iridescent, margins entire or occasionally with single broad tooth near base. **Leaves** lax-arching or pendent, closely spaced, 15–75 cm. **Petiole** 0.5–1.5 mm diam., glabrous, occasionally glaucous. **Blade** lanceolate, pinnate, 10–45 × 4–15 cm, glabrous, gradually reduced distally; proximal pinnae 3(–4)-pinnate; rachis straight to flexuous, glabrous, not glaucous. **Segment stalks** 0.5–3.5 mm, dark color extending into segment base. **Ultimate segments** various, generally cuneate or fan-shaped to irregularly rhombic (plants in American southwest occasionally with segments nearly round), about as long as broad; base broadly to narrowly cuneate; margins shallowly to deeply lobed, incisions 0.5–7 mm, occasionally ± laciniate, sharply denticulate in sterile segments; apex rounded to acute. **Indusia** transversely oblong or crescent-shaped, 1–3(–7) mm, glabrous. **Spores** mostly 40–50 μm diam. $2n = 120$.

Sporulating spring–summer. Moist calcareous cliffs, banks, and ledges along streams and rivers, walls of lime sinks, canyon walls (in the American southwest), around foundations, on mortar of storm drains; 0–2500 m; B.C.; Ala., Ariz., Ark., Calif., Colo., Fla., Ga., Ky., La., Miss., Mo., Nev., N.Mex., N.C., Okla., S.C., S.Dak., Tenn., Tex., Utah, Va.; Mexico; West Indies; Central America; South America in Venezuela, Peru; tropical to warm temperate regions in Eurasia and Africa.

No evident pattern to morphologic variation in the species is discernible, although a number of segregate species and infraspecific taxa have been recognized within North American *Adiantum capillus-veneris*. In the Eastern Hemisphere, the species is diploid, with $2n = 60$ (I. Manton 1950). Several tetraploid counts have been reported from North America (W. H. Wagner Jr. 1963). Spore-measurement data suggest, however, that the polyploid cytotype may not be widely distributed. Further investigation is needed to determine whether *Adiantum capillus-veneris* populations in North America are conspecific with those in Eurasia and Africa.

2. Adiantum tenerum Swartz, Prodr., 135. 1788
· Brittle maidenhair

Stems short-creeping; scales bicolored, centers dark reddish brown, margins pale tan, erose-ciliate. **Leaves** arching or sometimes pendent, closely spaced, 20–110 cm. **Petiole** 1–3 mm diam., glabrous, occasionally glaucous. **Blade** trowel-shaped, pinnate, 12–60 × 12–60 cm, gradually reduced distally, glabrous; proximal pinnae 3-pinnate; rachis straight, glabrous, not glaucous. **Segment stalks** 1–5 mm, with dark color ending abruptly at segment base, terminating in cupulelike swelling at base of segment (unlike any other species of *Adiantum* in the flora). **Ultimate segments** fan-shaped or rhombic, about as long as broad; base cuneate; apex rounded or acute, lobed, lobes separated by narrow incisions 0.5 mm wide. **Indusia** transversely oblong to crescent-shaped, 0.5–2 mm, glabrous. **Spores** mostly 40–58 μm diam. $2n = 60$.

Sporulating throughout the year. Restricted to moist, shaded, limestone ledges, sink walls, and grottoes in the flora; 0–50 m; Fla.; e,s Mexico; Central America in Guatemala, Honduras, Nicaragua, Costa Rica; South America in Venezuela.

Adiantum tenerum is readily distinguished from other species in the flora by the ultimate segments conspicuously articulate to the stalks.

3. Adiantum jordanii Müller Halle, Bot. Zeit. 1864: 26.
1864 · California maidenhair

Stems short-creeping; scales reddish brown, concolored, margins entire. **Leaves** arching or pendent, clustered, 30–45 cm. **Petiole** 1–1.5 mm diam., glabrous, not glaucous. **Blade** lanceolate, pinnate, 20–24 × 8–10 cm, gradually reduced distally, glabrous; proximal pinnae 3(–4)-pinnate; rachis straight, glabrous, not glaucous. **Segment stalks** 1–4 mm, with dark color ending abruptly at segment base. **Ultimate segments** fan-shaped, not quite as long as broad; base truncate or broadly cuneate; margins of fertile segments unlobed but very narrowly incised, sterile segments with margins lobed, denticulate; apex rounded. **Indusia** transversely oblong, 3–10 mm, glabrous. **Spores** mostly 40–50 μm diam. $2n = 60$.

Sporulating early spring–midsummer. Seasonally moist, shaded, rocky banks, canyons, and ravines; 0–1000 m; Calif., Oreg.; Mexico in Baja California.

Adiantum jordanii occasionally hybridizes with *A.*

aleuticum where their ranges overlap in northern California, yielding the sterile hybrid *Adiantum* × *tracyi* C. C. Hall ex W. H. Wagner. *Adiantum* × *tracyi*, morphologically intermediate between its parental species, can be distinguished from *A. jordanii* by its broadly deltate leaf blade that tapers abruptly from the 4(–5)-pinnate base to a 1-pinnate apex. It is best separated from *A. aleuticum* by leaf blades with a strong rachis, and by ultimate blade segments that are less than twice as long as broad. *Adiantum* × *tracyi* shows 59 univalents at metaphase; its spores are irregular and misshapen (W. H. Wagner Jr. 1962).

4. Adiantum tricholepis Fée, Mém. Foug. 8: 72. 1857 · Hairy maidenhair

Stems short-creeping to nearly erect; scales dark reddish brown, concolored, margins entire or minutely denticulate. **Leaves** arching or pendent, densely clustered, 20–62 cm. **Petiole** 0.8–1 mm diam., glabrous, occasionally glaucous. **Blade** ovate, pinnate, 15–38 × 8–26 cm, gradually reduced distally, hirsute; proximal pinnae 3–4-pinnate; rachis straight or becoming flexuous, glabrous, not glaucous. **Segment stalks** 1–4 mm, dark color ending ± abruptly at segment base. **Ultimate segments** transversely oblong, nearly round, or fan-shaped, about as long as broad; base truncate or cuneate; margins of fertile segments crenulate or entire, sterile segments with margins serrulate; apex rounded. **Indusia** transversely oblong or crescent-shaped, 0.5–4 mm, covered with whitish needlelike trichomes. **Spores** mostly 35–53 μm diam.

Sporulating late winter–early spring. Moist, shaded, limestone cliffs along streams and rivers, on boulders in creeks, and among rocks on steep slopes; 200–500 m; Tex.; Mexico; Central America in Guatemala, Belize.

Adiantum tricholepis occurs in the flora only in Bandera and Medina counties on the Edwards Plateau in central Texas. Collections identified as *A. tricholepis* from the mouth of the Pecos River are *Adiantum capillus-veneris*.

5. Adiantum hispidulum Swartz, J. Bot. (Schrader) 1800(2): 82. 1801 · Rosy maidenhair

Stems short-creeping; scales dark reddish brown, concolored, margins entire. **Leaves** arching, clustered, 20–37 cm. **Petiole** 1–2 mm diam., adaxially hispid, not glaucous. **Blade** lanceolate, pinnate or occasionally pseudopedate, 1-pinnate distally, 12–18 × 6.5–8 cm; proximal pinnae 1–4-pinnate; indument of light-colored, sparse, multicellular hairs; rachis straight, densely hispid, not glaucous. **Segment stalks** 0.2–0.3 mm, dark color generally entering into segment base. **Ultimate segments** oblong to long-triangular, ca. 2 times as long as broad, progressively reduced toward apex of penultimate divisions; basiscopic margin oblique; acroscopic margin of fertile segments crenulate, sterile segments sharply denticulate; apex obtuse or acute. False **indusia** ± round, 0.6–0.9 mm diam., covered with reddish brown, stiff, needlelike bristles. **Spores** mostly 40–60 μm diam.

Sporulating summer–fall. Banks and old walls; 0–100 m; introduced; Conn., Ga.; Asia in s India; e Africa; Pacific Islands.

Adiantum hispidulum is represented by sporadic escapes from cultivation in the flora, possibly naturalized locally. It also has been reported from Florida and Louisiana.

6. Adiantum melanoleucum Willdenow, Sp. Pl. 5(1): 443. 1810 · Fragrant maidenhair

Stems short-creeping; scales dark reddish brown, concolored, margins entire. **Leaves** arching to erect, clustered, (15–)35–80 cm. **Petiole** 1–2 mm diam., minutely rough, abaxially strigose, not glaucous. **Blade** ovate, pinnate, 1-pinnate distally (small leaves 1-pinnate throughout), (11–)15–35 × (3.5–)8–15 cm, glabrous; proximal pinnae (and sometimes also next 2–3 pairs) 1–2(–3)-pinnate; rachis straight, densely minutely rough. **Segment stalks** 0.1–0.8 mm, dark color entering into segment base. **Ultimate segments** oblong, about 2 times as long as broad; basiscopic margin straight or sometimes oblique; acroscopic margin shallowly lobed, lobes separated by narrow incisions; apex obtuse, shallowly lobed. False **indusia** crescent-shaped, 1–4.5 mm, glabrous. **Spores** mostly 40–50 μm diam. 2n = 60.

Sporulating throughout the year. Hammocks and limestone sinks in Everglades National Park; 0 m; Fla.; West Indies in Greater Antilles, Bahamas.

7. **Adiantum pedatum** Linnaeus, Sp. Pl. 2: 1095. 1753
· Northern maidenhair, adiante du Canada

Adiantum pedatum forma *billingsae* Kittredge; *A. pedatum* forma *laciniatum* (Hopkins) Weatherby **Stems** short-creeping; scales bronzy deep yellow, concolored, margins entire. **Leaves** lax-arching (rarely pendent), closely spaced, 40–75 cm. **Petiole** 1–2 mm diam., glabrous, occasionally glaucous. **Blade** fan-shaped, pseudopedate, 1-pinnate distally, 15–30 × 15–35 cm, glabrous; proximal pinnae 3–9-pinnate; rachis straight, glabrous, occasionally glaucous. **Segment stalks** 0.5–1.5(–1.7) mm, dark color entering into segment base. **Ultimate segments** oblong, ca. 3 times as long as broad; basiscopic margin straight; acroscopic margin lobed, lobes separated by narrow incisions 0–0.9(–1.1) mm wide; apex obtuse, divided into shallow, rounded lobes separated by shallow sinuses 0.1–2(–3.7) mm deep, margins of lobes crenulate or crenate-denticulate. **Indusia** transversely oblong, 1–3 mm, glabrous. **Spores** mostly 34–40 μm diam. $2n = 58$.

Sporulating summer–fall. Rich, deciduous woodlands, often on humus-covered talus slopes and moist lime soils; 0–700 m; N.B., N.S., Ont., Que.; Ala., Ark., Conn., Del., D.C., Ga., Ill., Ind., Iowa, Kans., Ky., La., Maine, Md., Mass., Mich., Minn., Miss., Mo., Nebr., N.H., N.J., N.Y., N.C., Ohio, Okla., Pa., R.I., S.C., Tenn., Vt., Va., W.Va., Wis.

Once considered a single species across its range in North America and eastern Asia, *Adiantum pedatum* is considered to be a complex of at least three vicariant species (*A. pedatum* and *A. aleuticum* occur in North America) and a derivative allopolyploid species (C. A. Paris 1991). *Adiantum pedatum* in the strict sense is restricted to deciduous woodlands in eastern North America.

8. **Adiantum aleuticum** (Ruprecht) Paris, Rhodora 93: 112. 1991 · Western maidenhair, Aleutian maidenhair, adiante des Aléoutiennes

Adiantum pedatum Linnaeus var. *aleuticum* Ruprecht, Distr. Crypt. Vasc. Ross., 49. 1845; *A. boreale* C. Presl; *A. pedatum* subsp. *aleuticum* (Ruprecht) Calder & Roy L. Taylor; *A. pedatum* subsp. *calderi* Cody; *A. pedatum* subsp. *subpumilum* (W. H. Wagner) Lellinger; *A. pedatum* var. *subpumilum* W. H. Wagner in W. H. Wagner & Boydston

Stems short-creeping or suberect; scales bronzy deep yellow, concolored, margins entire. **Leaves** lax-arching to stiffly erect or pendent, often densely clustered, 15–110 cm. **Petiole** 0.5–3 mm diam., glabrous, often glaucous. **Blade** fan-shaped to funnel-shaped, pseudopedate, 1-pinnate distally, 5–45 × 5–45 cm; proximal pinnae (1–)2–7-pinnate; rachis straight, glabrous, often with glaucous bloom. **Segment stalks** 0.2–0.9(–1.3) mm, dark color entering into segment base or not. **Ultimate segments** oblong, long-triangular, or occasionally reniform, ca. 2.5–4 times as long as broad; basiscopic margin straight to oblique, or occasionally excavate; acroscopic margin lobed, lobes separated by narrow to broad incisions 0.2–3 mm wide; apex acute to obtuse, obtuse apices divided into ± angular lobes separated by sinuses 0.6–4 mm deep, margins of lobes sharply denticulate. False **indusia** transversely oblong to crescent-shaped, 0.2–3.5(–6) mm, glabrous. **Spores** mostly 37–47 μm diam. $2n = 58$.

Sporulating summer–fall. Wooded ravines, shaded banks, talus slopes, serpentine barrens, and coastal headlands (uncommon); 0–3200 m; Alta., B.C., Nfld., Que.; Alaska, Ariz., Calif., Colo., Idaho, Maine, Md., Mont., Nev., Oreg., Pa., Utah, Vt., Wash., Wyo.; Mexico in Chihuahua.

Adiantum aleuticum is disjunct in wet rock fissures at high elevations in Arizona, Colorado, Montana, Nevada, Utah, Wyoming, and Mexico in Chihuahua, and it is disjunct on serpentine in Newfoundland, Quebec, Maine, Maryland, Pennsylvania, and Vermont.

Although the western maidenhair has traditionally been interpreted as an infraspecific variant of *Adiantum pedatum,* the two taxa are reproductively isolated and differ in an array of morphologic characteristics. Therefore, they are more appropriately considered separate species (C. A. Paris and M. D. Windham 1988). Morphologic differences between *A. pedatum* and *A. aleuticum* are subtle; the two may be separated, however, using characteristics in the key. *Adiantum aleuticum* occurs in a variety of habitats throughout its range, from moist, wooded ravines to stark serpentine barrens and from coastal cliffs to subalpine boulder fields. Although morphologic differences exist among populations in these diverse habitats, they are not consistent. Consequently, infraspecific taxa are not recognized here within *A. aleuticum.*

9. **Adiantum viridimontanum** Paris, Rhodora 93: 108. 1991 · Green Mountain maidenhair

Stems short-creeping; scales bronzy deep yellow, concolored, margins entire. **Leaves** arching to stiffly erect, often densely clustered, 38–75(–90) cm. **Petiole** 1–3 mm diam., glabrous, often glaucous. **Blade** fan-shaped to funnel-shaped, pseudopedate, 1-pinnate distally, 10–35 × 10–35(–45) cm, glabrous; proximal pinnae 2–7-pinnate; rachis straight, glabrous, often glaucous. **Segment stalks** (0.4–)0.6–1.5(–1.9) mm, dark color commonly entering into segment base. **Ultimate segments** long-triangular, ca. 2.5 times as long as broad; basiscopic margin oblique; acroscopic margin lobed, lobes separated by narrow (less than 1 mm) incisions; apex acute, usually entire. **False indusia** transversely oblong, mostly 2–5(–10) mm, glabrous. **Spores** mostly 45–58 μm diam. $2n = 116$.

Sporulating summer–fall. Restricted to serpentine sites where it occurs in rock clefts, on talus slopes, and in well-developed serpentine soils; 200–800 m; Vt.

Adiantum viridimontanum, an allopolyploid from a sterile hybrid between *A. pedatum* and *A. aleuticum*, is known only from north central Vermont (C. A. Paris and M. D. Windham 1988). Additional populations may eventually be located on serpentine in southern Quebec.

2. ACROSTICHUM Linnaeus, Sp. Pl. 2: 1067. 1753; Gen. Pl. ed. 5, 484. 1754 · Leather fern [Greek *acros*, at the end, tip, and *stichos*, row, referring to the distal spore-bearing pinnae]

Robert M. Lloyd

Plants terrestrial in fresh- or saltwater habitats. **Stems** erect or creeping, branched; scales dark brown, concolored, linear-lanceolate, margins entire. **Leaves** slightly dimorphic, clustered, 1–5 m. **Petiole** brown, with a single groove adaxially, glabrous, smooth or with scale scars, with several abaxial vascular bundles and 2 adaxial vascular bundles. **Blade** lanceolate, pinnate, leathery, abaxially glabrous or hispid, adaxially dull, not striate, glabrous; rachis straight. **Pinnae** stalked, free from rachis, narrowly oblong to lanceolate, 2–5 cm wide; base cuneate; stalk green; margins plane; fertile leaves bearing sporangia on most pinnae or on only more distal pinnae (fertile pinnae may be slightly smaller than sterile ones). **Veins** of pinnae conspicuous, strongly anastomosing. **False indusia** absent. **Sporangia** spread over abaxial surface, mixed with paraphyses (sori acrostichoid), containing 64 spores. **Spores** yellow, tetrahedral, minutely tuberculate or roughened, equatorial flange absent. $x = 30$.

Species 3, possibly more (2 in the flora): worldwide, warm and tropical regions.

SELECTED REFERENCES Adams, D. C. and P. B. Tomlinson. 1979. *Acrostichum* in Florida. Amer. Fern J. 69: 42–46. García de López, I. 1978. Revisión del género *Acrostichum* en la República Dominicana. Moscosoa 1: 64–70. Lloyd, R. M. 1980. Reproductive biology and gametophyte morphology of New World populations of *Acrostichum aureum*. Amer. Fern J. 70: 99–110. Small, J. K. 1938. Ferns of the Southeastern States. Lancaster, Pa. [Facsimile edition 1964, New York and London.]

1. Fertile leaves with only most distal pinnae fertile; pinnae often distant and not overlapping; costal areoles usually narrow, 3 or more times longer than wide; paraphyses terminating in isodiametric, irregularly lobed cell. . . . . .1. *Acrostichum aureum*
1. Fertile leaves with most pinnae fertile; pinnae distant to closely spaced and ± overlapping; costal areoles usually broad, less than 3 times longer than wide; paraphyses terminating in horizontally extended, smooth or lobed cell. . . . . 2. *Acrostichum danaeifolium*

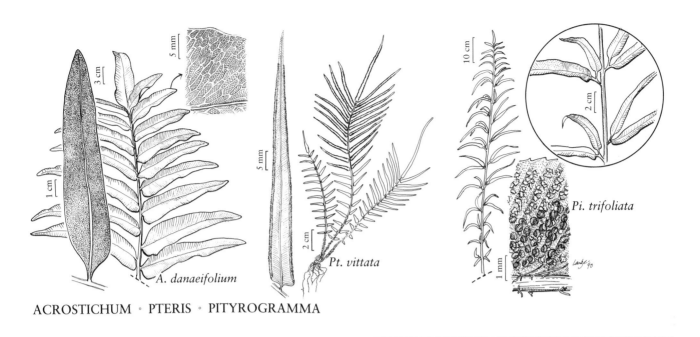

A. danaeifolium

Pt. vittata

Pi. trifoliata

ACROSTICHUM · PTERIS · PITYROGRAMMA

1. Acrostichum aureum Linnaeus, Sp. Pl. 2: 1069. 1753

Stems creeping or ascending, frequently branched. **Leaves** ± arching, 1–3 m × 12–50 cm. **Rachis** rounded abaxially, decidedly grooved adaxially. **Pinnae** 24–30(–40), usually not overlapping, 10–34 × 1.3–7 cm; proximal pinnae always distant, frequently rounded at tip; costal areoles 3 or more times longer than wide; distal 6–12 pinnae of fertile leaves bearing sporangia. **Sporangia** spread over abaxial surface of distal pinnae in fertile leaves; paraphyses stalked, ending with single isodiametric, irregularly lobed cell. **Spores** (37–)57(–72) μm diam., surface minutely tuberculate. $2n = 60$.

Sporulating all year. Coastal mangrove swamps, salt marshes, low hammocks, along lake and canal margins; 0–50 m; Fla.; worldwide in tropical and subtropical regions.

2. Acrostichum danaeifolium Langsdorff & Fischer, Pl. Voy. Russes Monde 1: 5, plate 1. 1810

Acrostichum lomarioides Jenman

Stems usually erect, infrequently branched. **Leaves** ascending or erect, 1.5–5 m × 15–60 cm. **Rachis** shallowly grooved abaxially, flat or shallowly grooved adaxially. **Pinnae** 20–32(–64), distant to closely spaced, usually overlapping, 7–37 × 1.5–5.5 cm, tapering toward apex, abruptly acute at tip; costal areoles less than 3 times longer than wide; most pinnae of fertile leaves bearing sporangia. **Sporangia** spread over abaxial surface of fertile pinnae; paraphyses stalked, ending with horizontally extended, smooth or little-lobed cell. **Spores** (44–)54(–72) μm diam., surface minutely roughened with small projecting papillae. $2n = 60$.

Somewhat saline to freshwater swamps, canal and pond margins, commonly in sinkholes in hammocks, disturbed marl sites, inland to coastal regions; 0–50 m; Fla.; Central America; South America.

These species frequently can be distinguished by the distribution of pinnae, the distribution of fertile pinnae, the shape of the costal areoles, and the structure of the paraphyses. In parts of Florida, their distributions are contiguous and abruptly separated by habitat. *Acrostichum aureum* is more frequently found in coastal shaded areas, in saline black-mangrove communities, and in the southern and southwestern parts of the state. *Acrostichum danaeifolium* grows vigorously in full sun and is common and widely distributed in Florida, where it has been collected in virtually every county throughout the southern two-thirds of the state. Hybrids have been produced in the laboratory, although these have not been analyzed cytologically. Hybrids are apparently rare in the field and have been reported only in the Dominican Republic (I. García de López 1978).

3. PTERIS Linnaeus, Sp. Pl. 2: 1073. 1753; Gen. Pl. ed 5, 484. 1754 · Brake [Greek *pteris*, fern, derived from *pteron*, wing or feather, for the closely spaced pinnae, which give the leaves a likeness to feathers]

Clifton E. Nauman

Plants terrestrial or on rock. **Stems** erect or creeping, branched; scales pale brown to black, concolored, elongate, margins entire. **Leaves** monomorphic, clustered or closely spaced, 1–20 dm. **Petiole** straw-colored, green, brownish red to purple black, longitudinally ridged, 2–3-grooved adaxially, scaly at base, glabrous or scaly distally, with 1 (less often 2 or more) vascular bundle. **Blade** oblong to lanceolate to deltate, 1–4-pinnate, herbaceous to leathery, abaxially and adaxially glabrous or sometimes pubescent or scaly, adaxially dull, not striate; rachis straight. **Ultimate segments** of blade sessile to short-stalked, linear to oblong-lanceolate, 1.5–8 mm wide; base truncate or narrowed to stalk, stalk when present green, not lustrous; margins plane or reflexed to form false indusia. **Veins** in leaves conspicuous, free (except in sori) and forking well above base of segment, or highly anastomosing. **False indusia** pale, scarious, covering sori. **Sporangia** intramarginal, sori usually continuous except at pinna or segment apex and sinuses, paraphyses present. **Spores** brown, trilete, tetrahedral, rugate and/or tuberculate, usually with prominent equatorial flange. $x = 29$.

Species ca. 300 (5 species and 1 hybrid in the flora): worldwide, warm and tropical regions.

SELECTED REFERENCES Kramer, K. U. 1990. *Pteris.* In: K. Kubitzki et al., eds. 1990 + . The Families and Genera of Vascular Plants. 1 + vol. Berlin etc. Vol. 1, pp. 250–252. Lakela, O. and R. W. Long. 1976. Ferns of Florida. Miami. Wagner, W. H. Jr. and C. E. Nauman. 1982. *Pteris* × *delchampsii*, a spontaneous fern hybrid from southern Florida. Amer. Fern J. 72: 97–102. Wunderlin, R. P. 1982. Guide to the Vascular Plants of Central Florida. Tampa.

1. Veins in leaves anastomosing except sometimes near margins of ultimate segments. . . . 1. *Pteris tripartita*
1. Veins in leaves entirely free.
  2. Leaves not strictly 1-pinnate, at least proximal pinnae pinnatifid-lobed or variously forked or divided.
    3. Pinnae of mature leaves decurrent to relatively broad-winged rachis in at least distal 1/2 of leaf. . . . . 4. *Pteris multifida*
    3. Pinnae of mature leaves not decurrent to relatively broad-winged rachis or only terminal pinna decurrent on rachis. . . . . 5. *Pteris cretica*
  2. Leaves strictly 1-pinnate, pinnae not lobed or divided.
    4. Petioles and often also rachises densely scaly, scales light to reddish, often grading into hairs on abaxial costae; pinnae appearing not articulate to rachis, apices long-attenuate or sharply acute; sori narrow, with most of abaxial blade surface exposed. . . . . 2. *Pteris vittata*
    4. Petioles often sparsely scaly or scaly only proximally, scales dark brown to nearly black, scales absent or few on rachises, abaxial costae with or without hairs; pinnae appearing articulate to rachis, apices acute; sori broad, little or no abaxial blade tissue exposed. . . . . 3. *Pteris bahamensis*

1. Pteris tripartita Swartz, J. Bot. (Schrader) 1800(2): 67. 1801 · Giant brake

Litobrochia tripartita (Swartz) C. Presl

Stems stout, short-creeping, densely and conspicuously scaly; scales pale brown. **Leaves** clustered, 1–2 m. **Petiole** straw-colored to brownish red, to more than 1 m, scaly proximally, otherwise glabrous at maturity. **Blade** deltate to pentagonal, pedate, ultimate divisions pinnately divided, 1–2 × 1–2 m; rachis not winged. **Pinnae** few, closely spaced, remaining green through winter, not decurrent on rachis, not articulate to rachis, oblong-lanceolate, 1–3-forked, to 7 × 6 dm; base asymmetrical, acute; apex acute; rachis and costae glabrate or with minute hairs, especially near axils of proximal pinnae; penultimate pinnules linear to linear-lanceolate, pinnatifid, separated, not remaining green through winter, not articulate to rachis. **Ultimate segments** of blade numerous, linear-oblong to linear-lanceolate, to 19 × 6 mm, margins entire or serrulate, apex obtuse and rounded to acute; terminal segments 3–4 cm longer and more tapering than lateral segments. **Veins** anastomosing near costae and costules, becoming forked and free near margins of ultimate segments. **Sori** narrow, blade tissue exposed abaxially.

Terrestrial in cypress, pond-apple, and other swamps or forested wet habitats, on constantly moist, circumneutral soils; 0–50 m; introduced, naturalized in scattered locations; Fla.; West Indies; Central America; South America; native to tropical Asia.

2. Pteris vittata Linnaeus, Sp. Pl. 2: 1074. 1753, not Schkuhr 1809 · Ladder brake, Chinese brake, Chinese ladder brake

Pycnodoria vittata (Linnaeus) Small

Stems stout, short-creeping, densely scaly; scales pale brown. **Leaves** clustered, 1–10 dm. **Petiole** green to pale brown, 1–30 cm, densely scaly; scales dense proximally, extending to and along rachis. **Blade** oblanceolate, 1-pinnate, (15–)25–50(–80) × (6–)13–25 cm; rachis not winged. **Pinnae** numerous, separated proximally, closely spaced to barely overlapping distally, not remaining green through winter, not decurrent on rachis, not articulate to rachis, linear-lanceolate to linear-attenuate, simple, 2–18 cm × 4–9 mm; base asymmetrically cordate to widened or truncate; margins serrulate, prominently so near apex; apex acuminate, attenuate, or acute; scales of rachis grading into uniseriate hairs on abaxial costae, or hairs absent on abaxial costae; proximal pinnae not divided or lobed. **Veins** free, forked. **Sori** narrow, blade tissue exposed abaxially. $2n = 116$.

Roadsides and other disturbed habitats; coastal plain; 0–50 m; introduced; Ala., Calif., D.C., Fla., Ga., La., Miss., S.C.; West Indies; South America; native to Asia.

Pteris vittata has escaped from cultivation. It is found on almost any calcareous substrate, such as old masonry, sidewalks, building crevices, and nearly every habitat in southern Florida with exposed limestone, notably pinelands. It is scattered throughout Florida and is sporadic, becoming less frequent to rare northward in the coastal plain.

Pteris vittata varies exceedingly in size, density of scales on the rachis, presence or absence of hairs on the abaxial costae, and overall color and aspect of the leaf. As a result, it may occasionally bear a resemblance to forms of *P.* × *delchampsii* W. H. Wagner & Nauman, the hybrid between *P. bahamensis* and *P. vittata*.

3. Pteris bahamensis (J. Agardh) Fée, Mém Foug. 5: 125. 1852 · Bahama ladder brake, plumy ladder brake, Bahama brake

Pteris diversifolia Swartz var. *bahamensis* J. Agardh, Recens. Spec. Pter., 6. 1839; *Pteris longifolia* Linnaeus var. *bahamensis* (J. Agardh) Hieronymus; *Pycnodoria bahamensis* (J. Agardh) Small; *Pycnodoria pinetorum* Small

Stems slender, short-creeping, sparsely scaly; scales dark brown to black. **Leaves** clustered, to ca. 1 m. **Petiole** green or straw-colored to purple-black proximally or medium brown with age, 10–25(–45) cm, glabrous or sparingly scaly at base, glabrous at maturity. **Blade** lanceolate, broadly linear or oblanceolate, 1-pinnate, 25–50(–60) × 3–16 cm; rachis not winged. **Pinnae** often numerous, well separated, mostly green over winter, not decurrent on rachis, articulate to rachis, narrowly linear, simple, 1.5–9 cm × 1.5–5 mm; base rounded or auriculate and widened but not cordate; margins obscurely dentate, often appearing entire; apex short-acute to obtuse; pinnae glabrous or rarely with a few scattered hairs abaxially on costa. **Veins** free, forked. **Sori** broad, little blade tissue exposed abaxially. $2n = 116$.

In crevices and pockets on oölitic limestone in rocky pinelands and infrequently on the edges of hammocks; 0–50 m; Fla.; West Indies in the Bahamas.

A form with dissected, deeply or completely 1–2-pinnate pinnae occurs throughout the range of *Pteris bahamensis* and is known in the flora from southern Florida.

Pteris bahamensis is often treated as a variety of *P. longifolia* Linnaeus, and some transition toward that species is evident. The primary differences are in the degree of rachis pubescence (denser in *P. longifolia*) and in pinna base shape (typically cordate in *P. longifolia*). The presence of transitional specimens and the quantitative nature of the differences suggest the taxa may be conspecific. Little is known, however, about the ranges and patterns of variation in both taxa. *Pteris bahamensis* is diploid and *P. longifolia* appears to be tetraploid. The two taxa are closely related, and further cytological and morphometric analyses will be needed before their relationships can be stated with confidence. *Pteris bahamensis* is maintained here at the species rank to emphasize the differences between the two taxa, though they are perhaps better treated as subspecies. Specimens identified as *P. longifolia* from the flora are *P. bahamensis*.

Pteris × *delchampsii* W. H. Wagner & Nauman is intermediate between *Pteris bahamensis* and *P. vittata*. Hybrid plants resemble a narrow, skeletonized form of *P. vittata* but have darker, shorter, and fewer stem scales, the petioles and rachises are less densely scaly, and pinnae are stiffer, farther apart, slender, and less ascending, with the margins less sharply dentate. The spores are largely misshapen. The chromosome number is $2n = 116$, with irregular pairing.

Pteris × *delchampsii* is terrestrial or on rock in disturbed calcareous habitats on limestone walls and ledges in Broward, Dade, and Monroe counties, Florida; it is also thought to occur in Collier County, Florida. Outside the flora it occurs in the West Indies in the Bahamas.

Plants of *Pteris* × *delchampsii* most often resemble one of the parent species, and this may confound identification. Hybrids can be distinguished by the high percentage of misshapen, collapsed, or empty spores and abortive sporangia.

4. **Pteris multifida** Poiret in Lamarck et al., Encycl. 5: 714. 1804 · Spider brake, spider fern, Chinese brake, Huguenot fern, saw-leaved bracken

Pycnodoria multifida (Poiret) Small

Stems slender, short-creeping, densely scaly; scales dark reddish brown to chestnut brown. **Leaves** clustered, 1–6 dm. **Petiole** pale or brownish, 5–30 cm, scaly proximally, otherwise glabrous. **Blade** oblong to oblanceolate, irregularly and pedately divided proximally (as in *Pteris cretica*) and pinnately divided distally, 10–35 × 13–25

cm; rachis slightly and evenly winged, wing constricted above each pinna pair. **Pinnae** 3–7 pairs, widely spaced, distal pinnae simple, adnate and decurrent to rachis; pinnae remaining green through winter, not articulate to rachis, lanceolate to linear; sterile pinnae wider than fertile pinnae (to ca. 1.2 cm), margins irregularly serrate to serrulate; fertile pinnae mostly less than 5 mm, margins entire to serrate at apex; adaxial costae with sparse, septate hairs; proximal pinnae with 1–4 elongate basal segments. **Veins** free, simple or forked. **Sori** narrow, blade tissue exposed abaxially. $2n = 116$.

Terrestrial or on rock in disturbed areas in circumneutral soils; primarily coastal plain; introduced; Ala., Ark., Fla., Ga., Ill., Ind., Ky., La., Md., Miss., N.Y., N.C., S.C., Tex.; West Indies; South America in Argentina, Brazil; native to e Asia.

Pteris multifida is found on old shady walls and masonry around cemeteries, dumps, and towns. It may no longer occur in Indiana. Juveniles of *Pteris multifida* may key to *Pteris cretica*.

5. **Pteris cretica** Linnaeus, Mant. Pl., 130. 1767 · Cretan brake

Pycnodoria cretica (Linnaeus) Small

Stems slender, creeping, sparingly scaly; scales dark brown to chestnut brown. **Leaves** clustered to closely spaced, to 1 m. **Petiole** straw-colored to light brown distally, darker proximally, 10–50 cm, base sparsely scaly. **Blade** irregularly ovate, primarily and irregularly pedately divided, 10–30 × 6–25 cm; rachis not winged; only terminal pinna decurrent on rachis. **Pinnae** 1–3 pairs, well separated, blade often 5-parted with terminal pinna and 2 lateral pairs of pinnae remaining green through winter, not articulate; sterile pinnae to 25 × 0.8–1.5 cm, serrulate; fertile pinnae narrower than sterile pinnae, to ca. 11 mm wide, spiny-serrate; base acute acroscopically and decurrent (sometimes narrowly and barely so) basiscopically, glabrous; proximal pinnae with 1 (rarely 2) basiscopic lobes. **Veins** free, simple or forked. **Sori** narrow, blade tissue exposed abaxially.

Varieties 2 (2 in the flora): widely scattered in tropical and subtropical regions worldwide.

Pteris cretica is almost pantropical in distribution (C. V. Morton 1957). Because this species is so commonly and widely cultivated and appears to escape easily in warmer regions, its native range is uncertain.

Young leaves of young plants of *Pteris multifida* may key to *P. cretica* because only the terminal pinnae may be decurrent on the rachis as in *P. cretica*. Juveniles of *P. multifida* can be separated by proximal pinnae with long-attenuate apices and thinner-textured leaves than *P. cretica*. Juveniles of *P. cretica* have proximal pinnae with acute to blunt or nearly rounded apices and thicker-textured leaves.

1. Pinnae green throughout. . . . .
. . . . 5a. *Pteris cretica* var. *cretica*
1. Pinnae with broad, white, central stripe. . . . .
. . . . 5b. *Pteris cretica* var. *albolineata*

5a. Pteris cretica Linnaeus var. **cretica**

Pinnae without white or pale green, longitudinal streak along middle; sterile pinnae less than 10 mm wide, fertile segments less than 8 mm wide; terminal pinna somewhat to barely decurrent on rachis.

Terrestrial or on rock in rocky hammocks, limesinks, grottoes, and river bluffs; introduced; Fla., La.; widely scattered worldwide in tropical and subtropical regions.

5b. Pteris cretica Linnaeus var. **albolineata** Hooker, Bot. Mag., plate 5194. 1860 · White-lined cretan brake

Pinnae with white or pale green, longitudinal streak along middle; sterile pinnae to ca. 25 mm wide, fertile pinnae to ca. 15 mm wide; terminal pinna usually not decurrent on rachis.

Terrestrial or on rock on wooded slopes or shaded limestone ledges and sink margins in circumneutral soil; introduced; Fla.; natural range uncertain (see discussion of species).

Two taxa are considered doubtful species and are therefore excluded. *Pteris ensiformis* Burman f. cv. *victoriae* Baker was reported for peninsular Florida by E. T. Wherry (1964), and as far as I can determine, this is the only report of the species for the flora. The source for the record is uncertain. *Pteris grandifolia* Linnaeus was reported from Dade County, Florida, by T. Darling Jr. (1961), the original find probably occurring in 1952 (E. T. Wherry 1964) and representing either an escape from cultivation or plants merely persistent from cultivation. Recent searches throughout the original location have failed to turn up extant populations; probably the plants have disappeared.

4. PITYROGRAMMA Link, Handbuch 3: 19. 1833 · Silverback fern, goldback fern [Greek *pityros,* bran, and *gramma,* lines (as in written characters), referring to the farina covering the abaxial leaf blade surface]

George Yatskievych

Michael D. Windham

Plants terrestrial [or on rock]. **Stems** erect or nearly so, unbranched; scales brown, concolored or nearly so, lanceolate, margins entire. **Leaves** monomorphic, closely spaced, 25–150 cm. **Petiole** black to purplish black or reddish brown, longitudinally grooved adaxially, glabrous, with 2(–3) vascular bundles. **Blade** linear-lanceolate to ovate or elongate-triangular, 1–4-pinnate, herbaceous to leathery, abaxially usually farinose, farina white or yellow, sometimes partially or completely replaced by trichomes, adaxially glabrous, dull, not striate; rachis straight. **Ultimate segments** stalked or sessile, free from costa or partially adnate to it, narrowly triangular to linear, entire or lobed; base narrowly cuneate, stalks when present green, not lustrous; margins not recurved to form false indusia. **Veins** of ultimate segments free, obscure, pinnately branched and divergent distally. **False indusia** absent. **Sporangia** scattered along veins, containing 32 or 64 spores, intermixed with farina-producing glands. **Spores** tan with dark brown ridges, tetrahedral-globose, perispore usually reticulate, with equatorial flange (except in *Pityrogramma trifoliata*). **Gametophytes** glabrous.

Species ca. 15 (2 in the flora): primarily neotropical, some in Africa, introduced elsewhere in Eastern Hemisphere.

The name *Nesoris bicolor* Rafinesque was based on a yellow-farinose plant from Florida and presumably represents a species of *Pityrogramma*. A type has not been located, and C. S.

Rafinesque's (1836[–1838]b, part 4) description is not precise enough to permit equation of this name with modern nomenclature. No yellow-farinose *Pityrogramma* has since been reported from Florida.

Pentagramma has been segregated from *Pityrogramma* (G. Yatskievych et al. 1990) and comprises what was the *Pityrogramma triangularis* (Kaulfuss) Maxon complex.

SELECTED REFERENCE   Tryon, R. M. 1962. Taxonomic fern notes. II. *Pityrogramma* (including *Trismeria*) and *Anogramma*. Contr. Gray Herb. 189: 52–76.

1. Distal pinnae pinnately lobed or divided, narrowly triangular. . . . .
    . . . . 1a. *Pityrogramma calomelanos* var. *calomelanos*
1. Distal pinnae entire or serrulate, linear to narrowly lanceolate. . . . . 2. *Pityrogramma trifoliata*

1. Pityrogramma calomelanos (Linnaeus) Link,
   Handbuch 3. 1833 · Silverback fern
Acrostichum calomelanos Linnaeus, Sp. Pl. 2: 1072. 1753
Varieties 3 (1 in the flora): primarily Neotropics, Africa.

1a. Pityrogramma calomelanos (Linnaeus) Link var.
   calomelanos

Leaves 25–120 cm. **Petiole** black, dull, glabrous or sparsely scaly proximally. **Blade** elongate-triangular to ovate-lanceolate, 2-pinnate-pinnatifid, sometimes 3-pinnate proximally, densely white-farinose abaxially. **Pinnae** narrowly triangular, pinnately lobed or divided, flat surfaces in plane of blade. **Pinnules** short-stalked or sessile, lanceolate, margins serrate to lobed near base. **Spores** with perispore reticulate and with equatorial flange.

Ditches in forested areas; 0–20 m; introduced; Fla.; Mexico; West Indies in the Antilles; Central America; South America; widely naturalized in tropical Africa.

Pityrogramma calomelanos var. *calomelanos* is commonly cultivated in greenhouses. In Florida, it is known from a small number of sites in Hillsborough and Polk counties, where it possibly escaped from cultivation. *Pityrogramma calomelanos* var. *austroamericana* (Domin) Farwell (Costa Rica to Brazil) is characterized by yellow farina; *P. calomelanos* var. *ochracea* (C. Presl) R. M. Tryon (Honduras to Bolivia) possesses scattered trichomes instead of farinose indument.

2. Pityrogramma trifoliata (Linnaeus) R. M. Tryon,
   Contr. Gray Herb. 189: 68. 1962 · Goldenrod fern

Acrostichum trifoliatum Linnaeus, Sp. Pl. 2: 1070. 1753; *Trismeria trifoliata* (Linnaeus) Diels

Leaves 50–150 cm, rarely longer. **Petiole** reddish brown, shiny, scaly. **Blade** linear-lanceolate, 1-pinnate apically to 2-pinnate proximally; fertile pinnae white-farinose abaxially. **Pinnae** palmate to pinnate, 2–3 (–7)-foliolate proximally, with flat surfaces parallel to ground, undivided and linear apically. **Pinnules** short-stalked, linear, occasionally lobed proximally, margins serrulate. **Spores** with perispore sparsely granular, lacking equatorial flange. $2n = 116$.

Roadside ditches and canal banks; 0–20 m; Fla.; Mexico; West Indies in Greater Antilles; Central America; South America.

The range of *Pityrogramma trifoliata* is apparently actively expanding in Florida, perhaps in response to increasing disturbance of wetland habitats. A good case can be made for segregation of this species into the monotypic *Trismeria* Fée, based on differences in dissection of leaves and spore ornamentation noted in the descriptions of this and the previous species. Because naturally occurring hybrids with four other *Pityrogramma* taxa are known to exist in the tropics, *P. trifoliata* is retained in *Pityrogramma* for the present, following R. M. Tryon (1962), who viewed *P. trifoliata* as a specialized offshoot of the core group of species in the genus. Elsewhere, plants are known with yellow farina on the fertile leaves, but these are nowhere common.

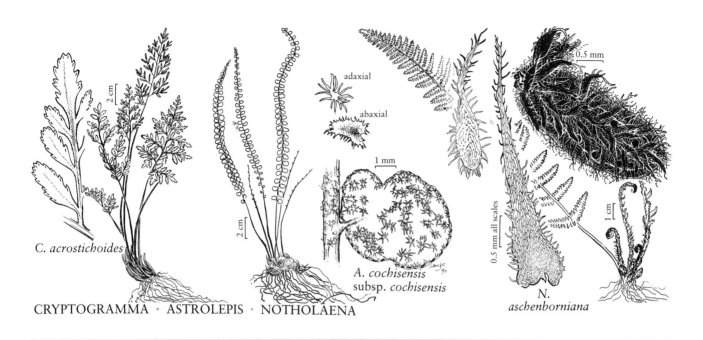

C. acrostichoides

adaxial

abaxial

1 mm

A. cochisensis
subsp. cochisensis

0.5 mm all scales

0.5 mm

N.
aschenborniana

CRYPTOGRAMMA · ASTROLEPIS · NOTHOLAENA

5. CRYPTOGRAMMA R. Brown in Franklin, Narr. Journey Polar Sea, 767. 1823

· Parsley fern, rock-brake, cliff-brake, cryptogramme [Greek *cryptos*, hidden, and *gramme*, line, referring to the ± marginal soral bands hidden by revolute margins]

Edward R. Alverson

Plants on rock. **Stems** decumbent to erect, or creeping, few to many branched; scales colorless or brownish, concolored or bicolored, ovate, lanceolate or linear, margins entire. **Leaves** dimorphic, scattered or densely tufted; fertile leaves 5–25 cm; sterile leaves 3–20 cm, shorter than fertile leaves. **Petiole** dark brown proximally, light brown to green distally, with single longitudinal groove adaxially, scaly, with single vascular bundle. **Blade** deltate, lanceolate to elliptic, 2–4-pinnate, somewhat leathery or herbaceous to membranaceous, abaxially glabrous, adaxially glabrous or sparsely pubescent, dull to somewhat lustrous, not striate; rachis straight. **Ultimate segments** of blade short-stalked or sessile, free or adnate to midrib; stalks dull, green; segments of sterile leaves ovate, elliptic, obovate, or fan-shaped, usually less than 4 mm wide, margins plane, dentate or shallowly to deeply cut; fertile segments strongly differentiated from sterile segments, lanceolate to linear, usually less than 2 mm wide, margins reflexed to form false indusia extending over entire length of segments, at first covering young sporangia, often becoming plane at maturity. **Veins** of ultimate segments usually obscure, free, pinnately branched and divergent distally. **False indusia** greenish to brown, broad, clearly marginal, usually concealing sporangia. **Sporangia** scattered along veins on abaxial leaf surface, often intermixed with farina-producing glands, containing 64 spores. **Spores** yellow, tetrahedral, trilete, verrucose, equatorial flange absent. **Gametophytes** glabrous. $x = 30$.

Species 8–11 (4 in the flora): temperate regions, North America, South America, Eurasia.

SELECTED REFERENCE Alverson, E. R. 1989. *Cryptogramma cascadensis*, a new parsley fern from western North America. Amer. Fern J. 79: 95–102.

1. Stems 1–1.5 mm diam., creeping, few-branched; leaves scattered along stems, delicate and ephemeral; petioles mostly dark brown in proximal 1/2–2/3, greenish distally; often in calcareous habits. . . . . 1. *Cryptogramma stelleri*
1. Stems 4–20 mm diam., decumbent to erect, many branched from base; leaves strongly tufted, herbaceous to somewhat leathery, present throughout growing season; petioles dark brown in proximal 1/8 or less; mostly of noncalcareous habitats.
  2. Blades herbaceous, thin, translucent when dried, shed in autumn, not persistent; hydathodes on dried leaves superficial; hairs absent from adaxial leaf surface. . . 2. *Cryptogramma cascadensis*
  2. Blades somewhat leathery, opaque when dried, green and persistent over winter; hydathodes on dried leaves sunken below surface; hairs small, appressed, cylindric, present in grooves of petiole, costae, and costules of adaxial leaf surface.
    3. Segments oblong to ovate-lanceolate, with 6–12 or more teeth or shallow lobes; sterile leaves 2–3-pinnate. . . . . 3. *Cryptogramma acrostichoides*
    3. Segments obovate, with 2–6 deep lobes; at least some sterile leaves 3–4-pinnate. . . . . 4. *Cryptogramma sitchensis*

1. Cryptogramma stelleri (S. G. Gmelin) Prantl in Engler, Bot. Jahrb. Syst. 3: 413. 1882 · Slender rock-brake or cliff-brake, Steller's rock-brake, fragile rock-brake, cryptogramme de Steller

Pteris stelleri S. G. Gmelin, Novi Comment. Acad. Sci. Imp. Petrop. 12: 519. 1768

Stems creeping, few branched, slender, 1–1.5 mm diam., succulent, brittle; scales colorless, sparse, transparent-reticulate, ovate, 0.4 × 0.3 mm; stems shriveling in 2d year following emergence of leaves. **Leaves** scattered along stems, ephemeral (dying by late summer), soon shed; sterile leaves erect, 3–15 cm; fertile leaves erect, 5–20 cm; petioles, costae, and costules glabrous. **Petiole** dark brown in proximal 1/2 or less, becoming greenish distally, ca. 1 mm wide when dry, only slightly furrowed, glabrous. **Blade** broadly lanceolate to ovate-lanceolate, all pinnate-pinnatifid to 2-pinnate, herbaceous to membranous, thin; hydathodes superficial, often poorly developed or absent. **Segments** of sterile leaves ovate-lanceolate to fan-shaped, distal 1/2–1/3 shallowly lobed; segments of fertile leaves horizontal to ascending, often only partially differentiated from sterile leaves, lanceolate to linear, 8–25 × 2–4 mm; margins reflexed, forming continuous false indusia. **Sporangia** often in discrete sori. $2n = 60$.

New growth produced in spring, dying by late summer. Sheltered calcareous cliff crevices and rock ledges, typically in coniferous forest or other boreal habitats; 0–3000 m; Alta., B.C., N.B., Nfld., N.W.T., N.S., Ont., P.E.I., Que., Yukon; Alaska, Colo., Conn., Ill., Iowa, Maine, Mass., Mich., Minn., Mont., Nev., N.H., N.J., N.Y., Oreg., Pa., Utah, Vt., Wash., W.Va., Wis., Wyo.; Europe in ne former Soviet republics; Asia.

2. Cryptogramma cascadensis E. R. Alverson, Amer. Fern J. 79: 95. 1989 · Cascade parsley fern

Stems decumbent to erect, much branched from base, stout, 4–8 mm diam. (including hardened, persistent leaf bases); scales often bicolored, dense, broadly lanceolate to linear, to 6 × 2 mm. **Leaves** strongly tufted, deciduous; sterile leaves spreading, 3–20 cm; fertile leaves erect, 5–25 cm; petioles, costae, and costules glabrous. **Petiole** green to straw-colored, dark brown only on proximal 1/8 or less, ca. 1 mm wide when dry, collapsing and strongly furrowed; scales bicolored or ± concolored, becoming sparse distally. **Blade** deltate to ovate-lanceolate, all 2–3-pinnate, herbaceous, thin and translucent when dried, hydathodes superficial. **Segments** of sterile leaves oblong to fan-shaped, bases cuneate, distal 1/2–1/3 of each segment regularly dentate and often more deeply incised every 2d and 4th tooth; segments of fertile leaves ascending to erect, strongly differentiated from those of sterile leaves, linear, 3–12 × 1–2 mm; fertile segments revolute, covering sporangia. **Sporangia** in sori that coalesce at maturity. $2n = 60$.

New growth produced in spring, spores maturing in late summer and autumn, leaves dying in autumn. Talus slopes and cliff crevices, often on igneous rocks, typically in relatively mesic subalpine habitats; 900–3500 m; B.C.; Calif., Idaho, Mont., Oreg., Wash.

Populations of *Cryptogramma cascadensis* were previously identified as *C. acrostichoides*.

3. Cryptogramma acrostichoides R. Brown in Franklin, Narr. Journey Polar Sea, 754, 767. 1823 · American parsley fern

Cryptogramma crispa (Linnaeus) R. Brown ex Hooker subsp. *acrostichoides* (R. Brown) Hultén; *C. crispa* var. *acrostichoides* (R. Brown) C. B. Clarke

Stems decumbent to erect, much branched from base, stout, 10–20 mm diam. (including hardened, persistent leaf bases); scales bicolored, dense, broadly lanceolate to linear, to 6 × 2 mm. **Leaves** densely tufted, green over winter, persistent after withering; sterile leaves spreading, 3–17 cm; fertile leaves erect, 5–25 cm; hairs small, appressed, cylindric, scattered along grooves of petiole and along costae and costules of adaxial blade surface. **Petiole** green to straw-colored, dark brown on proximal 1/8 or less, 1–2 mm wide, firm and strawlike, not collapsed; scales bicolored or ± concolored, becoming sparse distally. **Blade** deltate to ovate-lanceolate, all 2–3-pinnate, somewhat leathery, opaque; hydathodes sunken below leaf surface. **Segments** of sterile leaves oblong to ovate-lanceolate, bases cuneate, distal 2/3–1/2 of segments crenate to dentate, often somewhat more deeply incised every 2d tooth; segments of fertile leaves horizontal to ascending, strongly differentiated from those of sterile leaves, linear, 3–12 × 1–2 mm; margins of fertile segments revolute, covering sporangia. **Sporangia** in sori that coalesce at maturity. $2n = 60$.

New growth produced in spring, spores maturing in summer, sterile leaves green over winter, senescing 2d spring. Noncalcareous cliff crevices, rock outcrops, and talus, often in relatively dry habitats, typically montane but occurring in lowland to alpine habitats; 0–3700 m; Alta., B.C., Man., N.W.T., Ont., Sask., Yukon; Alaska, Ariz., Calif., Colo., Idaho, Mich., Minn., Mont., Nev., N.Mex., Oreg., Utah, Wash., Wyo.; reported from Mexico in Baja California; Asia.

Cryptogramma acrostichoides has often been treated as a variety or subspecies of the strictly European *Cryptogramma crispa* (Linneaus) R. Brown, which has a chromosome number of $2n = 120$.

4. Cryptogramma sitchensis (Ruprecht) T. Moore, Index Fil., 67. 1857 · Alaska parsley fern

Allosorus sitchensis Ruprecht, Distr. Crypt. Vasc. Ross. 3: 47. 1845; *Cryptogramma acrostichoides* R. Brown var. *sitchensis* (Ruprecht) C. Christensen; *C. crispa* (Linnaeus) R. Brown ex Hooker var. *sitchensis* (Ruprecht) C. Christensen

Stems decumbent to erect, much branched from base, stout, 10–20 mm diam. (including hardened, persistent leaf bases); scales bicolored, dense, broadly lanceolate to linear, to 7 × 2 mm. **Leaves** densely tufted, green over winter, persistent; fertile leaves erect, 5–25 cm; sterile leaves spreading, 3–17 cm; hairs small, appressed, cylindric, scattered along grooves of petioles and along costae and costules of adaxial blade surface. **Petiole** dark brown only on proximal 1/8 or less, green to straw-colored distally, 1–2 mm wide, firm and strawlike, not collapsed; scales bicolored or ± concolored, becoming sparse distally. **Blade** deltate to ovate-lanceolate, somewhat leathery, opaque; sterile blades dimorphic, 2–3-pinnate or 3–4-pinnate, hydathodes only slightly sunken below leaf surface. **Segments** of less dissected sterile leaves ovate-lanceolate, regularly dentate to incised with 8–16 teeth or lobes; segments of more finely dissected sterile leaves pinnatifid with 4–8 small, obovate lobes, lobe apices acute; segments of fertile leaves ascending, strongly differentiated, linear, 3–10 × 1–3 mm; margins of fertile segments revolute, covering sporangia. **Sporangia** in sori that coalesce at maturity.

New growth produced in spring, spores maturing in late summer; sterile leaves green over winter, senescing 2d spring. Cliff crevices and talus slopes, lowland to alpine; 0–1800 m; B.C., N.W.T., Yukon; Alaska.

Cryptogramma sitchensis is an allotetraploid species ($2n = 120$; E. Alverson, unpubl. data) that arose through hybridization between *C. acrostichoides* and another species, possibly the eastern Asian *C. raddeana* Fomin. Past difficulties in clearly distinguishing *C. sitchensis* from *C. acrostichoides* can be attributed to the frequent occurrence of sterile hybrids where the ranges of the two species overlap.

6. ASTROLEPIS D. M. Benham & Windham, Amer. Fern J. 82: 55. 1992

· Star-scaled cloak ferns [Greek *astro,* star, and *lepis,* scale, in reference to the starlike scales on the adaxial blade surface]

Dale M. Benham

Michael D. Windham

Plants usually on rock. **Stems** compact to short-creeping, erect to ascending, sparingly branched; scales tan to chestnut brown, concolored to weakly bicolored, linear-attenuate, margins ciliate-dentate to entire. **Leaves** monomorphic, densely clustered, 7–130 cm. **Petiole** dull chestnut brown or straw-colored, rounded adaxially, sparsely to densely covered with scales, with 2 vascular bundles. **Blade** linear to linear-oblong, 1-pinnate to pinnate-pinnatifid, leathery, abaxially covered with overlapping, lanceolate to ovate, ciliate scales with underlying layer of stellate scales, adaxially sparsely to densely covered with stellate or coarsely ciliate scales, often glabrescent when mature, dull, not striate; rachis straight. **Ultimate segments** (pinnae) stalked to subsessile, free from axis, ovate, oblong or elongate-deltate, cordate to subcordate or rarely truncate at base, usually more than 4 mm wide; segment margins plane, undifferentiated, not recurved to form false indusia. **Veins** of ultimate segments obscure, pinnately branched and divergent distally. **False indusium** absent. **Sporangia** scattered along veins near pinna margins (often clustered near notches between pinna lobes), containing 32 or 64 spores, not intermixed with farina-producing glands. **Spores** light to dark brown, tetrahedral-globose, rugose, lacking prominent equatorial ridge. **Gametophytes** glabrous. $x = 29$.

Species ca. 8 (4 in the flora): North America, Mexico, West Indies, Central America, South America.

The species of *Astrolepis* have traditionally been assigned to either *Notholaena* (R. M. Tryon 1956) or *Cheilanthes* (J. T. Mickel 1979b; R. M. Tryon and A. F. Tryon 1982). Recent biosystematic analyses by D. M. Benham and M. D. Windham (1992) indicate, however, that the star-scaled cloak ferns form a distinctive, monophyletic group worthy of generic recognition. The combination of a chromosome base number of $x = 29$, pinnate leaves, two vascular bundles in the petioles, unique stellate or coarsely ciliate scales on the adaxial blade surface, and other characteristics separate *Astrolepis* from related genera.

SELECTED REFERENCES   Benham, D. M. 1989. A Biosystematic Revision of the Fern Genus *Astrolepis* (Adiantaceae). Ph.D. dissertation. Northern Arizona University. Benham, D. M. 1992. Additional combinations in *Astrolepis*. Amer. Fern J. 82: 59–62. Benham, D. M. and M. D. Windham. 1992. Generic affinities of the star-scaled cloak ferns. Amer. Fern J. 82: 47–58. Hevly, R. H. 1965. Studies of the sinuous cloak fern (*Notholaena sinuata*) complex. J. Arizona Acad. Sci. 3: 205–208. Tryon, R. M. 1956. A revision of the American species of *Notholaena*. Contr. Gray Herb. 179: 1–106.

1. Largest pinnae usually 4–7 mm; most adaxial scales circular to elliptic, peltate; abaxial scales ovate, usually 0.5–1 mm. . . . . 1. *Astrolepis cochisensis*
1. Largest pinnae usually 7–35 mm; most adaxial scales elongate, attached at their base; abaxial scales lanceolate, 1–1.5 mm.
    2. Adaxial scales dense, usually persistent; largest pinnae asymmetrically lobed or entire; body of adaxial pinna scales 5–7 cells wide. . . . . 2. *Astrolepis integerrima*
    2. Adaxial scales sparse, often deciduous; largest pinnae usually symmetrically lobed; body of adaxial pinna scales 1–4 cells wide.
        3. Adaxial pinna surface sparsely scaly, at least some scales persistent; body of adaxial scales 2–4 cells wide; abaxial scales ciliate with coarse marginal projections; pinnae shallowly lobed, lobes usually broadly rounded. . . . . 3. *Astrolepis windhamii*

3. Adaxial pinna surface sparsely scaly to glabrescent, most scales deciduous with age; body of adaxial scales 1–2 cells wide; abaxial scales ciliate-dentate with delicate marginal projections; pinnae usually deeply lobed, lobes often acute. . . . . . . 4. *Astrolepis sinuata*

1. Astrolepis cochisensis (Goodding) D. M. Benham & Windham, Amer. Fern J. 82: 57. 1992

Notholaena cochisensis Goodding, Muhlenbergia 8: 93. 1912; *Cheilanthes cochisensis* (Goodding) Mickel; *C. sinuata* (Lagasca ex Swartz) Domin var. *cochisensis* (Goodding) Munz; *Notholaena sinuata* (Lagasca ex Swartz) Kaulfuss var. *cochisensis* (Goodding) Weatherby

Stems compact; stem scales uniformly tan or somewhat darker near base, to 10 mm, margins ciliate-dentate to entire. **Leaves** 7–40 cm. **Blade** 1-pinnate to pinnate-pinnatifid, pinna pairs 20–50. **Pinnae** oblong, largest usually 4–7 mm, entire or asymmetrically lobed, lobes 1–4, broadly rounded, separated by shallow sinuses; abaxial scales completely concealing surface, ovate, usually 0.5–1 mm, ciliate with coarse marginal projections; adaxial scales sparse, deciduous, stellate to coarsely ciliate, mostly circular to elliptic, peltate, body more than 5 cells wide. **Sporangia** containing 32 or 64 spores.

Subspecies 3 (3 in the flora): North America, Mexico.

Astrolepis cochisensis is reported to be toxic to sheep (F. P. Mathews 1945). Three cytotypes that occupy different ranges and/or habitats have been identified and are treated here as subspecies. These include a sexual diploid (subsp. *chihuahuensis*) found on calcareous substrates in the Chihuahuan Desert; an apogamous triploid (subsp. *cochisensis*), which inhabits primarily calcareous substrates in the Sonoran, Mojavean, and western Chihuahuan deserts; and an apogamous tetraploid (subsp. *arizonica*), which occupies primarily noncalcareous substrates in southern Arizona. Isozyme analyses suggest that subsp. *cochisensis* is an autotriploid derivative of the diploid subsp. *chihuahuensis* (D. M. Benham 1989). Both the isozymes and substrate preferences of subsp. *arizonica* indicate, however, that it is not a simple autotetraploid and that other taxa remain to be discovered within the *Astrolepis cochisensis* complex.

1. Sporangia containing 64 spores; spores averaging 39–46 μm diam.; plants of calcareous substrates in Texas, s New Mexico. . . . .
. . . . . 1a. *Astrolepis cochisensis* subsp. *chihuahuensis*
1. Sporangia containing 32 spores; spores averaging 59–86 μm diam.; plants of calcareous and noncalcareous substrates in California, Arizona, Nevada, w New Mexico, extreme w Texas, Oklahoma, Utah, n Mexico.

2. Spores averaging 59–70 μm diam.; plants primarily of calcareous substrates throughout Southwest. . . . .
. . . . 1b. *Astrolepis cochisensis* subsp. *cochisensis*
2. Spores averaging 72–86 μm diam.; plants primarily of noncalcareous substrates in s Arizona. . . . .
. . . . . 1c. *Astrolepis cochisensis* subsp. *arizonica*

1a. Astrolepis cochisensis (Goodding) D. M. Benham & Windham subsp. **chihuahuensis** D. M. Benham, Amer. Fern. J. 82: 59. 1992

Sporangia containing 64 spores; spores averaging 39–46 μm diam. $2n = 58$.

Sporulating summer–fall. Rocky calcareous slopes and cliffs; mostly on limestone; 1200–2600 m; N.Mex., Tex.; ne Mexico.

1b. Astrolepis cochisensis (Goodding) D. M. Benham & Windham subsp. **cochisensis**

Sporangia containing 32 spores; spores averaging 59–70 μm diam. $n = 2n = 87$, apogamous.

Sporulating summer–fall. Rocky slopes and cliffs; favoring limestone and other calcareous substrates; 400–2100 m; Ariz., Calif., N.Mex., Okla., Tex.; n Mexico.

1c. Astrolepis cochisensis (Goodding) D. M. Benham & Windham subsp. **arizonica** D. M. Benham, Amer. Fern J. 82: 60. 1992

Sporangia containing 32 spores; spores averaging 72–86 μm diam. $n = 2n = 116$, apogamous.

Sporulating summer–fall. Rocky slopes and cliffs; favoring granite, quartzite, and other noncalcareous substrates, occasionally found on limestone; 600–1200 m; Ariz.

2. Astrolepis integerrima (Hooker) D. M. Benham & Windham, Amer. Fern J. 82: 57. 1992

Notholaena sinuata (Lagasca ex Swartz) Kaulfuss var. *integerrima* Hooker, Sp. Fil. 5: 108. 1864; *Cheilanthes integerrima* (Hooker) Mickel; *Notholaena integerrima* (Hooker) Hevly

Stems compact; stem scales uniformly tan or somewhat darker near base, to 15 mm, margins ciliate-dentate to entire. **Leaves** 8–45 cm. **Blade** 1-pinnate to pinnate-pinnatifid, pinna pairs 20–45. **Pinnae** oblong to ovate, largest usually 7–15 mm, entire or asymmetrically lobed, lobes 2–7, broadly rounded, separated by shallow sinuses; abaxial scales concealing surface, lanceolate, usually 1–1.5 mm, ciliate with coarse marginal projections; adaxial scales abundant, mostly persistent, stellate to coarsely ciliate, elongate, attached at base, body mostly 5–7 cells wide. **Sporangia** containing 32 spores. $n = 2n = 87$, apogamous.

Sporulating summer–fall. Rocky hillsides and cliffs; usually on limestone or other calcareous substrates; 500–1800 m; Ariz., Nev., N.Mex., Okla., Tex.; n,c Mexico.

R. H. Hevly (1965) hypothesized that *Astrolepis integerrima* was produced by hybridization between *A. cochisensis* and *A. sinuata*. Recent isozyme analyses (D. M. Benham 1989) indicate, however, that *Astrolepis integerrima* is an apogamous allotriploid hybrid between *A. cochisensis* and an unnamed Mexican taxon related to *A. crassifolia* (Houlston & T. Moore) D. M. Benham & Windham. Two morphologic forms exist in this taxon: one with essentially entire pinnae, and one (more common in the United States) with larger, asymmetrically lobed pinnae. The former might be confused with *A. cochisensis* on occasion, but the abundance of adaxial scales and the larger pinnae of *A. integerrima* should serve to distinguish these species. The lobed form of *A. integerrima* is superficially similar to *A. windhamii*, from which it is distinguished by the abundance and greater width of adaxial scales and the asymmetrical lobing of the pinnae.

3. Astrolepis windhamii D. M. Benham, Amer. Fern J. 82: 60. 1992

Stems compact to short-creeping; stem scales uniformly tan or somewhat darker near base, to 15 mm, margins ciliate-dentate to entire. **Leaves** 10–50 cm. **Blade** pinnate-pinnatifid, pinna pairs 20–45. **Pinnae** ovate to deltate, largest 7–15 mm, usually symmetrically lobed, lobes 6–11,

broadly rounded, separated by shallow sinuses; abaxial scales concealing surface, lanceolate, usually 1–1.5 mm, ciliate with coarse marginal projections; adaxial scales sparse, mostly persistent, elongate, usually stellate, attached at base, body 2–4 cells wide. **Sporangia** containing 32 spores. $n = 2n = 87$, apogamous.

Sporulating summer–fall. Rocky hillsides and cliffs; occurring on calcareous and noncalcareous substrates; 1200–2100 m; Ariz., N.Mex., Tex.; n Mexico.

Recent isozyme analyses (D. M. Benham 1989) indicate that *Astrolepis windhamii* is an apogamous allotriploid that contains three different genomes, one each from *A. sinuata, A. cochisensis,* and an unnamed Mexican taxon related to *A. crassifolia.* Because of this genomic constitution, *Astrolepis windhamii* tends to bridge the morphologic gap between *A. sinuata* and *A. integerrima,* which is itself a hybrid between *A. cochisensis* and the unnamed Mexican species. Although the features that separate these taxa are subtle, the pinna lobing and scale characteristics of *A. windhamii* mentioned in the key adequately distinguish them in most cases.

4. Astrolepis sinuata (Lagasca ex Swartz) D. M. Benham & Windham, Amer. Fern J. 82: 56. 1992

Acrostichum sinuatum Lagasca ex Swartz, Syn. Fil., 14. 1806; *Cheilanthes sinuata* (Lagasca ex Swartz) Domin; *Notholaena sinuata* (Lagasca ex Swartz) Kaulfuss

Stems compact to short-creeping; stem scales uniformly chestnut brown or with lighter margin, to 6 mm, margins ciliate-dentate. **Leaves** 11–130 cm. **Blade** pinnate-pinnatifid, pinna pairs 30–60. **Pinnae** deltate to ovate, largest 7–35 mm, symmetrically lobed, lobes 6–14 often acute, separated by deep sinuses; abaxial scales concealing surface, lanceolate, usually 1–1.5 mm, ciliate-dentate with delicate marginal projections; adaxial scales sparse, deciduous, elongate, stellate, body 1–2 cells wide, attached at base. **Sporangia** containing 32 or 64 spores.

Subspecies 2 (2 in the flora): North America, Mexico, West Indies, Central America, South America.

Astrolepis sinuata comprises two cytotypes that tend to occupy different ranges and are treated here as subspecies. Sexual diploid populations (subsp. *mexicana*) are widely distributed in Mexico, but in the flora they are apparently confined to the Davis and Chisos mountains of Texas and to southeast New Mexico. The range of the apogamous triploid (subsp. *sinuata*) extends from Argentina to the southwestern United States, with a disjunct population in Georgia. Isozyme studies suggest that subsp. *sinuata* was derived from the diploid subsp. *mexicana* through autopolyploidy (D. M. Benham 1989).

1. Sporangia containing 32 spores; spores averaging 50–65 μm diam.; plants widespread in sw United States and disjunct in Georgia. . . . . .
. . . . 4a. *Astrolepis sinuata* subsp. *sinuata*

1. Sporangia containing 64 spores; spores averaging 37–44 μm diam.; plants apparently restricted to w Texas and se New Mexico. . . . . 4b. *Astrolepis sinuata* subsp. *mexicana*

4a. Astrolepis sinuata (Lagasca ex Swartz) D. M. Benham & Windham subsp. **sinuata**

Sporangia containing 32 spores; spores averaging 50–65 μm diam. $n = 2n = 87$, apogamous.

Sporulating summer–fall. Rocky slopes and cliffs; occurring on calcareous and noncalcareous substrates; 800–2100 m; Ariz., Ga., N.Mex., Tex.; Mexico; West Indies; Central America; South America.

4b. Astrolepis sinuata (Lagasca ex Swartz) D. M. Benham & Windham subsp. **mexicana** D. M. Benham, Amer. Fern J. 82: 59. 1992

Sporangia containing 64 spores; spores averaging 37–44 μm diam. $2n = 58$.

Sporulating summer–fall. Rocky slopes and cliffs; occurring on calcareous and noncalcareous substrates; 1300–1800 m; N.Mex., Tex.; Mexico; Central America.

7. NOTHOLAENA R. Brown, Prodr., 145. 1810 · Cloak fern [Greek *notho,* false, and *chlaena,* coat, in reference to the reflexed leaf segment margins that form false indusia]

Michael D. Windham

Plants usually on rock. **Stems** short-creeping to compact, ascending to horizontal, usually branched; scales black or often bicolored with dark central stripe and lighter margins, linear-subulate to lanceolate; margins ciliate, denticulate, or entire. **Leaves** monomorphic, clustered, 4–35 cm. **Petiole** brown or black, rounded, flattened, or with single longitudinal groove adaxially, often bearing scales, hairs, or farinose glands, with single vascular bundle. **Blade** linear-lanceolate, ovate, deltate, or pentagonal, pinnate-pinnatifid to 4-pinnate, leathery, abaxially covered by yellowish or whitish farina (completely obscured by stellate scales in *Notholaena aschenborniana*), adaxially often sparsely glandular, dull, not striate; rachis straight. **Ultimate segments** of blade sessile to subsessile, often adnate to costae, narrowly elliptic to oblong-ovate or deltate, usually less than 4 mm wide; base rounded to truncate or cuneate; stalks (when present) usually lustrous and dark-colored; segment margins recurved to form confluent, poorly defined false indusia extending entire length of segment. **Veins** of ultimate segments free, obscure, pinnately branched and divergent distally. **False indusia** greenish, narrow, clearly marginal, occasionally concealing the sporangia. **Sporangia** confined to submarginal vein tips, containing 64, 32, or 16 spores, intermixed with farina-producing glands. **Spores** black to dark brown, globose or tetrahedral-globose, granulate, lacking prominent equatorial ridge. **Gametophytes** glandular-farinose (gametophytes of all other Pteridaceae genera lack farina). $x = 30$.

Species ca. 25 (10 in the flora): North America, Mexico, West Indies, Central America, South America.

As pointed out by R. M. Tryon (1956), J. T. Mickel (1979), and many others, *Notholaena* in the broad sense is poorly defined and difficult to distinguish from either *Cheilanthes* or *Pellaea.* North American taxa traditionally assigned to *Notholaena* represent at least four distinct evolutionary lineages (M. D. Windham 1986). In order to clarify species relationships and generic boundaries among cheilanthoid ferns, *Notholaena* is defined here in a very restricted sense. The pubescent, nonfarinose species (such as *N. newberryi*) have been placed in *Cheilanthes* following R. M. Tryon and A. F. Tryon (1982). The scaly, nonfarinose taxa of the *N. sinuata* complex have been transferred to *Astrolepis* for reasons discussed by D. M. Benham and M. D. Windham (1992). The glabrous and farinose species related to *Pellaea* (e.g., *N. jonesii* and *N. dealbata*) have been placed in *Argyrochosma* (M. D. Windham 1987). The species retained in *Notholaena* following this reorganization (i.e., members of the farinose *N. grayi–N. standleyi* alliance) form a coherent, monophyletic group found only in the Western Hemisphere. The correct generic name for this group is in dispute because *Notholaena* has been lectotypified by several authors citing three different type species. The rules of priority favor the first typification (by J. Smith in 1875) based on *N. trichomanoides* (Linnaeus) Desvaux, which is definitely a member of the group here called *Notholaena*. R. E. G. Pichi-Sermolli (1989), however, urged acceptance of the typification by C. Christensen ([1905–] 1906). Both the second and third lectotypifications of *Notholaena* were based on species unrelated to the North American taxa discussed here. If Smith's typification is overturned, the correct generic name for our species will be *Chrysochosma* (J. Smith) Kümmerle.

SELECTED REFERENCES Tryon, R. M. 1956. A revision of the American species of *Notholaena*. Contr. Gray Herb. 179: 1–106. Windham, M. D. 1986. Reassessment of the phylogenetic relationships of *Notholaena*. Amer. J. Bot. 73: 742.

1. Blades with scales or multicelled hairs in addition to whitish or yellowish farina.
  2. Nonfarinose indument on blades primarily of deeply dissected, stellate scales; farina on abaxial surface completely concealed by overlapping scales. . . . . 1. *Notholaena aschenborniana*
  2. Nonfarinose indument on blades of lanceolate scales and/or multicelled hairs; farina on abaxial surface readily apparent.
    3. Nonfarinose indument on blades of shiny, dark brown, needlelike hairs; sporangia containing 64 spores. . . . . 2. *Notholaena nealleyi*
    3. Nonfarinose indument on blades primarily of dull, light brown, narrowly lanceolate scales; sporangia containing 16 or 32 spores.
      4. Blades sparsely villous adaxially with long, multicelled, whitish hairs; abaxial scales ciliate. . . . . 3. *Notholaena aliena*
      4. Blades glabrous adaxially except for scattered glandular hairs; abaxial scales entire.
      . . . . 4. *Notholaena grayi*
1. Blades with whitish or yellowish farina only, lacking scales or multicelled hairs.
  5. Petioles grooved or flattened adaxially; blades linear to narrowly deltate, 2–6 times longer than wide; basal pinnae not markedly larger than adjacent pair, ± equilateral, proximal basiscopic pinnules not greatly enlarged.
    6. Blades glabrous adaxially at maturity; ultimate segments adnate to costae for entire width and clearly sessile; segment margins slightly recurved. . . . . 5. *Notholaena lemmonii*
    6. Blades glandular adaxially at maturity; at least some ultimate segments constricted proximally and slightly stalked; segment margins strongly revolute. . . . . . . 6. *Notholaena greggii*
  5. Petioles rounded adaxially; blades pentagonal to ovate, 1–2 times longer than wide; basal pinnae much larger than adjacent pair, strongly inequilateral due to enlargement of proximal basiscopic pinnules.
    7. Blades clearly 3-pinnate proximally; at least some ultimate segments constricted proximally and slightly stalked.
      8. Stem scales weakly bicolored, brown margins very narrow; blades broadly pentagonal, adaxially glandular; segment margins recurved but not strongly revolute.
      . . . . 7. *Notholaena californica*
      8. Stem scales strongly bicolored, brown margins broad and conspicuous; blades narrowly deltate-pentagonal, adaxially glabrescent; segment margins strongly revolute. . . . . 8. *Notholaena neglecta*
    7. Blades pinnate-pinnatifid to 2-pinnate-pinnatifid; ultimate segments adnate to costae their entire width and clearly sessile.
      9. Blades ovate, fully pinnate beyond basal pinnae, distal pinnae not connected; abaxial farina white. . . . . 9. *Notholaena copelandii*

9. Blades broadly pentagonal, pinnatifid beyond basal pinnae, distal pinnae connected by narrow wing of green tissue; abaxial farina yellow. . . . . . . 10. *Notholaena standleyi*

1. Notholaena aschenborniana Klotzsch, Linnaea 20: 417. 1847

Cheilanthes aschenborniana (Klotzsch) Mettenius; *Chrysochosma aschenborniana* (Klotzsch) Pichi-Sermolli

Stem scales concolored, margins black, undifferentiated, thick, ciliate. **Leaves** 8–35 cm. **Petiole** black, much shorter than blade, rounded adaxially, covered with ciliate and stellate scales. **Blade** lanceolate, usually 2-pinnate, 3–6 times longer than wide, abaxially with whitish or yellowish farina completely concealed by overlapping ciliate and deeply dissected stellate scales, adaxially with scattered stellate scales; basal pinnae equal to or slightly smaller than adjacent pair, ± equilateral, proximal basiscopic pinnules not greatly enlarged. **Ultimate segments** sessile to subsessile, narrowly adnate to costae; segment margins slightly recurved, rarely concealing sporangia. **Sporangia** containing 32 spores. $n = 2n = 90$, apogamous.

Sporulating summer–fall. Rocky slopes and cliffs, apparently confined to limestone; 300–1900 m; Ariz., Tex.; Mexico.

Although *Notholaena aschenborniana* is often described as nonfarinose, the abaxial glandular farina characteristic of all species of *Notholaena* as defined here is present beneath the dense covering of scales. All individuals examined chromosomally by M. D. Windham (unpublished data) were apogamous triploids that apparently arose through autopolyploidy. Further investigation is necessary to determine whether or not 64-spored, diploid populations of *N. aschenborniana* are still extant.

2. Notholaena nealleyi Seaton ex J. M. Coulter, Contr. U.S. Natl. Herb. 1: 61. 1890

Cheilanthes nealleyi (Seaton ex J. M. Coulter) Domin; *Chrysochosma schaffneri* (E. Fournier) L. Underwood ex Davenport var. *nealleyi* (Seaton ex J. M. Coulter) Pichi-Sermolli; *Notholaena schaffneri* (E. Fournier) Davenport var. *nealleyi* (Seaton ex J. M. Coulter) Weatherby

Stem scales mostly bicolored, margins black, broad and poorly defined, thick, ciliate-denticulate. **Leaves** 5–15 cm. **Petiole** black to dark brown, much shorter than blade, rounded adaxially, sparsely glandular farinose, bearing needlelike, multicelled hairs and ciliate scales.

Blade lanceolate, 2-pinnate-pinnatifid, 3–6 times longer than wide, abaxially with conspicuous whitish farina and shiny, dark brown, needlelike hairs scattered along rachises and costae, adaxially distinctly glandular; basal pinnae equal to or slightly smaller than adjacent pair, ± equilateral, proximal basiscopic pinnules not greatly enlarged. **Ultimate segments** sessile to subsessile, narrowly adnate to costae; segment margins slightly recurved, rarely concealing sporangia. **Sporangia** containing 64 spores.

Sporulating summer–fall. Calcareous cliffs and ledges, usually on limestone; 200–1700 m; Tex.; Mexico.

Notholaena nealleyi often has been treated as a variety of the Mexican species *N. schaffneri* (Fournier) L. Underwood ex Davenport, but recent studies (Windham, unpublished data) suggest that it may represent an allotetraploid hybrid between diploid *N. schaffneri* and another, as yet unidentified, species. Results of these studies, combined with morphologic differences and geographic isolation, favor the recognition of *N. nealleyi* as a distinct species.

3. Notholaena aliena Maxon, Contr. U.S. Natl. Herb. 17: 605. 1916

Chrysochosma aliena (Maxon) Pichi-Sermolli

Stem scales concolored to weakly bicolored, margins usually brown, very narrow and poorly defined, thin, ciliate-denticulate. **Leaves** 5–15 cm. **Petiole** brown, equal to or shorter than blade, rounded adaxially, glandular-farinose, bearing scattered hairs and scales. **Blade** linear-lanceolate, 2-pinnate-pinnatifid, 3–6 times longer than wide, abaxially with conspicuous cream-colored or pale yellow farina and dull, light brown, narrowly lanceolate, ciliate scales scattered along rachises and costae, adaxially sparsely glandular and villous with long, multicelled, whitish hairs; basal pinnae equal to or slightly larger than adjacent pair, ± equilateral, proximal basiscopic pinnules not greatly enlarged. **Ultimate segments** sessile, broadly adnate to costae; segment margins slightly recurved, rarely concealing sporangia. **Sporangia** containing 16 spores.

Sporulating summer–fall. Rocky slopes and cliffs, usually on volcanic substrates; 700–1200 m; Tex.; Mexico.

Notholaena aliena is a rare species closely related to *N. grayi*. Preliminary studies indicate that *N. aliena* reproduces by apogamy and is probably triploid.

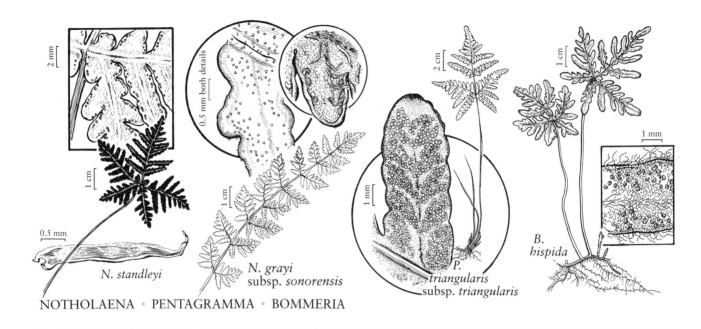

N. *standleyi*

N. *grayi*
subsp. *sonorensis*

P.
triangularis
subsp. *triangularis*

B.
hispida

NOTHOLAENA · PENTAGRAMMA · BOMMERIA

4. Notholaena grayi Davenport, Bull. Torrey Bot. Club 7: 50, plate 4. 1880

Cheilanthes grayi (Davenport) Domin; *Chrysochosma grayi* (Davenport) Pichi-Sermolli

Stem scales concolored to weakly bicolored, margins usually brown, very narrow and poorly defined, thin, ciliate-denticulate. **Leaves** 5–20 cm. **Petiole** brown, equal to or somewhat shorter than blade, rounded adaxially, glandular-farinose, bearing scattered hairs and scales. **Blade** linear-lanceolate, 2-pinnate-pinnatifid, 3–6 times longer than wide, abaxially with conspicuous whitish farina and dull, light brown, lanceolate, entire scales scattered along rachises and costae, adaxially distinctly glandular; basal pinnae equal to or slightly larger than adjacent pair, ± equilateral, proximal basiscopic pinnules not greatly enlarged. **Ultimate segments** sessile, broadly adnate to costae; segment margins slightly recurved, rarely concealing sporangia. **Sporangia** containing 16 or 32 spores.

Subspecies 2 (2 in the flora): North America, Mexico.

Notholaena grayi comprises two cytotypes here treated as subspecies. Sexually reproducing diploid populations (*N. grayi* subsp. *sonorensis*) are concentrated in southern Arizona and western Mexico. Apogamous triploids (*N. grayi* subsp. *grayi*) are more widespread, extending from Arizona to central Texas and northeastern Mexico. Isozyme analyses indicate that subsp. *grayi* is an autotriploid derivative of subsp. *sonorensis* (G. J. Gastony and M. D. Windham 1989).

1. Most sporangia containing 16 spores; spores generally more than 55 µm . . . . 4a. *Notholaena grayi* subsp. *grayi*
1. Most sporangia containing 32 spores; spores generally less than 55 µm . . . . 4b. *Notholaena grayi* subsp. *sonorensis*

4a. Notholaena grayi Davenport subsp. **grayi**

Most sporangia containing 16 spores; spores generally more than 55 µm. $n = 2n = 90$, apogamous.

Sporulating summer–fall. Rocky slopes and cliffs, on a variety of substrates including both granite and limestone; 300–1900 m; Ariz., N.Mex., Tex.; n Mexico.

4b. Notholaena grayi Davenport subsp. **sonorensis** Windham, Contr. Univ. Michigan Herb. 19: 36. 1993

Notholaena hypoleuca Goodding

Most sporangia containing 32 spores; spores generally less than 55 µm. $2n = 60$.

Sporulating summer–fall. Rocky slopes and cliffs, usually on granitic or volcanic substrates; 1200–1900 m; Ariz.; w Mexico.

5. Notholaena lemmonii D. C. Eaton, Bull. Torrey Bot. Club 7: 63. 1880

Cheilanthes lemmonii (D. C. Eaton) Domin; *Chrysochosma lemmonii* (D. C. Eaton) Pichi-Sermolli

Stem scales weakly bicolored, margins brown, narrow, poorly defined, thin, sparsely ciliate-denticulate. **Leaves** 7–30 cm. **Petiole** black to dark brown, much shorter than blade, grooved or flattened adaxially, bearing scattered glands and a few scales near base. **Blade** linear-lanceolate, 2-pinnate-pinnatifid, 3–6 times longer than wide, abaxially with conspicuous white or pale yellow farina, scales absent, adaxially glabrous at maturity; basal pinnae usually slightly smaller than adjacent pair, ± equilateral, proximal basiscopic pinnules not greatly enlarged. **Ultimate segments** sessile, broadly adnate to costae; segment margins slightly recurved, rarely concealing sporangia. **Sporangia** containing 64 spores. $2n = 60$.

Sporulating summer–fall. Rocky slopes and cliffs, usually on granitic or volcanic substrates; 1000–1500 m; Ariz.; w Mexico.

According to E. Wollenweber (1984), *Notholaena lemmonii* shows remarkable infraspecific variability in the chemical composition of the farina. R. M. Tryon (1956) recognized two varieties of *N. lemmonii*. The type collection of the species came from southern Arizona, and all specimens from the flora area are *N. lemmonii* var. *lemmonii*. The disjunct *N. lemmonii* var. *australis,* known only from the Mexican states of Puebla and Oaxaca, probably represents a distinct species.

6. Notholaena greggii (Mettenius ex Kuhn) Maxon, Contr. U.S. Natl. Herb. 17: 606. 1916

Pellaea greggii Mettenius ex Kuhn, Linnaea 36: 86. 1869; *Cheilanthes greggii* (Mettenius ex Kuhn) Mickel; *Chrysochosma greggii* (Mettenius ex Kuhn) Pichi-Sermolli

Stem scales strongly bicolored, margins brown, broad and well defined, thin, erose to denticulate. **Leaves** 4–20 cm. **Petiole** light brown, equal to or somewhat shorter than blade, grooved or flattened adaxially, bearing scattered glands and a few scales near base. **Blade** narrowly deltate, 2–3-pinnate, 2–4 times longer than wide, abaxially with conspicuous whitish farina, scales absent, adaxially glandular; basal pinnae slightly larger than adjacent pair, ± equilateral, proximal basiscopic pinnules not greatly enlarged. **Ultimate segments** sessile or subsessile, narrowly adnate to costae or free; segment margins strongly revolute, often concealing sporangia. **Sporangia** containing 64 spores. $2n = 60$.

Sporulating summer–fall. Calcareous slopes and ledges, usually on limestone or gypsum; 500–1000 m; Tex.; n Mexico.

Notholaena greggii is rarely collected in the flora area, and all known localities lie within 25 km of the Mexican border in the Big Bend region of Texas. It is most closely related to *N. bryopoda* Maxon, a gypsophile endemic to the southern Chihuahuan Desert.

7. Notholaena californica D. C. Eaton, Bull. Torrey Bot. Club 10: 27. 1883

Aleuritopteris cretacea (Liebmann) E. Fournier; *Cheilanthes deserti* Mickel; *Chrysochosma californica* (D. C. Eaton) Pichi-Sermolli

Stem scales weakly bicolored, margins brown, very narrow and poorly defined, thin, ciliate-denticulate. **Leaves** 4–20 cm. **Petiole** brown, equal to or somewhat longer than blade, rounded adaxially, bearing scattered glands and a few scales near base. **Blade** broadly pentagonal, 3-pinnate, 1–2 times longer than wide, abaxially with conspicuous white or yellow farina, scales absent, adaxially glandular; basal pinnae much larger than adjacent pair, strongly inequilateral, proximal basiscopic pinnules greatly enlarged. **Ultimate segments** sessile to subsessile, narrowly adnate to costae or free; segment margins recurved but rarely concealing sporangia. **Sporangia** containing 32 spores.

Subspecies 2 (2 in the flora): North America, Mexico.

Notholaena californica comprises two distinct chemotypes, recognizable by farina color, that are treated here as subspecies. These white and yellow forms are strikingly different in chemical composition of the farina (E. Wollenweber 1984), and they are rarely, if ever, found growing at the same locality. Gene flow between the chemotypes in the region of sympatry (southern California) is prevented because both are apogamous in this area. The recent discovery that plants of *N. californica* subsp. *californica* from Arizona are pentaploids (M. D. Windham, unpublished data) suggests that this species has a complex evolutionary history that cannot be resolved until sexually reproducing (64-spored) progenitors are found.

1. Farina on abaxial blade surface pale to bright yellow. . . . . 7a. *Notholaena californica* subsp. *californica*
1. Farina on abaxial blade surface white. . . . . 7b. *Notholaena californica* subsp. *leucophylla*

7a. Notholaena californica D. C. Eaton subsp. **californica**

Notholaena californica D. C. Eaton subsp. *nigrescens* Ewan

Blade with pale to bright yellow farina on abaxial surface. $n = 2n = 150$, apogamous.

Sporulating spring–fall. Rocky slopes and cliffs, usually on granitic or volcanic substrates; 100–1100 m; Ariz., Calif.; Mexico in Baja California, Sonora.

7b. Notholaena californica D. C. Eaton subsp. **leucophylla** Windham, Contr. Univ. Michigan Herb. 19: 35. 1993

Blade with white farina on abaxial surface.

Sporulating spring–early summer. Rocky slopes and cliffs, usually on granitic or volcanic substrates; 100–1100 m; Calif.; Mexico in Baja California.

8. Notholaena neglecta Maxon, Contr. U.S. Natl. Herb. 17: 602. 1916

Cheilanthes neglecta (Maxon) Mickel; *Chrysochosma neglecta* (Maxon) Pichi-Sermolli

Stem scales strongly bicolored, margins brown, broad and well defined, thin, ciliate-denticulate. **Leaves** 4–15 cm. **Petiole** black, equal to or somewhat longer than blade, rounded adaxially, bearing scattered glands and a few scales near base. **Blade** narrowly deltate-pentagonal, 3–4-pinnate, 1–2 times longer than wide, abaxially with conspicuous pale yellow farina, scales absent, adaxially glabrous to sparsely glandular; basal pinnae much larger than adjacent pair, strongly inequilateral, proximal basiscopic pinnules greatly enlarged. **Ultimate segments** sessile to subsessile, narrowly adnate to costae or free; segment margins strongly revolute, often concealing sporangia. **Sporangia** containing 32 spores. $n = 2n = 90$, apogamous.

Sporulating summer–fall. Rocky slopes and cliffs, apparently confined to limestone; 300–1900 m; Ariz., Tex.; Mexico.

Notholaena neglecta is closely related to *N. californica* and may have been involved in the origin of polyploids within that complex. Populations occurring in the flora are composed of apogamous triploids, but a sexually reproducing diploid cytotype has been found in Nuevo León, Mexico (M. D. Windham, unpublished data).

9. Notholaena copelandii C. C. Hall, Amer. Fern J. 40: 181, plate 16. 1950

Cheilanthes candida M. Martens & Galeotti var. *copelandii* (C. C. Hall) Mickel; *Chrysochosma candida* (M. Martens & Galeotti) Kümmerle var. *copelandii* (C. C. Hall) Pichi-Sermolli; *Notholaena candida* (M. Martens & Galeotti) Hooker var. *copelandii* (C. C. Hall) R. M. Tryon

Stem scales strongly bicolored, margins brown, broad and well defined, thin, weakly ciliate. **Leaves** 6–25 cm. **Petiole** black, equal to or somewhat longer than blade, rounded adaxially, glabrous except for a few scales and farinose glands near base. **Blade** ovate, 2-pinnate-pinnatifid, 1–2 times longer than wide, abaxially with conspicuous white farina, scales absent, adaxially glandular; basal pinnae much larger than adjacent pair, strongly inequilateral, proximal basiscopic pinnules greatly enlarged. **Ultimate segments** sessile, broadly adnate to costae; segment margins slightly recurved, rarely concealing sporangia. **Sporangia** containing 64 spores. $2n = 60$.

Sporulating summer–fall. Rocky slopes and cliffs, apparently confined to limestone; 300–1500 m; Tex.; Mexico.

Notholaena copelandii often has been treated as a variety of the Mexican species *N. candida* (M. Martens & Galeotti) Hooker. The two taxa show significant differences in stem scale and leaf morphology (R. M. Tryon 1956), density of glands on the adaxial leaf surface, and chemical composition of the farinose indument (E. Wollenweber 1984). Measures of genetic similarity based on isozyme data indicate that *N. copelandii* and *N. candida* are quite distinct and, at present, allopatry precludes gene flow. Past hybridization may account for the small number of intermediate specimens reported from San Luis Potosí, Mexico, by R. M. Tryon (1956).

10. Notholaena standleyi Maxon, Amer. Fern J. 5: 1. 1915

Cheilanthes standleyi (Maxon) Mickel; *Chrysochosma hookeri* (Domin) Kümmerle; *Notholaena hookeri* D. C. Eaton

Stem scales strongly bicolored, margins brown, broad and well defined, thin, entire. **Leaves** 5–30 cm. **Petiole** brown, equal to or somewhat longer than blade, rounded adaxially, glabrous except for a few scales at base. **Blade** broadly pentagonal, deeply pinnatifid but not fully pinnate above base (distal pinnae connected by narrow wing of green tissue), 1–2 times longer than wide, abaxially with conspicuous yellowish farina, scales absent, adaxially glabrous; basal pinnae much larger than adjacent pair, strongly inequilateral, proximal basiscopic pinnules greatly enlarged. **Ultimate segments** sessile, broadly adnate to costae; segment margins slightly recurved, rarely concealing sporangia. **Sporangia** containing 32 or 16 spores. $2n = 60$.

Sporulating late spring–fall. Rocky slopes and cliffs, on a variety of substrates including granite and limestone; 300–2100 m; Ariz., Colo., N.Mex., Okla., Tex.; Mexico.

The low numbers of spores per sporangium in *Notholaena standleyi* are apparently not associated with apogamy because all plants thus far analyzed are sexual diploids (M. D. Windham, unpublished data). D. S. Siegler and E. Wollenweber (1983) identified three chemotypes in this species correlated with substrate specificity and subtle variations in farina color. These three "races" occupy different portions of the geographic range, and further investigation may indicate that they deserve formal taxonomic recognition. A report of *Notholaena standleyi* from Nevada (D. B. Lellinger 1985) must be considered suspect because it is disjunct, and the closest populations in Arizona represent a different chemotype.

8. PENTAGRAMMA Yatskievych, Windham, & E. Wollenweber, Amer. Fern. J. 80: 15. 1990 · Silverback fern, goldback fern [Greek *penta*, five, and *gramma*, lines (as in written characters), for the pentagonal leaf blades]

George Yatskievych

Michael D. Windham

Plants on rock or terrestrial. **Stems** short-creeping, erect or ascending, sometimes branched; scales sharply bicolored with black, hard, central stripe and tan margins, narrowly lanceolate, margins entire. **Leaves** monomorphic, clustered, 5–40 cm. **Petiole** chestnut brown to black, rounded or nearly so adaxially, glabrous, farinose, or viscid-glandular, with single vascular bundle. **Blade** triangular-pentagonal, 1–2-pinnate-pinnatifid proximally, pinnatifid distally, herbaceous to leathery, abaxially usually farinose, farina white or yellow, adaxially glabrous or glandular, dull, not striate; rachis straight. **Ultimate segments** sessile, adnate to rachis and costae their full width, triangular to lanceolate, margins not recurved to form false indusia. **Veins** of ultimate segments free, obscured, pinnately branched and divergent distally. **False indusia** absent. **Sporangia** borne along veins, containing 64 spores, intermixed with farina-producing glands. **Spores** tan to brown, distinctly trigonal, coarsely tuberculate, tubercules usually somewhat fused, lacking prominent equatorial ridge. **Gametophytes** glabrous. $x = 30$.

Species 2 (2 in the flora): North America, nw Mexico.

Until recently, *Pentagramma* was included in *Pityrogramma*, based on superficial morphologic similarities between the two groups. Differences in leaf morphology, anatomy of the petioles, spore ornamentation, and presumed basal chromosome numbers suggest that the two genera are not closely related and are best placed in different tribes of the Pteridaceae. Within *Pentagramma*, species limits are not well understood, and workers have advocated recognition of one to three species, with several additional, morphologically cryptic taxa also indicated (but not named). A detailed examination of infra- and interpopulational variation in the group using modern techniques is needed.

SELECTED REFERENCES Alt, K. S. and V. Grant. 1960. Cytotaxonomic observations on the goldback fern. Brittonia 12: 153–170. Tryon, R. M. 1962. Taxonomic fern notes. II. *Pityrogramma* (including *Trismeria*) and *Anogramma*. Contr. Gray Herb. 189: 52–76. Yatskievych, G., M. D. Windham, and E. Wollenweber. 1990. A reconsideration of the genus *Pityrogramma* (Adiantaceae) in western North America. Amer. Fern J. 80: 9–17.

1. Petioles, especially in young leaves, and stem apices farinose; blades white-farinose abaxially and adaxially, appearing grayish adaxially when fresh. . . . . 1. *Pentagramma pallida*
1. Petioles and stem apices glabrous or somewhat glandular, petioles rarely farinose (in *P. triangularis* subsp. *maxonii*); blades yellow- or white-farinose abaxially, glabrous or with scattered, clear, nonfarinose glands adaxially, appearing bright green or sometimes yellowish green adaxially when fresh. 2. *Pentagramma triangularis*

1. Pentagramma pallida (Weatherby) Yatskievych, Windham, & E. Wollenweber, Amer. Fern J. 80: 15. 1990

Pityrogramma triangularis (Kaulfuss) Maxon var. *pallida* Weatherby, Rhodora 22: 119. 1920; *P. pallida* (Weatherby) K. S. Alt & V. E. Grant

Petiole black or nearly so, not shiny, farinose; farina white, especially in young leaves. **Blade** thin, herbaceous, white-farinose abaxially and adaxially, appearing grayish adaxially when fresh. $2n = 60$.

Rock crevices and at base of boulders, in drainages and on slopes, rarely on roadbanks, pine and oak woodlands; 100–400 m; Calif.

Pentagramma pallida is endemic to the foothills of the Sierra Nevada. C. A. Weatherby (1920), K. S. Alt and V. Grant (1960), and D. M. Smith (1980) agreed that *P. pallida* is among the most easily distinguishable taxa in the group, particularly in the field. In addition to the key characteristics, the usually blackish, nonlustrous petioles of this species are unique in the genus. Alt and Grant noted a case of apparent hybridization between this species and *Pentagramma triangularis* in Tuolumne County, California.

SELECTED REFERENCES Smith, D. M. 1980. Flavonoid analysis of the *Pityrogramma triangularis* complex. Bull. Torrey Bot. Club 107: 134–145. Weatherby, C. A. 1920. Varieties of *Pityrogramma triangularis*. Rhodora 22: 113–120.

2. Pentagramma triangularis (Kaulfuss) Yatskievych, Windham, & E. Wollenweber, Amer. Fern J. 80: 15. 1990

Gymnogramma triangularis Kaulfuss, Enum. Filic., 73. 1824; *Pityrogramma triangularis* (Kaulfuss) Maxon

Petiole chestnut brown to dark brown, somewhat shiny, glabrous or sometimes viscid-glandular or rarely somewhat white-farinose proximally. **Blade** thin and herbaceous to thick and leathery, abaxially densely farinose, farina white or yellow, adaxially bright green to yellowish green when fresh, glabrous to glandular or viscid.

Subspecies 4 (4 in the flora): North America, nw Mexico.

Pentagramma triangularis occurs in rock crevices and

at the base of overhanging boulders in drainages and on slopes and roadbanks. Occasional plants in which the farina is nearly absent may be encountered. These have been described as *Pityrogramma triangularis* var. *viridis* Hoover, a name of uncertain application that appears to refer to misshapen-spored hybrids of various parentage within the *Pentagramma triangularis* complex.

1. Leaves viscid-glandular adaxially; distal pinnae mostly entire; proximal basiscopic lobes of basal pinnae entire to undulate or crenate. . . . . . .
. . . . . . 2d. *Pentagramma triangularis* subsp. *viscosa*
1. Leaves glabrous or with scattered yellowish capitate glands adaxially, not viscid-glandular; distal pinnae mostly regularly lobed; proximal basiscopic lobes of basal pinnae pinnatifid, often deeply so.
  2. Leaves with scattered yellowish, capitate, nonfarinose glands adaxially, white-farinose abaxially. . . . .
. . . . 2a. *Pentagramma triangularis* subsp. *maxonii*
  2. Leaves glabrous adaxially, yellow- or white-farinose abaxially.
    3. Farina light to bright yellow. . . . . . .
. 2c. *Pentagramma triangularis* subsp. *triangularis*
    3. Farina white. . . . .
. 2b. *Pentagramma triangularis* subsp. *semipallida*

2a. Pentagramma triangularis (Kaulfuss) Yatskievych, Windham, & E. Wollenweber subsp. **maxonii** (Weatherby) Yatskievych, Windham, & E. Wollenweber, Amer. Fern J. 80: 16. 1990

Pityrogramma triangularis (Kaulfuss) Maxon var. *maxonii* Weatherby, Rhodora 22: 119. 1920

Petiole glabrous, not viscid-glandular, rarely white-farinose proximally. **Blades** thin and herbaceous (but not leathery), not viscid-glandular, abaxially densely white-farinose, adaxially with scattered yellow capitate glands. **Distal pinnae** mostly regularly lobed. **Proximal basiscopic lobes** of basal pinnae pinnatifid, often deeply so. $2n = 60, 120$.

Desert scrub and in pine and oak woodlands; 100–2500 m; Ariz., Calif., N.Mex.; Mexico in Baja California, Sonora.

2b. Pentagramma triangularis (Kaulfuss) Yatskievych, Windham, & E. Wollenweber subsp. **semipallida** (J. T. Howell) Yatskievych, Windham, & E. Wollenweber, Amer. Fern J. 80: 16. 1990

Pityrogramma triangularis (Kaulfuss) Maxon var. *semipallida* J. T. Howell, Leafl. W. Bot. 9: 223. 1962

Petiole glabrous, not viscid-glandular. **Blade** thin and herbaceous, sometimes thick (but not leathery), not viscid-glandular, abaxially densely white-farinose, adaxially glabrous. **Distal pinnae** mostly regularly lobed. **Proximal basiscopic lobes** of basal pinnae pinnatifid, often deeply so.

Chaparral, pine and oak woodlands; 100–900 m; Calif.

Pentagramma triangularis subsp. *semipallida,* as here treated, remains heterogeneous. Diploid (based on spore size) populations occur in the foothills of the Sierra Nevada in Butte County, California. Tetraploids (based on spore size) of similar morphology are apparently restricted to Santa Barbara County, California, including the adjacent Channel Islands. The relationship between these two variants has not been studied in detail.

2c. Pentagramma triangularis (Kaulfuss) Yatskievych, Windham, & E. Wollenweber subsp. **triangularis**

Petiole glabrous, not viscid-glandular. **Blade** thin and herbaceous to somewhat thick and leathery, not viscid-glandular, abaxially densely pale to bright yellow, adaxially glabrous. **Distal pinnae** mostly regularly lobed. **Proximal basiscopic lobes** of basal pinnae pinnatifid, often deeply so. $2n = 60$, ca. 90, 120.

Chaparral, pine and oak woodlands; 50–1800 m; B.C.; Calif., Idaho, Oreg., Wash.; Mexico in Baja California.

We here restrict *Pentagramma triangularis* subsp. *triangularis* to plants with yellow farina and glabrous adaxial leaf surfaces occurring throughout a large region in westernmost North America. This subspecies comprises a complex of morphological, cytological, and phytochemical variants, at least some of which may deserve formal taxonomic recognition, following more detailed studies. Plants with yellow farina reported from Arizona, Nevada, and Utah may represent tetraploid hybrids between *P. triangularis* subsp. *triangularis* and *P. triangularis* subsp. *maxonii* and are not mapped herein.

2d. Pentagramma triangularis (Kaulfuss) Yatskievych, Windham, & E. Wollenweber subsp. **viscosa** (Nuttall ex D. C. Eaton) Yatskievych, Windham, & E. Wollenweber, Amer. Fern J. 80: 15. 1990

Gymnogramma viscosa Nuttall ex D. C. Eaton, Ferns N. Amer. 2: 16. 1879; *Pityrogramma triangularis* (Kaulfuss) Maxon var. *viscosa* (Nuttall ex D. C. Eaton) Weatherby; *P. viscosa* (Nuttall ex D. C. Eaton) Maxon

Petiole sometimes viscid-glandular. **Blades** thick and leathery, abaxially densely white-farinose and viscid-glandular, adaxially viscid-glandular. **Distal pinnae** mostly entire. **Proximal basiscopic lobes** of basal pinnae entire to undulate or crenate. $2n = 60$.

Chaparral, pine and oak woodlands; 50–500 m; largely coastal; Calif.; Mexico in Baja California.

Pentagramma triangularis subsp. *viscosa* was said to introgress with *P. triangularis* subsp. *triangularis* by K. S. Alt and V. Grant (1960), who noted both diploid and tetraploid plants of intermediate morphology at some sites where these two occur together.

9. BOMMERIA E. Fournier in Baillon, Dict. Bot. 1: 448. 1877 [Named for the Belgian pteridologist Jean Edouard Bommer 1829–1895]

Christopher H. Haufler

Plants terrestrial. **Stems** prostrate, long-creeping, often branched [short, seldom branched]; scales pale brown to yellowish, lanceolate, concolored, margins entire. **Leaves** monomorphic, scattered, 4–30 cm. **Petiole** chestnut brown to dark purple, rounded or with single groove adaxially, indument of scales and/or trichomes, especially proximally and distally, with single vascular bundle. **Blade** pentagonal, pedately divided into 3 segments, deeply pinnate-pinnatifid, herbaceous, abaxially with scales, unicellular coiled trichomes, and unicellular needlelike trichomes, adaxially dull, not striate, with unicellular needlelike trichomes; rachis straight. **Segments** of blade sessile, 1–3(–5) mm wide, distal segment 1–2-pinnatifid in proximal portion, pinnatifid in distal portion; proximal segments usually connected to distal by narrow wing along rachis, inequilaterally elongate-deltate, proximal basiscopic portion elongate, 1–2-pinnatifid; margins not recurved to form false indusia. **Veins** in segments obscure, free to anastomosing, pinnately branched and divergent distally. **False indusia** absent. **Sporangia** borne along veins, covering most veins or somewhat marginally restricted, containing 64 [32] spores, paraphyses and glands absent. **Spores** brown, globose, trilete, exospore smooth, perispore surfaces crested or reticulate, equatorial flange absent. $x = 30$.

Species 5 (1 in the flora): North America, Mexico, Central America.

SELECTED REFERENCES Gastony, G. J. and C. H. Haufler. 1976. Chromosome numbers and apomixis in the fern genus *Bommeria*. Biotropica 8: 1–11. Haufler, C. H. 1979. A biosystematic revision of *Bommeria*. J. Arnold Arbor. 60: 445–476.

1. **Bommeria hispida** (Kuhn) L. Underwood, Bull. Torrey Bot. Club 29: 633. 1902

Gymnogramma hispida Kuhn, Linnaea 36: 72. 1869; *Bommeria schnafferi* E. Fournier; *Gymnogramma ehrenbergiana* Klotzsch var. *muralis* Pringle ex Davenport; *Gymnopteris hispida* (Kuhn) L. Underwood

Stems long-creeping and often branched. **Leaves** arising at 7–10 mm intervals. **Petiole** generally rounded (but distally grooved in some large leaves); indument of scales proximally, of scales and trichomes distally, central portion generally glabrous. **Blade** 1–7 cm, about as long as wide; ultimate segments rounded at apex; abaxial indument of scales (commonly with 10+ cells across base), unicellular needlelike trichomes (0.75–1.12 mm), and unicellular coiled trichomes (ca. 1–2 mm); adaxial indument of unicellular needlelike hairs 0.47–1.11 mm, arising from unspecialized basal cells; rachis chestnut brown. **Veins** free. **Sporangia** covering 2/3–3/4 distance from margin of blade to costa of each ultimate segment. **Spores** 64 per sporangium; perispore surface crested. $2n = 60$.

Sporulating summer–fall. At bases of large boulders on dry to moist slopes, primarily in mountainous, xeric regions; occasionally forming large mats; 1000–2500 m; Ariz., N.Mex., Tex.; Mexico.

Bommeria hispida is the only member of this genus to occur north of Mexico. It is the most morphologically distinct species in the genus, having a relatively small leaf size, dissected segments, and copious and diverse leaf indument. Natural hybrids involving *B. hispida* are unknown. The pedate blade shape, lack of colored farina abaxially, and presence of both needlelike and coiled trichomes abaxially serve to distinguish this species from sympatric members of *Cheilanthes*, *Notholaena*, and *Pentagramma*.

Reports of *Bommeria hispida* in California (C. H. Haufler 1979) are based on old specimens with questionable locality data.

10. **CHEILANTHES** Swartz, Syn. Fil., 126. 1806, name conserved · Lip fern

[Greek *cheilos*, margin, and *anthus*, flower, referring to the marginal sporangia]

Michael D. Windham
Eric W. Rabe

Plants usually on rock. **Stems** compact to long-creeping, ascending to horizontal, usually branched; scales brown to black or often bicolored with dark central stripe and lighter margins, linear-subulate to ovate-lanceolate, margins entire or denticulate. **Leaves** monomorphic, clustered to widely scattered, 4–60 cm. **Petiole** brown to black or straw-colored, rounded, flattened, or with single longitudinal groove adaxially, pubescent, scaly, or glabrous, with a single vascular bundle. **Blade** linear-oblong to lanceolate, ovate, or elongate-pentagonal, pinnate-pinnatifid to 4-pinnate at base, leathery or rarely somewhat herbaceous, abaxially

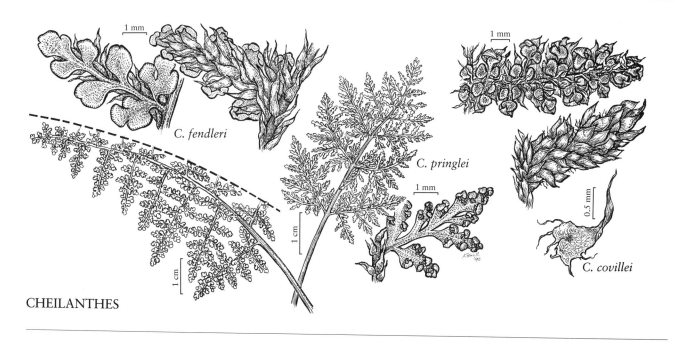

C. fendleri

C. pringlei

C. covillei

CHEILANTHES

pubescent and/or scaly, rarely glabrous, adaxially pubescent to glabrous, dull, not striate; rachis straight. **Ultimate segments** of blade stalked or sessile, usually free from costae, round to elongate or spatulate, usually less than 4 mm wide, base rounded, truncate, or cuneate; stalks (when present) often lustrous and dark-colored; segment margins usually recurved to form confluent, poorly defined false indusia, extending entire length of segment or discontinuous on apical or lateral lobes. **Veins** of ultimate segments free or rarely anastomosing, obscure, pinnately branched and divergent distally. **False indusia** greenish to whitish, usually narrow, clearly marginal or rarely inframarginal, often concealing sporangia. **Sporangia** confined to submarginal vein tips or scattered along veins near segment margins, containing 64 or 32 spores, not intermixed with farina-producing glands. **Spores** brown to black or gray (rarely yellowish), tetrahedral-globose, rugose or cristate, lacking prominent equatorial ridge. **Gametophytes** glabrous. $x = 30$ (29 in the *Cheilanthes alabamensis* complex).

Species ca. 150 (28 in the flora): mostly Western Hemisphere but a few in Europe, Asia, Africa, Pacific Islands, and Australia.

Cheilanthes is by far the largest and most diverse genus of xeric-adapted ferns. In its classic circumscription, the genus has been notoriously difficult to distinguish from other cheilanthoid genera, especially *Notholaena* and *Pellaea* (R. M. Tryon and A. F. Tryon 1982). This has led some authors (e.g., J. T. Mickel 1979b) to abandon several of the segregate genera and greatly expand the number of species assigned to *Cheilanthes*. Taxonomic problems in this group have motivated an ongoing series of biosystematic studies that offer hope for a stable classification through the identification of natural, monophyletic groups. The circumscription of *Cheilanthes* has been clarified recently by a redefinition of *Notholaena* and the transfer of several species (e.g., *N. parryi*, *N. newberryi*, and *N. aurea*) to *Cheilanthes* (R. M. Tryon and A. F. Tryon 1982). The boundaries of the genus have been further sharpened by the recognition of *Aspidotis* (A. R. Smith 1975), *Argyrochosma* (M. D. Windham 1987), and *Astrolepis* (D. M. Benham and M. D. Windham 1992) as distinct genera. Despite these efforts, *Chei-*

lanthes remains a very heterogeneous (and probably polyphyletic) genus in need of further critical study.

The circumscription of *Cheilanthes* used here closely parallels that proposed by T. Reeves (1979), who recognized four New World subgenera and a small group of species of uncertain placement. Among the North American species, *C. pringlei* and *C. wrightii* belong to the latter group, and *C. arizonica* is the sole representative of subgenus *Othonoloma* Link ex C. Christensen. The subgenus that Reeves called the *Cheilanthes alabamensis* group is represented in North America by *C. aemula, C. alabamensis, C. microphylla,* and *C. horridula.* It differs from other members of the genus in a number of critical features (e.g., blade indument, sporangial distribution, and chromosome base number) that suggest a relationship to the genus *Pellaea.* Although the group is somewhat anomalous in *Cheilanthes,* inclusion of the *C. alabamensis* complex in *Pellaea* (e.g., R. Cranfill 1980) would make that genus polyphyletic. The group is maintained here in *Cheilanthes* with the recognition that it may constitute a natural group worthy of consideration as a distinct genus.

The remaining 21 North American species of *Cheilanthes* are almost evenly divided between subgenus *Physapteris* (C. Presl) Baker in Hooker & Baker and subgenus *Cheilanthes.* The former subgenus is characterized by T. Reeves (1979) as having noncircinate vernation, usually scaly blades with small, beadlike ultimate segments, and hairs with cell walls that fit together in "tongue and groove" fashion. Species of subgenus *Cheilanthes* typically have circinate vernation, nonscaly blades with larger ultimate segments, and hairs with straight crosswalls. Despite the consistency of these differences, the two groups are closely related and linked by occasional intersubgeneric hybridization between *C. covillei* (subg. *Physapteris*) and *C. parryi* and *C. newberryi* (subg. *Cheilanthes*).

SELECTED REFERENCES Benham, D. M. and M. D. Windham. 1992. Generic affinities of the star-scaled cloak ferns. Amer. Fern. J. 82: 47–58. Mickel, J. T. 1979b. The fern genus *Cheilanthes* in the continental United States. Phytologia 41: 431–437. Reeves, T. 1979. A Monograph of the Fern Genus *Cheilanthes* Subgenus *Physapteris* (Adiantaceae). Ph.D. dissertation. Arizona State University. Smith, A. R. 1975. The California species of *Aspidotis*. Madroño 23: 15–24. Tryon, R. M. 1956. A revision of the American species of *Notholaena*. Contr. Gray Herb. 179: 1–106.

1. Costae with multiseriate scales abaxially (narrow and inconspicuous in *Cheilanthes gracillima* and *C. tomentosa*), intermixed with hairs in some species; vernation noncircinate, expanding leaves hooked but not coiled at tips.
    2. Rachises and distal portion of petioles grooved adaxially; ultimate segments spatulate; sori discontinuous, confined to apical and lateral lobes. . . . . 1. *Cheilanthes pringlei*
    2. Rachises and petioles rounded or slightly flattened adaxially; ultimate segments round, elliptic, or oblong; sori ± continuous around segment margins.
        3. Ultimate segments scabrous, covered with stiff, usually pustulose hairs; fertile ultimate segments narrowly elliptic, the largest 3–5 mm. . . . . 2. *Cheilanthes horridula*
        3. Ultimate segments smooth to touch, lacking stiff, pustulose hairs; fertile ultimate segments round to somewhat elliptic, the largest less than 3 mm.
            4. Costal scales linear, inconspicuous, the largest 0.1–0.4 mm wide.
                5. Ultimate segments pubescent adaxially with fine, unbranched hairs; costal scale margins entire; blades 3-pinnate for most of length. . . . . 3. *Cheilanthes tomentosa*
                5. Ultimate segments glabrescent adaxially or bearing scattered, branched hairs; costal scale margins ciliate at base; blades predominantly 2-pinnate (occasionally 3-pinnate at base). . . . . 4. *Cheilanthes gracillima*
            4. Costal scales lanceolate to ovate, conspicuous, the largest 0.4–1.5 mm wide.
                6. Costal scale margins entire to erose or denticulate (rarely with 1 or 2 cilia in *Cheilanthes eatonii*).

7. Ultimate segments completely glabrous adaxially; stems long-creeping, 1–3 mm diam.; stem scales mostly concolored, brown. . . . . 5. *Cheilanthes fendleri*
7. Ultimate segments densely to sparsely pubescent adaxially; stems compact, usually 5–10 mm diam.; stem scales mostly bicolored with dark, well-defined central stripe and light brown margins.
    8. Ultimate segments densely tomentose with fine hairs abaxially; costal scales usually linear-lanceolate, loosely imbricate, not concealing ultimate segments. . . . . 6. *Cheilanthes eatonii*
    8. Ultimate segments nearly glabrous abaxially except for a few coarse hairs; costal scales ovate-lanceolate, strongly imbricate, usually concealing ultimate segments. .7. *Cheilanthes villosa*
6. Costal scale margins ciliate, especially near base (inconspicuously so in *C. covillei*).
    9. Costal scales ovate-lanceolate with deeply cordate bases, basal lobes usually overlapping to close sinus, scales usually ciliate only in proximal 1/2; sporangia containing 64 spores; stems usually short-creeping, leaves clustered (separated to 10 mm in *C. clevelandii*).
        10. Costal scales ciliate on basal lobes only, thus often appearing to lack cilia; ultimate segments glabrous or with a few entire to weakly ciliate scales abaxially, lacking branched hairs; stem scales usually dark brown or black throughout, rarely with narrow, light brown margins. . . . . 8. *Cheilanthes covillei*
        10. Costal scales conspicuously ciliate over most of proximal 1/2; ultimate segments pubescent abaxially with branched hairs and ciliate scales; stem scales usually bicolored.
            11. Ultimate segments oblong to ovate, sparsely pubescent adaxially (often glabrescent when mature); stem scales linear-lanceolate with pale, narrow margins much narrower than width of dark, central stripe. . . . . 9. *Cheilanthes intertexta*
            11. Ultimate segments round to subcordate, glabrous adaxially; stem scales lanceolate with pale, broad margins approaching width of dark, central stripe. . . . . 10. *Cheilanthes clevelandii*
    9. Costal scales lanceolate with truncate or subcordate bases, basal lobes (when present) not overlapping, scales often ciliate throughout; sporangia containing 32 spores; stems usually long-creeping, leaves scattered.
        12. Pinnae appearing densely tomentose adaxially; costal scales with fine, curly cilia forming entangled mass; ultimate segments minute, 0.5–1 mm. . . . . 11. *Cheilanthes lindheimeri*
        12. Pinnae appearing glabrous or sparsely pubescent adaxially; costal scales with coarse cilia that are not strongly entangled; ultimate segments larger, at least some 1–3 mm.
            13. Pinnae appearing glabrous adaxially; costal scales often ciliate only in proximal 1/2; stem scales usually brown, concolored, loosely appressed and deciduous on older portions of stem. . . . . . 12. *Cheilanthes wootonii*
            13. Pinnae appearing sparsely pubescent adaxially; costal scales usually ciliate entire length; stem scales dark brown, often bicolored, usually strongly appressed and persistent. . . . . 13. *Cheilanthes yavapensis*
1. Costae lacking multiseriate scales, pubescent or glabrous; vernation circinate or noncircinate, expanding leaves tightly coiled at tip in most species.
  14. Rachises and pinnae essentially glabrous; petioles adaxially grooved for most of length.
    15. Basal pinnae conspicuously larger than adjacent pair; stems compact, usually 4–8 mm diam.; ultimate segments with scattered reddish glands abaxially. . . 14. *Cheilanthes arizonica*
    15. Basal pinnae slightly smaller than or equal to adjacent pair; stems long-creeping, 1–3 mm diam.; ultimate segments lacking reddish glands abaxially. . . . 15. *Cheilanthes wrightii*

14. Rachises (and often pinnae) pubescent or glandular; petioles rounded, flattened, or slightly grooved distally but never grooved below middle adaxially.

    16. Rachis pubescence dimorphic, abaxially sparsely hirsute with long, divergent hairs, adaxially densely covered with tortuous, appressed hairs; ultimate segments sparsely and inconspicuously pubescent abaxially, often appearing glabrous.

        17. Basal pinnae inequilateral, proximal basiscopic pinnules conspicuously enlarged; blades ovate to deltate, 5–15 cm wide. . . . . .16. *Cheilanthes aemula*

        17. Basal pinnae ± equilateral, proximal basiscopic pinnules not conspicuously enlarged; blades lanceolate to narrowly oblong, 1–7 cm wide.

            18. Costae green adaxially for most of length; most sporangia containing 32 spores; stems short-creeping to compact, usually 4–7 mm diam. . . . . 17. *Cheilanthes alabamensis*

            18. Costae black adaxially for most of length; most sporangia containing 64 spores; stems long-creeping, 1–3 mm diam. . . . . 18. *Cheilanthes microphylla*

    16. Rachis pubescence monomorphic, hairs similar in form and density abaxially and adaxially; ultimate segments conspicuously pubescent or glandular abaxially.

        19. False indusia inframarginal, strongly differentiated, 0.25–0.5 mm wide, forming pouch with constricted aperture abaxially on beadlike ultimate segments; ultimate segments nearly glabrous adaxially. . . . . 19. *Cheilanthes lendigera*

        19. False indusia marginal, weakly differentiated or absent, 0–0.25 mm wide, not forming pouch with constricted aperture abaxially on ultimate segments; ultimate segments pubescent or glandular adaxially (glabrescent in older leaves of *Cheilanthes feei*).

            20. Blades pinnate-pinnatifid throughout; pinnae clearly articulate, dark color of stalk stopping abruptly at swollen, hirsute node. . . . . . . 20. *Cheilanthes bonariensis*

            20. Blades 2–4-pinnate at base; pinnae not articulate, color of stalk continuing into pinna base.

                21. Blades elongate-pentagonal, usually 5–10 cm wide; basal pinnae conspicuously larger than adjacent pair, inequilateral, proximal basiscopic pinnules greatly enlarged.

                    22. Petioles reddish purple to dark brown; ultimate segments resinous-sticky, covered with short, capitate glands. . . . .21. *Cheilanthes kaulfussii*

                    22. Petioles straw-colored; ultimate segments not noticeably sticky, covered with long, noncapitate hairs. . . . . 22. *Cheilanthes leucopoda*

                21. Blades linear to ovate-lanceolate, usually 1–5 cm wide; basal pinnae about same size as adjacent pair, ± equilateral, proximal basiscopic pinnules not greatly enlarged.

                  23. Ultimate segments with very fine, cobwebby hairs; stem scales uniformly black or dark brown, strongly appressed. . . 23. *Cheilanthes newberryi*

                  23. Ultimate segments bearing coarse hairs, not at all cobwebby; stem scales uniformly brown or bicolored, loosely appressed.

                    24. Stem scales concolored, uniformly brown or slightly darker toward tip; rachises strongly flattened or shallowly grooved adaxially; costae green adaxially for most of length.

                        25. Ultimate segments conspicuously resinous-sticky, covered with short, capitate glands; stem scales strongly contorted. . . . . 24. *Cheilanthes viscida*

                        25. Ultimate segments not conspicuously sticky, covered with long, strongly flattened hairs; stem scales straight to slightly contorted. . . . . 25. *Cheilanthes cooperae*

                  24. At least some stem scales bicolored, darker toward center or base; rachises rounded or slightly flattened adaxially; costae brown adaxially for most of length.

26. Fertile ultimate segments nearly round, beadlike, 1–3 mm; sporangia containing 32 spores; blades 3-pinnate near base. . . . . 26. *Cheilanthes feei*
26. Fertile ultimate segments elongate, not beadlike, usually 3–5 mm; sporangia containing 64 spores; blades 2-pinnate-pinnatifid near base.
    27. Ultimate segments densely villous adaxially; many stem scales strongly bicolored with well-defined central stripe. . . . . 27. *Cheilanthes parryi*
    27. Ultimate segments sparsely hirsute adaxially; most stem scales weakly bicolored with poorly defined central stripe. . . . . 28. *Cheilanthes lanosa*

1. Cheilanthes pringlei Davenport, Bull. Torrey Bot. Club 10: 61. 1883 · Pringle's lip fern

Stems long-creeping, 1–3 mm diam.; scales uniformly brown or with poorly defined, dark, central stripe, linear-lanceolate, straight to slightly contorted, loosely appressed, usually persistent. **Leaves** clustered to somewhat scattered, 4–15 cm; vernation noncircinate. **Petiole** dark brown, grooved distally on adaxial surface. **Blade** ovate-deltate, 3-pinnate-pinnatifid at base, 1.5–5 cm wide; rachis grooved adaxially, with scattered, lanceolate scales, not pubescent. **Pinnae** not articulate, dark color of stalk continuing into pinna base, basal pair conspicuously larger than adjacent pair, inequilateral, basiscopic pinnules enlarged, appearing glabrous adaxially. **Costae** green adaxially for most of length; abaxial scales multiseriate, lanceolate, truncate or subcordate at base, without overlapping basal lobes, conspicuous, the largest 0.4–0.8 mm wide, loosely imbricate, not concealing ultimate segments, erose, not ciliate. **Ultimate segments** spatulate, not especially beadlike, the largest usually 2–3 mm, abaxially glabrous or with a few small scales near base, adaxially glabrous. **False indusia** marginal, weakly differentiated, 0.05–0.25 mm wide. **Sori** discontinuous, confined to apical or lateral lobes. **Sporangia** containing 64 spores. $2n = 60$.

Sporulating late spring–fall. Rocky slopes and ledges, usually on igneous substrates; 700–1200 m; Ariz.; n Mexico.

Cheilanthes pringlei is often confused with young, sterile plants of *C. fendleri*, but it is easily distinguished from the latter by having rachises that are grooved adaxially. This species appears to be restricted to the Sonoran Desert; records from Gila and Cochise counties, Arizona, and southern New Mexico are based on misidentifications.

2. Cheilanthes horridula Maxon, Amer. Fern J. 8: 94. 1918 · Prickly lip fern

Pellaea aspera (Hooker) Baker

Stems short-creeping, usually 4–7 mm diam.; scales uniformly brown or with poorly defined, dark central stripe, linear-lanceolate, straight to slightly contorted, loosely appressed, persistent. **Leaves** clustered, 5–30 cm; vernation noncircinate. **Petiole** black to dark brown, rounded adaxially. **Blade** linear-oblong to lanceolate, pinnate-pinnatifid to 2-pinnate at base, 1–4 cm wide; rachis rounded adaxially, with scattered linear-lanceolate scales and dimorphic pubescence, abaxially sparsely hirsute, adaxially covered with tortuous, appressed hairs. **Pinnae** not articulate, dark color of stalk continuing into pinna base, basal pair not conspicuously larger than adjacent pair, usually equilateral, appearing pustulose adaxially. **Costae** green adaxially for most of length; abaxial scales multiseriate, lanceolate, truncate to subcordate at base, without overlapping basal lobes, somewhat inconspicuous, the largest 0.4–0.6 mm wide, loosely imbricate, not concealing ultimate segments, erose, not ciliate. **Ultimate segments** narrowly elliptic to elongate-deltate, not beadlike, the largest 3–5 mm, abaxially and adaxially scabrous with stiff, usually pustulose hairs. **False indusia** marginal, slightly differentiated, 0.05–0.25 mm wide. **Sori** ± continuous around segment margins. **Sporangia** containing 64 spores. $2n = 58, 116$.

Sporulating summer–fall. Rocky slopes and ledges, usually on limestone; 100–1400 m; Okla., Tex.; n Mexico.

The scabrous, pustulose hairs of *Cheilanthes horridula* make it one of the most distinctive species of *Cheilanthes* in North America. As currently circumscribed, the species includes two sexually reproducing cytotypes that may be given formal recognition when their mor-

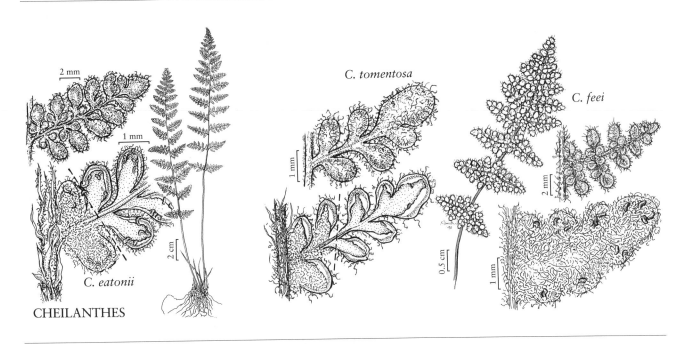

C. tomentosa

C. feei

C. eatonii

CHEILANTHES

phologic characteristics and distributions are sufficiently well known.

3. Cheilanthes tomentosa Link, Hort. Berol. 2: 42. 1833 · Woolly lip fern

Myriopteris tomentosa (Link) J. Smith

Stems compact, usually 4–8 mm diam.; scales mostly bicolored, with broad, well-defined, dark, central stripe and narrow, light brown margins, linear-lanceolate, straight to slightly contorted, loosely appressed, persistent. **Leaves** clustered, 8–45 cm; vernation noncircinate. **Petiole** usually dark brown, rounded adaxially. **Blade** oblong-lanceolate, usually 4-pinnate at base, 1.5–8 cm wide; rachis rounded abaxially, with scattered linear scales and monomorphic pubescence. **Pinnae** not articulate, dark color of stalk continuing into pinna base, basal pair not conspicuously larger than adjacent pair, usually equilateral, appearing tomentose adaxially. **Costae** green adaxially for most of length; abaxial scales multiseriate, linear, truncate at base, inconspicuous, the largest 0.1–0.4 mm wide, loosely imbricate, not concealing ultimate segments, usually entire, not ciliate. **Ultimate segments** oval or rarely oblong, beadlike, the largest 1–2 mm, abaxially densely tomentose, adaxially pubescent with fine, unbranched hairs. **False indusia** marginal to obscurely inframarginal, somewhat differ-

entiated, 0.05–0.25 mm wide. **Sori** ± continuous around segment margins. **Sporangia** containing 32 spores. $n = 2n = 90$, apogamous.

Sporulating summer–fall. Rocky slopes and ledges, on a variety of substrates including limestone and granite; 200–2400 m; Ala., Ariz., Ark., Ga., Kans., Mo., N.Mex., N.C., Okla., Pa., S.C., Tenn., Tex., Va., W.Va.; Mexico.

Cheilanthes tomentosa is an apogamous triploid of unknown parentage. It is closely related to *C. eatonii*, but it is distinguished by having narrower, less prominent costal scales. Natural hybrids between these two species have been reported (D. S. Correll 1956), but such hybrids are unlikely because both species are apogamous in the supposed region of hybridization.

4. Cheilanthes gracillima D. C. Eaton in Emory, Rep. U.S. Mex. Bound. 2: 234. 1859 · Lace fern

Myriopteris gracillima (D. C. Eaton) J. Smith

Stems short-creeping, 4–8 mm diam.; scales uniformly brown or with poorly defined, dark, central stripe, linear-lanceolate, straight to slightly contorted, loosely appressed, persistent. **Leaves** clustered, 5–25 cm; vernation noncircinate. **Petiole** dark brown, rounded adaxially. **Blade** linear-oblong, 2–3-pinnate at base, 1–2.5 cm wide; rachis rounded adaxially, with scattered

Cheilanthes

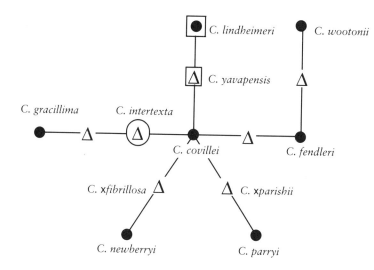

Relationships among some species of *Cheilanthes*. Solid circles represent parental taxa; boxed circles represent apogamous parental taxa; circled triangles represent fertile allotetraploids; boxed triangles represent apogamous tetraploids; and triangles represent sterile hybrids.

linear scales, not pubescent. **Pinnae** not articulate, dark color of stalk continuing into pinna base, basal pair not conspicuously larger than adjacent pair, usually equilateral, appearing sparsely pubescent or glabrous adaxially. **Costae** green adaxially for most of length; abaxial scales multiseriate, linear, truncate at base, inconspicuous, the largest 0.1–0.4 mm wide, loosely imbricate, not concealing ultimate segments, long-ciliate, cilia usually confined to base. **Ultimate segments** oblong or rarely oval, beadlike, the largest 1.5–3 mm, abaxially densely covered with branched hairs and small, ciliate scales, adaxially with scattered, branched hairs or glabrescent. **False indusia** marginal, slightly differentiated, 0.05–0.25 mm wide. **Sori** ± continuous around segment margins. **Sporangia** containing 64 spores.

Sporulating summer–fall. Cliffs and rocky slopes, usually on igneous substrates; 800–3000 m; Alta., B.C.; Calif., Idaho, Mont., Nev., Oreg., Utah, Wash.

Cheilanthes gracillima is a well-marked species, but it apparently hybridizes with *C. intertexta* (see reticulogram) to produce plants of intermediate morphology with malformed spores that have been called *C. gracillima* var. *aberrans* M. E. Jones (A. R. Smith 1974).

5. **Cheilanthes fendleri** Hooker, Sp. Fil. 2: 103, plate 107b. 1852 · Fendler's lip fern

Myriopteris fendleri (Hooker) E. Fournier

Stems long-creeping, 1–3 mm diam.; scales mostly uniformly brown, ovate-lanceolate, straight to slightly contorted, loosely appressed, often deciduous on older portions of stem. **Leaves** scattered, 7–30 cm; vernation noncircinate. **Petiole** usually dark brown, rounded adaxially. **Blade** lanceolate to ovate-deltate, 3–4-pinnate at base, 1.5–5 cm wide; rachis rounded adaxially, somewhat scaly, not pubescent. **Pinnae** not articulate, dark color of stalk continuing into pinna base, basal pair not conspicuously larger than adjacent pair, usually equilateral, appearing glabrous adaxially. **Costae** green adaxially for most of length; abaxial scales multiseriate, lanceolate-ovate, truncate or subcordate at base, without overlapping basal lobes, conspicuous, the largest 0.4–1.2 mm wide, strongly imbricate, often concealing ultimate segments, entire to denticulate, not ciliate. **Ultimate segments** round to oblong, beadlike, the largest 1.5–3 mm, abaxially glabrous or with a few small scales near base, adaxially glabrous. **False indusia** marginal, weakly differentiated, 0.05–0.25 mm wide. **Sori** ± continuous around segment margins. **Sporangia** containing 64 spores. $2n = 60$.

Sporulating summer–fall. Rocky slopes and ledges; found on a variety of acidic and mildly basic substrates; 1200–3100 m; Ariz., Colo., N.Mex., Tex.; n Mexico.

Young, sterile plants of *Cheilanthes fendleri* are occasionally misidentified as *C. pringlei*; they are distinguished from that species by having rachises that are rounded (not grooved) adaxially. *Cheilanthes fendleri* apparently hybridizes with both *C. wootonii* (T. Reeves 1979) and *C. covillei* (M. D. Windham, unpublished).

6. Cheilanthes eatonii Baker in Hooker & Baker, Syn. Fil. 4: 140. 1867 · Eaton's lip fern

Cheilanthes castanea Maxon; *C. eatonii* forma *castanea* (Maxon) Correll

Stems compact, 4–8 mm diam.; scales mostly bicolored, with broad, well-defined, dark, central stripe and narrow, light brown margins, linear-lanceolate, straight to slightly contorted, loosely appressed, persistent. **Leaves** clustered, 6–35 cm; vernation noncircinate. **Petiole** dark brown, rounded adaxially. **Blade** oblong-lanceolate, 3–4-pinnate at base, 1.5–5 cm wide; rachis rounded adaxially, with scattered linear-lanceolate scales and monomorphic pubescence. **Pinnae** not articulate, dark color of stalk continuing into pinna base, basal pair not conspicuously larger than adjacent pair, usually equilateral, appearing tomentose to glabrescent adaxially. **Costae** green adaxially for most of length; abaxial scales multiseriate, lanceolate to linear, truncate or subcordate at base, without overlapping basal lobes, conspicuous, the largest 0.4–0.7 mm wide, loosely imbricate, not concealing ultimate segments, erose-dentate, rarely with 1–2 cilia at base on a few scales. **Ultimate segments** oval to round, beadlike, the largest 1–3 mm, abaxially densely tomentose, adaxially pubescent with fine, unbranched hairs or glabrescent. **False indusia** marginal to obscurely inframarginal, somewhat differentiated, 0.05–0.25 mm wide. **Sori** ± continuous around segment margins. **Sporangia** containing 32 spores. $n = 2n = 90, 120$, apogamous.

Sporulating summer–fall. Rocky slopes and ledges, found on a variety of substrates including limestone and granite; 300–3000 m; Ariz., Ark., Colo., N.Mex., Okla., Tex., Utah, Va., W.Va.; Mexico; Central America in Costa Rica.

As here circumscribed, *Cheilanthes eatonii* is a variable species comprising apogamous triploid and tetraploid cytotypes of unknown parentage. It includes plants previously identified as *C. castanea* and *C. pinkavii* (ined.). Type specimens of *C. eatonii* and *C. castanea* are quite distinct morphologically, but most plants here

included within *C. eatonii* are intermediate between these two extremes (T. Reeves 1979). Because there is no clear morphologic break, *C. castanea* is placed here in synonymy under *C. eatonii* pending further study. Reports of hybridization between *C. eatonii* and *C. villosa* (D. B. Lellinger 1985) are based on specimens from western Texas and southern New Mexico that appear to be intermediate between these taxa in several characters. T. Reeves (1979) applied the name *C. pinkavii* to these specimens; that name has never been validly published. Formal recognition of this taxon is deferred pending completion of a biosystematic study of the *C. eatonii* complex as a whole.

7. Cheilanthes villosa Davenport ex Maxon, Proc. Biol. Soc. Wash. 31: 142. 1918 · Villous lip fern

Stems compact, 4–8 mm diam.; scales mostly bicolored, with broad, well-defined, dark, central stripe and narrow, light brown margins, linear-lanceolate, straight to slightly contorted, loosely appressed, persistent. **Leaves** clustered, 7–30 cm; vernation noncircinate. **Petiole** usually dark brown, rounded adaxially. **Blade** oblong-lanceolate to ovate, 3–4-pinnate at base, 1.5–5 cm wide; rachis rounded adaxially, with scattered filiform to lanceolate scales, not pubescent. **Pinnae** not articulate, dark color of stalk continuing into pinna base, basal pair not conspicuously larger than adjacent pair, usually equilateral, appearing villous adaxially. **Costae** green adaxially for most of length; abaxial scales multiseriate, ovate to lanceolate, shallowly cordate at base, often with overlapping basal lobes, conspicuous, the largest 0.4–1.5 mm wide, strongly imbricate, often concealing ultimate segments, erose-dentate, not ciliate. **Ultimate segments** round to oval, beadlike, the largest 1–2 mm, abaxially nearly glabrous except for a few coarse hairs, adaxially villous with coarse, unbranched hairs. **False indusia** marginal to obscurely inframarginal, slightly differentiated, 0.05–0.25 mm wide. **Sori** ± continuous around segment margins. **Sporangia** containing 32 spores. $n = 2n = 90$, apogamous.

Sporulating summer–fall. Cliffs and rocky slopes, usually on limestone; 400–2200 m; Ariz., N.Mex., Tex.; n Mexico.

Cheilanthes villosa is an apogamous triploid of unknown parentage. Although there are reports of hybridization between *C. villosa* and *C. eatonii* (D. B. Lellinger 1985), recent gene exchange is unlikely because both taxa are apogamous in North America. Morphologically intermediate specimens (tentatively called *C. pinkavii* in T. Reeves 1979) are included here in *C. eatonii,* pending further study and valid publication of Reeve's epithet.

8. Cheilanthes covillei Maxon, Proc. Biol. Soc. Wash. 31: 147. 1918 · Coville's lip fern

Myriopteris covillei (Maxon) A. Löve & D. Löve

Stems short-creeping, usually 2–4 mm diam.; scales usually uniformly dark brown to black or rarely with narrow, light brown margins, linear-lanceolate, straight to slightly contorted, strongly appressed, persistent. **Leaves** clustered, 5–30 cm; vernation noncircinate. **Petiole** dark brown, rounded adaxially. **Blade** lanceolate to ovate-deltate, 3–4-pinnate at base, 1.5–5 cm wide; rachis rounded adaxially, somewhat scaly, not pubescent. **Pinnae** not articulate, dark color of stalk continuing into pinna base, basal pair not conspicuously larger than adjacent pair, usually equilateral, appearing glabrous (or somewhat scaly) adaxially. **Costae** green adaxially for most of length; abaxial scales multiseriate, ovate-lanceolate, deeply cordate at base, with overlapping basal lobes, conspicuous, the largest 0.4–1.5 mm wide, strongly imbricate, usually concealing ultimate segments, ciliate only on basal lobes. **Ultimate segments** round to oblong, beadlike, the largest 1–3 mm, abaxially glabrous or with a few small scales near base, adaxially glabrous. **False indusia** marginal, weakly differentiated, 0.05–0.25 mm wide. **Sori** ± continuous around segment margins. **Sporangia** containing 64 spores. $2n = 60$.

Sporulating late spring–fall. Rocky slopes, cliffs, and ledges, usually on igneous substrates; 100–2500 m; Ariz., Calif., Nev., Utah; Mexico in Baja California.

Cheilanthes covillei can be difficult to distinguish from the closely related *C. intertexta* and *C. clevelandii;* it differs from these two species in having glabrous blades and costal scales ciliate only on the basal lobes. *Cheilanthes covillei* is occasionally misidentified as *C. fendleri* because the cilia of the scales are often obscure; it is distinguished from the latter species by having rigid, dark brown stem scales that are strongly appressed. *Cheilanthes covillei* hybridizes with *C. parryi* and *C. newberryi* to form rare, sterile diploids known as *C.* ×*parishii* Davenport and *C.* ×*fibrillosa* (Davenport) Davenport ex Underwood, respectively. A third sterile diploid hybrid with *C. fendleri* has recently been discovered in central Arizona (M. D. Windham, unpublished).

9. Cheilanthes intertexta (Maxon) Maxon in Abrams, Ill. Fl. Pacific States 1: 28. 1923 · Coastal lip fern

Cheilanthes covillei Maxon subsp. *intertexta* Maxon, Proc. Biol. Soc. Wash. 31: 149. 1918

Stems short-creeping, usually 3–7 mm diam.; scales usually bicolored, with broad, well-defined, dark, central stripe and narrow, light brown margins, linear-lanceolate, straight to slightly contorted, strongly appressed, persistent. **Leaves** clustered, 4–25 cm; vernation noncircinate. **Petiole** dark brown, rounded adaxially. **Blade** lanceolate to ovate-deltate, usually 3-pinnate at base, 1–4 cm wide; rachis rounded adaxially, with scattered scales and sparse monomorphic pubescence. **Pinnae** not articulate, dark color of stalk continuing into pinna base, basal pair not conspicuously larger than adjacent pair, usually equilateral, appearing glabrous to sparsely pubescent adaxially. **Costae** green adaxially for most of length; abaxial scales multiseriate, ovate-lanceolate, deeply cordate at base, with overlapping basal lobes, conspicuous, the longest 0.4–1 mm wide, imbricate, often concealing ultimate segments, long-ciliate, cilia usually confined to proximal 1/2. **Ultimate segments** oblong to ovate, beadlike, the largest 1–3 mm, abaxially densely covered with branched hairs and small, ciliate scales, adaxially with scattered branched hairs or glabrescent. **False indusia** marginal, weakly differentiated, 0.05–0.25 mm wide. **Sori** ± continuous around segment margins. **Sporangia** containing 64 spores.

Sporulating late spring–fall. Rocky slopes and ledges, usually on igneous substrates; 500–2800 m; Calif., Nev., Oreg.

Preliminary isozyme analyses support D. B. Lellinger's (1985) suggestion that *Cheilanthes intertexta* is a fertile allotetraploid hybrid between *C. gracillima* and *C. covillei.* It is morphologically most similar to the latter parent (see comments under *C. covillei*), but it is occasionally confused with *C. gracillima,* with which it apparently hybridizes to form sterile intermediates that have been called *C. gracillima* var. *aberrans* M. E. Jones. *Cheilanthes intertexta* may also be confused with *C. clevelandii,* with which it is partially sympatric. In addition to the characters given in the key, *C. intertexta* is distinguished from closely related sexual species by having larger spores averaging more than 55 μm in diameter.

10. **Cheilanthes clevelandii** D. C. Eaton, Bull. Torrey Bot. Club 6: 33. 1875 · Cleveland's lip fern

Stems usually short-creeping, 1–3 mm diam.; scales usually bicolored, with well-defined, dark, central stripe and broad, light brown margins, lanceolate, straight to slightly contorted, strongly appressed, persistent. **Leaves** clustered to somewhat scattered, 8–40 cm; vernation noncircinate. **Petiole** dark to light brown, rounded adaxially. **Blade** oblong-lanceolate to ovate, usually 4-pinnate at base, 2–8 cm wide; rachis rounded adaxially, with scattered scales and sparse monomorphic pubescence. **Pinnae** not articulate, dark color of stalk continuing into pinna base, basal pair not conspicuously larger than adjacent pair, usually equilateral, appearing glabrous adaxially. **Costae** green adaxially for most of length; abaxial scales multiseriate, ovate-lanceolate, deeply cordate at base, with overlapping basal lobes, conspicuous, 0.4–1 mm wide, imbricate, occasionally concealing ultimate segments, ciliate (usually only on proximal 1/2). **Ultimate segments** round to subcordate, beadlike, the largest 1–2 mm, abaxially with branched hairs and small ciliate scales, adaxially glabrous. **False indusia** marginal, weakly differentiated, 0.05–0.25 mm wide. **Sori** ± continuous around segment margins. **Sporangia** containing 64 spores.

Sporulating late spring–summer. Rocky slopes and ledges, usually on igneous substrates; 0–1600 m; Calif.; Mexico in Baja, California.

Although some specimens of *Cheilanthes clevelandii* can be difficult to distinguish from *C. covillei* and *C. intertexta* (see comments under *C. covillei*), the species is restricted to the coastal mountains of California and Baja California, and it rarely overlaps the ranges of these closely related species. In the region where *C. clevelandii* is sympatric with *C. intertexta*, the smaller spores of the former species (averaging less than 55 μm in diameter) are helpful in identification. T. Reeves (1979) tentatively identified two varieties of *C. clevelandii*, but formal recognition of these taxa must await further study.

11. **Cheilanthes lindheimeri** Hooker, Sp. Fil. 2: 101, plate 107a. 1852 · Fairy swords

Myriopteris lindheimeri (Hooker) J. Smith

Stems long-creeping, 0.7–3 mm diam.; scales uniformly brown, ovate-lanceolate, straight to slightly contorted, loosely appressed, often deciduous on older portions of stem. **Leaves** scattered, 7–30 cm; vernation noncircinate. **Petiole** usually dark brown, rounded adaxially. **Blade** oblong-lanceolate to ovate-deltate, 4-pinnate at base, 2–5 cm wide; rachis rounded adaxially, with scattered linear-lanceolate scales and sparse monomorphic pubescence. **Pinnae** not articulate, dark color of stalk continuing into pinna base, basal pair not conspicuously larger than adjacent pair, usually equilateral, appearing densely tomentose adaxially. **Costae** green adaxially for most of length; abaxial scales multiseriate, lanceolate-ovate, truncate to cordate at base, usually without overlapping basal lobes, conspicuous, the largest 0.4–1 mm wide, strongly imbricate, often concealing ultimate segments, long-ciliate throughout, cilia fine, curly, forming entangled mass. **Ultimate segments** round to slightly oblong, beadlike, the largest 0.7–1 mm, abaxially nearly glabrous, often with a few small scales or branched hairs, adaxially appearing tomentose but actually nearly glabrous. **False indusia** marginal, weakly differentiated, 0.05–0.25 mm wide. **Sori** ± continuous around segment margins. **Sporangia** containing 32 spores. $n = 2n = 90$, apogamous.

Sporulating summer–fall. Rocky slopes and ledges; on a variety of acidic to mildly basic substrates; 200–2500 m; Ariz., N.Mex., Tex.; Mexico.

Cheilanthes lindheimeri is an apogamous triploid of unknown parentage. It is occasionally misidentified as *C. wootonii* (actually the element here recognized as *C. yavapensis*); most specimens can be placed using the characteristics given in the key. The adaxial blade surface appears to be densely gray tomentose, but this is an illusion created by the fine, curly cilia of the abaxial costal scales that overtop the minute ultimate segments and form an entangled mass that prevents the easy removal of individual costal scales.

12. **Cheilanthes wootonii** Maxon, Proc. Biol. Soc. Wash. 3: 146. 1918 · Wooton's lip fern

Stems long-creeping, 1–3 mm diam.; scales uniformly brown or weakly bicolored with poorly defined, dark, central stripe, lanceolate-ovate, straight to slightly contorted, loosely appressed, often deciduous on older portions of stem. **Leaves** scattered, 7–35 cm; vernation noncircinate. **Petiole** usually dark brown, rounded adaxially. **Blade** oblong-lanceolate, 3–4-pinnate at base, 2–5 cm wide; rachis rounded adaxially, with scattered linear-lanceolate scales and sparse monomorphic pubescence. **Pinnae** not articulate, dark color of stalk continuing into pinna base, basal pair not conspicuously larger than adjacent pair, usually equilateral, appearing glabrous adaxially. **Costae** green adaxially for most of length; abaxial scales multiseriate, lanceolate-ovate, truncate or subcordate at base, without overlapping basal lobes, conspicuous, the largest 0.4–0.8 mm wide, strongly imbricate, often concealing ultimate segments, ciliate, with coarse cilia often confined to proximal 1/2. **Ultimate segments** round to oblong, beadlike, the largest 1–3 mm, abaxially glabrous or with a few small scales near base, adaxially glabrous. **False indusia** marginal, weakly differentiated, 0.05–0.25 mm wide. **Sori** ± continuous around segment margins. **Sporangia** containing 32 spores. $n = 2n = 90$, apogamous.

Sporulating summer–fall. Rocky slopes and ledges, usually on igneous substrates; 800–2900 m; Ariz., Calif., Colo., Nev., N.Mex., Okla., Tex., Utah; n Mexico.

Like its close relative *Cheilanthes lindheimeri*, *C. wootonii* is an apogamous triploid of unknown parentage. With the recognition of *C. yavapensis* as a distinct species, the name *C. wootonii* is restricted to populations with leaf blades that appear glabrous adaxially, costal scales that are often ciliate only in the proximal half, and stem scales that are usually brown and loosely appressed. In addition, *C. wootonii* is distinguished from *C. yavapensis* by having smaller spores, averaging less than 62 μm in diameter. These characteristics can be subtle, and some specimens will be difficult to place in either *C. wootonii* or *C. yavapensis*. T. Reeves (1979) identified several specimens from Arizona that he hypothesized were hybrids between *C. wootonii* and *C. fendleri*.

13. **Cheilanthes yavapensis** T. Reeves ex Windham, Contr. Univ. Michigan Herb. 19: 32. 1993 · Yavapai lip fern

Stems long-creeping, 1–3 mm diam.; scales often bicolored, with broad, poorly defined, dark, central stripe and narrow, brown margins, lanceolate, straight to slightly contorted, strongly appressed, persistent. **Leaves** scattered, 7–35 cm; vernation noncircinate. **Petiole** dark brown, rounded adaxially. **Blade** oblong-lanceolate to nearly ovate, 4-pinnate at base, 2–6 cm wide; rachis rounded adaxially, with scattered linear-lanceolate scales and sparse monomorphic pubescence. **Pinnae** not articulate, dark color of stalk continuing into pinna base, basal pair not conspicuously larger than adjacent pair, usually equilateral, appearing sparsely pubescent adaxially. **Costae** green adaxially for most of length; abaxial scales multiseriate, lanceolate, truncate to cordate at base, without overlapping basal lobes, conspicuous, the largest 0.4–1 mm wide, strongly imbricate, often concealing ultimate segments, ciliate, cilia coarse, usually distributed entire length of scale. **Ultimate segments** round to oblong, beadlike, the largest usually 1–2 mm, abaxially glabrous or with a few small scales near base, adaxially appearing sparsely pubescent but actually nearly glabrous. **False indusia** marginal, weakly differentiated, 0.05–0.25 mm wide. **Sori** ± continuous around segment margins. **Sporangia** containing 32 spores. $n = 2n = 120$, apogamous.

Sporulating summer–fall. Rocky slopes and ledges, usually on igneous substrates; 500–2400 m; Ariz., N.Mex., Tex.

Cheilanthes yavapensis is an apogamous tetraploid, apparently formed by hybridization between *C. lindheimeri* and *C. covillei* (G. J. Gastony and M. D. Windham 1989). Although *C. yavapensis* has long been included within the concept of *C. wootonii*, the discovery that the similarities resulted from hybrid convergence rather than common ancestry requires that they be recognized as two distinct species. Unfortunately, the morphologic characteristics that separate these taxa are subtle, and careful study will be necessary to determine the proper dispositon of problematic specimens. In addition to the characteristics mentioned in the key, *C. yavapensis* is distinguished from *C. wootonii* by having larger spores, averaging more than 62 μm in diameter.

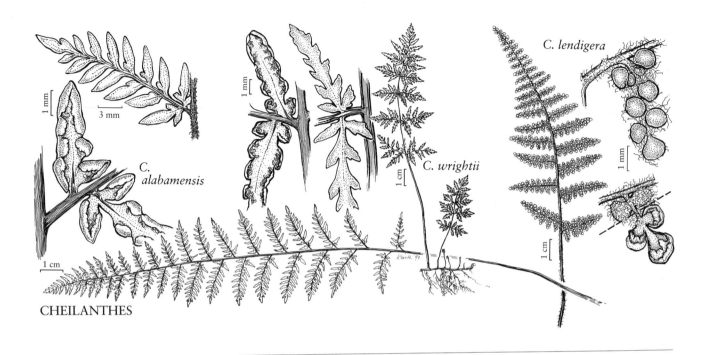

CHEILANTHES

14. Cheilanthes arizonica (Maxon) Mickel, Phytologia 41: 433. 1979 · Arizona lip fern

Cheilanthes pyramidalis Fée subsp. *arizonica* Maxon, Amer. Fern J. 8: 116. 1918; *C. pyramidalis* Fée var. *arizonica* (Maxon) Broun

Stems short-creeping, usually 4–8 mm diam.; scales uniformly dark brown or with poorly defined, dark, central stripe, narrowly lanceolate, straight to slightly contorted, loosely appressed, usually persistent. **Leaves** clustered, 8–25 cm; vernation circinate. **Petiole** dark brown, grooved most of length adaxially. **Blade** elongate-pentagonal, 3-pinnate-pinnatifid to 4-pinnate at base, 4–10 cm wide; rachis grooved adaxially, lacking scales, with minute, widely scattered hairs. **Pinnae** not articulate, dark color of stalk continuing into pinna base, basal pair conspicuously larger than adjacent pair, inequilateral, proximal basiscopic pinnules enlarged, glabrous adaxially. **Costae** green adaxially for most of length; abaxial scales absent. **Ultimate segments** elliptic to linear-oblong, not beadlike, the largest 4–10 mm, abaxially with reddish, scattered, sessile glands, adaxially glabrous; false indusia marginal to obscurely inframarginal, strongly differentiated, 0.25–0.50 mm wide. **Sori** ± continuous around segment margins. **Sporangia** containing 32 spores. $n = 2n = 90$, apogamous.

Sporulating summer–fall. Rocky slopes and ledges, on quartzite or igneous substrates; 1900–2400 m; Ariz.; Mexico; Central America in Guatemala, Honduras.

The distinctive *Cheilanthes arizonica* is an apogamous triploid of unknown parentage. In the flora, it is known from about seven localities, all of which are situated within 50 km of the Mexican border.

15. Cheilanthes wrightii Hooker, Sp. Fil. 2: 87, plate 110, fig. A. 1858 · Wright's lip fern

Stems long-creeping, 1–3 mm diam.; scales uniformly brown or slightly darker at base, linear-lanceolate, straight to slightly contorted, loosely appressed, often deciduous on older portions of stem. **Leaves** clustered to somewhat scattered, 4–25 cm; vernation circinate. **Petiole** brown, grooved adaxially. **Blade** lanceolate to ovate-deltate, 2-pinnate-pinnatifid at base, 1–4 cm wide; rachis grooved adaxially, not scaly or pubescent. **Pinnae** not articulate, dark color of stalk continuing into pinna base, basal pair often slightly larger than adjacent pair, ± equilateral, appearing glabrous adaxially. **Costae** green adaxially for most of length; abaxial scales absent. **Ultimate segments** oblong to linear, not beadlike, the largest 3–7 mm, abaxially and adaxially glabrous. **False indusia** marginal, slightly differentiated, 0.05–0.25 mm wide.

Sori discontinuous, concentrated on interrupted lateral lobes. **Sporangia** containing 64 spores. $2n = 60$.

Sporulating summer–fall. Rocky slopes and ledges usually on igneous substrates; 300–2000 m; Ariz., N.Mex., Tex.; n Mexico.

The glabrous *Cheilanthes wrightii* is occasionally confused with *C. alabamensis* and certain species of *Pellaea*. *Cheilanthes wrightii* is easily separated from *C. alabamensis* by its glabrous rachis, which is grooved on the adaxial surface. It is distinguished from all local members of *Pellaea* by having both a grooved rachis and a thin, long-creeping stem.

16. Cheilanthes aemula Maxon, Contr. U.S. Natl. Herb. 10: 495. 1908 · Texas lip fern

Stems short-creeping, usually 4–7 mm diam.; scales uniformly brown or slightly darker at base, linear-lanceolate, straight to slightly contorted, loosely appressed, persistent. **Leaves** clustered, 10–50 cm; vernation noncircinate. **Petiole** black to dark brown, rounded adaxially. **Blade** ovate-deltate, 3-pinnate to 3-pinnate-pinnatifid at base, 4–15 cm wide; rachis rounded adaxially, lacking scales, with dimorphic pubescence, abaxially sparsely hirsute, adaxially covered with tortuous, appressed hairs. **Pinnae** not articulate, dark color of stalk continuing into pinna base, basal pair slightly larger than adjacent pair, somewhat inequilateral, proximal basiscopic pinnules conspicuously enlarged, appearing glabrous or sparsely pubescent adaxially. **Costae** black adaxially for most of length; abaxial scales absent. **Ultimate segments** narrowly elliptic to elongate-deltate, not beadlike, the largest 3–6 mm, abaxially and adaxially sparsely hirsute to glabrescent. **False indusia** marginal, slightly differentiated, 0.05–0.25 mm wide. **Sori** somewhat discontinuous, often concentrated on interrupted lateral lobes. **Sporangia** containing 64 spores. $2n = 58$.

Sporulating summer–fall. Rocky slopes and ledges, apparently confined to limestone; 100–500 m; Tex.; n Mexico.

In addition to the characteristics mentioned in the key, *Cheilanthes aemula* is distinguished from North American populations of *C. alabamensis* by having 64 spores per sporangium rather than 32. In the flora, this species is known from about 10 localities in central and western Texas.

17. Cheilanthes alabamensis (Buckley) Kunze, Linnaea 20: 4. 1847 · Alabama lip fern

Pteris alabamensis Buckley, Amer. J. Sci. Arts 45: 177. 1843; *Pellaea alabamensis* (Buckley) Hooker

Stems short-creeping to compact, 3–7 mm diam.; scales uniformly brown or slightly darker at base, linear-lanceolate, straight to slightly contorted, loosely appressed, persistent. **Leaves** clustered, 6–50 cm; vernation noncircinate. **Petiole** black, rounded adaxially. **Blade** lanceolate to linear-oblong, 2-pinnate to 2-pinnate-pinnatifid at base, 1–7 cm wide; rachis rounded adaxially, lacking scales, with dimorphic pubescence, abaxially sparsely hirsute, adaxially covered with tortuous, appressed hairs. **Pinnae** not articulate, dark color of stalk continuing into pinna base, basal pair slightly smaller than adjacent pair, ± equilateral, appearing glabrous or sparsely pubescent adaxially. **Costae** green adaxially for most of length; abaxial scales absent. **Ultimate segments** narrowly elliptic to elongate-deltate, not beadlike, the largest 3–7 mm, abaxially and adaxially sparsely hirsute to glabrescent. **False indusia** marginal to obscurely inframarginal, somewhat differentiated, 0.1–0.4 mm wide. **Sori** ± continuous around segment margins. **Sporangia** containing 32 spores. $n = 2n = 87$, apogamous.

Sporulating summer–fall. Rocky slopes, cliffs, and ledges, usually on limestone; 100–2000 m; Ala., Ariz., Ark., Ga., Kans., Ky., La., Mo., N.Mex., N.C., Okla., Tenn., Tex., Va.; Mexico.

Plants of *Cheilanthes alabamensis* occurring in the flora are apogamous triploids; a sexual diploid cytotype has been found in Nuevo León, Mexico (M. D. Windham, unpublished). Given the high degree of morphologic similarity between the two cytotypes, the North American triploid probably was derived from the Mexican diploid through autopolyploidy. In the flora, *Cheilanthes alabamensis* is most often confused with *C. microphylla*, from which it is distinguished by having thicker stems, mostly green costae, and 32-spored sporangia.

18. Cheilanthes microphylla (Swartz) Swartz, Syn. Fil., 127. 1806 · Southern lip fern

Adiantum microphyllum Swartz, Prodr., 135. 1788

Stems long-creeping, 1–3 mm diam.; scales uniformly brown or slightly darker at base, linear-lanceolate, straight to slightly contorted, loosely appressed, persistent. **Leaves** usually scattered, 8–40 cm; vernation noncircinate. **Petiole** black, rounded adaxially. **Blade** lanceolate to linear-oblong, 2-pinnate-pinnatifid to 3-pinnate at base, 1.5–6 cm wide; rachis rounded adaxially, lacking scales, with dimorphic pubescence, abaxially sparsely hirsute, adaxially covered with tortuous, appressed hairs. **Pinnae** not articulate, dark color of stalk continuing into pinna base, basal pair often slightly larger than adjacent pair, ± equilateral, appearing glabrous or sparsely pubescent adaxially. **Costae** black adaxially for most of length; abaxial scales absent. **Ultimate segments** narrowly elliptic to elongate-deltate, not beadlike, the largest 3–7 mm, abaxially and adaxially sparsely hirsute to glabrescent. **False indusia** marginal to obscurely inframarginal, somewhat differentiated, 0.1–0.4 mm wide. **Sori** somewhat discontinuous, often concentrated on interrupted lateral lobes. **Sporangia** containing 64 spores. $2n = 116$.

Sporulating summer–fall. Calcareous rock outcrops and shell mounds; 0–100 m; Fla.; Mexico; West Indies; Central America; South America.

In the flora, the primarily Caribbean *Cheilanthes microphylla* is known from a small number of localities on the Florida peninsula. This restricted distribution, combined with its smaller stems, mostly black costae, and 64-spored sporangia, helps to separate *Cheilanthes microphylla* from the closely related *C. alabamensis*.

19. Cheilanthes lendigera (Cavanilles) Swartz, Syn. Fil., 328. 1806 · Beaded lip fern

Pteris lendigera Cavanilles, Descr. Pl., 268. 1802; *Myriopteris lendigera* (Cavanilles) J. Smith

Stems long-creeping, 1–3 mm diam.; scales uniformly brown or with poorly defined, dark, central stripe, linear-lanceolate, straight to slightly contorted, loosely appressed, usually persistent. **Leaves** scattered to clustered, 5–30 cm; vernation noncircinate. **Petiole** usually dark brown, rounded adaxially. **Blade** ovate-deltate to oblong-lanceolate, usually 4-pinnate at base, 1.5–8 cm wide; rachis rounded adaxially, with scattered linear scales and dense mono-

morphic pubescence. **Pinnae** not articulate, dark color of stalk continuing into pinna base, basal pair not conspicuously larger than adjacent pair, usually equilateral, appearing glabrous or sparsely pubescent adaxially. **Costae** green adaxially for most of length; abaxial scales uniseriate and hairlike. **Ultimate segments** round to slightly oblong, beadlike, the largest 1–3 mm, abaxially sparsely to moderately pubescent with coarse hairs, adaxially glabrous. **False indusia** inframarginal, strongly differentiated, 0.25–0.5 mm wide, forming pouch with constricted aperture on abaxial surface of ultimate segments. **Sori** ± continuous around segment margins. $2n = 120$.

Sporulating summer–fall. Rocky slopes and ledges, usually on igneous substrates; 1300–2400 m; Ariz., Tex.; Mexico; Central America; South America.

Cheilanthes lendigera has the small, beadlike ultimate segments characteristic of subgenus *Physapteris*; the prominent inframarginal false indusia and near absence of multiseriate costal scales serve to distinguish it from all other North American members of that group. T. Reeves (1979) suggested that *C. lendigera* is a fertile allotetraploid resulting from hybridization between the Mexican species *Cheilanthes mexicana* Davenport and *C. marsupianthes* (Fée) T. Reeves ex Windham (unpublished).

20. Cheilanthes bonariensis (Willdenow) Proctor, Bull. Inst. Jamaica, Sci. Ser. 5(1): 15. 1953 · Bonaire lip fern

Acrostichum bonariense Willdenow, Sp. Pl. 5: 114. 1810; *Notholaena aurea* (Poiret) Desvaux

Stems short-creeping to compact, usually 4–8 mm diam.; scales bicolored, with broad, well-defined, dark, central stripe and narrow, light brown margins, narrowly lanceolate, slightly contorted, strongly appressed, persistent. **Leaves** clustered, 10–60 cm; vernation noncircinate. **Petiole** dark brown, rounded adaxially. **Blade** linear, pinnate-pinnatifid throughout, 1–4 cm wide; rachis rounded adaxially, lacking scales, with dense monomorphic pubescence. **Pinnae** articulate at swollen, hirsute nodes, basal pair slightly smaller than adjacent pair, ± equilateral, appearing hirsute adaxially. **Costae** absent. **Ultimate segments** elongate-deltate to ovate, not especially beadlike, the largest 1–7 mm, abaxially densely tomentose, adaxially hirsute. **False indusia** marginal, weakly differentiated, 0.05–0.25 mm wide. **Sori** ± continuous around segment margins. **Sporangia** containing 32 spores. $n = 2n = 90$, apogamous.

Sporulating summer–fall. Rocky slopes and ledges;

found on a variety of substrates though rarely observed on limestone; 1200–2400 m; Ariz., N.Mex., Tex.; Mexico; West Indies; Central America; South America.

Cheilanthes bonariensis has been assigned to *Notholaena* in past treatments. It is distantly related (at best) to the species here included in *Notholaena*, however, and we concur with R. M. Tryon and A. F. Tryon (1982) that it should be transferred to *Cheilanthes*. Chromosomal studies (G. J. Gastony and M. D. Windham 1989) suggest that *C. bonariensis* is an apogamous triploid that arose through autopolyploidy. Further investigation is necessary to determine whether 64-spored, sexually reproducing populations of *C. bonariensis* are still extant.

21. Cheilanthes kaulfussii Kunze, Linnaea 13: 145. 1839 · Glandular lip fern

Stems short-creeping, usually 4–8 mm diam.; scales uniformly black or with narrow brown margins, linear-subulate, straight to slightly contorted, strongly appressed, persistent. **Leaves** clustered, 8–35 cm; vernation circinate. **Petiole** dark brown, flattened or slightly grooved distally on adaxial surface. **Blade** elongate-pentagonal, 3-pinnate-pinnatifid to 4-pinnate at base, 3–10 cm wide; rachis grooved adaxially, lacking scales, with monomorphic pubescence. **Pinnae** not articulate, dark color of stalk continuing into pinna base, basal pair larger than adjacent pair, strongly inequilateral, proximal basiscopic pinnules greatly enlarged, appearing glandular-pubescent adaxially. **Costae** green or straw-colored adaxially for most of length; abaxial scales absent. **Ultimate segments** linear-oblong, not especially beadlike, largest 3–8 mm, abaxially and adaxially glandular pubescent with short, sticky, capitate glands. **False indusia** marginal, weakly differentiated, 0.05–0.25 mm wide. **Sori** usually discontinuous, concentrated on apical and lateral lobes. **Sporangia** containing 32 spores.

Sporulating summer–fall. Rocky slopes and ledges, usually on igneous substrates; 300–2500 m; Tex.; Mexico; Central America; South America.

The few populations of *Cheilanthes kaulfussii* known in the flora produce 32 spores per sporangium and reproduce apogamously (D. M. Benham 1982). Although the chromosome number of North American specimens has not been established with certainty, the specimens appear to be polyploids that may have been derived from 64-spored Mexican populations through autopolyploidy. The species is quite distinctive and should not be confused with any other member of the flora.

22. Cheilanthes leucopoda Link, Fil. Spec., 66. 1841 · White-footed lip fern

Stems compact, usually 4–10 mm diam.; scales uniformly brown, linear-subulate, straight to slightly contorted, loosely appressed, persistent. **Leaves** clustered, 7–30 cm; vernation circinate. **Petiole** straw-colored, shallowly grooved distally on adaxial surface. **Blade** elongate-pentagonal, 4-pinnate at base, 3–10 cm wide; rachis grooved adaxially, lacking scales, with monomorphic pubescence. **Pinnae** not articulate, color of stalk continuing into pinna base, basal pair larger than adjacent pair, strongly inequilateral, proximal basiscopic pinnules greatly enlarged, appearing hirsute adaxially. **Costae** green or straw-colored adaxially for entire length; abaxial scales absent. **Ultimate segments** oblong to lanceolate, not especially beadlike, the largest 3–5 mm, abaxially and adaxially hirsute with long, noncapitate hairs. **False indusia** marginal, weakly differentiated, 0.05–0.25 mm wide. **Sori** usually discontinuous, concentrated on apical and lateral lobes. **Sporangia** containing 32 spores. $2n = 60$.

Sporulating summer–fall. Rocky slopes and ledges; apparently confined to limestone; 300–500 m; Tex.; n Mexico.

In North America, *Cheilanthes leucopoda* is known only from the Edwards Plateau in west central Texas. It is unique among local *Cheilanthes* species in being a sexual diploid that consistently produces 32 spores per sporangium.

23. Cheilanthes newberryi (D. C. Eaton) Domin, Biblioth. Bot. 20(85): 133. 1915 · Newberry's lip fern

Notholaena newberryi D. C. Eaton, Bull. Torrey Bot. Club 4: 12. 1873

Stems short-creeping to compact, usually 3–5 mm diam.; scales uniformly dark brown to black or with light brown, ephemeral tip, linear-subulate, slightly contorted, strongly appressed, persistent. **Leaves** clustered, 5–30 cm; vernation noncircinate. **Petiole** dark brown, rounded adaxially. **Blade** lanceolate to linear-oblong, 2-pinnate-pinnatifid to 3-pinnate at base, 1.5–5 cm wide; rachis rounded adaxially, lacking scales, with dense monomorphic pubescence. **Pinnae** not articulate, dark color of stalk continuing into pinna base, basal pair slightly smaller than adjacent pair, ± equilateral, proximal basiscopic pinnules slightly enlarged, appearing tomentose adaxially. **Costae** green or straw-colored adaxially for most of length; abaxial scales absent.

Ultimate segments oblong to lanceolate, not beadlike, the largest 3–5 mm, abaxially densely woolly with very fine, cobwebby hairs, tomentose adaxially. **False indusia** absent. **Sori** usually discontinuous, concentrated on small apical and lateral lobes. **Sporangia** containing 64 or 32 spores. $2n = 60$.

Sporulating late spring–summer. Rocky cliffs and slopes; usually on igneous substrates; 0–1200 m; Calif.; Mexico in Baja California.

As with *Cheilanthes bonariensis*, *C. newberryi* has traditionally been assigned to *Notholaena*; R. M. Tryon and A. F. Tryon (1982) argued for its placement in *Cheilanthes*. Hybridization of *C. newberryi* with *C. covillei* to form *C.* × *fibrillosa* supports placement in *Cheilanthes*. R. M. Tryon (1956) reported variations in spore number per sporangium in *C. newberryi* that may indicate cytologic variability within the species.

24. Cheilanthes viscida Davenport, Bull. Torrey Bot. Club 6: 191. 1877 · Viscid lip fern

Stems short-creeping, usually 4–8 mm diam.; scales uniformly brown, linear-subulate, strongly contorted, loosely appressed, persistent. **Leaves** clustered, 6–30 cm; vernation circinate. **Petiole** dark brown, flattened or slightly grooved distally on adaxial surface. **Blade** narrowly oblong to linear, 3-pinnate-pinnatifid at base, 1–4 cm wide; rachis flattened or slightly grooved adaxially, lacking scales, with monomorphic pubescence. **Pinnae** not articulate, dark color of stalk continuing into pinna base, basal pair slightly smaller than adjacent pair, ± equilateral, appearing glandular pubescent adaxially. **Costae** green adaxially for most of length; abaxial scales absent. **Ultimate segments** oblong to lanceolate, not beadlike, the largest 3–4 mm, abaxially and adaxially glandular-pubescent with short, sticky, capitate glands. **False indusia** marginal, weakly differentiated, 0.05–0.25 mm wide. **Sori** usually discontinuous, concentrated on apical and lateral lobes. **Sporangia** containing 64 spores.

Sporulating late spring–fall. Cliffs and rocky slopes, usually on igneous substrates; 200–1300 m; Calif.; Mexico in Baja California.

Cheilanthes viscida is confined to a relatively small region in the deserts of California. Variations in spore size among populations suggest that the species may include more than one cytotype.

25. Cheilanthes cooperae D. C. Eaton, Bull. Torrey Bot. Club 6: 33. 1875 · Mrs. Cooper's lip fern

Stems compact to short-creeping, usually 4–8 mm diam.; scales brown at base and darker toward tip, linear-subulate, straight to slightly contorted, loosely appressed, persistent. **Leaves** clustered, 5–30 cm; vernation circinate. **Petiole** dark brown, flattened or slightly grooved distally on adaxial surface. **Blade** linear-oblong to lanceolate-ovate, 3-pinnate at base, 1.5–5 cm wide; rachis flattened or slightly grooved adaxially, lacking scales, with monomorphic pubescence. **Pinnae** not articulate, dark color of stalk continuing into pinna base, basal pair slightly smaller than adjacent pair, ± equilateral, appearing hirsute adaxially. **Costae** green adaxially for most of length; abaxial scales absent. **Ultimate segments** linear-oblong to ovate, not beadlike, the largest 3–5 mm, abaxially and adaxially hirsute with long, strongly flattened hairs. **False indusia** marginal, weakly differentiated, 0.05–0.25 mm wide. **Sori** usually discontinuous, concentrated on apical and lateral lobes. **Sporangia** containing 64 spores. $2n = 60$.

Sporulating late spring–summer. Calcareous cliffs and ledges; usually on limestone; 100–700 m; Calif.

Although scattered throughout much of California, *Cheilanthes cooperae* is apparently rare and quite localized. It appears to be most closely related to *C. viscida*, from which it differs in lacking glandular pubescence. The ranges of the two species do not overlap, and they seem amply distinct.

26. Cheilanthes feei T. Moore, Index Fil., 38. 1857 · Slender lip fern

Stems compact to short-creeping, usually 4–8 mm diam.; scales often uniformly brown but at least some on each plant with well-defined, dark, central stripe, linear-lanceolate, slightly contorted, loosely appressed, persistent. **Leaves** clustered, 4–20 cm; vernation circinate. **Petiole** dark brown to black, rounded adaxially. **Blade** linear-oblong to lanceolate, 3-pinnate at base, 1–3 cm wide; rachis rounded adaxially, lacking scales, with dense monomorphic pubescence. **Pinnae** not articulate, dark color of stalk continuing into pinna base, basal pair usually smaller than adjacent pair, ± equilateral, appearing sparsely pubescent to glabrescent adaxially. **Costae** brown adaxially for most of length; abaxial scales absent. **Ultimate segments** round to slightly oblong, beadlike, the largest 1–3 mm, abaxially densely villous with long, segmented hairs, adaxially sparsely hirsute to glabrescent. **False indusia** marginal, weakly differentiated, 0.05–0.20 mm wide. **Sori** ± continuous around segment margins. **Sporangia** containing 32 spores. $n = 2n = 90$, apogamous.

Sporulating late spring–fall. Calcareous cliffs and ledges, usually on limestone or sandstone; 100–3800 m; Alta., B.C.; Ariz., Ark., Calif., Colo., Idaho, Ill., Iowa, Kans., Ky., Minn., Mo., Mont., Nebr., Nev., N.Mex., Okla., Oreg., S.Dak., Tex., Utah, Va., Wash., Wis., Wyo.; n Mexico.

Cheilanthes feei is an apogamous triploid of unknown parentage. It has small, beadlike blade segments similar to those of subg. *Physapteris,* but most morphological characteristics suggest a clear relationship to members of subg. *Cheilanthes* (T. Reeves 1979). The species is most often confused with *C. parryi,* from which it can be distinguished by its thinner, sparser pubescence and smaller ultimate segments.

27. Cheilanthes parryi (D. C. Eaton) Domin, Biblioth. Bot. 20(85): 133. 1915 · Parry's lip fern

Notholaena parryi D. C. Eaton, Amer. Naturalist 9: 351. 1875

Stems compact to short-creeping, 3–10 mm diam.; scales often uniformly brown but at least some on each plant with well-defined, dark, central stripe, linear-lanceolate, straight to slightly contorted, loosely appressed, persistent. **Leaves** clustered, 4–20 cm; vernation circinate. **Petiole** dark brown to black, rounded adaxially. **Blade** lanceolate to linear-oblong, usually 2-pinnate-pinnatifid at base, 1–4 cm wide; rachis rounded adaxially, lacking scales, with dense monomorphic pubescence. **Pinnae** not articulate, dark color of stalk continuing into pinna base, basal pair usually smaller than adjacent pair, ± equilateral, appearing densely villous adaxially. **Costae** brown adaxially for most of length; abaxial scales absent. **Ultimate segments** oblong to lanceolate, not beadlike, the largest 3–5 mm, abaxially and adaxially densely villous with long, segmented hairs. **False indusia** absent. **Sori** ± continuous around segment margins. **Sporangia** containing 64 spores. $2n = 60$.

Sporulating late spring–fall. Cliffs and ledges, on a variety of substrates including limestone and granite; 100–2300 m; Ariz., Calif., Nev., Utah; Mexico in Baja California, Sonora.

Many authors assign *Cheilanthes parryi* to *Notholaena,* but it is distantly related (at best) to the type species of that genus, and we concur with R. M. Tryon and A. F. Tryon (1982) that it should be tranferred to *Cheilanthes.* Further support for this generic placement is provided by the fact that *C. parryi* hybridizes with *C. covillei* to form the sterile diploid *C.* ×*parishii. Cheilanthes parryi* is most often confused with *C. feei,* from which it can be distinquished by its coarser, denser pubescence and larger ultimate segments.

28. Cheilanthes lanosa (Michaux) D. C. Eaton in Emory, Rep. U.S. Mex. Bound. 2(1): 234. 1859 · Hairy lip fern

Nephrodium lanosum Michaux, Fl. Bor.-Amer. 2: 270. 1803; *Cheilanthes vestita* (Sprengel) Swartz

Stems compact to short-creeping, usually 4–8 mm diam.; scales often uniformly brown but at least some on each plant with thin, poorly defined, dark, central stripe, linear-lanceolate, straight to slightly contorted, loosely appressed, persistent. **Leaves** clustered, 7–50 cm; vernation circinate. **Petiole** dark brown, rounded adaxially. **Blade** linear-oblong to lanceolate, usually 2-pinnate-pinnatifid at base, 1.5–5 cm wide; rachis rounded adaxially, lacking scales, with monomorphic pubescence. **Pinnae** not articulate, dark color of stalk continuing into pinna base, basal pair slightly smaller than adjacent pair, ± equilateral, appearing sparsely hirsute adaxially. **Costae** brown adaxially for most of length; abaxial scales absent. **Ultimate segments** oblong to lanceolate, not beadlike, the largest 3–5 mm, abaxially and adaxially sparsely hirsute with long, segmented hairs. **False indusia** marginal, weakly differentiated, 0.05–0.25 mm wide. **Sori** discontinuous, concentrated on small apical and lateral lobes. **Sporangia** containing 64 spores. $2n = 60$.

Sporulating summer–fall. Rocky slopes and ledges, on a variety of substrates including limestone and granite; 100–800 m; Ala., Ark., Conn., Fla., Ga., Ill., Ind., Kans., Ky., La., Md., Minn., Miss., Mo., N.J., N.Y., N.C., Ohio, Okla., Pa., S.C., Tenn., Tex., Va., W.Va., Wis.

Cheilanthes lanosa is apparently confined to the forests and prairies of eastern North America, and reports of this distinctive species from Arizona and New Mexico (A. J. Petrik-Ott 1979) have not been substantiated by herbarium specimens.

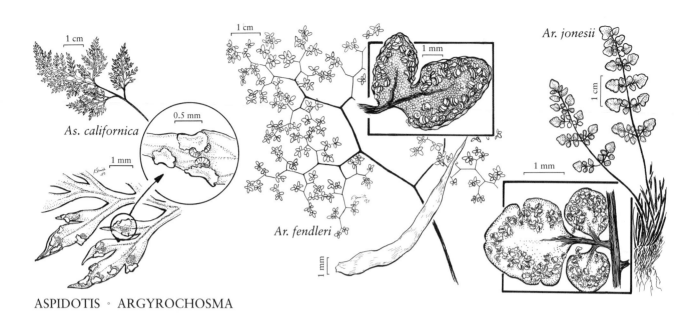

As. californica

Ar. fendleri

Ar. jonesii

ASPIDOTIS · ARGYROCHOSMA

11. ASPIDOTIS (Nuttall ex Hooker & Baker) Copeland, Gen. Fil., 68. 1947 · Lace ferns, aspidote [Greek *aspidotes,* shield-bearer, for the shieldlike false indusia]

Alan R. Smith

Hypolepis Bernhardi sect. *Aspidotis* Nuttall ex Hooker & Baker, Syn. Fil. 4: 131. 1867

Plants terrestrial or on rock. **Stems** ± compact, short-creeping, ascending at tip, branched; scales mostly dark brown, often with very narrow margin of lighter color, lanceolate, margins entire. **Leaves** monomorphic to somewhat dimorphic, crowded, 8–35 cm. **Petiole** usually dark reddish brown, with single groove adaxially, glabrous, with single vascular bundle. **Blade** ovate-triangular, deltate, or pentagonal, 3–4(–5)-pinnate, thick to thin, abaxially glabrous, adaxially lustrous, striate, glabrous; rachis straight. **Ultimate segments** of blades short-stalked or with base narrowed and decurrent onto costa or costule-bearing segments, linear to lanceolate, mostly 0.5–1.3 mm wide; stalks greenish, not darkened; fertile margins recurved. **Veins** of ultimate segments obscure, free, ± pinnate and unbranched. **False indusia** appearing inframarginal, scarious, whitish, broad, partly concealing sporangia. **Sporangia** in marginal, discrete or continuous sori on abaxial surface, containing 64 spores, lacking paraphyses and glands. **Spores** dark brown, tetrahedral-globose, trilete, reticulate, equatorial flange absent. $x = 30$.

Terrestrial, often at bases of boulders or in rock crevices, in dry to moist, montane areas, woodlands, or chaparral, sometimes on ultramafic rocks.

Species 4 (3 in the flora): North America, 1 in Mexico.

D. B. Lellinger (1968) recognized *Aspidotis* as separate from *Cheilanthes* based on its elongate, distantly dentate segments with striate shining surface and on its broad, scarious indusia.

SELECTED REFERENCE   Smith, A. R. 1975. The California species of *Aspidotis. Madroño* 23: 15–24.

1. Mature, fertile blades with continuous sori along length of segments (not at apex); indusia with 10–35 shallow, regular teeth or erose; fertile segments linear, margins ± entire. . . . 3. *Aspidotis densa*
1. Mature, fertile blades with sori discrete or partially discontinuous; indusia with coarse, irregular teeth or entire; fertile segments lanceolate to deltate, distantly dentate.
   2. Sori discrete, 1–3(–5) per blade segment; indusia margins with 2–6 coarse, irregular teeth or ± entire. . . . . 1. *Aspidotis californica*
   2. Sori partially discontinuous, connected by narrow indusial wings, 3–7(–9) per blade segment; indusia margins with 6–10 coarse, irregular teeth or lobes. . . . . . 2. *Aspidotis carlotta-halliae*

1. **Aspidotis californica** (Hooker) Nuttall ex Copeland, Gen. Fil., 68. 1947 · California lace fern

Hypolepis californica Hooker, Sp. Fil. 2: 71, plate 88a. 1852; *Cheilanthes californica* (Hooker) Mettenius

Leaves monomorphic, 10–35 cm. **Blade** 4(–5)-pinnate, 3–12 cm, nearly as wide as long, papery. **Ultimate segments** lanceolate or deltate, 1.5–4 mm; midrib usually obscure abaxially. **Sori** of mature blades short, discrete, 1–3(–5) per segment; indusia semicircular, margins with 2–6 coarse, irregular teeth or ± entire. $2n = 60, 120$.

Rocky outcrops and crevices (not serpentine), moist to dry, shaded slopes and cliffs; 20–1300 m; Calif.; Mexico in n Baja California.

2. **Aspidotis carlotta-halliae** (W. H. Wagner & E. F. Gilbert) Lellinger, Amer. Fern J. 58: 141. 1968

Cheilanthes carlotta-halliae W. H. Wagner & E. F. Gilbert, Amer. J. Bot. 44: 738, figs. 1b, 2b, 2e. 1957

Leaves monomorphic or weakly subdimorphic, 10–30 cm. **Blade** 4-pinnate, 3–12 cm, nearly as wide as long, thin to thick. **Ultimate segments** narrowly lanceolate to deltate, 2–6 mm; midrib obscure or evident abaxially. **Sori** of mature blades ± discrete to usually subcontinuous, 3–7(–9) per segment; indusia semicircular to usually elongate and connecting several adjacent sori, margins with 6–10 irregular and prominent teeth and/or lobes. $2n = 120$.

Generally on serpentine slopes, in crevices, and on rock outcrops; 100–1400 m; Calif.

Aspidotis carlotta-halliae is a fertile allotetraploid species derived from hybridization between *A. californica* and *A. densa*. Occasional sterile backcrosses are found.

3. **Aspidotis densa** (Brackenridge in Wilkes) Lellinger, Amer. Fern J. 58: 141. 1968 · Indian's dream, aspidote dense

Onychium densum Brackenridge in Wilkes, U.S. Expl. Exped. 16: 120, plate 13, fig. 2. 1854, 1855; *Cheilanthes siliquosa* Maxon; *Pellaea densa* (Brackenridge) Hooker

Leaves monomorphic or often somewhat dimorphic, 8–25 cm; fertile leaves more erect than sterile leaves, long-petioled, petioles often 2–5 times longer than blades, fertile blades with more ascending pinnae and narrower segments than sterile blades. **Blade** 3–4-pinnate, 2–10 cm, somewhat leathery. **Ultimate segments** linear, 3–8 mm; midrib prominent abaxially. **Sori** of mature blades continuous along length of segments except at apex; indusia linear, margins with 10–35, shallow, regular teeth or erose. $2n = 60$.

Slopes, crevices, rocky outcrops, often on serpentine, sometimes in chaparral; 300–3400 m; B.C., Que.; Calif., Idaho, Mont., Nev., Oreg., Utah, Wash., Wyo.

12. **ARGYROCHOSMA** (J. Smith) Windham, Amer. Fern J. 77: 38. 1987 [Greek *argyros*, silver, and *chosma*, powder, referring to whitish farina covering the abaxial surface of leaf blades in most species]

Michael D. Windham

Notholaena R. Brown sect. *Argyrochosma* J. Smith, J. Bot. (Hooker) 4: 50. 1841

Plants usually on rock. **Stems** compact, erect to ascending, usually unbranched; scales tan to brown, rarely black, concolored, subulate to narrowly lanceolate, margins entire. **Leaves** monomorphic, clustered, 3–30 cm. **Petiole** brown or black, rounded, flattened, or with single longitudinal groove adaxially, glabrous except for a few scales near base, with single vascular bundle. **Blade** lanceolate, ovate, or deltate, 2–6-pinnate, leathery to somewhat herbaceous, abaxially glabrous or covered by whitish farina, adaxially glabrous or sparsely glandular, dull, not striate; rachis straight or flexuous. **Ultimate segments** of blade stalked or subsessile, usually free from costae, elliptic to ovate or deltate, usually less than 4 mm wide; base often ± cordate, stalks lustrous and dark-colored; segment margins plane or often recurved and forming confluent, poorly defined false indusia extending entire length of segment. **Veins** of ultimate segments free, usually obscure, pinnately branched and divergent distally. **False indusia**, when present, greenish, narrow, clearly marginal, occasionally concealing sporangia. **Sporangia** scattered along veins on abaxial leaf surface, often submarginal, containing 64 or 32 spores, usually intermixed with farina-producing glands. **Spores** brown, tetrahedral-globose, with cristate or rugose surfaces, lacking prominent equatorial ridge. **Gametophytes** glabrous. $x = 27$.

Species ca. 20 (6 in the flora): North America, Mexico, West Indies, Central America, South America.

The species of *Argyrochosma* traditionally have been assigned to either *Notholaena* or *Pellaea*. Comparative studies (M. D. Windham 1987) have revealed that members of *Argyrochosma* are not closely related to *Notholaena* as typified by *N. trichomanoides* (Linnaeus) Desvaux. The two genera show consistent differences in stem and leaf morphology, sporangial distribution, spore color and ornamentation, chromosome base number, gametophyte morphology, and chemical composition of the farina.

Argyrochosma is more closely related to *Pellaea* (M. D. Windham 1987) and members of these genera are occasionally confused. These two groups are easily distinguished, however, based on the presence or absence of farina, leaf segment size and shape, and characteristics of the stem scales and leaf margins. In addition, all species of *Argyrochosma* thus far examined have a chromosome number based on $x = 27$, unique among cheilanthoid ferns. These differences suggest that *Argyrochosma* is monophyletic and worthy of recognition as a distinct genus.

SELECTED REFERENCES Tryon, R. M. 1956. A revision of the American species of *Notholaena*. Contr. Gray Herb. 179: 1–106. Windham, M. D. 1987. *Argyrochosma*, a new genus of cheilanthoid ferns. Amer. Fern J. 77: 37–41.

1. Abaxial surfaces of blades glabrous, lacking whitish farina.
    2. Ultimate leaf segments articulate, dark color of stalks stopping abruptly at segment bases; rachises flattened or shallowly grooved adaxially; margins of fertile ultimate segments usually revolute, often concealing mature sporangia. . . . . .5. *Argyrochosma microphylla*
    2. Ultimate leaf segments not articulate, dark color of stalks continuing into segment bases abaxially; rachises rounded adaxially; margins of fertile ultimate segments plane to somewhat recurved, not concealing mature sporangia. . . . . 6. *Argyrochosma jonesii*
1. Abaxial surfaces of blades obscured by whitish farina.
    3. Ultimate leaf segments articulate, dark color of stalks stopping abruptly at segment bases; margins of ultimate segments plane, not recurved or revolute; sporangia following secondary veins for most of length. 1. *Argyrochosma incana*
    3. Ultimate leaf segments not articulate, dark color of stalks continuing into segment bases abaxially; margins of ultimate segments usually recurved or revolute; sporangia following secondary veins for short distance near segment margin.
        4. Pinna costae distinctly flexuous; branches arising from prominent angles. . . 2. *Argyrochosma fendleri*
        4. Pinna costae straight or nearly so; branches not arising from prominent angles.
            5. Petioles chestnut brown, 0.50–0.75 mm diam.; blades somewhat herbaceous; veins often visible adaxially; sporangia containing 64 spores. . . . . . . 3. *Argyrochosma dealbata*

5. Petioles reddish brown to black, usually more than 0.75 mm diam.; blades leathery; veins obscure; sporangia containing 32 spores. . . . . 4. *Argyrochosma limitanea*

1. Argyrochosma incana (C. Presl) Windham, Amer. Fern J. 77: 40. 1987

Notholaena incana C. Presl, Reliq. Haenk. 1: 19, plate 1, fig. 2. 1825
Stem scales brown. **Leaves** 5–20 cm. **Petiole** black, 0.75–2 mm diam. **Blade** ovate, 3–4-pinnate proximally, leathery, abaxially covered by dense, white farina, adaxially glabrous; rachis rounded to slightly flattened adaxially. **Pinna costae** straight or nearly so, branches not arising from prominent angles. **Ultimate segments** articulate, dark color of stalks stopping abruptly at bases; segment margins plane, never concealing sporangia; veins obscure adaxially. **Sporangia** following secondary veins for most of length, containing 64 spores. $2n = 54$.

Sporulating summer–fall. Sheltered cracks and ledges on canyon walls; restricted to rocks of volcanic origin in the flora; 1200–1700 m; Ariz., N.Mex.; Mexico; West Indies in Hispaniola; Central America in Guatemala.

Argyrochosma incana has been collected at about 10 localities in the flora area, all within 50 km of the Mexican border. Most herbarium specimens were previously identified as *A. limitanea,* which is easily distinguished from *A. incana* by the characteristics given in the key. In addition, all known plants of *A. limitanea* have 32 spores per sporangium; those of *A. incana* have 64. Although *A. incana* shows considerable morphologic and biochemical diversity in Mexico, populations in the United States are relatively uniform.

2. Argyrochosma fendleri (Kunze) Windham, Amer. Fern J. 77: 40. 1987

Notholaena fendleri Kunze, Farrnkräuter 2: 87, plate 136. 1851; *Cheilanthes cancellata* Mickel; *Pellaea fendleri* (Kunze) Prantl
Stem scales brown. **Leaves** 5–25 cm. **Petiole** dark brown, 0.75–1.5 mm diam. **Blade** deltate, 4–6-pinnate proximally, leathery to somewhat herbaceous, abaxially covered by whitish farina, adaxially glabrous or glandular; rachis rounded adaxially. **Pinna costae** distinctly flexuous, branches arising from prominent angles. **Ultimate segments** not articulate, dark color of stalks continuing into segment bases abaxially; segment margins plane to recurved, often

partially concealing sporangia; veins usually obscure adaxially. **Sporangia** submarginal, borne on distal 1/4 of secondary veins, containing 64 spores. $2n = 54$.

Sporulating summer–fall. Rocky slopes and cliffs; usually on granitic or volcanic substrates; 1700–3000 m; Colo., N.Mex., Wyo.; Mexico in Sonora.

Argyrochosma fendleri is occasionally confused with *A. limitanea,* which can have slightly flexuous rachises and pinna costae. All *A. limitanea* specimens with slightly flexuous rachises and costae have 32 spores per sporangium, whereas specimens of *A. fendleri* consistently have 64. This southern Rocky Mountain species is the only member of the genus that is found on acidic substrates such as granite.

3. Argyrochosma dealbata (Pursh) Windham, Amer. Fern J. 77: 40. 1987

Cheilanthes dealbata Pursh, Fl. Amer. Sept. 2: 671. 1814; *Notholaena dealbata* (Pursh) Kunze; *Pellaea dealbata* (Pursh) Prantl
Stem scales brown. **Leaves** 3–15 cm. **Petiole** chestnut brown, 0.50–0.75 mm diam. **Blade** deltate, 3–5-pinnate proximally, somewhat herbaceous, abaxially covered by whitish farina, adaxially glabrous or sparsely glandular; rachis rounded to slightly flattened adaxially. **Pinna costae** straight or nearly so, branches not arising from prominent angles. **Ultimate segments** not articulate, dark color of stalks continuing into segment bases abaxially; segment margins recurved, often partially concealing sporangia; veins often visible adaxially. **Sporangia** submarginal, borne on distal 1/3 of secondary veins, containing 64 spores. $2n = 54$.

Sporulating summer–fall. Calcareous cliffs and ledges; 100–600 m; Ark., Ill., Kans., Ky., Mo., Nebr., Okla., Tex.

The distinctions between *Argyrochosma dealbata* and *A. limitanea* are subtle but apparently absolute. Although both occur in Texas, their ranges do not overlap. Despite the morphologic similarities, isozyme analyses indicate that *A. dealbata,* a diploid species, was not involved in the origin of polyploid *A. limitanea.*

4. Argyrochosma limitanea (Maxon) Windham, Amer. Fern J. 77: 40. 1987

Notholaena limitanea Maxon, Amer. Fern J. 9: 70. 1919; *Cheilanthes limitanea* (Maxon) Mickel; *Pellaea limitanea* (Maxon) C. V. Morton

Stem scales brown. **Leaves** 5–30 cm. **Petiole** reddish brown to black, 0.75–2 mm diam. **Blade** lanceolate to deltate, 3–5-pinnate proximally, leathery, abaxially covered by dense white farina, adaxially glabrous or sparsely glandular; rachis rounded to slightly flattened adaxially. **Pinna costae** straight to slightly flexuous, branches not arising from prominent angles. **Ultimate segments** not articulate, dark color of stalks continuing into segment bases abaxially; segment margins recurved, often concealing sporangia; veins obscure adaxially. **Sporangia** submarginal, borne on distal 1/2 of secondary veins, containing 32 spores.

Subspecies 2 (2 in the flora): North America, Mexico.

Populations of *Argyrochosma limitanea* represent two morphologically distinctive taxa treated here as subspecies. For the most part, these taxa have different geographic ranges, and genetic interaction between them is precluded because both are asexual triploids. Isozyme analyses indicate that the two subspecies have different polyploid origins, but evolutionary relationships (and proper taxonomic treatment) cannot be resolved until sexually reproducing (64-spored) progenitors are found.

1. Leaf blades broadly deltate-ovate, 4–5-pinnate, with proximal pinnae at least 1/2 as long as blades.
. . . . . 4a. *Argyrochosma limitanea* subsp. *limitanea*
1. Leaf blades lanceolate to oblong, 3–4-pinnate, with proximal pinnae 1/4–1/3 as long as blades.
. . . . . 4b. *Argyrochosma limitanea* subsp. *mexicana*

4a. Argyrochosma limitanea (Maxon) Windham subsp. **limitanea**

Leaf blade broadly deltate-ovate, 4–5-pinnate, with proximal pinnae at least 1/2 as long as blades. $n = 2n = 81$, apogamous.

Sporulating summer–fall. Rocky slopes and cliffs, usually on calcareous or volcanic substrates; 800–2300 m; Ariz., Calif., N.Mex., Utah; Mexico in Chihuahua, Sonora.

4b. Argyrochosma limitanea (Maxon) Windham subsp. **mexicana** (Maxon) Windham, Contr. Univ. Michigan Herb. 19: 32. 1993

Notholaena limitanea Maxon subsp. *mexicana* Maxon, Amer. Fern J. 9: 72. 1919; *N. limitanea* var. *mexicana* (Maxon) M. Broun

Leaf blade lanceolate to oblong, 3–4-pinnate, proximal pinnae 1/4–1/3 as long as blades. $n = 2n = 81$, apogamous.

Sporulating summer–fall.

Rocky slopes and cliffs; usually on calcareous or volcanic substrates; 1700–2500 m; Ariz., N.Mex., Tex.; n Mexico.

5. Argyrochosma microphylla (Mettenius ex Kuhn) Windham, Amer. Fern J. 77: 40. 1987

Pellaea microphylla Mettenius ex Kuhn, Linnaea 36: 86. 1869; *Cheilanthes parvifolia* (R. M. Tryon) Mickel; *Notholaena parvifolia* R. M. Tryon

Stem scales brown. **Leaves** 7–25 cm. **Petiole** brown, 0.75–1.5 mm diam. **Blade** deltate to ovate, 3–4-pinnate proximally, leathery, abaxially and adaxially glabrous; rachis flattened or shallowly grooved adaxially. **Pinna costae** straight to somewhat flexuous, branches rarely arising from prominent angles. **Ultimate segments** articulate, dark color of stalks stopping abruptly at segment bases; segment margins recurved to revolute, often concealing sporangia; veins obscure adaxially. **Sporangia** submarginal, borne on distal 1/3 of secondary veins, containing 64 spores. $2n = 54$.

Sporulating summer–fall. Rocky limestone hillsides and cliffs; 300–2100 m; N.Mex., Tex.; n Mexico.

Argyrochosma microphylla is probably the most distinctive species of *Argyrochosma* in the flora. Chromosome studies by I. W. Knobloch et al. (1973) suggest that it may include diploid and tetraploid cytotypes.

6. Argyrochosma jonesii (Maxon) Windham, Amer. Fern J. 77: 40. 1987

Notholaena jonesii Maxon, Amer. Fern J. 7: 108. 1917; *Cheilanthes jonesii* (Maxon) Munz; *Pellaea jonesii* (Maxon) C. V. Morton

Stem scales brown to nearly black. **Leaves** 4–15 cm. **Petiole** dark brown, 0.75–1.5 mm diam. **Blade** ovate-lanceolate, 2–3-pinnate proximally, leathery, abaxially and adaxially glabrous; rachis rounded to slightly flattened adaxially. **Pinna costae** straight or nearly so, branches not arising from prominent angles. **Ultimate segments** not articulate, dark color of stalks continuing into segment bases abaxially; segment margins plane to slightly recurved, not concealing sporangia; veins obscure adaxially. **Sporangia** submarginal, borne on distal 1/2 of secondary veins, containing 64 spores. $2n = 54, 108$.

Sporulating spring–fall. Calcareous cliffs and ledges; 600–1900 m; Ariz., Calif., Nev., Utah; Mexico in Sonora.

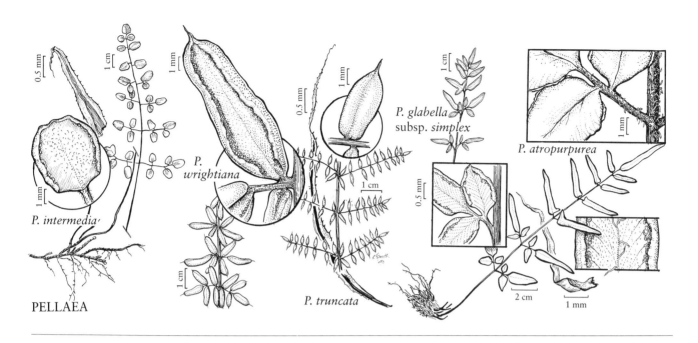

P. intermedia

P. wrightiana

P. truncata

P. glabella subsp. *simplex*

P. atropurpurea

PELLAEA

Argyrochosma jonesii includes two sexually reproducing cytotypes. The diploid is known from a few localities in the Sonoran and Mojave Deserts; the tetraploid is found throughout the Mojave Desert and cismontane southern California. Although subtle morphologic differences exist between these cytotypes, they are distinguished primarily by characteristics known to correlate with increases in ploidy level (such as spore size and the length of stomatal guard cells). Further investigation is necessary to determine whether the tetraploid arose through autopolyploidy or hybridization between cryptic species.

13. PELLAEA Link, Fil. Spec., 59. 1841, name conserved · Cliff-brake [Greek *pellos*, dark, possibly referring to bluish gray leaves]

Michael D. Windham

Plants usually on rock. **Stems** compact to long-creeping, ascending to horizontal, usually branched; scales brown to tan or often bicolored with dark, central stripe and lighter margins, linear-subulate to lanceolate (rarely ovate), margins dentate, erose, or entire. **Leaves** monomorphic to somewhat dimorphic, clustered to widely scattered, 2–100 cm. **Petiole** brown, black, straw-colored, or gray, rounded, flattened or with single longitudinal groove adaxially, glabrous or pubescent, usually with a few scales at base, with single vascular bundle. **Blade** linear to ovate-deltate, 1–4-pinnate proximally, leathery or rarely somewhat herbaceous, abaxially glabrous, pubescent, or with hairlike scales scattered along costae, adaxially usually glabrous, dull, not striate; rachis straight or flexuous. **Ultimate segments** of blade usually stalked and free from costae, elliptic, lanceolate to linear, usually more than 4 mm wide; base rounded, truncate, or cordate; stalks often lustrous and dark colored; segment margins reflexed to form confluent, poorly defined, false indusia extending entire length of segment. **Veins** of ultimate segments free or rarely anastomosing, usually obscure, pinnately branched and divergent distally. **False indusia** greenish to whitish, narrow, clearly marginal, often concealing the sporangia. **Sporangia** scattered along veins near segment margins, containing 32 or 64 spores, often intermixed with glands, farina-producing. **Spores** brown to tan (rarely yellow), tetrahedral-globose, rugose or cristate, lacking prominent equatorial ridge. $x = 29$.

Species ca. 40 (15 in the flora): most in the Western Hemisphere, a small number in Asia, Africa, the Pacific Islands, and Australia.

Pellaea in the broad sense is a diverse, poorly defined assemblage of xeric-adapted ferns (A. R. Smith 1981). Relationships among the North American, neotropical, and Eastern Hemisphere species are unclear, and it seems likely that the genus, as broadly construed by E. B. Copeland (1947) and R. M. Tryon and A. F. Tryon (1982), is polyphyletic. The species included here in *Pellaea* belong to a closely knit alliance that is usually recognized as a distinct section (sect. *Pellaea*). Although the inclusion of *P. bridgesii* in this group has been questioned (A. F. Tryon 1957), W. H. Wagner Jr. et al. (1983) have shown that the aberrant morphology of this species is simply an extreme expression of evolutionary trends commonly encountered in sect. *Pellaea*.

Among Western Hemisphere cheilanthoid ferns, species of *Pellaea* show clear morphologic, chromosomal, and biochemical affinities to *Argyrochosma* and members of the *Cheilanthes alabamensis* complex. In fact, the glabrous species of *Argyrochosma* (*A. jonesii* and *A. microphylla*) are commonly misidentified as *Pellaea*. These species are easily recognizable, however, because they have a combination of concolored stem scales and small ultimate segments (less than 4 mm wide).

SELECTED REFERENCES Gastony, G. J. 1988. The *Pellaea glabella* complex: Electrophoretic evidence for the derivations of the agamosporous taxa and a revised taxonomy. Amer. Fern J. 78: 44–67. Tryon, A. F. 1957. A revision of the fern genus *Pellaea* section *Pellaea*. Ann. Missouri Bot. Gard. 44: 125–193. Tryon, A. F. 1968. Comparisons of sexual and apogamous races in the fern genus *Pellaea*. Rhodora 70: 1–24.

1. Petioles and rachises straw-colored, tan, or gray, rarely lustrous; stem scales narrowly lanceolate to ovate, largest more than 0.3 mm wide.
  2. Rachises and costae strongly flexuous; pinnae retrorse, projecting downward toward base of leaf. . . . . 1. *Pellaea ovata*
  2. Rachises and costae straight or slightly flexuous; pinnae perpendicular to rachis or ascending.
    3. Stems stout and compact, more than 5 mm diam.; stem scales uniformly orange-brown and thin; ultimate segments rotund-cordate to deltate-cordate. . . . . . . 2. *Pellaea cordifolia*
    3. Stems slender and long-creeping, less than 5 mm diam.; stem scales mostly bicolored, with black, thick center and brown, thin margins; ultimate segments elliptic to ovate-deltate, not deeply cordate.
      4. Ultimate segments leathery, veins obscure abaxially; croziers only slightly scaly; blades usually 2-pinnate at base. . . . . 3. *Pellaea intermedia*
      4. Ultimate segments somewhat herbaceous, veins visible abaxially; croziers densely scaly; blades usually 3-pinnate at base. . . . . 4. *Pellaea andromedifolia*
1. Petioles and rachises dark brown to black, usually lustrous; stem scales linear-subulate, less than 0.3 mm wide.
  5. Some stem scales bicolored, with dark central region and lighter, brown margin.
    6. Leaf blades linear, 1-pinnate throughout, pinnae entire; fertile ultimate segments with rounded apices; segment margins not recurved, not concealing sporangia. . . . 5. *Pellaea bridgesii*
    6. Leaf blades linear-oblong to deltate, pinnate-pinnatifid to 3-pinnate, at least some pinnae lobed or divided; fertile ultimate segments with mucronate apices; segment margins recurved, usually concealing sporangia.
      7. Leaf blades deeply pinnate-pinnatifid at base, basal pinnae ternately lobed; petioles dark purple or black; sporangia not intermixed with farina-producing glands.
      . . . . 6. *Pellaea ternifolia*

7. Leaf blades 2–3-pinnate at base, basal pinnae fully pinnate (occasionally appearing ternate but with terminal segment on short, dark stalk); petioles chestnut brown to dark reddish brown; sporangia intermixed with glands producing yellowish farina (sparse in *Pellaea wrightiana*).
  8. Pinna costae usually shorter than or equal to ultimate segments; largest pinnae divided into 3–11 segments; blades linear-oblong, usually less than 4.5 cm wide.
    9. Ultimate segments linear, with greenish, strongly recurved margins covering more than 1/2 abaxial surface; sporangia with short stalks; pinnae strongly ascending. . . . . 7. *Pellaea brachyptera*
    9. Ultimate segments narrowly oblong, with white-bordered, recurved margins usually covering less than 1/2 abaxial surface; sporangia with long stalks; pinnae perpendicular to rachis or slightly ascending. . . . . . . 8. *Pellaea wrightiana*
  8. Pinna costae much longer than ultimate segments; largest pinnae divided into 11 or more segments; blades ovate-deltate to lanceolate, usually more than 4.5 cm wide.
    10. Blades usually 2-pinnate proximally, pinnae perpendicular to rachis or slightly ascending; margins of ultimate segments with whitish borders; sporangia with long stalks. . . . . 9. *Pellaea truncata*
    10. Blades 3-pinnate proximally or, if 2-pinnate, pinnae strongly ascending; margins of ultimate segments with greenish borders; sporangia with short stalks. . . . . 10. *Pellaea mucronata*
5. Stem scales uniformly reddish brown or tan.
  11. Ultimate segments glabrous abaxially or with isolated hairlike scales on a few segments or, if sparsely villous, then rachises nearly glabrous; pinnae or costae slightly decurrent on rachis; blades linear-oblong to lanceolate.
    12. Proximal pinnae usually bilobed and mitten-shaped; petioles with prominent articulation lines near base; rachises of mature leaves green in distal portion of blade; sporangia sessile or subsessile. . . . . 11. *Pellaea breweri*
    12. Proximal pinnae deeply divided into 3–7 lobes or segments (occasionally simple); petioles lacking articulation lines or, if present, then rachises of mature leaves brown to terminal pinna; some sporangia long-stalked. . . . . 12. *Pellaea glabella*
  11. Ultimate segments sparsely villous on abaxial costae; rachises variously pubescent; pinnae or costae not decurrent on rachis (obscurely so in *Pellaea gastonyi*); blades lanceolate, ovate, or deltate.
    13. Adaxial surface of rachis densely covered with short, curly, appressed hairs; largest ultimate segments (excluding terminal pinnae) usually more than 30 mm. . . . . 13. *Pellaea atropurpurea*
    13. Adaxial surface of rachis with sparse, long, divergent hairs; largest ultimate segments (excluding terminal pinnae) usually less than 30 mm.
      14. Proximal pinnae divided into 7–15 ultimate segments, segments shorter than longest pinna costae; fertile leaves usually more than 6 cm wide. . . . . 14. *Pellaea lyngholmii*
      14. Proximal pinnae divided into 3–7 ultimate segments, some segments longer than longest pinna costae; fertile leaves usually less than 6 cm wide. . . 15. *Pellaea gastonyi*

1. **Pellaea ovata** (Desvaux) Weatherby, Contr. Gray Herb. 114: 34. 1936

Pteris ovata Desvaux, Mém. Soc. Linn. Paris 6: 301. 1827; *Pellaea flexuosa* (Kaulfuss ex Schlechtendal & Chamisso) Link

Stems creeping, horizontal, slender, 2–5 mm diam.; scales mostly bicolored, lanceolate, largest scales 0.3–0.8 mm wide, centers black, thick, margins brown, thin, erose-dentate. **Leaves** monomorphic, clustered or scattered along stem, 15–100 cm; croziers pubescent, bearing a few scales. **Petiole** straw-colored, tan, or gray, not lustrous, rounded or slightly flattened adaxially, without prominent articulation lines. **Blade** elongate-deltate, usually 3-pinnate proximally, 5–25 cm wide; rachis tan throughout, strongly flexuous, rounded or flattened adaxially, usually glabrous. **Pinnae** retrorse, projecting downward toward base of leaf, not decurrent on rachis, with 5–40 ultimate segments; costae strongly flexuous, 25–120 mm, longer than ultimate segments. **Ultimate segments** lanceolate-deltate, 5–20 mm, leathery, glabrous or sparsely pubescent; margins recurved on fertile segments, covering less than 1/2 abaxial surface, borders whitish, entire; apex obtuse to truncate. **Veins** of ultimate segments obscure. **Sporangia** short-stalked, containing 64 spores, not intermixed with farina-producing glands. $2n = 58$.

Sporulating summer–fall. Rocky slopes and ledges, leaves often supported by associated vegetation, on a variety of substrates including granite and limestone; 300–1700 m; Tex.; Mexico; West Indies in Hispaniola; Central America; South America.

Populations of *Pellaea ovata* in the flora are composed of sexual diploids; an apogamous triploid cytotype predominates south of the United States. I have not seen herbarium specimens to substantiate reports of *P. ovata* from New Mexico (D. B. Lellinger 1985).

2. **Pellaea cordifolia** (Sessé & Mociño) A. R. Smith, Amer. Fern J. 70: 26. 1980

Adiantum cordifolium Sessé & Mociño, Naturaleza (Mexico City), ser. 2, 1(App.): 182. 1890; *Pellaea cardiomorpha* Weatherby; *P. sagittata* (Cavanilles) Link var. *cordata* (Cavanilles) A. F. Tryon

Stems compact, ascending, stout, 6–10 mm diam.; scales uniformly orange-brown and thin, lanceolate to ovate, largest scales 0.3–1 mm wide, margins dentate. **Leaves** somewhat dimorphic, sterile leaves shorter than fertile leaves, clustered on stem, 15–

50 cm; croziers not conspicuously pubescent, densely scaly. **Petiole** straw-colored, tan, or gray, not lustrous, rounded or slightly flattened adaxially, without prominent articulation lines. **Blade** ovate-deltate, 2-pinnate proximally, 5–20 cm wide; rachis tan throughout, straight to slightly flexuous, rounded or flattened adaxially, glabrous. **Pinnae** perpendicular to rachis or slightly ascending, not decurrent on rachis, usually with 3–15 ultimate segments; costae straight to slightly flexuous, 25–100 mm, longer than ultimate segments. **Ultimate segments** round-cordate to deltate-cordate, 5–15 mm, herbaceous to leathery, glabrous or puberulent; margins recurved on fertile segments, covering less than 1/2 abaxial surface, borders whitish, crenulate; apex rounded or retuse. **Veins** of ultimate segments usually evident. **Sporangia** short-stalked, containing 64 spores, not intermixed with farina-producing glands. $2n = 58$.

Sporulating summer–fall. Rocky slopes and ledges, usually on volcanic substrates; 1000–2500 m; Tex.; Mexico.

The diploid *Pellaea cordifolia* has often been treated as a variety of the Central American and South American apogamous triploid, *P. sagittata*. The two taxa are distinguished by a number of qualitative morphologic features (A. R. Smith 1980), and it seems unlikely that they represent cytotypes of a single species. A. F. Tryon (1957) suggested that *P. sagittata* may have originated through hybridization between *P. ovata* and *P. cordifolia* (as *P. sagittata* var. *cordata*).

3. **Pellaea intermedia** Mettenius ex Kuhn, Linnaea 38: 84. 1869

Pellaea intermedia var. *pubescens* Mettenius ex Kuhn

Stems creeping, horizontal, slender, 2–4 mm diam.; scales mostly bicolored, narrowly lanceolate, largest scales 0.3–0.8 mm wide, centers black, thick, margins brown, thin, irregularly dentate. **Leaves** monomorphic, widely scattered along stem, 12–50 cm; croziers pubescent and bearing a few scales. **Petiole** straw-colored, tan, or gray, not lustrous, rounded or slightly flattened adaxially, without prominent articulation lines. **Blade** ovate to elongate-deltate, usually 2-pinnate proximally, 4–20 cm wide; rachis tan throughout, straight to slightly flexuous, rounded or flattened adaxially, ± pubescent. **Pinnae** perpendicular to rachis or slightly ascending, not decurrent on rachis, usually with 7–21 ultimate segments; costae straight to slightly flexuous, 20–100 mm, longer than ultimate segments. **Ultimate segments** ovate to elliptic, 5–15 mm, leathery, glabrous or usually puberulent abaxially; margins recurved on fertile segments, usually covering less than 1/2 abaxial surface, borders whitish, nearly entire; apex obtuse to slightly mucronate. **Veins** of ultimate segments obscure. **Sporangia** short-stalked, containing 32 spores, not intermixed with farina-producing glands. $n = 2n = 87$, 116, apogamous.

Sporulating summer–fall. Rocky slopes and ledges, on a variety of substrates, including limestone and granite; 300–2400 m; Ariz., N.Mex., Tex.; n Mexico.

Plants of *Pellaea intermedia* in the flora are apogamous triploids and tetraploids; a sexual diploid cytotype has been found near Saltillo, Mexico (A. F. Tryon 1968). Given the high degree of morphologic similarity among the three cytotypes, the North American polyploids probably were derived from the Mexican diploid through autopolyploidy.

4. Pellaea andromedifolia (Kaulfuss) Fée, Mém. Foug. 5: 129. 1852

Pteris andromedifolia Kaulfuss, Enum. Filic., 188. 1824; *Pellaea andromedifolia* var. *pubescens* D. C. Eaton

Stems creeping, horizontal, slender, 2–4 mm diam.; scales mostly bicolored, narrowly lanceolate, largest scales 0.3–0.8 mm wide, centers black, thick, margins brown, thin, irregularly dentate. **Leaves** monomorphic, scattered along stem, 10–60 cm; croziers not conspicuously pubescent, densely scaly. **Petiole** straw-colored, tan, or gray, not lustrous, rounded or slightly flattened adaxially, without prominent articulation lines. **Blade** elongate-deltate, usually 3-pinnate proximally, 3–20 cm wide; rachis tan throughout, straight to slightly flexuous, rounded or flattened adaxially, glabrous or pubescent. **Pinnae** ascending or perpendicular to rachis, not decurrent on rachis, with 8–50 ultimate segments; costae usually straight, 15–140 mm, longer than ultimate segments. **Ultimate segments** elliptic to ovate, 3–15 mm, somewhat herbaceous, glabrous to sparsely pubescent abaxially; margins recurved on fertile segments, usually covering less than 1/2 abaxial surface, borders whitish, entire; apex retuse to rounded. **Veins** of ultimate segments evident. **Sporangia** short-stalked, containing 64 or 32 spores, not intermixed with farina-producing glands. $2n = 58$; $n = 2n = 87$, 116, apogamous.

Sporulating late spring–summer. Rocky slopes and ledges, usually on igneous substrates; 0–1500 m; Calif., Oreg.; Mexico in Baja California.

Pellaea andromedifolia comprises three cytotypes: a sexually reproducing diploid, an apogamous triploid, and an apogamous tetraploid. Isozyme studies by G. J.

Gastony and L. D. Gottlieb (1985) suggested that the apogamous triploid is an autopolyploid derived from sexual diploid populations. The apogamous tetraploid apparently resulted from hybridization between diploid and triploid individuals. These cytotypes have not been formally recognized as subspecies because their ranges seem to overlap extensively and because the ploidy level of the type collection of *P. andromedifolia* is not known.

5. Pellaea bridgesii Hooker, Sp. Fil. 2: 238, plate 142b. 1858

Stems compact, ascending, stout, 5–10 mm diam.; scales mostly weakly bicolored, linear-subulate, 0.1–0.3 mm wide, centers dark brown, thin, margins lighter, thin, denticulate to entire. **Leaves** monomorphic, clustered on stem, 7–30 cm; croziers nearly glabrous. **Petiole** dark brown, lustrous, rounded adaxially, without prominent articulation lines. **Blade** linear, 1-pinnate, 1.5–4 cm wide; rachis brown throughout, straight, rounded adaxially, glabrous. **Pinnae** perpendicular to slightly ascending, usually not decurrent on rachis, simple and unlobed; costae absent. **Ultimate segments** broadly ovate to elliptic, 7–20 mm, leathery, glabrous; margins plane, not recurved, not covering abaxial surface, borders whitish, entire; apex obtuse to rounded. **Veins** of ultimate segments obscure. **Sporangia** sessile or subsessile, containing 64 spores, intermixed with abundant farina-producing glands. $2n = 58$.

Sporulating summer–fall. Rocky slopes and cliffs, on granitic substrates; 1200–3600 m; Calif., Idaho, Nev., Oreg.

The morphology of *Pellaea bridgesii* is so distinctive that its sectional (and even generic) placement in *Pellaea* has long been a source of contention. W. H. Wagner Jr. et al. (1983) documented the existence of sterile diploid hybrids (called *P.* ×*glaciogena*) between *P. bridgesii* and *P. mucronata* (see reticulogram), suggesting that *P. bridgesii* is most closely related to members of sect. *Pellaea*. In addition to the more obvious characters mentioned above, *P. bridgesii* is distinguished from other North American species (except *P. ternifolia*) by its anastomosing veins.

6. Pellaea ternifolia (Cavanilles) Link, Fil. Spec., 59. 1841

Pteris ternifolia Cavanilles, Descr. Pl., 266. 1802

Stems compact, ascending, stout, 5–10 mm diam.; scales bicolored, linear-subulate, 0.1–0.3 mm wide, centers black, thick, margins brown, thin, erose-dentate. **Leaves** monomorphic, clustered on stem, 10–50 cm; croziers sparsely to densely villous. **Petiole** black or dark pur-

Pellaea

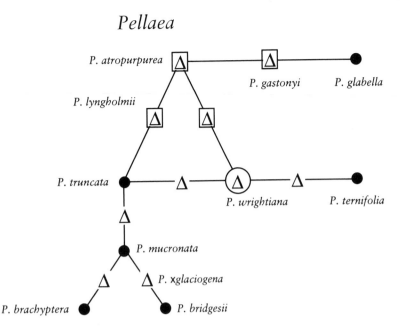

Relationships among some species of *Pellaea*. Solid circles represent parental taxa; circled triangles represent fertile allotetraploids; boxed triangles represent apogamous allopolyploids; and triangles represent sterile hybrids.

ple, lustrous, rounded or slightly flattened adaxially, without prominent articulation lines. **Blade** linear to ovate, deeply pinnate-pinnatifid proximally, 2.5–8 cm wide; rachis black or purple throughout, straight, often flattened adaxially, glabrous or villous. **Pinnae** perpendicular to rachis or slightly ascending, not decurrent on rachis, ternate at base of leaf; costae absent. **Ultimate segments** linear-oblong, 10–40 mm, leathery, glabrous to sparsely villous abaxially on midrib; margins recurved on fertile segments, rarely covering more than 1/2 abaxial surface, borders whitish, entire; apex mucronate. **Veins** of ultimate segments obscure. **Sporangia** long-stalked, containing 64 spores, not intermixed with farina-producing glands.

Subspecies 3 (3 in the flora): North America, Mexico, Central America, South America, Pacific Islands in Hawaii.

Pellaea ternifolia is represented in the flora by three morphologically and chromosomally distinct taxa. These discrete genetic entities also show a tendency toward geographic isolation and are treated here as subspecies. Diploid populations referred to *P. ternifolia* subsp. *ternifolia* are scattered from Texas through Mexico to South America. The pubescent tetraploid *(P. ternifolia* subsp. *villosa)* follows the Sierra Madre Oriental from Puebla, Mexico, north to Texas; the glabrous tetra-

ploid *(P. ternifolia* subsp. *arizonica)* occurs in Arizona, Texas, and northern Mexico. Isozyme and chromosome studies suggest that both tetraploids are segmental allopolyploids produced by hybridization between subsp. *ternifolia* and other (as yet unidentified) diploid elements within *P. ternifolia*.

1. Rachises villous, especially in axils of pinnae; pinnae with hairs scattered along main veins abaxially. . . . . . 6a. *Pellaea ternifolia* subsp. *villosa*
1. Rachises glabrous or with a few widely scattered hairs; pinnae completely glabrous.
    2. Largest ultimate segments (excluding terminal pinnae) usually less than 18 mm; distal portion of petioles grooved or flattened adaxially; spores usually 39–45 μm diam. . . . . 6b. *Pellaea ternifolia* subsp. *ternifolia*
    2. Largest ultimate segments (excluding terminal pinnae) usually more than 18 mm; distal portion of petioles rounded or slightly flattened adaxially; spores usually 46–53 μm diam. . . . . 6c. *Pellaea ternifolia* subsp. *arizonica*

6a. Pellaea ternifolia (Cavanilles) Link subsp. **villosa**
Windham, Contr. Univ. Michigan Herb. 19: 43. 1993

Petiole rounded or slightly flattened adaxially in distal portion; rachis villous, especially in axils of pinnae; pinnae with hairs scattered along main veins abaxially; largest ultimate segments (excluding terminal pinnae) usually more than 18 mm; spores averaging 46–53 μm diam. $2n = 116$.

Sporulating summer–fall. Rocky slopes and ledges, on igneous substrates; 2300–2700 m; Tex.; Mexico.

The only record of *Pellaea ternifolia* subsp. *villosa* from the flora is provided by a single plant collected in the Davis Mountains of western Texas.

6b. Pellaea ternifolia (Cavanilles) Link subsp. **ternifolia**

Petiole grooved or flattened adaxially in distal portion; rachis glabrous or with a few widely scattered hairs; pinnae completely glabrous; largest ultimate segments (excluding terminal pinnae) usually less than 18 mm; spores usually 39–45 μm diam. $2n = 58$.

Sporulating summer–fall. Rocky slopes and ledges on igneous substrates; 2200–2600 m; Tex.; Mexico; Central America; South America; Pacific Islands in Hawaii.

In the flora, this subspecies is known only from the Chisos Mountains in western Texas.

6c. Pellaea ternifolia (Cavanilles) Link subsp. **arizonica**
Windham, Contr. Univ. Michigan Herb. 19: 42. 1993

Petiole rounded or slightly flattened adaxially in distal portion; rachis glabrous or with a few widely scattered hairs; pinnae completely glabrous; largest ultimate segments (excluding terminal pinnae) usually more than 18 mm; spores averaging 46–53 μm diam. $2n = 116$.

Sporulating summer–fall. Cliffs and rocky slopes, on a variety of acidic substrates including quartzite and granite; 1700–2400 m; Ariz., Tex.; n Mexico.

Pellaea ternifolia subsp. *arizonica* and *P. wrightiana* hybridize; the hybrids are morphologically intermediate tetraploids and have malformed spores.

7. Pellaea brachyptera (T. Moore) Baker in Hooker & Baker, Syn. Fil. ed. 2, 477. 1874

Platyloma brachyptera T. Moore, Gard. Chron., 141. 1873

Stems compact, ascending, stout, 5–10 mm diam.; scales bicolored, linear-subulate, 0.1–0.3 mm wide, centers dark brown to black, thick, margins brown, thin, dentate. **Leaves** monomorphic, clustered on stem, 8–40 cm; croziers sparsely villous. **Petiole** dark brown, lustrous, flattened or slightly grooved adaxially, without prominent articulation lines. **Blade** linear-oblong, 2-pinnate proximally, 1–4 cm wide; rachis brown throughout, straight, shallowly grooved adaxially, usually glabrous. **Pinnae** strongly ascending, not decurrent on rachis, usually with 5–11 ultimate segments; costae straight, 5–20 mm, usually shorter than ultimate segments. **Ultimate segments** linear, 5–20 mm, leathery, glabrous; margins on fertile segments strongly revolute, covering more than 1/2 abaxial surface, borders greenish, crenate; apex mucronate. **Veins** of ultimate segments obscure. **Sporangia** short-stalked, containing 64 spores, intermixed with abundant farina-producing glands.

Sporulating summer–fall. Cliffs and rocky slopes, usually on igneous substrates, occasionally on serpentine; 900–2700 m; Calif., Oreg., Wash.

The distinctive *Pellaea brachyptera* reportedly hybridizes with *P. mucronata* (A. F. Tryon 1957; D. B. Lellinger 1985); the hybrids are morphologically intermediate plants with malformed spores.

8. Pellaea wrightiana Hooker, Sp. Fil. 2: 142. 1858

Pellaea ternifolia (Cavanilles) Link var. *wrightiana* (Hooker) A. F. Tryon

Stems compact, ascending, stout, 5–10 mm diam.; scales bicolored, linear-subulate, 0.1–0.3 mm wide, centers black, thick, margins brown, thin, erose-dentate. **Leaves** monomorphic, clustered on stem, 6–40 cm; croziers sparsely villous. **Petiole** dark brown, lustrous, flattened or slightly grooved adaxially, without prominent articulation lines. **Blade** linear-oblong, 2-pinnate proximally, 1.5–5 cm wide; rachis brown throughout, straight, shallowly grooved adaxially, usually glabrous. **Pinnae** perpendicular to rachis or slightly ascending, not decurrent on rachis, usually with 3–9 ultimate segments; costae straight, 2–20 mm, usually shorter than ultimate segments. **Ultimate segments** narrowly oblong, 5–20 mm, leathery, glabrous; margins recurved on fertile segments, usually covering less than

1/2 abaxial surface, borders whitish, crenulate; apex mucronate. **Veins** of ultimate segments obscure. **Sporangia** long-stalked, containing 64 spores, intermixed with sparse farina-producing glands. $2n = 116$.

Sporulating summer–fall. Cliffs and rocky slopes, on a variety of acidic to mildly basic substrates; 300–2900 m; Ariz., Colo., N.Mex., N.C., Okla., Tex., Utah; n Mexico.

W. H. Wagner Jr. (1965) suggested that *Pellaea wrightiana* was a fertile allotetraploid hybrid between *P. truncata* (as *P. longimucronata*) and *P. ternifolia*. This hypothesis has been confirmed by isozyme analyses (M. D. Windham 1988). *Pellaea wrightiana* is therefore treated as a distinct species rather than a variety of *P. ternifolia*. This tetraploid species hybridizes with *P. truncata* and *P. ternifolia* subsp. *arizonica* to produce sterile triploids and tetraploids with intermediate morphology and malformed spores. *Pellaea wrightiana* has also hybridized with *P. atropurpurea* to form a rare apogamous pentaploid known only from western Oklahoma.

9. **Pellaea truncata** Goodding, Muhlenbergia 8: 94. 1912

Pellaea longimucronata Hooker; *P. wrightiana* Hooker var. *longimucronata* (Hooker) Davenport

Stems compact, ascending, stout, 5–10 mm diam.; scales bicolored, linear-subulate, 0.1–0.3 mm wide, centers black, thick, margins brown, thin, erose-dentate. **Leaves** somewhat dimorphic, sterile leaves shorter and less divided than fertile leaves, clustered on stems, 8–40 cm; croziers sparsely villous. **Petiole** dark brown, lustrous, flattened or slightly grooved adaxially, without prominent articulation lines. **Blade** ovate-deltate, usually 2-pinnate proximally, 4–18 cm wide; rachis brown throughout, straight, shallowly grooved adaxially, usually glabrous. **Pinnae** perpendicular to rachis to slightly ascending, not decurrent on rachis, usually with 9–25 ultimate segments; costae straight, 20–70 mm, much longer than fertile ultimate segments. **Ultimate segments** narrowly oblong, 4–10 mm, leathery, glabrous; margins recurved on fertile segments, usually covering less than 1/2 abaxial surface, borders whitish, nearly entire; apex mucronate. **Veins** of ultimate segments obscure. **Sporangia** long-stalked, containing 64 spores, intermixed with abundant farina-producing glands. $2n = 58$.

Sporulating late spring–fall. Cliffs and rocky slopes, on various substrates but rarely observed on limestone; 600–2500 m; Ariz., Calif., Colo., Nev., N.Mex., Tex., Utah; n Mexico.

Most manuals refer to *Pellaea truncata* as *P. longimucronata*, a name shown to be invalid by A. Cronquist et al. (1972+, vol. 1). Populations located near the range of *P. mucronata* in the Mojave Desert are often difficult to identify because of the subtlety of the characters involved and an apparent tendency to produce sterile (and possibly fertile) hybrids. Morphologically intermediate hybrids between *P. truncata* and *P. wrightiana* are common in regions where the ranges of the two species overlap, but these are easily identified by their malformed spores.

10. **Pellaea mucronata** (D. C. Eaton) D. C. Eaton in Emory, Rep. U.S. Mex. Bound. 2(1): 233. 1859

Allosorus mucronatus D. C. Eaton, Amer. J. Sci. Arts 22: 138. 1856; *Pellaea ornithopus* Hooker

Stems compact, ascending, stout, 5–10 mm diam.; scales bicolored, linear-subulate, 0.1–0.3 mm wide, centers black, thick, margins brown, thin, erose-dentate. **Leaves** monomorphic, clustered on stem, 7–45 cm; croziers sparsely villous. **Petiole** dark brown, lustrous, flattened to slightly grooved adaxially, without prominent articulation lines. **Blade** ovate-deltate, (2–)3-pinnate proximally, 4–18 cm wide; rachis brown throughout, straight, shallowly grooved adaxially, usually glabrous. **Pinnae** perpendicular to rachis to strongly ascending, not decurrent on rachis, usually with 9–40 ultimate segments; costae straight, 10–70 mm, much longer than ultimate segments. **Ultimate segments** narrowly oblong, 2–12 mm, leathery, glabrous; margins recurved to strongly revolute on fertile segments, usually covering more than 1/2 abaxial surface, borders greenish, usually dentate; apex mucronate. **Veins** of ultimate segments obscure. **Sporangia** short-stalked, containing 64 spores, intermixed with abundant farina-producing glands.

Subspecies 2 (2 in the flora): North America, Mexico.

Pellaea mucronata encompasses two morphologic extremes that tend to occupy different habitats and are treated here as subspecies. The typical 3-pinnate form (*P. mucronata* subsp. *mucronata*) is scattered throughout California and southern Nevada, usually below 1800 m elevation. The 2-pinnate form with ascending, overlapping pinnae (*P. mucronata* subsp. *californica*) is apparently confined to the Sierra Nevada and Transverse Ranges of California at elevations greater than 1800 m. The taxonomic status of these entities remains in dispute, and they are often treated as mere ecological forms. W. H. Wagner Jr. et al. (1983) indicated that natural hybrids formed between *P. bridgesii* and these two taxa are morphologically distinct, suggesting that the differences observed between the subspecies of *P. mucronata* are genetically based. In addition to *P. bridgesii*, subsp. *mucronata* apparently hybridizes with both *P. truncata* and *P. brachyptera* (see comments under those species).

1. Blades 3-pinnate proximally; pinnae usually ± perpendicular to rachis, not overlapping; plants usually found below 1800 m. . . . .
. . . . . . 10a. *Pellaea mucronata* subsp. *mucronata*
1. Blades usually 2-pinnate proximally; pinnae ascending and overlapping, especially in distal portion of leaf; plants usually found above 1800 m.
. . . . . . 10b. *Pellaea mucronata* subsp. *californica*

10a. Pellaea mucronata (D. C. Eaton) D. C. Eaton subsp. mucronata

Leaf blade 3-pinnate proximally; pinnae usually ± perpendicular to rachis, not overlapping. $2n = 58$.

Sporulating late spring–summer. Cliffs and rocky slopes, on a variety of acidic to mildly basic substrates; 0–2300 m; Calif., Nev.; Mexico in Baja California.

10b. Pellaea mucronata (D. C. Eaton) D. C. Eaton subsp. californica (Lemmon) Windham, Contr. Univ. Michigan Herb. 19: 42. 1993

Pellaea wrightiana Hooker var. *californica* Lemmon, Ferns Pacif. Coast, 10. 1882; *P. compacta* (Davenport) Maxon; *P. mucronata* var. *californica* (Lemmon) Munz & I. M. Johnston

Leaf blade usually 2-pinnate proximally; pinnae ascending and overlapping, especially in distal portion of leaf.

Sporulating summer–fall. Cliffs and rocky slopes, usually on granitic substrates; 1800–3000 m; Calif.

11. Pellaea breweri D. C. Eaton, Proc. Amer. Acad. Arts 6: 555. 1865

Stems compact, ascending, stout, 5–10 mm diam.; scales uniformly reddish brown, linear-subulate, 0.1–0.3 mm wide, thin, margins sinuous, nearly entire. **Leaves** monomorphic, clustered on stem, 2.5–20 cm; croziers sparsely villous. **Petiole** brown, lustrous, rounded adaxially, with prominent articulation lines near base. **Blade** linear-oblong, pinnate-pinnatifid proximally, 1–4 cm wide; rachis brown proximally, green distally, straight, rounded adaxially, glabrous to sparsely villous. **Pinnae** ascending or perpendicular to rachis, decurrent on rachis, deeply 2-lobed (mitten-shaped) near base of leaf; costae absent. **Ultimate segments** lanceolate-deltate, 5–25 mm, herbaceous, glabrous; margins recurved on fertile segments, covering less than 1/2 abaxial surface, borders whitish, erose-denticulate; apex obtuse or rounded. **Veins** of ultimate segments evident. **Sporangia** sessile or subsessile, containing 64 spores, not intermixed with farina-producing glands. $2n = 58$.

Sporulating summer–fall. Cliffs and rocky slopes, on a variety of substrates including granite and limestone; 1600–3800 m; Calif., Colo., Idaho, Mont., Nev., Oreg., Utah, Wash., Wyo.

Pellaea breweri is distinguished from other North American taxa (except for some populations of *P. glabella*) by the presence of prominent articulation lines near the base of the petiole. The leaves are easily detached, and many herbarium specimens consist of separate leaves and stems, the latter covered with petiole bases of approximately equal length.

12. Pellaea glabella Mettenius ex Kuhn, Linnaea 36: 87. 1869 · Pelléa glabre

Stems compact, ascending, stout, 5–10 mm diam.; scales uniformly reddish brown, linear-subulate, 0.1–0.3 mm wide, thin, margins sinuous, entire to denticulate. **Leaves** monomorphic, clustered on stem, 2–40 cm; croziers sparsely villous. **Petiole** brown, lustrous, rounded adaxially, occasionally with prominent articulation lines near base. **Blade** linear-oblong to ovate-lanceolate, 1–2-pinnate proximally, 1–8 cm wide; rachis brown throughout, straight, rounded adaxially, nearly glabrous. **Pinnae** somewhat ascending, decurrent on rachis, usually with 3–7 lobes or ultimate segments; costae when present straight, 1–50 mm, often shorter than ultimate segments. **Ultimate segments** oblong-lanceolate, 5–20 mm, leathery to herbaceous, glabrous except for occasional hairlike scales abaxially near midrib; margins recurved on fertile segments, covering less than 1/2 abaxial surface, borders whitish, erose-denticulate; apex obtuse. **Veins** of ultimate segments usually obscure. **Sporangia** long-stalked, containing 32 or 64 spores, not intermixed with farina-producing glands.

Subspecies 4: only in the flora.

Pellaea glabella includes four geographically and genetically isolated taxa treated here as subspecies. D. B. Lellinger (1985) recognized three species in this difficult group, but isozyme analyses (G. J. Gastony 1988) showed that one of these (*P. suksdorfiana*) is an autotetraploid derivative of the diploid known as *P. occidentalis*. As a result, Gastony recognized just two species: *P. glabella* (with two varieties) and *P. occidentalis* (with two subspecies). The few morphologic features that distinguish these taxa, however, are subtle and environmentally plastic, and the isozyme data indicate that they are less divergent genetically than any other pair of *Pellaea* species in North America. Therefore, a more conservative taxonomic treatment seems warranted.

1. Sporangia containing 32 spores; spores averaging 60–72 μm diam.
    2. Some ultimate segments (especially terminal segments) with hairlike scales abaxially near midrib; e North America (with outlying station in Texas panhandle). . . . .
        . . . . 12a. *Pellaea glabella* subsp. *glabella*
    2. Ultimate segments essentially glabrous; w North America. . . . .
        . . . . 12b. *Pellaea glabella* subsp. *simplex*
1. Sporangia containing 64 spores; spores averaging 38–52 μm diam.
    3. Ultimate segments (especially terminal segments) with hairlike scales abaxially near midrib; Missouri. . . . .
        . . . . . 12c. *Pellaea glabella* subsp. *missouriensis*
    3. Ultimate segments glabrous; w North America. . . . 12d. *Pellaea glabella* subsp. *occidentalis*

12a. Pellaea glabella Mettenius ex Kuhn subsp. **glabella**

Pellaea atropurpurea (Linnaeus) Link var. *bushii* Mackenzie
Some ultimate segments (especially terminal segments) with hairlike scales abaxially near midrib; most sporangia containing 32 spores; spores averaging 60–72 μm diam. $n = 2n = 116$, apogamous.

Sporulating summer–fall. Calcareous cliffs and ledges, usually on limestone substrates; 0–1200 m; Ont., Que.; Ark., Conn., Ill., Ind., Iowa, Kans., Ky., Md., Mass., Mich., Minn., Mo., Nebr., N.Y., Ohio, Okla., Pa., Tenn., Tex., Vt., Va., W.Va., Wis.

G. J. Gastony (1988) has shown that this apogamous tetraploid was derived from *Pellaea glabella* subsp. *missouriensis* by an autopolyploid increase in chromosome number.

12b. Pellaea glabella Mettenius ex Kuhn subsp. **simplex** (Butters) A. Löve & D. Löve, Taxon 26: 325. 1977

Pellaea glabella var. *simplex* Butters, Amer. Fern J. 7: 84. 1917; *P. atropurpurea* (Linnaeus) Link var. *simplex* (Butters) C. V. Morton; *P. occidentalis* (E. E. Nelson) Rydberg subsp. *simplex* (Butters) Gastony; *P. suksdorfiana* Butters

Ultimate segments essentially glabrous; sporangia containing 32

spores; spores averaging 60–72 μm diam. $n = 2n = 116$, apogamous.

Sporulating summer–fall. Calcareous cliffs and ledges, usually on limestone; 900–3000 m; Alta., B.C.; Ariz., Colo., Idaho, Mont., N.Mex., Utah, Wash., Wyo.

This western counterpart of *Pellaea glabella* subsp. *glabella* is an apogamous tetraploid. A. F. Tryon (1957) and D. B. Lellinger (1985) hypothesized that it might have arisen as a hybrid between the western diploid member of the *P. glabella* complex (here called subsp. *occidentalis*) and *P. atropurpurea*. G. J. Gastony (1988) has shown conclusively, however, that *P. glabella* subsp. *simplex* is an autopolyploid derivative of subsp. *occidentalis* and does not contain genes contributed by *P. atropurpurea*.

12c. Pellaea glabella Mettenius ex Kuhn subsp. **missouriensis** (G. J. Gastony) Windham, Contr. Univ. Michigan Herb. 19: 39. 1993

Pellaea glabella var. *missouriensis* Gastony, Amer. Fern J. 78: 64. 1988
Some ultimate segments (especially terminal segments) with hairlike scales abaxially near midrib; sporangia containing 64 spores; spores averaging 38–52 μm diam. $2n = 58$.

Sporulating summer–fall. Cliffs and ledges, apparently confined to limestone; 100–300 m; Mo.

This diploid taxon is currently known only from southeastern Missouri in the Ozark region, but a thorough survey of spore number per sporangium in *Pellaea glabella* from eastern North America may yield additional localities. Plants of *P. glabella* subsp. *missouriensis* occasionally hybridize with *P. glabella* subsp. *glabella*; the hybrids are apogamous pentaploids resembling *P. glabella* subsp. *glabella* (G. J. Gastony 1988).

12d. Pellaea glabella Mettenius ex Kuhn subsp. **occidentalis** (E. E. Nelson) Windham, Contr. Univ. Michigan Herb. 19: 39. 1993

Pellaea atropurpurea (Linnaeus) Link var. *occidentalis* E. E. Nelson, Fern Bull. 7: 30. 1899; *P. glabella* var. *nana* (Richardson) Cody; *P. glabella* var. *occidentalis* (E. E. Nelson) Butters; *P. occidentalis* (E. E. Nelson) Rydberg; *P. pumila* Rydberg

Ultimate segments glabrous; sporangia containing 64 spores; spores averaging 38–52 μm diam. $2n = 58$.

Sporulating summer–fall. Calcareous cliffs and ledges,

usually on limestone; 500–2800 m; Man., Sask.; Mont., N.Dak., S.Dak., Utah, Wyo.

Most recent treatments refer to this taxon as *Pellaea glabella* var. *occidentalis*. It appears to be synonymous, however, with *P. glabella* var. *nana,* which was proposed 76 years earlier and has priority if the taxon is treated at varietal rank. This diploid (treated here as a subspecies) often shows prominent articulation lines near the base of the petiole, and plants with less divided leaves are occasionally misidentified as *P. breweri.* At least some of the sporangia of subsp. *occidentalis* are long-stalked, however, whereas those of *P. breweri* are sessile or subsessile.

13. Pellaea atropurpurea (Linnaeus) Link, Fil. Spec., 59. 1841 · Pelléa pourpre foncé

Pteris atropurpurea Linnaeus, Sp. Pl. 2: 1076. 1753; *Pellaea atropurpurea* var. *cristata* Trelease

Stems compact, ascending, stout, 5–10 mm diam.; scales uniformly reddish brown (or tan), linear-subulate, 0.1–0.3 mm wide, thin, margins entire to denticulate. **Leaves** somewhat dimorphic, sterile leaves shorter and less divided than fertile leaves, clustered on stems, 5–50 cm; croziers villous. **Petiole** reddish purple to nearly black, lustrous, rounded adaxially, without prominent articulation lines. **Blade** elongate-deltate, usually 2-pinnate proximally, 2–18 cm wide; rachis reddish purple throughout, straight, rounded adaxially, densely pubescent adaxially with short, curly, appressed hairs. **Pinnae** perpendicular to rachis or ascending, not decurrent on rachis, usually with 3–15 ultimate segments; costae straight, 10–100 mm, often longer than ultimate segments. **Ultimate segments** linear-oblong, 10–75 mm, leathery, sparsely villous abaxially near midrib; margins weakly recurved to plane on fertile segments, usually covering less than 1/2 abaxial surface, borders whitish, crenulate; apex obtuse to slightly mucronate. **Veins** of ultimate segments obscure. **Sporangia** long-stalked, containing 32 spores, not intermixed with farina-producing glands. $n = 2n = 87$, apogamous.

Sporulating summer–fall. Calcareous cliffs and rocky slopes, usually on limestone; 100–2500 m; Ont., Que.; Ala., Ariz., Ark., Colo., Conn., D.C., Fla., Ga., Ill., Ind., Iowa, Kans., Ky., La., Md., Mass., Mich., Minn., Miss., Mo., Nebr., Nev., N.J., N.Mex., N.Y., N.C., Ohio, Okla., Pa., R.I., S.C., S.Dak., Tenn., Tex., Utah, Vt., Va., W.Va., Wis., Wyo.; Mexico; Central America in Guatemala.

Contrary to D. B. Lellinger's (1985) hypothesis, isozyme data indicate that neither *Pellaea glabella* nor *P.*

ternifolia was involved in the origin of this apogamous triploid. Instead, it appears that *P. atropurpurea* is an autopolyploid derivative of a single diploid taxon that has not yet been located. A thorough survey of spore number per sporangium in this species should be undertaken to determine whether the diploid progenitor is still extant. Collections from western Canada identified as *P. atropurpurea* actually represent *P. gastonyi,* an apogamous tetraploid produced by hybridization between *P. atropurpurea* and diploid populations of *P. glabella. Pellaea atropurpurea* has also hybridized with *P. wrightiana;* the hybrid is a rare apogamous pentaploid known only from western Oklahoma. *Pellaea lyngholmii* is the apogamous tetraploid hybrid between *P. atropurpurea* and *P. truncata. Pellaea atropurpurea* is distinguished from all these hybrids by having rachises that are densely pubescent adaxially, larger ultimate segments, and spores averaging less than 62 µm in diameter.

14. Pellaea lyngholmii Windham, Contr. Univ. Michigan Herb. 19: 40. 1993

Stems compact, ascending, stout, 5–10 mm diam.; scales uniformly brown or tan, linear-subulate, 0.1–0.3 mm wide, thin, margins entire to denticulate. **Leaves** somewhat dimorphic, sterile leaves shorter and less divided than fertile leaves, clustered on stem, 10–30 cm; croziers villous. **Petiole** dark brown to reddish purple, lustrous, rounded adaxially, without prominent articulation lines. **Blade** elongate-deltate to ovate, 2-pinnate proximally, 5–15 cm wide; rachis brown or reddish purple throughout, straight, often slightly flattened adaxially, sparsely villous with long, divergent hairs. **Pinnae** perpendicular to rachis or slightly ascending, not decurrent on rachis, usually with 7–15 ultimate segments; costae straight, 25–80 mm, usually longer than ultimate segments. **Ultimate segments** oblong-lanceolate, to 7–25 mm, leathery, sparsely villous abaxially near midrib; margins usually recurved on fertile segments, covering less than 1/2 abaxial surface, borders whitish, crenulate; apex slightly mucronate. **Veins** of ultimate segments obscure. **Sporangia** long-stalked, containing 32 spores, not intermixed with farina-producing glands.

Sporulating summer–fall. Rocky slopes and ledges, usually on sandstone; 1200–1800 m; Ariz.

Pellaea lyngholmii is an apogamous tetraploid that arose through hybridization between *P. atropurpurea* and *P. truncata* (M. D. Windham 1993). It is most often confused with *P. atropurpurea,* from which *P. lyngholmii* differs in having sparsely villous rachises, smaller and more numerous ultimate segments, and spores usually more than 62 μm in diameter.

15. Pellaea gastonyi Windham, Contr. Univ. Michigan Herb. 19: 36. 1993

Stems compact, ascending, stout, 5–10 mm diam.; scales uniformly reddish brown, linear-subulate, 0.1–0.3 mm wide, thin, margins entire to denticulate. **Leaves** somewhat dimorphic, sterile leaves shorter than fertile leaves, clustered on stem, 8–25 cm; croziers villous. **Petiole** reddish purple to dark brown, lustrous, rounded adaxially, without prominent articulation lines. **Blade** elongate-deltate to lanceolate, 2-pinnate proximally, 3–6 cm wide; rachis purple or brown throughout, straight, rounded adaxially, sparsely villous with long, divergent hairs. **Pinnae** ascending or perpendicular to rachis, not decurrent on rachis or obscurely so, usually with 3–7 ultimate segments; costae straight, 2–30 mm, usually shorter than ultimate segments. **Ultimate segments** oblong-lanceolate, 7–30 mm, leathery, sparsely villous abaxially near midrib; margins usually recurved on fertile segments, covering less than 1/2 abaxial surface, borders whitish, crenulate; apex obtuse to slightly mucronate. **Veins** of ultimate segments obscure. **Sporangia** long-stalked, containing 32 spores, not intermixed with farina-producing glands.

Sporulating summer–fall. Calcareous cliffs and ledges, usually on limestone; 100–1500 m; Alta., B.C., Sask.; Mo., S.Dak., Wyo.

Pellaea gastonyi is an apogamous tetraploid that has originated through repeated hybridization between *P. atropurpurea* and *P. glabella*. Isozyme studies (G. J. Gastony 1988) indicate that *P. glabella* subsp. *missouriensis* was the diploid parent of plants found in Missouri, whereas diploid *P. glabella* subsp. *occidentalis* was involved in the origin of *P. gastonyi* populations occurring in western North America. *Pellaea gastonyi* is most often confused with *P. atropurpurea,* from which it differs in having sparsely villous rachises, smaller ultimate segments, and spores averaging more than 62 μm in diameter.

14. VITTARIACEAE Ching

• Shoestring Fern Family

Donald R. Farrar

Plants epiphytic or on rock. **Roots** abundant, covered with brown hairs. **Stems** short-creeping, branched, covered with scales with conspicuously thickened, clathrate cell walls. **Leaves** simple, entire, petioles indistinct. **Veins** (visible in cleared leaves) anastomosing on each side of midrib in row of long, polygonal areolae without included veinlets. **Epidermis** with spicular cells. **Sori** in submarginal groove on each side of midrib. **Indusia** absent, sporangia interspersed with branched soral paraphyses. **Spores** monolete or trilete. **Gametophytes** persistent, ribbonlike, much branched, and clone-forming by vegetative reproduction. **Gemmae** uniseriate, 2–5 mm, composed of 2–16 dark green body cells and 0–4 almost colorless, smaller rhizoid primordia cells.

Genera ca. 10, species ca. 100 (1 genus, 3 species in the flora): worldwide in tropics and subtropics.

In species outside the flora, leaves may be much shorter and broader with more than one row of areolae and several soral lines on each side of the midrib. Rarely veins are free, and sporangia are scattered over the abaxial surface. Gametophyte gemmae may be much longer than found in the flora, may be platelike, or may be absent.

Vegetative proliferation of gametophytes of this family allows the gametophyte generation to persist after sporophytes are produced. Although sporophytes of only one species, *Vittaria lineata,* are known in the flora, at least two additional species are represented by persistent gametophytes. Because sporophytes of *V. graminifolia* may occur in the flora, characteristics of its sporophyte are included in the key to species.

SELECTED REFERENCES Benedict, R. C. 1911. The genera of the fern tribe Vittarieae, their external morphology, venation, and relationships. Bull. Torrey Bot. Club 38: 153–190. Benedict, R. C. 1914. A revision of the genus *Vittaria* J. E. Smith. I. The species of the subgenus *Radiovittaria*. Bull. Torrey Bot. Club 41: 391–410. Farrar, D. R. 1974. Gemmiferous fern gametophytes—Vittariaceae. Amer. J. Bot. 61: 146–155. Tryon, R. M. 1964. Taxonomic fern notes. IV. Some American vittarioid ferns. Rhodora 66: 110–117.

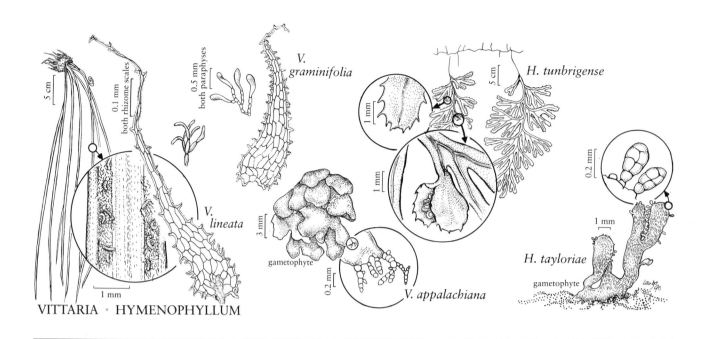

1. VITTARIA　Smith, Mém. Acad. Roy. Sci. (Turin) 5: 413. 1793 · Shoestring fern [Latin *vitt(a)*, fillet, ribbon, stripe]

Stems densely scaly. **Leaves** pendent, linear. **Veins** anastamosing in single row of areolae on each side of midrib. **Sori** in single submarginal groove on each side of midrib. **Gametophytes** 0.5–2 × 0.5–1 cm; gemmae uniseriate, 2–5 mm, borne at apices of aerial branches; body cells 2–16; rhizoid primordia cells 0–4.

Species ca. 50 (3 in the flora): worldwide in tropics and subtropics.

Key to Sporophytes

1. Stem scales tapering to long filiform tips; soral paraphyses without dilated terminal cells; spores monolete. . . . . 1. *Vittaria lineata*
1. Stem scales without long filiform tips; soral paraphyses with dilated terminal cells; spores trilete. . . . . 2. *Vittaria graminifolia*

Key to Gametophytes

1. Gemmae of uniform length; body cells 4, with rhizoid primordium on each end cell. . . . . 2. *Vittaria graminifolia*
1. Gemmae variable in length; body cells 2–16, rhizoid primordia present or absent.
   2. Gemmae with 4–16 body cells, end cells equal to or smaller than medial cells; rhizoid primordia on some medial cells and regularly on end cells; sporophytes frequently produced. . . . . 1. *Vittaria lineata*
   2. Gemmae with 2–12 body cells but at least some 2–3-celled gemmae present, end cells often swollen and larger than medial cells; rhizoid primordia often absent on 1 or both end cells, seldom present on medial cells; sporophytes not produced. . . . . . . 3. *Vittaria appalachiana*

1. **Vittaria lineata** (Linnaeus) Smith, Mém. Acad. Roy. Sci. (Turin) 5: 421. 1793 · Shoestring fern

Pteris lineata Linnaeus, Sp. Pl. 2: 1073. 1753

Plants epiphytic. **Stems** short-creeping, branched, densely scaly; scales brown, apex attenuate, filiform. **Leaves** 10–60 cm × 1–3 mm, petioles indistinct. **Sporangia** protected by soral paraphyses that lack dilated terminal cells. **Spores** monolete. **Gametophytes** much branched. **Gemmae** tapering at ends, end cells not swollen; body cells 4–16, rhizoid primordia on each end cell, often on 1–2 medial cells. $2n = 120$.

Epiphytic, most commonly on trunks of palms (*Sabal palmetto* Loddiges), in moist woods and especially along streams; 0–100 m; Fla., Ga.; Mexico; West Indies; Central America; South America.

Sporophytes, now extirpated, once occurred on rock cliffs at a single site in Lincoln County, east central Georgia. *Vittaria lineata* is now known outside of Florida only in Camden County, in southeastern Georgia. Gametophytes commonly form the dominant cover on moist logs and tree trunks, especially the bases of *Sabal palmetto* palms, within the range of the sporophyte. Such populations usually contain numerous small, sexually produced sporophytes.

2. **Vittaria graminifolia** Kaulfuss, Enum. Filic., 192. 1824 · Grass fern

Vittaria filifolia Fée, Mém. Foug. 3: 20. 1852

Plants epiphytic. **Sporophytes** absent in flora. **Gametophytes** much branched. **Gemmae** tapering at ends, end cells not swollen; body cells 4, rhizoid primordia on each end cell.

Epiphytic in dark, moist hollows formed by flaring root buttresses of beech (*Fagus grandifolia* Ehrhart); 0–50 m; La.; Mexico; West Indies; Central America; South America.

Louisiana plants, known only from St. Helena Parish, are identical to gametophytes grown from sporophyte plants from Central America both morphologically and in starch gel enzyme electrophoresis patterns. Gametophyte colonies and possibly sporophytes of *Vittaria graminifolia* should be expected in similar habitats at additional sites along the Gulf Coast and in peninsular Florida. Sporophytes of *V. graminifolia* differ from those of *V. lineata* in having trilete spores, dilated terminal cells of the soral paraphyses, stem scales with acute (not long filiform) apices, and a chromosome number of $2n = 60$. Earlier reports of sporophytes of *V. graminifolia* in Florida (as *V. filifolia*) have been shown to refer to aberrant forms of *V. lineata* (G. J. Gastony 1980).

3. **Vittaria appalachiana** Farrar & Mickel, Amer. Fern J. 81: 72. 1991 · Appalachian shoestring fern, Appalachian vittaria

Plants on rock. **Sporophytes** absent or abortive, rarely formed (see discussion). **Gametophytes** sparsely to much branched. **Gemmae** highly variable, often with end cells swollen; body cells 2–12, rhizoid primordia absent from medial cells, often lacking on 1 or both end cells.

In dark moist cavities and rock shelters in noncalcareous rocks. Occasionally epiphytic on tree bases in narrow ravines; 150–1800 m; Ala., Ga., Ind., Ky., Md., N.Y., N.C., Ohio, Pa., S.C., Tenn., Va., W.Va.

Dense colonies of *Vittaria appalachiana* coat rock surfaces in deeply sheltered habitats throughout the Appalachian Mountains and plateau. Abortive, apogamously produced embryos and small sporophytes with leaves less than 5 mm have been collected from one site in Ohio and have been produced from gametophytes in culture on two occasions. The largest of these produced simple, linear leaves and clathrate rhizome scales typical of Vittariaceae. Starch gel enzyme electrophoresis patterns, as well as morphology, distinguish these plants from other American species. Enzyme electrophoresis patterns and a somatic chromosome number of 120 (G. J. Gastony 1977) suggest that the plants are diploid and possibly of hybrid origin. Fixation of different genotypes in different sections of the range indicates an ancient origin of the independent gametophytes, possibly through Pleistocene elimination of the sporophyte generation (D. R. Farrar 1990).

A distinctive morphologic characteristic of *Vittaria appalachiana* is the variability displayed in gemma production, often including forms intermediate between gemmae and their supporting gemmifer cells and abortive "gemmae" arrested in early stages of development. This is in contrast to the remarkably regular pattern of gemma production in other species (D. R. Farrar 1978; E. S. Sheffield and D. R. Farrar 1988).

15. HYMENOPHYLLACEAE Link

• Filmy Fern Family

Donald R. Farrar

Plants epiphytic, terrestrial, or on rock. **Stems** long-creeping, often threadlike and intertwining, or short-erect, protostelic, bearing brown hairs of 1–2 types. **Roots** sparse or absent. **Leaves** small, 0.5–20 × 0.2–5 cm, often forming dense mats. **Petiole** short, threadlike to wiry, often winged part or entire length. **Blade** ovate or oblong to lanceolate, simple to decompound, usually 1 cell thick between veins (except *Trichomanes membranaceum* Linnaeus), entire or dentate; scales or simple and/or stellate hairs often borne on veins or leaf margins. **Veins** free and divergent, occasionally present as unattached "false" veins. **Sori** marginal on vein ends, enclosed by 2-valved or conic involucres. **Sporangia** borne on moundlike receptacle or on elongate "bristle," sessile or short-stalked; annulus oblique. **Spores** green, globose, trilete. **Gametophytes** filamentous or ribbonlike or a combination of both, much branched, 0.2–1 cm, often bearing gemmae, persistent, clone-forming by vegetative reproduction.

Genera 6, species ca. 650 (2 genera, 11 species in the flora): worldwide in wet tropics and subtropics, a few in temperate latitudes.

Species outside the flora display a wide range of morphologies and habits, and many are somewhat larger than North American species.

Some authors divide the Hymenophyllaceae into 30 or more genera. The subdivisions of these genera are treated here as subgenera and sections, following C. V. Morton (1968).

Although plants of the Hymenophyllaceae clearly have the capacity to withstand periodic desiccation and freezing, they have a delicate nature that requires they grow in deeply sheltered habitats of nearly continuous high moisture and humidity. This undoubtedly accounts for the relative rarity of all species in the flora. Possibly they are currently restricted from more widespread pre-Pleistocene occurrences. All owe their continuing existence largely or entirely to vegetative propagation by either the sporophyte or gametophyte generation. The capacity for vegetative reproduction and dispersal by gametophytes of the Hymenophyllaceae allows gametophyte colonies to persist indefinitely without completing a life cycle. In the flora, several species are maintained exclusively as gametophytes with sporophytes rarely or never produced.

SELECTED REFERENCES Farrar, D. R. 1967. Gametophytes of four tropical fern genera reproducing independently of their sporophytes in the southern Appalachians. Science 155: 1266–1267. Farrar, D. R. 1985. Independent fern gametophytes in the wild. Proc. Roy. Soc. Edinburgh, B 86: 361–369. Morton, C. V. 1968. The genera, subgenera, and sections of the Hymenophyllaceae. Contr. U.S. Natl. Herb. 38: 153–214. Stokey, A. G. 1940. Spore germination and vegetative stages of the gametophytes of *Hymenophyllum* and *Trichomanes*. Bot. Gaz. 101: 759–790.

1. Soral involucres 2-valved, the halves roundish to ovate, sporangial receptacle enclosed within involucre; gametophytes entirely ribbonlike, branched; gemmae platelike or absent. . . . . 1. *Hymenophyllum*, p. 191
1. Soral involucres conic, sporangial receptacle (bristle) becoming exserted beyond involucre; gametophytes entirely filamentous or proximally filamentous with aerial blades; gemmae uniseriate. . . . . 2. *Trichomanes*, p. 193

1. HYMENOPHYLLUM Smith, Mém. Acad. Roy. Sci. (Turin) 5: 418. 1793 · Filmy fern [Greek *hymen*, membrane, and *phyllon*, leaf]

Plants epiphytic or on rock. **Stems** long-creeping, intertwining, threadlike; hairs brown, sparse. **Roots** few, delicate. **Leaves** 1–3-pinnatifid, 2–6 × 0.5–1.5 cm. **Petiole** short, threadlike, not winged. **Blade** with inconspicuous glandular hairs or prominent stellate hairs; margins entire to distantly dentate. **Soral involucres** 2-valved, the halves roundish to ovate. **Sporangial receptacle** a low mound of tissue included within involucre. **Gametophytes** persistent, ribbonlike, much branched. **Gametophyte gemmae** platelike or absent.

Species ca. 310 (3 in the flora): worldwide in tropical regions.

Most species of *Hymenophllyum* occur in middle elevation rainforests; a few occur in continuously moist temperate habitats. Although all members of this genus are relatively small and delicate, leaves of many species outside the flora become considerably longer by indeterminate growth of the leaf apices.

SELECTED REFERENCES Taylor, M. S. 1938. Filmy-ferns in South Carolina. J. Elisha Mitchell Sci. Soc. 54: 345–348. Wagner, W. H. Jr., D. R. Farrar, and B. W. McAlpin. 1970. Pteridology of the Highlands Biological Station area, southern Appalachians. J. Elisha Mitchell Sci. Soc. 86: 1–27.

Key to Sporophytes
1. Leaf blades glabrous.
    2. Blade segments and involucres dentate. . . . . 1. *Hymenophyllum tunbrigense*
    2. Blade segments and involucres entire. . . . . 2. *Hymenophyllum wrightii*
1. Leaf blades with prominent stellate hairs. . . . . 3. *Hymenophyllum tayloriae*

Key to Gametophytes: Species commonly occurring independently of sporophytes
1. On rock in crevices and grottoes; s Appalachian Mountains. . . . . 3. *Hymenophyllum tayloriae*
1. On wet rock or epiphytic on bark and decaying wood of large conifers; coastal British Columbia, Alaska. . . . . 2. *Hymenophyllum wrightii*

1. **Hymenophyllum tunbrigense** (Linnaeus) Smith, Mém. Acad. Roy. Sci. (Turin) 5: 418. 1793 · Tunbridge filmy fern

Trichomanes tunbrigense Linnaeus, Sp. Pl. 2: 1098. 1753

Plants on rock. **Leaves** oblong, 2–3-pinnatifid, $2–6 \times 0.5–1.5$ cm, with minute, 2-celled, glandular hairs scattered on veins; margins distantly dentate. **Gametophyte gemmae** absent. $2n = 26$.

On rock, forming imbricate mats on vertical cliffs in narrow gorges usually near waterfalls and cascades; 350–500 m; S.C.; Mexico; West Indies; Central America; South America; Europe; Asia; in tropical and temperate regions.

About two dozen small populations of *Hymenophyllum tunbrigense* exist in a single river gorge in Pickens County, South Carolina. It is slow to recover from disturbance, and its numbers have been substantially reduced by collecting since its initial discovery in 1936. Gametophytes characteristic of the genus but lacking gemmae have been described from Great Britain, where populations are more vigorous and where spore production and sexual reproduction via gametophytes are more common (F. J. Rumsey et al. 1990; C. A. Raine et al. 1991). In plants in the flora, spore production is relatively rare, and gametophytes have not been observed.

2. **Hymenophyllum wrightii** Bosch, Ned. Kruidk. Arch. 4: 391. 1859 · Wright's filmy fern

Mecodium wrightii (Bosch) Copeland

Plants on rock or epiphytic. **Leaves** triangular-ovate, 2–3-pinnatifid, $2–5 \times 1–1.5$ cm, nearly glabrous with a few multicellular, gland-tipped hairs; margins entire. **Gametophyte gemmae** platelike, abundant. $2n = 56, 84$.

Shady, wet, vertical cliffs (gametophytes on wet rock or epiphytic on bark and decaying wood of large conifers); 0–20 m; B.C.; Alaska; e Asia.

Although sporophytes of *Hymenophyllum wrightii* are known from only a single locality on Queen Charlotte Islands, British Columbia, gametophytes are more generously distributed along the coasts of Alaska and British Columbia (T. M. C. Taylor 1967). The gametophytes reproduce vegetatively and are capable of persisting and dispersing via gemmae without the intervention of the sporophyte generation.

3. **Hymenophyllum tayloriae** Farrar & Raine in Raine et al., Amer. Fern J. 81: 116, figs. 4–6, 11–13, 16–17. 1991 · Taylor's filmy fern

Plants on rock. **Leaves** (single juvenile specimen known less than 1 cm), with prominent stellate hairs on midrib and margins. **Gametophyte gemmae** platelike, abundant.

On rock in deeply shaded, moist crevices in narrow gorges and behind waterfalls; 350–1200 m; n Ala., N.C., S.C.

This species is described from gametophyte plants. A juvenile sporophyte collected in 1936 by M. S. Taylor is presumed to be this species.* It consists of a short stem with 4 leaves, the largest of which is less than 1 cm. The plant lacks mature characteristics including sori, but the leaves bear stellate hairs typical of subg. *Leptocionium* sect. *Sphaerocionium* of C. V. Morton (1968). Gametophytes collected with the sporophyte occur commonly in the area and differ from those of *H. tunbrigense* both morphologically, especially in bearing copious gemmae, and in enzyme electrophoretic patterns. Therefore, they are here considered to be a distinct species, *H. tayloriae* (C. A. Raine et al. 1991).

D. B. Lellinger (1985) and G. R. Proctor (1985) have considered the South Carolina sporophyte to be *Hymenophyllum hirsutum* (Linnaeus) Swartz. Although the characters of the single sporophyte do not exclude this possibility, they are insufficient to permit definite assignment of the plant to this species. Furthermore, adaptations of the gametophytes to independent existence in temperate habitats of the southern Appalachian Mountains suggest genetic differentiation sufficient to warrant species recognition.

* As this volume goes to press, additional juvenile sporophytes identical to those collected by Taylor were found growing with gametophytes of *H. tayloriae* in Lawrence County, Alabama.

2. TRICHOMANES Linnaeus, Sp. Pl. 2: 1097. 1753; Gen. Pl. ed. 5, 485. 1754 · Bristle fern [Greek *thrix,* hair, and *manes,* cup, alluding to the hairlike receptacle extending from the cuplike involucre]

Plants epiphytic or on rock. **Stems** long-creeping or short and erect, clothed in masses of dark brown hairs of 2 or more types, including multicellular gland-tipped hairs and elongate, sometimes branched and often multicellular, rhizoidlike hairs. **Roots** sparse or absent on creeping stems, numerous and wiry on erect stems. **Leaves** entire, lobed, or compound, 0.5–20 × 0.2–5 cm. **Petiole** short, wiry, often partially or wholly winged. **Blade** glabrous or with scattered, multicellular, gland-tipped hairs on veins; margins entire or minutely lobed, sometimes bearing dark stellate hairs (or orbicular scales, *Trichomanes membranaceum*). **Soral involucres** conic. **Sporangia** sessile, formed at base of exserted bristle and carried outward by intercalary growth of bristle base. **Gametophytes** persistent, entirely filamentous or with proximal filamentous net producing aerial blades with gemmiferous apices. **Gametophyte gemmae** uniseriate.

Species ca. 320 (8 in the flora): nearly worldwide, mostly tropical, a few temperate.

Trichomanes occurs primarily in tropical lowland and montane rainforests, a few species occurring in continuously moist, deeply sheltered habitats in temperate latitudes. Species outside the flora display a wide range of morphologies and habits. Some are terrestrial, some attain considerably larger size, and some have dimorphic fertile and sterile leaves.

Filamentous gametophytes of *Trichomanes* can be distinguished from algae and from moss protonemata by their short cells with numerous discoid chloroplasts, by the presence of short, brown, unicellular rhizoids, and by their production of specialized gemmifer cells and gemmae.

SELECTED REFERENCES Farrar, D. R., J. C. Parks, and B. W. McAlpin. 1982. The fern genera *Vittaria* and *Trichomanes* in the northeastern United States. Rhodora 85: 83–92. Wessels Boer, J. G. 1962. The New World species of *Trichomanes* sect. *Didymoglossum* and *Microgonium.* Acta Bot. Neerl. 11: 277–330.

Key to Gametophytes: Species commonly occurring independently of Sporophytes
1. Plants entirely filamentous; on rock in temperate uplands of e U.S.. . . . . 8. *Trichomanes intricatum*
1. Plants composed of proximal filaments and aerial blades; epiphytic on decomposing cypress along Gulf Coast. . . . . 2. *Trichomanes holopterum*

Key to sporophytes
1. Leaves pinnately lobed or compound with strongly pinnate venation.
  2. Leaves 2–4 cm, bearing dark, large, branched hairs on margins between lobes. . . 3. *Trichomanes krausii*
  2. Leaves 4–20 cm, without hairs on margins.
    3. Leaves 1–2-pinnate-pinnatifid, widely spaced on long-creeping stems. . . 1. *Trichomanes boschianum*
    3. Leaves pinnatifid, clustered on short erect stems. . . . . 2. *Trichomanes holopterum*
1. Leaves entire to irregularly palmately lobed with veins repeatedly forking from base to weakly pinnate.
  4. Leaf margins fringed with paired disclike scales. . . . . 7. *Trichomanes membranaceum*
  4. Leaf margins fringed with dark stellate hairs.
    5. Soral involucres 1 per leaf; involucral lips not dark-edged; venation weakly pinnate.
    . . . . 4. *Trichomanes petersii*
    5. Soral involucres 1–6 per leaf; involucral lips dark-edged; veins repeatedly forking from base.
      6. Leaves 0.5–1 cm; soral involucres flaring at mouth. . . . .
      . . . . 5. *Trichomanes punctatum* subsp. *floridanum*
      6. Leaves 1–3 cm; soral involucres not flaring at mouth. . . . . 6. *Trichomanes lineolatum*

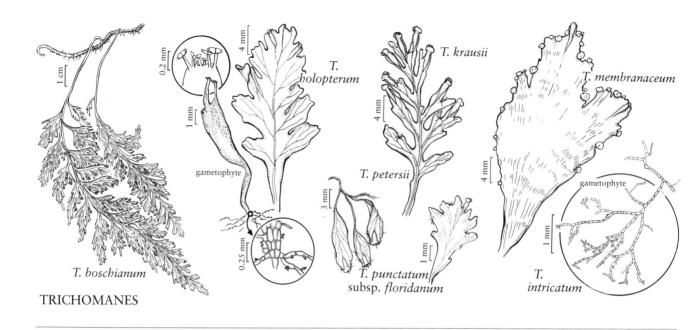

TRICHOMANES

1. Trichomanes boschianum Sturm, Ned. Kruidk. Arch. 5(2): 160. 1861 · Appalachian filmy fern, Appalachian bristle fern

Plants on rock. **Stems** long-creeping, slender, bearing widely spaced leaves; stems covered with dark multicellular hairs of 2 kinds, unbranched gland-tipped hairs and branched or unbranched rhizoidlike hairs, sparsely rooted. **Leaves** lanceolate, 1–2-pinnate-pinnatifid, 4–20 × 1–4 cm, bearing scattered short, unbranched, glandular hairs on principal veins; petioles shorter than blades. **Venation** pinnate, without unconnected false veins. **Soral involucres** terminal on lateral veins at base of lobes, conic, not flaring at mouth; involucral lips not dark edged. **Gametophytes** composed entirely of branching filaments. **Gemmae** composed of short filaments of undifferentiated cells. 2*n* = 72, 108, 144.

In deeply sheltered grottoes on noncalcareous rocks; 150–800 m; Ala., Ark., Ga., Ill., Ind., Ky., N.C., Ohio, S.C., Tenn., Va., W.Va.; Mexico in Chihuahua.

Although earlier treated as synonymous with the tropical American *Trichomanes radicans* Swartz, recent authors have agreed that *Trichomanes boschianum* is a distinct taxon endemic to eastern North America. It exists as fertile diploids and tetraploids with occasional sterile triploids. Diploid cytotypes are prevalent in western localities, and polyploids are more common to the east. Although occurring in climatically moderated

habitats, most populations suffer heavy mortality from sporadic droughts. The plants are very slow to regrow, and many populations are currently but a fraction of their size of 20 years ago. They seldom show evidence of sexual reproduction although gametophyte colonies of this species may be found in the vicinity of fertile sporophytes. Identity of these gametophytes has been confirmed by enzyme electrophoresis, but most occurrences of independent *Trichomanes* gametophytes in the eastern United States have been shown by this method to be those of *T. intricatum* (D. R. Farrar 1985).

2. Trichomanes holopterum Kunze, Farrnkräuter 1: 185. 1845 · Entire-winged bristle fern

Plants epiphytic. **Stems** erect, slender, bearing clustered leaves and numerous, slender, wiry roots, clothed at apices with dark, multicellular, gland-tipped hairs. **Leaves** lanceolate-oblong, pinnatifid, 2.5–10 × 0.5–2 cm, with sparse, 2-celled glandular hairs on veins; petioles shorter than blades. **Venation** pinnate, without unconnected false veins. **Soral involucres** terminal on vein ends at pinna apices, broadly conic and flaring at mouth. **Gametophytes** composed of proximal filamentous network supporting aerial blades tipped by gemmae. **Gemmae** regularly 4-celled with 2 green body cells and 2 nongreen rhizoid primordia. 2*n* = 128.

Epiphytic on decaying stumps and logs of bald cypress in dense cypress swamps; below 10 m; Fla.; West Indies.

Although mature sporophytes of *Trichomanes holopterum* are very rare, gametophyte colonies are relatively common and more extensive and usually bear numerous young sporophytes, most of which seem unable to grow beyond a height of 1–2 cm. Persistence and vegetative reproduction by the gametophyte generation appear primarily responsible for maintenance of the species in the flora (C. E. Delchamps 1966; D. R. Farrar and W. H. Wagner Jr. 1968). The species probably also occurs elsewhere in cypress swamps along the Gulf Coast.

3. **Trichomanes krausii** Hooker & Greville, Icon. Filic. 2: plate 149. 1830 · Kraus's bristle fern

Didymoglossum krausii (Hooker & Greville) C. Presl

Plants epiphytic or on rock. **Stems** long-creeping, threadlike, bearing scattered leaves; stems covered with dark hairs of 2 types: 2-celled glandular hairs and elongate rhizoidlike hairs; roots absent. **Leaves** oblong, 1–2-pinnatifid, 1–5 × 0.5–1.5 cm, with dark, stellate marginal hairs between lobes, 2-celled glandular hairs on petioles and veins, and dark rhizoidlike hairs on petioles and sometimes abaxially on blades; petioles shorter than blades. **Venation** pinnate with unconnected false veins. **Soral involucres** terminal on lobes near leaf apices, conic, flaring at mouth; involucre lips narrowly dark edged. **Gametophytes** composed entirely of branching filaments. **Gemmae** composed of short filaments of undifferentiated cells. $2n = 136$.

On rock walls of limestone sinks or epiphytic on trunks and roots of trees growing in and around limestone sinks; below 10 m; Fla.; Mexico; West Indies; Central America; South America.

Trichomanes krausii is fairly common in and around limestone sinks in hardwood forests in Dade County, Florida. Gametophytes may be found occasionally in the vicinity of sporophytes, but they do not form large independent colonies.

4. **Trichomanes petersii** A. Gray, Amer. J. Sci. Arts, ser. 2, 15: 326. 1853 · Dwarf bristle fern, Peters's bristle fern

Didymoglossum petersii (A. Gray) Copeland

Plants on rock or epiphytic. **Stems** long-creeping, threadlike, and intertwining, bearing scattered leaves, covered with dark hairs of 2 types, 2-celled glandular hairs and elongate rhizoidlike hairs; roots absent. **Leaves** elliptic to oblanceolate, simple, 0.5–2 cm × 2–5 mm, ± entire, bearing simple or 2-cleft dark hairs on margin and 2-celled glandular hairs on petioles and veins; petioles nearly as long as blades. **Venation** weakly pinnate with numerous unconnected false veins. **Soral involucres** usually 1 per leaf, terminal on blades, short-conic, flaring widely at mouth; involucre lips not dark edged. **Gametophytes** composed entirely of branched filaments. **Gemmae** composed of short filaments of undifferentiated cells. $2n = $ ca. 102.

On tree trunks and noncalcareous rocks in deep narrow gorges; 0–500 m; Ala., Ark., Fla., Ga., La., Miss., N.C., S.C., Tenn.; Mexico; Central America in Guatemala.

Trichomanes petersii has irregular meiosis and generally misshapen spores, but it produces some large and viable spores, presumably unreduced in chromosome number. Gametophytes, identified as this taxon by enzyme electrophoresis, have been observed to produce apogamous sporophytes. Therefore, the species seems capable of some reproduction by an apogamous life cycle. Sporophytes also reproduce vegetatively by dispersible buds formed on the leaves. Although some gametophytes in the vicinity of sporophytes have been shown to be *Trichomanes petersii*, most of the independent *Trichomanes* gametophyte populations of the eastern United States are *T. intricatum*.

Sporophytes of *Trichomanes petersii* and other species of subg. *Didymoglossum* form dense mats of imbricated leaves, often excluding all other vegetation. In this habit, as well as in their reduced size and absence of roots, they have adopted a growth form mimicking and successfully competing with bryophytes. In Louisiana and Mississippi, *T. petersii* occurs on trunks of *Fagus* and *Magnolia*. Elsewhere it is on noncalcareous rocks, but in Florida, *T. petersii* also occurs on chert boulders in limestone sinks and cliffs.

5. **Trichomanes punctatum** Poiret in Lamarck et al., Encycl. 8: 64. 1808

Didymoglossum punctatum (Poiret) Desvaux

Subspecies 4 (1 in the flora): tropical and subtropical

regions, North America, Mexico, West Indies, Central America, South America.

5a. Trichomanes punctatum Poiret subsp. floridanum
Wessels Boer, Acta Bot. Neerl. 11: 299. 1962
· Florida bristle fern

Plants on rock or epiphytic. **Stems** long-creeping, threadlike, bearing scattered leaves, covered with dark hairs of 2 types, 2-celled glandular hairs and elongate rhizoidlike hairs; roots absent. **Leaves** round to oblanceolate, simple or irregularly lobed at apices, 5–10 × 2–9 mm, with dark stellate hairs on margin and 2-celled glandular hairs on petioles and veins; petioles shorter than blades. **Venation** repeatedly forking from the base with few unconnected false veins. **Soral involucres** 1–6 per blade, terminal on blades, long-conic, flaring at mouth; involucre lips inconspicuously dark edged. **Gametophytes** composed entirely of branched filaments. **Gemmae** composed of short filaments of undifferentiated cells. $2n = 68$.

On rock in limestone sinks, rarely epiphytic on trunks and roots of trees in limestone sinks; 0–100 m; Fla.

Some early authors listed *Trichomanes sphenoides* Kunze as occurring in Florida. J. G. Wessels Boer (1962), however, reduced *T. sphenoides* to *T. punctatum* subsp. *sphenoides*, which occurs in the Greater Antilles, Central America, and western South America, and considered all the Florida material to be the endemic *T. punctatum* subsp. *floridanum.*

6. Trichomanes lineolatum (Bosch) Hooker in Hooker
& Baker, Syn. Fil. 3: 73. 1867 · Lined bristle fern

Didymoglossum lineolatum Bosch, Ned. Kruidk. Arch. 5: 136. 1863
Plants on rock. **Stems** long-creeping, threadlike, bearing scattered leaves, covered with dark hairs of 2 types, 2-celled glandular hairs and elongate rhizoidlike hairs; roots absent. **Leaves** round to obovate, simple to irregularly lobed, 1–3 × 0.5–1.5 cm, with dark stellate hairs on margin, 2-celled glandular hairs on petioles and veins, and dark rhizoidlike hairs on petioles; petioles shorter than blades. **Venation** repeatedly forking from the base; unconnected false veins few or absent. **Soral involucres** 1–5 per leaf, terminal on blades, narrowly conic, not flaring at mouth; involucre lips with conspicuous dark marginal band 2–5 cells wide. **Gametophytes** unknown, presumed to be as others of subgenus. $2n = 68$.

On rock in limestone sinks; below 10 m; Fla.; Mexico; West Indies; Central America; South America.

Trichomanes lineolatum leaves have unusually thick veins that are enlarged toward the margin and conspicuous in dried specimens.

7. Trichomanes membranaceum Linnaeus, Sp. Pl. 2:
1097. 1753 · Scale-edged bristle fern

Lecanium membranaceum (Linnaeus) C. Presl
Plants epiphytic, on rock, or terrestrial. **Stems** long-creeping, threadlike, bearing scattered leaves, covered with dark hairs of 2 types, multicellular gland-tipped hairs and elongate rhizoidlike hairs. **Leaves** subsessile, irregularly ovate to oblong and often irregularly cleft, 2.5–6.5 × 1–3 cm, bearing multicellular gland-tipped hairs and elongate rhizoidlike hairs on petiole; margin fringed with minute, paired, roundish scales. **Blades** 2 cell layers thick between veins. **Venation** flabellate with many unconnected false veins. **Soral involucres** 5–15 per leaf, marginal on leaf apices, narrowly conic with flaring lips. **Gametophytes** unknown. $2n = 68$.

Terrestrial in acid humus; below 10 m; Miss.; epiphytic and on rock, Mexico; West Indies; Central America; South America.

Trichomanes membranaceum resembles species of subg. *Didymoglossum* in its growth form and chromosome number, and it has been considered by some authors to be of that subgenus. Its marginal scales, absence of stellate hairs, and leaf blades 2 cell layers thick set it apart from other species of subg. *Didymoglossum* and all other American *Trichomanes.*

The single population reported from Harrison County, Mississippi, in 1929 (E. T. Wherry 1964) may be extirpated, and no other occurrences of *Trichomanes membranaceum* are currently known in the flora. Because it is a common and adaptable species of varied habitats throughout tropical America, occurrences along the Gulf Coast and in Florida are not unlikely.

8. Trichomanes intricatum Farrar, Amer. Fern J. 82: 68.
1992 · Appalachian trichomanes, weft fern

Plants on rock, occasionally epiphytic. **Sporophytes** not known. **Gametophytes** entirely filamentous, much branched, persistent. **Gemmae** composed of short filaments of undifferentiated cells.

On noncalcareous rocks in deeply sheltered crevices and grottoes; 150–1800 m; Ala.,

Conn., Ga., Ill., Ind., Ky., Md., Mass., N.H., N.J., N.Y., N.C., Ohio, Pa., S.C., Tenn., Vt., Va., W.Va.

Throughout the eastern uplands of the United States, gametophytes of *Trichomanes intricatum* form feltlike populations covering up to a square meter or more of rock surface in climatically moderated rock shelters and narrow canyons. Sporophytes are not produced, and reproduction is by gemmae and by perennial gametophyte growth and branching.

Filamentous gametophytes of the various *Trichomanes* species have not been distinguished morphologically. Enzyme electrophoresis, however, has shown the vast majority of independent *Trichomanes* gametophyte populations, as well as all of those existing beyond the range of sporophytes of *T. boschianum* and *T. petersii*, to be *T. intricatum*. (All populations of gametophytes tested in Arkansas and several populations in the immediate vicinity of sporophytes of *T. petersii* and *T. boschianum* in eastern states have enzyme banding patterns identical to one or the other of those species.) The adaptation of *T. intricatum* to far northern habitats and its inability to produce sporophytes suggest that this is a distinct taxon, possibly derived from a pre-Pleistocene North American species possessing a normal alternation of generations (D. R. Farrar 1985, 1992).

16. DENNSTAEDTIACEAE Ching

Raymond B. Cranfill

Plants perennial, mostly terrestrial, rarely epiphytic, generally in mesic, forested habitats. **Stems** short- to long-creeping, solenostelic [protostelic], bearing hairs (or less often scales), often branching by means of buds on proximal part of petiole. **Leaves** monomorphic, circinate in bud. **Petiole** not articulate, with 1–many vascular bundles, hairy or glabrous [scaly]. **Blade** 1-pinnate to decompound (rarely simple), glabrous or hairy or with mixture of hairs and glands; rachis and costae grooved adaxially [not grooved in some genera]. **Veins** free or sometimes joined at margin in fertile segments, pinnate or forking in ultimate segments. **Sori** near or at blade margin on vein tips or submarginal commissural vein; true (inner) indusia present, free or fused with portion of blade margin to form cup or pouch, or obscured by revolute and usually modified portion of blade margin [indusia rarely absent]; sporangial stalk of 1–3 rows of cells. **Spores** not green, tetrahedral or bilateral, monolete or trilete. **Gametophytes** green, cordate, with archegonia and antheridia borne on lower surface.

Genera ca. 20, species perhaps 400 (4 genera, 6 species in the flora): worldwide, mostly tropical.

The family is variously circumscribed, in the strict sense including only eight genera, while in the broadest sense encompassing about half the recognized genera of higher ferns (R. E. Holttum 1947). Here it is delimited in the sense of J. T. Mickel (1973). Characteristics that define the family include submarginal or marginal sori with generally two indusia, an inner true indusium and an outer false indusium formed by the revolute, often modified segment margin (although either type may be reduced or absent in some genera); indument usually of hairs rather than scales; and long-creeping protostelic or solenostelic stems with stem buds on the bases of the petioles. Not all genera in the family share all the characteristics.

SELECTED REFERENCES Holttum, R. E. 1947. A revised classification of the leptosporangiate ferns. J. Linn. Soc., Bot. 51: 123–158. Mickel, J. T. 1973. Position of and classification within the Dennstaedtiaceae. In: A. C. Jermy et al., eds. 1973. The Phylogeny and Classification of the Ferns. London. Pp. 135–144. [Bot. J. Linn. Soc., Suppl. 1.]

1. Sori continuous along margins of segments; inner indusium, if present, hidden by reflexed margin of blade and maturing sporangia; blades usually broadly triangular. . . . . . . 3. *Pteridium*, p. 201
1. Sori distinct, not continuous along margins of ultimate segments; inner indusium present or absent; blades narrowly deltate to lanceolate to ovate.
   2. Sori at tips of ultimate segments; veins dichotomously forked. . . . . 4. *Odontosoria*, p. 204
   2. Sori lateral on ultimate segments; veins pinnately branched.
      3. Inner indusium absent, outer indusium recurved over sori. . . . . 2. *Hypolepis*, p. 201
      3. Inner indusium present, outer indusium fused with inner one to form cup or tubular structure containing sporangia. . . . . 1. *Dennstaedtia*, p. 199

1. DENNSTAEDTIA Bernhardi, J. Bot. (Schrader) 1800(2): 124. 1802 · [Named after A. W. Dennstaedt, 1776–1826, German botanist]

Clifton E. Nauman

A. Murray Evans

Plants terrestrial, often forming colonies. **Stems** subterranean, long- to short-creeping; hairs dark reddish brown, jointed. **Leaves** clustered or scattered, erect to arching, ovate to lanceolate to deltate, 0.4–3 m. **Petiole** glabrous to pubescent, usually without prickles, often with stem buds near base; vascular bundles 1–2, arranged in U- or Ω-shape in cross section. **Blade** 2–4-pinnate; rachis without prickles; nectaries absent. **Segments** pinnately divided, ultimate segments ovate to lanceolate, margins dentate or lobed. **Veins** free, pinnately branched. **Sori** marginal at vein tips, distinct, round or cylindric; indusia formed by fusion of true indusium and minute blade tooth to form circular or slightly 2-valvate cup. **Spores** tetrahedral-globose, trilete (rarely monolete), tuberculate or ridged. $x = 34, 46, 47$.

Species ca. 70 (3 in the flora): worldwide, mostly tropical.

SELECTED REFERENCE Tryon, R. M. 1960. A review of the genus *Dennstaedtia* in America. Contr. Gray Herb. 187: 23–52.

1. Blades yellow-green or pale green, mostly 2-pinnate-pinnatifid, leaves usually less than 1 m. . . . . 1. *Dennstaedtia punctilobula*
1. Blades dark green, 3-pinnate, leaves 1–2(–3) m.
   2. Basal segments of pinnules alternate; blades dull; indusia globose; Texas. . . . 2. *Dennstaedtia globulifera*
   2. Basal segments of pinnules opposite; blades lustrous; indusia tubular or cylindric; Florida . . . . 3. *Dennstaedtia bipinnata*

1. **Dennstaedtia punctilobula** (Michaux) T. Moore, Index Fil., 97. 1857 · Hay-scented fern, dennstaedtie à lobules ponctués

Nephrodium punctilobulum Michaux, Fl. Bor.-Amer. 2: 268. 1803

Stems long-creeping, 2–3 mm diam. **Leaves** clustered, erect, 0.4–1(–1.3) × 0.1–0.3 m. **Petiole** straw-colored to brown, darker at base, dull, ca. 1/2 length of blade, pubescent with soft, jointed hairs. **Blade** yellow-green or pale green, dull, lanceolate, 2-pinnate-pinnatifid, ca. 3 times as long as wide, base slightly narrowed but truncate, apex acuminate, with soft, silver-gray, jointed hairs on both surfaces. **Basal segments** of pinnules opposite; ultimate segments ovate to lanceolate, base equilateral, truncate, margins deeply lobed, serrate-crenate. **Sori** globose to almost cylindric; indusia tubular to cylindric. **Spores** trilete, globose with low, tuberculate, distal face and equatorial flange. $2n = 68$.

Sporulates in summer. Rocky slopes, meadows, woods, stream banks, and roadsides, in acid soils; 0–1200 m; N.B., Nfld., N.S., Ont., P.E.I., Que.; Ala., Ark., Conn., Del., Ga., Ill., Ind., Ky., Maine, Md., Mass., Mich., Mo., N.H., N.J., N.Y., N.C., Ohio, Pa., R.I., S.C., Tenn., Vt., Va., W.Va., Wis.

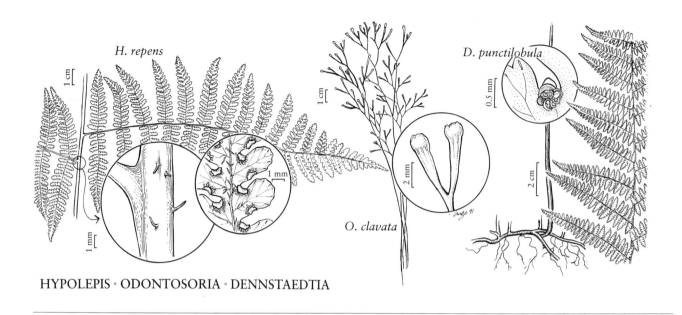

H. repens

D. punctilobula

O. clavata

HYPOLEPIS · ODONTOSORIA · DENNSTAEDTIA

Reports of occurrences of *Dennstaedtia punctilobula* in Iowa are based on incorrectly labeled specimens collected in Massachusetts (T. S. Cooperrider 1968). *Dennstaedtia punctilobula* spreads aggressively in open woods and clearings.

2. Dennstaedtia globulifera (Poiret) Hieronymus, Bot. Jahrb. Syst. 34: 455. 1904 · Beaded cuplet fern

Polypodium globuliferum Poiret in Lamarck et al., Encycl. 5: 554. 1804

Stems long-creeping, 3–10 mm diam. **Leaves** clustered to well separated to distant, erect or arching, 2–3 × (1–)2(–3) m. **Petiole** yellowish to brown throughout, lustrous, ca. 1/3 length of blade, sparsely pubescent with soft, jointed hairs. **Blade** green, dull, ovate to deltate, 3(–4)-pinnate, nearly as wide as long, base truncate, apex acute, with reddish jointed hairs on veins abaxially, glabrous adaxially. **Basal segments** of pinnules alternate; ultimate segments ovate to elliptic, base ± equilateral, margins lobed ca. 1/2 distance to midrib. **Sori** globose; indusia globose. **Spores** trilete, strongly 3-lobed, surface coarsely verrucose. $2n$ = ca. 94.

Sporulates summer. Moist caves or sinks; 300–500 m; Tex.; Mexico; West Indies; Central America; South America.

The tropical *Dennstaedtia globulifera* and *D. bipinnata* are among the largest ferns in the flora. In the flora *Dennstaedtia globulifera* is found only in Val Verde County, Texas.

3. Dennstaedtia bipinnata (Cavanilles) Maxon, Proc. Biol. Soc. Wash. 61: 39. 1938 · Bipinnate cuplet fern, cuplet fern

Dicksonia bipinnata Cavanilles, Descr. Pl., 174. 1802

Stems long-creeping, 5–6 mm diam. **Leaves** clustered to well separated, arching, 1.5–2.5 × ca. 1 m. **Petiole** straw-colored to brown, darker at base, lustrous, 1/2 to equal length of blade, sparsely pubescent with soft, jointed hairs at base when young. **Blade** bright green, lustrous, ovate, 2–4-pinnate, 1/2 to nearly as wide as long, base obtuse, apex acute, sparsely pubescent throughout to nearly glabrous abaxially. **Basal segments** of pinnules opposite; ultimate segments mostly oblong-ovate, base inequilaterally cuneate, margins incised-dentate. **Sori** globose to almost cylindric; indusia tubular or cylindric. **Spores** trilete, prominently 3-lobed, surface irregularly tuberculate. $2n$ = 188.

Sporulates spring and early summer. Moist to wet, forested habitats in acid soils; 0 m; Fla.; Mexico; West Indies; Central America; South America to Bolivia.

Dennstaedtia bipinnata was first found in Florida in 1926 around Lake Okeechobee (J. K. Small 1938). These populations have been extirpated and have been replaced by sugar cane and other agriculture. Populations in Seminole County may be naturalized. The species has been reported as occurring in Okeechobee County. Collections made in the 1920s from Palm Beach County suggest that this species may be native, although *Pteris tripartita* Swartz, which is definitely naturalized, was also collected in that region at that time.

2. HYPOLEPIS Bernhardi, Neues J. Bot. 1(2): 34. 1806 · Bramble ferns [Greek *hypo*, below, and *lepis*, scale, in reference to position of sori under the revolute leaf margin]

Clifton E. Nauman

Plants terrestrial, often forming colonies. **Stems** subterranean, long-creeping; hairs reddish. **Leaves** scattered, arching, deltate, 45–160 cm [to 7 m]. **Petiole** glabrescent or pubescent, often with prickles, sometimes with stem buds near base; vascular bundles more than 3, forming Ω-shaped pattern in cross section. **Blade** 1–4-pinnate; rachises with prickles; nectaries absent. **Segments** pinnatifid, ultimate segments oblong, margins lobed. **Veins** free, simple or pinnately branched. **Sori** ± marginal at vein tips, discrete, mostly round, protected by revolute blade tooth, rarely inframarginal and unprotected. **Spores** ellipsoid, monolete, tuberculate or papillate. $x = 26, 29$.

Species ca. 45 (1 in the flora): worldwide in tropical regions.

SELECTED REFERENCE Brownsey, P. J. 1983. Polyploidy and aneuploidy in *Hypolepis* and the evolution of the Dennstaedtiales. Amer. Fern J. 73: 97–108.

1. Hypolepis repens (Linnaeus) C. Presl, Tent. Pterid., 162. 1836 · Creeping bramble fern, flakelet fern

Lonchitis repens Linnaeus, Sp. Pl. 2: 1078. 1753

Stems long-creeping, branching, 2.5–4 mm diam., pubescence brown. **Leaves** 9–60 dm. **Petiole** straw-colored to reddish brown, 2–4(–6) dm, glabrous or pubescent, bearing prickles; primary pinnae similarly pubescent. **Blade** deltate, 3–4-pinnate-pinnatifid, lateral divisions opposite or nearly so, (2.5–)4–15 × 1.2–6 dm. **Rachis** nearly glabrous or glandular-pubescent, with prickles. **Ultimate segments** spreading, elliptic to lanceolate or oblong-elliptic, pinnatifid, 4–9 × 7 mm, abaxially paler, somewhat papery, nearly glabrous to glandular-pubescent, lobes rounded. **Veins** simple to 1-forked, ending just short of margins. **Sori** in sinuses of ultimate segment divisions, reniform to semicircular or appearing circular at maturity. **Indusia** formed from thin, recurved flap of blade tissue, sometimes obscured by sporangia when mature. $2n = 208$.

Sporulates essentially all year; low hammocks and swamps, generally wet to moist wooded areas in circumneutral to subacid soils; 0 m; Fla.; West Indies; Mexico; Central America; South America.

Within the flora *Hypolepis repens* is found primarily in central peninsular Florida with disjunct populations in southernmost Florida and in northeastern Florida.

3. PTERIDIUM Gleditsch ex Scopoli, Fl. Carniol., 169. 1760, name conserved · Bracken ferns [Greek *pteridion*, a small fern]

Carol A. Jacobs

James H. Peck

Plants terrestrial, often forming colonies or thickets. **Stems** subterranean, slender, long-creeping; hairs pale to dark, jointed; scales absent; true vessels present (absent in other Dennstaedtiaceae genera in the flora). **Leaves** widely spaced, broadly deltate, 0.5–4.5 m. **Petiole** glabrous to short-hairy, without prickles, with stem buds near base, vascular bundles numerous, U- or Ω-shaped in cross section. **Blade** 2–4-pinnate, rachis and costae grooved adaxially; rachis without prickles; nectaries at base of proximal and sometimes distal pinnae. **Segments** pinnately divided, ultimate segments ovate to oblong to linear, base extending proximally on costae (decurrent) or proximally (surcurrent), margins entire. **Veins** free or joined at margin by commissural vein beneath sori, pinnately 2–3-forked. **Sori** ± continuous, covered by recurved, outer false indusium and obscure, extrorse, inner true indusium. **Spores** tetrahedral-globose, trilete, very finely granulate. $x = 26$.

Species 1 (1 species, 4 varieties in the flora): almost worldwide.

SELECTED REFERENCES Page, C. N. 1976. The taxonomy and phytogeography of bracken—A review. Bot. J. Linn. Soc. 73: 1–34. Perring, F. H. and B. G. Gardener, eds. 1976. The biology of bracken. [Symposium.] Bot. J. Linn. Soc. 73(1–3): i–vi, 1–302. Tryon, R. M. 1941. A revision of the genus *Pteridium*. Rhodora 43: 1–31, 37–67.

1. **Pteridium aquilinum** (Linnaeus) Kuhn in Decken, Reisen Ost-Afrika 3(3): 11. 1879 · Bracken, fougère des aigles

Pteris aquilina Linnaeus, Sp. Pl. 2: 1075. 1753

Petioles scattered along creeping stems, 0.3–3.5 m, shallowly to deeply grooved adaxially, base not strongly distinct from stem. **Blades** broadly deltate, papery to leathery, sparsely to densely hairy abaxially, rarely glabrous. **Pinnae** often opposite to subopposite [alternate]; proximal pinnae often prolonged basiscopically, each proximal pinna nearly equal to distal part of leaf in size and dissection (except in var. *caudata*). **Segments** alternate, numerous.

Varieties 12 (4 in the flora): almost worldwide.

In accord with the most recent revision (R. M. Tryon 1941) of the genus, *Pteridium* is treated here as a single widespread species composed of two subspecies with 12 varieties. So treated, it is probably the most widespread species of all vascular plants, with the exception of a few annual weeds (F. H. Perring and B. G. Gardner 1976). The plants are generally aggressive, invading disturbed areas as weeds in pastures, cultivated fields, and roadsides. In Europe, it was harvested and burned to produce potash. Although croziers are eaten in many temperate cultures, bracken has been shown to contain thiaminase (and other compounds with mutagenic and carcinogenic properties).

Disagreement exists among taxonomists regarding the rank that should be accorded to the taxa treated herein as varieties. In a survey of the genus, C. N. Page (1976) noted uniform chromosome numbers and flavonoid compositions of the varieties. D. B. Lellinger (1985) separated the genus into at least two species based on morphology, recognizing as species the subspecies of R. M. Tryon (1941). J. T. Mickel and J. M. Beitel (1988) reported sympatric occurrence in Mexico of three taxa that maintained consistent characteristics and only rarely produced plants with combined characteristics. They suggested that these three taxa should be considered as species that occasionally hybridize. P. J. Brownsey (1989)

reported that two different brackens in Australia formed sterile hybrids and should be treated as species. Modern systematic studies are needed to evaluate the status and rank of the four North American varieties. As treated below, *Pteridium aquilinum* var. *pubescens*, var. *latiusculum*, and var. *pseudocaudatum* are in subsp. *aquilinum*, and var. *caudatum* is in subsp. *caudatum* (Linnaeus) Bonaparte.

1. Fertile ultimate segments only decurrent or more decurrent than surcurrent, mostly 1–2 mm wide; hairs on abaxial surface of blades abundant, straight, stiff, subappressed to spreading. . . . . 1a. *Pteridium aquilinum* var. *caudatum*
1. Fertile ultimate segments adnate or equally decurrent and surcurrent, mostly 3–6 mm wide; hairs on abaxial surface of blades abundant to sparse, twisted and flexible, if abundant then lax, spreading.
    2. Pinnules at nearly 90° angle to costa; outer indusium pilose on margin and often on surface; hairs on abaxial surface of blades abundant, lax, and spreading. . . . . 1b. *Pteridium aquilinum* var. *pubescens*
    2. Pinnules at 45°–60° angle to costa; outer indusium glabrous; hairs on abaxial surface of blades sparse or blades nearly glabrous.
        3. Terminal segments of pinnules 2–4 times longer than wide; segment margins and abaxial surface of blade midrib and costae shaggy. . . . . 1c. *Pteridium aquilinum* var. *latiusculum*
        3. Terminal segments of pinnules ca. 6–15 times longer than wide; segment margins and abaxial surface of blade midrib and costae sparsely pilose to glabrous. . . . . 1d. *Pteridium aquilinum* var. *pseudocaudatum*

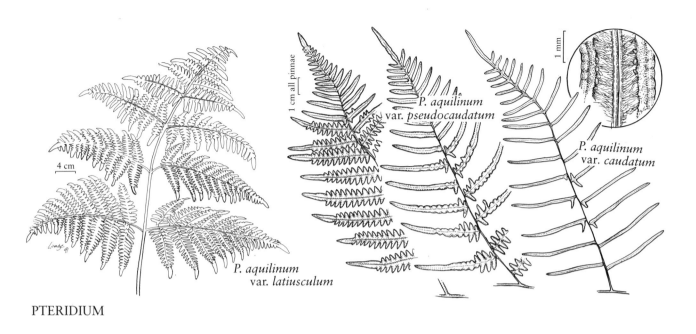

P. aquilinum
var. *pseudocaudatum*

P. aquilinum
var. *caudatum*

1 cm all pinnae

1 mm

P. aquilinum
var. *latiusculum*

PTERIDIUM

1a. Pteridium aquilinum (Linnaeus) Kuhn var. **caudatum** (Linnaeus) Sadebeck, Jahrb. Hamburg. Wiss. Anst. Beih. 3: 5. 1897 · Lacy bracken

Pteris caudata Linnaeus, Sp. Pl. 2: 1075. 1753; *Pteridium caudatum* (Linnaeus) Maxon

Petiole 20–75 cm. **Blade** broadly ovate to deltate, 2–3-pinnate-pinnatifid, 30–100 × 20–80 cm; blades, rachises, and costae usually densely covered abaxially with abundant, straight, stiff, subappressed to spreading hairs. **Pinnae** all narrowly to broadly triangular; terminal segment of each pinna ca. 10 times longer than wide, longer ultimate segments several times their width apart, ca. 1–2.5 mm wide. **Pinnules** at nearly 90° angle to costa; fertile ultimate segments only decurrent, or more decurrent than surcurrent. **Outer indusia** entire, glabrous.

In barrens, pine woodlands, and edges of deciduous woods in strongly acid to circumneutral soil, forming large colonies in exposed sites; 0 m; Fla.; Mexico; West Indies; Bermuda; Central America; South America from Colombia to Peru.

In Florida and West Indies material, the abaxial surfaces are quite hairy; in Central American material they are often much less hairy.

1b. Pteridium aquilinum (Linnaeus) Kuhn var. **pubescens** L. Underwood, Native Ferns ed. 6, 91. 1900 · Western bracken

Pteris aquilina Linnaeus var. *pubescens* (L. Underwood) Clute, Fern Bull. 8: 37. 1900

Petiole 10–100 cm. **Blade** ovate-triangular to nearly pentagonal, 3-pinnate to 3-pinnate-pinnatifid, 30–200 × 15–100 cm; blades, rachises, and costae usually densely covered abaxially with abundant, contorted, lax, spreading hairs. **Pinnae** (proximal) triangular, distal pinnae oblong; terminal segment of each pinna ca. 4 times longer than wide, longer ultimate segments less than their width apart, ca. 1.5–5 mm wide. **Pinnules** at nearly 90° angle to costa; fertile ultimate segments adnate or equally decurrent and surcurrent. **Outer indusia** entire, pilose on margin and surface, hairs like those of axes.

In dry to moist woods and open areas in partial to full sun, forming abundant colonies; 0–3000 m; Alta., B.C.; Alaska, Ariz., Calif., Colo., Idaho, Mont., Nev., N.Mex., Oreg., S.Dak., Tex., Utah, Wash., Wyo.; Mexico in Baja California, Chihuahua, and Durango.

1c. Pteridium aquilinum (Linnaeus) Kuhn var. **latiusculum** (Desvaux) L. Underwood ex A. Heller, Cat. N. Amer. Pl. ed. 3, 17. 1909 · Eastern bracken

Pteris latiuscula Desvaux, Mém. Soc. Linn. Paris 6. 2: 303. 1827; *Pteridium latiusculum* (Desvaux) Fries

Petiole 15–100 cm. **Blade** broadly triangular to sometimes ovate, 3-pinnate or 3-pinnate-pinnatifid at base, 20–80 × 25–50 cm; blade margins and abaxial surface shaggy, rachises and costae glabrous or sparsely pilose abaxially. **Pinnae** (proximal) broadly triangular, distal pinnae narrowly triangular or oblong; terminal segment of each pinna ca. 2–4 times longer than wide, longer ultimate segments less than their width apart, ca. 3–6 mm wide. **Pinnules** at 45°–60° angle to costa; fertile ultimate segments adnate or equally decurrent and surcurrent. **Outer indusia** entire or somewhat erose, glabrous. $2n = 104$.

In barrens, pastures, open woodlands in moderately to strong acid soil, abundant, forming large colonies; 0–1500 m; St. Pierre and Miquelon; Alta., B.C., Man., N.B., Nfld., N.S., P.E.I., Ont., Que.; Ala., Ark., Conn., Del., D.C., Fla., Ga., Ill., Ind., Iowa, Ky., La., Maine, Md., Mass., Mich., Minn., Miss., Mo., N.H., N.J., N.Y., N.C., N.Dak., Ohio, Pa., R.I., S.C., S.Dak., Tenn., Vt., Va., W.Va., Wis.; Mexico in Nuevo León; Europe; Asia.

Colonies are more frequent in the northern part of the range. Fertile colonies, however, are more frequent in the southern and eastern portion of the range. Outliers in British Columbia and Alberta, which we have not seen, are documented in W. J. Cody and D. M. Britton (1989).

1d. Pteridium aquilinum (Linnaeus) Kuhn var. **pseudocaudatum** (Clute) A. Heller, Cat. N. Amer. Pl. ed. 2, 12. 1900 · Tailed bracken

Pteris aquilina Linnaeus var. *pseudocaudata* Clute, Fern Bull. 8: 39. 1900

Petiole 10–70 cm. **Blade** broadly triangular to sometimes ovate, 2–3-pinnate-pinnatifid at base, 20–80 × 20–70 cm; blades, rachises, and costae sparsely pilose to glabrous abaxially. **Pinnae** (proximal) ovate-lanceolate or triangular, distal pinnae oblong; terminal segment of each pinna ca. 6–15 times longer than wide, longer ultimate segments 1–2 times their width apart, ca. 2–5 mm wide. **Pinnules** at 45–60° angle to costa; fertile ultimate segments adnate or equally decurrent and surcurrent. **Outer indusia** entire to somewhat erose, glabrous.

In barrens and open pine or oak woods in acid, often sandy soil, abundant, forming large colonies; 0–1000 m; Ala., Ark., Conn., Del., D.C., Fla., Ga., Ill., Ind., Ky., La., Md., Mass., Miss., Mo., N.J., N.Y., N.C., Ohio, Okla., Pa., S.C., Tenn., Tex., Va., W.Va.

Pteridium aquilinum var. *pseudocaudatum* is more common in the southern portion of the range.

4. ODONTOSORIA Fée, Mém. Foug. 5: 325. 1852 · Wedgelet fern [Greek *odous*, tooth, and *soros*; the sori are at the tips of toothed segments]

Karl U. Kramer

Sphenomeris Maxon, J. Wash. Acad. Sci. 3: 144. 1913

Plants terrestrial or on rock, not forming colonies. **Stems** creeping on substrate surface, mostly short; hairs brown, jointed (grading into long, narrow scales with uniseriate tip), not clathrate. **Leaves** closely spaced to quite distant, erect to arching, narrowly oblong [triangular, ovate, to irregularly shaped], 10–50 cm [to 6 m]. **Petiole** glabrous or with a few basal scales, stem buds absent, vascular bundles 1. **Blade** 3–4-pinnate [2–5-pinnate, sometimes scandent or climbing], ± glabrous [some species with spines on petiole, midrib, or costae]; rachises without prickles; nectaries absent. **Segments** nearly dichotomously divided; ultimate segments linear to cuneate, margins entire or bifid. **Veins** free, simple or forked. **Sori** terminal near blade margin on single vein or on commissure joining 2–8 veins, ± spheric to transversely elongate, indusia attached at base and sides, opening toward margin. **Spores** tetrahedral to nearly globose, or oblong, trilete or monolete, smooth or granulate. $x = 38, 47$.

Species 22 (1 in the flora): mostly tropical.

SELECTED REFERENCE   Kramer, K. U. 1957. A revision of the genus *Lindsaea* in the New World with notes on allied genera. Acta Bot. Neerl. 6: 97–290.

1. **Odontosoria clavata** (Linnaeus) J. Smith, Hist. Fil., 264. 1875 · Wedgelet fern

Adiantum clavatum Linnaeus, Sp. Pl. 2: 1096. 1753; *Sphenomeris clavata* (Linnaeus) Maxon; *Stenoloma clavatum* (Linnaeus) Fée

Stems short-creeping, scales very narrow, dark brown, leaves closely spaced. **Blade** oblong, ca. 10–25 cm; ultimate segments narrowly elongate-cuneate, often 1–1.5 cm × 0.5–2 mm, papery, apically denticulate especially when sterile. **Sori** on 1–8 vein endings; indusia erose at distal margin. **Spores** trilete. $2n = 76$.

Calcareous soil, sinks and pinelands; 0 m; Fla.; s Mexico; West Indies in Bahamas, Greater Antilles.

17. THELYPTERIDACEAE Ching ex Pichi-Sermolli · Marsh Fern Family

Alan R. Smith

Plants terrestrial or on rock [epiphytic]. **Stems** creeping to erect, scaly at apex. **Leaves** monomorphic or somewhat dimorphic [dimorphic]. **Petiole** in cross section with 2 crescent-shaped vascular bundles at base. **Blade** pinnate to pinnate-pinnatifid, rarely more than 2-pinnate [simple]; rachis grooved adaxially or not, grooves not continuous with grooves of next order. **Veins** free or anastomosing, running to margin, areoles with or without included free veinlets. **Indument** of transparent, needlelike, hooked, septate, or stellate hairs, or rarely hairs lacking. **Sori** inframedial to supramedial, occasionally nearly marginal, round or oblong, rarely elongate along veins; indusia reniform or sometimes absent. **Spores** bilateral, monolete [rarely globose-tetrahedral and trilete], usually with a prominent, crested, echinate, or reticulate perispore. **Gametophytes** green, cordate, usually hairy or glandular; antheridia 3-celled.

Genera 1 to ca. 30, depending on circumscription, species ca. 900 (as circumscribed here, 3 genera and 25 species in the flora): mostly tropical.

Members of Thelypteridaceae have historically been associated with Dryopteridaceae (in particular, *Dryopteris*) but in fact have no close relationship with that family. *Thelypteris* and allies differ from *Dryopteris* and allies by their indument of transparent needlelike hairs (versus needlelike hairs absent in Dryopteridaceae); general absence of blade scales (versus blade scales often present); petiole vasculature in cross section with two crescent-shaped bundles (versus many round bundles arranged in an arc, *Athyrium* and allies exceptional); generally 1-pinnate to pinnate-pinnatifid blades (versus often more divided); veins usually not forking in the ultimate segments (versus often forking); adaxial grooves discontinuous from rachis to costae, or grooves lacking (versus grooves often continuous); and chromosome base numbers from 27–36 (versus generally 40, 41).

1. Blades 1-pinnate to deeply pinnate-pinnatifid; costae grooved adaxially; veins meeting margin at or above sinus or united below sinus. . . . . 1. *Thelypteris*, p. 207
1. Blades 2-pinnatifid, with pinnae at least in distal 1/2 of blade connected by wings along rachis, or blades 2-pinnate-pinnatifid; costae not grooved adaxially; veins commonly meeting margin above sinus.
   2. Pinnae free, rachis not winged; blades 2-pinnate or more divided; costal hairs septate, often longer than 1 mm; indusia small, less than 0.3 mm diam. . . . . 2. *Macrothelypteris*, p. 220
   2. Pinnae mostly connected by wings along rachis, the wings often forming semicircular lobes between pinnae; blades 2-pinnatifid; costal hairs not septate, shorter than 0.5 mm; indusia absent. . . . . 3. *Phegopteris*, p. 221

1. THELYPTERIS Schmidel, Icon. Pl. ed. Keller, 3, 45, plates 11, 13. 1763 (Oct.), name conserved · Female fern [Greek *thelys*, female, and *pteris*, fern]

Stems long-creeping to ascending to erect, 1.5–12 mm or more diam. **Blades** 1-pinnate to pinnate-pinnatifid, rarely 2-pinnate, proximal pinnae reduced or not, apex commonly gradually reduced, infrequently abruptly reduced and pinnalike; pinnae entire to deeply pinnatifid, sessile or short-stalked; costae grooved adaxially; buds absent or uncommonly present in axils of pinnae; veins free to regularly anastomosing, commonly simple (1-forked in a few species, e.g., *Thelypteris palustris*) and reaching margin; indument various abaxially, often of simple or branched hairs on blades, rachises and costae with or usually without scales. **Sori** round, oblong, or elongate along veins, commonly medial to supramedial; indusia round-reniform, large (ca. 1 mm diam.) and persistent or sometimes small (less than 0.3 mm diam.), occasionally ephemeral, sometimes absent; sporangial capsules glabrous or occasionally hairy. $x = 27$, 29, 31, 32, 33, 34, 35, 36.

Species ca. 875 (21 in the flora): nearly worldwide.

In the broadest sense, *Thelypteris* is a very large and complex genus of about 900 species and constitutes the only genus in the family. It has been divided into ca. 30 genera by R. E. Holttum (1971, 1982); these are treated as subgenera and/or sections by various workers. In the treatment adopted here, the genus is broadly circumscribed but excludes the small segregate genera *Phegopteris* and *Macrothelypteris*, two of the most distinctive elements. The subgroups of *Thelypteris* (treated as genera by some workers) are indicated in the key to species and by their subgeneric names preceding the treatment of species groups. The name to be used if a narrowly circumscribed segregate genus is adopted is included in the synonymy.

SELECTED REFERENCES Christensen, C. 1913. A monograph of the genus *Dryopteris*. Part I. The tropical American pinnatifid-bipinnatifid species. Kongel. Danske Vidensk. Selsk. Skr., Naturvidensk. Math. Afd., ser. 7, 10: 55–282. Holttum, R. E. 1971. Studies in the family Thelypteridaceae III. A new system of genera in the Old World. Blumea 19: 17–52. Holttum, R. E. 1981. The genus *Oreopteris* (Thelypteridaceae). Kew Bull. 36: 223–226. Holttum, R. E. 1982. Thelypteridaceae. In: C. G. G. J. van Steenis and R. E. Holttum, eds. 1959–1982. Flora Malesiana. Series II. Pteridophyta. Vol. 1, part 5. Iwatsuki, K. 1964. An American species of *Stegnogramma*. Amer. Fern J. 54: 141–153. Smith, A. R. 1971. Systematics of the neotropical species of *Thelypteris* section *Cyclosorus*. Univ. Calif. Publ. Bot. 59: 1–143. Tryon, A. F., R. M. Tryon, and F. Badré. 1980. Classification, spores, and nomenclature of the marsh fern. Rhodora 82: 461–474.

1. Pinnae entire or merely serrate; veins regularly united in pairs to form parallel rows of more than 5 areoles between costa and pinna margin, areoles each with single excurrent, usually free veinlet (subg. *Meniscium*).
   2. Pinna margins hooked-serrate. . . . . 21. *Thelypteris serrata*
   2. Pinna margins entire to undulate or crenulate. . . . . 20. *Thelypteris reticulata*

1. Pinnae shallowly to deeply lobed; veins free, connivent at sinuses, or 1–2 pairs between primary veins uniting to form common vein extending to sinus.
   3. Stellate and/or forked hairs present, especially on scales at apex of stems, on rachises, especially in adaxial grooves, and often on other parts of blades (subg. *Goniopteris*).
      4. Blades mostly 15–25 cm wide, proximal pinnae not reduced, blade abruptly narrowed distally, apical pinna similar to lateral pinnae; pinnae usually 6–8 pairs. . . . . 19. *Thelypteris tetragona*
      4. Blades mostly 2–8(–10) cm wide, with or without reduced proximal pinnae, blade gradually narrowed distally to pinnatifid apex, apical pinna not similar to lateral pinnae; pinnae 6–25(–30) pairs.
         5. Pinnae with acute lobes; blade hairs subsessile, stellate; blades erect, not rooting at apex. . . . . 18. *Thelypteris sclerophylla*
         5. Pinnae entire or with shallow, rounded lobes; blade hairs stalked, stellate, mixed with needlelike hairs; blades commonly arching or pendent and rooting at apex. . . . . 17. *Thelypteris reptans*
   3. Stellate or forked hairs absent.
      6. Sori elongate; sporangia minutely hairy (subg. *Stegnogramma*). . . . . 16. *Thelypteris pilosa* var. *alabamensis*
      6. Sori round or slightly oblong; sporangia glabrous.
         7. Blades with all veins extending to margin above sinus; proximal pinnae greatly reduced or not.
            8. Proximal pinnae not reduced or only slightly so, not much shorter than middle pinnae.
               9. Segment veins often forked; blades lacking glands; abaxial surface of costae with tan, ovate scales (subg. *Thelypteris*). . . . . 6a. *Thelypteris palustris* var. *pubescens*
               9. Segment veins not forked; blades bearing sessile glands abaxially; costae lacking scales (subg. *Parathelypteris*). . . . . 3. *Thelypteris simulata*
            8. Proximal pinnae greatly reduced, much shorter than middle pinnae.
               10. Petioles (and often rachises and costae) with persistent tan to straw-colored, linear-lanceolate scales abaxially (subg. *Lastrea*). . . . . 5. *Thelypteris quelpaertensis*
               10. Petioles persistently scaly only toward base, rachises and costae lacking scales abaxially.
                  11. Stems erect, trunklike with age; leaves evergreen; blades abaxially with dense reddish, hemispheric, sessile glands (subg. *Amauropelta*). . . . . 4. *Thelypteris resinifera*
                  11. Stems creeping (apices sometimes upturned with leaves clustered); leaves dying back in winter; blades glandular or not abaxially (subg. *Parathelypteris*).
                     12. Costae conspicuously hairy abaxially, hairs ca. 1 mm; blades abaxially lacking glands or with sparse, mostly sessile glands. . . . . 2. *Thelypteris noveboracensis*
                     12. Costae glabrous or sparsely hairy abaxially, hairs 0.2–0.7 mm; blades abaxially with numerous yellow to orange, mostly short-stipitate glands. . . . . 1. *Thelypteris nevadensis*
         7. Blades with at least some basal veins extending to, connivent at, or united below sinus; proximal pinnae usually not reduced (subg. *Cyclosorus*), or if reduced, then veins united below sinus.
            13. Basal veins of adjacent segments united below sinus, excurrent vein leading toward sinus.
               14. Stems long-creeping; costae with tan ovate scales abaxially; veins, costules, and costae adaxially glabrous or sparsely hairy with hairs less than 0.2 mm. . . . . 15. *Thelypteris interrupta*

14. Stems short-creeping or suberect; costae without scales abaxially; veins, costules, and costae moderately to rather densely hairy adaxially with hairs greater than 0.3 mm.

    15. Costae with predominantly short hairs uniform in length (less than 0.2 mm and often less than 0.1 mm) on abaxial surface; excurrent veins mostly greater than 2 mm; petioles purplish; leaves with usually more than 2 pairs of greatly reduced proximal pinnae. . . . . 14. *Thelypteris dentata*

    15. Costae with most hairs greater than 0.3 mm (some exceeding 0.5 mm) on abaxial surface, hairs not uniform; excurrent veins less than 2 mm; petioles straw-colored; leaves with 0–2 pairs of slightly reduced proximal pinnae. . . . . 13a. *Thelypteris hispidula* var. *versicolor*

13. Basal veins of adjacent segments free, or connivent at sinuses.

    16. Stems suberect to erect; scales at base of petioles ovate, glabrous. . . . . 11a. *Thelypteris patens* var. *patens*

    16. Stems long- to short-creeping; scales at base of petioles lanceolate, usually hairy.

        17. Costules and veins adaxially with at least a few rather stout hairs mostly longer than 0.3 mm; blades adaxially often rather glandular; scales absent on abaxial surface of rachises and costae of mature leaves.

            18. Proximal pinnae (1–2 pairs) somewhat reduced; stems short-creeping, sometimes appearing suberect; venation variable, even on same leaf, veins united and giving rise to short-excurrent vein or veins connivent at sinus; blades adaxially often somewhat hairy; veins adaxially always with stout hairs, many longer than 0.4 mm. . . . . . 13a. *Thelypteris hispidula* var. *versicolor*

            18. Proximal pinnae usually not at all reduced; stems short- to long-creeping; veins connivent at sinus or distal vein of each pair meeting margin slightly above sinus; blades glabrous or sparsely hairy adaxially; veins adaxially with or without stout hairs. . . . . 12. *Thelypteris kunthii*

        17. Costules and veins adaxially glabrous; blades adaxially lacking glands; a few scales often persistent on rachises and/or costae abaxially.

            19. Apical portion of blades somewhat similar to lateral pinnae; pinnae less than 2 cm wide. . . . . 7. *Thelypteris augescens*

            19. Apical portion of blades ± attenuated, different from lateral pinnae; pinnae 0.8–4.8 cm wide.

                20. Basal segments of pinnae near base of blade slightly elongate and parallel to rachis; costae abaxially lacking scales or scales sparse. . . . . 9. *Thelypteris ovata*

                20. Basal segments of pinnae near base of blade not elongate and not parallel to rachis, or only those facing apex of rachis elongate and enlarged; costae abaxially bearing few to many scales.

                    21. Pinnae more than 2 cm wide; costae abaxially with hairs ca. 0.1 mm. . . . . 8a. *Thelypteris grandis* var. *grandis*

                    21. Pinnae less than 2 cm wide; costae abaxially with hairs at least 0.2 mm. . . . 10a. *Thelypteris puberula* var. *sonorensis*

1a. THELYPTERIS Schmidel subg. **PARATHELYPTERIS** (H. Itô) R. M. Tryon & A. F. Tryon, Rhodora 84: 128. 1982

Thelypteris sect. *Parathelypteris*, H. Itô in Nakai & Honda, Nov. Fl. Jap. 4: 127. 1939; *Parathelypteris* (H. Itô) Ching

Distinguished from other subgenera by the combination of seasonal leaves, narrow, long-creeping stems 1–3 mm diam., pinnate-pinnatifid blades with generally greatly reduced proximal pinnae (except *Thelypteris simulata*), and veins meeting margin above the sinuses. $x = 27, 32$.

Species 1–3 are included in *Parathelypteris* by some workers (e.g., R. E. Holttum 1976).

1. **Thelypteris nevadensis** (Baker) Clute ex C. V. Morton, Amer. Fern J. 48: 139. 1958 · Nevada marsh fern

Nephrodium nevadense Baker, Ann. Bot. (London) 5: 320. 1891; *Dryopteris nevadensis* (D. C. Eaton) L. Underwood; *D. oregana* C. Christensen; *Parathelypteris nevadensis* (Baker) Holttum

Stems creeping for 2–5 cm, then ascending or suberect, 1.5–3 mm diam. **Leaves** monomorphic, dying back in winter, tightly clustered, (25–)40–105 cm. **Petiole** straw-colored, 3–20(–35) cm × 1–3 mm, at base with scales tan to reddish brown, ovate, glabrous. **Blade** elliptic, 20–70 cm, proximal 4–10 pinna pairs gradually reduced (smallest 5–20 mm), blade tapering gradually to pinnatifid apex. **Pinnae** 3–10 × (0.6–)1–2 cm, deeply pinnatifid to within 1 mm of costa; segments oblong to linear, oblique (sides slanted, not perpendicular to costa), entire to crenulate; proximal pair of veins from adjacent segments meeting margin above sinus. **Indument** abaxially of sparsely set hairs 0.2–0.7 mm on rachises, costae, and sometimes veins, also of numerous orangish, sessile to usually short-stalked glands on blade tissue; blades adaxially glabrous except along costae. **Sori** round, supramedial; indusia tan, glabrous or short-ciliate, sometimes also with glands; sporangia glabrous. $2n = 54$.

Terrestrial in woods and meadows, especially near springs, seepage areas, and streams; 0–1800 m; B.C.; Calif., Oreg., Wash.

Thelypteris nevadensis is named for the Sierra Nevada and, contrary to its common name, is not found in Nevada.

2. **Thelypteris noveboracensis** (Linnaeus) Nieuwland, Amer. Midl. Naturalist 1: 225. 1910 · New York fern, thélyptéride de New York, fougère de New York

Polypodium noveboracense Linnaeus, Sp. Pl. 2: 1091. 1753; *Dryopteris noveboracensis* (Linnaeus) A. Gray; *Parathelypteris noveboracensis* (Linnaeus) Ching; *Thelypteris thelypterioides* (Michaux) Holub

Stems usually long-creeping, 1.5–2.5 mm diam. **Leaves** monomorphic, dying back in winter, mostly evenly spaced 1 cm or more (sun-gathering leaves in loose cluster), (25–)40–85 cm. **Petiole** straw-colored, 4–25 cm × 1–3 mm, at base with scales tan to reddish brown, ovate, glabrous. **Blade** elliptic, 15–60 cm, proximal 4–10 pinna pairs gradually smaller toward base (smallest often less than 5 mm), blade tapering gradually to pinnatifid apex. **Pinnae** deeply pinnatifid to within 1 mm of costa, 3–9(–13) × 1–2(–2.5) cm; segments oblong to linear, somewhat oblique, entire to crenulate; proximal pair of veins from adjacent segments meeting margin above sinus. **Indument** abaxially of moderately to densely set hairs to 1 mm on rachises, costae, and veins, glands lacking or yellowish to light orangish, mostly sessile on blade tissue; blades adaxially often with hairs on veins. **Sori** round, supramedial; indusia tan, often ciliate; sporangia glabrous. $2n = 54$.

Terrestrial in moist woods, especially near swamps, streams, and in vernal seeps of ravines, often in slightly disturbed secondary forests, frequently forming large colonies; 0–1100 m; St. Pierre and Miquelon; N.B., Nfld., N.S., Ont., P.E.I., Que.; Ala., Ark., Conn., Del., D.C., Ga., Ill., Ind., Ky., La., Maine, Md., Mass., Mich., Miss., N.H., N.J., N.Y., N.C., Ohio, Okla., Pa., R.I., S.C., Tenn., Vt., Va., W.Va.

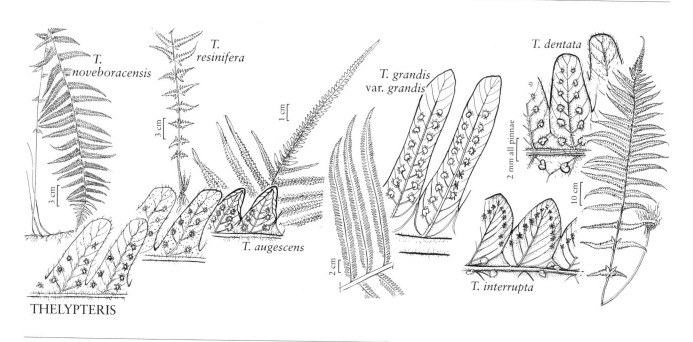

3. **Thelypteris simulata** (Davenport) Nieuwland, Amer. Midl. Naturalist 1: 226. 1910 · Massachusetts fern, thélyptère simulée

Aspidium simulatum Davenport, Bot. Gaz. 19: 495. 1894; *Parathelypteris simulata* (Davenport) Holttum

Stems long-creeping, 1.5–3 mm diam. **Leaves** monomorphic, dying back in winter, often 1 cm or more apart, fertile leaves often somewhat longer than sterile leaves, 25–80 cm. **Petiole** straw-colored above base, 12–45 cm × 1–3 mm, at base sparsely set with tan, ovate, glabrous scales. **Blade** lanceolate, 14–35 cm, proximal pinnae largest, or lowest pair slightly smaller, blade ta-pered gradually to pinnatifid apex. **Pinnae** deeply pinnatifid to ca. 1 mm from costa, 3–10 × 1–2 cm; segments oblong, somewhat oblique, entire; proximal pair of veins from adjacent segments meeting margin above sinus. **Indument** abaxially of sparsely set hairs 0.2–0.4 mm primarily on costae, also of yellowish short-stalked glands less than 0.1 mm, blade tissue with numerous reddish to orangish, resinous, shiny, sessile, hemispheric glands; blades adaxially with scattered hairs on veins. **Sori** round, medial; indusia tan, often glandular; sporangia glabrous. $2n = 128$.

Terrestrial in acid soils of shaded swamps and bogs, frequently associated with sphagnum; 0–100 m; N.B., N.S., Ont., Que.; Conn., Del., Maine, Md., Mass., N.H., N.J., N.Y., Pa., R.I., Vt., Va., W.Va., Wis.

1b. THELYPTERIS Schmidel subg. AMAUROPELTA (Kunze) A. R. Smith, Amer. Fern J. 63: 121. 1973

Amauropelta Kunze, Farrnkräuter 1: 109. 1843

Distinguished from other subgenera by the combination of evergreen leaves, suberect to erect stems hidden by old petiole bases, veins meeting margin above the sinuses, and the generally greatly reduced proximal pinnae. $x = 29$.

Thelypteris resinifera is included in the genus *Amauropelta* by some workers.

4. Thelypteris resinifera (Desvaux) Proctor, Bull. Inst. Jamaica, Sci. Ser. 5: 63. 1953 · Glandular maiden fern, wax-dot maiden fern

Polypodium resiniferum Desvaux, Ges. Naturf. Freunde Berlin Mag. Neuesten Entdeck. Gesammten Naturk. 5: 317. 1811; *Amauropelta resinifera* (Desvaux) Pichi-Sermolli; *Dryopteris panamensis* (C. Presl) C. Christensen; *Thelypteris panamensis* (C. Presl) E. P. St. John.

Stems erect, (3–)5–12 mm diam. **Leaves** monomorphic, evergreen, clustered, (15–)25–100(–135) cm; croziers often mucilaginous. **Petiole** straw-colored, 2–15(–25) cm × 1–4(–6) mm, at base with brownish, ovate-lanceolate, ± glabrous scales. **Blade** to 110 cm, proximally with 6–12 pairs of smaller pinnae, distally tapering gradually to a pinnatifid apex. **Proximal pinnae** hastate or auriculate; longest (medial) pinnae 2–14(–20) × 0.4–2.5 cm, incised to within 1 mm of costa; segments 2–3 mm wide, strongly oblique and somewhat curved; veins meeting margin above sinus. **Indument** abaxially of hairs 0.2–0.5 mm on rachises, costae, and sometimes veins and blade tissue, also of numerous glands, these yellowish to often reddish, resinous, shiny, sessile, hemispheric; blade tissue adaxially glabrous or sparsely hairy. **Sori** round, medial to submarginal; indusia tan, glandular, and sparsely hairy. $2n = 58$.

Damp woods and swamps in subacid soil; 0–100 m; Fla.; Mexico; West Indies in the Greater Antilles; Central America; nw South America.

1c. THELYPTERIS Schmidel subg. LASTREA (Hooker) Alston, J. Wash. Acad. Sci. 48: 232. 1958

Nephrodium Marthe ex Michaux subg. *Lastrea* Hooker, Sp. Fil. 4: 5, 84. 1862

Distinguished from other subgenera by the combination of seasonal leaves, short-creeping to suberect stems hidden by old petiole bases, persistently scaly petioles, pinnate-pinnatifid blades with greatly reduced proximal pinnae, and veins meeting margin above the sinuses. $x = 34$.

Thelypteris quelpaertensis, *T. limbosperma* (Allioni) H. P. Fuchs from Europe, and *T. elwesii* (Baker) Ching from Asia constitute the genus *Oreopteris* Holub (R. E. Holttum 1981).

5. Thelypteris quelpaertensis (H. Christ) Ching, Bull. Fan Mem. Inst. Biol. 6: 328. 1936

Dryopteris quelpaertensis H. Christ in A. Léveillé, Bull. Acad. Int. Géogr. Bot. 20: 7. 1910; *Oreopteris quelpaertensis* (H. Christ) Holub

Stems short-creeping to suberect, 5–10 mm diam. **Leaves** monomorphic, dying back in winter, crowded, (15–)25–100 cm. **Petiole** straw-colored to tan above base, 3–20 cm × 2–5 mm, scales on petioles and rachises tan to straw-colored, persistent, ovate to lanceolate. **Blade** elliptic, 25–80 cm, 5–10 pairs of proximal pinnae gradually smaller toward base, lowest pinnae ca. 1 cm, blade tapering gradually to pinnatifid apex. **Pinnae** deeply pinnatifid to ca. 1 mm or less from costa, 3–12 × 1–2 cm; segments linear to oblong, somewhat oblique and often somewhat curved, entire or crenulate, basal segments of proximal pinnae more often crenulate; proximal pair of veins from adjacent segments meeting margin above sinus. **Indument** abaxially of tan to whitish linear scales along costae, hairs lacking or sparse along costae, blade tissue lacking glands or sparsely glandular. **Sori** round, submarginal; indusia tan, glabrous; sporangia glabrous. $2n = 68$.

Terrestrial in open, rocky woods and subalpine meadows in acid soils; 30–1300 m; B.C., Nfld.; Alaska, Wash.; e Asia.

Although the name *Thelypteris limbosperma* (Allioni) H. P. Fuchs, type from Europe, has usually been applied to plants in the flora, specimens from western North America match more closely those from eastern Asia; therefore, a name based on a Korean type is used here. The single collection from the coast of Newfoundland (reported by A. Bouchard and S. G. Hay 1976) is remarkably disjunct but matches collections from western North America rather than those of the European species.

1d. THELYPTERIS Schmidel subg. THELYPTERIS

Distinguished from other subgenera by the combination of seasonal leaves, long-creeping stems 1–3 mm diam., pinnate-pinnatifid blades, veins often forked and meeting margins above the sinuses, scales along the abaxial costae, and by the proximal pinnae not or only slightly shortened. $x = 35$.

Thelypteris palustris and *T. confluens* (Thunberg) C. V. Morton from the Southern Hemisphere constitute the genus *Thelypteris,* according to R. E. Holttum (1971).

6. Thelypteris palustris Schott, Gen. Fil., plate 10. 1834 · Marsh fern, thélyptère des marais

Acrostichum thelypteris Linnaeus; *Dryopteris thelypteris* (Linnaeus) A. Gray

Varieties 2 (1 in the flora): North America, Europe, Asia.

Thelypteris palustris var. *palustris* occurs in Eurasia.

The name *Thelypteris thelypterioides* (Michaux) Holub has been applied to *T. palustris,* but A. F. Tryon et al. (1980) have shown that this was a result of an incorrect typification by C. V. Morton (1967).

6a. Thelypteris palustris Schott var. **pubescens** (Lawson) Fernald, Rhodora 31: 34. 1929

Lastrea thelypteris (Linnaeus) Bory var. *pubescens* Lawson, Edinburgh New Philos. J., n.s. 19: 277. 1864; *Dryopteris thelypteris* (Linnaeus) A. Gray var. *pubescens* (Lawson) Weatherby; *Thelypteris palustris* var. *haleana* Fernald

Stems long-creeping, 1–3 mm diam. **Leaves** monomorphic or slightly dimorphic, dying back in winter, often 1–3 cm apart, fertile leaves more erect, narrower, and with somewhat contracted pinnae and segments, 20–90 cm. **Petiole** straw-colored above base, 9–45(–60) cm × 1–3 mm, at base sparsely set with tan, ovate, glabrous scales. **Blade** lanceolate, 10–40(–55) cm, proximal pinnae commonly slightly shorter, blade tapering gradually to pinnatifid apex. **Pinnae** pinnatifid to within 1 mm of costa, 2–10 × 0.5–2 cm; segments oblong, somewhat oblique, entire; proximal pair of veins from adjacent segments meeting margin above sinus, veins frequently forked. **Indument** abaxially of sparsely to densely set hairs on costae and sometimes veins, costae also commonly with a few small, tan scales, blade tissue glabrous on both sides. **Sori** round, medial; indusia tan, often hairy; sporangia glabrous. $2n = 70$.

Terrestrial in swamps, bogs, and marshes, also along riverbanks and roadside ditches, and in wet woods; 0–1000 m; St. Pierre and Miquelon; Man., N.B., Nfld., N.S., Ont., P.E.I., Que.; Ala., Ark., Conn., Del., D.C., Fla., Ga., Ill., Ind., Iowa, Kans., Ky., La., Maine, Md., Mass., Mich., Minn., Miss., Mo., Nebr., N.H., N.J., N.Y., N.C., N.Dak., Ohio, Okla., Pa., R.I., S.C., S.Dak., Tenn., Tex., Vt., Va., W.Va., Wis.; perhaps Mexico (as *Dryopteris tremula* H. Christ); West Indies in Bermuda, Cuba.

1e. THELYPTERIS Schmidel subg. CYCLOSORUS (Link) C. V. Morton, Amer. Fern J. 53: 153. 1963

Cyclosorus Link, Hort. Berol. 2: 128. 1833

Distinguished from other subgenera by the combination of evergreen leaves, usually long- to short-creeping stems more than 5 mm diam., pinnate-pinnatifid blades with proximal pinnae usually not shortened (except *Thelypteris hispidula* and *T. dentata*), and some veins meeting margin at or below the sinuses. $x = 36$.

Species 7–12 were monographed by A. R. Smith (1971), who treated them in *Thelypteris* sect. *Cyclosorus.* These same species were placed in *Christella* sect. *Pelazoneuron* by R. E. Holttum (1982, p. 553), who restricted use of *Cyclosorus* to *Thelypteris interrupta* (species 15 here) and two other species.

7. Thelypteris augescens (Link) Munz & I. M. Johnston, Amer. Fern J. 12: 75. 1922 · Abrupt-tipped maiden fern

Aspidium augescens Link, Fil. Spec., 103. 1841; *Christella augescens* (Link) Pichi-Sermolli; *Dryopteris augescens* (Link) C. Christensen

Stems creeping, 4–8 mm diam. **Leaves** monomorphic, evergreen, (0.5–)1–3.5 cm apart, (30–)65–140 cm. **Petiole** straw-colored, 15–70 cm × 2–7(–9) mm, at base with scales tan to brownish, linear-lanceolate, hairy at margin. **Blade** 30–70 cm, broadest at or near base, abruptly narrowed distally, apical pinna ± similar to lateral pinnae, 5–17 × 1–3(–5) cm. **Pinnae** (4–)10–22(–28) × (0.3–)0.7–1.5 cm, incised 1/2–3/4 of width; segments somewhat curved, margins revolute, those at base of proximal pinnae slightly elongate; proximal pair of veins from adjacent segments running to sinus. **Indument** abaxially of hairs 0.2–0.4 mm on costae, veins, and blade tissue, also of brownish scales 0.6–1.2 mm on costae; veins and blade tissue glabrous adaxially. **Sori** round, medial to supramedial; indusia tan, bearing hairs 0.2–0.4 mm; sporangia glabrous. 2n = 144.

Limestone banks, in sun or partial shade; 0–50 m; Fla.; s Mexico; West Indies in the Bahamas, Cuba; Central America in Guatemala.

Thelypteris augescens occasionally hybridizes with *T. kunthii* and *T. ovata* var. *ovata* in southern Florida.

8. Thelypteris grandis A. R. Smith, Univ. Calif. Publ. Bot. 59: 96. 1971 · Stately maiden fern

Nephrodium paucijugum Jenman

Varieties 4 (1 in the flora): North America, Mexico, West Indies in the Greater Antilles, Central America, South America in Andes to nw Argentina and s Brazil.

8a. Thelypteris grandis A. R. Smith var. **grandis**

Stems long-creeping, ca. 1 cm diam. **Leaves** monomorphic, evergreen, 4–8 cm apart, to ca. 3 m. **Petiole** straw-colored, to 1.3 m × 0.8–1.2 cm, at base sparsely set with brown, lanceolate, hairy scales. **Blade** to ca. 1.7 m, broadest at or near base, gradually to somewhat abruptly tapered to pinnatifid apex. **Pinnae** (15–)20–45 × 2–3.5(–4.8) cm, incised mostly 3/4–9/10 of width; segments oblique, curved, 2–4 basal basiscopic ones (those facing base of petiole) on proximal pinnae greatly reduced or wanting; proximal 1–2(–3) pairs of veins from adjacent

segments connivent at sinus. **Indument** abaxially of hairs 0.1 mm on costae and veins, also of brownish, hairy scales on costae; blade tissue glabrous on both sides. **Sori** round, medial to supramedial; indusia pinkish to reddish brown, glabrous; sporangia glabrous. 2n = 72.

Mixed swamps and old logging roads; 0–50 m; Fla.; West Indies in the Greater Antilles, St. Christopher (St. Kitts).

In the flora *Thelypteris grandis* var. *grandis* is known only from one population in Collier County, Florida.

9. Thelypteris ovata R. P. St. John in Small, Ferns S.E. States, 230, with plate. 1938 · Ovate maiden fern

Christella ovata (R. P. St. John) A. Löve & D. Löve; *Dryopteris ovata* (R. P. St. John) Broun

Stems usually long-creeping, 3–6 mm diam. **Leaves** monomorphic, evergreen, (0.5–)1–4 cm apart, (30–)55–135(–165) cm. **Petiole** straw-colored, 15–80 cm × 2–6 mm, at base with tan to brownish, linear-lanceolate, hairy scales. **Blade** about equaling petiole length, broadest at base, gradually to somewhat abruptly tapered to pinnatifid apex. **Pinnae** (5–)10–25 × 0.8–2.2 cm, incised 4/5 of width; segments oblique, somewhat curved, basal pair from middle pinnae often elongate parallel to rachis; proximal pair of veins from adjacent segments reaching margin at or just above sinus. **Indument** abaxially of hairs mostly 0.2–0.5 mm on costae, veins, and blade tissue, also sometimes of a few tan scales on costae and rachises; blades adaxially glabrous except along rachises and costae. **Sori** round, supramedial to inframarginal; indusia tan, hairy, hairs mostly 0.2–0.4 mm; sporangia glabrous. 2n = 72.

Varieties 2 (2 in the flora): North America, Mexico, West Indies, Central America.

1. Scales usually absent on costae abaxially; blade tissue glabrous adaxially; petiole base and stem scales brownish. . . . 9a. *Thelypteris ovata* var. *ovata*
1. Scales few, very narrow, on costae abaxially; blade tissue minutely pubescent or glabrous adaxially; petiole base and stem scales tan. . . . . 9b. *Thelypteris ovata* var. *lindheimeri*

9a. Thelypteris ovata R. P. St. John var. **ovata**

Scales usually absent on costae abaxially; blade tissue glabrous adaxially. 2n = 72.

Hammocks and limestone sinks, occasionally on rocks; 0–50 m; Ala., Fla., Ga., S.C.; West Indies in the Bahamas.

This variety occasionally hybridizes with *Thelypteris augescens* and *T. kunthii* in Florida.

9b. Thelypteris ovata R. P. St. John var. **lindheimeri** (C. Christensen) A. R. Smith, Amer. Fern J. 61: 30. 1971 · Lindheimer's maiden fern

Dryopteris normalis C. Christensen var. *lindheimeri* C. Christensen, Kongel. Danske Vidensk. Selsk. Skr., Naturvidensk. Math. Afd., ser. 7, 10: 182. 1913

Scales few, very narrow, usually present on costae abaxially; blades minutely pubescent or glabrous adaxially.

Riverbanks and moist canyons; 0–100 m; Tex.; Mexico; West Indies in Jamaica; Central America in Belize, Guatemala.

10. Thelypteris puberula (Baker) C. V. Morton, Amer. Fern J. 48: 138. 1958

Nephrodium puberulum Baker in Hooker & Baker, Syn. Fil. ed. 2, 495. 1874; *Christella puberula* (Baker) A. Löve & D. Löve; *Dryopteris feei* C. Christensen; *D. puberula* (Baker) Kuntze; *Thelypteris feei* (C. Christensen) Moxley

Varieties 2 (1 in the flora): North America, Mexico, Central America.

Thelypteris puberula var. *puberula* occurs throughout Mexico and Central America to Costa Rica.

10a. Thelypteris puberula (Baker) C. V. Morton var. **sonorensis** A. R. Smith, Univ. Calif. Publ. Bot. 59: 91. 1971 · Sonoran maiden fern

Stems creeping, 4–8 mm diam. **Leaves** monomorphic, evergreen, (0.5–)1–3 cm apart, 35–120 (–165) cm. **Petiole** straw-colored, 15–60(–80) cm × 2–7 mm, at base sparsely set with brownish, lanceolate, hairy scales. **Blade** 20–55(–85) cm, broadest at or near base, gradually to somewhat abruptly tapered to pinnatifid apex. **Pinnae** 7–20(–26) × 1–2(–3) cm, incised 1/2–4/5 of width, proximal pinnae often narrowed toward base except for slightly enlarged acroscopic basal segment; segments oblique, somewhat curved; proximal 1–3 pairs of veins from adjacent segments connivent at sinus, or meeting margin just above sinus. **Indument** abaxially of often irregularly crimped hairs ca. 0.2–0.3 mm on costae, veins, and blade tissue, costae also with a few brownish scales less than 0.5 mm; blade tissue abaxially with scattered hairs ca. 0.2 mm. **Sori** round, medial to supramedial; indusia tan to brownish, usually densely hairy, hairs irregularly crimped, ca. 0.2 mm; sporangia glabrous.

In canyons, especially along streams and seepage areas, sometimes on calcareous substrates; 100–1300 m; Ariz., Calif.; w Mexico.

11. Thelypteris patens (Swartz) Small, Ferns S.E. States, 243. 1938 · Grid-scale maiden fern

Polypodium patens Swartz, Prodr., 133. 1788; *Christella patens* (Swartz) Holttum; *Cyclosorus arcuatus* (Poiret) Alston; *Dryopteris patens* (Swartz) Kuntze

Varieties 3 (1 in the flora): North America, s Mexico, West Indies in the Antilles, Central America, South America to s Brazil and n Argentina.

11a. Thelypteris patens (Swartz) Small var. **patens**

Stems erect, 4–12 mm diam. **Leaves** monomorphic, evergreen, clustered, (15–)75–125(–200) cm. **Petiole** (5–)15–50(–100) cm × (1–)2–9(–12) mm, at base with tan (whitish when young), ovate, glabrous scales. **Blade** (8–)25–75(–100) cm, broadest at or near base, tapering gradually to pinnatifid apex. **Pinnae** (3–)10–32 × (0.5–)1–3(–4) cm, incised 3/4 or more of width; segments oblique, somewhat curved, basal segments elongate parallel to rachises or acroscopic segment of a pinna enlarged and dentate; proximal pair of veins from adjacent segments running to sinus or meeting margin just above sinus. **Indument** abaxially of hairs 0.2–0.8 mm and stalked glands 0.1 mm on costae, veins, and sometimes blade tissue; veins and costules adaxially glabrous, lacking glands. **Sori** round, medial to supramedial; indusia tan, usually hairy; sporangia glabrous or with stalked pear-shaped glands borne on sporangial stalks. $2n = 144$.

In limestone crevices in hammocks; 0–50 m; Fla.; s Mexico; West Indies in the Antilles; Central America; South America to s Brazil, nw Argentina.

Thelypteris patens var. *patens* is known in the flora from one collection made in 1905 from Dade County, Florida.

12. Thelypteris kunthii (Desvaux) C. V. Morton, Contr. U.S. Natl. Herb. 38: 53. 1967 · Widespread maiden fern, southern shield fern

Nephrodium kunthii Desvaux, Mém. Soc. Linn. Paris 6: 258. 1827; *Christella normalis* (C. Christensen) Holttum; *Dryopteris normalis* C. Christensen; *Thelypteris macrorhizoma* R. P. St. John; *T. normalis* (C. Christensen) Moxley; *T. saxatilis* R. P. St. John; *T. unca* R. P. St. John

Stems short- to long-creeping, 4–8 mm diam. **Leaves** monomorphic, evergreen, up to 2(–3) cm apart, (15–) 50–160 cm. **Petiole** straw-colored, (5–)20–80 × (1–) 3–6 mm, at base with brown, linear-lanceolate, hairy scales. **Blade** (9–)30–80 cm, broadest at base, gradually tapered to pinnatifid apex. **Pinnae** (2–)8–15(–20) × (0.6–)1–2.5 cm, incised 3/5–4/5 of width; segments oblong, rounded to acute at apex; proximal pair of veins from adjacent segments running to sinus, or nearly so. **Indument** abaxially of hairs mostly 0.3–1 mm on costae, veins, and blade tissue; veins adaxially with similar hairs but blade tissue usually without hairs, often with scattered yellowish, stalked glands 0.1 mm. **Sori** round, medial to supramedial; indusia tan, hairy, hairs 0.2–0.4 mm; sporangial glands obscure, yellowish, stalked, arising from sporangial stalks. $2n = 144$.

Roadsides, ditches, riverbanks, woodlands, limestone sinks; 0–100 m; Ala., Ark., Fla., Ga., La., Miss., S.C., Tex.; e,s Mexico; West Indies; Bermuda; Central America to Costa Rica; South America from Colombia to n Brazil.

Thelypteris kunthii occasionally hybridizes with *T. augescens* and *T. ovata* in Florida; hybrids with *T. hispidula* may also occur.

13. **Thelypteris hispidula** (Decaisne) C. F. Reed, Phytologia 17: 283. 1968 · Hairy maiden fern

Aspidium hispidulum Decaisne, Nouv. Ann. Mus. Hist. Nat. 3: 346. 1834; *Christella hispidula* (Fée) Holttum; *Cyclosorus quadrangularis* (Fée) Tardieu-Blot; *Dryopteris hispidula* (Decaisne) Kuntze; *D. quadrangularis* (Fée) Alston; *Thelypteris quadrangularis* (Fée) Schelpe

Varieties 4 (1 in the flora): tropical and subtropical, North America, Mexico, West Indies in the Antilles, Central America, South America, Asia, Africa.

The relationship between Old World and New World varieties is unstudied.

This species and the next are included in *Christella* subg. *Christella* by R. E. Holttum (1982).

13a. **Thelypteris hispidula** (Decaisne) C. F. Reed var. **versicolor** (R. P. St. John) Lellinger, Amer. Fern J. 71: 94. 1981 · Variable maiden fern, St. John's shield fern

Thelypteris versicolor R. P. St. John in Small, Ferns S.E. States, 250, plate. 1938; *Dryopteris versicolor* (R. P. St. John) Broun; *Thelypteris macilenta* E. P. St. John; *T. quadrangularis* (Fée) Schelpe var. *versicolor* (R. P. St. John) A. R. Smith

Stems short-creeping to ascending at apex, 3–5 mm diam. **Leaves** monomorphic, evergreen, closely placed, (20–)40–95 cm. **Petiole** straw-colored, (5–)10–40 cm × 1.5–3 mm, at base with brown, lanceolate, hairy scales. **Blade** 14–55 cm, proximal 0–2(–4) pairs of pinnae slightly to greatly reduced (lowest pinnae to 1/3 length of longest), blade gradually tapered to pinnatifid apex. **Pinnae** 4–14 × 0.8–2 cm, incised 3/4–4/5 of width; segments spreading or somewhat oblique, rounded at tip, basal acroscopic segment of proximal pinnae sometimes auriculate and crenate; proximal pair of veins from adjacent segments united just below sinus or approaching each other and turning abruptly toward sinus, or connivent at sinus, excurrent vein, if any, 1 mm or less. **Indument** on both surfaces of densely set hairs 0.2–0.8 mm on costae, veins, and often blade tissue, also often of yellowish, stalked glands 0.1 mm. **Sori** round, medial to supramedial; indusia hairy; sporangial stalks with inconspicuous stalked, pear-shaped glands. $2n = 72$.

Woodlands and limestone sinks, especially in seepage areas and along streams; 0–100 m; Ala., Fla., Ga., La., Miss., S.C., Tex.; West Indies in Cuba.

14. **Thelypteris dentata** (Forsskål) E. P. St. John, Amer. Fern J. 26: 44. 1936 · Downy maiden fern, downy shield fern, tapering tri-vein fern

Polypodium dentatum Forsskål, Fl. Aegypt.-Arab., 185. 1775; *Christella dentata* (Forsskål) Brownsey & Jermy; *Cyclosorus dentatus* (Forsskål) Ching; *Dryopteris dentata* (Forsskål) C. Christensen; *D. mollis* (Swartz) Hieronymus; *Thelypteris reducta* Small

Stems short-creeping, 4–6 mm diam. **Leaves** often somewhat dimorphic, evergreen, often closely placed, 50–150 cm, fertile leaves with longer petioles and more contracted pinnae. **Petiole** often purplish brown, 15–50 cm × 3–6 mm, at base with brown, linear-lanceolate, hairy scales. **Blade** (25–)40–100 cm, 1–4 (–6) proximal pairs of pinnae reduced, blade gradually tapered to pinnatifid apex. **Pinnae** 7–17 × 1–3 cm, incised 1/2–3/4 of width; segments rounded at apex, basal acroscopic segment of proximal pinnae often auriculate; proximal pair of veins from adjacent segments united at obtuse angle below sinus with excurrent vein 2–4 mm. **Indument** abaxially of uniformly short hairs 0.1–0.2 mm on costae, veins, and blade tissue; veins adaxially with stouter hairs, also with hairs 0.1–0.2 mm on blade tissue. **Sori** round, medial to supramedial; indusia tan, pubescent, hairs 0.1–0.3 mm; sporangial stalks with orangish, stalked glands. $2n = 144$.

Damp woods; 0–100 m; introduced; Ala., Fla., Ga., Ky., La.; s Mexico; West Indies in the Antilles; South America to n Argentina; native to tropical and subtropical Asia, Africa.

Thelypteris dentata probably does not persist northward in areas (such as Kentucky) where winters are sometimes severe (R. Cranfill 1980).

15. Thelypteris interrupta (Willdenow) K. Iwatsuki, Jap. J. Bot. 38: 314. 1963 · Hottentot fern, Willdenow's fern, spready tri-vein fern

Pteris interrupta Willdenow, Phytographia, 13. 1794; *Cyclosorus gongylodes* (Schkuhr) Link; *C. interruptus* (Willdenow) H. Itô; *C. tottus* (Thunberg) Pichi-Sermolli; *Dryopteris gongylodes* (Schkuhr) Kuntze; *Thelypteris gongylodes* (Schkuhr) Small; *T. totta* (Thunberg) Schelpe

Stems long-creeping, cordlike, 3–6 mm diam. **Leaves** monomorphic, evergreen, 3–6 cm apart, 50–150(–250) cm. **Petiole** straw-colored to tan, 20–125 cm × 3–6 mm, scaleless. **Blade** 30–125 cm, broadest at base, gradually narrowed distally to pinnatifid apex. **Pinnae** 7–30 × 1–2 cm, incised 1/3–1/2(–3/5) of width; segments deltate, rounded to acute; proximal pair of veins from adjacent segments united at acute or obtuse angle below sinus, with excurrent vein 2–4 mm. **Indument** abaxially of hairs 0.1–0.3 mm on costae and veins, or hairs often lacking, costae also with tan, ovate scales; veins, costules, and costae adaxially glabrous or sparsely pubescent; blade tissue without hairs on both sides, or hairy abaxially, usually with red to orange, shiny, sessile, hemispheric glands abaxially. **Sori** round, medial to supramedial; indusia tan, glabrous to hairy; sporangia with red- or orange-capped, stalked, globose glands arising from sporangial stalks. $2n = 144$.

Wet roadside ditches, riverbanks, marshes, and cypress swamps; 0–50 m; Fla., La.; Mexico; West Indies in the Antilles; Central America; South America to Argentina; tropical and subtropical Asia, Africa.

D. B. Lellinger (1985) applied the name *Thelypteris interrupta* to specimens from India, while using *T. totta* (type from South Africa) for North American and South American specimens. Diploid cytotypes are known from Africa and Asia, whereas all counts from the Neotropics are tetraploid. Until more counts are available and the morphologic variation (chiefly in glands, pubescence, and leaf size) in this species complex is better understood, I prefer to circumscribe the species broadly.

R. E. Holttum (1982) circumscribed *Cyclosorus* (as a genus) to include this species and one or two others.

1f. THELYPTERIS Schmidel subg. STEGNOGRAMMA (Blume) C. F. Reed, Phytologia 17:254.1968

Stegnogramma Blume, Enum. Pl. Javae, 172. 1828

Distinguished from other subgenera by the combination of evergreen leaves, broadly adnate, shallowly lobed distal pinnae, elongate exindusiate sori, and sporangial capsules bearing hairs. $x = 36$.

16. Thelypteris pilosa (M. Martens & Galeotti) Crawford, Amer. Fern J. 41: 16. 1951 · Streak-sorus fern

Gymnogramma pilosa M. Martens & Galeotti, Nouv. Mém. Acad. Roy. Sci. Bruxelles 15(5): 27, plate 4, fig. 1. 1842; *Dryopteris pilosa* (M. Martens & Galeotti) C. Christensen; *Stegnogramma pilosa* (M. Martens & Galeotti) K. Iwatsuki

Varieties 2 (1 in the flora): North America, Mexico, Central America.

Thelypteris pilosa var. *pilosa* ranges from northwestern Mexico to Central America in the Honduras.

16a. Thelypteris pilosa (M. Martens & Galeotti) Crawford var. **alabamensis** Crawford, Amer. Fern J. 41: 19. 1951 · Alabama streak-sorus fern

Leptogramma pilosa (M. Martens & Galeotti) L. Underwood var. *alabamensis* (Crawford) Wherry; *Stegnogramma pilosa* (M. Martens & Galeotti) K. Iwatsuki var. *alabamensis* (Crawford) K. Iwatsuki

Stems short-creeping, 1.5–2.5 mm diam. **Leaves** monomorphic, evergreen, clustered, 0.1–1 cm apart, 10–45 cm. **Petiole** straw-colored, (1.5–)5–20 cm × ca. 1 mm, at base very sparsely set with ovate-lanceolate, hairy scales; blades linear-lanceolate, to ca. 30(–40) cm, slightly narrowed at base, proximal 1–3 pinna pairs slightly shortened, gradually tapered to pinnatifid apex. **Pinnae** spreading, rounded at tip, 0.5–3 × 0.5–1.3 cm, proximal pinnae stalked to 3 mm, distal pinnae broadly adnate and basiscopically decurrent, crenate or incised 1/4–1/2 width or single basal acro-scopic segment on proximal pinnae nearly or quite free; segments somewhat oblique, rounded at tip; proximal pair of veins from adjacent segments running to sinus or nearly so. **Indument** on both sides of thin to stout hairs mostly 0.2–1.5 mm on costae, veins, and blade tissue. **Sori** elongate along veins, lacking indusia; sporangia often minutely hairy on capsule, hairs 0.1–0.2 mm.

On sandstone cliffs in river gorges; 150 m; Ala.; Mexico in Chihuahua, Sonora.

Thelypteris pilosa is included in the Asian and African genus *Stegnogramma* by K. Iwatsuki (1964). Variety *alabamensis* differs from var. *pilosa* in Mexico and Central America by the much narrower blades, spreading (vs. ascending) pinnae with rounded tips, and free or nearly free basal acroscopic segment on the proximal pinnae. The Winston County, Alabama, populations are remarkably disjunct, about 2000 km from the nearest Mexican populations of both varieties.

1g. THELYPTERIS Schmidel subg. GONIOPTERIS (C. Presl) Duek, Adansonia, n. s. 11: 720. 1971

Goniopteris C. Presl, Tent. Pterid., 181. 1836

Distinguished from other subgenera by the combination of evergreen leaves, shallowly to deeply lobed pinnae, and stellate or furcate hairs on the stem scales, along the petiole and rachis, especially in the adaxial grooves, and sometimes on the blades. *x* = 36.

Species 17–19 are included in the genus *Goniopteris* by some workers.

17. **Thelypteris reptans** (J. F. Gmelin) C. V. Morton, Fieldiana, Bot. 28: 12. 1951 · Creeping star-hair fern

Polypodium reptans J. F. Gmelin, Syst. Nat. 2: 1309. 1791; *Dryopteris reptans* (J. F. Gmelin) C. Christensen; *Goniopteris reptans* (J. F. Gmelin) C. Presl

Stems creeping to suberect, 2–3 mm diam. **Leaves** somewhat dimorphic, evergreen, laxly arching or prostrate, sterile leaves often rooting at attenuate apices or along rachises, mostly (10–)15–55 cm, fertile leaves more erect and with longer petioles, not rooting, with more contracted pinnules. **Petiole** green, 1–25 cm × 0.5–1 mm, at base sparsely set with brown, lanceolate, stellate-hairy scales. **Blade** usually 10–30 cm, pinnate in proximal half only or throughout, narrowed distally to pinnatifid apex. **Pinnae** entire to crenate to shallowly lobed ca. 1/3 of width, 1–2.5(–5) × 0.3–1(–1.5) cm, sometimes subcordate at base; proximal pairs from adjacent segments usually united with excurrent vein or veins free. **Indument** abaxially of stellate, forked, and needlelike hairs on rachises, costae, veins, and blade tissue; blade tissue adaxially also with stellate hairs. **Sori** round, medial to supramedial; indusia minute or lacking; sporangia with stellate hairs. 2*n* = 144.

Limestone rocks and grottoes, damp woods; 0–50 m; Fla.; s Mexico; West Indies; Central America in Guatemala; South America in n Venezuela.

18. **Thelypteris sclerophylla** (Poeppig ex Sprengel) C. V. Morton, Amer. Fern J. 41: 87. 1951 · Stiff star-hair fern

Aspidium sclerophyllum Poeppig ex Sprengel, Syst. Veg. 4(1): 99. 1827; *Dryopteris sclerophylla* (Poeppig ex Sprengel) C. Christensen; *Goniopteris sclerophylla* (Poeppig ex Sprengel) Wherry

Stems short-creeping to suberect, 5–8 mm diam. **Leaves** monomorphic, evergreen, clustered, 20–55(–80) cm. **Petiole** straw-colored, 5–18(–25) cm × 1–3 mm, at base with brown, lanceolate, stellate-hairy scales. **Blade** 15–40(–55) × 5–10 cm, proximal 2–5(–10) pairs of pinnae gradually reduced, blade gradually narrowed distally to a pinnatifid apex. **Pinnae** 18–25 pairs, 2–5(–8) × 1–1.5(–2) cm, deeply serrate to incised nearly 3/4 of width, distal pinnae often strongly adnate; segments of rather harsh texture, somewhat oblique, rounded-deltate to often acute; proximal pair of veins from adjacent segments united below sinus with excurrent vein. **Indument** on both surfaces of numerous, sessile or short-stalked, stellate hairs 0.1–0.2 mm on costae, veins, and blade tissue; rachises and costae sometimes with longer simple hairs to 0.8 mm abaxially. **Sori** round, medial to supramedial; indusia tan, densely stellate-hairy; sporangia glabrous. 2*n* = 144.

Terrestrial or on rock in limestone hammocks; 0–50 m; Fla.; West Indies in the Greater Antilles.

In the flora *Thelypteris sclerophylla* is known only from Dade County, Florida, where it is rare.

C. Christensen (1913), C. V. Morton (1951), and

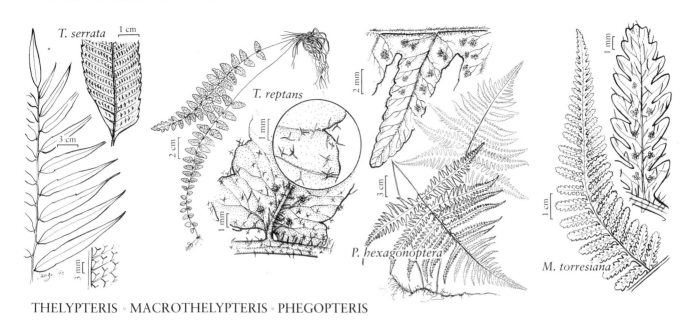

THELYPTERIS · MACROTHELYPTERIS · PHEGOPTERIS

D. B. Lellinger (1985) have attributed the basionym to Kunze in Sprengel, but Sprengel clearly credited Poeppig, rightly or wrongly. Sprengel's original description also differs in a number of details from that by G. Kunze (Linnaea 9: 92. 1834), so that Kunze's later attribution of the basionym to himself cannot be accepted.

19. **Thelypteris tetragona** (Swartz) Small, Ferns S.E. States, 256. 1938 · Free-tip star-hair fern

Polypodium tetragonum Swartz, Prodr., 132. 1788; *Dryopteris tetragona* (Swartz) Urban; *Goniopteris tetragona* (Swartz) C. Presl

Stems short-creeping, 3–5 mm diam. **Leaves** somewhat dimorphic, evergreen, somewhat spaced, fertile leaves long-peti-oled, more erect, and with more contracted pinnae, to ca. 1.1 m. **Petiole** straw-colored, to 60 cm × 2–5 mm, at base sparsely set with brown, lanceolate, stellate-hairy scales. **Blade** 30–45(–55) × 15–25 cm, broadest at base, with apical pinna similar to lateral pinnae. **Pinnae** 6–8(–12) pairs, 7–15(–18) × 2–3 cm (fertile 1–2 cm), incised 1/2–3/4 of width; segments 3–5 mm wide, rounded at apex; veins 6–10 pairs, proximal pair from adjacent segments united at obtuse angle below sinus with excurrent vein ca. 0.5–1 mm to sinus. **Indument** abaxially of mostly needlelike hairs 0.1–0.3 mm on costae and veins, blade tissue glabrous on both sides. **Sori** round, medial to supramedial; indusia lacking; sporangia with numerous hairs 0.1 mm. $2n = 144$.

Damp woods; 0–50 m; Fla.; Mexico; West Indies in the Antilles; Central America to Panama; n South America.

1h. THELYPTERIS Schmidel subg. **MENISCIUM** (Schreber) C. F. Reed, Phytologia 17: 254. 1968

Meniscium Schreber, Gen. Pl. 2: 757. 1791

Distinguished from other subgenera by the combination of evergreen leaves, pinnae entire or merely serrate, cross-veins regularly uniting in pairs to form parallel rows of usually more than 10 areoles between costa and pinna margin, areoles each with a single excurrent, usually free, vein produced at the point of union of the cross-veins, and elongate, exindusiate sori. $x = 36$.

Species 20 and 21 are included in the genus *Meniscium* by some workers.

20. Thelypteris reticulata (Linnaeus) Proctor, Bull. Inst. Jamaica, Sci. Ser. 5: 63. 1953 · Lattice-vein fern

Polypodium reticulatum Linnaeus, Syst. Nat. ed. 10, 2: 1325. 1759; *Dryopteris reticulata* (Linnaeus) Urban; *Meniscium reticulatum* (Linnaeus) Swartz

Stems short- to long-creeping, 3–10 mm diam. **Leaves** monomorphic to somewhat dimorphic, evergreen, fertile leaves with longer petioles, to 2 m or more. **Petiole** tan, to ca. 1 m × 10 mm, at base with scales absent or ephemeral. **Blade** 50–120 cm, with apical pinna similar to lateral pinnae and 1/2 to nearly equaling longest lateral pinna; buds often present at base of proximal pinna. **Pinnae** to 20 pairs or more, widest at or near rounded or broadly cuneate base, 20–30 × 2–6 cm, margin entire to crenulate; main lateral veins of fertile pinnae 2–4 per cm, with 12–20 rows of areoles between costa and margin, secondary veins somewhat curved. **Indument** abaxially of mostly adpressed, curved hairs 0.2–0.3 mm on costae; blades glabrous adaxially. **Sori** oblong-arcuate, uniseriate between lateral veins on cross-veins; indusia lacking; sporangia glabrous. $2n = 144$.

Hammocks in subacid, swampy soil in full shade; 0–50 m; Fla.; s Mexico; West Indies in the Antilles; South America in Colombia, n Venezuela.

Thelypteris reticulata is very rare in the flora.

21. Thelypteris serrata (Cavanilles) Alston, Kew Bull. 1932: 309. 1932 · Dentate lattice-vein fern

Meniscium serratum Cavanilles, Descr. Pl., 548. 1802; *Dryopteris serrata* (Cavanilles) C. Christensen

Stems short-creeping, 4–10 mm diam. **Leaves** monomorphic, evergreen, ca. 65–200 cm or more. **Petiole** tan, (25–)40–120 cm × 4–10 mm, at base with scales absent or ephemeral. **Blade** 40–100 cm or more, gradually reduced distally with lanceolate apical pinna that is 1/4 to equaling longest lateral pinna; buds sometimes present at base of proximal pinna. **Pinnae** lanceolate, 15–25 pairs, (10–)15–25 × (1.5–)2.5–3.5(–4.5) cm, rounded to truncate at base, margin hooked-serrate; main lateral veins of fertile pinnae 3–5 per cm, with rows of 10–18 areoles between costa and margin, cross-veins somewhat curved. **Indument** abaxially of spreading, irregularly crimped hairs ca. 0.2–0.3 mm on costae and sometimes veins adaxially; blades glabrous adaxially. **Sori** oblong-arcuate, uniseriate between main lateral veins on cross-veins, often appearing confluent at maturity; indusia lacking; sporangia glabrous. $2n = 72$.

Cypress sloughs and swamps; 0–50 m; Fla.; s Mexico; West Indies in the Antilles; Central America; South America to n Argentina.

2. MACROTHELYPTERIS (H. Itô) Ching, Acta Phytotax. Sin. 8: 308. 1963 · [Greek *makros*, large, *thelys*, female, and *pteris*, fern]

Thelypteris Schmidel sect. *Macrothelypteris* H. Itô in Nakai & Honda, Nov. Fl. Jap. 4: 141. 1938; *Thelypteris* Schmidel subg. *Macrothelypteris* (H. Itô) A. R. Smith

Stems short-creeping, thick, 1 cm diam. **Blades** 2-pinnate-pinnatifid nearly throughout, broadest at base, apex gradually reduced; pinnae pinnate-pinnatifid, sessile or stalked, not connected by wing along rachis; costae not grooved adaxially; buds absent; veins free, often forked, tips not reaching margin; rachises and costae lacking scales; indument abaxially of unbranched, septate hairs mostly over 1 mm. **Sori** round, medial to supramedial; indusia small, less than 0.3 mm diam., often obscured in mature sori; sporangial capsules bearing short-stalked glands. $x = 31$.

Species ca. 10 (1 introduced in the flora): tropical and subtropical regions, North America, Asia, Africa, Pacific Islands, Australia in Queensland.

SELECTED REFERENCES Holttum, R. E. 1969. Studies in the family Thelypteridaceae. The genera *Phegopteris*, *Pseudophegopteris*, and *Macrothelypteris*. Blumea 17: 5–32. Leonard, S. W. 1972. The distribution of *Thelypteris torresiana* in the southeastern United States. Amer. Fern J. 62: 97–99.

1. Macrothelypteris torresiana (Gaudichaud-Beaupré) Ching, Acta Phytotax. Sin. 8: 310. 1963 · Torres's fern, Mariana maiden fern

Polystichum torresianum Gaudichaud-Beaupré, Voy. Uranie 8: 333. 1828; *Dryopteris uliginosa* (Kunze) C. Christensen; *Thelypteris torresiana* (Gaudichaud-Beaupré) Alston; *T. uliginosa* (Kunze) Ching **Stems** short-creeping, thick, to 10 mm diam. **Leaves** monomorphic, evergreen, 60–150 cm. **Petiole** to 75 cm × 3–12 mm, glaucous when living. **Blade** 2–3-pinnate, to ca. 85 cm. **Pinnae** to 35 × 10(–17) cm; pinnules sessile to adnate, oblique, 2–8 × 0.8–2.5 cm, incised almost to costule into oblique segments 2–4 mm wide, segments entire to dentate or pinnatifid; veins forked or simple. **Indument** abaxially of septate hairs 1–2 mm on costae and costules; rachis, costae, and blade tissue with capitate glands 0.05 mm. **Sori** round, indusia small, less than 0.3 mm diam., glabrous or glandular; sporangia with minute capitate glands near annulus. $2n = 124$.

Terrestrial in damp woods and along stream banks; 0–100 m; introduced; Ala., Ark., Fla., Ga., La., Miss., S.C., Tex.; s Mexico; West Indies in the Antilles; Central America; South America to n Argentina; native to tropical and subtropical Asia, Africa.

The name *Dryopteris setigera* (Blume) Kuntze has been misapplied to plants in the flora.

3. PHEGOPTERIS (C. Presl) Fée, Mém Foug. 5: 242. 1852 · Beech fern [Greek *phegos,* beech, and *pteris,* fern]

Polypodium Linnaeus sect. *Phegopteris* C. Presl, Tent. Pterid., 179. 1836; *Thelypteris* Schmidel subg. *Phegopteris* (C. Presl) Ching

Stems long-creeping, 1–4 mm diam. **Blades** 2–3-pinnatifid in proximal part, broadest at base, apex gradually reduced; pinnae deeply lobed, mostly strongly adnate, connected by wing along rachis, wing sometimes forming lobe between pinnae and served by vein arising from rachis; costae not grooved adaxially; buds absent; veins free, simple or often forked, distal veins of segment reaching margin or nearly so; indument abaxially of unbranched, unicellular hairs, rachises and costae also with spreading, ovate-lanceolate scales. **Sori** round to oblong, supramedial to inframarginal, lacking indusia; sporangial capsule often bearing stalked glands or hairs. $x = 30$.

Species 3 (2 in the flora): north temperate and boreal, 1 in temperate e Asia.

SELECTED REFERENCE Holttum, R. E. 1969. Studies in the family Thelypteridaceae. The genera *Phegopteris, Pseudophegopteris,* and *Macrothelypteris.* Blumea 17: 5–32.

1. Proximal pair of pinnae (7–)10–20 cm, connected to those next above by wing along rachis; scales on costae abaxially whitish to light tan, narrowly lanceolate, mostly 3–5 cells wide at base; hairs of costae abaxially mostly less than 0.25 mm, a few sometimes to 0.5 mm; segments of larger pinnae deeply lobed; veins always forked to pinnate. . . . . .1. *Phegopteris hexagonoptera*
1. Proximal pair of pinnae 3–10(–15) cm, sessile or slightly adnate to rachis; scales on costae abaxially tan to often shiny brown, ovate-lanceolate, usually 6–12 or more cells wide at base; hairs of costae abaxially mostly more than 0.3 mm with many ca. 0.5 mm; segments of pinna entire or crenate, rarely shallowly lobed; veins in segments of middle pinnae mostly simple. . . . . .2. *Phegopteris connectilis*

1. Phegopteris hexagonoptera (Michaux) Fée, Mém. Foug. 5: 242. 1852 · Broad beech fern, southern beech fern, phégoptère á hexagones, grand fougère du hêtre

Polypodium hexagonopterum Michaux, Fl. Bor.-Amer. 2: 271. 1803; *Dryopteris hexagonoptera* (Michaux) C. Christensen; *Thelypteris hexagonoptera* (Michaux) Nieuwland

Stems long-creeping, 2–4 mm diam. **Leaves** monomorphic, dying back in winter, often 1–2 cm apart, ca. 25–75 cm. **Petiole** straw-colored, (7–)20–45 cm × 1.5–3 mm, at base with scales tan, lanceolate, glabrous or marginally hairy. **Blade** broadly deltate, about as broad as long, (8–)15–33 cm, proximal pinnae longest and narrowed at base, usually spreading or slightly ascending. **Pinnae** 7–20 × 2–6(–8) cm, all connected by wing along rachis, deeply pinnatifid; segments entire or largest pinnatifid about halfway to costule; proximal pair of veins from adjacent segments meeting margin above sinus, veins often forked. **Indument** abaxially of moderately to densely set hairs mostly 0.1–0.25 mm along costae and veins, also of yellowish stalked glands 0.1 mm on veins and blade tissue, costae with whitish to light tan, narrowly lanceolate, spreading, marginally hairy scales to ca. 1.5 mm. **Sori** subterminal on veins. $2n = 60$.

In moist woods, usually in full shade, often in moderately acid soils; 0–1000 m; Ont., Que.; Ala., Ark., Conn., Del., D.C., Fla., Ga., Ill., Ind., Iowa, Kans., Ky., La., Maine, Md., Mass., Mich., Minn., Miss., Mo., N.H., N.J., N.Y., N.C., Ohio, Okla., Pa., R.I., S.C., Tenn., Tex., Vt., Va., W.Va., Wis.

G. A. Mulligan and W. J. Cody (1979) reported hybrids between *Phegopteris hexagonoptera* and *P. connectilis* from a few localities in Quebec, New Brunswick, and Nova Scotia. These hybrids are apogamous and have a chromosome number of $2n = 120$.

2. Phegopteris connectilis (Michaux) Watt, Canad. Naturalist & Quart. J. Sci., n. s. 3: 29. 1866 · Northern beech fern, narrow beech fern, phégoptère à segments joint, fougère du hêtre

Polypodium connectile Michaux, Fl. Bor.-Amer. 2: 271. 1803; *Dryopteris phegopteris* (Linnaeus) C. Christensen; *Phegopteris polypodioides* Fée; *Polypodium phegopteris* Linnaeus; *Thelypteris phegopteris* (Linnaeus) Slosson

Stems long-creeping, 1–2(–3) mm diam. **Leaves** monomorphic, dying back in winter, often 1–2 cm apart, 15–60 cm. **Petiole** straw-colored, (8–)15–36 cm × 1–3 mm, at base with scales brownish, lanceolate, glabrous or sparingly hairy on margin. **Blade** narrowly to broadly deltate, usually somewhat longer than broad, (6–)12–25 cm, proximal pinnae longest and slightly narrowed at base, spreading or reflexed. **Pinnae** deeply pinnatifid, (3–)6–12 × 1–3.3 cm, lowermost 1–2 pairs separate, sessile, more distal pairs strongly adnate and connected by narrow rachis wing; segments entire, or those of proximal pinna pair sometimes crenate, uncommonly shallowly lobed; proximal pair of veins from adjacent segments meeting margin above sinus, veins simple or sometimes forked in lowermost pinnae. **Indument** abaxially of moderately to densely set hairs 0.3–1 mm along costae, veins, and blade tissue, costae also with scales light tan to shiny brown, ovate-lanceolate, spreading, to ca. 3 mm, scales sometimes sparingly hairy on margin. **Sori** subterminal on veins. $n = 2n = 90$, apogamous.

In moist, strongly to moderately acid soil, or on rocks in shaded rock crevices; 0–2200 m; Greenland; St. Pierre and Miquelon; Alta., B.C., Man., N.B., Nfld., N.W.T., N.S., Ont., P.E.I., Que., Sask., Yukon; Alaska, Conn., Del., Idaho, Ill., Indiana, Iowa, Maine, Md., Mass., Mich., Minn., Mo., Mont., N.H., N.J., N.Y., N.C., Ohio, Oreg., Pa., R.I., Tenn., Vt., Va., Wash., W.Va., Wis.; Eurasia.

18. BLECHNACEAE C. Presl

• Chain Fern Family

Raymond B. Cranfill

Plants perennial, mostly terrestrial, occasionally on rock or epiphytic. **Stems** creeping to sub-erect or ascending, sometimes climbing [rarely arborescent], slender to stout, dictyostelic, scaly. **Leaves** monomorphic or dimorphic, large and coarse, generally greater than 30 cm, often exceeding 1 m. **Petiole** not articulate, generally more than 2 vascular bundles arranged in arc, generally scaly at least at base. **Blade** often anthocyanic (reddish) when young, pinnatifid [rarely simple] to pinnate-pinnatifid or 2-pinnate [rarely decompound], glabrous or occasion-ally bearing scales or capitate glands. **Rachis** frequently grooved adaxially. **Veins** of sterile leaves generally free, rarely anastomosing, veins of fertile leaves united to form sorus-bearing secondary vein parallel to costa or costule (vascular commisure), sometimes anastomosing further. **Sori** elongate along secondary vein; indusia present [rarely absent], opening along costal side of fertile vein, frequently hidden by dehisced sporangia; sporangial stalk of 3 rows of cells. **Spores** monolete, reniform; perine present, variously ornamented. **Gametophytes** green, cordate, sometimes bearing capitate hairs, antheridia and archegonia borne on lower surface.

Genera ca. 10, species ca. 250 (2 genera, 6 species in the flora): mostly tropical and south temperate (except *Woodwardia*, which is north temperate).

Circumscription of genera is controversial, especially as to placement of those species now included in *Blechnum*. Characteristics holding the family together include the anastomoses of veins along the axes of the blade to form a series of areoles or a single continuous vein along which the sorus is borne, elongate sori with indusia opening toward midvein, bilateral spores, and chromosome base numbers of generally $x = 28–36$. Relationships of the family with both dryopteroid and athyrioid ferns have been suggested.

Stenochlaena tenuifolia (Desvaux) T. Moore, native to the Old World, was reported as escaped from cultivation in the 1930s in southern Florida; it has not been collected there recently. It is distinguished by having climbing stems and by having contracted, 2-pinnate fertile leaves with sporangia covering the abaxial surface.

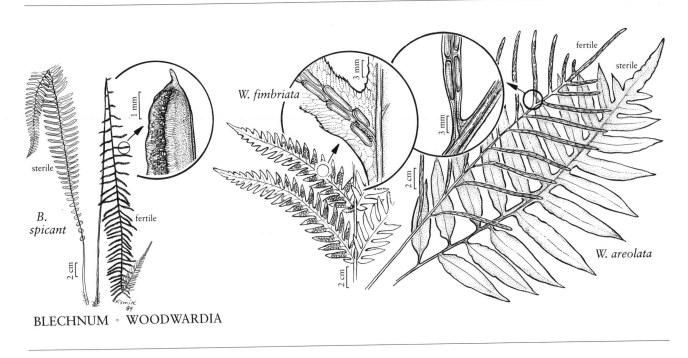

BLECHNUM · WOODWARDIA

1. Sori continuous along length of costa or costule; veins of sterile leaves free. . . . . . . 1. *Blechnum*, p. 224
1. Sori distinct, in chainlike rows along costa or costule; veins of sterile leaves anastomosing
    at least along costae and costules. . . . . 2. *Woodwardia*, p. 226

1. BLECHNUM Linnaeus, Sp. Pl. 2: 1077. 1753; Gen. Pl. ed. 5, 485. 1754 · Deer ferns

[Greek *blechnon*, an ancient name for ferns in general]

Clifton E. Nauman

Plants terrestrial or rarely on rock. **Stems** creeping to ascending or erect, slender to stout, sometimes climbing [rarely subarborescent]; scales brown or black. **Leaves** monomorphic or variously dimorphic, cespitose to scattered. **Blades** pinnatifid to l-pinnate, rarely simple or 2-pinnate. **Rachis and costae** glabrous, scaly, or hairy abaxially. **Veins** free, often forked. **Sori** borne on vascular commissures parallel to costae, 1 per side, normally uninterrupted, linear, continuous along length of costa. **Spores** with perine smooth to variously winged or rugose. x = 28, 29, 31, 32, 34, 36.

Species ca. 220 (3 in the flora): mostly tropical, especially Southern Hemisphere.

1. Leaves strongly dimorphic, fertile leaves notably more erect, longer, and with narrower
    pinnae than sterile leaves; sterile blades tapering at base. . . . . 1. *Blechnum spicant*
1. Leaves ± monomorphic, fertile leaves only slightly longer and somewhat contracted rela-
    tive to sterile leaves; sterile blades ± truncate at base.
      2. Blades 1-pinnate proximally and pinnatifid distally or pinnatifid nearly throughout, usually
         less than 50 cm; pinna margins ± entire. . . . . 2a. *Blechnum occidentale* var. *minor*
      2. Blades 1-pinnate throughout, usually more than 50 cm; pinna margins ± evenly serrulate.
      . . . . 3. *Blechnum serrulatum*

1. **Blechnum spicant** (Linnaeus) Smith, Mém. Acad. Roy. Sci. (Turin) 5: 411. 1793 · Deer fern

Osmunda spicant Linnaeus, Sp. Pl. 2: 1066. 1753; *Struthiopteris spicant* (Linnaeus) Weis

Stems slender, short-creeping or ascending, not climbing. **Leaves** dimorphic, cespitose, erect or spreading, fertile leaves more erect and longer than sterile leaves. **Petioles of fertile leaves** reddish brown to purple-black, 15–60 cm, coarsely scaly proximally. **Petioles of sterile leaves** reddish brown, 2–30 cm, coarsely scaly proximally. **Fertile blades** erect, narrowly rhombic, 1-pinnate, without conform terminal pinna, 25–65 × 3–15 cm, glabrous. **Sterile blades** spreading, narrowly oblanceolate to linear-lanceolate, pinnatifid to apex and so without conform terminal pinna, 20–75 × 3–14 cm, tapering at base, glabrous. **Rachises** of fertile and sterile leaves with indument of filiform, spreading scales abaxially. **Fertile pinnae** not articulate to rachis, sessile, decurrent and surcurrent to rachis; larger pinnae slightly curved, linear, 25–32 × 1.5–2 mm; margins entire, slightly to moderately revolute; costae with indument of a few small scales abaxially, often concealed by sori. **Sterile pinnae** closely spaced, not articulate to rachis, base fully adnate; larger pinnae curved, linear to oblong-linear or barely wider beyond middle, 15–35 × 3.5–5 mm; margins entire. $2n = 68$.

Wet coniferous woods and swamps; 0–1400 m; B.C.; Alaska, Calif., Idaho, Oreg., Wash.

Blechnum spicant is found primarily along the coast and in coastal mountain ranges.

2. **Blechnum occidentale** Linnaeus, Sp. Pl. ed. 2, 2: 1524. 1763

Varieties 2 (1 in the flora): North America, Mexico, West Indies, Central America, South America.

2a. **Blechnum occidentale** Linnaeus var. **minor** Hooker, Sp. Fil. 3: 51. 1860 · Hammock fern, New World midsorus fern

Blechnum occidentale forma *pubirachis* (Rosenstock) Lellinger; *Blechnum occidentale* var. *pubirachis* Rosenstock

Stems slender, creeping, elongate, branched, ascending to erect at tip, not climbing. **Leaves** ± monomorphic, cespitose to widely spaced, erect to arching, fertile leaves only slightly contracted and longer than sterile leaves. **Petiole** straw-colored to light brown, (4–)8–34 cm, coarsely scaly proximally. **Blade** narrowly to broadly lanceolate, 1-pinnate proximally, becoming pinnatifid distally, or pinnatifid throughout, without conform terminal pinna, 10–30 × 3–12 cm, base truncate, pubescent abaxially. **Rachis** with indument of spreading, ± gland-tipped hairs abaxially. **Pinnae** not articulate to rachis, proximal pinnae sessile to subsessile, distal pinnae adnate; larger pinnae curved, lanceolate, 2–7 × 0.5–1.5 cm. **Fertile pinnae** slightly contracted; margins minutely serrulate to nearly entire; costae with indument of hairs abaxially. $2n = 124$.

Rocky and clayey places near seasonally dry streams, shady hammocks or open woods, over limestone, soil nearly neutral; 0 m; Fla., Ga., La., Tex.; West Indies; Central America; South America to Bolivia.

Blechnum occidentale var. *minor* differs from var. *occidentale* in having rachises slightly pubescent or puberulous abaxially. Both varieties are found throughout the New World tropics; *B. occidentale* var. *minor* grows at higher elevations (D. B. Lellinger 1985). The nomenclature of these taxa is complicated, and other names may apply. Systematic problems involving different ploidy levels and apparent geographic clines in *Blechnum occidentale* sensu lato remain to be solved.

Plants of *Blechnum occidentale* often reproduce extensively by stolons as well as by spores.

3. **Blechnum serrulatum** Richard, Actes Soc. Hist. Nat. Paris 1: 114. 1792 · Swamp fern, marsh fern, dentate midsorus fern

Stems stout, horizontal and long-creeping, branched, partly erect at tip, rarely climbing tree trunks. **Leaves** ± monomorphic, widely spaced, erect to arching. **Petiole** dull yellow or grayish brown or light brown, 10–55 cm, finely scaly proximally. **Blade** broadly linear to elliptic-lanceolate, 1-pinnate throughout, with conform terminal pinnae, 25–70 × 5–28 cm, base truncate, glabrous. **Rachis** lacking indument abaxially. **Pinnae** articulate to rachis except for terminal pinna, subsessile to short-stalked; larger pinnae ± straight, linear to linear-elliptic or linear-lanceolate, 3–15 × 0.5–1.8 cm, fertile pinnae often slightly smaller and contracted; margins serrulate; costae with indument of scales abaxially. $2n = 72$.

Swamps, marshes, wet prairies, and adjacent moist pine woods or hammocks; 0 m; Fla.; Central America; South America.

Plants of *Blechnum serrulatum* occurring in open sun are often dwarfed and stiffly erect. Those occurring in brackish conditions or perennially flooded areas may become hemiepiphytes.

2. WOODWARDIA   Smith, Mém. Acad. Roy. Sci. (Turin) 5: 411. 1793 · Chain fern [in honor of Thomas Jenkinson Woodward, 1745–1820, English botanist]

Raymond B. Cranfill

Anchistea C. Presl; *Lorinseria* C. Presl

Plants terrestrial or rarely on rock. **Stems** long-creeping to erect, slender to stout, not climbing; scales brown. **Leaves** monomorphic (dimorphic in 1 species), clustered or well separated. **Blades** pinnate or pinnatifid. **Rachises and costae** scaly. **Veins** anastomosing in both sterile and fertile leaves, forming a regular series of areoles along costae and costules, further anastomosing in 1 species. **Sori** discrete, in chainlike rows along costae or costules, extending only the length of individual areolar veins. **Spores** with perine irregularly folded. $x = 34, 35$.

Species 14 (3 in the flora): North America, Central America, Mediterranean Europe, e Asia.

Woodwardia radicans (Linnaeus) Smith has been reported as an escape from cultivation in Florida and in the Sierra Nevada in California; it has not persisted. It and the commonly cultivated *Woodwardia unigemmata* Makino resemble *Woodwardia fimbriata* Smith, but both *W. radicans* and *W. unigemmata* are distinguished by having a scaly bulblet near the apex of the leaf.

1. Leaves strongly dimorphic; sterile blades ± pinnatifid, with 2 or more rows of areoles between costae and margin, veins free only at margin. . . . . 1. *Woodwardia areolata*
1. Leaves ± monomorphic; sterile blades pinnate, with 1 row of areoles adjacent to costae or costules, veins free to margin.
    2. Stems forming stout caudex covered with petiole bases, suberect; petioles straw-colored and densely covered with orangish scales at base; pinnae not articulate to rachis. . . . . 3. *Woodwardia fimbriata*
    2. Stems relatively slender to ca. 1 cm diam., long-creeping; petioles blackish and glabrate at base; pinnae articulate to rachis. . . . . 2. *Woodwardia virginica*

1. Woodwardia areolata (Linnaeus) T. Moore, Index Fil. 45. 1857

Acrostichum areolatum Linnaeus, Sp. Pl. 2: 1069. 1753; *Lorinseria areolata* (Linnaeus) C. Presl

Stems long-creeping, slender; scales brown, many, broadly lanceolate. **Leaves** dimorphic, deciduous, few, well separated; sterile leaves 40–58 cm, fertile leaves 49–70 cm. **Petiole** reddish brown proximally, straw-colored distally; base not swollen, with sparsely set brown scales. **Blade** bright green, generally lanceolate, scaly-glandular upon emergence but soon glabrate; sterile leaves pinnatifid, 13–26 cm; fertile leaves pinnate, sharply contracted, 20–27 cm. **Pinnae** not articulate to rachis, arranged in 7–12 alternate pairs; sterile pinnae lanceolate, 3–11 × 1–2.5 cm; fertile pinnae contracted, linear, 3–11 × 0.2–0.5 cm. **Veins** anastomosing into 2 or more rows of areoles between costae and margin, free only at blade margin. **Sori** linear-oblong, deeply sunken into blades, nearly occupying full breadth of blade. **Indusia** ± membranous, lacking thickened cells, tucked under sporangia, not recurving but mostly disintegrating with age. $2n = 70$.

Acidic bogs, seeps, and wet woods, rarely on rock of siliceous cliffs and ledges on northern edge of range; 0–600 m; N.S.; Ala., Ark., Conn., Del., Fla., Ga., Ill., Ind., Ky., La., Maine, Md., Mass., Miss., Mo., N.H., N.J., N.Y., N.C., Ohio, Okla., Pa., R.I., S.C., Tenn., Tex., Va., W.Va.

Woodwardia areolata is most abundant on the coastal plain of the eastern United States, scattered in the Ouachita and Boston mountains, Ozark and Cumberland plateaus, and the Piedmont, but not in the high Appalachians, the heavy gumbo soils of the Mississippi Valley, or the limestone regions of the Interior Low Plateaus. It apparently has been extirpated in Maine where it is known only from specimens collected in the 1860s.

Features such as extreme leaf dimorphism, sunken sori, and expanded persistent indusia set *Woodwardia areolata* apart from all others in the genus. The existence of closely related transitional species in Asia, however, makes generic segregation uncertain. Those who wish to recognize a monotypic generic segregate based on *Woodwardia areolata* must coin a new name because *Lorinseria* C. Presl (1849) is an orthographic variant of *Lorinsera* Opiz (1839). For a detailed discussion of the ecology and geography of this species, see R. Cranfill (1983). Sterile specimens of this species are sometimes confused with *Onoclea sensibilis*.

2. Woodwardia virginica (Linnaeus) Smith, Mém. Acad. Roy. Sci. (Turin) 5: 412. 1793 · Virginia chain fern, Woodwardie de Virginie

Blechnum virginicum Linnaeus, Mant. Pl. [2]: 307. 1771; *Anchistea virginica* (Linnaeus) C. Presl

Stems long-creeping, ropelike; scales dark brown, few, triangular. **Leaves** monomorphic, deciduous, numerous, well separated, 50–100 cm. **Petiole** dark purple to black proximally, straw-colored distally, base lustrous and swollen, glabrate. **Blade** greenish, lanceolate, 28–60 cm, bearing mixture of glands and scales upon emergence with glands persisting. **Pinnae** articulate to rachis, arranged in 12–23 evenly distributed pairs, linear to narrowly lanceolate, deeply pinnatifid; middle pinnae 6–16 × 1–3.5 cm. **Veins** anastomosing to form a single row of areoles, then free to margin. **Sori** elongate, linear, superficial, often appearing confluent upon dehiscence of sporangia. **Indusia** ± membranous, lacking thickened cells, freely spreading, often hidden by dehisced sporangia. $2n = 70$.

Swamps, marshes, bogs, and roadside ditches over noncalcareous substrates; 0–300 m; N.B., N.S., Ont., P.E.I., Que.; Ala., Ark., Conn., Del., Fla., Ga., Ill., Ind., La., Maine, Md., Mass., Mich., Miss., N.H., N.J., N.Y., N.C., Ohio, Pa., R.I., S.C., Tenn., Tex., Vt., Va., W.Va.; Bermuda.

Woodwardia virginica is primarily confined to the coastal plain of eastern North America. It is not likely to be confused with any other species of the genus, but it is sometimes mistaken for *Thelypteris palustris* (Linnaeus) Schott, *T. interrupta* (Willdenow) K. Iwatsuki, or *Osmunda cinnamomea*. It may be distinguished from the first by having netted venation, from the second by having linear sori, and from *Osmunda* by having blackish petiole bases.

3. Woodwardia fimbriata Smith in Rees, Cycl. 38(76). 1818 · Giant chain fern

Woodwardia chamissoi Brackenridge; *Woodwardia paradoxa* C. H. Wright

Stems forming a stout caudex covered with petiole bases, suberect; scales light brown, many, lanceolate-attenuate. **Leaves** monomorphic, evergreen, numerous in vaselike cluster, 40–170 cm. **Petiole** straw-colored, sometimes reddish at base; base thickened, with densely set orange scales. **Blade** pale green, elliptic-lanceolate, 25–100 cm, scaly-glandular upon emergence but soon glabrate. **Pinnae** not articulate to rachis, in 8–24 pairs, narrowly deltate to lanceolate, pinnatifid; proximal to middle pinnae 12–42 × 2.5–8 cm. **Veins** anastomosing to form single row of areoles, then free to margin. **Sori** short and broad, mostly curved and confined to costular areoles, deeply sunken into blades. **Indusia** cartilaginous and vaulted; cells thickened, retaining configuration after dehiscence of sporangia. $2n = 68$.

Redwood forests, mixed conifer forests, and mixed conifer-hardwood forests, always where moisture is present, such as stream banks or springs; 0–1000 m; B.C.; Ariz., Calif., Nev., Oreg., Wash.; Mexico in n Baja California.

Woodwardia fimbriata is confined primarily to the California floristic province; it is disjunct and local elsewhere.

19. ASPLENIACEAE Newman

• Spleenwort Family

Warren H. Wagner Jr.

Robbin C. Moran

Charles R. Werth

Plants terrestrial, on rock, or rarely epiphytic. **Stems** erect or nearly erect, rarely long-creeping, scaly. **Steles** radially symmetric or dorsiventral (with structurally distinct abaxial and adaxial aspects) dictyosteles. **Leaves** monomorphic, rarely almost dimorphic with fertile leaves taller and more erect than sterile ones. **Petioles** with 1 vascular bundle X-shaped in cross section or with 2 vascular bundles back to back and C-shaped. **Blades** extremely diverse, simple to 4-pinnate, commonly with tiny glandular hairs and a few linear scales, rarely with spreading hairs. **Veins** free to anastomosing. **Sori** borne on veins, ± lunate to linear. **Indusia** usually present, shape conforming to sorus and originating along 1 side of sorus. **Sporangia** with stalk of 1 row of cells, annulus vertical, interrupted by sporangial stalk. **Spores** monolete; perispore typically winged, spiny, reticulate, or perforate. **Gametophytes** surficial, green, cordate.

Genera 1, species ca. 700 (1 genus, 28 species, and 3 nothospecies in the flora): worldwide.

Members of this family can usually be identified by the combination of clathrate stem scales and indusiate linear sori. Supporting anatomic characteristics include the two vascular bundles in the petiole that unite distally in the petiole to form an X-shaped petiolar strand, and the single row of cells in the sporangial stalk. The scales consist of cells with dark, thick, radial walls and clear, thin, tangential walls, giving the scales a clathrate (laticelike) appearance reminiscent of lead moldings between plates of stained glass.

As construed here, Aspleniaceae comprise a single, huge, extremely diverse genus, *Asplenium*. A satisfactory taxonomic division into subgenera or satellite genera has not been possible because of the absence of any significant gaps. Various segregates have been proposed (e.g., *Camptosorus, Phyllitis, Ceterach, Pleurosorus*), but numerous "intergeneric" hybrids occur.

The members of *Asplenium* are popular with plant evolutionists, field naturalists, and fern gardeners, not only because of the interesting morphology of the plants but also because of their remarkable ability to form spectacular hybrids, often combining dramatically different leaf shapes. In North America, 23 diploid hybrids and allopolyploids have been recorded. At least two of these hybrid combinations occur as both sterile diploids and their fertile allotetraploid derivatives. Only those hybrids that are reproductively competent (through vigorous clone-forming by root proliferations or apogamy, or rarely through sexual reproduction) are treated in the key and fully described below.

Only about two-fifths of the reproductively competent species are believed to be cladistically divergent species; the other three-fifths are of hybrid origin (allopolyploids). The most unusual allopolyploid phytogeographically is *Asplenium adiantum-nigrum,* the parents of which are known only in the Old World. These reticulate relationships are summarized in the reticulogram.

Polyploidy is widespread in *Asplenium,* and the chromosome numbers vary from $2x$ to $6x$. Two species, *Asplenium trichomanes* and *A. heterochroum,* occur in different levels of ploidy—$2x$ and $4x$, and $4x$ and $6x$, respectively. The highest chromosome number known for *Asplenium* in North America is $2n = 216$ (in *A. trichomanes-dentatum* and the hexaploid form of *A. heterochroum*). The only three apogamous taxa are *A. monanthes* ($3x$), *A. resiliens* ($3x$), and *A.* ×*heteroresiliens* ($5x$).

SELECTED REFERENCES Gastony, G. J. 1986. Electrophoretic evidence for the origin of a fern species by unreduced spores. Amer. J. Bot. 73: 1563–1569. Kramer, K. U. and R. Viane. 1990. Aspleniaceae. In: K. Kubitzki et al., eds. 1990+. The Families and Genera of Vascular Plants. 1+ vol. Berlin etc. Vol. 1, pp. 52–56. Reichstein, T. 1981. Hybrids in European Aspleniaceae (Pteridophyta). Bot. Helv. 91: 89–139.

1. ASPLENIUM Linnaeus, Sp. Pl. 2: 1078. 1753; Gen. Pl. ed. 5, 485. 1754 · Spleenwort [Greek *splen,* spleen; thought by Dioscorides to be useful for treating spleen diseases]

Roots fibrous, not proliferous or proliferous and producing tiny plantlets. **Stems** erect, rarely long-creeping; scales basally attached, clathrate. **Petioles** not articulate. **Blades** 1–4-pinnate, of diverse size and shape. **Indusia** present. $x = 36$.

Species ca. 700 (28 species, 3 nothospecies in the flora): worldwide.

1. Blades simple, pinnatifid, or forked (sometimes with free pinnae at base), not pinnate throughout.
    2. Blades less than 2–3 mm wide, linear, frequently forking or with 1–3 small, narrow projections. . . . . 7. *Asplenium septentrionale*
    2. Blades more than 10 mm wide, linear-lanceolate to lanceolate, if lobed, segments resembling adnate pinnae.
        3. Veins anastomosing to form areoles; blades rooting at tip. . . . . 4. *Asplenium rhizophyllum*
        3. Veins free; blades with apical buds or not but not rooting at tip.
            4. Blades simple, linear to oblanceolate.
                5. Blades cordate at base; roots not proliferous. . . 5a. *Asplenium scolopendrium* var. *americanum*
                5. Blades tapered at base; roots proliferous. . . . . 6. *Asplenium serratum*
            4. Blades pinnatifid, elliptic, lanceolate, to deltate.
                6. Petioles 1 cm or less, dull; stem scales 0.6–1 mm wide, sparsely denticulate.
                    . . . . 1. *Asplenium dalhousiae*
                6. Petioles 1–10 cm, lustrous; stem scales less than 0.5 mm wide, entire.
                    7. Petioles dark-pigmented only at base; blade narrowly deltate. . . 3. *Asplenium pinnatifidum*
                    7. Petioles and proximal midrib darkly pigmented abaxially; blade oblong-lanceolate. . . . . 2. *Asplenium ebenoides*
1. Blades pinnate throughout or pinnatifid only in distal 1/3, pinnae undivided to 1–3-divided.
    8. Blades 2–4-pinnate, with only largest pinnae divided to all pinnae divided.

Asplenium

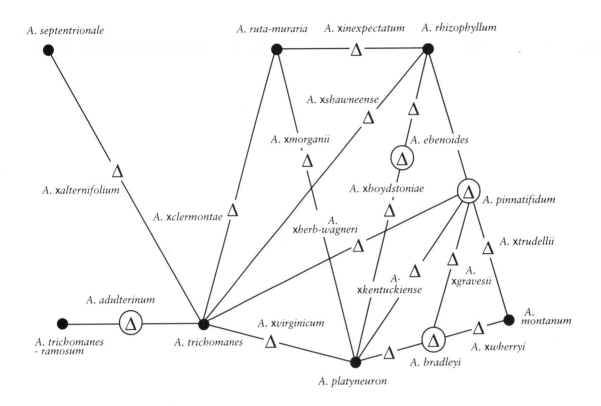

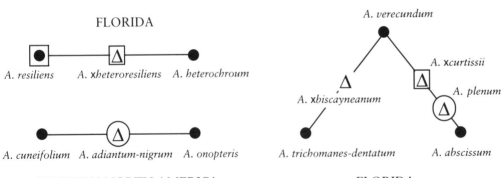

Relationships among some species of *Asplenium*. Solid circles represent parental taxa; boxed circles represent apogamous parental taxa; circled triangles represent allopolyploids; boxed triangles represent apogamous hybrids; and triangles represent sterile hybrids. The sterile, diploid forms of *A. bradleyi* and *A. ebenoides* are not here distinguished from their allotetraploid counterparts.

9. Blades deltate-ovate, with 2–5 pairs of pinnae; pinnae generally deltate-ovate to obdeltate, rarely lanceolate to narrowly elliptic; color in living plants dull bluish green. . . . . 30. *Asplenium ruta-muraria*
9. Blades ovate-lanceolate to linear, with 5–20 pairs of pinnae; pinnae linear to oblong; color in living plants shiny green.
   10. Delicate root proliferations usually present (commonly buried in mossy soil in crevices, but enabling clone formation); damp, shaded limestone; Florida.
       11. Spores uniform and normal.
           12. Blades oblong-lanceolate, mostly 2-pinnate; pinnae lanceolate to oblong-lanceolate. . . . . 25. *Asplenium cristatum*
           12. Blades narrowly lanceolate, 2–3-pinnate; pinnae ovate to ovate-deltate. . . . . 31. *Asplenium verecundum*
       11. Spores of different sizes and shapes, malformed (irregular in size and shape).
           13. Blades linear-lanceolate, s Florida. . . . . 23. *Asplenium* × *biscayneanum*
           13. Blades oblong-lanceolate, c Florida.
               14. Medial pinnae pinnate throughout; pinnules mostly deeply notched apically. . . . . 26. *Asplenium* × *curtissii*
               14. Medial pinnae pinnate only in proximal 1/2; pinnules mostly shallowly notched apically. . . . . 29. *Asplenium plenum*
   10. Delicate root proliferations absent or rare; mainly rock crevices; plants not in Florida.
       15. Buds at blade, pinna, and pinnule apices minute, scaly, dormant; Arizona. . . . . 27. *Asplenium exiguum*
       15. Buds absent; foothills and mountains in e United States and s Rocky Mountains.
           16. Petioles all pale green except at base. . . . . 28. *Asplenium montanum*
           16. Petioles dark reddish brown.
               17. Blades lanceolate to linear-lanceolate, 1–2-pinnate; proximal pinnae often somewhat reduced; e United States foothills and mountains. . . . . 24. *Asplenium bradleyi*
               17. Blades deltate-ovate to deltate-lanceolate, 2–3-pinnate; proximal pinnae usually largest; s Rocky Mountains. . . . . 22. *Asplenium adiantum-nigrum*
8. Blades 1-pinnate, pinnae undivided (except in some individuals of *A. pumilum*).
   18. Pinna pairs 1–2(–4); blades ovate-deltate, with fine nonglandular hairs on veins. . . . . 17. *Asplenium pumilum*
   18. Pinna pairs 5–35; blades mainly oblong to lanceolate, usually with glandular hairs or narrow hairlike scales on veins or blades glabrous.
       19. Pinnae with conspicuous basal auricles overlapping rachis; sori inframedial; leaves ± dimorphic, upright and tall, or spreading and short; plants often terrestrial. . . . . 16. *Asplenium platyneuron*
       19. Pinnae bases not overlapping rachis; sori medial to supramedial; leaves monomorphic; plants mainly on rock (epiphytic in *A. auritum*).
           20. Blades mostly 4–9 cm wide, linear-lanceolate to oblong-lanceolate.
               21. Blades linear-lanceolate, to 35 cm; pinna pairs 10–22, base with conspicuous acroscopic auricle; epiphytic on tree trunks. . . . . . . 10. *Asplenium auritum*
               21. Blades oblong to oblong-lanceolate, to 20 cm; pinna pairs 4–8, base with or without small acroscopic auricle; on mossy limestone. . . . . 8. *Asplenium abscissum*
           20. Blades mostly less than 3 cm wide, mainly linear.
               22. Sori 1(–3), mainly confined to basiscopic side of strongly asymmetric pinnae. . . . . 14. *Asplenium monanthes*
               22. Sori 4–10, on both sides of costa of nearly symmetric pinnae.
                   23. Blade apex prolonged into whiplike extension of rachis terminating in bud. . . . . 15. *Asplenium palmeri*

23. Blade apex not prolonged, terminal bud absent.
    24. Pinnae ovate, oblong-ovate, to rhombic, mainly 4–7 mm, length 1–2 times width.
        25. Rachises dark reddish brown throughout. . . . . 19. *Asplenium trichomanes*
        25. Rachises dark in proximal 1/3–2/3 or only at base.
            26. Rachises dark in proximal 1/3–2/3; pinna margins entire to shallowly crenate. . . . . 9. *Asplenium adulterinum*
            26. Rachises dark only at base; pinna margins dentate-crenate.
                27. Pinnae with basiscopic 1/2 much reduced; blades mainly less than 1 cm wide; n North America. . . . . 21. *Asplenium trichomanes-ramosum*
                27. Pinnae nearly symmetric; blades mainly more than 1 cm wide; s Florida. . . . . 11. *Asplenium trichomanes-dentatum*
    24. Pinnae oblong-lanceolate to oblong, mainly 6–10 mm, length 3–5 times width.
        28. Pinna margins deeply lobed, especially acroscopic and distal margins, cut 1/3 or more distance from margin to costa; s California. . . . . 20. *Asplenium vespertinum*
        28. Pinna margins nearly entire or shallowly crenate-dentate; se United States to e Arizona.
            29. Pinna margins nearly entire; sori supramedial; spores 32; widespread in s United States. . . . . . . 18. *Asplenium resiliens*
            29. Pinna margins ± shallowly crenate-dentate; sori medial to inframedial; spores 32 or 64; local in se United States.
                30. Pinna margins shallowly crenulate; auricle present at pinna base; veins obscure; spores 32 per sporangium. . . . . 13. *Asplenium ×heteroresiliens*
                30. Pinna margins serrate to crenate; auricle rudimentary or absent at pinna base; veins evident; spores 64 per sporangium. . . . . 12. *Asplenium heterochroum*

1. **Asplenium dalhousiae** Hooker, Icon. Pl., plate 105. 1837 · Countess Dalhousie's spleenwort

Ceterach dalhousiae (Hooker) C. Christensen; *Ceterachopsis dalhousiae* (Hooker) Ching

Roots not proliferous. **Stems** erect, unbranched; scales black with brown margins, lanceolate, 2–5 × 0.6–1 mm, sparsely denticulate. **Leaves** monomorphic. **Petiole** dark to light brown throughout, dull, to 1 cm, 1/10–1/15 length of blade, indument of scales throughout. **Blade** narrowly elliptic to narrowly lanceolate, pinnatifid, 4–15 × 1.5–6 cm, thick, sparsely puberulent to glabrescent; base gradually tapered; apex obtuse, not rooting. **Rachis** light brown to tan, dull-scaly; scales brown, lanceolate. **Veins** free, obscure. **Sori** 3–7 pairs per pinna, on both basiscopic and acroscopic sides of lobes. **Spores** 64 per sporangium. $2n = 72$.

Moist, rocky ravines, terrestrial among and at bases of rocks; 1300–2000 m; Ariz.; n Mexico; Asia in the Himalayas.

In the flora, *Asplenium dalhousiae* is found only in the Mule, Huachuca, and Baboquivari mountains of southern Arizona. The pattern of disjunction in the worldwide range of this species is highly unusual.

Asplenium dalhousiae is sometimes placed in the genus *Ceterach* on the basis of its thick, pinnatifid leaves. Most pteridologists, however, restrict *Ceterach* to species with densely scaly, pinnatifid leaves. *Asplenium dalhousiae* is placed in *Ceterachopsis* by pteridologists who believe it merits its own genus.

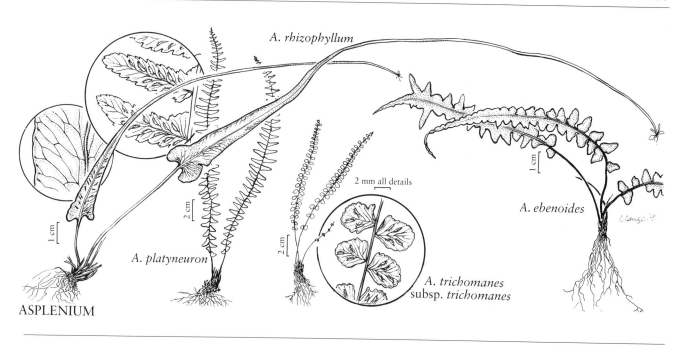

A. rhizophyllum

A. platyneuron

2 mm all details

A. ebenoides

A. trichomanes
subsp. trichomanes

ASPLENIUM

2. Asplenium ebenoides R. R. Scott, Gard. Monthly & Hort. Advertiser 7: 267. 1865 · Scott's spleenwort

× Asplenosorus ebenoides (R. R. Scott) Wherry

Roots not proliferous. **Stems** ascending to erect, rarely branched; scales dark brown to blackish throughout, narrowly deltate, 2–4 × 0.25–0.45 mm, margins entire. **Leaves** weakly subdimorphic, fertile leaves taller and more erect than sterile leaves. **Petiole** reddish or purplish brown throughout, lustrous, 1–10 cm, 1/5–1 times length of blade; indument of dark brown to black scales, narrowly deltate at very base, grading into hairs distally. **Blade** highly variable and typically irregularly shaped, narrowly deltate to lanceolate, pinnatifid or 1-pinnate in proximal 1/3, 2–20 × 1–6(–13) cm, medium thick, sparsely pubescent adaxially only; base ± truncate; apex acute to long-attenuate, apical buds borne occasionally but not known to root in nature. **Rachis** reddish or purplish brown abaxially, fading to green distally, lustrous, glabrous. **Pinnae** in 0–3 pairs, often irregular in size and shape, deltate to narrowly deltate; proximal pinnae 5–30(–80) × 3–10(–15) mm; base truncate to obtuse, auriculate on both sides; margins entire to finely serrate or crenulate; apex obtuse to acute or occasionally attenuate. **Veins** somewhat evident, mostly free, rarely anastomosing. **Sori** 1–10(–15 +) pairs per pinna, on both acroscopic and basiscopic lobes. **Spores** malformed (sterile form) or 64 per sporangium (fertile form). $2n = 72, 144$.

Conglomerate boulders; 70 m; Ala.

The above description applies to the sterile hybrid *Asplenium platyneuron × rhizophyllum* and its allopolyploid derivative. The allotetraploid form is known only from Hale County, Alabama, where it occurs with *A. platyneuron* (but not with *A. rhizophyllum*) on conglomerate boulders (K. S. Walter et al. 1982). The sterile diploid form of *A. ebenoides* occurs at elevations of 70 to 500 m within the region where the ranges of the parental species overlap, always occurring with both parents on limestone, sandstone, or other rock strata. A hybrid between the allopolyploid and *A. platyneuron* [*A. × boydstoniae* (K. S. Walter) J. W. Short] was discovered at Havana Glen. An unnamed hybrid between the sterile diploid (presumably via unreduced spores) and *A. rhizophyllum* is known from West Virginia and Missouri (K. S. Walter et al. 1982).

This fern has been pivotal in the study of fern hybridization. Called the "most famous hybrid fern," it was one of the first crosses to be synthesized deliberately in culture (M. Slosson 1902) and the first to be converted from the sterile diploid state to the fertile tetraploid state experimentally (W. H. Wagner Jr. and R. S. Whitmire 1957).

3. Asplenium pinnatifidum Nuttall, Gen. N. Amer. Pl. 2: 251. 1818 · Lobed spleenwort

× *Asplenosorus pinnatifidus* (Nuttall) Mickel

Roots not proliferous. **Stems** short-creeping to erect, frequently branched; scales dark reddish brown, narrowly deltate, 3–5 × 0.3–0.5 mm, margins entire. **Leaves** monomorphic. **Petiole** dark reddish brown at base, fading to green in distal 1/3–1/2, lustrous, 1–10 cm, 1/5–1 times length of blade; indument of dark reddish brown, narrowly deltate scales at very base, grading distally into hairs. **Blade** narrowly deltate, often irregular in outline, pinnatifid or often with single pair of pinnae proximally, 2–17(–20) × 1–4(–13) cm, thick, pubescent abaxially only; base truncate, cordate, or auriculate; apex acute to long-attenuate, proliferous bud very rare, not known to root in nature. **Rachis** green, sometimes drying to tan, dull; hairs on abaxial surface only, scattered, minute. **Pinnae** 0–1 pair, ovate to deltate, sometimes narrowly so, 5–20(–90) × 0.4–1(–1.2) mm; base truncate to acute; margins crenate to serrate; apex rounded to attenuate. **Veins** free (rarely anastomosing), obscure. **Sori** 1–6(–40+) per segment, usually confluent with age. **Spores** 64 per sporangium. $2n = 144$.

Cliffs, ledges, and boulders of sandstone and other acidic rocks; 0–1000 m; Ala., Ark., D.C., Ga., Ill., Ind., Ky., Md., Miss., Mo., N.J., N.C., Ohio, Okla., Pa., S.C., Tenn., Va., W.Va., Wis.

Asplenium pinnatifidum is an allotetraploid derived from the hybrid *A. montanum* × *rhizophyllum*. Although isozyme studies indicate that this species originated at more than one site (C. R. Werth et al. 1985b), the sterile diploid hybrid is unknown. The species is uncommon in the eastern part of the Appalachian region and becomes much more frequent in the Cumberland and Interior Low plateaus, extending westward into the Ozarks and Ouachitas. It is disjunct in the Driftless Area of Wisconsin in Iowa County (M. G. and R. P. Hanson 1979). It crosses frequently with *A. montanum* (producing *A.* × *trudellii* Wherry), with *A. bradleyi* (producing *A.* × *gravesii* Maxon), with *A. platyneuron* (producing *A.* × *kentuckiense* McCoy), and with *A. trichomanes* (producing *A.* × *herb-wagneri* W. C. Taylor & Mohlenbrock).

4. Asplenium rhizophyllum Linnaeus, Sp. Pl. 2: 1078. 1753 (as *rhizophylla*) · Walking fern, doradille à feuilles radicantes

Camptosorus rhizophyllus (Linnaeus) Link

Roots not proliferous. **Stems** erect or ascending, usually unbranched; scales dark brown throughout, narrowly deltate, 2–3 × (0.2–)0.5–1 mm, margins entire. **Leaves** monomorphic but fertile leaves generally larger than sterile leaves. **Petiole** reddish brown at base, becoming green distally, dull but sometimes lustrous at base, 0.5–12 cm, 0.1–1.5 times length of blade; indument of dark brown, narrowly deltate scales at base, of minute, club-shaped hairs distally. **Blades** highly variable in size and shape, even on 1 plant, narrowly deltate to linear-lanceolate, simple, 1–30 × 0.5–5 cm, leathery, sparsely pubescent, hairs more numerous abaxially than adaxially; blade base cordate, auriculate, or occasionally hastate, auricles rarely attenuate and radicant; margins entire to sinuate, rarely irregularly incised; apex rounded to very long-attenuate and, if attenuate, generally rooting at tip. **Rachis** green, dull, nearly glabrous. **Veins** obscure, anastomosing to form areoles near midrib. **Sori** numerous, scattered somewhat irregularly over blade, often joined at vein junctures. **Spores** 64 per sporangium. $2n = 72$.

Shaded, usually moss-covered boulders and ledges, usually on limestone or other basic rocks, but occasionally on sandstone or other acidic rocks, rarely on fallen tree trunks; 0–1000 m; Ont., Que.; Ala., Ark., Conn., D.C., Ga., Ill., Ind., Iowa, Kans., Ky., Maine, Md., Mass., Mich., Minn., Miss., Mo., N.H., N.J., N.Y., N.C., Ohio, Okla., Pa., R.I., Tenn., Vt., Va., W.Va., Wis.

Asplenium rhizophyllum, a diploid species, is morphologically very distinctive within *Asplenium* and is segregated by many authors, along with its sister species *A. ruprechtii* Kurata of eastern Asia, into the genus *Camptosorus* Link. Proliferations arising from leaf tips result in the formation of clonal patches, often dense and extensive, on the mossy boulders and ledges where it typically grows. Naturally occurring sterile hybrids are known with *A. platyneuron* (these and their fertile allotetraploid derivatives are both referred to *A. ebenoides*), *A. ruta-muraria* (*A.* × *inexpectatum* E. L. Braun ex C. V. Morton), *A. trichomanes* subsp. *trichomanes* [*A.* × *shawneense* (R. C. Moran) H. E. Ballard], and *A. ebenoides*. In addition, the allotetraploid *A. pinnatifidum* is derived from the hybrid *A. montanum* × *rhizophyllum*; the sterile diploid hybrid is unknown.

5. **Asplenium scolopendrium** Linnaeus, Sp. Pl. 2: 1079. 1753

Asplenium altajense (Komarov) Grubov; *Phyllitis scolopendrium* (Linnaeus) Newman

Varieties 2 or 3 (1 in the flora): North America, Mexico, West Indies in Hispaniola, Europe.

5a. **Asplenium scolopendrium** Linnaeus var. **americanum** (Fernald) Kartesz & Gandhi, Phytologia 70: 196. 1991 · American hart's-tongue

Phyllitis scolopendrium (Linnaeus) Newman var. *americanum* Fernald, Rhodora 37: 220. 1935; *P. fernaldiana* A. Löve; *P. japonica* Komarov subsp. *americana* (Fernald) A. Löve & D. Löve

Roots not proliferous. **Stems** erect, unbranched; scales brown, lanceolate, 3–6 × 1–1.5 mm, margins entire. **Leaves** monomorphic. **Petiole** brown to straw-colored, dull, 3–10 cm, 1/8–1/4 length of blade; indument of brown, narrowly lanceolate scales. **Blade** linear, simple, 8–35 × 2–4.5 cm, thick, papery, nearly glabrous; base cordate; margins entire; apex acuminate, not rooting. **Rachis** brown proximally, straw-colored distally, dull, glabrous. **Veins** free, obscure. **Sori** numerous, perpendicular to rachis, or nearly so, usually restricted to distal 1/2 of blade. **Spores** 64 per sporangium. Usually $2n = 144$.

On calcareous rocks in sinkholes, at cave entrances, and on cool, moist talus, always in deep shade; 0–100 m; Ont.; Ala., Mich., N.Y., Tenn.; Mexico in Nuevo León.

Asplenium scolopendrium is rare and has a spotty distribution in North America. Consequently, when new populations are found, they receive considerable attention from pteridologists.

The American variety was distinguished from the European plants by M. L. Fernald (1935, pp. 220–221) on the basis of their smaller leaves, narrower scales with longer, attenuate tips, more promptly glabrate midribs, and blades tending to bear sori in the distal half rather than the entire length. In addition, M. L. Arreguín-Sánchez and R. Aguirre-Claverán (1986, pp. 400, 402) listed two differences in perispore morphology of the two varieties; they did not say how many specimens were studied. The most clear-cut difference is that the American plants are tetraploid and the European plants are diploid (D. M. Britton 1953), a difference insufficient for recognition of this taxon at specific rank (W. H. Wagner Jr. 1955b). More work is needed to determine if *Asplenium scolopendrium* var. *lindenii* (Hooker) Viane, Rasbach, & Reichstein, which occurs in southern Mexico (Oaxaca and Chiapas) and Hispaniola, is distinguishable from var. *americanum*.

6. **Asplenium serratum** Linnaeus, Sp. Pl. 2: 1079. 1753 · New World bird's-nest fern, American bird's-nest fern

Roots proliferous. **Stems** erect, unbranched; scales brown throughout, narrowly lanceolate, 5–10 × 1–1.5 mm, margins entire. **Leaves** monomorphic. **Petiole** vestigial. **Blade** linear, oblanceolate, simple, (10–)20–40(–70) × 3–8 cm, thick, glabrous; base gradually tapered; margins entire to irregularly crenate; apex attenuate, not rooting. **Rachis** green throughout, dull, glabrous. **Veins** numerous, free, mostly immersed. **Sori** parallel to each other, nearly perpendicular to midrib. **Spores** 64 per sporangium. $2n = 144$.

Rotten logs and stumps; 0–50 m; Fla.; s Mexico; West Indies; Central America; South America.

Asplenium serratum is found rarely in southern peninsular Florida, where it is at the extreme edge of its tropical American range. This large simple-leaved spleenwort is called "American bird's-nest fern" because of its superficial resemblance to the Old World *A. nidus* Linnaeus, which is regularly grown in temperate conservatories. *Asplenium serratum* is unusual in having roots with abundant, matted hairs rather than scattered hairs as found in other species.

7. **Asplenium septentrionale** (Linnaeus) Hoffmann, Deutschl. Fl. 2: 12. 1796 · Forked spleenwort

Acrostichum septentrionale Linnaeus, Sp. Pl. 2: 1068. 1753

Roots not proliferous. **Stems** erect, much branched to produce dense many-stemmed tufts or mats bearing numerous crowded leaves; scales dark reddish brown to black throughout, narrowly deltate, 2–4 × 0.3–0.6 mm, margins entire. **Leaves** monomorphic. **Petiole** dark reddish brown proximally, fading to green distally, 2–13 cm, 2–5 times length of blade; indument absent. **Blade** linear, simple or more often 1-pinnate, 0.5–4 × 0.1–0.4 cm, occasionally wider when pinnae strongly diverge, leathery, glabrous; base acute; apex acute, not rooting at tip. **Rachis** green, lustrous, glabrous. **Pinnae** of pinnate leaves 2(–4), strongly ascending to give forked appearance, linear, (5–)10–30 × 0.75–3 mm; base acute; margins remotely lacerate; apex acute. **Veins** free, obscure. **Sori** usually 2+ per pinna, parallel to margins. **Spores** 64 per sporangium. $2n = 144$.

Cliffs of various substrates; 700–2900 m; Ariz., Calif., Colo., D.C., N.Mex., Okla., Oreg., S.Dak., Tex., Utah, W.Va., Wyo.; Mexico in Baja California; Europe; Asia.

In North America *Asplenium septentrionale* is principally a western species with isolated disjunct populations in Monroe and Hardy counties, West Virginia. Because of its close resemblance to a tuft of grass, it is easily overlooked, and discoveries of additional localities are to be expected. In Europe *A. septentrionale* is known to hybridize with several species, but in North America only the hybrid with *A. trichomanes* (*A. ×alternifolium* Wulfen) is known.

8. Asplenium abscissum Willdenow, Sp. Pl. 5(1): 321. 1810 · Abscised spleenwort

Roots proliferous. **Stems** erect, unbranched; scales brown throughout, linear-deltate, 1.2 × 0.1–0.3 mm, margins entire. **Leaves** monomorphic. **Petiole** green throughout, dull, 3–15 (–20) cm, 1/3–2/3 length of blade; indument absent. **Blade** deltate, 1-pinnate, 7–12(–20) × 3–6(–9) cm, thick, papery, glabrous; base not tapered; apex attenuate, not rooting. **Rachis** green throughout, dull, glabrous. **Pinnae** in 4–8 pairs, linear-lanceolate; medial pinnae 30–70 × 7–15 mm, slightly curved, strongly excavated in proximal basiscopic 1/3; base with small acroscopic auricle; margins entire to dentate; apex attenuate. **Veins** free, obscure. **Sori** 2–9 pairs, medial, commonly 1–2 less on basiscopic side than on acroscopic side. **Spores** 64 per sporangium. 2n = 72.

Shaded limestone boulders, cliff ledges, grottoes, sinkholes; 0–50 m; Fla.; Mexico; West Indies in the Antilles; Central America; South America to Bolivia and Brazil.

Asplenium abscissum hybridizes with *A. verecundum* to produce the triploid hybrid, *A. ×curtissii.*

9. Asplenium adulterinum J. Milde, Höh. Sporenpfl. Deutschl., 40. 1865 · Adulterated spleenwort

Roots not proliferous. **Stems** short-creeping, mainly unbranched; scales black or with narrow pale borders, narrowly lanceolate, 1.5–3 × 0.2–0.4 mm, margins entire. **Leaves** monomorphic. **Petiole** dark reddish brown throughout, 1–4 mm; indument of black linear scales at base. **Blade** linear, 1-pinnate, 2.5–14 × 0.5–1.2 cm, thick (open habitat) to herbaceous (shaded, moist habitat), essentially glabrous; base somewhat tapered; apex obtuse, not rooting. **Rachis** reddish brown in proximal 1/2–4/5, green distally, lustrous, glabrous. **Pinnae** in 10–30 pairs, ovate to rhombic to ovate-oblong, 2.5–11 ×

2–6 mm; base truncate to shortly acute; margins shallowly crenate (shade forms) to essentially entire (exposed forms); apex obtuse, broadly rounded. **Veins** free, evident to obscure. **Sori** 1–3 pairs per pinna on both basiscopic and acroscopic sides. **Spores** 64 per sporangium. 2n = 144.

Crevices in limestone; 1250 m; B.C.; Europe.

In North America *Asplenium adulterinum* is known to occur on Vancouver Island, British Columbia, where only the fertile allotetraploids are known. It is likely to occur in areas where the two parents, *A. trichomanes* and *A. trichomanes-ramosum*, grow together. The genetics of the American plants should be compared with that of the European, among which two nothosubspecies occur (F. Mokry et al. 1986).

10. Asplenium auritum Swartz, J. Bot. (Schrader) 1800(2): 52. 1801 · Eared spleenwort

Roots proliferous. **Stems** erect, unbranched; scales brown throughout, broadly linear, 1–2 × 0.7–1.1 mm, margins shallowly and widely dentate. **Leaves** monomorphic. **Petiole** green to black, dull, 2–10(–12) cm, 1/3–1/2 length of blade; indument absent. **Blade** narrowly deltate, 1–2-pinnate, 4–20(–30) × 1.8–12(–18) cm, thick, nearly glabrous; base not tapered; apex gradually tapered, not rooting. **Rachis** green to black, dull, abaxially glabrous. **Pinnae** in 10–22 pairs, linear-deltate, medial pinnae 1–4(–9) × 0.3–1(–2.5) cm; base with acroscopic auricle or pinnule enlarged, excavated in proximal 1/5–1/4; margins mostly 1–2-dentate-serrate, or lobed or pinnate proximally or in proximal 2/3; apex blunt in some 1-pinnate forms, gradually reduced to attenuate in strongly 2-pinnate forms; pinnules narrow, not auriculate. **Veins** free, evident. **Sori** 4–9(–10) pairs per pinna, subcostal, nearly parallel to costae on both basiscopic and acroscopic sides. **Spores** 64 per sporangium.

Mainly epiphytic on old sloping tree trunks in shady forests; 0–50 m; Fla.; Mexico; West Indies in the Antilles; Central America; South America.

In the flora *Asplenium auritum* is evidently confined to Florida, where it is rare, occurring primarily on live oaks (*Quercus virginiana* Miller). The species is highly variable. Juvenile plants, less than 2 cm, tend to be 2-pinnate. In mature plants all stages between 1-pinnate and 2-pinnate leaves are found, but 1-pinnate are more common.

11. Asplenium trichomanes-dentatum Linnaeus, Sp. Pl.

2: 1080. 1753 *Asplendium dentatum* Linnaeus

Roots proliferous. **Stems** erect, unbranched; scales blackish throughout, linear, 0.9–1.1 × 0.2–0.3 mm, margins entire. **Leaves** dimorphic; sterile leaves short with small crowded pinnae; fertile leaves 9–20 cm with long, widely spaced pinnae. **Petiole** pale green except darkish proximally, dull, (1–)4–6(–10) cm, 1/5–1/2 length of blade; indument absent. **Blade** linear, 1-pinnate, 3–10 (–15) × 1–2.5 cm, papery, slightly glandular; base not tapered; apex with elongate terminal pinna similar in size to subtending lateral pinnae, not rooting. **Rachis** green throughout, dull, glabrous. **Pinnae** in 5–8(–12) pairs, ovate, strongly asymmetric, lanceolate; medial pinnae 3–11 × (2–)4–6 mm; base acute; margins crenate-dentate; apex truncate. **Veins** free, somewhat evident. **Sori** (1–)2–5(–8), mostly on acroscopic side. **Spores** 64 per sporangium. $2n = 108$.

Shady hammocks, on walls of limesinks, moist limestone rocks, rocky banks of streams; 0–50 m; Fla.; s Mexico; West Indies in the Antilles; Central America; South America in Colombia, Venezuela.

In the flora *Asplenium trichomanes-dentatum* is known only from shady hammocks in extreme southern peninsular Florida, where it grows in colonies. This delicate spleenwort should be studied in detail and compared with its similar and apparently conspecific counterparts in tropical America.

12. Asplenium heterochroum Kunze, Linnaea 9: 67.

1834 · Varicolored spleenwort

Asplenium muticum Gilbert

Roots not proliferous. **Stems** erect or ascending, rarely branched; scales black throughout, linear-lanceolate, 2–3 × 0.2–0.4 mm, margin entire, apex attenuate. **Leaves** monomorphic. **Petiole** black or purplish black throughout, lustrous, 0.3–5 cm, 1/4– 1/15 length of blade; indument of black filiform scales at base. **Blade** linear or narrowly oblanceolate, 1-pinnate throughout, 6–22(–37) × 1–1.8(–2.3) cm, thin, glabrous; base tapered; apex acute, not rooting. **Rachis** black throughout, lustrous, glabrous. **Pinnae** in 15–40 pairs, oblong; medial pinnae somewhat asymmetric, oblong, 4–10 × 2–4(–5) mm; base acute, acroscopically enlarged; margins serrate to crenate, lobed; apex obtuse. **Veins** free, evident. **Sori** 3–6 pairs per pinna, on both basiscopic and acroscopic sides of pinnae. **Spores** 64 per sporangium. $2n = 144, 216$.

Sinkholes, limestone rocks in shady hammocks, masonry; 0–50 m; Fla., Ga.; Mexico; West Indies in Cuba, Puerto Rico; Bermuda; Central America in Belize.

The tetraploid cytotype (4x) of *Asplenium heterochroum* hybridizes with *A. resiliens* (3x) to produce *A. ×heteroresiliens* (5x). It is extremely rare and local, known in northern peninsular Florida from Alachua, Citrus, Columbia, Jackson, and Marion counties.

13. Asplenium ×heteroresiliens W. H. Wagner, Amer.

Fern J. 56: 12, plate 3. 1966 · Morzenti's spleenwort

Roots not proliferous. **Stems** erect, rarely branched; scales black throughout, linear-lanceolate, 1– 3.5 × 0.1–0.4 mm, margins entire. **Leaves** monomorphic. **Petiole** black or purplish black throughout, lustrous, 1–2 cm, 1/5–1/10 length of blade; indument of black filiform scales at base. **Blade** linear to narrowly oblanceolate, 1-pinnate throughout, 5–15 × 1–2 cm, thick, glabrous; base gradually tapered, apex acute, not rooting. **Rachis** black or purplish black throughout, lustrous, glabrous or with a few scattered scales. **Pinnae** in 12–30 pairs, oblong to quadrangular; medial pinnae 5–10 × 2–3 mm; base with slight or prominent acroscopic auricle and often with basiscopic auricle; margins usually shallowly crenulate, rarely ± entire; apex acute to obtuse. **Veins** free, obscure. **Sori** 1–5 pairs per pinna, on both basiscopic and acroscopic sides. **Spores** 32 per sporangium. $n = 2n = 180$ (apogamous).

Limestone rocks; 0–100 m; Fla., N.C., S.C.

V. M. Morzenti (1966) studied the cytology and morphology of *Asplenium* ×*heteroresiliens* and found it to be a pentaploid (5x) of hybrid origin between *A. resiliens* (3x) and *A. heterochroum* (4x). The hybrid is intermediate in morphology between the parents, and it is difficult to separate from them. Misshapen spores mixed with large globose spores can usually be found on the same specimen.

14. Asplenium monanthes Linnaeus, Mant. Pl. 1: 130.

1767 · Single-sorus fern

Roots not proliferous. **Stems** erect, unbranched; scales black with lighter margins, linear-lanceolate, 3–6 × 0.4–0.8 mm, margins entire. **Leaves** monomorphic. **Petiole** reddish brown throughout, lustrous, 1–12(–20) cm, 1/3–1/10 length of blade; indument of black filiform scales. **Blade** linear, 1-pinnate throughout, 5–25(–40) × 1– 2.5(–3) cm, thick, glabrous; base gradually tapered; apex acute, not rooting. **Rachis** reddish brown throughout, lustrous, glabrous. **Pinnae** in 10–40 pairs, oblong to quadrangular, somewhat asymmetric; medial pinnae 4–15 × 2–5 mm; base rounded to cuneate; margins crenulate or ± entire; apex obtuse. **Veins** free, obscure. **Sori** 1(–3) per pinna, only on basiscopic side. **Spores** 32 per sporangium. $n = 2n = 108$ (apogamous).

Rock; 50–1000 m; Ala., Ariz., N.C., S.C.; Mexico; West Indies in Hispaniola, Jamaica; Central America; South America to n Argentina; Africa including Madagascar, Madeira, Réunion, Tristan da Cunha; Pacific Islands in Hawaii.

15. Asplenium palmeri Maxon, Contr. U.S. Natl. Herb. 13: 39. 1909 · Palmer's spleenwort

Roots not proliferous. **Stems** short-creeping, unbranched; scales black with lighter margins, linear-lanceolate, 1.5–3 × 0.1–0.4 mm, margins entire. **Leaves** monomorphic. **Petiole** purplish black, lustrous, 0.5–3 cm, 1/3–1/20 length of blade; indument of black filiform scales at base. **Blade** linear, 1-pinnate throughout, 7–17.5 × 0.9–1.8 cm, thick, glabrous; base gradually reduced; apex gradually reduced to whiplike rooting tip. **Rachis** purplish black throughout, lustrous, glabrous or nearly so. **Pinnae** in (12–)20–40 pairs, oblong; medial pinnae 6–9 × 3–4 mm; base broadly cuneate or auriculate; margins crenate-serrate; apex obtuse. **Veins** free, obscure. **Sori** 3–7 pairs per pinna, on both basiscopic and acroscopic sides. **Spores** 64 per sporangium.

Shaded rocky slopes, wet ledges, often in protected places; 900–2000(–2750) m; Ariz., N.Mex., Tex.; Mexico; Central America in Guatemala, Belize.

16. Asplenium platyneuron (Linnaeus) Britton, Sterns, & Poggenburg, Prelim. Cat., 3. 1888 · Ebony spleenwort, doradille à nervures plates

Acrostichum platyneuron Linnaeus, Sp. Pl. 2: 1069. 1753; *Asplenium platyneuron* var. *bacculum-rubrum* (Fernald) Fernald; *A. platyneuron* var. *incisum* (E. C. Howe) Robinson

Roots not proliferous. **Stems** short-creeping, unbranched; scales dark brown to black throughout, narrowly linear-deltate, 2–4 × 0.3–0.6 mm, margins entire. **Leaves** ± dimorphic; fertile leaves taller and more erect than sterile leaves. **Petiole** reddish brown throughout, lustrous, 1–10 cm, 1/4–1/3 length of blade; indument of dark brown to black, filiform scales at base. **Blade** lustrous, linear to narrowly oblanceolate, 1-pinnate throughout, 4–50 × 2–5(–7) cm, thin, glabrous, or occasionally sparsely pubescent; base gradually tapered; apex acute, not rooting. **Rachis** reddish or purplish brown throughout, lustrous, glabrous. **Pinnae** in 15–45 pairs, oblong to quadrangular; medial pinnae 1–2.5 × 0.3–0.5 cm; base with conspicuous acroscopic and sometimes basiscopic auricle, this overlapping rachis; margins crenate to serrulate, sometimes more deeply incised in robust specimens; apex acute to obtuse. **Veins** free, evident. **Sori** 1–12 pairs per pinna, on both basiscopic and acroscopic sides. **Spores** 64 per sporangium. $2n = 72$.

Forest floor or on rocks, often invading masonry and disturbed soils; 0–1300 m; Ont., Que.; Ala., Ark., Ariz., Colo., Conn., Del., D.C., Fla., Ga., Ill., Ind., Iowa, Kans., Ky., La., Maine, Md., Mass., Mich., Minn., Miss., Mo., Nebr., N.H., N.Mex., N.J., N.Y., N.C., Ohio, Okla., Pa., R.I., S.C., Tenn., Tex., Vt., Va., W.Va., Wis.; s Africa.

The combining author for *Asplenium platyneuron* is often given as Oakes ex D. C. Eaton; see D. B. Lellinger (1981) for justification of the authorship employed here.

Asplenium platyneuron is remarkable in that it occurs in southern Africa as well as in North America. No other North American fern has this distribution. *Asplenium platyneuron* is an ecological generalist and is particularly characteristic of disturbed woodlands. This species is migrating northward on the northern portions of its range in the upper Great Lake states (W. H. Wagner Jr. and D. M. Johnson 1981). Proliferous buds on the lowest pinnae allow formation of clumps with stems at several layers in the litter. *Asplenium platyneuron* hybridizes with *A. rhizophyllum*, *A. trichomanes* (producing *A. ×virginicum* Maxon), *A. pinnatifidum*, *A. ruta-muraria* (producing *A. ×morganii* W. H. Wagner & F. S. Wagner), *A. bradleyi*, and *A. montanum* (producing sterile *A. bradleyi*).

17. Asplenium pumilum Swartz, Prodr., 129. 1788 · Triangle spleenwort, hairy spleenwort

Asplenium pumilum Swartz var. *anthriscifolium* (Jacquin) Wherry

Roots proliferous. **Stems** erect, unbranched; scales black with pale margins, linear, extremely narrow, 2–3 mm, only several cells wide. **Leaves** monomorphic. **Petiole** green in small leaves, black abaxially and green adaxially in large leaves, (1–)2–7(–16) cm, 1–2 times length of blade; indument of fine, nonglandular hairs on veins. **Blade** deltate, simple to 2-pinnate, 1–8(–12) × 1–6(–8) cm, thin, papery with scattered hairs on both surfaces; base truncate; margins crenate-dentate; apex pointed, not rooting. **Rachis** green, dull, glabrous. **Pinnae** in 0–5 pairs, ovate to deltate, simple to lobed to pinnate proximally, 1–6 × 1–3.5 cm, proximal pinna pair largest; base broadly cuneate to truncate; margins irregularly crenate; apex rounded to pointed. **Veins** free, evident. **Sori** 1–15(–35) per pinna, on both basiscopic and acroscopic sides. **Spores** 64 per sporangium. $2n = 72$.

Shaded limestone boulders; 0–50 m; Fla.; Mexico; West Indies; Central America; South America.

Asplenium pumilum is a widespread tropical American fern known only from a few spots in north central Florida. It is a very distinct species, readily recognized by its hairy blades and deltate leaves. Fertile forms vary from simple and only 2 cm to 2-pinnate and 28 cm, and all stages between the two extremes exist. Extreme forms are different enough to suggest that two species might be present.

18. Asplenium resiliens Kunze, Linnaea 18: 331. 1844

· Black-stemmed spleenwort

Roots not proliferous. **Stems** erect, unbranched; scales black throughout, linear-lanceolate, 4–5 × 0.2–0.6 mm, margins entire. **Leaves** monomorphic. **Petiole** black throughout, lustrous, 1.5–3(–5) cm, 1/4–1/10 length of blade; indument of blackish brown, filiform scales. **Blade** linear to narrowly oblanceolate, 1-pinnate throughout, 9–20(–30) × 1–2(–2.5) cm, thick, glabrous; base gradually tapered; apex acute, not rooting. **Rachis** black throughout, lustrous, glabrous. **Pinnae** in 20–40 pairs, oblong; medial pinnae (5–)10–20 × (2–)3–5 mm; base usually with an acroscopic auricle; margins ± entire to shallowly crenate; apex obtuse. **Veins** free, obscure. **Sori** 2–5 pairs per pinna, on both basiscopic and acroscopic sides, often confluent with age. **Spores** 32 per sporangium. $n = 2n = 108$ (apogamous).

Cliffs, sinkholes, on limestone or other basic rocks; 100–1500 m; Ala., Ariz., Ark., Del., Fla., Ga., Ill., Kans., Ky., La., Md., Miss., Mo., N.Mex., Nev., N.C., Okla., Pa., Tenn., Tex., Utah, Va., W.Va.; Mexico; West Indies in Hispaniola, Jamaica; Central America in Guatemala; South America.

Asplenium parvulum M. Martens & Galeotti is an older, but illegitimate, name because it is a later homonym of *A. parvulum* Hooker.

In Florida *Asplenium resiliens* hybridizes with *A. heterochroum* Kunze (4x), producing *A.* ×*heteroresiliens* (5x).

19. Asplenium trichomanes Linnaeus, Sp. Pl. 2: 1080.

1753 · Maidenhair spleenwort, doradille chevelue

Roots not proliferous. **Stems** short-creeping, often branched; scales black throughout or with brown borders, lanceolate, 2–5 × 0.2–0.5 mm, margins entire to denticulate. **Leaves** monomorphic. **Petiole** reddish brown or blackish brown throughout, lustrous, 1–4(–7) cm, 1/6–1/4 length of blade; indument absent or of black, linear-lanceolate or filiform scales at base. **Blade** linear, 1-pinnate, 3–22 × 0.5–1.5 cm, thin, glabrous or sparsely pubescent; base gradually tapered; apex narrowly acute, not rooting. **Rachis** reddish brown throughout, lustrous, glabrous or nearly so. **Pinnae** in 15–35 pairs, oblong to oval; medial pinnae 2.5–8 × 2.5–4 mm; base broadly cuneate, with or without low, rounded acroscopic auricle; margins shallowly crenate to serrate or ± entire; apex obtuse. **Veins** free, evident. **Sori** 2–4 pairs per pinna, on both basiscopic and acroscopic sides. **Spores** 64 per sporangium. $2n = 72, 144$.

Subspecies 4 (2 in the flora): worldwide.

In North America, as in Europe, *Asplenium trichomanes* consists of diploid and tetraploid cytotypes, treated here as subspecies. *Asplenium trichomanes* subsp. *trichomanes*, the diploid, is found on noncalcareous rocks. In the southwestern United States it occurs at high elevations. *Asplenium trichomanes* subsp. *quadrivalens*, the tetraploid, grows on calcareous substrates and has a more northern distribution (R. C. Moran 1982). Triploid hybrids are known between the diploids and tetraploids (R. C. Moran 1982; W. H. Wagner Jr. and F. S. Wagner 1966).

1. Spores 27–32 μm; on acidic substrates. . . .
. . . . 19a. *Asplenium trichomanes* subsp. *trichomanes*
1. Spores 37–43 μm; on limestone. . . . .
. . . . 19b. *Asplenium trichomanes* subsp. *quadrivalens*

19a. Asplenium trichomanes Linnaeus subsp. **trichomanes**

Asplenium melanocaulon Willdenow

Spores (as measured in Hoyer's Solution) 27–32 μm. $2n = 72$.

Acidic rocks such as sandstone, basalt, and granite, very rarely on calcareous rocks; 0–3000 m; B.C., N.B., Nfld., N.S., Ont., Que.; Ala., Alaska, Ariz., Ark., Calif., Colo., Conn., Del., Ga., Ill., Ind., Kans., Ky., La., Maine, Md., Mass., Mich., Minn., Mo., Nebr., N.H., N.J., N.Mex., N.Y., N.C., Ohio, Okla., Oreg., Pa., R.I., S.C., S.Dak., Tenn., Tex., Vt., Va., Wash., W.Va., Wis., Wyo.; Mexico in Chihuahua; Europe; Asia; Africa; Australia.

In southern Illinois *Asplenium trichomanes* subsp. *trichomanes* hybridizes with *A. rhizophyllum* to produce *A.* ×*shawneense* (R. C. Moran) H. E. Ballard (R. C. Moran 1981).

19b. Asplenium trichomanes Linnaeus subsp. **quadrivalens** D. E. Meyer, Ber. Deutsch. Bot. Ges. 74: 456. 1962

Spores (as measured in Hoyer's Solution) 37–43 μm. $2n = 144$.

Calcareous rocks such as limestone and dolomite; 0–3000 m; B.C., Ont., Que.; Conn., Del., Ill., Maine, Md., Mass., Mich., N.H., N.J., N.Y., Ohio, Oreg., Pa., Vt., Va., Wash., W.Va., Wis.; Europe; Asia; Africa; Pacific Islands in New Zealand; Australia.

20. Asplenium vespertinum Maxon, Bull. Torrey Bot. Club 27: 197. 1900 · Western spleenwort

Asplenium trichomanes Linnaeus var. *vespertinum* (Maxon) Jepson

Roots not proliferous. **Stems** short-creeping, rarely branched; scales black with lighter margins, linear-lanceolate, 2–3 × 0.2–0.4 mm, margins entire, or denticulate distally. **Leaves** monomorphic. **Petiole** reddish brown or purplish black throughout, lustrous, 2–5 cm, 1/4–1/6 length of blade, sometimes curved like a J at base; indument of blackish brown, filiform scales at base. **Blade** linear, 1-pinnate throughout, 5–15(–28) × 1–2.5 cm, thick, glabrous; base gradually tapered; apex acute, not rooting. **Rachis** reddish brown or purplish black throughout, shiny, glabrous. **Pinnae** 15–30 pairs, oblong; medial pinnae 5–10 × 2–3 mm; base rounded to cuneate; margins lobed or serrate; apex obtuse. **Veins** free, obscure. **Sori** 2–6 pairs per pinna, on both basiscopic and acroscopic sides. **Spores** 64 per sporangium.

Moist, shaded canyon walls and at base of overhanging rocks; 0–1000 m; Calif.; Mexico in Baja California.

21. Asplenium trichomanes-ramosum Linnaeus, Sp. Pl. 2: 1082. 1753 · Green spleenwort, doradille verte

Asplenium viride Hudson

Roots not proliferous. **Stems** short-creeping or ascending, frequently branched; scales dark reddish brown to blackish throughout, narrowly deltate, 2–4 × 0.2–0.4 mm, margins entire to undulate or with widely spaced shallow teeth. **Leaves** monomorphic. **Petiole** reddish brown at base, green distally, lustrous, 1–5(–6) cm, 1/4–1/2(–1) times length of blade; indument of dark reddish brown to black, narrowly deltate scales grading into glandular hairs. **Blade** linear, 1-pinnate throughout, 2–13 × 0.6–1.2 cm, thin, glabrous or with sparse minute hairs; base slightly tapering or truncate; apex acute, not rooting. **Rachis** green throughout, dull, glabrous or with scattered hairs as on petioles. **Pinnae** in 6–21 pairs, deltate to rhombic; medial pinnae 5–6 × 4–5 mm; base obtuse and often inequilateral; distal margins crenate; apex rounded to acute. **Veins** free, evident. **Sori** 2–4 pairs per pinna, on both basiscopic and acroscopic sides. **Spores** 64 per sporangium. $2n = 72$.

Limestone and other basic rocks; 0–4000 m; Greenland; Alta., B.C., N.B., Nfld., N.W.T., N.S., Ont., P.E.I., Que., Yukon; Alaska, Calif., Colo., Idaho, Maine, Mich., Mont., Nev., N.Y., Oreg., S.Dak., Utah, Vt., Wash., Wis., Wyo.; Europe; Asia.

Hybridization between *Asplenium trichomanes-ramosum* and *A. trichomanes* produces the fertile allotetraploid *A. adulterinum*, which occurs on Vancouver Island.

22. Asplenium adiantum-nigrum Linnaeus, Sp. Pl. 2: 1081. 1753 · Black spleenwort

Asplenium andrewsii A. Nelson;
A. chihuahuense J. G. Baker;
A. dubiosum Davenport

Roots not proliferous. **Stems** ascending or short-creeping, infrequently branched; scales dark brown to blackish throughout, narrowly deltate, 2–4(–5) × 0.2–0.5 mm, margins entire or shallowly denticulate to serrulate. **Leaves** monomorphic. **Petiole** dark reddish brown proximally, often fading to green distally, lustrous, 2–20 cm, 2/3–2 times length of blade; indument of black filiform scales and minute hairs. **Blade** deltate, 2–3-pinnate, 2.5–10 × 2–6.5 cm, thick, hairs dark, scattered, minute; base truncate; apex acute to acuminate, not rooting. **Rachis** greenish throughout or sometimes reddish brown proximally, lustrous, sparsely pubescent. **Pinnae** in 4–10 pairs, deltate to lanceolate; most proximal (largest) pinnae 1.5–4 × 1–2.5 cm; base obliquely obtuse; segment margins coarsely incised; apex acute. **Veins** free, evident. **Sori** 1–numerous pairs per pinna [1–6 pairs per segment], on both basiscopic and acroscopic sides. **Spores** 64 per sporangium. $2n = 144$.

Cliffs; 1675–2300 m; Ariz., Colo., Utah; Eurasia; Africa; Mexico in Chihuahua; Pacific Islands.

Asplenium adiantum-nigrum is principally a Eurasian species and occurs extremely rarely in North America (see M. G. Shivas 1969 and M. D. Windham 1983 for discussions of the conspecificity of Western Hemisphere and Eastern Hemisphere material). It is an allotetraploid derived from hybridization of two European taxa, *A. cuneifolium* Viviani and *A. onopteris* Linnaeus (M. G. Shivas 1969). Hybrids involving *A. adiantum-nigrum* and other *Asplenium* species occur in Europe but are unknown in North America.

23. Asplenium ×biscayneanum (D. C. Eaton) A. A.
Eaton, Fern Bull. 12: 45. 1904 (as species) · Biscayne spleenwort

Asplenium rhizophyllum Linnaeus var. *biscayneanum* D. C. Eaton, Bull. Torrey Bot. Club 14: 97, plate 168. 1887

Roots proliferous. **Stems** erect, not branched; scales blackish throughout, linear-deltate, sparse, 1–1.3 × 0.1–2.4 mm, margins entire. **Leaves** nearly monomorphic. **Petiole** green, becoming blackish in older leaves, 1–5(–12) cm, 1/8–1/3 length of blade; indument of black, narrowly lanceolate scales. **Blade** dull, linear, 2-pinnate, (4–)12–22 × 1–3.5 cm, papery, glabrous; base slightly tapered; apex narrowing gradually, not rooting. **Rachis** mostly blackish, green distally and in juvenile leaves, shiny, sparsely scaly. **Pinnae** in 9–20 pairs, oblong, 0.5–2 × 0.4–1.3 cm; apex blunt to truncate. **Pinnules** of 1–2 segments; segments linear-oblong, 3–6 × 1–3 mm; margins dentate; apex notched, pointed, rounded, or blunt. **Veins** free, evident. **Sori** 1–2 per segment, on both basiscopic and acroscopic sides. **Spores** abortive. $2n = $ ca. 180.

Hammocks, on limestone rock faces; 0–20 m; Fla.

Asplenium ×*biscayneanum* is the hybrid of *A. trichomanes-dentatum* and *A. verecundum*, with which it occurs. This peculiar spleenwort may most readily be separated from *A. trichomanes-dentatum* by its deeply cut pinnae, and from *A. verecundum* by being only 2-pinnate and having long petioles. Chromosome pairing in *A.* ×*biscayneanum* is irregular. Judging from herbarium collections, it shows considerable hybrid vigor.

All the collections are from Dade County, Florida, where it may now be extirpated.

24. Asplenium bradleyi D. C. Eaton, Bull. Torrey Bot.
Club 4: 11. 1873 · Bradley's spleenwort

Asplenium stotleri Wherry

Roots not proliferous. **Stems** short-creeping to ascending, occasionally branched; scales dark reddish to brown throughout, narrowly deltate, (2–)3–5 × 0.2–0.4 mm, margins entire or shallowly dentate. **Leaves** monomorphic. **Petiole** reddish or purplish brown throughout, lustrous, 1–10(–13) cm, 1/3–3/4 length of blade; indument of brown, narrowly lanceolate scales at base, grading into hairs. **Blade** narrowly oblong to lanceolate, pinnate-pinnatifid to 2-pinnate, 2–17(–20) × 1–6 cm, thin to moderately thick, sparsely pubescent; base truncate or obtuse; apex acute, not rooting. **Rachis** reddish or purplish brown proximally, fading to green in distal 1/3–2/3, lustrous, sparsely pubescent. **Pinnae** in (3–)5–15(–25) pairs, ovate, obovate to lanceolate or deltate-lanceolate; medial pinnae 6–40 × 3–10 mm; base truncate to obliquely obtuse; margins dentate to denticulate; apex acute or rounded. **Veins** free, barely evident. **Sori** 3 to numerous pairs per pinna, on both basiscopic and acroscopic sides. **Spores** 64 per sporangium. $2n = 144$.

Acidic rocks, usually on steep ledges; 0–1000 m; Ala., Ark., Ga., Ill., Ky., Md., Mo., N.J., N.Y., N.C., Ohio, Okla., Pa., S.C., Tenn., Va., W.Va.

Asplenium bradleyi is a morphologically variable species, the allotetraploid derivative of *A. montanum* × *platyneuron* (W. H. Wagner Jr. 1954; D. M. Smith and D. A. Levin 1963; C. R. Werth et al. 1985). The sterile diploid form of *A. bradleyi* has been collected twice in nature (W. H. Wagner Jr. et al. 1973; A. M. Evans 1988), and isozyme studies indicate that the allotetraploid has had a polytopic origin (C. R. Werth et al. 1985b). Occurring rarely to locally in the Appalachian region, *A. bradleyi* overlaps with both progenitor taxa, but it is fairly frequent in the Ozark and Ouachita region where *A. montanum* is absent. Sterile hybrids with *A. pinnatifidum* (*A.* ×*gravesii*), *A. montanum* (*A.* ×*wherryi* D. M. Smith et al.), and *A. platyneuron* are known from nature.

25. Asplenium cristatum Lamarck in Lamarck et al., Encycl. 2: 310. 1786 · Hemlock spleenwort

Roots proliferous. **Stems** erect, not branched; scales blackish throughout or rarely with narrow pale margins, narrowly linear, mostly only 8–12 cells wide proximally, 2–3 × 0.5–0.8 mm. **Leaves** monomorphic. **Petiole** blackish to greenish, dull, 1–7 (–17) cm, 1/5–1/2 length of blade; glabrous with few scales at base. **Blade** oblong-lanceolate, 2–3-pinnate, (2–)6–12(–18) × 2.7–10 cm, thin, glabrous; basal pinna pairs unreduced to reduced 1/4; apex gradually reduced, not rooting. **Rachis** blackish, dull, glabrous. **Pinnae** in (7–)10–15(–20) pairs, oblong, 2-pinnate throughout; medial pinnae (1–)2–3(–6) cm, all ± curved upward; proximal pinnae descending; base cuneate; apex attenuate. **Pinnules** ± asymmetric, 1–5(–10) × 1–4 mm; margins strongly dentate, cut 1/3–1/2 to axis. **Veins** free, evident. **Sori** 1–4(–6) per pinnule, usually on both basiscopic and acroscopic sides. **Spores** 64 per sporangium. $2n = 72$.

Low limestone boulders and ledges in deep moist woods; 0–20 m; Fla.; Mexico; West Indies; Central America; South America.

A widespread tropical American species, *Asplenium cristatum* is local in west central Florida.

26. Asplenium ×curtissii L. Underwood, Bull. Torrey Bot. Club 33: 194. 1906 · Curtiss's spleenwort

Roots proliferous. **Stems** erect, unbranched; scales blackish throughout, narrowly deltate, 1 × 0.2 mm, margins entire. **Leaves** monomorphic. **Petiole** brownish black, 3–10(–15) cm, 1/3–2/5 length of blade; indument of blackish, narrowly lanceolate scales at base. **Blade** oblong-lanceolate, 2-pinnate, 10–30 × (1.5–)5–10 cm, thin, glabrous; base not or only slightly tapered; apex gradually narrowing, not rooting. **Rachis** blackish to green, dull, nearly glabrous. **Pinnae** in (7–)14–22 pairs, oblong; medial pinnae 1–6 × 0.5–1.5 cm; base truncate; apex pointed. **Pinnules** linear to fan-shaped to unequally pinnate, 3–9 × 1–7 mm, mostly notched apically. **Veins** free, evident. **Sori** 1–4 per segment, usually more on acroscopic side. **Spores** abortive. $2n = 108$.

Shaded damp limestone rocks; 0–50 m; Fla.

Asplenium ×curtissii, sterile and with irregular meiosis, is the product of hybridization between *A.*

abscissum and *A. verecundum* and occurs with them in central Florida, sometimes forming large colonies by root proliferation. It can readily be separated from *A. abscissum* by its pinnate blades. From *A. verecundum* it can be distinguished by its relatively long petioles and less divided blades. *Asplenium ×curtissi* is known only from several localities in north central Florida. It makes a showy conservatory plant.

27. Asplenium exiguum Beddome, Ferns S. India, plate 146. 1864 · Little spleenwort

Roots not proliferous. **Stems** erect or ascending, unbranched; scales black throughout, narrowly deltate, 2–3 × 0.2–0.3 mm, margins mostly with widely spaced, shallow teeth. **Leaves** monomorphic. **Petiole** dark reddish brown throughout, dull, 1–3 cm, 1/10–1/6 length of blade; indument of black filiform scales. **Blade** lanceolate, 2-pinnate to 2-pinnate-pinnatifid, 4–10 × 1–3 cm, thin, sparsely pubescent; base tapering; apex acute to acuminate, often bearing minute, scaly, proliferous bud. **Rachis** basally reddish brown, fading to green in distal 1/2 to 3/4, dull, sparsely pubescent and with a few filiform scales. **Pinnae** 10–20 pairs, narrow, oblong; medial pinnae 5–12 × 4–7 mm; base acute to obtuse; margins coarsely incised; apex notched, bearing proliferous bud. **Veins** free, obscure. **Sori** 1–4 pairs per pinna, on both basiscopic and acroscopic sides. **Spores** 64 per sporangium. $2n = 72$.

Cliffs; 1200 m; Ariz.; Mexico; Asia in the Himalayas; Pacific Islands in the Philippines.

Asplenium exiguum has an interesting disjunct distribution, its range barely extending into the United States. Its vegetative propagation by buds scattered on the blade was reported by J. T. Mickel (1976).

28. Asplenium montanum Willdenow, Sp. Pl. 5(1): 342. 1810 · Mountain spleenwort

Roots proliferous. **Stems** horizontal, often arching upward, unbranched (although clusters of stems often form from root proliferations, giving false appearance of single much-branched stem); scales dark brown throughout, narrowly deltate, 2–4 × 0.2–0.4 mm, margins entire. **Leaves** monomorphic. **Petiole** dark brown to purplish black, lustrous proximally, fading to green distally, 2–11 cm, 1/2–1 1/2 length of blade; indument of

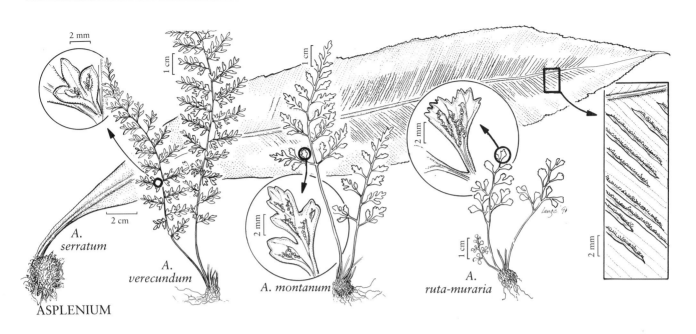

blackish, narrowly lanceolate scales only at very base and of minute hairs. **Blade** deltate to lanceolate, 1–2-pinnate-pinnatifid, 2–11 × 1–7(–10) cm, thick, essentially glabrous; base truncate or obtuse; apex acuminate to acute, not rooting. **Rachis** green throughout, dull, sparsely pubescent. **Pinnae** in 4–10 pairs, deltate to lanceolate; proximal (longest) pinnae 6–35 × 4–20 mm; base obtuse to acute; margins coarsely incised; apex acute to rounded. **Veins** free, obscure. **Sori** 1–15 per pinna, on both basiscopic and acroscopic sides. **Spores** 64 per sporangium. $2n = 72$.

Crevices in sandstone or other acidic rocks; 0–2000 m; Ala., Conn., Ga., Ind., Ky., Md., Mass., Mo., N.J., N.Y., N.C., Ohio, Pa., R.I., S.C., Tenn., Va., W.Va.

Asplenium montanum occurs principally in the Appalachian region, with outlying localities in the Shawnee Hills of western Kentucky (R. Cranfill 1980) and adjacent Indiana (D. M. Smith 1956). A report of its disjunct occurrence on the northern edge of the Ozarks is based on a single specimen whose label indicates the collection locality near Graham Cave, Montgomery County, Missouri. Efforts by several botanists to relocate the population have failed. Reports of a disjunct station in the upper peninsula of Michigan are doubtful.

Asplenium montanum is an ecological specialist. It is typically the sole vascular plant species in the siliceous rock crevices in which it is found. It may occur, however, with two allotetraploid species, *A. bradleyi* and *A. pinnatifidum*, which were derived from hybrids of *A. montanum* with *A. platyneuron* and *A. rhizophyllum*, respectively. In addition, *A. montanum* crosses frequently with *A. pinnatifidum* producing *A.* ×*trudellii* and rarely with allotetraploid individuals of *A. bradleyi* producing *A.* ×*wherryi*.

29. **Asplenium plenum** E. P. St. John ex Small, Ferns S.E. States, 173. 1938 · Ruffled spleenwort

Roots proliferous. **Stems** erect, unbranched; scales blackish throughout, narrowly deltate, 0.4–1.1 × 0.3–0.7 mm, margins entire to denticulate. **Leaves** monomorphic. **Petiole** blackish throughout, dull, 2–6(–10) cm, 1/4–2/5 length of blade; indument absent. **Blade** lanceolate, 1–2-pinnate, (4–)8–12(–15) × 1.5–5 cm, thin, glabrous; base not tapered; apex gradually narrowing. **Rachis** mostly green except occasionally blackish at base, dull, glabrous. **Pinnae** in (5–)10–20(–25) pairs, oblong-deltate, 1–3.5 × 0.5–1.8 cm; base excavate on basiscopic side; apex pointed. **Pinnules** linear to oblong, 4–10 mm; apex mostly notched. **Veins** free, not conspicuous. **Sori** mostly 1 per segment, 1–3 mm. **Spores** mostly abortive, some viable. $2n = 144$.

Limestone rocks in shaded forests; 0–50 m; Fla.

Asplenium plenum occurs with its parents on limestone rocks in shaded forests and is known only from Florida, although it could occur in the Antilles, Central

America, and South America (D. B. Lellinger 1981). It is noteworthy for constituting one of the first known examples of backcrossing and formation of a new taxon by unreduced spores from a sterile hybrid. According to V. M. Morzenti (1967) and G. J. Gastony (1986), hybridization between *A. abscissum* and *A. verecundum* produced *A.* ×*curtissii*. An unreduced spore of the hybrid gave rise to a 3*x* gametophyte. This gametophyte produced a 3*x* sperm that backcrossed with an *x* egg of *A. abscissum* producing the 4*x* allotetraploids *A. plenum*, that is not only capable of propagation by minute root proliferations like those of the parents but also to some extent by spores. This complex hypothesis was confirmed by electrophoretic comparisons of the plants involved (G. J. Gastony 1986).

30. **Asplenium ruta-muraria** Linnaeus, Sp. Pl. 2: 1081.
1753 · Wall rue, doradille rue-des-murailles

Asplenium cryptolepis Fernald; *A. cryptolepis* Fernald var. *ohionis* Fernald; *A. ruta-muraria* var. *cryptolepis* (Fernald) Wherry

Roots not proliferous. **Stems** short-creeping to erect, often branched; scales very dark brown throughout, narrowly deltate, 1–3 × 0.1–0.25 mm, margins with widely spaced teeth. **Leaves** monomorphic. **Petiole** reddish brown proximally, green distally, dull, 1–9 cm, (1/2–)1–2 times length of blade; indument of dark brown, narrowly deltate scales proximally grading into multicellular hairs. **Blade** deltate-ovate to obovate or oblanceolate, 1–2(–3)-pinnate to 2-pinnate-pinnatifid, 2–6 × 1–4 cm, somewhat thick, glabrous; base obtuse; apex acute to rounded, not rooting. **Rachis** green, dull, glabrous except for very sparse, minute hairs. **Pinnae** in 2–4 pairs, deltate-ovate to obdeltate; proximal (largest) pinnae 7–30 × 5–20 mm; base truncate to acute; margins finely (sometimes coarsely) incised; apex rounded to acute. **Veins** free, evident. **Sori** as many as 30 or more per pinna, usually 1–5 per segment, on both basiscopic and acroscopic sides. **Spores** 64 per sporangium. 2*n* = 144.

Limestone (or calcareous shale) cliffs and boulders, rarely invading masonry; 0–1000 m; Ont., Que.; Ala., Ark., Conn., Ind., Ky., Md., Mass., Mich., Mo., N.H., N.J., N.Y., N.C., Ohio, Pa., R.I., Tenn., Vt., Va., W.Va.; Europe; e Asia.

The relationship of North American *Asplenium ruta-muraria* to its European counterparts is incompletely understood and bears further investigation. Based on features of the stems, M. L. Fernald (1928) segregated

the North American taxon as *A. cryptolepis*, but most current authors agree that morphologic differentiation of North American and European material is too slight and inconsistent for recognition at the specific level. In Europe, two ploidy levels are treated as subspecies, diploid *A. ruta-muraria* subsp. *dolomiticum* Lovis & Reichstein and tetraploid *A. ruta-muraria* subsp. *ruta-muraria*, the latter representing the most compelling case for true autopolyploidy (i.e., based on chromosomal homology) known in ferns (G. Vida 1970). Chromosome counts of North American plants are consistently tetraploid; whether or not these plants are referable to subsp. *ruta-muraria* will remain unclear until additional evidence (e.g., isozymes) is obtained. Meanwhile, North American material should be designated simply as *A. ruta-muraria*, the convention used in most current manuals.

Although M. L. Fernald (1928) recognized *Asplenium cryptolepis* var. *ohionis* (= *A. ruta-muraria* var. *subtenuifolium* Christ), based on its acute rather than rounded segment apices, leaves assignable to this variety may occur on plants also bearing leaves more similar to those of the type variety (R. Cranfill 1980). The former is not recognized taxonomically here.

Numerous hybrids of *Asplenium ruta-muraria* with various taxa are known from Europe (T. Reichstein 1981), but only three are known from North America, all exceedingly rare. These are the hybrids with *A. rhizophyllum* (*A.* ×*inexpectatum*), with *A. trichomanes* (*A.* ×*clermontae* Syme), and with *A. platyneuron* (*A.* ×*morganii*).

31. **Asplenium verecundum** Chapman ex L. Underwood, Bull. Torrey Bot. Club 33: 193. 1906

Asplenium scalifolium E. P. St. John; *A. suare* E. P. St. John; *A. subtile* E. P. St. John.

Roots proliferous. **Stems** erect, unbranched; scales blackish throughout, linear to narrowly triangular, 0.9–1.3 × 0.1–0.4 mm, margins entire. **Leaves** monomorphic. **Petiole** brownish black, 0.5–3(–5) cm, 1/9–1/7 length of blade; indument of blackish, linear-lanceolate scales at base. **Blade** narrowly lanceolate, 2–3-pinnate, (4–)10–20(–)30 × 1–3(–5) cm, thin, glabrous; base tapered somewhat; apex formed by gradual reduction, not rooting. **Rachis** blackish brown, dull, essentially glabrous. **Pinnae** in (8–)12–16(–22) pairs, oblong; medial pinnae 0.5–2.5 × 3–10 mm; base cuneate; apex obtuse. **Pinnules** of 1–5 segments; segments oblanceolate, 2–3 × 1–2 mm, entire; apex mostly round. **Veins** free, evident, 1 per segment. **Sori** 1 per segment. **Spores** 64 per sporangium. $2n = 144$.

Limestone outcrops in grottoes, on cliffs, and on boulders in shaded woods; 0–50 m; Fla.

This delicate spleenwort, *Asplenium verecundum,* occurs in both southern and central peninsular Florida where it is very local. Sterile hybrids are known to result from crossing with *A. trichomanes-dentatum (A. × biscayneanum)* and *A. abscissum (A. × curtissii).*

Asplenium verecundum may be a variety or cytotype of the similar *A. myriophyllum* (Swartz) C. Presl of the West Indies (D. B. Lellinger 1985).

20. DRYOPTERIDACEAE Herter

· Wood Fern Family

Alan R. Smith

Plants perennial, terrestrial or on rock, occasionally hemiepiphytic or epiphytic. **Stems** creeping to erect, rarely arborescent, sometimes climbing, branched or unbranched, dictyostelic, bearing scales. **Leaves** circinate in bud, monomorphic or dimorphic. **Petiole** usually not articulate to stem, scales usually persistent at base, in cross section with 2–many roundish bundles, or bundles 2 and lunate. **Blade** simple to commonly 1–5-pinnate or more divided, leaf buds absent or present. **Veins** pinnate or parallel in ultimate segments, simple or forked, free or anastomosing, areoles sometimes with included free veinlets. **Indument** on blade commonly of glands, hairs, and/or scales, especially on rachis and costae abaxially. **Sori** borne abaxially on veins or at vein tips (but usually not marginal), or sporangia acrostichoid and covering abaxial surface, if in discrete sori then variously shaped (round, oblong, or elongate); receptacle not or only slightly elevated, with or without indusium, indusium variously linear, falcate, or reniform, sometimes hoodlike, cuplike, or round. **Sporangia** with stalk of 2–3 rows of cells; annulus vertical, interrupted by stalk. **Spores** all of 1 kind, usually not green (except *Matteuccia, Onoclea*), oblong or reniform in outline, monolete, variously ornamented (often broadly winged), 64 per sporangium (32 in apogamous spp.). **Gametophytes** green, aboveground, cordate, glabrous or often bearing glands or hairs; archegonia and antheridia borne on lower surface, antheridia 3-celled.

Genera ca. 60, species perhaps exceeding 3000 (18 genera, 79 species in the flora): worldwide.

The family Dryopteridaceae has been variously circumscribed; it is here delimited in a manner similar to that of R. M. Tryon and A. F. Tryon (1982) but with the inclusion of *Nephrolepis*. In many works, the family has gone under the illegitimate name Aspidiaceae. Some authorities define Dryopteridaceae more narrowly, to exclude *Athyrium, Deparia, Diplazium, Cystopteris,* and *Gymnocarpium* (Athyriaceae or Woodsiaceae), *Woodsia* (Woodsiaceae), *Lomariopsis* (Lomariopsidaceae), *Nephrolepis* (Nephrolepidaceae or Davalliaceae), *Onoclea* and *Matteuccia* (Onocleaceae), and *Ctenitis* and *Tectaria* (Tectariaceae). Characteristics holding Dryopteridaceae (as circumscribed here) together include the bilateral, monolete spores, often broadly winged perispore, absence of needlelike hairs, scaly stem and petiole bases, abaxial (nonmarginal) sori, base chromosome number of 40 or 41 (also 38 and 39 in *Woodsia,* 37 in *Onoclea,* 42 in *Cystopteris*), and usually indusiate sori. Loss of indusium, dimorphism, areolate venation, and reduced blade dissection have occurred repeatedly along many evolutionary lines in Dryopteridaceae, and in general these characteristics are often not very useful in delimiting genera or assessing intergeneric relationships.

In some genera, especially *Phanerophlebia* and *Polystichum,* the blade bears very narrow scales (sometimes called microscales) that resemble uniseriate hairs. These scales may be only one or two cells wide. Every intergradation exists between these filiform microscales and more typical, wider scales, and the two types are the same color, generally tan to brownish. Microscales are probably not homologous with true hairs, which may be either unicellular or multicellular, uncolored or sometimes reddish (as in *Tectaria* and *Ctenitis*), glandular (as in *Woodsia*) or not. Hairs in Dryopteridaceae, if present at all, are generally readily distinguishable from the needlelike, transparent ones found in Thelypteridaceae.

1. Leaves strongly dimorphic, either fertile or sterile, the 2 types very dissimilar.
   2. Sporangia completely covering abaxial surface of blade, not in discrete sori or hidden by revolute segment margins; sterile blades 1-pinnate, pinnae serrate. . . . . . . 17. *Lomariopsis,* p. 304
   2. Sporangia not covering abaxial surface of blade, in discrete sori or hidden by revolute segment margins; sterile blades variously divided, not 1-pinnate with pinnae serrate.
      3. Plants hemiepiphytic, rooted in ground and with stems climbing trees; sterile blades 3–4-pinnate; indusia thick, conspicuous. . . . . 11. *Maxonia,* p. 289
      3. Plants terrestrial, not climbing; sterile blades deeply pinnatifid to 1-pinnate-pinnatifid; indusia thin, fragile, hidden by revolute segment margins.
         4. Sterile blades pinnatifid to 1-pinnate at base; venation areolate; fertile blades 2-pinnate, sori enclosed in small, globose, hardened pinnules. . . . . 2. *Onoclea,* p. 251
         4. Sterile blades 1-pinnate-pinnatifid; veins free; fertile blades 1-pinnate, sori on linear pinnae and enclosed by hardened pinna margin. . . . . 1. *Matteuccia,* p. 249
1. Leaves monomorphic, fertile and sterile similar in size and dissection, occasionally with somewhat contracted fertile pinnae on same leaf as sterile pinnae (as in *Polystichum acrostichoides*).
   5. Stolons present, wiry, arising from stem; blades 1-pinnate; pinnae articulate to rachis, sometimes deciduous with age; indusia lunate to reniform or circular with narrow sinus. . . . . 18. *Nephrolepis,* p. 305
   5. Stolons absent; blades variously divided; pinnae not articulate to rachis, or rarely proximal pinnae weakly articulate but not deciduous; indusia various or absent.
      6. Indusia completely surrounding receptacle and composed of filaments or scalelike segments arranged in cuplike fashion around sorus; petiole base with 2 vascular bundles; scales absent on costae abaxially. . . . . 8. *Woodsia,* p. 270
      6. Indusia attached centrally or laterally, not completely surrounding receptacle, or indusia absent; petiole base with 2 or more vascular bundles; scales absent or present on costae abaxially.
         7. Veins areolate or copiously anastomosing.
            8. Sori and indusia linear; petiole base with 2 vascular bundles. . . . . . . 3. *Diplazium,* p. 252
            8. Sori round, with indusia round or round-reniform; petiole base with many vascular bundles.
               9. Costae adaxially rounded or flattened, bearing multicellular reddish hairs; margins of pinnae lacking spinules or teeth. . . . . 16. *Tectaria,* p. 302
               9. Costae adaxially grooved, lacking reddish multicellular hairs but sometimes with tan, very reduced, filiform scales; margins of pinnae spinulose to denticulate or crenate. . . . . 13. *Cyrtomium,* p. 299

7. Veins free or only casually and sparingly anastomosing.
  10. Indusia round, attached at center (peltate); sori round; petiole base with 3 or more vascular bundles.
    11. Blades 1-pinnate with terminal pinna similar to lateral pinnae; sori in 2–4 rows between costa and margin; s Arizona to w Texas. . . 14. *Phanerophlebia*, p. 300
    11. Blades 1–3-pinnate with gradually reduced and pinnatifid apex; sori in 1(–2) rows between costa and margin or between costule and margin; widespread. . . . . 12. *Polystichum*, p. 290
  10. Indusia round-reniform, reniform, linear, or absent, attached laterally at sinus; sori round or elongate; petiole base with 2 or more vascular bundles.
    12. Sori elongate, straight or hooked at one end, indusiate; petiole base with 2 vascular bundles.
      13. Adaxial grooves of costae shallow, not decurrent into rachis groove; multicellular hairs borne along costae, especially adaxially; stems moderately long-creeping; blades 1-pinnate-pinnatifid. . . . . . . 4. *Deparia*, p. 254
      13. Adaxial grooves of costae deep, decurrent into rachis groove; multicellular hairs absent on costae; stems short-creeping to erect; blades 1-pinnate to 2-pinnate-pinnatifid.
        14. Blades commonly 2-pinnate or more divided; proximal pinnae often slightly to greatly reduced; sori usually hooked at distal end. . . . . 5. *Athyrium*, p. 255
        14. Blades 1-pinnate, 1-pinnate-pinnatifid, or 2-pinnate (if 2-pinnate then veins anastomosing); proximal pinnae not or slightly reduced; sori ± straight, not hooked at distal end. . . . . . 3. *Diplazium*, p. 252
    12. Sori round or nearly so, indusia present or absent; petiole with 2 or more vascular bundles.
      15. Costae rounded or flat adaxially, bearing dense, obviously multicellular hairs with reddish crosswalls. . . . . .15. *Ctenitis*, p. 301
      15. Costae grooved adaxially, lacking hairs.
        16. Indusia attached at distinct sinus, round-reniform; petiole base with 3 or more vascular bundles.
          17. Stems short-creeping, nearly erect or erect; blades lanceolate to ovate, not pentagonal; widespread. . . . . . . .9. *Dryopteris*, p. 280
          17. Stems moderately long-creeping; blades pentagonal, with basal basiscopic pinnules decidedly longer than next pair; South Carolina. . . . . 10. *Arachniodes*, p. 288
        16. Indusia absent or laterally attached and hoodlike, arching over sori; petiole base with 2 vascular bundles.
          18. Stems long-creeping; blades deltate to pentagonal, proximal pinnae by far the largest; petioles mostly 1/3–3 times length of blades.
            19. Indusia present but often inconspicuous in mature leaves, laterally attached and arching over sori; segment margins serrate-dentate (*Cystopteris montana*). . . . . 7. *Cystopteris*, p. 263
            19. Indusia always absent; segment margins entire or crenate but never serrate-dentate. . . . . . . 6. *Gymnocarpium*, p. 258
          18. Stems short-creeping to ascending; blades ovate to lanceolate, proximal pinnae occasionally slightly longer than adjacent pair; petioles mostly shorter than blades.
            20. Indusia absent from all sori; stems erect to ascending, surface obscured by petiole bases and scales, scales more than 5 mm. . . . . 5. *Athyrium*, p. 255

20. Indusia present but often inconspicuous in mature leaves, laterally attached and arching over sori; stems decumbent, usually creeping, surface often visible through petiole bases and scales, scales less than 5 mm. . . . . 7. *Cystopteris*, p. 263

1. MATTEUCCIA Todaro, Giorn. Sci. Nat. Econ. Palermo 1: 235. 1866, name conserved · Ostrich fern [for Carlo Matteucci, 1800–1863, physicist at the University of Florence, Italy]

David M. Johnson

Plants terrestrial. **Stems** obliquely ascending to erect, stolons present or absent, not wiry. **Leaves** strongly dimorphic, sterile leaves with longer and wider pinnae than fertile ones, sterile dying back in winter, fertile persistent, brownish, hardened. **Petiole** of sterile leaf ca. 1/10–1/5 length of blade, petiole of fertile leaf ± equaling length of blade, bases swollen, persisting as trophopods over winter; vascular bundles 2, lateral, lunate in cross section. **Blade** elliptic (sterile) or narrowly oblong to oblanceolate (fertile), pinnate-pinnatifid or sometimes pinnate with lacerate pinnae in fertile leaves, gradually to nearly abruptly reduced distally to pinnatifid apex, sterile blades herbaceous. **Pinnae** not articulate to rachis, segment margins entire (sterile); proximal pinnae (several pairs) greatly reduced, sessile, equilateral; costae shallowly grooved adaxially, grooves not continuous from rachis to costae; indument absent on both surfaces or deciduously hairy abaxially. **Veins** free, simple. **Sori** in 1 row between midrib and margin, ± round, covered by hardened revolute pinna margin; indusia vestigial, triangular, persistent but not easily seen in mature leaves. **Spores** greenish, with low folds and minute echinate-cristate elements. $x = 40$.

Species 3 (1 in the flora): north temperate regions.

Matteuccia is one of several genera known to store starch grains in long-persistent petiole bases (trophopods) (W. H. Wagner Jr. and D. M. Johnson 1983). *Struthiopteris* Willdenow is a later homonym of *Struthiopteris* Scopoli and hence illegitimate. *Pteretis* Rafinesque, even though it is the oldest legitimate name, has been rejected in favor of *Matteuccia*.

SELECTED REFERENCES Fernald, M. L. 1935. Critical plants of the upper Great Lakes region of Ontario and Michigan. Rhodora 37: 197–222. Hill, R. W. and W. H. Wagner Jr. 1974. Seasonality and spore type of pteridophytes in Michigan. Michigan Bot. 13: 40–44. Lloyd, R. M. 1971. Systematics of the onocleoid ferns. Univ. Calif. Publ. Bot. 61: 1–86, 5 plates. Lloyd, R. M. and E. J. Klekowski Jr. 1970. Spore germination and viability in Pteridophyta: Evolutionary significance of chlorophyllous spores. Biotropica 2: 129–137. Von Aderkas, P. 1984. Economic history of ostrich fern, *Matteuccia struthiopteris*, the edible fiddlehead. Econ. Bot. 38: 14–23. Wagner, W. H. Jr. and D. M. Johnson. 1983. Trophopod, a commonly overlooked storage structure of potential systematic value in ferns. Taxon 32: 268–269.

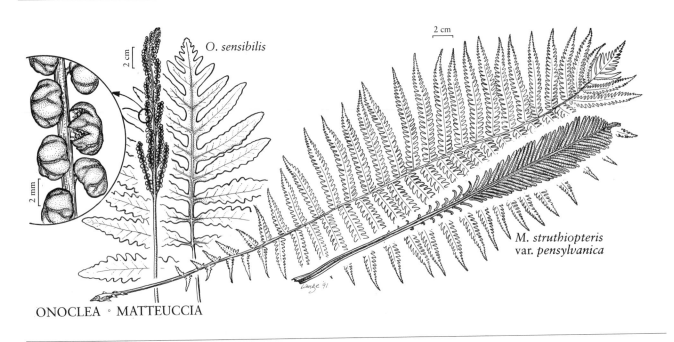

O. sensibilis

M. struthiopteris
var. *pensylvanica*

ONOCLEA · MATTEUCCIA

1. Matteuccia struthiopteris (Linnaeus) Todaro, Giorn. Sci. Nat. Econ. Palermo 1: 235. 1866

Osmunda struthiopteris Linnaeus, Sp. Pl. 2: 1066. 1753

Varieties 2 (1 in the flora): North America, Eurasia.

1a. Matteuccia struthiopteris (Linnaeus) Todaro var. **pensylvanica** (Willdenow) C. V. Morton, Amer. Fern J. 40: 247. 1950 · Ostrich fern, fougère-à-l'autriche

Struthiopteris pensylvanica Willdenow, Sp. Pl. 5(1): 289. 1810; *Matteuccia pensylvanica* (Willdenow) Raymond; *M. struthiopteris* var. *pubescens* (Terry) Clute; *Pteretis pensylvanica* (Willdenow) Fernald; *P. struthiopteris* (Linnaeus) Nieuwland

Leaves in vaselike cluster to 1.75 m. **Sterile leaves** oblanceolate, 30–130 × 12–25 cm. **Petiole** of sterile leaf black, 4.5–46 cm, flattened at base, becoming deeply grooved distally, scales pale orange-brown; rachis grooved; petiole and rachis occasionally puberulent. **Pinnae** linear, 20–60 per side, longest 6.5–13.5 cm, gradually decreasing in length toward base; segments 20–40 pairs per pinna. **Petiole** of sporophyll 11–24 cm, base scaly. **Blade** oblong to oblanceolate, 15–40 × 2.5–6.5 cm. **Pinnae** greenish, becoming dark brown at maturity, linear, 30–45 per side, 3–5.6 cm, constricted at regular intervals. $2n = 80$.

Sporophylls produced in mid to late summer, persist-ing through winter. Rich woods, often in alluvial or mucky swamp soils; 0–1500 m; Alta., B.C., Man., N.B., Nfld., N.W.T., N.S., Ont., P.E.I., Que., Sask., Yukon; Alaska, Conn., Del., Ill., Ind., Iowa, Maine, Md., Mass., Mich., Minn., Mo., N.H., N.J., N.Y., N.Dak., Ohio, Pa., R.I., S.Dak., Vt., Va., W.Va., Wis.

The name *Pteretis nodulosa* (Michaux) Nieuwland has been misapplied to this species.

Matteuccia struthiopteris is most common in north-eastern North America, primarily north of the limit of Wisconsin glaciation. The sporangia dehisce in the spring before the new sterile leaves have expanded, thus re-leasing the spores into an unimpeded airstream (R. W. Hill and W. H. Wagner Jr. 1974). The green spores germinate in two to five days (R. M. Lloyd and E. J. Klekowski Jr. 1970).

Matteuccia struthiopteris var. *struthiopteris*, which differs in its bicolored petiole scales and more truncate pinna lobes, occurs in temperate Eurasia. As in *Ono-clea sensibilis*, leaf forms intermediate between sterile leaves and sporophylls are sometimes found (M. L. Fernald 1935).

Matteuccia struthiopteris has been used as a land-scaping plant in the United States and Canada, where it is frequently planted as a border along house foun-dations. It is also the source of edible fiddleheads, the canning of which is a local industry in New England and adjacent Canada. The fiddlehead of *M. struthio-pteris* is the state vegetable of Vermont.

2. ONOCLEA Linnaeus, Sp. Pl. 2: 1062. 1753; Gen. Pl. ed. 5, 484. 1754 · Sensitive fern [Greek *onos*, vessel, and *kleiein*, to close, in reference to the sori, which are enclosed by the revolute fertile leaf margins]

David M. Johnson

Plants terrestrial. **Stems** creeping, stolons absent. **Leaves** strongly dimorphic, fertile leaves usually shorter, greatly contracted, persistent 2–3 years, sterile leaves dying back in winter. **Petiole** of sterile leaf ca. 1–1.5 times length of blade, petiole of fertile leaf 2–6 times length of blade, bases swollen and persisting as trophopods over winter; vascular bundles 2, lateral, lunate in cross section. **Blade** of sterile leaf deltate, pinnatifid to pinnate-pinnatifid proximally, reduced and shallowly pinnatifid distally, herbaceous to papery, blade of fertile leaf linear-oblong, 2-pinnate, leathery. **Pinnae** not articulate to rachis, segment margins of sterile blades entire to sinuate or shallowly lobed, margins of fertile pinnules strongly revolute and forming hardened beadlike structures; proximal pinnae largest or nearly so, sessile or adnate, equilateral; costae adaxially flat; indument on both sides of linear to lanceolate scales and/or multicellular hairs on rachis and costae. **Veins** reticulate with areoles lacking included veinlets in sterile leaves, veins free in fertile leaves. **Sori** covered by strongly revolute margins of pinnae, ± round; indusia vestigial, triangular, persistent but not easily seen in mature leaves. **Spores** greenish, with a few low folds and numerous, minute, echinate-cristate elements. $x = 37$.

Species 1 (1 in the flora): temperate regions in Northern Hemisphere, Asia.

Onoclea is one of several genera known to store starch grains in long-persistent petiole bases (trophopods) (W. H. Wagner Jr. and D. M. Johnson 1983).

1. Onoclea sensibilis Linnaeus, Sp. Pl. 2: 1062. 1753
· Sensitive fern, onoclée sensible

Onoclea sensibilis forma *hemiphyllodes* (Kiss & Kümmerle) Gilbert; *O. sensibilis* forma *obtusilobata* (Schkuhr) Gilbert; *O. sensibilis* var. *obtusilobata* (Schkuhr) Torrey

Leaves irregularly spaced along stem. **Sterile leaves** yellow-green, deltate, coarsely divided, 13–34 × 15–30 cm. **Petiole** of sterile leaf black, 22–58 cm, flattened at base; rachis winged, becoming broader toward apex. **Pinnae** 5–11 per side, lanceolate; proximal pinnae 9–18 cm, margins entire, sinuate, or laciniate. **Sporophyll leaves** green, becoming black at maturity, oblong, 7–17 × 1–4 cm. **Petiole** 19–40 cm, base sparsely scaly. **Pinnae** linear, 5–11 per side, 2.5–5 cm; ultimate segments revolute to form beadlike structures, 2–4 mm diam. **Sori** borne on free veins, enclosed by ultimate segments. $2n = 74$.

Sporophylls produced May–October. Open swamps, thickets, marshes, or low woods, in sunny or shaded locations, often forming thick stands; 0–1500 m; St. Pierre and Miquelon; Man., N.B., Nfld., N.S., Ont., P.E.I., Que.; Ala., Ark., Colo., Conn., Del., Fla., Ga., Ill., Ind., Iowa, Kans., Ky., La., Maine, Md., Mass., Mich., Minn., Miss., Mo., Nebr., N.H., N.J., N.Y., N.C., N.Dak., Ohio, Okla., Pa., R.I., S.C., S.Dak., Tenn., Tex., Vt., Va., W.Va., Wis., Wyo.; e Asia.

Onoclea sensibilis occurs in eastern North America, principally east of the Great Plains. Leaf forms with pinnae intermediate between those of sporophylls and sterile leaves, or with pinnae fertile only on one side of the blade, can occur on plants that also bear normal leaf forms. These do not merit taxonomic recognition (J. M. Beitel et al. 1981).

Onoclea sensibilis resembles *Woodwardia areolata* (Linnaeus) T. Moore, with which it often grows. *Onoclea* has entire pinna margins and nearly opposite basal pinnae whereas *Woodwardia areolata* has serrate pinna margins and alternate pinnae.

As in *Matteuccia struthiopteris* (Linnaeus) Todaro, sporophylls of *Onoclea sensibilis* persist through the winter and release the green spores in spring before the sterile leaves expand (R. W. Hill and W. H. Wagner Jr. 1974; L. G. Labouriau 1958; R. M. Lloyd and E. J. Klekowski Jr. 1970). *Onoclea sensibilis* is occasionally cultivated; it has a tendency to spread rapidly and become weedy. The name "sensitive fern" refers to the susceptibility of the leaves to even a light frost.

3. DIPLAZIUM Swartz, J. Bot. (Schrader) 1800(2): 4, 61. 1801 · Twin-sorus fern

[Greek *diplazein*, double, or *di*, two, and *plasion*, oblong, referring to a double sorus]

Masahiro Kato

Homalosorus Small ex Pichi-Sermolli

Plants terrestrial or on rock. **Stems** creeping, ascending, or erect, stolons absent. **Leaves** monomorphic, evergreen or dying back in winter. **Petiole** ca. 1/2 to equaling length of blade, base swollen and persisting as trophopod over winter or not; vascular bundles 2, lateral, lunate in cross section. **Blade** oblong-lanceolate to deltate, 1-pinnate to 2-pinnate-pinnatifid [simple to 4-pinnate-pinnatifid], gradually reduced distally to pinnatifid apex or apical pinna similar to (conform) adjacent pinnae, herbaceous to papery. **Pinnae** not articulate to rachis, segment margins entire, crenulate, or serrate; proximal pinnae not reduced, sessile, equilateral or inequilateral; costae adaxially deeply grooved, grooves continuous with that of rachis; indument abaxially absent or of linear to ovate scales, adaxially absent. **Veins** free, simple or forked, or basal pairs of adjacent segments anastomosing. **Sori** single or paired back-to-back on veins, oblong to linear, straight or slightly falcate; indusia linear, laterally attached, persistent. **Spores** brownish, usually broadly winged. $x = 40, 41$.

In a few species outside the flora, rachises and costae bear multicellular hairs like those of *Deparia*, which differs from *Diplazium* in having grooves of costae not decurrent onto rachis groove, veins free or anastomosing, sori long or short and costular, and indusia present or absent. Many species of *Diplazium* are known to reproduce apogamously.

Species about 400 (3 in the flora): worldwide.

SELECTED REFERENCE Johnson, D. M. 1986. Trophopods in North American species of *Athyrium* (Aspleniaceae). Syst. Bot. 11: 26–31. Kato, M. 1977. Classification of *Athyrium* and allied genera of Japan. Bot. Mag. (Tokyo) 90: 23–40.

1. Leaves 2-pinnate; veins anastomosing. . . . . 3. *Diplazium esculentum*
1. Leaves 1-pinnate or 1-pinnate-pinnatifid; veins free.
    2. Scales brown, entire; pinnae nearly entire. . . . . 1. *Diplazium pycnocarpon*
    2. Scales dark brown, dentate; pinnae lobed. . . . . 2. *Diplazium lonchophyllum*

1. Diplazium pycnocarpon (Sprengel) M. Broun, Index N. Amer. Ferns, 60. 1938 · Narrow-leaved glade fern, narrow-leaved-spleenwort, glade fern, diplazie à sores denses

Asplenium pycnocarpon Sprengel, Anleit. Kenntn. Gew. 3: 112. 1804; *Athyrium pycnocarpon* (Sprengel) Tidestrom; *Homalosorus pycnocarpos* (Sprengel) Pichi-Sermolli

Stems creeping; scales brown, broadly lanceolate, margins entire. **Petiole** 15–40(–50) cm. **Blade** oblong-lanceolate, 1-pinnate, 30–75 × 8–25 cm, ± narrowed to base with reduced proximal pinnae, broadest above base, abruptly acuminate to apex. Pinnae linear, ± entire to shallowly crenulate, base truncate or acroscopically auriculate, apex acuminate. **Veins** usually 1–2-forked, nearly reaching sinuses between crenations. **Sori** elongate, straight or slightly falcate, single or rarely double; indusia vaulted, ± thick. $2n = 80$.

Moist woods and slopes in neutral soil; 150–1000 m; Ont., Que.; Ala., Ark., Conn., Del., D.C., Fla., Ga., Ill., Ind., Iowa, Kans., Ky., La., Md., Mass., Mich., Minn., Miss., Mo., N.H., N.J., N.Y., N.C., Ohio, Pa., R.I., S.C., Tenn., Vt., Va., W.Va., Wis.

Diplazium pycnocarpon has commonly been placed in *Athyrium*, but it is closely related to the east Malesian *Diplazium flavoviride* Alston (M. Kato and D. Darnaedi 1988).

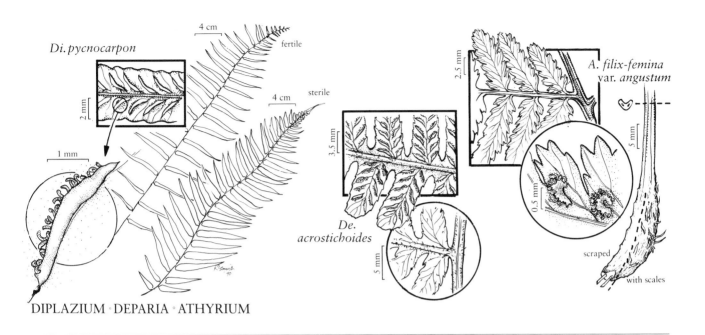

DIPLAZIUM · DEPARIA · ATHYRIUM

2. Diplazium lonchophyllum Kunze, Linnaea 13: 141. 1839 · Diplazie à feuilles allongées

Stems ascending to erect; scales dark brown, ovate to lanceolate, margins dentate. **Petiole** 15–45 cm. **Blade** deltate-lanceolate, pinnate-pinnatifid, 20–36 × 8–22 cm, broadest at or just above base, acuminate at apex. **Pinnae** lanceolate-oblong, inequilateral, base cuneate basiscopically, truncate acroscopically, apex acuminate, lobed halfway or more toward costa; basal acroscopic segments of basal pinnae free, margins serrate. **Veins** pinnate, lateral veins simple or sometimes forked. **Sori** elongate, straight, single or double, indusiate; indusia vaulted, thin, erose.

Moist, wooded slopes; lowland; very rare; 0–100 m; La.; Mexico; Central America; n South America.

Central and South American species closely related to *Diplazium lonchophyllum*, including *D. cristatum* (Desrousseaux) Alston, *D. drepanolobium* A. R. Smith, and *D. werckleanum* H. Christ, are in need of monographic work (R. G. Stolze 1981; A. R. Smith 1981; J. T. Mickel and J. M. Beitel 1988).

3. Diplazium esculentum (Retzius) Swartz, J. Bot. (Schrader) 1801(2): 312. 1803 · Vegetable fern

Hemionitis esculenta Retzius, Observ. Bot. 6: 38. 1791

Stems erect; scales brown, linear-lanceolate, margins dentate. **Petiole** 30–60 cm. **Blade** ovate, 2-pinnate to 2-pinnate-pinnatifid, 50–100 × 15–50 cm, base ± narrowed, apex abruptly acuminate. **Pinnae** 1-pinnate to 1-pinnate-pinnatifid. **Pinnules** oblong, base ± truncate, ± auriculate, apex acuminate, incised or lobed halfway to costule. **Veins** pinnate, anastomosing. **Sori** elongate, single or double, indusiate; indusia vaulted, thin, erose. $2n = 82$.

Moist soil near stream; 0 m; introduced; Fla., La.; se Asia; Africa.

Originally a tropical Eastern Hemisphere species, *Diplazium esculentum* is introduced in North America. This fern is used as a vegetable in eastern and southeastern Asia.

4. DEPARIA Hooker & Greville, Icon. Filic. 2(8). 1829–1830 · [Greek *depas*, saucer, referring to the saucerlike indusium of the type species, *Deparia prolifera,* which is aberrant in the genus]

Masahiro Kato

Plants terrestrial. **Stems** creeping, stolons absent. **Leaves** monomorphic, dying back in winter. **Petiole** 1/3–2/3 length of blade, base swollen and persisting as trophopod over winter or not; vascular bundles 2, lateral, lunate in cross section. **Blade** elliptic to ovate-lanceolate, 1-pinnate-pinnatifid [pinnatifid to 3-pinnate-pinnatifid], gradually reduced distally to pinnatifid apex, herbaceous. **Pinnae** not articulate to rachis, segment margins entire, crenulate, or serrate; proximal pinnae (several pairs) reduced or not, sessile, equilateral; costae adaxially shallowly grooved, grooves not continuous with that of rachis; indument on rachis and costae (both sides) of multicellular hairs. **Veins** free, simple or forked. **Sori** on veins, elongate, ± straight, or hooked at distal end; indusia linear, laterally attached, persistent. **Spores** brownish, broadly winged. $x = 40$.

Species ca. 50 (2 in the flora): North America, e Asia, se and tropical Africa including Madagascar, Pacific Islands, Australia.

Petiole bases are swollen and toothed in sect. *Lunathyrium* (Koidzumi) M. Kato but not or only slightly thickened and without teeth in sects. *Athyriopsis* (Ching) M. Kato, *Deparia,* and *Dryoathyrium* (Ching) M. Kato.

Two American species, one native and the other introduced, are usually placed in *Athyrium* or *Diplazium.* The genus *Deparia,* however, including these two species, is sufficiently distinct to warrant generic separation because of its nondecurrent costal grooves and the presence of multicellular hairs on blades (M. Kato 1984).

SELECTED REFERENCE   Kato, M. 1984. A taxonomic study of the athyrioid fern genus *Deparia* with main reference to the Pacific species. J. Fac. Sci. Univ. Tokyo, Sect. 3, Bot. 13: 375–430.

1. Leaves markedly narrowed to base; petiole bases swollen and dentate. . . . . .1. *Deparia acrostichoides*
1. Leaves not or only slightly narrowed to base; petiole bases neither markedly swollen nor
   dentate. . . . . 2. *Deparia petersenii*

1. Deparia acrostichoides (Swartz) M. Kato, Ann. Carnegie Mus. 49: 177. 1980 · Silvery-spleenwort, silvery glade fern, déparie fausse-acrostiche

Asplenium acrostichoides Swartz, J. Bot. (Schrader) 1800(1): 54. 1800; *Athyrium acrostichoides* (Swartz) Diels; *A. thelypterioides* (Michaux) Desvaux; *Diplazium acrostichoides* (Swartz) Butters

Stems short-creeping. **Petiole** dark red-brown at base, straw-colored distally, 10–45 cm, swollen, with 2 rows of teeth; scales at base light brown, linear-lanceolate to lanceolate. **Blade** oblong-lanceolate, pinnate-pinnatifid, 30–80 × 12–25(–30) cm, narrowed to base, broadest near middle, acuminate at apex. **Pinnae** linear-oblong, base truncate, apex acuminate;

segments oblong, margins entire to slightly lobed, apex round to slightly pointed. **Costae and veins** with multicellular hairs. **Veins** pinnate, lateral veins simple or 1-forked. **Sori** elongate, straight or hooked; indusia ± thick, margin ± entire. $2n = 80$.

Damp woods, often on slopes; 30–1500 m; N.B., N.S., Ont., P.E.I., Que.; Ala., Ark., Conn., Del., Ga., Ill., Ind., Iowa, Ky., Maine, Md., Mass., Mich., Minn., Mo., N.H., N.J., N.Y., N.C., Ohio, Pa., R.I., S.C., Tenn., Vt., Va., W.Va., Wis.

Deparia acrostichoides belongs to sect. *Lunathyrium.* Closely related Asian ferns have been treated as conspecific with *Deparia acrostichoides,* but *D. acrostichoides* differs from them in having creeping stems with rather distant leaves and pinnate-pinnatifid leaves. *Deparia acrostichoides* and Asian species such as *D. pycnosora* (H. Christ) M. Kato and *D. allantodioides*

(Beddome) M. Kato are examples of vicariant species pairs with amphipacific disjunct distributions.

2. Deparia petersenii (Kunze) M. Kato, Bot. Mag. (Tokyo) 90: 37. 1977

Asplenium petersenii Kunze, Analecta Pteridogr., 24. 1837; *Athyrium petersenii* (Kunze) Copeland; *Diplazium petersenii* (Kunze) H. Christ

Stems creeping. **Petiole** dark brown or blackish at base, straw-colored distally, 10–30 cm, not swollen, teeth absent; scales pale brown, linear-lanceolate. **Blade** ovate-lanceolate, deeply pinnate-pinnatifid, 15–40 × 6–28 cm, moderately or slightly narrowed to base, broadest above base, abruptly narrowed to acuminate, pinnatifid apex. **Pinnae** oblong to linear-lanceolate, base ± truncate or broadly cuneate, apex acuminate to caudate; segments oblong, margins ± entire to serrate, apex obtuse or ± acute. **Costae and veins** with multicellular hairs. **Veins** pinnate, lateral veins usually simple, sometimes forked in larger leaves. **Sori** elongate, straight or rarely hooked; indusia membraneous, margins laciniate.

Moist ravines, lowlands; 0–100 m; introduced; Ala., Fla., Ga.; Asia; Pacific Islands; Australia.

Deparia petersenii belongs to sect. *Athyriopsis*. It is an edible fern native to southeastern Asia, and it persists or escapes from cultivation in southeastern North America (C. V. Morton and R. K. Godfrey 1958) but does not seem to be truly naturalized (O. Lakela and R. W. Long 1976).

Specimens of *D. petersenii* from North America are often misidentified as the Asian species, *Deparia japonica* (Thunberg) M. Kato [*Diplazium japonicum* (Thunberg) Beddome].

5. ATHYRIUM Roth, Tent. Fl. Germ. 3(1,1): 31, 58. 1799 · Lady fern [Greek *athyros,* doorless; the sporangia only tardily push back the outer edge of the indusium]

Masahiro Kato

Plants generally terrestrial. **Stems** short-creeping or ascending, stolons absent. **Leaves** monomorphic, usually dying back in winter. **Petiole** ± 0.5 times length of blade or less, base swollen and dentate, persisting as trophopod over winter or not; vascular bundles 2, lateral, lunate in cross section. **Blade** lanceolate to elliptic or oblanceolate, 1–3-pinnate-pinnatifid, gradually reduced distally to confluent, pinnatifid apex, herbaceous. **Pinnae** not articulate to rachis, segment margins serrulate or crenate; proximal pinnae often reduced, sessile to short-petiolulate, ± equilateral; costae adaxially grooved, grooves continuous from rachis to costae to costules; indument absent or of linear to lanceolate scales or 1-celled glands abaxially. **Veins** free, simple or forked. **Sori** in 1 row between midrib and margin, round to elongate, straight or hooked at distal end, or horseshoe-shaped; indusia shaped like sori, persistent, attached laterally or with narrow sinus, or indusia absent. **Spores** brownish, rugose. $x = 40$.

Species about 180 (2 in the flora): worldwide.

In species outside the flora stems are sometimes long-creeping to erect, with leaves radially or dorsiventrally arranged.

SELECTED REFERENCES Johnson, D. M. 1986b. Trophopods in North American species of *Athyrium* (Aspleniaceae). Syst. Bot. 11: 26–31. Kato, M. 1977. Classification of *Athyrium* and allied genera of Japan. Bot. Mag. (Tokyo) 90: 23–40. Liew, F. S. 1972. Numerical taxonomic studies on North American lady ferns and their allies. Taiwania 17: 190–221.

1. Sori round, submarginal; indusia much reduced or usually absent. . . 1a. *Athyrium alpestre* var. *americanum*
1. Sori elongate or hooked, medial; indusia well developed. . . . . 2. *Athyrium filix-femina*

1. Athyrium alpestre (Hoppe) Clairville, Man. Herbor. Suisse, 301. 1811

Aspidium alpestre Hoppe, Bot. Taschenb. 1805: 216. 1805; *Athyrium distentifolium* Tausch ex Opiz

Varieties ca. 3 (1 in the flora): North America, Europe, Asia.

1a. Athyrium alpestre (Hoppe) Clairville var. **americanum** Butters, Rhodora 19: 204. 1917
· American alpine lady fern, athyrie alpestre américaine

Athyrium alpestre subsp. *americanum* (Butters) Lellinger; *A. alpestre* var. *gaspense* Fernald; *A. americanum* (Butters) Maxon; *A. distentifolium* Tausch ex Opiz subsp. *americanum* (Butters) Hultén; *A. distentifolium* var. *americanum* (Butters) Cronquist

Stems ascending or short-creeping. **Petiole** straw-colored or red-brown distally, (7–)10–30 cm, base dark red-brown to black with 2 rows of teeth, swollen; scales at base brown to dark brown, lanceolate or broadly lanceolate, 13 × 3(–5) mm. **Blade** narrowly elliptic or lanceolate, 2–3-pinnate-pinnatifid, 15–55(–65) × 3–25 cm, moderately narrowed proximally, broadest below middle, apex acuminate. **Pinnae** short-stalked, narrowly deltate to deltate-oblong, apex acute. **Pinnules** deeply pinnatifid, segments oblong, crenulate. **Rachis, costae, and costules** with small, pale brown scales. **Veins** pinnate. **Sori** round to elliptic; indusia absent or very minute, scalelike. $2n = 80$.

Wet talus slopes, rocky hillsides, alpine meadows; 600–3100 m; Greenland; Alta., B.C., Nfld., Que., Yukon; Alaska, Calif., Colo., Idaho, Mont., Nev., Oreg., Utah, Wash., Wyo.

Athyrium alpestre var. *americanum* differs from var. *distentifolium* of Europe in its more finely dissected leaves with crenulate pinnule-segments, relatively broader pinnae with abruptly larger basal pinnules, and much more rudimentary indusia, if any. Japanese plants are more similar to var. *americanum* than to var. *distentifolium,* and they need further study.

2. Athyrium filix-femina (Linnaeus) Roth ex Mertens, Arch. Bot. (Leipzig) 2(1): 106. 1799 · Lady fern, fougère femelle

Polypodium filix-femina Linnaeus, Sp. Pl. 2: 1090. 1753

Stems short-creeping or ascending. **Petiole** straw-colored distally, 7–60 cm, base dark red-brown or black, swollen, with 2 rows of teeth; scales light to dark brown, linear- to ovate-lanceolate, 7–20 × 1–5 mm. **Blade** elliptic, lanceolate to oblanceolate, 2-pinnate to 2-pinnate-pinnatifid, 18–30 × 5–50 cm, herbaceous but with cartilaginous margin, narrowed to base, apex acuminate. **Pinnae** sessile to short-stalked, linear-oblong to lanceolate, apex acuminate. **Pinnules** pinnatifid, segments oblong-linear to narrowly deltate, margins serrate. **Rachis, costae, and costules** glabrous or with glands or hairs. **Veins** pinnate. **Sori** straight, hooked at distal end, or horseshoe-shaped; indusia dentate or ciliate.

Varieties ca. 5 (4 in the flora): North America, Mexico, Central America, South America, Europe, Asia.

Athyrium filix-femina is circumboreal, and this or closely related species extend into Mexico, Central America, and South America. The delimitation and infraspecific classification of *A. filix-femina* need detailed study.

1. Petiole scales more than 1 cm; blades ca. 2 times length of petioles, elliptic to oblanceolate.
   2. Pinnules narrowly deltate or oblong-lanceolate, nearly equilateral at base; indusia long-ciliate; spores yellow. . . . .
   . . . . . 2a. *Athyrium filix-femina* var. *cyclosorum*
   2. Pinnules linear-oblong or linear-lanceolate, inequilateral at base; indusia dentate or long-ciliate; spores brown. . . . .
   . . . . . 2b. *Athyrium filix-femina* var. *californicum*
1. Petiole scales 1 cm or less; blades 1–1.5 times length of petioles, elliptic to lanceolate.
   3. Petiole scales brown to dark brown; blades elliptic, narrowed to base, broadest near or just below middle; pinnae sessile or short-stalked; pinnules linear to oblong; indusia not glandular; spores yellow. . . . .
   . . . . . 2c. *Athyrium filix-femina* var. *angustum*
   3. Petiole scales light brown to brown; blades ovate-lanceolate to lanceolate, slightly narrowed to base, broadest just above base; pinnae usually stalked; pinnules oblong-lanceolate to narrowly deltate; indusia glandular or not; spores dark brown. . . . .
   . . . . . 2d. *Athyrium filix-femina* var. *asplenioides*

2a. Athyrium filix-femina (Linnaeus) Mertens var. **cyclosorum** Ruprecht, Distr. Crypt. Vasc. Ross., 41. 1845 · Northwestern lady fern

Athyrium alpestre (Hoppe) Clairville ex T. Moore var. *cyclosorum* (Ruprecht) T. Moore; *A. filix-femina* subsp. *cyclosorum* (Ruprecht) C. Christensen in Hultén; *A. filix-femina* var. *sitchense* Ruprecht

Petiole 15–45(–60) cm, base densely scaly; scales brown or dark brown, lanceolate or ovate-lanceolate, (7–)10–20 × (1–)2–5 mm, ± crisped. **Blade** elliptic to oblanceolate, 1–2-pinnate-pinnatifid, 25–120 × 10–50 cm, gradually narrowed proximally, broadest at or just above middle, apex acuminate. **Pinnae** very short-stalked or sessile, linear-oblong or oblong, apex acuminate. **Pinnules** sessile, narrowly deltate or oblong-lanceolate, base basiscopically broadly cuneate, acroscopically truncate, ± auriculate, apex obtuse to ± acute. **Rachis, costae, and costules** glabrous or with scales or pale glands. **Sori** round to elliptic, hooked at distal end, or horseshoe-shaped, medial to supramedial; indusia long-ciliate with nonglandular marginal hairs as long as or longer than width of indusia. **Spores** yellow. $2n = 80$.

Moist woods, swamps, streambanks; 10–1600 m; Alta., B.C., Man., N.W.T., Ont., Que., Sask., Yukon; Alaska, Calif., Idaho, Mont., Oreg., S.Dak., Wash., Wyo.

Athyrium filix-femina var. *cyclosorum* is most similar to the European var. *filix-femina*; it differs in having broader, nearly equilateral pinnules and medial to supramedial sori. The variety is distributed in northwestern North America with disjunct populations in northwestern Quebec and Ontario.

2b. Athyrium filix-femina (Linnaeus) Mertens var. **californicum** Butters, Rhodora 19: 201. 1918 · Southwestern lady fern

Athyrium filix-femina subsp. *californicum* (Butters) Hultén **Petiole** 13–60 cm; scales brown or black, lanceolate or ovate-lanceolate, 7–15 × 1–4 mm. **Blade** narrowly elliptic to oblanceolate, 1–2-pinnate-pinnatifid, 25–130 × 8–25 cm, gradually narrowed proximally, broadest at or just above middle, apex acuminate. **Pinnae** sessile or very short-stalked, linear-oblong, apex acuminate. **Pinnules** linear-oblong or linear-lanceolate, base basiscopically cuneate, acroscopically truncate, ± auriculate, apex ± acute to acuminate. **Rachis, costae, and costules** with scales and usually with pale glands. **Sori** elliptic or oblong, straight, hooked at distal end, or horseshoe-shaped, medial; indusia dentate or ciliate with nonglandular marginal hairs as long as width of indusia. **Spores** brown.

Moist woods, meadows, streambanks; 1000–3500 m; Ariz., Calif., Colo., Idaho, Nev., N.Mex., Oreg., S.Dak., Utah, Wyo.

This southwestern variety, *Athyrium filix-femina* var. *californicum*, occurs at higher elevations than var. *cyclosorum*. *Athyrium filix-femina* var. *californicum* and var. *cyclosorum* are more closely related to each other than to eastern varieties. Distinctness between western varieties was shown by F. K. Butters (1917); F. S. Liew (1972) treated them as consubspecific.

2c. Athyrium filix-femina (Linnaeus) Mertens var. **angustum** (Willdenow) G. Lawson, Edinburgh New Philos. J., n.s. 19: 115. 1864 · Northern lady fern

Aspidium angustum Willdenow, Sp. Pl. 5(1): 277. 1810; *Athyrium angustum* (Willdenow) C. Presl; *A. filix-femina* subsp. *angustum* (Willdenow) Clausen; *A. filix-femina* var. *michauxii* (Sprengel) Farwell **Petiole** (7–)15–55 cm; scales brown to dark brown, linear-lanceolate, 8(–10) × 2 mm. **Blade** elliptic, 2-pinnatifid to 2-pinnate, (20–)30–75 × (5–)10–35 cm, moderately narrowed proximally, broadest near or just below middle, apex acuminate. **Pinnae** short-stalked or sessile, oblong-lanceolate. **Pinnules** decurrent onto costal wing or sessile, linear to oblong, base unequally cuneate, apex obtuse to acute or in larger leaves acuminate, lobed halfway or more to costules. **Rachis, costae, and costules** glabrous or with pale glands. **Sori** straight, less frequently hooked at distal end or horseshoe-shaped; sporangial stalks bearing glandular hairs; indusia irregularly dentate, ± ciliate. **Spores** yellow or brown. $2n = 80$.

Moist woods, swamps, thickets; 0–1100 m; Greenland; Man., N.B., Nfld., N.S., Ont., P.E.I., Que., Sask.; Conn., Del., Ill., Ind., Iowa, Maine, Md., Mass., Mich., Minn., Mo., Nebr., N.H., N.J., N.Y., N.C., N.Dak., Ohio, Pa., R.I., S.Dak., Vt., Va., W.Va., Wis.

2d. Athyrium filix-femina (Linnaeus) Mertens var. **asplenioides** (Michaux) Farwell, Pap. Michigan Acad. Sci. 2: 13. 1923 · Southern lady fern, lowland lady fern

Nephrodium asplenioides Michaux, Fl. Bor.-Amer. 2: 268. 1803; *Athyrium asplenioides* (Michaux) Desvaux; *A. filix-femina* subsp. *asplenioides* (Michaux) Hultén **Petiole** 13–55 cm; scales light brown or brown, lanceolate, 6–9 × 2 mm, ± crisped. **Blade** ovate-lanceolate to lanceolate, 2-pinnate-pinnatifid, (18–)25–60 × (5–)10–30 cm, slightly narrowed proximally, broadest just above base to just below middle, apex acuminate or ± caudate. **Pinnae** usually stalked, oblong-lanceolate to lanceolate, base truncate, apex acuminate. **Pinnules** oblong-lanceolate to narrowly deltate, base unequally cuneate, apex ± acute. **Rachis, costae, and costules** glabrous or with scales or pale glands. **Sori** elongate, straight or hooked at distal end or horseshoe-shaped; sporangial stalks with glandular hairs; indusia ciliate, hairs glandular or nonglandular and ± as long as indusial width. **Spores** brown or dark brown. $2n = 80$.

Moist woods, thickets, swamps; 10–2000 m; Ala., Ark., Conn., Del., Fla., Ga., Ill., Ind., Kans., Ky., La., Md., Mass., Miss., Mo., N.J., N.Y., N.C., Okla., Pa., R.I., S.C., Tenn., Tex., Va., W.Va.

6. GYMNOCARPIUM Newman, Phytologist 4: 371. 1851 · Oak fern [Greek *gymnos*, naked, and *karpos,* fruit, referring to the absence of indusia]

Kathleen M. Pryer

Plants terrestrial. **Stems** long-creeping, stolons absent. **Leaves** monomorphic, dying back in winter. **Petiole** ca. 1.5–3 times length of blade, base not swollen; vascular bundles 2, lateral, ± oblong in cross section. **Blade** broadly deltate, ternate, or ovate, 2–3-pinnate-pinnatifid, reduced distally to pinnatifid apex, herbaceous. **Pinnae** weakly articulate to rachis but persistent, segment margins entire to crenate; proximal pinnae longest, petiolulate, usually ± inequilateral with pinnules on basiscopic side longer than those on acroscopic side; costae adaxially grooved, grooves not continuous from rachis to costae; indument lacking or of minute (0.1 mm) glands abaxially and sometimes along costae adaxially. **Veins** free, simple or forked. **Sori** in 1 row between midrib and margin, ± round; indusia absent. **Spores** brownish, rugose. $x = 40$.

Species 8 (5 in the flora): north temperate regions, North America, Eurasia.

SELECTED REFERENCES Pryer, K. M. 1992. The status of *Gymnocarpium heterosporum* and *G. robertianum* in Pennsylvania. Amer. Fern J. 82: 34–39. Pryer, K. M. and D. M. Britton. 1983. Spore studies in the genus *Gymnocarpium*. Canad. J. Bot. 61: 377–388. Pryer, K. M., D. M. Britton, and J. McNeill. 1983. A numerical analysis of chromatographic profiles in North American taxa of the fern genus *Gymnocarpium*. Canad. J. Bot. 61: 2592–2602. Pryer, K. M. and C. H. Haufler. 1993. Isozymic and chromosomal evidence for the allotetraploid origin of *Gymnocarpium dryopteris* (Dryopteridaceae). Syst. Bot. 18: 150–172. Sarvela, J. 1978. A synopsis of the fern genus *Gymnocarpium*. Ann. Bot. Fenn. 15: 101–106. Sarvela, J., D. M. Britton, and K. M. Pryer. 1981. Studies on the *Gymnocarpium robertianum* complex in North America. Rhodora 83: 421–431. Wagner, W. H. Jr. 1966b. New data on North American oak ferns, *Gymnocarpium*. Rhodora 68: 121–138.

1. Adaxial blade surface glabrous or moderately glandular, abaxial blade surface and rachis moderately or densely glandular.
   2. Blades glabrous on adaxial surface; proximal pinnae and basiscopic pinnules of proximal pinnae curving toward apex of leaf and apex of pinna, respectively; pinnae of 2d pair almost always sessile with basal pinnules ± equal in length to adjacent pinnules. . . . . 4a. *Gymnocarpium jessoense* subsp. *parvulum*
   2. Blades moderately glandular on adaxial surface; proximal pinnae and basiscopic pinnules of proximal pinnae ± perpendicular to rachis and costa, respectively; pinnae of 2d pair usually stalked, or if sessile with basal pinnules shorter than adjacent pinnules. . . . . 5. *Gymnocarpium robertianum*
1. Adaxial and abaxial blade surfaces and rachis essentially glabrous.
   3. Pinnae of 2d pair and basal basiscopic pinnule of proximal pinnae stalked. . . . . . .
. . . . 1. *Gymnocarpium appalachianum*
   3. Pinnae of 2d pair sessile or rarely stalked; basal basiscopic pinnule of proximal pinnae sessile.
      4. Pinnae of 2d pair sessile with basal pinnules unequal in length (basiscopic markedly longer); ultimate segments of proximal pinnae slightly lobed to crenate, apex often crenulate, acute; blades 8–24 cm. . . . . 2. *Gymnocarpium disjunctum*
      4. Pinnae of 2d pair rarely stalked, if sessile with basal pinnules ± equal in length (basiscopic = acroscopic); ultimate segments of proximal pinnae crenate to entire, apex entire, rounded; blades 3–14 cm.

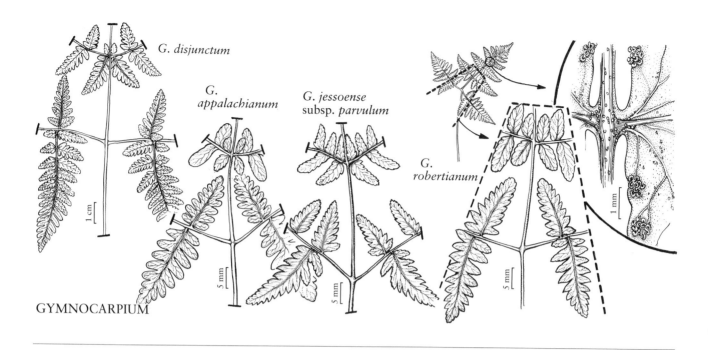

G. disjunctum

G. appalachianum

G. jessoense subsp. *parvulum*

G. robertianum

GYMNOCARPIUM

5. Sessile basal basiscopic pinnule of proximal pinnae with basal basiscopic pinnulet (division of pinnule) ± equal in length to adjacent pinnulet; pinnae of 2d pair usually sessile, with basal pinnules ± equal in length to adjacent basal pinnule; spores 34–39 μm. . . . . 3. *Gymnocarpium dryopteris*

5. Sessile basal basiscopic pinnule of proximal pinnae with basal basiscopic pinnulet shorter than adjacent pinnulet; pinnae of 2d pair sessile, with basal pinnules shorter than adjacent pinnule, or 2d basal pinnae rarely stalked; spores 27–31 μm. . . . . 1. *Gymnocarpium appalachianum*

1. Gymnocarpium appalachianum Pryer & Haufler, Syst. Bot. 18: 161. 1993 · Appalachian oak fern

Stems 0.5–1.5 mm diam.; scales 1.5–3 mm. **Fertile leaves** usually 10–32 cm. **Petiole** 6–20 cm, with sparse glandular hairs distally; scales 2–5 mm. **Blade** broadly deltate, 2–3-pinnate-pinnatifid, 4–12 cm, lax and delicate, abaxial surface and rachis glabrous or with occasional glandular hairs, adaxial surface glabrous. **Pinna apex** entire, rounded. **Proximal pinnae** 3–10 cm, ± perpendicular to rachis, with basiscopic pinnules ± perpendicular to costa; basal basiscopic pinnules stalked or sessile, pinnate-pinnatifid or pinnatifid, if sessile then with basal basiscopic pinnulet (division of pinnule) always shorter than adjacent pinnulet; 2d basal basiscopic pinnule infrequently stalked, if sessile then with basal basiscopic pinnulet shorter than adjacent pinnulet; basal acroscopic pinnule occasionally stalked, if sessile then with basal basiscopic pinnulet shorter than adjacent pinnulet. **Pinnae of 2d pair** usually stalked, if sessile then with basal basiscopic pinnule shorter than adjacent pinnule and equaling basal acroscopic pinnule; basal acroscopic pinnule shorter than adjacent pinnule, often with entire, rounded apex. **Pinnae of 3d pair** usually sessile with basal basiscopic pinnule shorter than adjacent pinnule and equaling or shorter than basal acroscopic pinnule; basal acroscopic pinnule equaling or shorter than adjacent pinnule. **Ultimate segments** of proximal pinnae oblong, entire to crenate, apex entire, rounded. Spores 27–31 μm. $2n = 80$.

Maple-birch-hemlock (*Acer-Betula-Tsuga*) woods on mountain slopes and summits, on moist sandstone or talus slopes with cold air seepage (algific); of conservation concern; 200–1400 m; Md., N.C., Ohio, Pa., Va., W.Va.

Gymnocarpium appalachianum, restricted to the Appalachian region, is a very local endemic.

2. Gymnocarpium disjunctum (Ruprecht) Ching, Acta Phytotax. Sin. 10: 304. 1965 · Western oak fern

Polypodium dryopteris Linnaeus var. *disjunctum* Ruprecht, Distr. Crypt. Vasc. Ross., 52. 1845; *Dryopteris disjuncta* (Ruprecht) C. V. Morton; *Gymnocarpium dryopteris* (Linnaeus) Newman subsp. *disjunctum* (Ruprecht) Sarvela; *G. dryopteris* var. *disjunctum* (Ruprecht) Ching

Stems 1–3 mm diam.; scales 2–4 mm. **Fertile leaves** usually 20–68 cm. **Petiole** 12–44 cm with sparse glandular hairs distally; scales 2–6 mm. **Blade** broadly deltate, 3-pinnate-pinnatifid, 8–24 cm, lax and delicate, abaxial surface and rachis glabrous or with sparse glandular hairs, adaxial surface glabrous. **Pinna apex** acuminate. **Proximal pinnae** 5–18 cm, ± perpendicular to rachis, with basiscopic pinnules ± perpendicular to costa; basal basiscopic pinnule sessile, pinnate-pinnatifid (with basal pinnulets, and sometimes 2 adjacent pinnulets, separate), basal basiscopic pinnulet usually longer (sometimes equaling or shorter) than adjacent pinnulet; 2d basal basiscopic pinnule sessile with basal basiscopic pinnulet usually longer than or equaling adjacent pinnulet; basal acroscopic pinnule sessile, with basal basiscopic pinnulet usually longer than or equaling adjacent pinnulet. **Pinnae of 2d pair** usually sessile with basal basiscopic pinnule longer than or equaling adjacent pinnule and markedly longer than basal acroscopic pinnule; basal acroscopic pinnule distinctly shorter than adjacent pinnule or rarely absent, apex often crenulate, obtuse. **Pinnae of 3d pair** usually sessile with basal basiscopic pinnule longer than or equaling adjacent pinnule and longer than basal acroscopic pinnule; basal acroscopic pinnule shorter than adjacent pinnule. **Ultimate segments** of proximal pinnae oblong, crenate to slightly lobed, apex crenulate, acute. Spores 27–31 μm. $2n = 80$.

Shaded, rocky slopes and ravines, mixed coniferous woods, moist stream and creek banks; of conservation concern; 0–2400 m; Alta., B.C.; Alaska, Idaho, Mont., Oreg., Wash., Wyo.; Asia in ne former Soviet republics.

In addition to the west coast of North America, *Gymnocarpium disjunctum* is found on Sakhalin Island in southern Kamchatka, in the former Soviet republics.

3. Gymnocarpium dryopteris (Linnaeus) Newman, Phytologist 4: app. 24. 1851 · Common oak fern, fougère-du-chêne

Polypodium dryopteris Linnaeus, Sp. Pl. 2: 1093. 1753; *Dryopteris linnaeana* C. Christensen; *Lastrea dryopteris* (Linnaeus) Bory; *Phegopteris dryopteris* (Linnaeus) Fée; *Thelypteris dryopteris* (Linnaeus) Slosson

Stems 0.5–1.5 mm diam.; scales 1–4 mm. **Fertile leaves** usually 12–42 cm. **Petiole** 9–28 cm, with sparse glandular hairs distally; scales 2–6 mm. **Blade** broadly deltate, 2-pinnate-pinnatifid, 3–14 cm, lax and delicate, abaxial surface and rachis glabrous or with sparse glandular hairs, adaxial surface glabrous. **Pinna apex** entire, rounded. **Proximal pinnae** 2–12 cm, ± perpendicular to rachis, with basiscopic pinnules ± perpendicular to costa; basal basiscopic pinnule usually sessile, pinnatifid or rarely pinnate-pinnatifid, if sessile then with basal basiscopic pinnulet often equaling or longer than adjacent pinnulet; 2d basal basiscopic pinnule sessile, with basal basiscopic pinnulet equaling or longer than adjacent pinnulet; basal acroscopic pinnule sessile, with basal basiscopic pinnulet longer than or equaling adjacent pinnulet. **Pinnae of 2d pair** usually sessile with basal basiscopic pinnule longer than or equaling adjacent pinnule and about equal to basal acroscopic pinnule; basal acroscopic pinnule equaling or slightly shorter than adjacent pinnule, often with entire, rounded apex. **Pinnae of 3d pair** sessile with basal basiscopic pinnule equaling adjacent pinnule and equaling basal acroscopic pinnules; basal acroscopic pinnule equaling or slightly shorter than adjacent pinnule. **Ultimate segments** of proximal pinnae oblong, entire to crenate, apex entire, rounded. Spores 34–39 μm. $2n = 160$.

Cool, coniferous and mixed woods and at base of shale talus slopes; 0–3000 m; Greenland; St. Pierre and Miquelon; Alta., B.C., Man., N.B., Nfld., N.W.T., N.S., Ont., P.E.I., Que., Sask., Yukon; Alaska, Ariz., Colo., Conn., Idaho, Iowa, Maine, Mass., Mich., Minn., Mont., N.H., N.J., N.Mex., N.Y., Ohio, Oreg., Pa., R.I., S.Dak., Vt., Wash., W.Va., Wis., Wyo.; n,c Europe; n Asia to China, Japan.

Gymnocarpium dryopteris is a fertile allotetraploid species that arose following hybridization between *G. appalachianum* and *G. disjunctum* (see reticulogram). Its wide distribution over much of the north temperate zone has provided ample opportunity for secondary contact between *G. dryopteris* and each of its diploid parents, thereby resulting in a wide-ranging composite of abortive-spored triploid crosses (*G. disjunctum* ×

Gymnocarpium

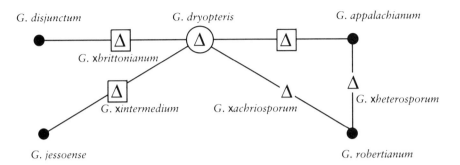

Relationships in *Gymnocarpium*. Solid circles represent parental taxa; circled triangle represents an allopolyploid; boxed triangles represent hybrids presumed to be apogamous; and triangles represent other sterile hybrids.

G. dryopteris and *G. appalachianum* × *G. dryopteris*). These relationships are shown on the diagram. Sterile triploid plants are not restricted only to areas where the range of the tetraploid overlaps with that of either diploid. Their broad distribution could be explained in part by their spores, which are of two types: malformed, black, and with very exaggerated perispores, or round with extensive netted perispores (K. M. Pryer and D. M. Britton 1983). The latter spore type is capable of germination and presumably permits the plants to reproduce apogamously. The name *G.* × *brittonianum* (Sarvela) Pryer & Haufler has been applied to the *G. disjunctum* × *G. dryopteris* hybrid formula (K. M. Pryer and C. H. Haufler 1993). The type of *G.* × *brittonianum* has aborted and round spores, and leaves that strongly resemble those of *G. disjunctum*. They are large, 3-pinnate-pinnatifid, and the second and third pairs of pinnae are sessile with basal basiscopic pinnules markedly longer than the basal acroscopic pinnules. Sterile triploid plants with a morphology similar to the type of *G.* × *brittonianum* are frequent. The biology of both of these cryptic hybrid taxa needs further study, which should lead to detailed morphologic descriptions and distribution maps.

Gymnocarpium dryopteris also hybridizes with both *G. jessoense* subsp. *parvulum* and *G. robertianum*.

4. Gymnocarpium jessoense (Koidzumi) Koidzumi, Acta Phytotax. Geobot. 5: 40. 1936

Dryopteris jessoensis Koidzumi, Bot. Mag. (Tokyo) 38: 104. 1924; *Aspidium dryopteris* (Linnaeus) Baumgarten var. *longulum* H. Christ; *Gymnocarpium longulum* (H. Christ) Ki-

tagawa; *G. robertianum* (Hoffman) Newman subsp. *longulum* (H. Christ) Toyokuni

Subspecies 2 (1 in the flora): North America, Asia.

4a. Gymnocarpium jessoense (Koidzumi) Koidzumi subsp. parvulum Sarvela, Ann. Bot. Fenn. 15: 103. 1978 · Nahanni oak fern

Gymnocarpium continentale (Petrov) Pojark

Stems 0.5–1.5 mm diam.; scales 1–4 mm. **Fertile leaves** usually 8–39 cm. **Petiole** 5–25 cm, with moderately abundant glandular hairs distally; scales 2–6 mm. **Blade** narrowly deltate to narrowly ovate, 2-pinnate-pinnatifid, 3–14 cm, firm and robust or lax and delicate, abaxial surface moderately glandular, rachis moderately to densely glandular, adaxial surface glabrous. **Pinna apex** acute. **Proximal pinnae** 2–9 cm, strongly curved toward apex of leaf, basiscopic pinnules strongly curved toward apex of pinna; basal basiscopic pinnule usually sessile, pinnatifid or rarely pinnate-pinnatifid, if sessile then with basal basiscopic pinnulet often equaling adjacent pinnulet; 2d basal basiscopic pinnule sessile, with basal basiscopic pinnulet equaling adjacent pinnulet; basal acroscopic pinnule sessile with basal basiscopic pinnulet longer than or equaling adjacent pinnulet. **Pinnae of 2d pair** almost always sessile with basal basiscopic pinnule usually equaling or slightly shorter than adjacent pinnule and equaling basal acroscopic pinnule; basal acroscopic pinnule equaling or slightly shorter than adjacent pinnule, apex often entire, rounded. **Pinnae of 3d pair** sessile with basal basiscopic pinnule equaling adjacent pinnule and equaling basal acroscopic pinnule; basal acroscopic pinnule equaling or slightly shorter than adjacent pinnule. **Ultimate segments** of proximal pinnae oblong, entire to slightly crenate, apex entire, rounded. **Spores** 32–37 μm. $2n = 160$.

Acid or neutral substrates at summit of cool, shale talus slopes, and on granitic cliffs and outcrops; 0–2000 m; Alta., B.C., Man., N.B., N.W.T., Ont., Que., Sask., Yukon; Alaska, Conn., Iowa, Maine, Mich., Minn., Vt., Wis.; Europe in Finland; Asia in Siberia, Kazakhstan.

Hybrids between *Gymnocarpium jessoense* subsp. *parvulum* and *G. dryopteris* (*G.* × *intermedium* Sarvela) are usually found wherever these two taxa occur together (Finland; Manitoba, Northwest Territories, Ontario, Quebec, Saskatchewan, Yukon; Alaska, Michigan, Minnesota, Wisconsin), and they are particularly abundant in the Great Lakes region. These hybrids have sometimes been referred to as *G.* × *heterosporum*, a name that is, however, correctly restricted to unique hybrids between *G. robertianum* and *G. appalachianum* (see discussion under *G. robertianum*; K. M. Pryer 1992). *Gymnocarpium* × *intermedium* is intermediate between the two parental species in its leaf morphology and glandularity, and it can be readily distinguished by its small, blackish, malformed, abortive spores, as well as large, brown, round spores that may allow this taxon to reproduce apogamously. Of the *Gymnocarpium* sterile hybrids, *G.* × *intermedium* is the easiest to distinguish morphologically.

5. **Gymnocarpium robertianum** (Hoffmann) Newman, Phytologist 4: app. 24. 1851 · Limestone oak fern, gymnocarpe du Robert

Polypodium robertianum Hoffmann, Deutschl. Fl. 2: add. et emend. 10. 1795; *Dryopteris robertiana* (Hoffmann) C. Christensen; *Phegopteris robertianum* (Hoffmann) Fée; *Thelypteris robertiana* (Hoffmann) Slosson

Stems 1–2 mm diam.; scales 2–4 mm. **Fertile leaves** usually 10–52 cm. **Petiole** 5–33 cm, with numerous glandular hairs distally; scales 2–6 mm. **Blade** broadly deltate, 2–3-pinnate-pinnatifid, 5–19 cm, usually firm and robust, abaxial surface moderately to densely glandular, rachis densely glandular, adaxial surface moderately glandular. **Pinna apex** acute. **Proximal pinnae** 3–13 cm, ± perpendicular to rachis, basiscopic pinnules ± perpendicular to costa; basal basiscopic pinnules either sessile or stalked, pinnate-pinnatifid or pinnatifid, if sessile then with basal basiscopic pinnulet usually shorter than adjacent pinnulet; 2d basal basiscopic pinnule sometimes stalked, if sessile then with basal basiscopic pinnulet shorter than or equaling adjacent pinnulet; basal acroscopic pinnule sometimes stalked, if sessile then with basal basiscopic pinnulet shorter than or equaling adjacent pinnulet. **Pinnae of 2d pair** usually stalked, if sessile then with basal basiscopic pinnule usually shorter than adjacent pinnule and equaling basal acroscopic pinnule; basal acroscopic pinnule shorter than adjacent pinnule, apex often entire, rounded. **Pinnae of 3d pair** usually sessile with basal basiscopic pinnule shorter than adjacent pinnule and equaling basal acroscopic pinnule; basal acroscopic pinnule equaling or shorter than adjacent pinnule. **Ultimate segments** of proximal pinnae oblong, entire to slightly crenate, apex entire, rounded. **Spores** 34–39 μm. $2n = 160$.

Calcareous substrates; limestone pavement, outcrops, and cliffs; *Thuja* swamps; 0–1000 m; Man., N.B., Nfld., Ont., Que.; Iowa, Mich., Minn., Wis.; Europe; Asia in Caucasus Mountains.

Gymnocarpium robertianum occurs in numerous localities in eastern Canada, especially in Ontario and Quebec where it is widely distributed; populations are small. Hybrids with *G. robertianum* are extremely rare. *Gymnocarpium* × *heterosporum* W. H. Wagner, a putative triploid hybrid between *G. robertianum* and *G. appalachianum*, is known only from one county in Pennsylvania (plants now extirpated, K. M. Pryer 1992). *Gymnocarpium* × *achriosporum* Sarvela, a putative tetraploid hybrid between *G. robertianum* and *G. dryopteris*, is known only from Sweden and two localities in Quebec. Both hybrids resemble *G. robertianum* in their leaf morphology and dense glandularity but have black, malformed spores.

Cystopteris

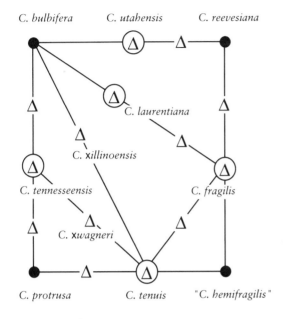

Relationships among some species of *Cystopteris*. Solid circles represent parental taxa; circled triangles represent allopolyploids; and triangles represent sterile hybrids. *Cystopteris hemifragilis* is a hypothetical ancestor.

7. **CYSTOPTERIS** Bernhardi, Neues J. Bot. 1(2): 26. 1806, name conserved · Bladder fern, brittle fern [Greek *kystos,* bladder, and *pteris,* fern, alluding to the indusium, which is inflated when young]

Christopher H. Haufler

Robbin C. Moran

Michael D. Windham

Plants terrestrial or on rock. **Stems** short- to long-creeping, stolons absent. **Leaves** monomorphic, dying back in winter. **Petiole** 1/3–3 times length of blades, base often swollen and persisting as trophopod over winter; vascular bundles 2, lateral, round or oblong in cross section. **Blade** ovate-lanceolate to deltate, 1–3-pinnate-pinnatifid, gradually reduced distally to a pinnatifid apex, membranaceous to herbaceous. **Pinnae** not articulate to rachis, segment margins crenulate, dentate, or serrate; proximal pinnae not reduced or 1 pair slightly reduced, sessile or petiolulate, equilateral or ± inequilateral, if inequilateral basiscopic side more narrowly cuneate; costae adaxially grooved, grooves continuous from rachis to costae; indument absent or of uniseriate, multicellular hairs in pinnae axils or of unicellular, gland-tipped hairs abaxially, absent adaxially. **Veins** free, simple or forked. **Sori** in 1 row between midrib and margin on ultimate segments, round; indusia ovate to lanceolate, hoodlike and arching over sorus toward margin, attached to receptacle base on costal side, persistent to ephemeral or often obscure at maturity. **Spores** brownish, echinate, or verrucate. $x = 42$.

Species ca. 20 (9 in the flora): worldwide.

Cystopteris is a taxonomically difficult genus at the species level. Especially troublesome is the worldwide and polymorphic species *C. fragilis* sensu lato. To maintain it as a single species with several varieties would be easiest (and least controversial). This approach, however, may not accurately reflect true evolutionary history.

Although *Cystopteris* species are found in temperate climates worldwide at tetraploid to octaploid ploidy levels, extant diploid species are concentrated in North America. The diploid species are *relatively* distinct from one another and are the progenitors of numerous allopolyploid derivatives (see reticulogram). In addition, an extinct (or undiscovered) diploid may have been involved in the origin of some polyploids (shown as "*C. hemifragilis*" on the reticulogram).

Considerable overlap exists among the leaf morphologies in the species of *Cystopteris*, even among the diploid taxa. Consequently, the key requires observation of subtle and sometimes overlapping characteristics.

Several general recommendations can be made for identifying *Cystopteris*. (1) Field workers should be aware that whenever *Cystopteris* species occur together, hybridization is likely; hybrids usually have shriveled and malformed spores. (2) Species of *Cystopteris* frequently occur as highly reduced plants, especially in stressful habitats such as high elevations, high latitudes, and cold and/or dry climates. Such stunted plants can be fertile, but leaf and stem characters required to distinguish species can be obscured. (3) Because of the importance of examining stem and spore features in distinguishing species, collectors should always attempt to obtain complete, fertile specimens.

SELECTED REFERENCES   Blasdell, R. F. 1963. A monographic study of the fern genus *Cystopteris*. Mem. Torrey Bot. Club 21: 1–102. Haufler, C. H., M. D. Windham, D. M. Britton, and S. J. Robinson. 1985. Triploidy and its evolutionary significance in *Cystopteris protrusa*. Canad. J. Bot. 63: 1855–1863. Haufler, C. H. and M. D. Windham. 1991. New species of North American *Cystopteris* and *Polypodium*, with comments on their reticulate relationships. Amer. Fern J. 81: 7–23. Haufler, C. H., M. D. Windham, and T. A. Ranker. 1990. Biosystematic analysis of the *Cystopteris tennesseensis* (Dryopteridaceae) complex. Ann. Missouri Bot. Gard. 77: 314–329.

1. Leaf blades elongate-pentagonal; proximal pinnae inequilateral, with enlarged basiscopic pinnules; stems cordlike, long-creeping, leaf bases more than 1 cm apart. . . . . . . 1. *Cystopteris montana*
1. Leaf blades elliptic to deltate; proximal pinnae equilateral or nearly so; stems not cordlike, short-creeping (leaf bases less than 0.5 cm apart) or if long-creeping, leaf bases generally less than 1 cm apart.
    2. Rachises, costae, indusia, and midribs of ultimate segments sparsely to densely covered by gland-tipped hairs; leaf blades deltate to ovate, usually widest at or near base; rachises and costae often with bulblets.
        3. Rachises and costae frequently with bulblets; rachises, costae, indusia, and midribs of ultimate segments usually densely covered by gland-tipped hairs; leaf blades broadly to narrowly deltate, almost always widest at base, apex long-attenuate; leaves seasonally bearing sori (earliest leaves lack sori, subsequent leaves with sori); petioles reddish when young, green or straw-colored in mature specimens; spores usually 33–38 μm. . . . . .2. *Cystopteris bulbifera*
        3. Rachises and costae occasionally with bulblets (often misshapen); rachises, costae, indusia, and midribs of ultimate segments usually sparsely covered by glandular hairs; leaf blades narrowly deltate to ovate-lanceolate, widest at or near base, apex short-attenuate; nearly all leaves bearing sori; petioles dark brown to straw-colored or green; spores usually 38–60 μm.
            4. Blades ovate to lanceolate, usually widest above base; spores usually 49–60 μm; ne North America. . . . . 3. *Cystopteris laurentiana*
            4. Blades deltate to narrowly deltate, usually widest at or near base; spores usually 38–48 μm; e,c to sw North America.
                5. Stem scales usually dark brown, ± clathrate, cell walls dark brown, thick, luminae prominent; leaves usually with multicellular, gland-tipped hairs in axils of pinnae; sw United States. . . . . 4. *Cystopteris utahensis*

5. Stem scales usually tan to light brown, cell walls brown, thin, luminae not obvious; leaves rarely with multicellular, gland-tipped hairs in axils of pinnae; e United States (e Kansas, s Minnesota throughout East Coast area). . . . . 5. *Cystopteris tennesseensis*
2. Rachises, costae, indusia, and midribs of ultimate segments without glandular hairs; leaf blades elliptic to lanceolate, generally widest at or just below middle of blade; rachises and costae without bulblets.
    6. Leaves clustered 1–4 cm behind protruding stem apex; stems pubescent, hairs yellow; spores usually 28–34 μm. 6. *Cystopteris protrusa*
    6. Leaves clustered at stem apex; stems lacking hairs; spores usually 33–60 μm.
        7. Proximal pinnae pinnate-pinnatifid to 2-pinnate; stems usually long-creeping; spores usually 33–41 μm. . . . . 7. *Cystopteris reevesiana*
        7. Proximal pinnae pinnatifid to pinnate-pinnatifid; stems short-creeping; spores usually 39–60 μm.
            8. Pinnae typically at acute angle to rachis, often curving toward blade apex; pinnae along distal 1/3 of blades ovate to narrowly elliptic; margins of pinnae usually crenulate or with rounded teeth; basal basiscopic pinnules of proximal pinnae cuneate to rounded at base. . . . . 8. *Cystopteris tenuis*
            8. Pinnae typically perpendicular to rachis, not curving toward blade apex; pinnae along distal 1/3 of blade deltate to ovate; margins of pinnae with sharp teeth; basal basiscopic pinnules of proximal pinnae truncate to rounded at base. . . . . 9. *Cystopteris fragilis*

1. **Cystopteris montana** (Lamarck) Bernhardi ex Desvaux, Neues J. Bot. 1(2): 26. 1806 · Mountain bladder fern, cystoptère des montagnes

Polypodium montanum Lamarck, Fl. Franç. 1: 23. 1779

Stems long-creeping, cordlike, internodes 1–2(–4) cm, old petiole bases few, hairs absent; scales usually tan to light brown, ovate-lanceolate, radial walls tan to brown, thin, luminae tan. **Leaves** monomorphic, at stem apex but not tightly clustered, to 45 cm, sori production about equal on all leaves (fairly independent of season). **Petiole** dark brown to black at base, gradually becoming green or straw-colored distally, (1–)2–3 times length of blades, sparsely scaly throughout. **Blade** elongate-pentagonal, 3(–4)-pinnate-pinnatifid; rachis and costae lacking gland-tipped hairs or bulblets; axils of pinnae with occasional multicellular gland-tipped hairs. **Pinnae** ascending, typically at acute angle to rachis, only proximal pinnae occasionally curving toward blade apex, margins serrate; proximal pinnae pinnate-pinnatifid, inequilateral, basal basiscopic pinnule stalked, enlarged, base truncate to obtuse; distal pinnae deltate to ovate. **Veins** directed into notches. **Indusia** cup-shaped, apex truncate, hairs gland-tipped only along margin. **Spores** spiny, usually 37–42 μm. $2n = 168$.

Sporulating summer–fall. Terrestrial in wet woods or along water courses; rare; 0–3500 m; Greenland; Alta., B.C., N.B., Nfld., N.W.T., N.S., Ont., Que., Sask., Yukon; Alaska, Colo., Mont.; Eurasia.

Cystopteris montana, the most distinctive of the *Cystopteris* in the flora, probably is allied to Asian species. Although this boreal species is restricted primarily to high latitudes, it occurs disjunctly at high elevations in Colorado, where its habitats are being threatened by development. *Cystopteris montana* does not hybridize with any other *Cystopteris* in the flora, but it has been implicated in the origin of the European allopolyploid *C. alpina* (Roth) Desvaux.

2. **Cystopteris bulbifera** (Linnaeus) Bernhardi, Neues J. Bot. 1(2): 10. 1806 · Bulblet bladder fern, cystoptère bulbifère

Polypodium bulbiferum Linnaeus, Sp. Pl. 2: 1091. 1753

Stems creeping, not cordlike, internodes very short, less than 0.5 cm, heavily beset with old petiole bases, hairs absent; scales uniformly brown to somewhat clathrate, lanceolate, radial walls brown, luminae clear. **Leaves** monomorphic, clustered at stem apex, to 75 cm, seasonally bearing sori (earliest leaves lack sori, subsequent leaves with sori). **Petiole** reddish when young, usually green or straw-colored throughout (occasionally darker) in mature specimens, shorter than blade, base sparsely scaly. **Blade** broadly to narrowly deltate, 2-pinnate to 2-pinnate-pinnatifid, almost always widest at base, apex long-attenuate; rachis and costae usually densely covered by unicellular, gland-tipped hairs, often with bulblets; axils of pinnae occasionally with multi-

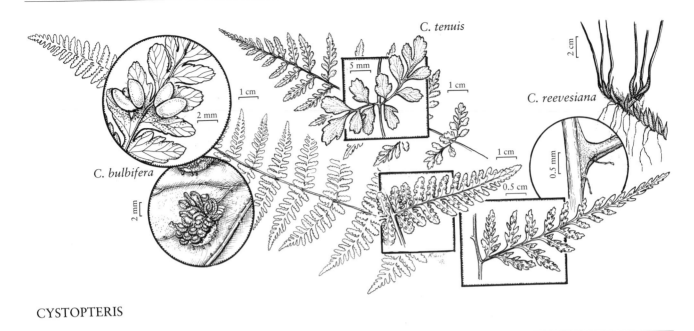

CYSTOPTERIS

cellular, gland-tipped hairs. **Pinnae** mostly perpendicular to rachis, not curving toward blade apex, margins serrate; proximal pinnae pinnate-pinnatifid to pinnatifid, ± equilateral, basiscopic pinnules not enlarged, basal basiscopic pinnules sessile to short-stalked, bases truncate to obtuse; distal pinnae ovate to oblong. **Veins** directed into notches. **Indusia** cup-shaped, apex truncate, typically invested with unicellular, gland-tipped hairs. **Spores** spiny, usually 33–38 μm. $2n = 84$.

Sporulating summer–fall. Cracks and ledges on cliffs, rarely terrestrial; usually on calcareous substrates; 0–2500 m; N.B., Nfld., N.S., Ont., Que.; Ala., Ariz., Ark., Conn., Del., Ga., Ill., Ind., Iowa, Ky., Maine, Md., Mass., Mich., Minn., Mo., N.H., N.J., N.Mex., N.Y., N.C., Ohio., Pa., R.I., S.C., Tenn., Tex., Utah, Vt., Va., W.Va., Wis.

Cystopteris bulbifera usually occurs on moist calcareous cliffs, but it also grows on rock in dense woods and occasionally occurs terrestrially in northern swamps. Blades on most individuals are narrowly deltate and distinctively long-attenuate. The rachises, costae, and indusia are densely beset with gland-tipped, unicellular hairs. Mature specimens often have deciduous bulblets. These characteristics readily distinguish *C. bulbifera* from other diploid species.

Hybridization and allopolyploidy involving *Cystopteris bulbifera* and other North American *Cystopteris* species have generated several species. In the eastern portion of its range, *C. bulbifera* and *C. protrusa* are the diploid progenitors of the tetraploid *C. tennesseensis* (C. H. Haufler et al. 1990). In northeastern North America, *C. bulbifera* has hybridized with tetraploid *C.*

fragilis, ultimately resulting in the hexaploid *C. laurentiana* (R. F. Blasdell 1963). In the southwest, the diploid *C. reevesiana* and disjunct representatives of *C. bulbifera* are the progenitors of the tetraploid *C. utahensis* (C. H. Haufler and M. D. Windham 1991). In addition to these fertile allopolyploids, sterile hybrids are also possible when *C. bulbifera* is sympatric with its polyploid derivatives. Sterile hybrids between *C. bulbifera* and *C. tennesseensis* have been identified from several localities. *Cystopteris bulbifera* may hybridize with *C. tenuis* to form *C.* ×*illinoensis* (C. H. Haufler et al. 1990; R. C. Moran 1982b). Diploid sexual *C. bulbifera* may be distinguished from these allopolyploid species and sterile hybrids because the hybrid-derived taxa (1) will normally have less prominent glandular hairs, (2) will have misshapen bulblets, (3) will more likely have blades that are widest above the base, and (4) will have large spores (in sexual allopolyploids) or malformed spores (in sterile hybrids).

3. **Cystopteris laurentiana** (Weatherby) Blasdell, Mem. Torrey Bot. Club 21(4): 51. 1963 · Laurentian bladder fern, cystoptère laurentienne

Cystopteris fragilis (Linnaeus) Bernhardi var. *laurentiana* Weatherby, Rhodora 28: 129. 1926 **Stems** creeping, not cordlike, internodes very short, less than 5 mm, heavily beset with old petiole bases, hairs absent; scales uniformly brown to ± clathrate, radial walls brown, luminae clear.

Leaves monomorphic, clustered at stem apex, to 45 cm, nearly all bearing sori. **Petiole** usually dark at base, grading to straw-colored distally, shorter than blade, sparsely scaly at base. **Blade** ovate to narrowly ovate, 2-pinnate to 2-pinnate-pinnatifid, widest above base, apex short-attenuate; rachis and costae usually sparsely invested with unicellular, gland-tipped hairs, occasionally with misshapen bulblets; axils of pinnae with occasional multicellular, gland-tipped hairs. **Pinnae** typically perpendicular to rachis, not curving toward blade apex, margins serrate; proximal pinnae pinnate-pinnatifid to pinnatifid, ± equilateral, basiscopic pinnules not enlarged, basal basiscopic pinnules sessile to short-stalked, base truncate to obtuse; distal pinnae ovate to oblong. **Veins** directed into teeth and notches. **Indusia** cup-shaped, apex truncate, typically sparsely invested with unicellular, gland-tipped hairs. **Spores** spiny, usually 49–60 μm. $2n = 252$.

Sporulating summer–fall. Cracks and ledges on cliffs, often on calcareous substrates; 0–1000 m; N.B., Nfld., N.S., Ont., Que.; Ill., Iowa, Mass., Mich., Minn., Pa., Vt., Wis.

Cystopteris laurentiana is a sexual allohexaploid species with *C. bulbifera* as the diploid parent and *C. fragilis* as the tetraploid. *Cystopteris laurentiana* was previously thought to be common only in the Great Lakes region (R. F. Blasdell 1963); it is now known to occur frequently in the Driftless Area of the Midwest. Because *C. laurentiana* can be difficult to distinguish from *C. fragilis,* specimens with ovate leaves having unusually large spores and growing on moist cliffs should be checked carefully for occasional glandular hairs, the distinguishing feature of *C. laurentiana.* Sterile pentaploid hybrids between *C. laurentiana* and *C. fragilis* have been discovered where the two species are sympatric.

4. **Cystopteris utahensis** Windham & Haufler in Haufler & Windham, Amer. Fern J. 81: 13. 1991

Stems creeping, not cordlike, internodes short, heavily beset with old petiole bases, hairs absent; scales lanceolate, ± clathrate, radial walls dark brown, thick, luminae clear. **Leaves** monomorphic, clustered at stem apex, to 45 cm, nearly all bearing sori. **Petiole** green to straw-colored throughout or darker near base, shorter than blade, base sparsely scaly. **Blade** deltate to narrowly deltate, 2-pinnate-pinnatifid, usually widest at or near base, apex short-attenuate; rachis and costae with unicellular, gland-tipped hairs, misshapen bulblets present or absent; axils of pinnae usually with multicellular, gland-tipped hairs. **Pinnae** typically perpendicular to rachis, not curving toward blade apex, margins serrate; proximal

pinnae pinnatifid to pinnate-pinnatifid, ± equilateral, basiscopic pinnules not enlarged; basal basiscopic pinnules sessile or short-stalked, base truncate to obtuse, distal pinnae ovate to oblong. **Veins** directed into teeth and notches. **Indusia** cup-shaped, apex truncate, with scattered, unicellular, gland-tipped hairs. **Spores** spiny, usually 39–48 μm. $2n = 168$.

Sporulating summer–fall. Cracks and ledges on cliffs; on calcareous substrates including sandstone, limestone, and dacite; 1300–2700 m; Ariz., Colo., Tex., Utah.

Cystopteris utahensis is an allopolyploid derived from the diploid species *C. bulbifera* and *C. reevesiana* (C. H. Haufler and M. D. Windham 1991). Because *C. utahensis* shares one parent (*C. bulbifera*) with *C. tennesseensis* and because of morphologic similarities between *C. reevesiana* and *C. protrusa* (the second diploid parent of *C. tennesseensis*), populations of *C. utahensis* were previously considered to have originated by long-distance dispersal from eastern populations of *C. tennesseensis*. Genetic studies using isozyme markers, however, indicated that *C. utahensis* was a distinct species and stimulated the discovery of morphologic criteria for distinguishing it from its eastern cousin. When combined with the geographic separation of the two tetraploids, the minor differences in indument features provide a means of circumscribing this genetically distinct species. Potential confusion in identifying *C. utahensis* arises because sterile triploid hybrids may form when it is sympatric with the more common diploid *C. reevesiana.*

5. **Cystopteris tennesseensis** Shaver, J. Tennessee Acad. Sci. 25(2): 107. 1950 · Tennessee bladder fern

Cystopteris fragilis (Linnaeus) Bernhardi forma *simulans* Weatherby; *C. fragilis* var. *tennesseensis* (Shaver) McGregor

Stems creeping, not cordlike, internodes short, heavily beset with old petiole bases, hairs absent; scales usually tan to light brown, lanceolate, radial walls tan to brown, thin, luminae tan. **Leaves** monomorphic, crowded near stem apex, to 45 cm, nearly all bearing sori. **Petiole** variable in color but mostly dark brown at base, gradually becoming straw-colored distally, shorter than blade, sparsely scaly at base. **Blade** deltate to narrowly deltate, 2-pinnate-pinnatifid, usually widest at or near base, apex short-attenuate; rachis and costae with occasional unicellular, gland-tipped hairs, with or without bulblets (usually misshapen); axils of pinnae with infrequent multicellular, gland-tipped hairs. **Pinnae** usually perpendicular to rachis, not curving toward blade apex, margins serrate; proximal pinnae pinnatifid to pinnate-pinnatifid, ± equilateral, basiscopic pinnules not enlarged, basal basiscopic pinnules sessile to short-stalked, base truncate to obtuse; distal pinnae ovate to oblong. **Veins** directed into teeth and notches. **Indusia** cup-shaped, apex truncate, with scattered, unicellular, gland-tipped hairs. **Spores** spiny, usually 38–42 μm. $2n = 168$.

Sporulating summer–fall. Cracks and ledges on cliffs, rarely terrestrial; often on calcareous substrates or associated with man-made habitats such as rock walls or bridge abutments; 100–500 m; Ala., Ark., Ga., Ill., Ind., Iowa, Kans., Ky., Md., Mo., N.C., Ohio., Okla., Pa., Tenn., Va., W.Va., Wis.

Cystopteris tennesseensis, an allotetraploid species, has *C. bulbifera* and *C. protrusa* as diploid progenitors. The relative distinctiveness of these diploids suggests that identification of *C. tennesseensis* individuals should be straightforward. As with other members of *Cystopteris*, however, a series of features makes reliable recognition of this tetraploid challenging. For some characteristics (occasional unicellular, gland-tipped hairs and bulblets; short-attenuate, narrowly deltate blades), it is intermediate between its parents; for others (very short internodes and crowded leaves; occurrence on rock), it tends toward *C. bulbifera*. This unequal intermediacy, the multiple origins from genetically different individuals (C. H. Haufler et al. 1990), and the occurrence of sterile backcross triploids with its diploid progenitors in zones of sympatry has blurred the already subtle features distinguishing this allopolyploid. For example, some individuals of *C. bulbifera* may have very few glandular hairs, and some *C. tennesseensis* appear to lack glandular hairs entirely (R. F. Blasdell 1963). Further, sterile tetraploid hybrids (called *C.* ×*wagneri* R. C. Moran) between *C. tennesseensis* and *C. tenuis* have been reported (R. C. Moran 1983) and verified through isozyme analyses (C. H. Haufler, unpubl. data). Finally, as discussed above, the recently recognized *C. utahensis* (C. H. Haufler and M. D. Windham 1991) is extremely similar morphologically to *C. tennesseensis*.

6. **Cystopteris protrusa** (Weatherby) Blasdell, Mem. Torrey Bot. Club 21(4): 41. 1963 · Southern bladder fern, lowland brittle fern

Cystopteris fragilis (Linnaeus) Bernhardi var. *protrusa* Weatherby, Rhodora 37: 373. 1935

Stems creeping, not cordlike, internodes long, 0.5–1 cm, with persistent petiole bases, covered with tan to golden hairs, especially toward apex; scales tan to light brown, ovate-lanceolate to lanceolate, radial walls thin, luminae tan, mostly

crowded at stem apex. **Leaves** seasonally dimorphic, clustered 1 to several cm beyond persistent old petiole bases, especially in late spring and early summer, to 45 cm, bearing sori (earliest leaves smaller, sterile, coarsely divided, margins with rounded teeth; subsequent leaves larger, fertile, more finely divided, margins with sharply pointed teeth). **Petiole** mostly green to straw-colored throughout, shorter than or nearly equaling blade, base sparsely scaly. **Blade** ovate to elliptic, 1-pinnate-pinnatifid to 2-pinnate, widest at or just below middle, apex broadly acute; rachis and costae lacking gland-tipped hairs or bulblets; axils of pinnae lacking multicellular, gland-tipped hairs. **Pinnae** usually perpendicular to rachis, not curving toward blade apex, margins dentate to serrate; proximal pinnae pinnatifid to pinnate-pinnatifid, ± equilateral, basiscopic pinnules not enlarged, basal basiscopic pinnules stalked, base truncate to obtuse; distal pinnae deltate to ovate. **Veins** mostly directed into teeth. **Indusia** ovate to cup-shaped, lacking gland-tipped hairs. **Spores** spiny, usually 28–34 μm. $2n = 84$.

Sporulating spring–summer. In soil of moist, deciduous forests; 0–1500 m; Ont.; Ala., Ark., Conn., Del., Ga., Ill., Ind., Iowa, Kans., Ky., La., Md., Mass., Mich., Minn., Miss., Mo., Nebr., N.J., N.Y., N.C., Ohio., Okla., Pa., S.C., Tenn., Va., W.Va., Wis.

The terrestrial habit and characteristic stem features (golden hairs and protruding apex) readily distinguish *Cystopteris protrusa* from other *Cystopteris* species with which it may be sympatric (*C. tenuis, C. tennesseensis,* and *C. bulbifera*). *Cystopteris tennesseensis* and *C. tenuis* are allotetraploids that have *C. protrusa* as one parent. The other progenitor for *C. tennesseensis* is *C. bulbifera,* and a presumably extinct diploid is proposed here as the progenitor for *C. tenuis*. When *C. protrusa* is sympatric with either of these derived tetraploid species, sterile triploids are often produced. In addition, there are sterile autotriploids within *C. protrusa* (C. H. Haufler et al. 1985).

7. **Cystopteris reevesiana** Lellinger, Amer. Fern J. 71: 92. 1981 · Southwestern brittle fern

Cystopteris fragilis (Linnaeus) Bernhardi subsp. *tenuifolia* Clute

Stems creeping, not cordlike, internodes usually long, with scattered persistent petiole bases, hairs absent; scales tan to brown, ovate to lanceolate, radial walls thin, luminae tan. **Leaves** monomorphic, clustered at stem apex, to 45 cm, bearing sori throughout year. **Petiole** highly variable in color, from uniformly dark purple to uniformly straw-colored, but mostly dark purple at base, grading to straw-colored at junction with blade, shorter than blade, base sparsely scaly. **Blade** ovate to elliptic, 2–3-pinnate, widest at or just below middle, apex short-attenuate; rachis and costae lacking gland-tipped hairs or bulblets; axils of pinnae with occasional multicellular, gland-tipped hairs. **Pinnae** usually perpendicular to rachis, not curving toward blade apex, margins dentate to crenate; proximal pinnae pinnate-pinnatifid to 2-pinnate, ± equilateral, basiscopic pinnules not enlarged; basal basiscopic pinnules mostly short-stalked, base truncate to obtuse, distal pinnae deltate to ovate. **Veins** directed into teeth and notches. **Indusia** cup-shaped to lanceolate, gland-tipped hairs absent. **Spores** spiny, usually averaging 33–41 μm. $2n = 84$.

Sporulating summer–fall. Terrestrial or on rock on variety of substrates; 1500–4000 m; Ariz., Colo., N.Mex., Tex., Utah; Mexico.

The finely dissected leaves, dark petioles, creeping stems, smaller spores, and terrestrial habit distinguish *Cystopteris reevesiana* from *C. fragilis* in the southwest. On rock and at high elevations, however, *C. reevesiana* can have stems with short internodes and leaves that are reduced in size and dissection (resembling *C. fragilis*). In southern Colorado, the two species are sympatric in some areas and form triploid hybrids. *Cystopteris reevesiana* and *C. bulbifera* are the diploid progenitors of *C. utahensis*, which occasionally crosses with *C. reevesiana* to produce sterile triploid hybrids of intermediate morphology.

8. **Cystopteris tenuis** (Michaux) Desvaux, Mém. Soc. Linn. Paris 6: 264. 1827

· Mackay's brittle fern, cystoptère ténue

Nephrodium tenue Michaux, Fl. Bor.-Amer. 2: 269. 1803; *Cystopteris fragilis* (Linnaeus) Bernhardi var. *mackayi* G. Lawson

Stems creeping, not cordlike, internodes short, beset with old petiole bases, hairs absent; scales tan to light brown, lanceolate, radial walls thin, luminae tan. **Leaves** monomorphic, clustered at stem apex, to 40 cm, nearly all bearing sori. **Petiole** dark at base, mostly green to straw-colored distally, shorter than or nearly equaling blade, base sparsely scaly. **Blade** lanceolate to narrowly elliptic, 1(–2)-pinnate-pinnatifid, widest at or just below middle, apex short-attenuate; rachis and costae lacking gland-tipped hairs or bulblets; axils of pinnae lacking multicellular, gland-tipped hairs. **Pinnae** typically at acute angle to rachis, often curving toward blade apex, margins crenulate; proximal pinnae pinnatifid to pinnate-pinnatifid, ± equilateral, basiscopic pinnules not enlarged; basal basiscopic pinnules sessile, base cuneate to obtuse, distal pinnae ovate to narrowly elliptic. **Veins** directed into teeth and notches. **Indusia** ovate to cup-shaped, without gland-tipped hairs. **Spores** spiny, usually 39–50 μm. $2n = 168$.

Sporulating summer–fall. Mostly on shaded rock and cliff faces but also occasionally on forest floors; 0–2800 m; N.B., N.S., Ont., Que.; Ark., Ariz., Conn., Del., Ill., Ind., Iowa, Kans., Ky., Maine, Md., Mass., Mich., Minn., Mo., Nebr., Nev., N.H., N.J., N.Y., N.C., Ohio, Okla., Pa., R.I., Tenn., Utah, Vt., Va., W.Va., Wis.

Long recognized as *Cystopteris fragilis* var. *mackayi*, *C. tenuis* was returned to species status by R. C. Moran (1983b). It is probably an allotetraploid originating from *C. protrusa* and an extinct diploid related to *C. fragilis* (C. H. Haufler 1985; C. H. Haufler and M. D. Windham 1991).

Cystopteris tenuis is common in eastern North America and less frequent at the northern and western perimeter of its range. In the center of its distribution (Minnesota, Iowa, Illinois, Wisconsin, Indiana, Ohio, Pennsylvania), the narrow, elliptic pinnae angled toward the blade apex and the rounded teeth make *C. tenuis* relatively distinct from *C. fragilis* and *C. protrusa* (although the early season, sterile leaves of *C. protrusa* often resemble those of *C. tenuis*). In the west and especially in the northeast, *C. tenuis* and *C. fragilis* are difficult to distinguish. For the most part, *C. fragilis* is confined to higher latitudes and elevations than *C. tenuis*, but the two species can be sympatric and occasionally form sterile tetraploid hybrids. *Cystopteris protrusa* and *C. tenuis* are infrequently sympatric, but where they are, sterile triploid hybrids can occur. Hybrids between *C. tenuis* and *C. tennesseensis* are recognized as *C.* ×*wagneri* (R. C. Moran 1983). Hybridization between *C. tenuis* and *C. bulbifera* has also been reported (R. C. Moran 1982b). This hybrid, *C.* ×*illinoensis* R. C. Moran, is known only from the type and needs to be studied further.

9. **Cystopteris fragilis** (Linnaeus) Bernhardi, Neues J. Bot. 1(2): 26, plate 2, fig. 9. 1806 [1805]

· Brittle fern, fragile fern, cystoptère fragile

Polypodium fragile Linnaeus, Sp. Pl. 2: 1091. 1753; *Cystopteris dickieana* Sim; *C. fragilis* subsp. *dickieana* (Sim) Hylander

Stems creeping, not cordlike, internodes short, beset with old petiole bases, hairs absent; scales tan to light brown, lanceolate, radial walls thin, luminae tan. **Leaves** monomorphic, clustered at stem apex, to 40 cm, nearly all bearing sori. **Petiole** dark at base, mostly green to straw-colored distally, shorter than or nearly equaling blade, base sparsely scaly. **Blade** lanceolate to narrowly elliptic, 1(–2)-pinnate-pinnatifid, widest at or just below middle, apex acute; rachis and costae lacking gland-tipped hairs or bulblets; axils of pinnae lacking multicellular, gland-tipped hairs. **Pinnae** usually perpendicular to rachis, not curving toward blade apex, margins serrate to sharply dentate; proximal pinnae pinnatifid to pinnate-pinnatifid, ± equilateral, basiscopic pinnules not enlarged; basal basiscopic pinnules sessile, base truncate to obtuse, distal pinnae deltate to ovate. **Veins** directed mostly into teeth. **Indusia** ovate to lanceolate, without gland-tipped hairs. **Spores** spiny or verrucate, usually 39–60 μm. $2n = 168, 252$.

Sporulating summer–fall. Mostly on cliff faces, also in thin soil over rock; 0–4500 m; Greenland; St. Pierre and Miquelon; Alta., B.C., Man., N.B., Nfld., N.W.T., N.S., Ont., P.E.I., Que., Sask., Yukon; Alaska, Calif., Colo., Conn., Idaho, Ill., Ind., Iowa, Kans., Maine, Mass., Mich., Minn., Mont., Nebr., Nev., N.H., N.Mex., N.Y., N.Dak., Ohio, Oreg., Pa., S.Dak., Tex., Utah, Vt., Wash., Wis., Wyo.; worldwide.

Cystopteris fragilis is most often confused with *C. tenuis* in the east and *C. reevesiana* in the southwest. Habitat and geography, as well as the morphologic features discussed in the key, usually serve to separate these taxa. For instance, *C. fragilis* is more likely to be found on cliffs whereas the other species prefer boulders and soil (*C. fragilis* occurs at higher elevations and/or latitudes than the other species). These distinctions can be confounded when *C. fragilis* forms hybrids with sympatric species. Sterile pentaploid plants have been discovered where *C. fragilis* overlaps with *C. laurentiana*, tetraploid hybrids are likely where *C. fragilis* occurs with *C. tenuis*, and triploids may form where *C. fragilis* is found with *C. reevesiana*. Even after segregating relatively distinct elements such as *Cystopteris protrusa, C. reevesiana*, and *C. tenuis*, and identifying sterile hybrids, *C. fragilis* still remains a polymorphic and complex taxon that probably contains a number of natural, cryptic evolutionary units. For example, morphologically distinct hexaploid cytotypes have been reported

(C. H. Haufler and M. D. Windham 1991). These occur as isolated and disjunct populations in Ontario, Alaska, Arizona, Colorado, Montana, and Wyoming. Isozymic profiles of each of these populations indicate that the hexaploids are polyphyletic and should not be accorded species status.

The presence of verrucate spores (as opposed to the normal spiny spores) has been used to circumscribe *Cystopteris dickieana*. Although genetic analyses have not been undertaken, we think the verrucate spore is probably a recessive feature controlled by one or a few genes. While present at low frequency in much of the range of *C. fragilis*, verrucate spores are particularly prominent in the Great Plains. Perhaps in this region the genetic combinations specifying verrucate spores have been fixed. Following R. F. Blasdell (1963), *C. dickieana* is also considered here to be conspecific with *C. fragilis* because (1) early stages in the development of spiny spores can appear verrucate (A. C. Jermy and L. Harper 1971), (2) the hexaploid cytotypes discussed above always have verrucate spores, regardless of their parentage, (3) individuals with verrucate spores can be found in populations that are otherwise uniformly spiny-spored, and (4) individuals and populations that have verrucate spores are not otherwise (morphologically, ecologically, or genetically) distinct from those that have spiny spores.

Especially in the western portion of its North American range (British Columbia, Washington, Montana, Idaho, Oregon, California), *Cystopteris fragilis* appears to be developing morphologically and ecologically distinctive variants. Hybrid individuals with aborted spores have been discovered, and plants from these areas increasingly tend to grow on both soil and rock and to have slightly different morphologies on the two substrates. These variants intergrade, however, and are not sufficiently distinct to warrant species status. This polymorphic polyploid is probably actively speciating at the tetraploid level, perhaps through gene silencing (C. R. Werth and M. D. Windham 1991).

8. WOODSIA R. Brown, Prodr., 158. 1810 · Cliff fern [for English botanist Joseph Woods]

Michael D. Windham

Plants usually on rock. **Stems** compact to creeping; ascending or erect (rarely horizontal), stolons absent. **Leaves** monomorphic, dying back over winter or sometimes persistent into the next season. **Petiole** 1/5–3/4 length of blade, base not conspicuously swollen; vascular bundles 2, arranged laterally, ± round or oblong in cross section. **Blade** linear to lanceolate or ovate,

1–2-pinnate-pinnatifid, gradually reduced distally to pinnatifid apex, herbaceous. **Pinnae** not articulate to rachis, segment margins entire to dentate, not spiny; proximal pinnae somewhat reduced, sessile, bases usually ± equilateral; costae often shallowly grooved adaxially, grooves ± continuous from rachis to costae; indument of glandular (occasionally nonglandular) hairs on both surfaces, rarely absent. **Veins** free, simple or forked. **Sori** in 1 row between midrib and margin on ultimate segments, round; indusia basal, dissected into several to numerous filamentous or scalelike segments encircling sorus, persistent but often obscure in mature sori. **Spores** brownish, cristate, rarely rugose. $x = 38, 39, 41$.

Species ca. 30 (10 in the flora): mostly north temperate regions and higher elevations in the tropics.

Woodsia is a well-marked genus; its morphology and chromosome base number ($x = 41$) provide evidence of relationships to the dryopteroid ferns. Most authors consider *Cystopteris* to be its closest ally, and the two genera are often confused in herbarium collections. The resemblance is superficial in many ways, however, and *Woodsia* is easily distinguished from *Cystopteris* by its persistent petiole bases, multilobed indusia, and obscure veins that end in hydathodes before reaching the leaf margin. The North American species of *Woodsia* fall into two natural groups that might be recognized as subgenera. *Woodsia ilvensis, W. glabella,* and *W. alpina* have articulate petioles, indusial segments that are uniseriate throughout and composed of cells that are much longer than wide, entire or crenate pinnules, strictly concolored stem scales, and chromosome base numbers of 39–41. They are circumboreal in distribution and show clear affinities to species found only in Eurasia. The remainder of the North American taxa have petioles that are not articulate, indusial segments that are multiseriate at the base and composed of cells that are isodiametric or slightly longer than wide, dentate pinnules, often bicolored stem scales, and a chromosome base number of 38. All of these species are endemic to the New World and probably represent a distinct lineage within the genus. Hybridization is common within these natural groups, but intergroup hybrids are relatively rare.

SELECTED REFERENCES Brown, D. F. M. 1964. A monographic study of the fern genus *Woodsia*. Nova Hedwigia 16: 1–154. Taylor, T. M. C. 1947. New species and combinations in *Woodsia* section *Perrinia*. Amer. Fern J. 37: 84–88. Wagner, F. S. 1987. Evidence for the origin of the hybrid cliff fern, *Woodsia* × *abbeae* (Aspleniaceae: Athyrioideae). Syst. Bot. 12: 116–124. Windham, M. D. 1987b. Chromosomal and electrophoretic studies of the genus *Woodsia* in North America. Amer. J. Bot. 74: 715.

1. Blades and rachises completely glabrous except for occasional sessile glands; proximal pinnae distinctly fan-shaped, usually wider than long; leaves 0.5–1.2 cm wide, mature petioles green or straw-colored throughout. . . . . 1. *Woodsia glabella*
1. Blades and/or rachises with scattered hairs, scales, or stalked glands, rarely glabrescent; proximal pinnae ovate-lanceolate to deltate, usually longer than wide; leaves 1.2–12 cm wide or, if less, then proximal portion of mature petioles reddish brown or dark purple.
   2. Petioles articulate well above base, abscission zone visible as swollen node; indusial segments uniseriate throughout, composed of cells that are many times longer than wide; pinnules entire or crenate, without acute teeth on margins.
      3. Linear-lanceolate scales absent or very rare on abaxial pinnae surfaces; rachises with widely scattered hairs and scales, sometimes nearly glabrous; largest pinnae with 1–3 pairs of pinnules.. . . . . 2. *Woodsia alpina*
      3. Linear-lanceolate scales common on abaxial pinnae surfaces; rachises with abundant hairs and scales; largest pinnae with 4–9 pairs of pinnules. . . . . 3. *Woodsia ilvensis*
   2. Petioles not articulate above base, swollen abscission zone absent; indusial segments usually multiseriate at base, composed of cells that are isodiametric or slightly longer than wide; pinnules dentate, with acute teeth on margins.

4. Pinnae with flattened, multicellular hairs concentrated along midrib on both surfaces; mature petioles usually reddish brown or dark purple, relatively brittle and easily shattered. . . . . 4. *Woodsia scopulina*

4. Pinnae lacking flattened, multicellular hairs along midrib; mature petioles light brown to straw-colored or, if reddish brown/dark purple (in *Woodsia plummerae* and *W. oregana*), somewhat pliable and resistant to shattering.

    5. Indusia composed of relatively broad segments, these multiseriate for most of length but often branched or divided distally.

        6. Proximal portion of mature petioles reddish brown or dark purple; blades densely glandular, often somewhat viscid; vein tips not enlarged, barely visible on adaxial surface. . . . . 5. *Woodsia plummerae*

        6. Proximal portion of mature petioles light brown or straw-colored (sometimes darker at very base); blades sparsely to moderately glandular, rarely viscid; vein tips usually enlarged to form whitish hydathodes visible on adaxial surface.

            7. Indusial segments branched or divided distally to form narrow, filamentous lobes; glandular hairs of blade with thin stalks and slightly expanded tips; pinnule margins usually thickened, lustrous on adaxial surface. . . . . 6. *Woodsia cochisensis*

            7. Indusial segments often glandular along distal edge but otherwise nearly entire, not divided into narrow, filamentous lobes; many glandular hairs of blade with thick stalks and distinctly bulbous tips; pinnule margins not noticeably thickened or lustrous. . . . . 7. *Woodsia obtusa*

    5. Indusia composed of narrow, usually filamentous segments, these uniseriate for most of length.

        8. Pinnule margins (viewed from abaxial surface) smooth to somewhat ragged but usually lacking translucent projections or filaments; proximal portion of mature petioles reddish brown or dark purple; indusial filaments generally inconspicuous, concealed by or slightly surpassing mature sporangia. . . . 8. *Woodsia oregana*

        8. Pinnule margins (viewed from abaxial surface) with translucent projections or filaments on teeth; proximal portion of mature petioles usually light brown or straw-colored (sometimes darker at very base); indusial filaments generally apparent, often greatly surpassing mature sporangia.

            9. Translucent projections on pinnule margins mostly 1–2-celled, occasionally filamentous; spores averaging 44–52 μm; largest pinnae divided into 3–7 pairs of closely spaced pinnules, pinna apices usually abruptly tapered to rounded. . . . . 9. *Woodsia neomexicana*

            9. Translucent projections on pinnule margins mostly multicellular, often prolonged to form twisted filaments; spores averaging 37–44 μm; largest pinnae with 7–18 pairs of discrete, widely spaced pinnules, pinna apices often attenuate to narrowly acute. . . . . 10. *Woodsia phillipsii*

1. **Woodsia glabella** R. Brown ex Richardson in Franklin, Narr. Journey Polar Sea, 754. 1823 · Smooth cliff fern, woodsie glabre

Woodsia alpina (Bolton) Gray var. *glabella* (R. Brown ex Richardson) D. C. Eaton; *W. hyperborea* (Liljeblad) R. Brown var. *glabella* (R. Brown ex Richardson) Watt

Stems compact, erect to ascending, with cluster of persistent petiole bases of ± equal length; scales uniformly brown, lanceolate. **Leaves** 3.5–15 × 0.5–1.2 cm. **Petiole** green or straw-colored throughout, articulate above base at swollen node, somewhat pliable and resistant to shattering. **Blade** linear to linear-lanceolate, pinnate-pinnatifid proximally, glabrous or with occasional sessile glands, never viscid; rachis glabrous. **Proximal pinnae** fan-shaped, wider than long; distal pinnae ovate-lanceolate, longer than wide, abruptly tapered to a rounded or broadly acute apex; largest pinnae with 1–3 pairs of pinnules, abaxial and adaxial surfaces glabrous. **Pinnules** entire or broadly crenate; margins nonlustrous, thin, lacking cilia or translucent projections. **Vein tips** slightly (if at all) enlarged, barely visible adaxially. **Indusia** of narrow hairlike segments, these

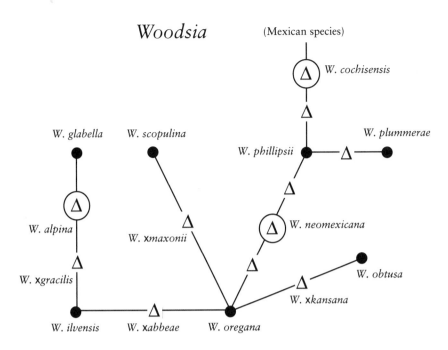

Relationships in *Woodsia*. Solid circles represent parental taxa; circled triangles represent allotetraploids; and triangles represent sterile hybrids.

uniseriate throughout, composed of cells many times longer than wide, usually surpassing mature sporangia. **Spores** averaging 39–45 μm. $2n = 78$.

Sporulating summer–early fall. Shaded cracks and ledges on cliffs; mostly calcareous rocks, especially limestone; 0–1500 m; Greenland; Alta., B.C., Man., N.B., Nfld., N.W.T., N.S., Ont., Que., Sask., Yukon; Alaska, Maine, Minn., N.H., N.Y., Vt.; n Eurasia.

Woodsia glabella is a well-marked species occasionally confused with narrow, glabrescent forms of *W. alpina* and *W. oregana* subsp. *oregana*. These taxa are readily distinguished from *W. glabella* by their petioles, which are reddish brown or dark purple near the base.

2. Woodsia alpina (Bolton) Gray, Nat. Arr. Brit. Pl. 2: 17. 1821 · Alpine cliff fern, woodsie alpine

Acrostichum alpinum Bolton, Fil. Brit. 2: 76, plate 42. 1790; *Woodsia alpina* var. *bellii* Lawson; *W. bellii* (Lawson) Porsild; *W. hyperborea* (Liljeblad) R. Brown; *W. ilvensis* (Linnaeus) R. Brown var. *alpina* (Bolton) Watt

Stems compact, erect to ascending, with cluster of persistent petiole bases of ± equal length; scales uniformly brown, lanceolate. **Leaves** 2.5–20 × 0.5–2.5 cm. **Petiole** red-

dish brown or dark purple when mature, articulate above base at swollen node, relatively brittle and easily shattered. **Blade** linear to narrowly lanceolate, usually pinnate-pinnatifid proximally, lacking glands, never viscid; rachis with widely scattered hairs and scales. **Pinnae** ovate-lanceolate to deltate, longer than wide, abruptly tapered to a rounded or broadly acute apex; largest pinnae with 1–3 pairs of pinnules; abaxial surface with isolated hairs and linear scales, adaxial surface glabrous. **Pinnules** entire or broadly crenate; margins nonlustrous, thin, with occasional isolated cilia, lacking translucent projections. **Vein tips** often enlarged to form whitish hydathodes visible adaxially. **Indusia** of narrow, hairlike segments, these uniseriate throughout, composed of cells many times longer than wide, usually surpassing mature sporangia. **Spores** averaging 46–53 μm.

Sporulating summer–early fall. Crevices and ledges on cliffs (occasionally on rocky slopes); mostly slaty and calcareous rocks; 0–1500 m; Greenland; B.C., Man., N.B., Nfld., N.W.T., N.S., Ont., Que., Sask., Yukon; Alaska, Maine, Mich., Minn., N.H., N.Y., Vt.; n Eurasia.

Isozyme studies confirm the longstanding hypothesis that *Woodsia alpina* is an allotetraploid derived from hybridization between *W. glabella* and *W. ilvensis* (see reticulogram). Considerable disagreement exists con-

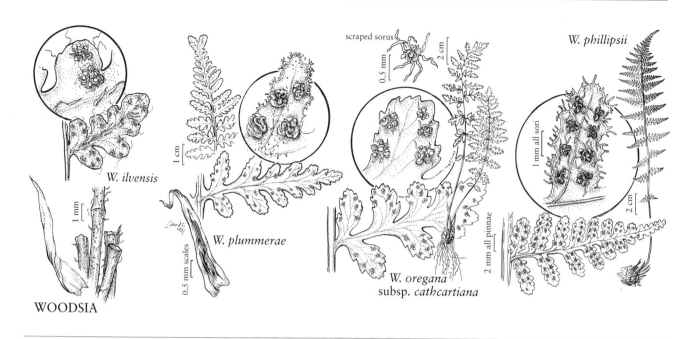

W. ilvensis

W. plummerae

scraped sorus

W. oregana
subsp. *cathcartiana*

W. phillipsii

WOODSIA

cerning the chromosome number of *W. alpina*, but 2*n* = 160 seems most likely, given the numbers reported for the two parental species. Hybrids between *W. alpina* and *W. ilvensis* have been reported from both Europe and North America. These morphologically intermediate triploids with malformed spores have been called *W.* × *gracilis* (Lawson) Butters.

3. **Woodsia ilvensis** (Linnaeus) R. Brown, Trans. Linn. Soc. London 11: 173. 1813 · Rusty cliff fern, woodsie de l'île d'Elbe

Acrostichum ilvense Linnaeus, Sp. Pl. 2: 1071. 1753

Stems compact, erect to ascending, with abundant persistent petiole bases of ± equal length; scales uniformly brown, lanceolate. **Leaves** 4.5–25 × 1.2–3.5 cm. **Petiole** usually brown or dark purple when mature, articulate above base at swollen node, relatively brittle and easily shattered. **Blade** narrowly lanceolate, usually 2-pinnate proximally, lacking glands, never viscid; rachis usually with abundant hairs and scales. **Pinnae** ovate-lanceolate to deltate, longer than wide, abruptly tapered to a rounded or broadly acute apex; largest pinnae with 4–9 pairs of pinnules; abaxial surface with mixture of hairs and linear-lanceolate scales, adaxial surface with multicellular hairs concentrated along midrib. **Pinnules** entire or crenate, rarely shallowly lobed; margins nonlus-

trous, thin, ciliate with multicellular hairs, lacking translucent projections. **Vein tips** frequently enlarged to form whitish hydathodes visible adaxially. **Indusia** of narrow, hairlike segments, these uniseriate throughout, composed of cells many times longer than wide, usually surpassing mature sporangia. **Spores** averaging 39–46 μm. 2*n* = 82.

Sporulating summer–early fall. Cliffs and rocky slopes; found on variety of substrates including serpentine; 0–1500 m; Greenland; Alta., B.C., Man., N.B., Nfld., N.W.T., N.S., Ont., Que., Sask., Yukon; Alaska, Conn., Ill., Iowa, Maine, Md., Mass., Mich., Minn., N.H., N.J., N.Y., N.C., Ohio, Pa., R.I., Vt., Va., W.Va., Wis.; n Eurasia.

Although generally separable by the characters given in the key, shade forms of *Woodsia ilvensis* with a reduced number of scales and hairs are occasionally misidentified as *W. alpina*. The morphologic distinctions between these species are further blurred by natural hybridization, which produces the intermediate triploid known as *W.* × *gracilis*. Some of the best characters for distinguishing these taxa are spore size and morphology. Spores average less than 46 μm in *W. ilvensis*, more than 46 μm in *W. alpina*, and are malformed and abortive in *W.* × *gracilis*. *Woodsia ilvensis* also hybridizes with *W. oregana* subsp. *cathcartiana* to form the sterile triploid *W.* × *abbeae* (F. S. Wagner 1987).

4. Woodsia scopulina D. C. Eaton, Canad. Naturalist &
Quart. J. Sci., n. s. 2: 91. 1865 · Mountain cliff fern,
woodsie des falaises

Woodsia obtusa (Sprengel) Torrey var. *lyallii* Hooker; *W.*
oregana D. C. Eaton var. *lyallii* (Hooker) Boivin

Stems compact, erect to ascending, with few to many
persistent petiole bases of unequal lengths; scales uniformly brown or bicolored with dark central stripe and
pale brown margins, ovate to narrowly lanceolate.
Leaves 9–35 × 1–8 cm. **Petiole** usually reddish brown
to dark purple proximally when mature, not articulate
above base, relatively brittle and easily shattered. **Blade**
lanceolate to linear-lanceolate, 2-pinnate proximally,
moderately glandular, rarely somewhat viscid; most
glandular hairs with thick stalks and distinctly bulbous
tips; rachis usually with abundant glandular and nonglandular hairs. **Pinnae** lanceolate-deltate to ovate, longer
than wide, abruptly tapered to a rounded or broadly
acute apex, occasionally attenuate; largest pinnae with
5–14 pairs of pinnules; abaxial and adaxial surfaces
glandular and sparsely villous, with flattened, multicellular hairs concentrated along midribs. **Pinnules** dentate, often shallowly lobed; margins nonlustrous, thin,
slightly glandular and occasionally ciliate with isolated,
multicellular hairs, lacking translucent projections. **Vein
tips** slightly (if at all) enlarged, barely visible adaxially.
Indusia of filamentous or nonfilamentous segments, these
multiseriate proximally, often uniseriate distally, composed of ± isodiametric cells, concealed by or slightly
surpassing mature sporangia. **Spores** averaging 39–57
μm.

Subspecies 3: only in the flora.

Woodsia scopulina shows substantial variation in leaf
size, shape, and dissection, and in the abundance of
multicellular hairs on the pinnae. Although much of
this variation seems to be environmentally induced, recent studies (M. D. Windham 1993) have identified three
chromosomal/morphologic variants that are treated here
as subspecies. Diploid populations of *W. scopulina* are
divisible into two groups, one of which (subsp. *scopulina*) is scattered throughout the mountainous regions
of western North America while the other (subsp. *appalachiana*) is confined to montane habitats in the
southeastern United States. These taxa seem amply distinct (T. M. C. Taylor 1947) and might be considered
separate species if not for the existence of populations
in the Great Lakes region and western cordillera that
tend to bridge the morphologic and geographic gap between them. These intermediate populations (subsp.
laurentiana) appear to be uniformly tetraploid and may
have arisen through ancient hybridization between subsp.
scopulina and subsp. *appalachiana*. In regions where
subsp. *laurentiana* is sympatric with subsp. *scopulina*,
the two taxa are rarely found growing together, sug-

gesting that they differ in their ecological tolerances and/
or habitat requirements.

1. Scales of stems and petiole bases narrowly lanceolate, mostly bicolored with broad, usually
continuous, dark central stripe; longest hairs on
pinnae composed of 5–8 cells; indusial segments
broad, not at all filamentous. . . . .
. . . . . . 4a. *Woodsia scopulina* subsp. *appalachiana*
1. Scales of stem and petiole bases ovate-lanceolate,
mostly concolored or weakly bicolored with narrow, often discontinuous, dark central stripe;
longest hairs on pinnae composed of 2–5 cells;
indusial segments narrow, often filamentous distally.
  2. Spores averaging 42–50 μm; stem and petiole
base scales usually concolored or with a few
isolated, dark, occluded cells. . . . .
. . . . . . .4b. *Woodsia scopulina* subsp. *scopulina*
  2. Spores averaging 50–57 μm; at least some
stem and petiole base scales with clusters of
dark, occluded cells near center forming narrow, usually discontinuous stripe. . . . . .
. . . . . . 4c. *Woodsia scopulina* subsp. *laurentiana*

4a. Woodsia scopulina D. C. Eaton subsp.
appalachiana (T. M. C. Taylor) Windham, Contr.
Univ. Michigan Herb. 19: 58. 1993

Woodsia appalachiana T. M. C.
Taylor, Amer. Fern J. 37: 88. 1947;
W. scopulina var. *appalachiana*
(T. M. C. Taylor) C. V. Morton

Scales of stems and petiole bases
mostly bicolored with broad,
usually continuous, dark central
stripe, scales narrowly lanceolate. **Pinnae** with longest hairs
composed of 5–8 cells. **Indusial segments** broad, not at
all filamentous. **Spores** averaging 39–46 μm. 2n = 76.

Sporulating summer–fall. Shaded cracks and ledges
on cliffs; mostly subacidic shale and sandstone; 300–
1700 m; Ark., Ky., N.C., Tenn., Va., W.Va.

4b. Woodsia scopulina D. C. Eaton subsp. **scopulina**

Scales of stems and petiole bases
usually concolored or with a few
isolated, dark, occluded cells,
scales ovate-lanceolate. **Pinnae**
with longest hairs composed of
2–5 cells. **Indusial segments** narrow, often filamentous distally.
Spores averaging 42–50 μm. 2n
= 76.

Sporulating summer–fall. Cliffs and rocky slopes;
found on variety of substrates including both granite
and limestone; 100–4000 m; Alta., B.C., Sask., Yu-

kon; Alaska, Calif., Colo., Idaho, Mont., Nev., Oreg., Utah, Wash., Wyo.

Woodsia scopulina subsp. *scopulina* is known to hybridize with subsp. *laurentiana* at localities where the two grow in close proximity. The resultant triploids have malformed spores and appear to be sterile.

4c. Woodsia scopulina D. C. Eaton subsp. laurentiana Windham, Contr. Univ. Michigan Herb. 19: 59. 1993

Scales of stems and petiole bases (at least some) with clusters of dark, occluded cells near center forming narrow, usually discontinuous stripe, scales ovate-lanceolate. **Pinnae** with longest hairs composed of 2–5 cells. **Indusial segments** narrow, often filamentous distally. **Spores** averaging 50–57 μm. 2n = 152.

Sporulating summer–fall. Cliffs and rocky slopes; found on a variety of substrates including both granite and limestone; 0–3000 m; Alta., B.C., Ont., Que.; Ariz., Calif., Colo., Idaho, Mont., Nev., Oreg., S.Dak., Utah, Wash., Wyo.

In addition to hybridizing with *Woodsia scopulina* subsp. *scopulina* (see comments above), subsp. *laurentiana* may have crossed with *Woodsia oregana* subsp. *cathcartiana* to form *W.* ×*maxonii* R. M. Tryon. With very few collections and no biosystematic data available, however, the origin of this putative hybrid remains in doubt. Contrary to previous hypotheses (D. F. M. Brown 1964; D. B. Lellinger 1985), Great Lakes populations of *W. scopulina* were not involved in the origin of the local hybrid known as *W.* ×*abbeae* (F. S. Wagner 1987).

5. Woodsia plummerae Lemmon, Bot. Gaz. 7: 6. 1882 · Plummer's cliff fern

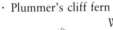

Woodsia obtusa (Sprengel) Torrey var. *glandulosa* D. C. Eaton & M. Faxon; *W. obtusa* var. *plummerae* (Lemmon) Maxon; *W. pusilla* E. Fournier var. *glandulosa* (D. C. Eaton & M. Faxon) T. M. C. Taylor

Stems compact, erect to ascending, with a few persistent petiole bases of unequal lengths; scales often uniformly brown but at least some bicolored with dark central stripe and pale brown margins, narrowly lanceolate. **Leaves** 5–25 × 1.5–6 cm. **Petiole** reddish brown to dark purple when mature, not articulate above base, somewhat pliable and resistant to shattering. **Blade** lanceolate to ovate, usually 2-pinnate proximally, densely glandular, often somewhat viscid; most glandular hairs with thick stalks

and distinctly bulbous tips; rachis with abundant glandular hairs and a few narrow scales. **Pinnae** ovate-deltate to elliptic, longer than wide, abruptly tapered to a rounded or broadly acute apex, occasionally attenuate; largest pinnae with 5–11 pairs of pinnules; abaxial and adaxial surfaces glandular, lacking nonglandular hairs or scales. **Pinnules** dentate, often shallowly lobed; margins nonlustrous, thin, densely glandular, lacking cilia but with occasional 1–2-celled translucent projections. **Vein tips** slightly (if at all) enlarged, barely visible adaxially. **Indusia** of relatively broad segments; segments multiseriate for most of length, often divided and uniseriate distally, composed of ± isodiametric cells, often surpassing mature sporangia. **Spores** averaging 44–50 μm. 2n = 152.

Sporulating late spring–fall. Cliffs and rocky slopes; usually on granite or volcanic substrates; 700–3100 m; Ariz., Calif., Colo., N.Mex., Okla., Tex.; n Mexico.

The origin and phylogenetic affinities of the tetraploid *Woodsia plummerae* have not been established with certainty. The hypothesis that it arose as a hybrid between the Mexican species *W. mollis* (Kaulfuss) J. Smith and *W. mexicana* Fée (D. F. M. Brown 1964; D. B. Lellinger 1985) seems untenable in light of recent chromosome studies indicating that the latter species is also tetraploid (M. D. Windham 1993). On the basis of sporophyte morphology and spore ornamentation, *W. plummerae* appears most closely related to the *W. mexicana* complex and *W. oregana*. In fact, *W. oregana* can be difficult to separate from *W. plummerae* in western New Mexico and northern Arizona. Intermediate plants occurring in this region may represent stable allotetraploids resulting from hybridization between the diploid progenitors of *W. plummerae* and *W. oregana* subsp. *cathcartiana*. Considering the available evidence, populations of *W. plummerae* in the United States probably originated through autopolyploidy from a recently discovered, but as yet unnamed, Mexican diploid of similar morphology. *Woodsia plummerae* occasionally hybridizes with *W. phillipsii* to produce sterile, morphologically intermediate triploids.

6. Woodsia cochisensis Windham, Contr. Univ. Michigan Herb. 19: 54. 1993 · Cochise cliff fern

Stems compact, erect to ascending, with a few persistent petiole bases of unequal lengths; scales often uniformly brown but at least some bicolored with dark central stripe and pale brown margins, narrowly lanceolate. **Leaves** 5–25 × 1.5–6 cm. **Petiole** light brown or straw-colored throughout when mature, occasionally darker at very base, not articulate above base, relatively brittle and easily shattered. **Blade** narrowly lanceolate to ovate, pinnate-pinnatifid to 2-pinnate proximally, sparsely to moderately glandular, never viscid; glandular hairs with thin stalks and slightly expanded tips; rachis with glandular hairs and occasional hairlike scales. **Pinnae** ovate-deltate to elliptic, longer than wide, abruptly tapered to a rounded or broadly acute apex, occasionally attenuate; largest pinnae with 4–9 pairs of pinnules; abaxial and adaxial surfaces glandular, lacking nonglandular hairs or scales. **Pinnules** dentate, often shallowly lobed; margins lustrous adaxially, usually thickened, lacking cilia but sparsely glandular, with occasional 1–2-celled translucent projections. **Vein tips** enlarged to form whitish hydathodes visible adaxially. **Indusia** of relatively broad segments; segments multiseriate most of length, usually divided and uniseriate distally, composed of ± isodiametric cells, often surpassing mature sporangia. **Spores** averaging 43–49 µm. $2n = 152$.

Sporulating late spring–fall. Shaded ledges and alcoves near springs and seeps; usually on granitic or volcanic substrates; 1000–2200 m; Ariz., N.Mex.; n Mexico.

Woodsia cochisensis traditionally has been identified as *W. plummerae* or (rarely) *W. mexicana* Fée. It is readily separated from *W. plummerae* by the characteristics given in the key, and from North American members of the *mexicana* group (*W. phillipsii* and *W. neomexicana*) by having indusial segments that are broad and nonfilamentous at the base. *Woodsia cochisensis* is less glandular than typical *W. mexicana* from northeastern Mexico and is further distinguished from that species by the thickened, lustrous pinnule margins and well-developed hydathodes. Isozyme and chromosome studies suggest that *W. cochisensis* is an allotetraploid that may have originated through hybridization between *W. phillipsii* and an undescribed Mexican diploid (M. D. Windham 1993). It crosses with the former species to produce sterile triploids of intermediate morphology.

7. **Woodsia obtusa** (Sprengel) Torrey, in New York State, Rep. Geol. Surv., 195. 1840 · Blunt-lobed cliff fern

Polypodium obtusum Sprengel, Anleit. Kenntn. Gew. 3: 93. 1804; *Woodsia perriniana* (Sprengel) Hooker & Greville

Stems compact to creeping, erect to horizontal, with few to many persistent petiole bases of unequal lengths; scales often uniformly brown but at least some bicolored with dark central stripe and pale brown margins, narrowly lanceolate. **Leaves** 8–60 × 2.5–12 cm. **Petiole** light brown or straw-colored when mature, occasionally darker at very base, not articulate above base, relatively brittle and easily shattered. **Blade** lanceolate to ovate, 2-pinnate to 2-pinnate-pinnatifid proximally, moderately glandular, rarely somewhat viscid; many glandular hairs with thick stalks and distinctly bulbous

tips; rachis with glandular hairs and scattered, often hairlike scales. **Pinnae** ovate-deltate to elliptic, longer than wide, abruptly tapered to a rounded or broadly acute apex, occasionally attenuate; largest pinnae with 5–14 pairs of pinnules; abaxial and adaxial surfaces glandular, lacking nonglandular hairs or scales. **Pinnules** dentate, sometimes deeply lobed; margins nonlustrous, thin, with occasional glands, lacking cilia or translucent projections. **Vein tips** usually enlarged to form whitish hydathodes visible adaxially. **Indusia** of relatively broad, nonfilamentous segments, these multiseriate throughout, composed of ± isodiametric cells, entire or glandular along distal edge, concealed by or slightly surpassing mature sporangia. **Spores** averaging 35–47 µm.

Subspecies 2: only in the flora.

Woodsia obtusa comprises two cytotypes that are treated here as subspecies because they show subtle morphologic and ecological distinctions and tend to have different distributions. Tetraploid populations (subsp. *obtusa*) are found throughout the eastern flora, commonly occurring on limestone. The diploid (subsp. *occidentalis*) is found near the western edge of the species range, usually on sandstone and granitic substrates. Isozyme studies suggest that subsp. *obtusa* may have been derived from subsp. *occidentalis* through autopolyploidy (M. D. Windham 1993). The westernmost collections of *Woodsia obtusa* (all subsp. *occidentalis*) come from the Wichita Mountains of Oklahoma and the Edwards Plateau of Texas. Reports of this species from the trans-Pecos region of western Texas are apparently based on misidentifications.

1. Spores averaging 42–47 µm; proximal pinnules of lower pinnae usually shallowly lobed or merely dentate; blades coarsely cut and evidently 2-pinnate; stems compact to short-creeping, individual branches usually 5–10 mm diam. . . . . 7a. *Woodsia obtusa* subsp. *obtusa*
1. Spores averaging 35–42 µm; proximal pinnules of lower pinnae usually deeply lobed or pinnatifid; blades finely cut, 2-pinnate-pinnatifid; stems short- to long-creeping, individual branches 3–5 mm diam. . . . 7b. *Woodsia obtusa* subsp. *occidentalis*

7a. Woodsia obtusa (Sprengel) Torrey subsp. **obtusa**

Stems compact to short-creeping, individual branches usually 5–10 mm diam. **Blade** coarsely cut and evidently 2-pinnate. **Proximal pinnules** of lower pinnae usually shallowly lobed or merely dentate. **Spores** averaging 42–47 µm. $2n = 152$.

Sporulating summer–fall. Cliffs and rocky slopes (rarely terrestrial); found on a variety of substrates including both granite and limestone; 0–1000 m; Ont., Que.; Ala., Ark., Conn., Del., Fla., Ga., Ill., Ind., Iowa, Kans., Ky., La., Maine, Md., Mass., Mich., Minn., Miss., Mo., Nebr., N.H., N.J., N.Y., N.C., Ohio, Okla., Pa., R.I., S.C., Tenn., Tex., Vt., Va., W.Va., Wis.

D. F. M. Brown (1964) hypothesized that tetraploid *Woodsia obtusa* might be an autopolyploid derived from *W. oregana*. Recent isozyme and spore ornamentation studies indicate, however, that these species are not closely related, and the discovery of a diploid cytotype of *W. obtusa* suggests a different (albeit autopolyploid) origin for this taxon (M. D. Windham 1993). Tetraploid subsp. *obtusa* crosses with diploid subsp. *occidentalis*; the resulting triploids are sterile and have malformed spores. It also hybridizes with *W. oregana* subsp. *cathcartiana* to form the sterile tetraploid hybrid known as *W. × kansana* Brooks.

7b. Woodsia obtusa (Sprengel) Torrey subsp. **occidentalis** Windham, Contr. Univ. Michigan Herb. 19: 56. 1993

Stems short- to long-creeping, individual branches usually 3–5 mm diam. **Blade** finely cut, 2-pinnate-pinnatifid. **Proximal pinnules** of lower pinnae usually deeply lobed or pinnatifid. **Spores** averaging 35–42 μm. $2n = 76$.

Sporulating summer–fall. Cliffs and rocky slopes (rarely terrestrial); found on a variety of substrates but mostly sandstone and granite; 200–500 m; Ark., Kans., Mo., Okla., Tex.

Woodsia obtusa subsp. *occidentalis* hybridizes with subsp. *obtusa* sporadically throughout the region of sympatry; the hybrids are sterile triploids with malformed spores.

8. Woodsia oregana D. C. Eaton, Canad. Naturalist & Quart. J. Sci., n. s. 2: 90. 1865 · Oregon cliff fern, woodsie de l'Orégon

Stems compact, erect to ascending, with few to many persistent petiole bases of unequal lengths; scales often uniformly brown but at least some bicolored with dark central stripe and pale brown margins, narrowly lanceolate. **Leaves** 4–25 × 1–4 cm. **Petiole** reddish brown to dark purple proximally when mature, not articulate above base, somewhat pliable and resistant to shattering. **Blade** linear-lanceolate to narrowly ovate, pinnate-pinnatifid or 2-pinnate proximally, sparsely to moderately glandular, never viscid; glandular hairs with thin stalks and slightly expanded tips; rachis with scattered glandular hairs and occasional hairlike scales. **Pinnae** ovate-deltate to elliptic, longer than wide, abruptly tapered to a rounded or broadly acute apex; largest pinnae with 3–9 pairs of pinnules; abaxial and adaxial surfaces glabrescent to moderately glandular, lacking nonglandular hairs or scales. **Pinnules** dentate, often shallowly lobed; margins nonlustrous, thin, with occasional glands, lacking cilia, rarely with 1–2-celled translucent projections. **Vein tips** slightly (if at all) enlarged, barely visible adaxially. **Indusia** of narrow, usually filamentous segments, these uniseriate for most of length, composed of ± isodiametric cells, concealed by or slightly surpassing mature sporangia. **Spores** averaging 39–50 μm.

Subspecies 2: only in the flora.

The variability and promiscuity of *Woodsia oregana* have been major sources of taxonomic difficulties in *Woodsia*, and more work will be necessary before relationships in this complex are fully resolved. As defined here, *W. oregana* comprises two subspecies that are chromosomally and biochemically distinct. In addition, the two taxa are nearly allopatric, with the diploid (subsp. *oregana*) confined to the Pacific Northwest and the tetraploid (subsp. *cathcartiana*) extending from the southwestern United States to eastern Canada. Isozyme studies indicate that subsp. *cathcartiana* is not an autotetraploid derived from known diploid populations of subsp. *oregana*, as was hypothesized by D. F. M. Brown (1964), and it may be more appropriate to recognize these taxa as distinct species. The morphologic features that distinguish these subspecies are very subtle, however, and they are associated primarily with differences in chromosome number. Until further systematic analyses are undertaken, these cytotypes should be maintained as subspecies of *W. oregana*.

1. Spores averaging 39–45 μm; cells on pinnule margins regular in shape, margins appearing entire; adaxial epidermal cells averaging less than 120 μm. . . . . 8a. *Woodsia oregana* subsp. *oregana*
1. Spores averaging 45–50 μm; cells on pinnule margins irregular in shape, margins usually minutely dentate and appearing ragged; adaxial epidermal cells averaging more than 120 μm. . . . . 8b. *Woodsia oregana* subsp. *cathcartiana*

8a. Woodsia oregana D. C. Eaton subsp. **oregana**

Cells on pinnule margins regular in shape, margins appearing entire; adaxial epidermal cells averaging less than 120 μm. **Spores** averaging 39–45 μm. $2n = 76$.

Sporulating summer–fall. Cliffs and rocky slopes; usually on granitic or volcanic substrates; 100–2800 m; Alta., B.C.,

Sask.; Calif., Idaho, Mont., Nev., Oreg., Utah, Wash., Wyo.

The leaves of *Woodsia oregana* subsp. *oregana* tend to be narrower and less glandular than those of subsp. *cathcartiana*. The two subspecies hybridize in the narrow region of sympatry; hybrids are sterile triploids with malformed spores.

8b. Woodsia oregana D. C. Eaton subsp. **cathcartiana** (B. L. Robinson) Windham, Contr. Univ. Michigan Herb. 19: 58. 1993

Woodsia cathcartiana B. L. Robinson, Rhodora 10: 30. 1908; *W. oregana* forma *cathcartiana* (B. L. Robinson) Boivin; *W. oregana* forma *glandulosa* T. M. C. Taylor; *W. oregana* var. *cathcartiana* (B. L. Robinson) C. V. Morton; *W. pusilla* Fournier var. *cathcartiana* (B. L. Robinson) T. M. C. Taylor

Cells on pinnule margins irregular in shape, margins usually minutely dentate and appearing ragged; adaxial epidermal cells averaging more than 120 μm. **Spores** averaging 45–50 μm. $2n = 152$.

Sporulating summer–fall. Cliffs and rocky slopes; found on a variety of substrates including both granite and limestone; 0–4000 m; Man., Ont., Que., Sask.; Ariz., Calif., Colo., Idaho, Iowa, Kans., Mich., Minn., Mont., Nebr., Nev., N.Mex., N.Y., N.Dak., Okla., S.Dak., Utah, Wis., Wyo.

D. F. M. Brown (1964) believed that *Woodsia oregana* subsp. *cathcartiana* was confined to a single locality on the Minnesota-Wisconsin border. Recent chromosome counts, however, indicate that the tetraploid cytotype of *Woodsia oregana* is actually more widespread than the diploid subsp. *oregana* (M. D. Windham 1993). The inclusion of western U.S. collections within the definition of this taxon is supported by isozyme data that indicate some plants from Arizona and New Mexico are identical to those collected at the type locality of subsp. *cathcartiana*. In addition to crossing with subsp. *oregana* (see comments above), *W. oregana* subsp. *cathcartiana* hybridizes with *W. neomexicana* to produce sterile tetraploids of intermediate morphology. It also crosses with *W. obtusa* subsp. *obtusa*, resulting in the sterile tetraploid *W.* ×*kansana* Brooks. F. S. Wagner (1987) has shown that *W. oregana* subsp. *cathcartiana*, not *W. scopulina*, hybridizes with *W. ilvensis* to form the sterile triploid *W.* ×*abbeae*. Some morphologic evidence suggests that *W.* ×*maxonii* may be a hybrid between subsp. *cathcartiana* and *W. scopulina* subsp. *laurentiana*; this hypothesis requires further testing. The difficulties involved with separating subsp. *cathcartiana* from certain plants of *W. plummerae* are discussed under that species.

9. Woodsia neomexicana Windham, Contr. Univ. Michigan Herb. 19: 52. 1993 · New Mexican cliff fern

Stems compact, erect to ascending, with few to many persistent petiole bases of unequal lengths; scales mostly uniformly brown but at least some bicolored with dark central stripe and pale brown margins, narrowly lanceolate. **Leaves** 4–30 × 1.5–6 cm. **Petiole** light brown or straw-colored when mature, occasionally darker at very base, not articulate above base, relatively brittle and easily shattered. **Blade** linear to lanceolate, usually pinnate-pinnatifid proximally, glabrescent to sparsely glandular, never viscid; glandular hairs with thin stalks and slightly expanded tips; rachis with scattered glandular hairs and rare, hairlike scales. **Pinnae** ovate-deltate to elliptic, longer than wide, abruptly tapered to a rounded or broadly acute apex; largest pinnae with 3–7 pairs of closely spaced pinnules; abaxial and adaxial surfaces glabrescent to sparsely glandular, lacking nonglandular hairs or scales. **Pinnules** dentate, often shallowly lobed; margins nonlustrous, thin, with occasional glands, lacking cilia, with 1–2-celled translucent projections on teeth. **Vein tips** occasionally enlarged to form whitish hydathodes visible adaxially. **Indusia** of narrow, filamentous segments, these uniseriate for most of length, composed of ± isodiametric cells, usually surpassing mature sporangia. **Spores** averaging 44–52 μm. $2n = 152$.

Sporulating summer–fall. Cliffs and rocky slopes; usually on sandstone or igneous substrates; 300–3500 m; Ariz., Colo., N.Mex., Okla., S.Dak., Tex.

Woodsia neomexicana traditionally has been identified as *W. mexicana*. Both taxa are tetraploid and may share one parent (M. D. Windham 1993); *W. neomexicana* is separated from typical *W. mexicana* by its completely filamentous indusial segments, reduced glandularity, and more northerly distribution. Isozyme data suggest that *W. neomexicana* is an allotetraploid hybrid between *W. phillipsii* and the diploid progenitor of *W. oregana* subsp. *cathcartiana* (M. D. Windham 1993). As with all allopolyploids, *W. neomexicana* can vary in the direction of either parent, and some plants (especially those resembling *W. phillipsii*) can be difficult to identify. All characters except those controlled directly by ploidy level show this tendency, and spore size remains the most dependable character for distinguishing *W. phillipsii* and *W. neomexicana*. This species hybridizes with *W. oregana* subsp. *cathcartiana* and *W. phillipsii* to produce sterile tetraploids and triploids, respectively.

10. Woodsia phillipsii Windham, Contr. Univ. Michigan Herb. 19: 50. 1993 · Phillips's cliff fern

Stems compact to short-creeping, erect to horizontal, with few to many persistent petiole bases of unequal lengths; scales mostly uniformly brown but at least some bicolored with dark central stripe and pale brown margins, narrowly lanceolate. **Leaves** 5–35 × 1.5–6 cm. **Petiole** light brown or straw-colored when mature, occasionally darker at very base, not articulate above base, relatively brittle and easily shattered. **Blade** lanceolate, usually 2-pinnate proximally, sparsely to moderately glandular, never viscid; glandular hairs with thin stalks and slightly expanded tips; rachis with scattered glandular hairs and hairlike scales. **Pinnae** elongate-deltate to elliptic, longer than wide, often attenuate to a narrowly acute apex; largest pinnae with 7–18 pairs of widely spaced pinnules; abaxial and adaxial surfaces somewhat glandular, lacking nonglandular hairs or scales. **Pinnules** dentate, often shallowly lobed; margins often lustrous adaxially, somewhat thickened, with occasional glands, appearing ciliate due to presence of multicellular translucent projections on teeth that are often prolonged to form twisted filaments. **Vein tips** usually enlarged to form whitish hydathodes visible adaxially. **Indusia** of narrow, filamentous segments, these uniseriate for most of length, composed of ± isodiametric cells, often greatly surpassing mature sporangia. **Spores** averaging 37–44 μm. $2n = 76$.

Sporulating summer–fall. Cliffs and rocky slopes; usually on granitic or volcanic substrates; 1600–3200 m; Ariz., N.Mex., Tex.; n Mexico.

Woodsia phillipsii traditionally has been identified as *W. mexicana*. It differs from typical *W. mexicana*, however, in having completely filamentous indusial segments, multicellular (often filamentous) translucent projections on the pinnule margins, a greater number of pinnules per pinna, and a diploid chromosome number. *Woodsia phillipsii* is the only diploid species currently recognized in the *W. mexicana* complex, and it was probably involved in the hybrid origins of both *W. mexicana* and *W. neomexicana*. Some individuals of the latter species are difficult to distinguish from *W. phillipsii* (see comments under *W. neomexicana*), and the two taxa occasionally hybridize to produce sterile triploids of intermediate morphology. *Woodsia phillipsii* is also known to hybridize with *W. plummerae* (see comments under that species) and *W. cochisensis*.

9. DRYOPTERIS Adanson, Fam. Pl. 2: 20, 551. 1763, name conserved · Wood fern, shield fern, dryoptère [Greek *drys*, tree, and *pteris*, fern]

James D. Montgomery

Warren H. Wagner Jr.

Plants terrestrial, rarely on rock. **Stems** short-creeping to erect, stolons absent. **Leaves** monomorphic, green through winter or dying back in winter. **Petiole** ca. 1/4–2/3 blade length, bases swollen or not; vascular bundles more than 3, arranged in an arc, ± round in cross section. **Blade** deltate-ovate to lanceolate, 1–3-pinnate-pinnatifid, gradually reduced distally to pinnatifid apex, herbaceous to somewhat leathery. **Pinnae** not articulate to rachis, segment margins entire, crenate, or serrate, spinulose or not; proximal pinnae reduced (several pairs), same size as or enlarged relative to more distal pinnae, sessile to petiolulate, equilateral or often inequilateral with pinnules on basiscopic side longer than those on acroscopic side; costae adaxially grooved, grooves continuous from rachis to costae to costules; indument of linear to ovate scales abaxially, also sometimes with glands, blades ± glabrous adaxially. **Veins** free, forked. **Sori** in 1 row between margin and midrib, round; indusia round-reniform, attached at narrow sinus, persistent or caducous. **Spores** brownish, coarsely rugose or with folded wings. $x = 41$.

Species ca. 250 (14 in the flora): mostly in temperate Asia.

The relationships of the North American species are reasonably well understood, but species identifications are complicated by the frequent presence of hybrids in field populations. Sterile hybrids can be distinguished from fertile species by their misshapen spores and intermediate morphology. They are not included in the key, but they may be identified as to parentage by combinations of characters in the key (e.g., marginal sori for *Dryopteris marginalis*, narrow blades for *D. cristata*). Relationships are shown in the accompanying reticulogram.

Dryopteris

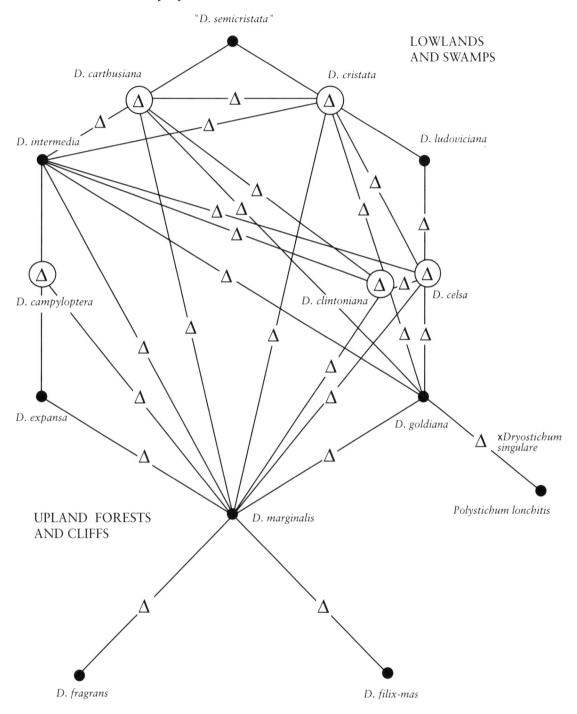

Relationships among some species of *Dryopteris*. Solid circles represent parental taxa; circled triangles represent allopolyploids; and triangles represent sterile hybrids. *Dryopteris semicristata* is a hypothetical ancestor.

SELECTED REFERENCES Carlson, T. J. and W. H. Wagner Jr. 1982. The North American distribution of the genus *Dryopteris*. Contr. Univ. Michigan Herb. 15: 141–162. Fraser-Jenkins, C. R. 1989. A classification of the genus *Dryopteris* (Pteridophyta: Dryopteridaceae). Bull. Brit. Mus. (Nat. Hist.), Bot. 18: 323–477. Montgomery, J. D. 1982. *Dryopteris* in North America. Part II. The hybrids. Fiddlehead Forum 9: 23–30. Montgomery, J. D. and E. M. Paulton. 1981. *Dryopteris* in North America. Fiddlehead Forum 8: 25–31. Petersen, R. L. and D. E. Fairbrothers. 1983. Flavonols of the fern genus *Dryopteris*: Systematic and morphological implications. Bot. Gaz. 144: 104–109. Viane, R. L. 1986. Taxonomical significance of the leaf indument in *Dryopteris* (Pteridophyta): I. Some North American, Macronesian and European taxa. Pl. Syst. Evol. 153: 77–105. Wagner, W. H. Jr. 1971. Evolution of *Dryopteris* in relation to the Appalachians. In: P. C. Holt, ed. 1971. The Distributional History of the Biota of the Southern Appalachians. Part 2. Flora. Blacksburg, Va. Pp. 147–192. [Virginia Polytechnic Inst. and State Univ., Res. Div. Monogr. 2.] Werth, C. R. 1991. Isozyme studies on the *Dryopteris* "spinulosa" complex. I: The origin of the log fern *Dryopteris celsa*. Syst. Bot. 16(3): 446–461.

1. Blades densely scaly abaxially, aromatic-glandular; old leaves forming conspicuous gray or brown clumps; leaves 6–25(–40) cm. . . . . 1. *Dryopteris fragrans*
1. Blades glabrous to sparsely scaly abaxially, not aromatic-glandular; old leaves not persisting in conspicuous gray or brown clumps; leaves usually more than 25 cm.
  2. Blades 2-pinnate to 3-pinnate-pinnatifid at base.
    3. Basal pinnules of basal pinnae shorter than adjacent pinnules.
      4. Pinnae with elongate serrate tip; blades and indusia lacking glands or sparsely glandular. 10. *Dryopteris cinnamomea*
      4. Pinnae lacking elongate serrate tip; blades, at least the midrib of segments and indusia, finely glandular. . . . . 11. *Dryopteris intermedia*
    3. Basal pinnules of basal pinnae longer than adjacent pinnules.
      5. First basal basiscopic pinnule not much wider than 1st acroscopic pinnule on basal pinnae; blades ovate-lanceolate. . . . . 12. *Dryopteris carthusiana*
      5. First basal basiscopic pinnule 2 times width of 1st acroscopic pinnule on basal pinnae; blades ovate-deltate.
        6. Petiole scales tan, with dark central stripe; leaves erect to slightly arching (Rocky Mountains, n Great Lakes, ne Canada). . . . . 13. *Dryopteris expansa*
        6. Petiole scales tan or dark at base, lacking distinct dark stripe; leaves widely spreading (Appalachian Mountains north to ne Canada). . . . . . 14. *Dryopteris campyloptera*
  2. Blades pinnate-pinnatifid to 2-pinnate at base.
    7. Sori at or near margins of segments; petioles with dense tuft of pale tawny scales at base. . . . . 2. *Dryopteris marginalis*
    7. Sori midway between margin and midrib or closer to midribs of segments; petioles with scattered tan to dark brown scales at base.
      8. Pinnules finely spiny with spreading teeth (Arizona and west coast of North America). . . . . 3. *Dryopteris arguta*
      8. Pinnules not finely spiny, teeth blunt or incurved (e, nw North America).
        9. Petioles less than 1/4 length of leaves, scales of 2 kinds, mixed, broad and hairlike. . . . . 4. *Dryopteris filix-mas*
        9. Petioles 1/4–1/3 length of leaves, scales broad to narrow, but not hairlike.
          10. Fertile pinnae narrower than vegetative pinnae, restricted to distal 1/2 of blade. . . . . 5. *Dryopteris ludoviciana*
          10. Fertile pinnae same width as vegetative pinnae, occupying distal 1/2 of blade to nearly entire blade.
            11. Basal pinnae ovate; blades ovate to ovate-lanceolate; scales at base of petioles dark brown or with dark brown stripe.
              12. Blades ovate, tapering abruptly to tip; sori nearer midvein than margin of segments. . . . . 6. *Dryopteris goldiana*
              12. Blades ovate-lanceolate, gradually narrowed to tip; sori about midway between midvein and margin of segments. . . . . . . 7. *Dryopteris celsa*
            11. Basal pinnae deltate; blades lanceolate or with parallel sides; scales at base of petioles tan.

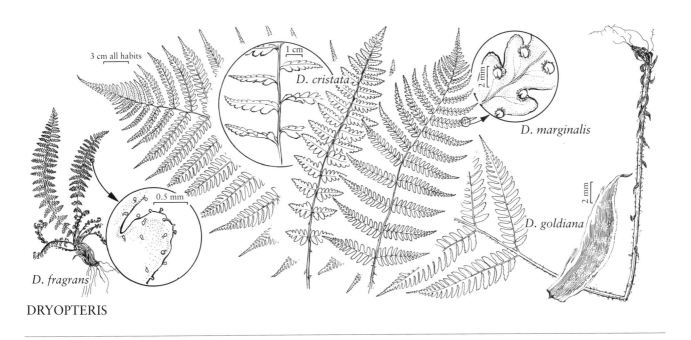

D. cristata

D. marginalis

D. goldiana

D. fragrans

DRYOPTERIS

1. **Dryopteris fragrans** (Linnaeus) Schott, Gen. Fil., plate 9. 1834 · Fragrant wood fern, dryoptère odorante

Polypodium fragrans Linnaeus, Sp. Pl. 2: 1089. 1753; *Dryopteris fragrans* var. *remotiuscula* Komarov

Leaves monomorphic, green through winter, 6–40 × 1–6 cm, old leaves persistent as gray or brown conspicuous clump at plant base (only in this species). **Petiole** 1/3 length of leaf, scaly throughout; scales dense, brown to red-brown. **Blades** green, linear-lanceolate, pinnate-pinnatifid to 2-pinnate, papery, densely scaly (only in this species) abaxially, glandular, aromatic when handled (only in this species). **Pinnae** ± in plane of blade, linear-oblong; basal pinnae linear-oblong, much reduced, basal pinnules longer than adjacent pinnules, basal basiscopic pinnule and basal acroscopic pinnule equal; pinnule margins crenately toothed. **Sori** midway between midvein and margin of segments. **Indusia** glandular. $2n = 82$.

Shaded cliffs and talus, often of limestone; 50–1800 m; Greenland; Alta., B.C., Man., N.B., Nfld., N.W.T., N.S., Ont., Que., Sask., Yukon; Alaska, Maine, Mich., Minn., N.H., N.Y., Vt., Wis.; Europe in n Finland; Asia in n, nw former Soviet republics.

Dryopteris fragrans is a northern species and is not closely related to the other species in North America. The only hybrid known to involve *D. fragrans* is with *D. marginalis*, producing *D.* ×*algonquinensis* D. Britton.

2. **Dryopteris marginalis** (Linnaeus) A. Gray, Manual, 632. 1848 · Marginal wood fern, dryoptère à sores marginaux

Polypodium marginale Linnaeus, Sp. Pl. 2: 1091. 1753

Leaves monomorphic, green through winter, 30–100 × 10–25 cm. **Petiole** 1/4–1/3 length of leaf, scaly at base; scales in dense tuft, pale tawny. **Blade** bluish green, ovate-lanceolate, pinnate-pinnatifid to 2-pinnate at base, leathery, not glandular. **Pinnae** ± in plane of blade, lanceolate; basal pinnae lanceolate, slightly reduced, basal pinnules longer than adjacent pinnules, basal basiscopic pinnule longer than basal acroscopic pinnule; pinnule margins shallowly crenate to nearly entire. **Sori** near margin of segments. **Indusia** lacking glands. $2n = 82$.

Rocky, wooded slopes and ravines, edges of woods,

stream banks and roadbanks, and rock walls; 50–1500 m; Greenland; N.B., Nfld, N.S., Ont., Que.; Ala., Ark., Conn., Del., Ga., Ill., Ind., Iowa, Kans., Ky., Maine, Md., Mass., Mich., Miss., Mo., N.H., N.J., N.Y., N.C., Ohio, Okla., Pa., R.I., S.C., Tenn., Vt., Va., W.Va., Wis.

Dryopteris marginalis is an eastern North America endemic. Even though this species hybridizes with 10 other species, and some of these hybrids are fairly common, *D. marginalis* is not known to be involved in the formation of any fertile polyploid. Hybrids can be detected by malformed spores and the nearly marginal sorus position.

3. **Dryopteris arguta** (Kaulfuss) Maxon, Amer. Fern J. 11: 3. 1921 · Marginal wood fern, coastal wood fern

Aspidium argutum Kaulfuss, Enum. Filic., 242. 1824

Leaves monomorphic, green through winter, 25–90 × 8–30 cm. **Petiole** 1/4–1/3 length of leaf, scaly at least at base; scales scattered, light brown. **Blade** green to yellow-green, ovate-lanceolate, pinnate-pinnatifid to 2-pinnate at base, herbaceous, glandular. **Pinnae** ± in plane of blade, lance-oblong; basal pinnae deltate, not much reduced, basal pinnules ± same length as adjacent pinnules, basal basiscopic pinnule and basal acroscopic pinnule ± equal, pinnule margins serrate with spreading, spinelike teeth. **Sori** midway between midvein and margin of segments. **Indusia** lacking glands. 2*n* = 82.

Shaded slopes and open woods; 0–2100 m; B.C.; Ariz., Calif., Oreg., Wash.; Mexico in Baja California.

Dryopteris arguta is somewhat variable. It has been suggested that more than one taxon is involved. No hybrids involving *D. arguta* are known.

4. **Dryopteris filix-mas** (Linnaeus) Schott, Gen. Fil., plate 67. 1834 · Male fern, fougère mâle

Polypodium filix-mas Linnaeus, Sp. Pl. 2: 1090. 1753

Leaves monomorphic, dying back in winter, 28–120 × 10–30 cm. **Petiole** less than 1/4 length of leaf, scaly at least at base; scales scattered, brown, of 2 distinct kinds, 1 broad, 1 hairlike (only this species has 2 distinct forms of scales without intermediates). **Blade** green, ovate-lanceolate, pinnate-pinnatifid to 2-pinnate at base, firm but not leathery, not glandular. **Pinnae** ± in plane of blade, lanceolate; basal pinnae ovate-lanceolate, much reduced, basal pinnules or segments ± same length as adjacent pinnules, basal basiscopic pinnule and basal acroscopic pinnule equal; pinnule margins serrate to lobed. **Sori** midway between midvein and margin of segments. **Indusia** lacking glands. 2*n* = 164.

Dense woods and talus slopes on limestone (ne North America); open woods among boulders and talus of granite or igneous rock (Rocky Mountains); 200–2500 m; Greenland; Alta., B.C., Nfld., N.S., Ont., Que., Sask.; Ariz., Calif., Colo., Idaho, Mich., Mont., N.Mex., Nev., Okla., Oreg., Tex., Utah, Wash., Wis., Wyo.; Europe; Asia.

The taxonomy of *Dryopteris filix-mas* is not well understood. In North America, this fern has been considered both an auto- and an allopolyploid and may be composed of at least two closely related taxa. Plants in the northeast and northwest are tetraploid. These differ morphologically and ecologically from a taxon of unknown chromosome number in the southwestern Rocky Mountains. The Rocky Mountain taxon closely resembles the Mexican *D. pseudofilix-mas* (Fée) Rothmaler. *Dryopteris filix-mas* also occurs in Europe, and it is known to be an allopolyploid of *D. caucasica* (A. Braun) Fraser-Jenkins & Corley × *oreades* Fomin.

5. **Dryopteris ludoviciana** (Kunze) Small, Ferns S.E. States, 281. 1938 · Southern wood fern

Aspidium ludovicianum Kunze, Amer. J. Sci. Arts, ser. 2, 6: 84. 1848; *Dryopteris floridana* (Hooker) Kunze

Leaves somewhat dimorphic, green through winter, 35–120 × 10–30 cm. **Petiole** more than 1/4 length of leaf, scaly at base; scales scattered, brown. **Blade** dark green, lanceolate, pinnate-pinnatifid, herbaceous, not glandular. **Pinnae** nearly in plane of blade, lance-oblong; fertile pinnae in distal 1/2 of leaf, distinctly narrower than proximal vegetative pinnae (only in this species); basal pinnae lanceolate-oblong, much reduced, basal pinnules slightly shorter than adjacent pinnules, basal basiscopic pinnule slightly longer than basal acroscopic pinnule; pinnule margins distantly serrate. **Sori** midway between midvein and margin of segments. **Indusia** lacking glands. 2*n* = 82.

Swamps and wet woods; 0–100 m; Ala., Ark., Fla., Ga., La., N.C., S.C.

Dryopteris ludoviciana is endemic to southeastern United States. This diploid is one of the parents of *D. celsa* and *D. cristata*. It crosses with *D. celsa* to produce sterile hybrids.

a

6. Dryopteris goldiana (Hooker ex Goldie) A. Gray, Manual, 631. 1848 · Goldie's wood fern, dryoptère de Goldie

Aspidium goldianum Hooker ex Goldie, Edinburgh Philos. J. 6: 333. 1822

Leaves monomorphic, dying back in winter, 35–120 × 15–40 cm. **Petiole** 1/3 length of leaf, scaly at base; scales scattered, dark, glossy brown to nearly black, with pale border. **Blade** green, often white-mottled at tip, ovate, tapering abruptly at apex, pinnate-pinnatifid to 2-pinnate at base, herbaceous, not glandular. **Pinnae** parallel to plane of blade, ovate-lanceolate, broadest above base; basal pinnae broadly oblong-lanceolate, slightly reduced, basal pinnule equal to adjacent pinnules, basal basiscopic pinnule and basal acroscopic pinnule equal; pinnule margins crenulate or serrate. **Sori** nearer midvein than margin. **Indusia** and axes lacking glands. $2n = 82$.

Dense, moist woods, especially ravines, limey seeps, or at the edge of swamps; 50–1500 m; N.B., Ont., Que.; Ala., Conn., Del., Ga., Ill., Ind., Iowa, Ky., Maine, Md., Mass., Mich., Minn., Mo., N.H., N.J., N.Y., N.C., Ohio, Pa., R.I., Tenn., Vt., Va., W.Va., Wis.

Dryopteris goldiana is diploid and is one of the parents of *D. celsa* and of *D. clintoniana*. *Dryopteris goldiana* hybridizes with five species. Hybrids can be identified by the glossy dark scales and large blade size. A remarkable additional hybrid (× *Dryostichum singulare* W. H. Wagner), involving this species and *Polystichum lonchitis*, is known from Gray and Simcoe counties, Ontario. It is intermediate between the parents and is sterile (W. H. Wagner Jr., F. S. Wagner et al. 1992).

7. Dryopteris celsa (W. Palmer) Knowlton, W. Palmer, & Pollard, Proc. Biol. Soc. Wash. 13: 202. 1900 · Log fern

Dryopteris goldiana (Hooker ex Goldie) A. Gray subsp. *celsa* W. Palmer, Proc. Biol. Soc. Wash. 13: 65. 1899

Leaves monomorphic, dying back in winter, 65–120 × 15–30 cm. **Petiole** 1/3 length of leaf, scaly at least at base; scales scattered, dark brown or tan with dark central stripe. **Blade** green, ovate-lanceolate, gradually tapering to tip, pinnate-pinnatifid, herbaceous, not glandular. **Pinnae** ± in plane of blade, lanceolate-ovate; basal pinnae linear-oblong, much reduced, basal pinnules longer than adjacent pinnules, basal basiscopic pinnule and basal acroscopic pinnule equal; pinnule margins crenately toothed. **Sori** midway between midvein and margin of segments. **Indusia** lacking glands. $2n = 164$.

Seepage slopes, hammocks and logs in swamps, mostly on the Piedmont and Coastal Plain; 50–800 m; Ala., Ark., Del., Ga., Ill., Ky., La., Md., Mich., Mo., N.J., N.Y., N.C., Pa., S.C., Tenn., Va., W.Va.

Dryopteris celsa is a fertile allotetraploid derived from hybridization between *D. goldiana* and *D. ludoviciana*. *Dryopteris celsa* hybridizes with six species; hybrids can usually be identified by the dark-striped scales.

8. Dryopteris clintoniana (D. C. Eaton) Dowell, Proc. Staten Island Assoc. Arts 1: 64. 1906 · Clinton's wood fern, dryoptère de Clinton

Aspidium cristatum (Linnaeus) Swartz var. *clintonianum* D. C. Eaton in A. Gray, Manual ed. 5, 665. 1867; *Dryopteris cristata* (Linnaeus) A. Gray var. *clintoniana* (D. C. Eaton) L. Underwood

Leaves dimorphic, 45–100 × 12–20 cm; fertile leaves dying back in winter; sterile leaves 1–several, smaller, green through winter. **Petiole** 1/4–1/3 length of leaf, scaly at least at base; scales scattered, tan, sometimes with dark brown center. **Blade** green, lanceolate, with nearly parallel sides, pinnate-pinnatifid, herbaceous, not glandular. **Pinnae** of fertile leaves twisted out of plane of blade but not fully perpendicular to it, narrowly elongate-deltate; basal pinnae narrowly elongate-deltate, much reduced; basal pinnules longer than or equal to adjacent pinnules, basal basiscopic pinnule and basal acroscopic pinnule equal; pinnule margins serrate or biserrate, with spiny teeth. **Sori** midway between midvein and margin of segments. **Indusia** lacking glands. $2n = 246$.

Swampy woods; 50–600 m; N.B., Ont., Que.; Conn., Ind., Maine, Mass., Mich., N.H., N.J., N.Y., Ohio, Pa., R.I., Vt.

Dryopteris clintoniana is a North American endemic and an allohexaploid derived from *D. cristata* and *D. goldiana*. *Dryopteris clintoniana* hybridizes with six species. Hybrids can be identified by the fairly narrow blades and elongate-deltate proximal pinnae.

9. Dryopteris cristata (Linnaeus) A. Gray, Manual, 631. 1848 · Crested wood fern, dryoptère à crêtes

Polypodium cristatum Linnaeus, Sp. Pl. 2: 1090. 1753

Leaves dimorphic, 35–70 × 8–12 cm; fertile leaves dying back in winter; sterile leaves several, small, green through winter, forming "rosette." **Petiole** 1/4–1/3 length of leaf, scaly at least at base; scales scattered, tan. **Blade** green, narrowly lanceolate or with parallel sides, pinnate-pinnatifid, not glandular. **Pinnae** of fertile leaves twisted out of plane of blade and perpendicular to it, deltate; basal pinnae deltate, somewhat reduced, basal pinnules longer than adjacent pinnules, basal basiscopic pinnule and basal acroscopic pinnule equal; pinnule margins distantly serrate, with spiny teeth. **Sori** midway between midvein and margin of segments. **Indusia** lacking glands. $2n = 164$.

Swamps, swampy woods, or open shrubby wetlands; 0–1200 m; Alta., B.C., Man., N.B., Nfld., N.S., Ont., P.E.I., Que., Sask.; Ala., Conn., Del., Ill., Ind., Iowa, Maine, Md., Mass., Mich., Minn., Mont., Nebr., N.H., N.J., N.Y., N.C., N.Dak., Ohio, Pa., R.I., Tenn., Vt., Va., W.Va., Wis.; Europe.

Dryopteris cristata is believed to be an allotetraploid derived from *D. ludoviciana* and an unknown diploid called "*D. semicristata*" by W. H. Wagner Jr. (1971). This ancestral taxon could have been either North American or Eurasian and may have become extinct during the last glaciation (T. J. Carlson and W. H. Wagner Jr. 1982). *Dryopteris cristata* hybridizes with five species; these hybrids can be identified by the narrow blades and deltate proximal pinnae.

10. Dryopteris cinnamomea (Cavanilles) C. Christensen, Amer. Fern J. 1: 95. 1911 · Cinnamon wood fern

Tectaria cinnamomea Cavanilles, Descr. Pl., 252. 1802

Leaves monomorphic, green through winter, 22–50 × 6–12 cm. **Petiole** 1/4 length of leaf, scaly at least at base; scales scattered, cinnamon-colored. **Blade** light green, deltate-ovate, 3-pinnate-pinnatifid, herbaceous, not or sparsely glandular. **Pinnae** in plane of blade, narrowly deltate-lanceolate to deltate-oblong, narrowed to elongate, serrate tip; basal pinnae deltate-oblong, somewhat reduced, basal pinnules shorter than adjacent pinnules, basal basiscopic pinnule longer than basal acroscopic pinnule; pinnule margins serrate. **Sori** near sinus. **Indusia** lacking glands.

Rock outcrops; 400–2600 m; Ariz., Tex.; Mexico.

Dryopteris cinnamomea belongs to the *D. patula* complex of Mexico and Central America, which is poorly understood. Arizona material of *D. cinnamomea* has been misidentified as *D. patula*, according to J. T. Mickel and J. M. Beitel (1988).

11. Dryopteris intermedia (Muhlenberg ex Willdenow) A. Gray, Manual, 630. 1848 · Evergreen wood fern, fancy fern, dryoptère intermédiaire

Aspidium intermedium Muhlenberg ex Willdenow, Sp. Pl. 5(1): 262. 1810; *Dryopteris austriaca* (Jacquin) Woynar var. *intermedia* (Muhlenberg ex Willdenow) C. V. Morton; *D. spinulosa* (O. F. Mueller) Watt var. *intermedia* (Muhlenberg ex Willdenow) L. Underwood

Leaves monomorphic, green through winter, 32–90 × 10–20 cm. **Petiole** 1/3 length of leaf, scaly at least at base; scales scattered, tan. **Blade** green, ovate, 3-pinnate-pinnatifid, herbaceous, glandular. **Pinnae** ± in plane of blade, lanceolate-oblong; basal pinnae lanceolate, not reduced, basal pinnules longer than adjacent pinnules, basal basiscopic pinnule longer than basal acroscopic pinnule; pinnule margins serrate, teeth spiny. **Sori** midway between midvein and margin of segments. **Indusia** with minute glandular hairs. $2n = 82$.

Moist rocky woods, especially hemlock hardwoods, ravines, and edges of swamps; 0–2000 m; N.B., Nfld., N.S., Ont., P.E.I., Que.; Ala., Conn., Del., Ga., Ill., Ind., Iowa, Ky., Maine, Md., Mass., Mich., Minn., Mo., N.H., N.J., N.Y., N.C., Ohio, Pa., R.I., Tenn., Vt., Va., W.Va., Wis.

A related taxon, *Dryopteris intermedia* subsp. *maderensis* (J. Milde ex Alston) Fraser-Jenkins, occurs on eastern Atlantic islands.

Dryopteris intermedia and the other taxa in the "*D. spinulosa* complex" have long confounded taxonomists. *Dryopteris intermedia* is diploid and is one of the parents of the allotetraploids *D. carthusiana* and *D. campyloptera*. *Dryopteris intermedia* hybridizes with eight species. All hybrids are easily detected by the distinctive glandular hairs on the indusia and, usually, on the costae and costules.

12. Dryopteris carthusiana (Villars) H. P. Fuchs, Bull. Soc. Bot. France 105: 339. 1959 · Spinulose wood fern, toothed wood fern, dryoptère de cartheuser

Polypodium carthusianum Villars, Hist. Pl. Dauphiné 1: 292. 1786; *Dryopteris austriaca* (Jacquin) Schinz & Thellung var. *spinulosa* (O. F. Mueller) Fiori; *D. spinulosa* (O. F. Mueller) Watt; *Polypodium spinulosum* O. F. Mueller

Leaves monomorphic, dying in winter, 15–75 × 10–30 cm.

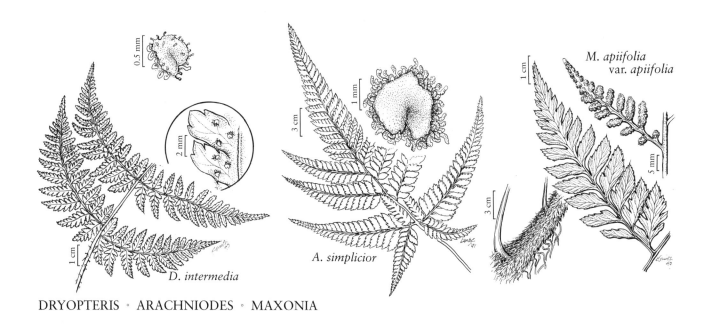

DRYOPTERIS · ARACHNIODES · MAXONIA

Petiole 1/4–1/3 length of leaf, scaly at least at base; scales scattered, tan. **Blade** light green, ovate-lanceolate, 2–3-pinnate-pinnatifid, herbaceous, not glandular. **Pinnae** ± in plane of blade, lance-oblong; basal pinnae lanceolate-deltate, slightly reduced, basal pinnules usually longer than adjacent pinnules, basal basiscopic pinnule longer than basal acroscopic pinnule; pinnule margins serrate, teeth spiny. **Sori** midway between midvein and margin of segments. **Indusia** lacking glands. $2n = 164$.

Swampy woods, moist wooded slopes, stream banks, and conifer plantations; 0–1200 m; Alta., B.C., Man., N.B., Nfld., N.W.T., N.S., Ont., P.E.I., Que., Sask., Yukon; Ark., Conn., Del., Idaho, Ill., Ind., Iowa, Ky., Maine, Md., Mass., Mich., Minn., Mo., Mont., Nebr., N.H., N.J., N.Y., N.C., N.Dak., Ohio, Pa., R.I., S.C., Tenn., Vt., Va., Wash., W.Va., Wis.; Eurasia.

Dryopteris carthusiana is tetraploid. *Dryopteris intermedia* is one parent, as indicated by chromosome pairing in their hybrid *D.* ×*triploidea* Wherry. The other parent is the hypothetical missing ancestral species "*D. semicristata*" (see discussion for *D. cristata*). *Dryopteris carthusiana* hybridizes with five species; hybrids can be separated from *D. intermedia* by the lack of glandular hairs and by having 2-pinnate leaves.

13. **Dryopteris expansa** (C. Presl) Fraser-Jenkins & Jermy, Brit. Fern Gaz. 11: 338. 1977 · Northern wood fern, spreading wood fern, dryoptère dressée

Nephrodium expansum C. Presl, Reliq. Haenk. 1: 38. 1825; *Dryopteris assimilis* S. Walker; *D. dilatata* (Hoffmann) A. Gray subsp. *americana* (Fischer) Hultén

Leaves monomorphic, tardily dying back in winter, to 90×30 cm. **Petiole** 1/3 length of leaf, scaly at least at base; scales scattered, brown with dark brown stripe. **Blade** green, deltate-ovate, 3-pinnate-pinnatifid, herbaceous, usually not glandular, occasionally finely and densely glandular. **Pinnae** ± in plane of blade, lanceolate-oblong; basal pinnae deltate, slightly reduced, basal pinnules equal to or longer than adjacent pinnules, basal basiscopic pinnule longer than basal acroscopic pinnule; pinnule margins serrate. **Sori** midway between midvein and margin of segments. **Indusia** lacking glands or sparsely glandular. $2n = 82$.

Cool moist woods and rocky slopes; 50–1500 m; Greenland; Alta., B.C., Nfld., N.W.T., Ont., Que., Yukon; Alaska, Calif., Idaho, Mich., Minn., Mont., Oreg., Wash., Wis., Wyo.; Europe.

Dryopteris expansa is diploid and is one of the parents of *D. campyloptera*. Where their ranges overlap in eastern Canada, these two species are very difficult to distinguish except by chromosome number. The growth habit (*D. expansa* leaves are more erect) is useful in the field. Three hybrids involving *D. expansa* are known; all are very rare.

14. **Dryopteris campyloptera** (Kunze) Clarkson, Amer. Fern J. 20: 118. 1930 · Mountain wood fern, eastern spreading wood fern, dryoptère arquée

Aspidium campylopterum Kunze, Amer. J. Sci. Arts, ser. 2, 6: 84. 1848; *Dryopteris austriaca* (Jacquin) Schinz & Thellung; *D. spinulosa* (Swartz) Watt var. *americana* (Fischer ex Kunze) Fernald; *D. spinulosa* var. *concordiana* (Davenport) Eastman

Leaves monomorphic, dying back in winter, 25–90 × 15–30 cm. **Petiole** 1/3 length of leaf, scaly at least at base; scales scattered, brown, sometimes with darker patch at base. **Blade** light green, deltate-ovate, 3-pinnate-pinnatifid, herbaceous, not glandular. **Pinnae** in plane of blade, lanceolate; basal pinnae deltate to broadly lanceolate, not reduced, basal pinnules equal to adjacent pinnules, basal basiscopic pinnule much longer and 2 times width of basal acroscopic pinnule (only in this species); pinnule margins toothed, teeth spine-tipped. **Sori** midway between midvein and margin of segments. **Indusia** usually lacking glands. $2n = 164$.

Cool, moist woods at increasing elevation southward; frequently only at summits of mountains; 0–1500 m; N.B., Nfld., N.S., P.E.I., Que.; Conn., Maine, Md., Mass., N.H., N.Y., N.C., R.I., Tenn., Vt., Va., W.Va.

10. ARACHNIODES Blume, Enum. Pl. Javae 2: 241. 1828 · East Indian holly fern [Greek *arachnion,* spider's web, and *-odes,* having the form or nature of; it has been suggested that Blume saw fungal hyphae or spider webs on his original material]

Alan R. Smith

Plants terrestrial. **Stems** moderately long- to short-creeping, stolons absent. **Leaves** monomorphic, evergreen. **Petiole** ± as long as blade, base not swollen; vascular bundles more than 3, arranged in an arc, ± round in cross section. **Blade** broadly deltate or pentagonal, 2–3-pinnate-pinnatifid, gradually to abruptly reduced distally to pinnate or pinnatifid apex, papery to somewhat leathery. **Pinnae** not articulate to rachis, segment margins and especially apex spinulose; proximal pinnae largest, petiolulate, inequilateral with basal basiscopic pinnule much larger and more elongate than more distal pinnules; costae adaxially grooved, grooves continuous from rachis to costae to costules; indument of hairlike scales abaxially, absent adaxially. **Veins** free, forked. **Sori** in 1 row between midrib and margin, round; indusia round-reniform, attached at narrow sinus, persistent. **Spores** brownish, rugate or tuberculate, sometimes spiny. $x = 41$.

Species ca. 50 (1 in the flora naturalized from Asia): tropics and subtropics, mostly in e Asia and Pacific Islands, a few in Africa, ca. 4 in Mexico, Central America, South America.

SELECTED REFERENCES Ching, R. C. 1934. A revision of the compound leaved Polysticha and other related species in the continental Asia including Japan and Formosa. Sinensia 5: 23–91. Gordon, J. E. 1981. *Arachniodes simplicior* new to South Carolina and the United States. Amer. Fern J. 71: 65–68. Tindale, M. D. 1960. Pteridophyta of south eastern Australia. Contr. New South Wales Natl. Herb., Fl. Ser. 211: 47–78.

1. Arachniodes simplicior (Makino) Ohwi, J. Jap. Bot. 37: 76. 1962 · Simpler East Indian holly fern

Aspidium aristatum (G. Forster) Swartz var. *simplicius* Makino, Bot. Mag. (Tokyo) 15: 65. 1901; *Byrsopteris simplicior* (Makino) Kurata; *Polystichopsis simplicior* (Makino) Tagawa; *Rumohra simplicior* (Makino) Ching

Stems creeping, 5–8 mm diam., densely scaly; scales tan to brownish, lanceolate. **Leaves** monomorphic, evergreen, 40–85 cm. **Petiole** straw-colored to tan, spaced 0.5–4 cm apart, 15–46 cm × 4–6 mm, base scaly; scales lanceolate. **Blade** deltate-pentagonal, 2–3-pinnate, 24–40 cm, broadest at base, abruptly tapering distally, apex pinnate, ultimately pinnatifid, as long as or longer than rest of blade. **Pinnae** 3–5 pairs, 1-pinnate, proximal pair basally 2-pinnate, 8–18 × 2–4 cm (excluding expanded base of proximal pinnae). **Basal pinnules** of proximal pair of pinnae elongate and pinnalike; basiscopic pinnule to 10 cm, acroscopic pinnule 1/3–1/2 length of basiscopic. **Ultimate segments** undivided to pinnatifid, finely spiny along margins and at tip. **Sori** with indusia 0.5–1 mm diam., indusia with deep sinus and often overlapping lobes, glabrous.

Terrestrial along stream banks in woods; 0 m; introduced; S.C.; Asia in Japan, China.

Arachniodes simplicior was introduced and naturalized in South Carolina and is known from a single population there (J. E. Gordon 1981). It was still locally common in April 1990 and apparently poses no threat to the native flora.

11. MAXONIA C. Christensen, Smithsonian Misc. Collect. 66(9): 3. 1916 · Climbing wood fern [for William R. Maxon, (1877–1948), American pteridologist]

Robbin C. Moran

Plants hemiepiphytic. **Stems** long-creeping and climbing trees, stolons absent. **Leaves** strongly dimorphic, fertile leaves greatly contracted, evergreen. **Petiole** ± equaling length of blade, base not swollen; vascular bundles more than 3, arranged in an arc, ± round in cross section. **Blade** deltate, 3–4-pinnate-pinnatifid, gradually reduced distally to pinnatifid apex, somewhat leathery. **Pinnae** not articulate to rachis, segment margins dentate to lobed; proximal pinnae largest or nearly so, petiolulate, ± equilateral or inequilateral, basiscopic side with pinnules longer than on acroscopic side; costae adaxially deeply grooved, grooves continuous from rachis to costae to costules; indument of transparent hairs along costae on both sides, also with a few linear scales abaxially on costae. **Veins** free, simple or forked. **Sori** in 1 row between midrib and margin, round; indusia round-reniform with shallow sinus, persistent. **Spores** brownish, spiny to broadly rugose. $x = 41$.

Species 1 (1 in the flora): tropical.

SELECTED REFERENCES Christensen, C. 1916. *Maxonia*, a new genus of tropical American ferns. Smithsonian Misc. Collect. 66(9): 1–4. Moran, R. C. 1987. Monograph of the neotropical fern genus *Polybotrya* (Dryopteridaceae). Bull. Illinois Nat. Hist. Surv. 34: 1–138. Walker, T. G. 1972. The anatomy of *Maxonia apiifolia*: A climbing fern. Brit. Fern Gaz. 10: 241–250.

1. Maxonia apiifolia (Swartz) C. Christensen, Smithsonian Misc. Collect. 66(9): 3. 1916

Dicksonia apiifolia Swartz, J. Bot (Schrader) 1800(2): 91. 1801

Varieties 2 (1 in the flora): North America, West Indies in Greater Antilles, Central America, South America in Ecuador.

1a. Maxonia apiifolia (Swartz) C. Christensen var. **apiifolia** · Climbing wood fern

Stems 1.5–4 cm diam., internodes 3–10 cm; scales orange, brownish, or golden, denticulate. **Roots** from under surface. **Petiole** 30–60 cm, essentially glabrous, grooved adaxially. **Blade** broadly ovate, 3–4-pinnate, base slightly reduced, somewhat leathery, glabrous. **Pinnae** stalked, closely spaced to overlapping, lanceolate. **Pinnules** anadromous throughout. **Tertiary segments** obliquely ascending.

Climbing on trees in wooded limestone hills; 0 m; Fla.; West Indies in Cuba, Jamaica.

Maxonia apiifolia, known in the flora from one collection made in 1921 in southern Florida, has not been relocated and may be extirpated. J. T. Mickel (1979)

suggested that it may represent a horticultural escape, but the species is not known in cultivation. Its occurrence in Florida may indicate long-distance dispersal from the West Indies. *Maxonia apiifolia* var. *dualis* (J. D. Smith) C. Christensen occurs in Central America in Guatemala, Honduras, and Panama, and in South America in Ecuador. It differs from var. *apiifolia* only by its entire, rather than denticulate, scales.

12. POLYSTICHUM Roth, Tent. Fl. Germ. 3: 31, 69. 1799, name conserved · Sword fern, Christmas fern, holly fern [Greek *poly*, many, and *stichos*, row, presumably in reference to the rows of sori on each pinna]

David H. Wagner

Plants terrestrial. **Stems** decumbent to erect, stolons absent. **Leaves** monomorphic (dimorphic in *P. acrostichoides*), evergreen. **Petiole** 1/9–1 times length of blade, bases swollen or not; vascular bundles more than 3, arranged in an arc, ± round in cross section. **Blade** linear-lanceolate to broadly lanceolate, 1–3-pinnate, gradually reduced distally to pinnatifid apex, somewhat leathery to leathery. **Pinnae** not articulate to rachis, segment or pinna margins spinulose-toothed (except *P. lemmonii*); proximal pinnae (several pairs) usually gradually reduced, sessile to short-petiolulate, bases usually inequilateral with acroscopic lobe; costae adaxially grooved, grooves continuous from rachis to costae; indument of linear to lanceolate scales on costae and sometimes between veins abaxially (microscales), ± glabrous or similarly scaly adaxially (scales forming loosely tangled network over blade and sori in *P. dudleyi*). **Veins** free, forked, rarely (*P. imbricans*) anastomosing. **Sori** in 1 row (to several) between midrib and margins, round (confluent, covering abaxial surface in *P. acrostichoides*); indusia peltate, persistent or caducous [absent]. **Spores** yellow or brownish to black, with inflated folds. $x = 41$.

Species ca. 180 (15 in the flora): worldwide.

The mating systems of *Polystichum* seem to be highly outcrossing (P. S. Soltis and D. E. Soltis 1987; P. S. Soltis et al. 1989); hybrids are frequent where two or more species occur. Sterile hybrids are discussed under one of their putative parents.

Sterile hybrids are best recognized by their misshapen sporangia, which produce little black dots at the end of the season instead of forming the fuzzy brown bump typical of sori after spores have been expelled. In many cases the intermediacy and robustness of hybrids make them stand out as odd. At least one or two hybrid plants are to be expected in large, mixed populations. The allopolyploids, having hybrid origins, present particular problems. They exhibit the Vavilov effect: allopolyploids tend to resemble one of their parental species when they grow with, or in the habitat typical of, that species (D. S. Barrington et al. 1989).

In the flora there are six diploids, five tetraploids, one hexaploid, and three species whose chomosome number is unknown. Relationships among the diploids are generally not very close; that is, each is probably more closely related to a species outside the flora than to one of the other species in the flora. The exception to this is the group composed of *Polystichum acrostichoides*, *P. imbricans*, and *P. munitum*. *Polystichum acrostichoides* appears to share a Tertiary common ancestor with *P. munitum*, and *P. imbricans* is more recently derived from *P. munitum*. All of the polyploid species are fertile allopolyploids. One of these species (*P. braunii*) is also involved in the formation of the hexaploid *P. setigerum* (see below).

Relationships among *Polystichum* Species

Allopolyploid	Presumed Originating Crosses
andersonii	*kwakiutlii* × *munitum*
californicum	*dudleyi* × *imbricans* or *dudleyi* × *munitum*
kruckebergii	*lemmonii* × *lonchitis*
scopulinum	*lemmonii* × *imbricans* or *lemmonii* × *munitum*
setigerum	*braunii* × *munitum*

The morphological similarity among *Polystichum* species may make identification difficult, particularly among the species with more divided leaves. The keys presented here are designed for mature, typical individuals. Some of the characters mentioned in the keys and descriptions require the use of a microscope. The microscales (small trichomes that occur on the abaxial leaf surface of all species and adaxially in some) are best observed by peeling them off with cellophane tape and mounting the tape on a slide, sticky side up, under a coverslip. The tape can also be used to lift off the components of the sori. *Polystichum acrostichoides, P. andersonii, P. lemmonii,* and *P. munitum* are known to have sclereid clusters in their pith. *Polystichum imbricans* lacks such clusters, and data are not available for the other species.

SELECTED REFERENCES Wagner, D. H. 1979. Systematics of *Polystichum* in western North America north of Mexico. Pteridologia 1: 1–64. Wagner, W. H. Jr. 1973. Reticulation of holly ferns (*Polystichum*) in the western United States and adjacent Canada. Amer. Fern J. 63: 99–115. Welsh, S. L. 1974. Anderson's Flora of Alaska and Adjacent Parts of Canada. Provo. Yatskievych, G., D. B. Stein, and G. J. Gastony. 1988. Chloroplast DNA evolution and systematics of *Phanerophlebia* (Dryopteridaceae) and related fern genera. Proc. Natl. Acad. Sci. U.S.A. 85: 2589–2593.

1. Fertile pinnae contracted, sori confluent, completely covering abaxial surface. . . . . . .
   . . . . 1. *Polystichum acrostichoides*
1. Fertile pinnae not contracted, sori often distinct.
   2. Leaves 1-pinnate.
      3. Pinnae denticulate but not spiny; pinna apex rounded; microscales dense on both leaf surfaces; restricted to the Aleutian Islands. . . . . 2. *Polystichum aleuticum*
      3. Pinnae serrulate-spiny; pinna apex acute to cuspidate; microscales prominent on abaxial surface only; widespread.
         4. Petioles mostly less than 1/6 length of leaf; blades narrowing toward base; proximal pinnae ± deltate; pinnae spreading-spinulose. . . . . 11. *Polystichum lonchitis*
         4. Petioles usually greater than 1/5 length of leaf; blades narrowing slightly, if at all, toward base; proximal pinnae auriculate-ovate to falcate; pinnae incurved-spinulose.
            5. Indusia ciliate; pinna apex acuminate, base cuneate. . . . . 13. *Polystichum munitum*
            5. Indusia entire to sharply dentate; pinna apex apiculate or cuspidate, base oblique.
               . . . . 7. *Polystichum imbricans*
   2. Leaves 1-pinnate-pinnatifid or 2-pinnate.
      6. Leaves 2-pinnate, pinnules petiolate.
         7. Pinnules rounded at tip, margins not spiny. . . . . 10. *Polystichum lemmonii*
         7. Pinnules apiculate at tip, margins spiny.
            8. Proliferous bulblets present on distal portion of leaves. . . . . .9. *Polystichum kwakiutlii*
            8. Proliferous bulblets absent.
               9. Blades narrowed toward base; microscales on abaxial surface dense but not forming tangled network. . . . . 4. *Polystichum braunii*
               9. Blades not narrowed toward base; microscales on abaxial surface forming loosely tangled network over blades and sori. . . . . 6. *Polystichum dudleyi*

6. Leaves 1-pinnate-pinnatifid but in some species with deeply incised pinna margins appearing 2-pinnate, segments (pinnules) sessile, adnate to costa for at least 2 mm.

    10. Microscales lanceolate to linear-lanceolate, on abaxial leaf surface only; leaves often smaller than 3 dm; to 3500 m.

        11. Pinna apex with spreading teeth (visible without magnification), subapical teeth nearly equal to apical tooth; apex of at least proximal pinnae acute. . . . . 8. *Polystichum kruckebergii*

        11. Pinna apex with incurved teeth, subapical teeth much smaller than apical tooth; pinnae apices obtuse. . . . . 14. *Polystichum scopulinum*

    10. Microscales filiform on both leaf surfaces; leaves often larger than 3 dm; to 1700 m.

        12. Rachis with 1 or more bulblets at pinna base(s) on distal 1/3 of blade. . . . . 3. *Polystichum andersonii*

        12. Rachis without bulblets.

            13. Pinnae not incised to costa; California to s British Columbia. . . . . 5. *Polystichum californicum*

            13. Pinnae incised to costa; s British Columbia northward.

                14. Pinnule margins spinulose-dentate but not incised; Alaska to British Columbia. 15. *Polystichum setigerum*

                14. Pinnule margins deeply incised; w tip of Aleutian Islands. . . . . 12. *Polystichum microchlamys*

1. Polystichum acrostichoides (Michaux) Schott, Gen. Fil., plate 9. 1834 · Christmas fern, polystic faux-acrostiche

Nephrodium acrostichoides Michaux, Fl. Bor.-Amer. 2: 267. 1803

Stems erect. **Leaves** dimorphic (only in this species); fertile pinnae distal, much contracted; sterile leaves arching, 3–8 dm; bulblets absent. **Petiole** 1/4–1/3 length of leaf, densely scaly; scales light brown, diminishing in size distally. **Blade** linear-lanceolate, 1-pinnate; base narrowed. **Pinnae** oblong to falcate, not overlapping, in 1 plane, 2–6 cm; base oblique, acroscopic auricles well developed; margins serrulate-spiny with teeth ascending; apex acute or blunt with subapical and apical teeth same size; microscales filiform, lacking projections, dense, on abaxial surface only. **Sori** confluent, completely covering abaxial surface of pinnae (only in this species); indusia entire. **Spores** light brown. $2n = 82$.

Forest floor and shady, rocky slopes; 0–1500 m; N.B., N.S., Ont., P.E.I., Que.; Ala., Ark., Conn., Del., D.C., Fla., Ga., Ill., Ind., Iowa, Kans., Ky., La., Maine, Md., Mass., Mich., Minn., Miss., Mo., Nebr., N.H., N.J., N.Y., N.C., Ohio, Okla., Pa., R.I., S.C., Tenn., Tex., Vt., Va., W.Va., Wis.; Mexico; naturalized in Europe.

Polystichum acrostichoides is a common species most closely related to *P. munitum* (G. Yatskievych et al. 1988), which also occurs extensively on forest floors. The dimorphic pinnae of *Polystichum acrostichoides*

are not unique to the genus; they are found also in some Asian species. Numerous variants have been named, mostly as forms, but none are of taxonomic consequence. Hybrids are known with *P. braunii* (*P.* × *potteri* Barrington) and *P. lonchitis* (*P.* × *hagenahii* Cody). The latter hybrid is rare, known only from its type locality in Ontario, where it grows with both parents. It is recognized by its intermediate morphology (leaves wider than *P. lonchitis*, narrower than *P. acrostichoides*, with slightly contracted sorus-bearing pinnae) and malformed sporangia and spores. *Polystichum* × *potteri* is much more widespread, from Nova Scotia, New Brunswick, and Quebec through New England to Pennsylvania. It resembles *P. braunii* but has narrower leaves bearing malformed sporangia.

2. Polystichum aleuticum C. Christensen in Hultén, Svensk Bot. Tidskr. 30: 515, fig. 1a. 1936

Stems erect. **Leaves** monomorphic, erect, 1–1.5 dm; bulblets absent. **Petiole** 1/6–1/4 length of leaf; scales tan, sparse or falling off early. **Blade** linear-lanceolate, 1-pinnate, gradually tapered to base. **Pinnae** ± deltate to ovate, slightly overlapping, in 1 plane, 4–8 mm; base truncate, acroscopic auricle well developed; margins denticulate, not spiny; apex rounded, not dentate; microscales linear, lacking projections, dense on both surfaces. **Indusia** entire to minutely erose-dentate. **Spores** brown.

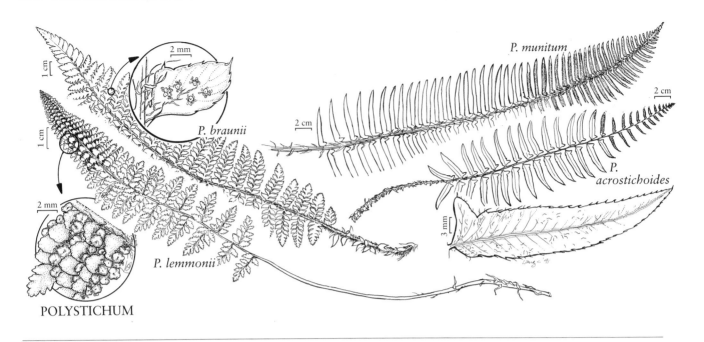

P. munitum

P. braunii

P. acrostichoides

P. lemmonii

POLYSTICHUM

Growing in crevices of rocks; of conservation concern; 0–400 m; Alaska.

Polystichum aleuticum is known only from Atka and Adak islands, Alaska. Its nearest relatives are in southwest China. *Polystichum aleuticum* resembles the dwarfed plants of *P. kruckebergii,* but it differs, especially, in its smaller, more rounded pinna apices and more abundant scales.

3. **Polystichum andersonii** M. Hopkins, Amer. Fern J. 3: 116, plate 9. 1913 · Anderson's sword fern

Polystichum braunii (Spenner) Fée subsp. *andersonii* (M. Hopkins) Calder & Roy L. Taylor; *P. braunii* var. *andersonii* (M. Hopkins) Hultén

Stems erect. **Leaves** monomorphic, arching, 3–10 dm; bulblets 1 or more, on distal 1/3 of rachis. **Petiole** 1/8–1/4 length of leaf, densely scaly; scales light brown, diminishing in size distally. **Blade** lanceolate, 1-pinnate-pinnatifid; base narrowed. **Pinnae** lanceolate-falcate, proximal pinnae ± triangular, not overlapping, in 1 plane, 2–10 cm; base oblique, acroscopic auricle well developed; margins incised to costae, segments adnate to costa for at least 2 mm, serrulate-spiny with teeth ascending; apex acute with subapical and apical teeth same size; microscales filiform, with contorted projections, dense abaxially, sparse adaxially. **Indusia** sparsely ciliate. **Spores** light brown to brown. $2n = 164$.

Lowland coastal to midmontane forests; interior moist forests; 100–1700 m; B.C.; Alaska, Idaho, Mont., Oreg., Wash.

Polystichum andersonii is an allotetraploid (D. H. Wagner 1979); its diploid parents are *P. munitum* and *P. kwakiutlii.* The triploid cross, *P. munitum × andersonii,* has been analyzed cytologically (W. H. Wagner Jr. 1973). It is the only sterile hybrid in the genus that develops large colonies through vegetative propagation by its bulblets. Hybrids look very much like some of the more deeply incised forms of *Polystichum munitum* except that they have abundant filiform scales, abortive sori, and nearly triangular lowermost pinnae with ± equally incised acroscopic and basiscopic auricles.

4. **Polystichum braunii** (Spenner) Fée, Mém. Foug. 5: 278. 1852 · Braun's holly fern, polystic de Braun

Aspidium braunii Spenner, Fl. Friburg. 1: 9, plate 2. 1825; *Polystichum braunii* subsp. *purshii* (Fernald) Calder & Roy L. Taylor; *P. braunii* var. *purshii* Fernald

Stems erect. **Leaves** monomorphic, arching, 2–10 dm; bulblets absent. **Petiole** 1/8–1/6 length of leaf, densely scaly; scales light brown, gradually diminishing in size distally. **Blade** broadly lanceolate, 2-pinnate; base narrowed. **Pinnae** oblong-lanceolate or falcate, proximal pinnae ± rectangular, not overlapping, in 1 plane, 2–10 cm; base oblique except proximal 3–4 pinnae, where auricles not developed; apex acute, subapical and apical teeth same size; microscales filiform to linear, lacking projections, dense abaxially, sparse adaxially. **Pinnules** ± stalked, short-falcate to oblique-rhombic, acroscopic auricle well developed on proximal pinnules; margins dentate, with slender bristle tips; apex broadly acute. **Indusia** ciliate. **Spores** brown. $2n = 164$.

Moist places in boreal forests; interior moist forests; 0–300 m; St. Pierre and Miquelon; B.C., N.B., Nfld., N.S., Ont., Que., Yukon; Alaska, Conn., Idaho, Maine, Mass., Mich., Minn., N.H., N.Y., Vt., Wis.; Eurasia.

Because no diploid ancestors have been found, *Polystichum braunii* is thought to be an ancient tetraploid. Its sterile hybrid with *P. acrostichoides*, *P. ×potteri*, is discussed under *P. acrostichoides*. *Polystichum braunii* × *lonchitis* has been reported from southeast Alaska (S. L. Welsh 1974). This hybrid has been cytologically confirmed in Europe and named *P. ×meyeri* Sleep & Reichstein (A. Sleep and T. Reichstein 1967). It has narrower and less divided leaves than *P. braunii* and poorly developed auricles. *Polystichum braunii* × *lonchitis* was described by J. Ewan (1944), but the type (from British Columbia) is *P. braunii* × *munitum* (A. Sleep and T. Reichstein 1967). This latter hybrid is the postulated progenitor of *P. setigerum* (D. H. Wagner 1979). North American *P. braunii* has been segregated as var. *purshii* Fernald, distinguished from European populations (var. *braunii*) by having broader microscales.

5. **Polystichum californicum** (D. C. Eaton) Diels in Engler & Prantl, Nat. Pflanzenfam. 1(4): 191. 1899 · California sword fern

Aspidium californicum D. C. Eaton, Proc. Amer. Acad. Arts 6: 555. 1865; *Polystichum aculeatum* (Linnaeus) Roth var. *californicum* (D. C. Eaton) Jepson

Stems erect or ascending. **Leaves** monomorphic, arching or erect, 2–8 dm; bulblets absent. **Petiole** 1/5–1/3 length of leaf; scales light brown, abruptly diminishing in size distally, falling off early distally. **Blade** lanceolate to linear-lanceolate, 1-pinnate-pinnatifid, base slightly narrowed. **Pinnae** oblong to lanceolate to falcate, shallowly to deeply divided, pinnae overlapping or not, in 1 plane, 2–10 cm; base oblique, acroscopic auricle lobed; margins not incised to costae, serrulate-spiny with teeth ascending; apex acute-attenuate, subapical and apical teeth same size (southern form) or obtuse and cuspidate with subapical teeth smaller than apical tooth (northern form); microscales filiform, dense abaxially, sparse adaxially. **Indusia** ciliate. **Spores** brown. $2n = 164$.

On forest floor in southern part of range and in rock crevices at cliff bottoms (most commonly andesite) to north; 100–850 m; B.C.; Calif., Oreg., Wash.

Polystichum californicum is restricted to the Coast Ranges and the Sierra-Cascade axis. It is most abundant in the Coast Range north of San Francisco.

Polystichum californicum is an allopolyploid, the evolutionary roots of which include *P. dudleyi* as the 2-pinnate ancestor. Morphologic and ecological data indicate *P. imbricans* is ancestor to the northern forms and *P. munitum* is ancestor to southern forms, suggesting *P. californicum* is an amalgam of interfertile tetraploids with polyphyletic origins (D. H. Wagner 1979). Cytological analysis corroborates this (A. D. Callan 1972; W. H. Wagner Jr. 1973), but chloroplast DNA studies have detected only the involvement of *P. imbricans* in the ancestry of *P. californicum* (P. S. Soltis et al. 1991).

The more xeric, rock-inhabiting members of the complex (showing the parental influence of *P. imbricans*) occupy the northern half of the range whereas plants of more mesic habitats are found to the south. Hybrids with both *P. dudleyi* and *P. munitum* are found frequently, because these three species are often sympatric (W. H. Wagner 1973). The hybrid with *P. dudleyi* (a triploid) will key to that species. The hybrid with *P. munitum* resembles a less-incised form of *P. californicum* with aborted sporangia. *Polystichum californicum* × *imbricans* has been found only once, in Oregon (A. D. Callan 1972). Another hybrid that will key here, based on its overall appearance, is *P. munitum* × *scopulinum*. It lacks filiform microscales and also has malformed sporangia. Such a specimen was the basis of the report of *Polystichum californicum* in eastern Washington (C. L. Hitchcock et al. 1955–1969, vol. 1). The sterile diploid hybrid between *P. dudleyi* and *P. munitum* is indistinguishable from *P. californicum* except for aborted sporangia and chromosome number (W. H. Wagner Jr. 1973).

6. **Polystichum dudleyi** Maxon, J. Wash. Acad. Sci. 8: 620. 1918 · Dudley's sword fern

Polystichum aculeatum (Linnaeus) Roth var. *dudleyi* (Maxon) Jepson

Stems erect. **Leaves** monomorphic, arching, 2–10 dm; bulblets absent. **Petiole** 1/5–1/3 length of leaf, densely scaly; scales light brown, gradually diminishing in size distally. **Blade** broadly lanceolate, 2-pinnate, base not narrowed. **Pinnae** narrowly lanceolate, not overlapping, in 1 plane, 3–13 cm; base oblique, apex acute with subapical and apical teeth same size; microscales filiform, lacking projections, sparse abaxially, but longer than in other *Polystichum* species, forming loosely tangled network over blade and sori (such network only in this species), sparse adaxially. **Pinnules** ± stalked, linear-falcate to oblique-rhombic, acroscopic auricle well developed on proximal pinnules; margins spinulose-dentate; apex acute. **Indusia** ciliate. **Spores** brown. $2n = 82$.

Moist forests; 0–100 m; Calif.

Polystichum dudleyi is confined to coastal central California. Hybrids with *P. californicum* are relatively frequent where these species occur together. These hybrids would key here but, unlike *P. dudleyi*, they are less divided and have aborted sporangia. The sterile diploid hybrid with *P. munitum* is also frequent in areas of sympatry. It is indistinguishable from *P. californicum* except for malformed sporangia and chromosome number (W. H. Wagner Jr. 1973).

7. Polystichum imbricans (D. C. Eaton) D. H. Wagner, Pteridologia 1: 50. 1979 · Imbricate sword fern

Aspidium munitum Kaulfuss var. *imbricans* D. C. Eaton, Ferns N. Amer. 1: 188, plate 25, fig. 3. 1878; *Polystichum munitum* (Kaulfuss) C. Presl subsp. *imbricans* (D. C. Eaton) Munz; *P. munitum* var. *imbricans* (D. C. Eaton) Maxon

Stems ascending to erect. **Leaves** erect to arching back at tip, 2–8 dm; bulblets absent. **Petiole** 1/4–1/3 length of leaf; scales abruptly diminishing in size distally and falling off early but retaining conspicuous tuft of brown scales at base. **Blade** linear-lanceolate to linear, 1-pinnate, base not or slightly narrowed. **Pinnae** oblong, slenderly lanceolate, or falcate, usually overlapping, in 1 plane or twisted out of plane of blade, 2–4 cm; base oblique, auricles well developed; margins serrulate-spiny with teeth ascending; apex cuspidate or apiculate with subapical teeth smaller than apical tooth; microscales lanceolate to linear with straight or sharply angular projections, sparse, on abaxial surface only. **Indusia** entire to sharply dentate. **Spores** dark brown.

Subspecies 2: only in the flora.

Polystichum imbricans is one of the postulated ancestors of two allopolyploids, *P. californicum* and *P. scopulinum* (D. H. Wagner 1979). Relationships to *P. munitum* are discussed under that species.

1. Pinnae oblong, less than 5 times longer than wide, adaxial surfaces facing upward and twisted out of plane of blade; leaves stiffly erect to tip; at base of boulders or in cliff crevices in exposed sites; California to British Columbia. . . . . 7a. *Polystichum imbricans* subsp. *imbricans*
1. Pinnae narrowly lanceolate, more than 5 times longer than wide, generally in 1 plane; leaf tips arching back; on forest floor in shade; only in California. . . 7b. *Polystichum imbricans* subsp. *curtum*

7a. Polystichum imbricans (D. C. Eaton) D. H. Wagner subsp. **imbricans**

Polystichum munitum (Kaulfuss) C. Presl subsp. *nudatum* (D. C. Eaton) Ewan

Stems decumbent to ascending. **Leaves** stiffly erect, 2.5–5 dm. **Pinnae** lanceolate-oblong, less than 5 times longer than wide, twisted out of plane of blade (adaxial surfaces facing upward like Venetian blinds), base usually cupped, apex cuspidate. **Indusia** entire.

Rooted at base of boulders or in cliff crevices in drier sites of mesic montane forests; 0–2500 m; B.C.; Calif., Oreg., Wash.

Polystichum imbricans subsp. *imbricans* grows in the Coast Ranges and the Sierra-Cascade axis. It is isolated in the Wallowa Mountains of eastern Oregon.

Sun forms of *Polystichum munitum* are often mistaken for *P. imbricans*; characteristics of the distal petiolar scales and indusial margins are more reliable than gross morphologic features for distinguishing them. *Polystichum imbricans* has narrow distal petiolar scales that fall off early; *P. munitum* has wide distal petiolar scales (the largest more than 1 mm wide) that are persistent. *Polystichum imbricans* hybridizes readily with *P. munitum*, the hybrids usually being sterile but in some places forming hybrid swarms because of partial fertility of the hybrids (D. H. Wagner 1979). The hybrids with *P. californicum* are discussed under that species.

7b. Polystichum imbricans (D. C. Eaton) D. H. Wagner subsp. **curtum** (Ewan) D. H. Wagner, Pteridologia 1: 51. 1979

Polystichum munitum (Kaulfuss) C. Presl subsp. *curtum* Ewan, Amer. Fern J. 32: 99. 1942

Stems erect. **Leaves** arching at tip, planar, 4–8 dm. **Pinnae** narrowly lanceolate, straight to falcate, in same plane as blade, more than 5 times longer than wide, apex acuminate-apiculate. **Indusia** entire or margin short-dentate.

On floor of mixed evergreen forests in light shade; 400–1400 m; Calif.

Polystichum imbricans var. *curtum* occurs mostly inland in California, and occasionally in the central Sierra Nevada.

8. Polystichum kruckebergii W. H. Wagner, Amer. Fern J. 56: 4. 1966 · Kruckeberg's holly fern

Stems ascending. **Leaves** erect, 1–2.5 dm; bulblets absent. **Petiole** 1/10–1/5 length of leaf, sparsely scaly; scales light brown, gradually diminishing in size distally. **Blade** linear, 1-pinnate-pinnatifid, base narrowed. **Pinnae** rhombic-ovate to short-falcate, proximal pinnae ± triangular; pinnae overlapping, twisted somewhat out of plane of blade, 0.5–1.5 cm; base oblique, acroscopic auricle well developed; margins shallowly incised to merely dentate or serrulate, teeth spreading and spiny at tip; apex acute with subapical and apical teeth same size; microscales lanceolate with few projections, confined to costa, on abaxial surface only. **Indusia** entire. **Spores** dark brown. 2*n* = 164.

Rocks and cliffs in subalpine to alpine habitats; 1500–3200 m; B.C.; Calif., Idaho, Mont., Nev., Oreg., Utah, Wash.

Polystichum kruckebergii is widely but sporadically distributed in small numbers in both the Sierra-Cascade and Rocky Mountain systems. Populations sometimes consist of only two or three dwarfed plants that are difficult to distinguish from *P. scopulinum,* with which they may occur. The spreading teeth of equal size at the pinna apex will usually distinguish this species. *Polystichum kruckebergii* is a tetraploid presumed to be of hybrid origin, with *P. lonchitis* and *P. lemmonii* as its diploid progenitors (W. H. Wagner Jr. 1973), although this hypothesis has not been confirmed. The hybrid with *P. munitum* has been found in Washington (P. S. Soltis et al. 1987) with both parents, and it is distinguished by intermediate morphology and abortive sporangia.

9. Polystichum kwakiutlii D. H. Wagner, Amer. Fern J. 80: 50, fig. 1. 1990

Stems unknown. **Leaves** (only distal portion known) with bulblets present. **Blade** lanceolate, 2-pinnate, base probably narrowed. **Pinnae** narrowly lanceolate, ca. 2–7 cm, base truncate to oblique, acroscopic proximal pinnule enlarged, apex obtuse. **Pinnules** short-stalked, ovate-rhombic, acroscopic auricle ± well developed, margins finely spiny-dentate; apex acuminate; microscales filiform, lacking projections, dense abaxially, sparse adaxially. **Indusia** entire.

Habitat unknown; of conservation concern; B.C.

Polystichum kwakiutlii is known only from the type specimen, collected at Alice Arm, British Columbia (whether referring to inlet or town is unknown). This species is presumed to be one of the diploid progenitors of *P. andersonii.* It should be sought among the boreal 2-pinnate polystichums, from which it can be distinguished by the presence of bulblets. *Polystichum kwakiutlii* differs from *P. andersonii* in its completely divided pinnae and entire indusia.

10. Polystichum lemmonii L. Underwood, Native Ferns ed. 6, 116. 1900 · Shasta fern

Polystichum mohrioides (Bory) C. Presl var. *lemmonii* (L. Underwood) Fernald

Stems decumbent to ascending. **Leaves** erect, 1–3.5 dm; bulblets absent. **Petiole** 1/5–1/4 length of leaf, sparsely scaly; scales pale tan, abruptly diminishing in size distally. **Blade** narrowly lanceolate, 2-pinnate, scarcely narrowed at base. **Pinnae** ovate, overlapping, folded inward and twisted horizontally, 0.5–2 cm; base truncate to oblique, proximal acroscopic pinnules not enlarged; apex broadly acute; microscales narrowly lanceolate, with few projections, sparse, ± confined to costa of both surfaces. **Pinnules** ± stalked, rounded, acroscopic auricle not well developed, margins entire to weakly dentate, apex rounded. **Indusia** entire or minutely dentate-erose. **Spores** dark brown to blackish. 2*n* = 82.

On rocky serpentine slopes; 1200–2400 m; B.C.; Calif., Oreg., Wash.

Polystichum lemmonii forms sterile hybrids with *P. scopulinum* and *P. munitum.* The first hybrid may be abundant where the two parents grow together, which they frequently do in the Wenatchee Mountains of Washington and Siskiyou Mountains of northern California and southwest Oregon. The hybrid is very similar to *P. lemmonii* but has malformed sporangia and slightly less divided pinnae than *P. lemmonii.* The *P. lemmonii* × *munitum* hybrid is morphologically indistinguishable from *P. scopulinum;* it is a sterile diploid reported only twice from the Wenatchee Mountains of Washington (W. H. Wagner Jr. 1973; P. S. Soltis et al. 1989). It is possible that this hybrid involves *P. imbricans* and not *P. munitum;* neither study distinguished between them.

American authors have misapplied the name *Polystichum mohrioides* (Bory) C. Presl, a South American species, to *P. lemmonii.*

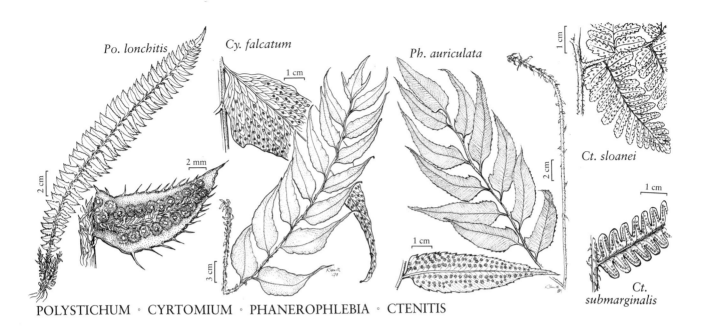

Po. lonchitis   *Cy. falcatum*   *Ph. auriculata*   *Ct. sloanei*

Ct. submarginalis

POLYSTICHUM · CYRTOMIUM · PHANEROPHLEBIA · CTENITIS

11. Polystichum lonchitis (Linnaeus) Roth, Tent. Fl. Germ. 3(1): 71. 1799 · Holly fern, polystic à feuilles de houx

Polypodium lonchitis Linnaeus, Sp. Pl. 2: 1088. 1753

Stems erect to occasionally ascending. **Leaves** erect, not arching except at tip, 1–6 dm; bulblets absent. **Petiole** 1/10–1/6 of blade, densely scaly; scales light brown, gradually diminishing in size distally. **Blade** linear, often widest above middle, 1-pinnate, base narrowed. **Pinnae** oblong to lanceolate to falcate, proximal pinnae ± deltate, rarely overlapping, in 1 plane, 0.5–3 cm, base truncate to oblique, acroscopic auricle well developed; margins serrulate-spiny with teeth spreading; apex acute, subapical tooth hardly smaller than apical tooth; microscales dense, on abaxial surface only. **Indusia** entire or minutely dentate-erose. **Spores** dark brown. $2n = 82$.

In rock crevices or at base of boulders, mostly in boreal and subalpine coniferous forests or alpine regions; 0–3200 m; Greenland; Alta., B.C., Nfld., N.S., Ont., Que., Yukon; Alaska, Ariz., Calif., Colo., Idaho, Mich., Minn., Mont., Nev., Oreg., Utah, Wash., Wis., Wyo.

The hybrid between *Polystichum lonchitis* and *P. acrostichoides* (= *P.* ×*hagenahii* Cody) is discussed under *P. acrostichoides*. The hybrid with *P. braunii* (= *P.* ×*meyeri* Sleep & Reichstein) is discussed under *P.*

braunii. In the Georgian Bay area of Ontario, *P. lonchitis* hybridizes with *Dryopteris goldiana* to produce the peculiar ×*Dryostichum singulare* W. H. Wagner (W. H. Wagner Jr., F. S. Wagner et al. 1992).

The spiny spores of *P. lonchitis* are distinctive and distinguish this from dwarfed forms of other 1-pinnate species.

12. Polystichum microchlamys (H. Christ) Matsumura, Index Pl. Jap. 1: 343. 1904

Aspidium microchlamys H. Christ, Bull. Herb. Boissier 7: 820. 1899

Stems erect. **Leaves** arching, 3–8 dm; bulblets absent. **Petiole** 1/8–1/4 length of leaf, densely scaly; scales brown, diminishing in size distally. **Blade** broadly lanceolate, 1-pinnate-pinnatifid, base slightly narrowed. **Pinnae** narrowly lanceolate, not overlapping, in 1 plane, 3–13 cm; base oblique, proximal acroscopic segments enlarged; margins incised to costa but segments sessile and adnate to costa for at least 2 mm, segments excised and decurrent, serrulate-spiny with teeth spreading to ascending; apex acute with subapical and apical teeth same size; microscales filiform, dense abaxially, sparse adaxially. **Indusia** erose-dentate. **Spore** color unknown.

Terrestrial; 0–100 m; Alaska; Asia in Kamtchatka and Japan.

Polystichum microchlamys is found in the flora only on Attu, at the western tip of the Aleutian Archipelago.

13. Polystichum munitum (Kaulfuss) C. Presl, Tent.
Pterid., 83. 1836 · Common sword fern

Aspidium munitum Kaulfuss, Enum. Filic., 236. 1824

Stems erect or ascending. **Leaves** arching, 5–18 dm; bulblets absent. **Petiole** 1/8–1/4 length of leaf, densely scaly; scales redbrown to dark brown or nearly black, gradually diminishing in size distally. **Blade** linear-lanceolate, 1-pinnate, base slightly narrowed. **Pinnae** narrowly lanceolate, straight to falcate, not overlapping, pinnae of shade-growing plants in 1 plane, those of sun-growing plants twisted or contorted, 1–15 cm; base ± cuneate, auricles well developed; margins serrulate-spiny with teeth ascending; apex acuminate with subapical teeth same size as apical tooth; microscales ovate-lanceolate to linear-lanceolate, with contorted projections, dense, on abaxial surface only. **Indusia** ciliate. **Spores** light yellow. $2n = 82$.

Terrestrial, forest floor, only occasionally on rock, in mesic coniferous to moist, mixed evergreen forests; 0–2200 m; B.C.; Calif., Idaho, Mont., Oreg., S.Dak., Wash.; Mexico on Guadalupe Island; naturalized in Europe.

One of the most abundant ferns in the western flora (rivaled only by *Pteridium*), *Polystichum munitum* also is of significant economic importance. Enormous quantities of leaves are gathered for backgrounds in funeral wreaths and other floral displays; the evergreen leaves keep well in cold storage and are exported to Europe. It is extensively used in landscaping, the trade being mainly in wild-collected plants.

Polystichum munitum appears to be most closely related to *P. imbricans* based on morphologic (D. H. Wagner 1979) and electrophoretic (P. S. Soltis et al. 1990) analyses. The chloroplast DNA of *P. imbricans*, however, is divergent (G. Yatskievych et al. 1988), suggesting a chloroplast origin independent of the nuclear genome. That *Polystichum munitum* is related to *P. acrostichoides* is supported by data from chloroplast DNA analysis (G. Yatskievych et al. 1988) but contradicted by data from electrophoretic studies (P. S. Soltis et al. 1990).

Polystichum munitum can be distinguished from *P. imbricans* by its persistent, wide (the largest wider than 1 mm) distal petiolar scales; such scales of *P. imbricans* are less than 1 mm wide and fall off early.

From an evolutionary standpoint, *Polystichum munitum* is a diploid progenitor of *P. andersonii*, *P. californicum*, *P. setigerum*, and, perhaps, *P. scopulinum*. Hybrids with all except *P. setigerum* have been reported, all triploid, attesting to its parental role in the tetraploids (see discussion under each). Hybrids with *P. braunii* (A. Sleep and T. Reichstein 1967), *P. kruckebergii* (P. S. Soltis et al. 1987), *P. dudleyi* (W. H. Wagner Jr. 1973), and *P. lemmonii* (P. S. Soltis et al. 1989) also have been reported.

The population on Guadalupe Island has been called *Polystichum solitarium* Maxon.

14. Polystichum scopulinum (D. C. Eaton) Maxon, Fern
Bull. 8: 29. 1900 · Rock sword fern, polystic des falaises

Aspidium aculeatum (Linnaeus) Swartz var. *scopulinum* D. C. Eaton, Ferns N. Amer. 2: 125. 1880; *Polystichum mohrioides* (Bory) C. Presl var. *scopulinum* (D. C. Eaton) Fernald

Stems ascending. **Leaves** erect, 1–3(–5) dm; bulblets absent. **Petiole** 1/5–1/3 length of leaf, densely scaly but scales falling off distally; scales light brown, abruptly diminishing in size distally. **Blade** narrowly lanceolate, 1-pinnate-pinnatifid, base narrowed. **Pinnae** oblong-lanceolate, overlapping, folded inward and twisted horizontally, 1–3 cm; base oblique; margins serrulate with teeth curved inward; apex obtuse to cuspidate with subapical teeth smaller than apical tooth; microscales narrowly lanceolate, with stout projections, sparse, on abaxial surface only. **Indusia** entire-ciliate. **Spores** brown. $2n = 164$.

Rock crevices and at base of boulders, serpentine to acidic substrates, usually exposed to full sun; 0–3500 m; B.C., Nfld., Que.; Ariz., Calif., Colo., Idaho, Mont., Nev., Oreg., Utah, Wash., Wyo.

Polystichum scopulinum is widely distributed in the United States west of the 110th meridian, where it occurs in sporadic, usually small populations. The species is abundant only on montane serpentine outcrops. The populations in Newfoundland and Quebec are dramatically disjunct.

Polystichum scopulinum is an allopolyploid, believed on morphologic grounds to be derived from *P. imbricans* × *lemmonii* (D. H. Wagner 1979). Based on putative hybridization between *P. scopulinum* and *P. munitum* (P. S. Soltis et al. 1989; W. H. Wagner Jr. 1973), however, *P. munitum* may also be involved. This hybrid is discussed under *P. californicum*.

15. Polystichum setigerum (C. Presl) C. Presl, Tent. Pterid, 83. 1836 · Alaska sword fern

Nephrodium setigerum C. Presl, Reliq. Haenk. 1: 37. 1825; *P. braunii* (Spenner) Fée subsp. *alaskense* (Maxon) Calder & R. L. Taylor; *P. braunii* var. *alaskense* (Maxon) Hultén

Stems erect. **Leaves** arching, 4–10 dm; bulblets absent. **Petiole** 1/8–1/5 length of leaf, densely scaly; scales light brown, gradually diminishing in size distally. **Blade** lanceolate, deeply 1-pinnate-pinnatifid to 2-pinnate, base narrowed. **Pinnae** lanceolate, not overlapping, in 1 plane, 4–8 cm, base oblique, margins incised to costa on middle pinnae, serrulate-spiny with teeth spreading-ascending, apex acute-apiculate with subapical and apical teeth same size; microscales filiform, sparse abaxially, confined to costa adaxially. **Indusia** erose-ciliate. **Spores** brown. $2n = 246$.

Forest floor in lowland coastal forests; 0–250 m; B.C.; Alaska.

Polystichum setigerum is disjunct on Attu Island at the western tip of the Aleutian Archipelago. It is presumed to be of hybrid origin, the result of a cross between *P. munitum* and *P. braunii* (D. H. Wagner 1979). This hybrid has been produced experimentally (A. Sleep and T. Reichstein 1967) and is reported from British Columbia (see discussion under *P. braunii*).

13. CYRTOMIUM C. Presl, Tent. Pterid., 86. 1836 · Asiatic holly fern [Greek *cyrtoma*, arch, for the arched veins]

George Yatskievych

Plants terrestrial or on rock. **Stems** erect or ascending, stolons absent. **Leaves** monomorphic, evergreen. **Petiole** ± 1/2–3/4 length of blade, base not swollen; vascular bundles more than 3, arranged in arc, ± round in cross section. **Blade** oblong-lanceolate, 1-pinnate with pinnae not lobed, distal pinnae only slightly smaller, blade ending in basally lobed pinna ± similar to lateral pinnae, papery or somewhat leathery. **Pinnae** not articulate to rachis, segment margins crenate to spinulose, sometimes also coarsely dentate; proximal pinnae not reduced, petiolulate, equilateral or somewhat inequilateral with acroscopic base more developed; costae adaxially grooved, grooves continuous from rachis to costae; indument of filiform scales abaxially, absent adaxially. **Veins** elaborately anastomosing, areoles formed with 1–3 included veinlets. **Sori** in 2 or more rows between midrib and margin, round. **Indusia** peltate, persistent or caducous. **Spores** brown, with inflated folds or wings. $x = 41$.

Species ca. 15 (2 in the flora): North America, Asia, Africa including Madagascar, Pacific Islands in Hawaii.

This difficult genus requires further systematic study. Estimates of species numbers have ranged from 9 (C. Christensen 1930) to 59 (Shing K. H. 1965). The group might better be considered a subgenus of *Polystichum*, from which it is poorly differentiated morphologically.

SELECTED REFERENCES Christensen, C. 1930. The genus *Cyrtomium*. Amer. Fern J. 20: 41–52. Ching, R. C. 1936. On the genus *Cyrtomium* Presl. Bull. Chin. Bot. Soc. 2: 85–106. Shing, K. H. 1965. A taxonomical study of the genus *Cyrtomium* Presl. Acta Phytotax. Sin., suppl. 1: 1–48.

1. Pinnae 4–10(–12) pairs, leathery, shiny adaxially, margins sometimes undulate or coarsely dentate, but not minutely crenulate or denticulate. . . . . .1. *Cyrtomium falcatum*
1. Pinnae (8–)10–25 pairs, papery, not shiny adaxially, margins minutely crenulate-denticulate. . . . . .2a. *Cyrtomium fortunei* var. *fortunei*

1. Cyrtomium falcatum (Linnaeus f.) C. Presl, Tent. Pterid., 86. 1836

Polypodium falcatum Linnaeus f., Suppl. Pl., 446. 1782

Stem scales orange-brown, ovate with attenuate apices. **Leaves** 30–60(–100) cm. **Pinnae** bright green and shiny adaxially, 4–10(–12) pairs, ovate-attenuate, usually falcate, 4–8.5 cm, leathery, sometimes with short, basal, acroscopic lobe, margins often undulate or coarsely and irregularly dentate. **Indusia** brown, sometimes with blackish centers, not shriveled at maturity. $n = 2n = 123$, apogamous.

Brick or stone walls, rocky areas, mesic forests, and coastal bluffs; 0–100 m; introduced; Calif., Fla., Ga., La., Miss., S.C.; Europe; Asia.

Cyrtomium falcatum is native to east Asia and widely escaped from cultivation. All plants in the flora appear to be the 32-spored, apogamous triploid.

2. Cyrtomium fortunei J. Smith, Ferns Brit. For., 286. 1866

Varieties 3 (1 in the flora): North America, Asia.

2a. Cyrtomium fortunei J. Smith var. **fortunei**

Stem scales brown, ovate to lanceolate with attenuate apices. **Leaves** to 90 cm. **Pinnae** dark green and not shiny adaxially, (8–)10–25 pairs, lanceolate-attenuate, usually falcate, 5–9 cm, papery, sometimes with short, basal, acroscopic lobe, margins minutely crenulate-denticulate. **Indusia** pale brown to tan, thin, shriveled at maturity. $n = 2n = 123$, apogamous.

Brick or stone walls, clay banks, mesic ravines; 0–100 m; introduced; Ga., La., Miss., Oreg., S.C.; Europe; Asia.

Cyrtomium fortunei is native to east Asia and widely escaped from cultivation. Two other varieties, var. *intermedium* Tagawa and var. *clivicola* (Makino) Tagawa, occur in Asia. *Cyrtomium caryotideum* (Wallich ex Hooker & Greville) C. Presl is another commonly cultivated species of Asiatic holly fern. It is native in Hawaii, but it is not yet known to have become established outside of cultivation in the flora area. It is characterized by having only 3–5 pairs of pinnae, with finely and sharply serrulate margins.

14. PHANEROPHLEBIA C. Presl, Tent. Pterid., 84. 1836 · Mexican holly fern [Greek *phaneros*, free, and *phlebium*, vein, for the nonanastomosing venation found in the type species, *P. nobilis*]

George Yatskievych

Plants terrestrial, less commonly on rock. **Stems** short-creeping to erect, stolons absent. **Leaves** monomorphic, evergreen. **Petiole** shorter than or ± equaling length of blade, base not swollen; vascular bundles more than 3, arranged in an arc, ± round in cross section. **Blade** ovate-lanceolate, 1-pinnate, with a ± similar apical pinna, papery. **Pinnae** not articulate to rachis, segment margins serrulate to spinulose; proximal pinnae largest or nearly so, short-petiolulate, ± equilateral or inequilateral with acroscopic lobe; costae adaxially grooved, grooves continuous from rachis to costae; indument of filiform scales on costae and veins abaxially, ± glabrous adaxially. **Veins** free [anastomosing], forked. **Sori** in 2 or more rows between midrib and margin, round; indusia persistent or caducous [absent]. **Spores** brown, with inflated folds or wings. $x = 41$.

Species 8 (2 in the flora): North America, Mexico, Central America, n South America, West Indies in Hispaniola.

Phanerophlebia has sometimes been included in *Cyrtomium*, which it resembles superficially. As with that genus, *Phanerophlebia* might better be considered a subgenus of *Polystichum*, from which it is poorly differentiated morphologically. Elsewhere, some species have bicolored stem scales, netted venation, and/or lack indusia.

SELECTED REFERENCES Maxon, W. R. 1912. Notes on the North American species of *Phanerophlebia*. Bull. Torrey Bot. Club 39: 23–28. Underwood, L. M. 1899. North American ferns—II. The genus *Phanerophlebia*. Bull. Torrey Bot. Club 26: 205–216.

1. Pinnae, at least some on each plant, with basal, acroscopic lobe; indusia flat to concave centrally, not umbonate, shriveled at maturity. . . . . 1. *Phanerophlebia auriculata*
1. Pinnae lacking basal, acroscopic lobe; indusia with raised, central umbo, not shriveled at maturity. . . . . 2. *Phanerophlebia umbonata*

1. Phanerophlebia auriculata L. Underwood, Bull. Torrey Bot. Club 26: 212. 1899

Cyrtomium auriculatum (L. Underwood) C. V. Morton

Leaves 10–60(–75) cm, very short leaves sometimes fertile. **Pinnae** (2–)5–12 pairs, ovate to lanceolate, usually falcate, 2–9 cm, base obliquely cuneate to nearly cordate, apex usually attenuate; acroscopic lobe usually present, prominent (at least some present on every leaf); pinnae rarely irregularly incised. **Indusia** membranous, flat or concave centrally, not umbonate, shriveled at maturity. $2n = 164$.

Soil pockets in sheltered rock crevices of canyons and ravines; 600–2100 m.; Ariz., N.Mex., Tex.; n Mexico.

2. Phanerophlebia umbonata L. Underwood, Bull. Torrey Bot. Club 26: 211. 1899

Cyrtomium umbonatum (L. Underwood) C. V. Morton

Leaves 30–90 cm. **Pinnae** 10–18 pairs, lanceolate, usually falcate, 5–15 cm, base cuneate to ± truncate, apex attenuate; acroscopic lobe absent. **Indusia** scarious, convex with darker, raised umbo centrally, persistent and not shriveled at maturity. $2n = 82$.

Moist, sheltered canyons and ravines; 1800–1900 m; Tex.; n Mexico.

In the flora *Phanerophlebia umbonata* is known only from the Chisos Mountains, in Brewster County, Texas.

15. CTENITIS (C. Christensen) C. Christensen in Verdoorn et al., Man. Pteridol., 543. 1938 · Comb fern [Greek *kteis*, comb]

Robbin C. Moran

C. Christensen in L. K. Rosenvinge, Biol. Arb. Tilegn

Plants generally terrestrial. **Stems** erect to obliquely ascending, stolons absent. **Leaves** monomorphic, evergreen or dying back in winter. **Petiole** 2/3 to equaling length of blade, base not swollen; vascular bundles more than 3, arranged in an arc, ± round in cross section. **Blade** lanceolate to deltate, 1–4-pinnate-pinnatifid, gradually reduced distally to confluent, pinnatifid apex, herbaceous. **Pinnae** not articulate to rachis, segment margins nearly entire to crenulate, ciliate; proximal pinnae not reduced, sometimes basal pair much the longest, sessile to petiolulate, equilateral or inequilateral with basiscopic side more developed (pinnules noticeably longer); costae adaxially rounded or flat, not grooved; indument of linear to lanceolate scales and often multicellular glandular hairs abaxially, of multicellular reddish hairs adaxially. **Veins** free, simple or forked. **Sori** in 1 row between midrib and margin, round; indusia round-reniform, attached at narrow sinus, sometimes small or seemingly absent, persistent or caducous. **Spores** brownish, usually spiny, sometimes prominently cristate, rarely finely reticulate. $x = 41$.

Species ca. 100 (2 in the flora): nearly worldwide in the tropics.

1. Blades ovate-lanceolate, 2–4-pinnate-pinnatifid; basal pinnae inequilateral, elongate basiscopically. . . . . 1. *Ctenitis sloanei*
1. Blades oblong or narrowly lanceolate, 1-pinnate-pinnatifid; basal pinnae equilateral. . . . . 2. *Ctenitis submarginalis*

1. Ctenitis sloanei (Poeppig ex Sprengel) C. V. Morton, Amer. Fern J. 59: 66. 1969 · Red-hair comb fern, Florida tree fern

Polypodium sloanei Poeppig ex Sprengel, Syst. Veg. 4(1): 59. 1827 **Petiole scales** orangish, rarely brown, linear or filiform, 20–40 × 0.4–1.5 mm, densely tangled and woollike. **Blade** 2–4-pinnate-pinnatifid, glabrous or glandular on both surfaces; glands pale yellow, ca. 0.1 mm, appressed. **Basal pinnae** 22–50 × 10–23 cm, inequilateral, elongate basiscopically. **Ultimate segments** 2–4 mm wide; margins ciliate. **Veins** 3–6 pairs per segment, unbranched or 1-forked. **Sori** medial to inframedial; indusia present but soon deciduous and therefore appearing absent. $2n = 82$.

Wooded limestone ledges, hammocks, cypress swamps; 0 m; Fla.; s Mexico; West Indies in Antilles, Trinidad; Central America; South America in Colombia, Ecuador, Peru, and Venezuela.

The names *Ctenitis ampla* (Humboldt & Bonpland ex Willdenow) Ching and *Dryopteris ampla* (Humboldt & Bonpland ex Willdenow) Kuntze have been misapplied to this taxon. *Ctenitis sloanei* and *C. submarginalis* both have numerous scales at the base of the petiole; in *C. sloanei*, however, the scales form a large, conspicuous, tangled tuft.

2. Ctenitis submarginalis (Langsdorff & Fischer) Ching, Sunyatsenia 5: 250. 1940

Polypodium submarginale Langsdorff & Fischer, Pl. Voy. Russes Monde, 12, plate 13. 1818; *Dryopteris submarginalis* (Langsdorff & Fischer) C. Christensen C. Christensen **Petiole scales** brown, linear, 10–20 × 0.8–1.5 mm, lax, not densely tangled or woollike. **Blade** 1-pinnate-pinnatifid, glabrous or pubescent on both surfaces, glandular abaxially and occasionally adaxially, glands pale yellow, ca. 0.5 mm. **Basal pinnae** 8–18 × 2–3 cm, equilateral, incised more than 3/4 distance to costae. **Ultimate segments** 4–7 mm wide, margins ciliate. **Veins** 6–10(–15) pairs per segment, unbranched. **Sori** medial to supramedial; indusia present but soon deciduous or completely absent.

Cypress swamps, hammocks, old forested spoil banks; 0 m; Fla., La.; e, s Mexico; West Indies in Hispaniola; Central America; South America to Uruguay.

Combining authorship of the accepted name sometimes has been incorrectly attributed to E. B. Copeland (1947). The Louisiana population of *Ctenitis submarginalis* occurs more than 960 km from populations in Florida and represents the northernmost locality for the species (G. P. Landry and W. D. Reese 1991). Unlike most ferns in North America, the plants in the Louisiana population are nonseasonal, producing leaves and sori throughout the year.

SELECTED REFERENCE Landrey, G. P. and W. D. Reese. 1991. *Ctenitis submarginalis* (Langsd. & Fisch.) Copel. new to Louisiana: First record in the U.S. outside of Florida. Amer. Fern J. 81: 105–106.

16. TECTARIA Cavanilles, Anales Hist. Nat. 1(2): 115. 1799 · Halberd fern [Latin *tectum,* roof, and *aria,* a substantive suffix, alluding to the rooflike indusium of some species]

Robbin C. Moran

Plants terrestrial or on rock. **Stems** short-creeping to erect, stolons absent. **Leaves** monomorphic, evergreen. **Petiole** 3/4–3 times length of blade, base not swollen; vascular bundles more than 3, arranged in an arc, ± round in cross section. **Blade** lanceolate to deltate or pentagonal, entire to 1-pinnate-pinnatifid [3-pinnate-pinnatifid], reduced distally to shallowly lobed or hastate apex, herbaceous to papery. **Pinnae** not articulate to rachis, segment margins entire to sinuate or shallowly lobed; proximal pinnae not or only slightly reduced, sessile to short-petiolulate, base equilateral or often inequilateral with prominent basiscopic lobe(s); costae adaxially rounded or shallowly grooved, grooves not continuous from rachis to costae; indument lacking or of multicellular hairs on costae abaxially, of multicellular hairs on costae adaxially. **Veins** reticulate, areoles with or without included veinlets. **Sori** in 1–several rows

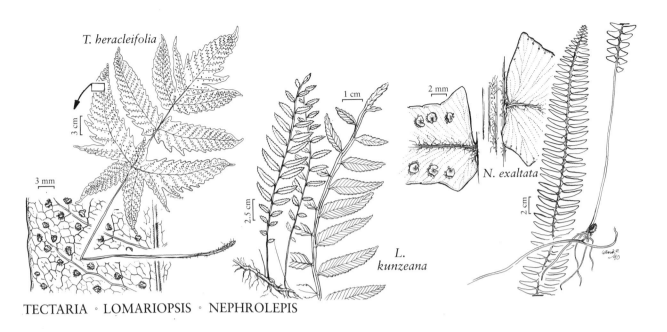

T. heracleifolia

L. kunzeana

N. exaltata

TECTARIA · LOMARIOPSIS · NEPHROLEPIS

between midrib and margin, round; indusia peltate to round-reniform and with narrow sinus, persistent or caducous. **Spores** brownish, with inflated folds or wings. $x = 40$.

Species ca. 200 (4 in the flora): mostly tropical.

SELECTED REFERENCE   Morton, C. V. 1966. The Mexican species of *Tectaria*. Amer. Fern J. 56: 120–137.

1. Blades 15–50 cm; areoles with included veinlets.
  2. Indusia peltate, attached centrally; pinnae cordate at base; rachises and costae glabrous abaxially. . . . . 1. *Tectaria heracleifolia*
  2. Indusia round-reniform, attached at sinus; pinnae oblique at base; rachises and costae puberulent abaxially. . . . . 2. *Tectaria incisa*
1. Blades 5–15 cm; areoles rarely with included veinlets.
  3. Pinnae 1–8 pairs, often with proliferous buds in axils; petiole shorter than or rarely equaling blade, pubescent on both surfaces; indusia round-reniform, attached at sinus.
  . . . . 3. *Tectaria coriandrifolia*
  3. Pinnae absent or 1(–2) pair, without proliferous buds in axils; petiole 1–3 times length of blade, pubescent adaxially, lacking hairs abaxially; indusia peltate, attached centrally.
  . . . . 4. *Tectaria fimbriata*

1. Tectaria heracleifolia (Willdenow) L. Underwood, Bull. Torrey Bot. Club 33: 200. 1906 · Broad halberd fern

Aspidium heracleifolium Willdenow, Sp. Pl. 5(1): 217. 1810

Stems erect, compact. **Petiole** straw-colored or rarely brown, 1–2 times length of blade, glabrous on both surfaces, base scaly; scales linear to lanceolate, 3–10 × 0.5–2 mm. **Blade** ovate to pentagonal, 1-pinnate or (in small plants) ternately lobed, (12–)20–40(–50) × 14–40 (–45) cm. **Rachis and costae** glabrous abaxially. **Pinnae** 1–2(–3) pairs, 12–22 cm, base cordate, margins shallowly lobed, with pair of basal lobes and without proliferous buds in axils. **Areoles** with included veinlets. **Indusia** peltate. $2n = 160$.

Rocky hammocks, limestone outcrops in shade, cave entrances; 0–100 m; Fla., Tex.; Mexico; West Indies in Antilles; Central America; South America to Peru.

2. Tectaria incisa Cavanilles, Descr. Pl., 249. 1802 · Incised halberd fern

Stems erect, compact. **Petiole** straw-colored to tan or chestnut brown, equaling or slightly longer than blade, pubescent on both surfaces, base scaly; scales narrowly lanceolate, 3–8 × 0.2–0.7 mm. **Blade** oblong or ovate-oblong, 1-pinnate, 0.15–1.2 × 0.2–0.5 m. **Rachis and costae** puberulent abaxially. **Pinnae** (2–)3–6 pairs, 12–25 cm, base oblique, with 1–2 proximal basiscopic lobes, margins crenate without proliferous buds in axils. **Areoles** with included veinlets. **Indusia** attached laterally. $2n = 160$.

Hammocks; 0 m; Fla.; Mexico; West Indies in Antilles; Central America; South America to n Argentina.

3. Tectaria coriandrifolia (Swartz) L. Underwood, Bull. Torrey Bot. Club 33: 200. 1906 · Hairy halberd fern

Aspidium coriandrifolium Swartz, J. Bot. (Schrader) 1800(2): 36. 1801

Stems decumbent, compact. **Petiole** tan to reddish brown, often shorter than to rarely equaling blade, pubescent on both surfaces, scaly; scales narrowly deltate or lanceolate, 2–5 × 0.1–0.2 mm. **Blade** lanceolate to oblong, 1-pinnate-pinnatifid, 5–12 × 2.5–10 cm. **Rachis and costae** pubescent abaxially. **Pinnae** 1–8 pairs, margins crenate to pinnatifid, 1.2–5 cm, base rounded, often with proliferous buds in axils. **Areoles** lacking included veinlets. **Indusia** round-reniform, attached at sinus. $2n = 80$ (Jamaica).

Sinkholes, rock, ledges, crevices; 0 m; Fla.; West Indies in Cuba, Jamaica.

Tectaria coriandrifolia is known in North America only from southernmost Florida; it has not been seen there for several years and is perhaps extirpated. It hybridizes with *Tectaria fimbriata* to form *T.* × *amesiana* A. A. Eaton [*Aspidium trifoliatum* (Linnaeus) Swartz var. *amesianum* (A. A. Eaton) Clute], which may be distinguished by its misshapen spores, fewer pinnae pairs, and dense row of hairs on the adaxial surface of the petioles. The hybrid is known only from Florida.

4. Tectaria fimbriata (Willdenow) Proctor & Lourteig, Bradea 5: 386. 1990 · Least halberd fern

Aspidium fimbriatum Willdenow, Sp. Pl. 5(1): 213. 1810; *A. minimum* (L. Underwood) Clute; *A. trifoliatum* (Linnaeus) Swartz var. *minimum* Clute; *Sagenia lobata* C. Presl; *Tectaria lobata* (C. Presl) C. V. Morton; *T. minima* L. Underwood

Stems horizontal, short-creeping. **Petiole** straw-colored, 1–3 times length of blade, hairs absent abaxially, pubescent adaxially, base scaly; scales lanceolate to linear, 1–3 × 0.3–0.5 mm. **Blade** deltate to pentagonal, 1-pinnate or ternately lobed, 5–10 × 3–7 cm. **Rachis and costae** pubescent abaxially. **Pinnae** absent or 1(–2) pair, or blade shallowly pinnatifid, 4–7 cm, base rounded, without proliferous buds in axils. **Areoles** rarely with included veinlets. **Indusia** peltate. $2n = 80$.

Sinkholes, on shaded ledges or occasionally in sun; 0 m; Fla.; Mexico in Yucatán; West Indies in Bahamas, Greater Antilles.

The hybrid between *Tectaria fimbriata* and *T. coriandrifolia* is *T.* × *amesiana* A. A. Eaton. Although intermediate between the two parents, the hybrid will lead to *T. fimbriata* in the key. The hybrid may be distinguished from *T. fimbriata* by scattered hairs on abaxial surface of petioles and misshapen spores.

17. LOMARIOPSIS Fée, Mém. Foug. 2: 10, 66. 1845 · Climbing holly fern [*Lomaria*, a subgenus of *Blechnum* (Blechnaceae), plus Greek -*opsis*, like]

Robbin C. Moran

Plants terrestrial [hemiepiphytic]. **Stems** long-creeping or climbing, stolons absent. **Leaves** strongly dimorphic, sterile ones longer and with wider pinnae than fertile ones, evergreen. **Petiole** ca. 1/2 length of blade, base not swollen; vascular bundles more than 3, arranged in an arc, ± round in cross section. **Blade** ovate-lanceolate, 1-pinnate, apex similar to lateral pinnae, papery. **Pinnae** articulate to rachis, sometimes deciduous, segment margins (pinnae) entire to serrate; proximal pinnae slightly reduced, sessile, equilateral; costae adaxially shallowly grooved, grooves not continuous from rachis to costae; indument of narrow scales abaxially, blades glabrous adaxially. **Veins** free, simple or forked, ± parallel perpendicular to costae. **Sori** absent, sporangia covering abaxial surface of linear, entire pinnae; indusia absent. **Spores** brownish, with spiny or with prominent crested wings. $x = 41$.

Species ca. 45 (1 in the flora): mostly tropical.

SELECTED REFERENCES   Holttum, R. E. 1940. New species of *Lomariopsis*. Bull. Misc. Inform. Kew 1939: 613–628. Underwood, L. M. 1906. American ferns VII A. The American species of *Stenochlaena*. Bull. Torrey Bot. Club 33: 591–603.

1. **Lomariopsis kunzeana** (C. Presl ex L. Underwood) Holttum, Bull. Misc. Inform. Kew 1939: 617. 1940 · Holly fern, climbing holly fern

Stenochlaena kunzeana C. Presl ex L. Underwood, Bull. Torrey Bot. Club 33: 196. 1906

Petiole straw-colored, 3–12 cm, narrowly winged, base scaly. **Blade** oblanceolate, 7–25 × 3–6 cm, tapered toward base. **Rachis** winged throughout; wing ca. 0.5 mm wide. **Sterile pinnae** narrowly oblong to lanceolate, 3–6 × 1–1.5 cm, coarsely serrate, base cuneate. **Fertile pinnae** linear, ca. 3 mm wide.

Hammocks, limestone sinkholes; 0 m; Fla.; West Indies in Cuba, Hispaniola.

Plants from Florida differ from those of Cuba and Hispaniola by their smaller size and more deeply serrate sterile pinnae. Unlike all other species in the genus, the Florida plants rarely climb trees and freely produce fertile leaves from the terrestrial stem. According to James H. Peck (pers. comm.), the gametophytes of *Lomariopsis kunzeana* can often be found among the stem scales.

18. **NEPHROLEPIS** Schott, Gen. Fil., plate 3. 1834 · Boston fern [Greek *nephros,* kidney, and *lepis,* scale, in reference to shape of the indusia]

Clifton E. Nauman

Plants terrestrial, epiphytic, or on rock. **Stems** ascending to erect, bearing wiry stolons and sometimes underground tubers. **Leaves** monomorphic, evergreen. **Petiole** ca. 1/10–1/2 length of blade, base not swollen; vascular bundles more than 3, arranged in an arc, ± round in cross section. **Blade** narrowly elliptic to linear-lanceolate, 1-pinnate (to 4–5-pinnate in various cultivated forms), very gradually reduced distally to minute pinnatifid apex, often seemingly indeterminate with apex never expanded, herbaceous to papery. **Pinnae** articulate to rachis, sometimes deciduous, segment (pinna) margins entire, crenulate, or biserrate; proximal pinnae (usually several pairs) slightly to greatly reduced, sessile, equilateral or inequilateral with basiscopic base excised and often an acroscopic basal auricle; costae adaxially grooved, grooves not continuous from rachis to costae; indument of linear-lanceolate scales and sometimes multicellular hairs on abaxial and sometimes adaxial surfaces. **Veins** free, forked. **Sori** ± round; indusia round-reniform and with deep sinus to semicircular with broad sinus or lunate without sinus and seemingly laterally attached, persistent. **Spores** brownish, tuberculate to rugose. $x = 41$.

Species 25–30 (4 in the flora): widespread in tropical areas.

Nephrolepis often has veins ending in hydathodes and whitish lime-dots adaxially.

Cultivars of *Nephrolepis* occasionally are found in the wild, where they persist for some time. Numerous forms of *N. exaltata* cv. 'Bostoniensis' and its derivatives are widely cultivated, and the following are known from Florida: *N. exaltata* cv. 'Bostoniensis', *N. exaltata* cv. 'Elegantissima' complex, *N. exaltata* cv. 'Florida Ruffles', *N. exaltata* cv. 'M. P. Mills'.

Nephrolepis falcata forma *furcans* (T. Moore in Nicholson) Proctor resembles *N. biserrata* in size, pinna shape, and sori, but it differs characteristically in having forking pinnae and rachises. It is widely cultivated and persists when escaped; it is not known to spread from spores. It is known in the literature under the following names: *Aspidium biserratum* Swartz var. *furcans* (T. Moore in Nicholson) Farwell, *Nephrolepis biserrata* (Swartz) Schott var. *furcans* (T. Moore in Nicholson) Hortus ex Bailey, and *Nephrolepis davallioides* var. *furcans* T. Moore in Nicholson.

Nephrolepis hirsutula (G. Forster) C. Presl cv. 'Superba' has irregularly pinnatisect, elliptic pinnae and a dense covering of reddish orange scales over most of the leaf surfaces.

The report of *Nephrolepis pectinata* (Willdenow) Schott for Florida by E. T. Wherry (1964) was based on a misdetermination (T. Darling Jr. 1982).

SELECTED REFERENCES   Darling, T. Jr. 1982. The deletion of *Nephrolepis pectinata* from the flora of Florida. Amer. Fern J. 72: 63. Nauman, C. E. 1981. The genus *Nephrolepis* in Florida. Amer. Fern J. 71: 35–40. Nauman, C. E. 1985. A Systematic Revision of the Neotropical Species of *Nephrolepis* Schott. Ph.D. dissertation. University of Tennessee. Wherry, E. T. 1964. The Southern Fern Guide. Garden City, N.Y.

1. Adaxial costae of central pinnae sparsely to densely covered with short, erect hairs (often also with scales).
   2. Mature petioles at base covered moderately to densely with appressed, dark brown scales with pale margins. . . . . 1. *Nephrolepis multiflora*
   2. Mature petioles at base often with a few loose, reddish to light brown, concolored scales.
      3. Adaxial costae sparsely hairy, hairs ca. 0.5 mm; pinnae mostly falcate. . . . . 2. *Nephrolepis* ×*averyi*
      3. Adaxial costae densely hairy to tomentose, hairs 0.2–0.4 mm; pinnae not falcate or only slightly so. . . . . 3. *Nephrolepis biserrata*
1. Adaxial costae of central pinnae glabrous, with or without scales.
   4. Indusia circular and peltate or horseshoe-shaped and attached at narrow sinus, largest ca. 1 mm wide; pinnae usually more than 5 cm, often with conspicuous hairs 0.3–0.4 mm on blade surface. . . . . 3. *Nephrolepis biserrata*
   4. Indusia reniform, horseshoe-shaped, or lunate to deltate-rounded and attached by narrow to broad sinus, 1.1–1.7 mm wide or wider; pinnae usually less than 5 cm, without hairs or hairs less than 0.3 mm and inconspicuous.
      5. Plants with or without tubers; adaxial rachis scales distinctly bicolored (pale with darker point of attachment), often dense; points of pinna attachment 5–12 mm apart; pinnae glabrous; indusia lunate to deltate-rounded or reniform. . . . 4. *Nephrolepis cordifolia*
      5. Plants never bearing tubers; adaxial rachis scales concolored or indistinctly bicolored, dense to sparse; points of pinna attachment 7–21 mm apart; pinnae with a few scales near costae; indusia usually reniform to horseshoe-shaped. . . . . . . 5. *Nephrolepis exaltata*

1. **Nephrolepis multiflora** (Roxburgh) F. M. Jarrett ex C. V. Morton, Contr. U.S. Natl. Herb. 38: 309. 1974 · Asian sword fern

Davallia multiflora Roxburgh, Calcutta J. Nat. Hist. 4: 515, plate 31 (left). 1844

Stem scales appressed, bicolored with margins transparent. **Tubers** absent. **Leaves** 3–25 × 0.3–1.6 dm. **Petiole** 0.4–4.4 dm, moderately to densely scaly; scales appressed, dark brown with pale margins. **Blade** sparsely to moderately scaly, hairy abaxially, hairs pale brown, 0.1–0.3 mm. **Rachis** 2.7–20 dm, points of pinna attachment 8–24 mm apart; scales scattered to dense, brown, margins pale. **Central pinnae** narrowly deltate, sometimes elliptic, 3.4–12.3 × 0.6–1.8 cm, base rounded basiscopically, slightly auriculate to truncate acroscopically (latter more common in sterile pinnae), acroscopic lobe acute to oblong, margins biserrate to irregularly serrate to serrulate, apex attenuate and occasionally slightly falcate; costae adaxially densely hairy, hairs pale, erect, 0.1–0.5 mm. **Indusia** circular to horseshoe-shaped, peltate or attached at narrow sinus, 1.1–1.3 mm wide. $2n = 82$.

Terrestrial or epiphytic in open waste places and roadsides; 0 m; introduced; Fla.; s Mexico; West Indies; Central America; South America; Africa; Asia.

Nephrolepis multiflora is native to the Old World tropics and is widely scattered and naturalized in the New World tropics as an escaped cultigen.

2. **Nephrolepis ×averyi** Nauman, Amer. Fern J. 69: 69. 1979 · Avery's sword fern

Stem scales loosely appressed to spreading, essentially concolored. **Tubers** absent. **Leaves** 8–30 × 0.5–1.3 dm. **Petiole** 1.2–5 dm, moderately to densely scaly; scales spreading, reddish to light brown throughout. **Blade** moderately scaly, glabrous to occasionally pubescent abaxially, glabrous adaxially, hairs pale to light brown, 0.2–0.7 mm. **Rachis** 5–26 dm, points of pinna attachment 9–25 mm apart; scales scattered, brown throughout. **Central pinnae** narrowly deltate to oblong-deltate, 3.6–9 × 0.7–2 cm, base truncate, sometimes rounded basiscopically, barely auriculate acroscopically, acroscopic lobe obtuse to acute, margins serrulate, apex narrowly acute to attenuate, falcate to slightly so; costae adaxially sparsely pubescent, hairs pale, erect, 0.1–0.5 mm. **Indusia** horseshoe-shaped to circular, attached at narrow sinus or peltate, 0.9–1.1 mm wide. $2n = 164$.

Terrestrial or epiphytic in forested, often moist habitats, e.g., swamps, hammocks, or relatively open, disturbed habitats; 0 m; Fla.; Mexico; West Indies.

Nephrolepis ×averyi is an allotetraploid derived from hybridization between *N. biserrata* and *N. exaltata* and is known to occur only in mixed populations of both parent species. It is distinguished from *N. exaltata* by its sparsely hairy adaxial costae, larger size, and misshapen spores, and from *N. biserrata* by its falcate pinnae and narrower leaves.

3. **Nephrolepis biserrata** (Swartz) Schott, Gen. Fil., plate 3. 1834 · Giant sword fern

Aspidium biserratum Swartz, J. Bot. (Schrader) 1800(2): 32. 1801
Stem scales loosely appressed to spreading, concolored or bicolored with pale margins. **Tubers** absent. **Leaves** 2–22 × 0.3–3.5 dm. **Petiole** 0.2–5.4 dm, sparsely to moderately scaly; scales spreading, reddish to light brown throughout (rarely with pale margins). **Blade** sparsely to densely scaly, glabrous or pubescent, hairs mostly on veins and abaxial, pale to light brown, 0.2–0.7 mm. **Rachis** 1.8–17 dm, points of pinna attachment 7.5–35 mm apart; scales moderately spaced, pale brown throughout. **Central pinnae** narrowly deltate to narrowly elliptic-lanceolate, 2.5–23 × 0.5–2 cm, base cu-

neate, truncate to auriculate-cordate acroscopically, rounded basiscopically, acroscopic lobe small and oblong or absent, margins biserrate to serrulate, apex attenuate; costae adaxially glabrous or densely hairy, hairs erect, pale, 0.3 mm. **Indusia** circular to horseshoe-shaped, peltate or attached at narrow sinus, 0.8–1.1 mm wide. $2n = 82$.

Terrestrial or less commonly epiphytic in forested, relatively wet habitats, e.g., swamps, but occasionally thickets, roadsides, or clearings; 0 m; Fla.; Mexico; West Indies; Central America; South America; Africa; se Asia.

Some forms of *Nephrolepis biserrata* closely resemble *N. multiflora* in pinna shape and indument but lack the distinctively transparent-margined (i.e., bicolored) and persistent petiole scales of the latter species. *Nephrolepis multiflora* also has more appressed and darker-colored stem scales.

4. **Nephrolepis cordifolia** (Linnaeus) C. Presl, Tent. Pterid., 79. 1836 · Tuberous sword fern, tuber sword fern

Polypodium cordifolium Linnaeus, Sp. Pl. 2: 1089. 1753; *Aspidium cordifolium* (Linnaeus) Swartz
Stem scales spreading, concolored. **Tubers** present or absent. **Leaves** 2.5–10.7 × 0.3–0.7 dm. **Petiole** 0.3–2 dm, moderately to densely scaly; scales spreading, pale brown throughout. **Blade** lacking scales, glabrous (rarely with a few branched hairs abaxially). **Rachis** 2.2–9 dm, points of pinna attachment 5–12 mm apart; scales moderately spaced to dense, pale to dark brown, point of attachment distinctly darker. **Central pinnae** oblong to lanceolate-oblong, straight to slightly falcate, 0.9–5 × 0.4–0.9 cm, base auriculate-cordate, acroscopically overlapping rachis, acroscopic lobe deltate, margins entire to serrulate to smoothly crenate, apex acute to bluntly rounded; costae adaxially glabrous. **Indusia** reniform to lunate or deltate-rounded, attached along broad sinus, 1.1–1.7 mm wide. $2n = 82$.

Terrestrial or epiphytic in wet, shady places, limestone ledges, cliffs, rock, roadsides, and often old homesites or waste places; widely escaped from cultivation and only questionably native to any particular region; 0 m; Fla.; Mexico; West Indies; Central America; South America; Africa; se Asia; Pacific Islands in Hawaii.

5. Nephrolepis exaltata (Linnaeus) Schott, Gen. Fil., plate 3. 1834 · Sword fern, wild Boston fern

Polypodium exaltatum Linnaeus, Syst. Nat. ed. 10, 2: 1326. 1759

Stem scales spreading, concolored. **Tubers** absent. **Leaves** 4–15 × 0.5–1.2 dm. **Petiole** 0.2–4 dm, sparsely to moderately scaly; scales spreading, pale brown to reddish brown, concolored. **Blade** glabrous, sparsely to moderately scaly abaxially near costae and adaxially. **Rachis** 2.4–16.3 dm, points of pinna attachment 7.3–21 mm apart; scales moderately spaced, pale to dark brown, essentially concolored or margin indistinctly paler; hairs absent. **Central pinnae** deltate-oblong, slightly to distinctly falcate, 2.3–7.4 × 0.6–1.8 cm, base truncate to truncate-auriculate or auriculate, occasionally overlapping rachis, acroscopic lobe deltate to acute, margins serrulate, apex acute to deltate; costae adaxially glabrous. **Indusia** reniform to horseshoe-shaped, attached at narrow or broad sinus, 1–1.7 mm wide.

Terrestrial or epiphytic in forested to open habitats, most often as an epiphyte; 0 m; Fla.; West Indies; Pacific Islands in scattered locations.

Nephrolepis exaltata is occasionally found farther north in the flora, but only as an escape from cultivation. *Nephrolepis exaltata* is usually confused with *N. cordifolia* when sterile; the latter species can be distinguished by its distinctly bicolored, adaxial rachis scales. These bicolored scales will distinguish *N. cordifolia* from all of the other species, even in the absence of other key features.

21. GRAMMITIDACEAE Newman

Alan R. Smith

Plants perennial, mostly small, on rock or commonly epiphytic [rarely terrestrial]. **Stems** long- to short-creeping or suberect, usually unbranched, bearing scales [rarely scales absent], solenostelic (having phloem on both sides of xylem) to dictyostelic (having complex nets of xylem). **Leaves** erect, arching, or pendent, monomorphic [rarely with specialized fertile areas], less than 50 cm [rarely longer], usually scaleless throughout. **Petioles** often dark-colored and wiry, commonly terete, usually less than 2 mm diam., articulate or not articulate, with 1 or 2 vascular strands. **Blades** simple and entire to commonly pinnatifid or 1-pinnate, rarely 2-pinnate or more divided, glabrous or commonly bearing hairs, especially on petioles and rachises; hairs tan to dark reddish brown [or transparent], unicellular to multicellular; rachises often dark-colored, not grooved adaxially. **Veins** free [to anastomosing in simple patterns]; hydathodes present or absent, sometimes obscured by lime dots adaxially. **Sori** abaxial on veins, round to oblong [occasionally elongate]; paraphyses present or absent, these glandular or hairlike; sporangia with stalk of 1 row of cells; indusia absent. **Spores** greenish, tetrahedral-globose, trilete, surface commonly papillate. **Gametophytes** greenish, borne aboveground, ribbon-shaped, sometimes bearing multicellular gemmae.

Genera ca. 10 or more, species ca. 500 or more (1 genus, 1 species in the flora): mostly tropical and subtropical, North America, ca. 250 in Mexico, West Indies, Central America, South America.

Circumscription of genera is unsettled and in some cases arbitrarily based on blade dissection. Generic limits need redefinition and are the subject of current study. The family description above encompasses the worldwide variation. Relationships of the family have generally been thought to be with the Polypodiaceae with which the family is sometimes combined, but that is open to question. Most species are epiphytic, and many occur in cloud forests at middle and high elevations.

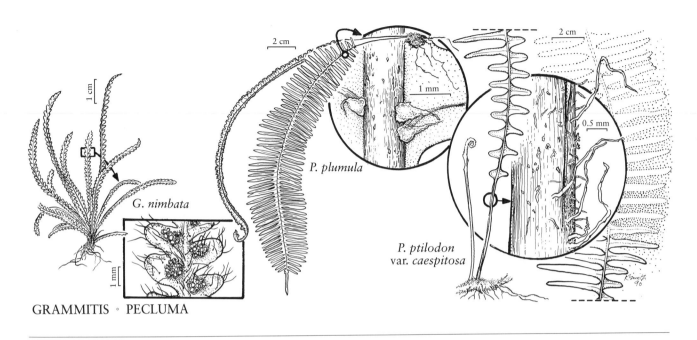

G. nimbata

P. plumula

P. ptilodon
var. *caespitosa*

GRAMMITIS · PECLUMA

1. GRAMMITIS   Swartz, J. Bot. (Schrader) 1800(2): 3, 17. 1802 · [Greek *gramme*, line, alluding to the elongate sori in a few species]

Plants often less than 15 cm [rarely more than 50 cm]. **Blade** pinnatifid [simple to 1-pinnate or rarely more divided], bristly [or glabrous], glandless [or glandular]; setae dark reddish brown [to transparent]. **Veins** simple or 1-forked [to pinnately veined] in each segment, free, ending in hydathodes adaxially. **Sori** round [to oval or elongate], not forming a coenosorus, not sunken; paraphyses absent [or present]. $x = 32, 33, 36, 37$.

Species ca. 300 or more (1 in the flora): tropical and subtropical, North America, South America, Asia, Africa, Pacific Islands, Australia.

The total number of species is based on acceptance of a relatively broad circumscription. Our sole species belongs to the group (genus or subgenus) formerly called *Xiphopteris,* a name that is now treated as a synonym of *Cochlidium* (L. E. Bishop 1978). This group comprises perhaps 30 species and is defined by the following characteristics: stem scales not clathrate, often marginally bristly; veins simple or with a single acroscopic branch in each segment; blades linear, usually pinnatifid or pinnatisect; lack of a marginal black border around blade; one sorus per segment; and hydathodes present. It probably warrants generic status under the name *Micropolypodium* Hayata, a primarily neotropical genus with representatives in eastern Asia (Malaysia, China, Sikkim, Taiwan, and Japan).

SELECTED REFERENCES   Bishop, L. E. 1978. Revision of the genus *Cochlidium* (Grammitidaceae). Amer. Fern J. 68: 76–94. Farrar, D. R. 1967. Gametophytes of four tropical fern genera reproducing independently of their sporophytes in the southern Appalachians. Science 155: 1266–1267.

1. **Grammitis nimbata** (Jenman) Proctor, Bull. Inst. Jamaica, Sci. Ser. 5: 34. 1953

Polypodium nimbatum Jenman, J. Bot. 24: 271. 1886; *Xiphopteris nimbata* (Jenman) Copeland

Stems erect, apices bearing scales; scales lustrous yellow to brownish, concolored, 0.5–1 mm, margins stiffly ciliate. **Leaves** densely fasciculate, 3–7 cm. **Petiole** 2–7 × 0.5 mm, densely hairy; hairs dark reddish or brownish, 0.8–1 mm. **Blade** linear, pinnatifid ca. 2/3 to rachis, narrowed and short-decurrent proximally, 3–5 mm wide; rachises and blade tissue on both sides with crowded, spreading setae like those of petiole or longer. **Segments** mostly in 20–30 pairs, oblique-spreading, oblong, 0.8–1.2 mm wide with entire margins. **Veins** concealed abaxially, unbranched, or fertile segments sometimes with single acroscopic branch, veins ending in hydathodes on adaxial surface. **Sori** round, 1 per segment, arising near base of acroscopic veinlet; sporangia glabrous.

On moist cliffs behind falls; ca. 500–1000 m; N.C.; West Indies in Cuba, Jamaica, Hispaniola.

Grammitis nimbata was discovered in the United States in Macon County, North Carolina, at a single locality, in 1966 (D. R. Farrar 1967) and has persisted to the present (D. R. Farrar, in litt. 1989). Primarily, the colony is gametophytic and reproduces by microscopic gemmae borne on the elongate-cordate thallus, but sterile sporophytes have been found in favorable years. The species description and illustration are based on mature spore-bearing plants from the Greater Antilles. The largest plants found from North Carolina are less than 3 cm.

22. POLYPODIACEAE Berchtold & J. Presl
• Polypody Family

Alan R. Smith

Plants perennial, terrestrial, on rock, or often epiphytic, erect, arching, or occasionally pendent. **Stems** long- to short-creeping, branched or not, bearing scales and few to numerous roots, usually dictyostelic. **Leaves** monomorphic to dimorphic, circinate in bud. **Petiole** usually articulate at base [rarely nonarticulate, as in *Loxogramme*], lacking scales or sometimes scaly, with usually 3 vascular bundles. **Blade** simple to often pinnatifid, pinnatisect, or pinnate, infrequently more divided; rachis grooved or not adaxially. **Veins** free (and simple to several times forked) to often anastomosing in complex systems, areoles with or without included veinlets. **Indument** on blade absent, or petiole, rachis, costae, and sometimes blade tissue usually bearing hairs (these often septate and with reddish crosswalls) and/or scales. **Sori** borne abaxially on veins, round to oblong, occasionally elongate, rarely marginal, rarely covering surface; paraphyses present or absent; sporangia with stalk of 2 or 3 rows of cells; indusia absent. **Spores** usually transparent or yellowish (rarely greenish), all 1 kind, bilateral, monolete [rarely trilete, as in some *Loxogramme*], surface most often smooth, tuberculate, verrucose, or granulate, occasionally spiny, 64 per sporangium (spores globose and 32 per sporangium in apogamous spp.). **Gametophytes** green, aboveground, cordate or elliptic, glabrous or sometimes glandular; archegonia and antheridia borne on lower surface, antheridia 3-celled.

Genera ca. 40, species perhaps 500 (7 genera, 25 species in the flora): worldwide, especially tropics and subtropics.

Phymatosorus scolopendria (Burman f.) Pichi-Sermolli, native to the Old World, is a rare escape in southern Florida.

Genera in this family are variously circumscribed, and the New World species historically were placed in the single genus *Polypodium*. Many of the segregates recognized here are still placed in *Polypodium* in recent floristic accounts. Limits of genera in both Old World and New World are controversial and are currently under study by several workers.

1. Blades simple, undivided.
    2. Sporangia confined to marginal or nearly marginal bands in distal 1/2 of blade. . .   7. *Neurodium*, p. 330
    2. Sporangia in discrete, round to oblong or slightly elongate sori on abaxial surface, not in marginal bands.
        3. Blades abaxially with peltate scales. . . . . 4. *Pleopeltis*, p. 324
        3. Blades abaxially glabrous, except for scattered scales on midrib.
            4. Stems 2–10 mm diam.; sori in 1–10 or more rows between midrib and margin; petioles clustered, proximate; main lateral veins often prominent, ± parallel. . .
            . . . . 5. *Campyloneurum*, p. 327
            4. Stems 0.5–1.5 mm wide; sori in 1 row between midrib and margin; petioles well separated, often 1–2 cm apart; main lateral veins obscure, not parallel. . . . .
            . . . . 6. *Microgramma*, p. 329
1. Blades pinnatifid or pinnatisect, rarely 1-pinnate.
    5. Blades with numerous peltate or ovate scales abaxially. . . . . 4. *Pleopeltis*, p. 324
    5. Blades lacking scales abaxially except along midrib.
        6. Blades pectinate, usually with more than (20–)25 pairs of segments; segments narrow, linear, 1.5–5(–8) mm wide; veins free; stems short-creeping; Florida. . . . . 1. *Pecluma*, p. 313
        6. Blades pinnatifid, rarely 1-pinnate, with fewer than 20(–25) pairs of segments; segments broad, generally (3–)5–20(–30) mm wide; veins free or anastomosing; stems moderately to widely creeping.
            7. Blades 1-pinnate; Florida. . . . . 2. *Polypodium triseriale*, p. 318
            7. Blades pinnatifid.
                8. Venation free or with 1 row of areoles between costa and margin; sori at end of 1 included veinlet or on forked free vein; widespread but not Florida. . .
                . . . . 2. *Polypodium*, p. 315
                8. Venation highly reticulate, with 3–4 rows of areoles between costa and margin; sori at end of usually 2 included veinlets; Florida. . . . . 3. *Phlebodium*, p. 323

1. PECLUMA   M. G. Price, Amer. Fern J. 73: 109. 1983 · [Latin *pectinatus*, in the form of a comb, and *plumula*, feathery, for the leaf blades]

A. Murray Evans

Polypodium Linnaeus subg. *Pectinatum* Lellinger

Plants terrestrial, on rock, or epiphytic. **Stems** short-creeping, unbranched, 4–8 mm diam., not whitish pruinose; scales concolored, ovate to linear-lanceolate, not clathrate, glabrous or pubescent, margins entire. **Leaves** monomorphic, crowded, narrowed toward tip, to 90 cm. **Petiole** dark brown to black, round in cross section except for narrow lateral ridge decurrent from base of blade. **Blade** narrowly oblong to linear, usually pectinate, with more than (20–) 25 pairs of segments, not glaucous, pubescent, scales absent or on midrib only; rachis scaly and/or pubescent or glabrous abaxially, pubescent adaxially, scales basally attached and linear to cordate. **Segments** usually linear, 1.5–5(–8) mm wide, adnate to costa, closely spaced, margins entire, apex acute. **Veins** 1–2-forked [simple], occasionally anastomosing. **Sori** terminal on veins, round; indument of branched glandular hairs. **Spores** tuberculate. $x = 37$.

Species ca. 30 (3 in the flora): North America, Mexico, West Indies, Central America, South America.

Within Polypodiaceae, *Pecluma* is distinctive in its short-creeping stems, pectinate blades (usually with more than 30 pairs of segments), and ungrooved, adaxially pubescent rachises.

Many species of *Pecluma* are well adapted to dry seasons. As the leaves dry out, the segments curl inwardly, presumably retarding moisture loss. After rains, the segments uncurl, apparently undamaged by the period of dryness.

SELECTED REFERENCES Evans, A. M. 1969. Interspecific relationships in the *Polypodium pectinatum–plumula* complex. Ann. Missouri Bot. Gard. 55: 193–293. Price, M. G. 1983. *Pecluma,* a new tropical American fern genus. Amer. Fern J. 73: 109–116.

1. Plants terrestrial or on logs or tree bases; segments at base of blade gradually reduced to auricles; petiole scales threadlike and inconspicuous or absent; veins 2–3-forked. . . . . 1a. *Pecluma ptilodon* var. *caespitosa*
1. Plants epiphytic or on rock; segments at base of blade abruptly reduced, or base narrowed to truncate; petiole scales cordate and appearing inflated or hastate and flat; veins 1–2-forked.
    2. Leaves linear-elliptic; segments linear, proximal segments not deflexed, reduced abruptly to mere lobes; petiole scales cordate and appearing inflated; veins 1-forked; spores 64 per sporangium. . . . . 2. *Pecluma plumula*
    2. Leaves narrowly ovate; segments narrowly ovate to linear, proximal segments deflexed, occasionally reduced to auricles; petiole scales hastate, flat; veins 2-forked; spores 32 per sporangium. . . . . 3. *Pecluma dispersa*

1. **Pecluma ptilodon** (Kunze) M. G. Price, Amer. Fern J. 73: 115. 1983

Polypodium ptilodon Kunze, Linnaea 9: 42. 1834

Varieties 4 (1 in the flora): North America, Mexico, West Indies, Central America, South America.

1a. **Pecluma ptilodon** (Kunze) M. G. Price var. **caespitosa** (Jenman) Lellinger, Proc. Biol. Soc. Wash. 98: 387. 1985

Polypodium pectinatum Linnaeus var. *caespitosum* Jenman, Bull. Bot. Dept., Jamaica, n.s. 4: 125. 1897; *P. ptilodon* Kunze var. *caespitosum* (Jenman) A. M. Evans

Stems 5–8 mm diam.; scales blackish, linear-lanceolate. **Leaves** erect, or arching in large plants. **Petiole** dark brown, less than 1/10 length of blade, pubescent with both simple and branched, short, multicellular hairs; scales threadlike, linear, scattered, inconspicuous. **Blade** lanceolate, 25–90 × 6.5–18 cm; base narrowly cuneate; apex acute. **Segments** lanceolate, 4–8 mm wide, sparsely pubescent, at base of blade reduced to auricles. **Veins** 2–3-forked. **Sori** oval, blade pubescence denser and longer around sorus; sporangia with 64 spores. $2n = 148$.

Sporulating all year. Terrestrial, or on logs or tree bases, in hammocks, swamps, and wet woods; 0 m; Fla.; Mexico; West Indies; Central America.

2. **Pecluma plumula** (Humboldt & Bonpland ex Willdenow) M. G. Price, Amer. Fern J. 73: 115. 1983 · Plume polypody

Polypodium plumula Humboldt & Bonpland ex Willdenow, Sp. Pl. 5(1): 178. 1810

Stems ca. 5 mm diam.; scales dark brown, linear-lanceolate. **Leaves** pendent, sometimes erect in small plants. **Petiole** black, 1/10–1/5 length of blade, hairs few, unbranched; scales ovate to cordate, conspicuous, inflated. **Blade** linear-elliptic, 15–50 × 3–7 cm; base abruptly cuneate; apex tapering to acute. **Segments** linear, 1.5–2.5 mm wide, sparsely short-pubescent; segments at base of blade abruptly reduced to lobes, not deflexed. **Veins** 1-forked. **Sori** round; sporangia with 64 spores. $2n = 148$.

Sporulating all year. Epiphytic or occasionally on rock, wet woods, river banks, hammocks, and limesinks; 0 m; Fla.; Mexico; West Indies; Central America; South America to s Brazil.

3. **Pecluma dispersa** (A. M. Evans) M. G. Price, Amer. Fern J. 73: 114. 1983 · Widespread polypody

Polypodium dispersum A. M. Evans, Amer. Fern J. 58: 173, plate 27. 1968

Stems 4–6 mm diam.; scales blackish, linear-lanceolate. **Leaves** erect or arching. **Petiole** black, 1/4–1/3 length of blade, hairs short, mostly simple, multicellular; scales deltate to linear, base broadly hastate-lacerate. **Blade** narrowly ovate, 20–70 × 5–11 cm; base narrowly truncate; apex acute. **Segments** narrowly ovate to linear, 3–5 mm wide; segments at base of blade abruptly reduced, usually reflexed. **Veins** 1–2-forked. **Sori** round or oval, sporangia with 32 spores. $n = 2n = 111$, apogamous.

Sporulating all year. Usually on limestone outcrops, occasionally epiphytic in hammocks; 0 m; Fla.; Mexico; West Indies; Central America; South America to s Brazil.

Pecluma dispersa frequently occurs as widely scattered clusters of small juveniles on mossy limestone, arising vegetatively from exposed roots of older plants.

2. POLYPODIUM Linnaeus, Sp. Pl. 2: 1082. 1753; Gen. Pl. ed. 5, 485. 1754 · Polypody [Greek *poly*, many, and *pous, podion*, little foot, in allusion to numerous knoblike prominences of the stem]

Christopher H. Haufler
Michael D. Windham
Frank A. Lang
S. A. Whitmore

Plants on rock, occasionally terrestrial or epiphytic. **Stems** creeping, usually branched, 3–15 mm diam., sometimes whitish pruinose; scales concolored to bicolored, lanceolate to ovate-acuminate, not clathrate to strongly clathrate, glabrous, margins entire to denticulate. **Leaves** monomorphic, closely spaced to distant, not conspicuously narrowed at tip, to 90 cm. **Petiole** articulate to stem, straw-colored, somewhat flattened or grooved to nearly terete, winged distally. **Blade** broadly ovate to deltate, pinnatifid to 1-pinnate at base, not pectinate, usually with fewer than 25 pairs of pinnae, not glaucous or conspicuously scaly; rachis sparsely scaly to glabrescent abaxially, puberulent to glabrous adaxially; scales ovate-lanceolate to linear, not peltate or clathrate. **Segments** linear to oblong; margins entire to serrate; apex rounded to attenuate. **Venation** free to anastomosing, if strongly anastomosing, then never with more than 1 included veinlet in fertile areoles. **Sori** often confined to distal region of leaf, discrete, circular to oval when immature, borne at tips of single veins, in 1–3 rows on either side of midrib; indument absent or of modified sporangia (sporangiasters), often bearing glandular hairs on bulbous head. **Spores** monolete, rugose to tuberculate. $x = 37$.

Species ca. 100 (11 in the flora): worldwide.

Some species traditionally included in *Polypodium* are treated here in other genera, for example, *Pleopeltis* and *Pecluma*.

Except for the tropical species *Polypodium triseriale*, North American *Polypodium* is a complex assemblage of interactive species. The North American species have ties to European taxa (e.g., *P. vulgare* sensu stricto, which probably originated by allopolyploidy between *P. glycyrrhiza* and *P. sibiricum*) but are quite distinct from them. Morphologic comparisons and continuing biochemical and molecular studies indicate that two groups of diploid species occur within the North American *P. vulgare* complex. One group includes *P. glycyrrhiza* and *P. californicum*; the second, *P. amorphum*, *P. appalachianum*, and *P. sibiricum*. Allopolyploid species have originated following hybridizations within a species group (i.e., *P. calirhiza* from *P. glycyrrhiza* × *californicum*, *P. saximontanum* from *P. amorphum* × *sibiricum*, and *P. virginianum* from *P. appalachianum* × *sibiricum*) as well as between members of the two groups (i.e., *P. hesperium* from *P. amorphum* × *glycyrrhiza*). These reticulate relationships

Polypodium

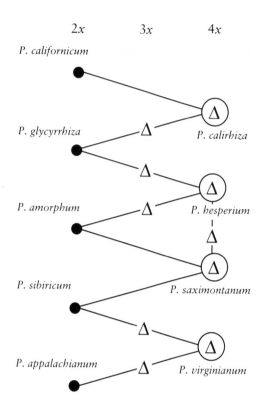

Relationships among some species of ***Polypodium***. Solid circles represent diploids; circled triangles represent allopolyploids; and triangles represent sterile hybrids.

are summarized in the reticulogram. We consider *P. scouleri* to be peripheral to the "core" diploids even though hybrids have been reported.

We have not included the European *Polypodium cambricum* Linnaeus [*P. australe* Fée], reported from San Clemente Island, California (R. M. Lloyd and J. E. Hohn 1969), in the North American flora because, since the single, original collection, efforts to relocate specimens in nature have failed (R. M. Lloyd et al. 1992).

Because taste is a characteristic used in the descriptions, the reader is cautioned to taste clean rhizomes from uncontaminated soils.

SELECTED REFERENCES   Cranfill, R. and D. M. Britton. 1983. Typification within the *Polypodium virginianum* complex (Polypodiaceae). Taxon 32: 557–560. Evans, A. M. 1971. *Polypodium.* In: A. M. Evans, ed. 1971. A Review of Systematic Studies of the Pteridophytes of the Southern Appalachians. Blacksburg. Haufler, C. H. and M. D. Windham. 1991. New species of North American *Cystopteris* and *Polypodium,* with comments on their reticulate relationships. Amer. Fern J. 81: 7–23. Haufler, C. H. and Wang Z. R. 1991. Chromosomal analyses and the origin of allopolyploid *Polypodium virginianum.* Amer. J. Bot. 78: 624–629. Lang, F. A. 1971. The *Polypodium vulgare* complex in the Pacific Northwest. Madroño 21: 235–254. Whitmore, S. A. and A. R. Smith. 1991. Recognition of the tetraploid, *Polypodium calirhiza* (Polypodiaceae), in western North America. Madroño 38: 233–248.

1. Leaves 1-pinnate at base; venation anastomosing, forming 2–3 or more rows of areoles on both sides of costae; sori in 2–3 rows on both sides of costae. . . . . 1. *Polypodium triseriale*
1. Leaves pinnatifid; venation free or, if anastomosing, only occasionally forming more than 1 row of areoles on both sides of costae; sori in 1 row on both sides of costae.
    2. Mature blades leathery, stiff; sori more than 3 mm diam.; venation always anastomosing, forming 1 row of areoles; leaf segments more than 12 mm wide. . . . . 2. *Polypodium scouleri*
    2. Mature blades herbaceous to leathery, rarely stiff; sori less than 3 mm diam.; venation free to anastomosing, occasionally forming 1 or more rows of areoles; leaf segments usually less than 12 mm wide.
        3. Rachises and segment midribs puberulent adaxially; segment margins usually serrate.
            4. Scales on abaxial surface of rachises linear and hairlike, less than 3 cells wide; venation entirely free; stems intensely sweet, licorice-flavored. . . . . . 3. *Polypodium glycyrrhiza*
            4. Scales on abaxial surface of rachises narrowly lanceolate to ovate, usually more than 3 cells wide; venation often anastomosing; stems acrid, sweet, or bland, but not licorice-flavored.
                5. Blade widest near base, proximal 1–3 pinnae equal to or longer than more distal pinnae; several areoles present on leaf segments; spores usually less than 58 μm; coastal California s of Humboldt County. . . . . 4. *Polypodium californicum*
                5. Blade widest above base, proximal 1–3 pinnae shorter than more distal pinnae; areoles often absent from many leaf segments; spores usually more than 58 μm; California n of San Luis Obispo County to s Oregon. . . . . 5. *Polypodium calirhiza*
        3. Rachises and segment midribs essentially glabrous adaxially; segment margins entire or crenulate, but rarely serrate.
            6. Sporangiasters absent (although occasional misshapen sporangia may be present); stem scales usually entire and symmetric; scales on abaxial surface of rachis linear-lanceolate, less than 6 cells wide; immature sori oval. . . . . 6. *Polypodium hesperium*
            6. Sporangiasters (with bulbous heads with or without glandular hairs) present; stem scales usually coarsely toothed and contorted distally; scales on abaxial surface of rachis lanceolate-ovate, more than 6 cells wide; immature sori circular.
                7. Sporangiasters bearing glandular hairs more than 40 per sorus; stem scales golden brown, nearly concolored; blades typically elongate-deltate, widest at or near base; tips of leaf segments acute to narrowly rounded. . . . .
                . . . . 7. *Polypodium appalachianum*
                7. Sporangiasters (with or without glandular hairs) less common, fewer than 40 per sorus; at least a portion of each stem scale dark brown, often distinctly bicolored; blades typically oblong, widest near middle; tips of leaf segments rounded to obtuse.
                    8. Sporangiasters predominantly without glandular hairs; spores usually less than 52 μm; stem scales dark brown throughout or obscurely bicolored with lighter margins. 8. *Polypodium sibiricum*
                    8. Sporangiasters mostly bearing glandular hairs; spores usually more than 52 μm; some stem scales light brown or prominently bicolored.
                        9. Leaf blades more than 3.5 cm wide; e North America. . . . . 9. *Polypodium virginianum*
                        9. Leaf blades less than 3.5 cm wide; w North America.
                            10. Sporangiasters consistently bearing glandular hairs; spores rugose with small surface projections less than 3 μm tall; Cascade Ranges.
                            . . . . 10. *Polypodium amorphum*
                            10. Sporangiasters bearing a few glandular hairs or occasionally lacking glands; spores with large tubercles, usually more than 3 μm tall; Rocky Mountains. . . . . 11. *Polypodium saximontanum*

1. Polypodium triseriale Swartz, J. Bot. (Schrader) 1800(2): 26. 1801

Goniophlebium triseriale (Swartz) Pichi-Sermolli; *Polypodium brasiliense* Poiret

Stems not whitish pruinose, slender to stout, 5–15 mm diam., taste unknown; scales brown, ovate-acuminate, symmetric, somewhat to strongly clathrate, margins somewhat lighter, entire. **Leaves** to 90 cm. **Petiole** slender to stout, to 7 mm diam. **Blade** broadly ovate, 1-pinnate at base, widest at or near base, to 60 cm wide, papery to almost leathery; rachis glabrous abaxially and adaxially. **Segments** (pinnae) linear to oblong, apex acuminate; proximal segments stalked to nearly sessile, distal ones slightly narrowed but broadly adnate at base, less than 35 mm wide; margins entire or slightly wavy; apex acute; midrib glabrous adaxially. **Venation** anastomosing with a regular series of 2–5 rows of areoles on both sides of costae. **Sori** in 1–3 parallel rows on both sides of costa, 0.5–3 mm diam., circular when immature. **Sporangiasters** absent. **Spores** less than 58 μm, verrucose, with surface projections less than 3 μm. $2n = 148$.

Epiphytic; 0 m; Fla.; s Mexico; West Indies; Central America; South America to s Brazil, Bolivia.

Commonly found in montane tropical rainforests, the epiphytic *Polypodium triseriale* is quite distinct from and probably only distantly related to other North American members of *Polypodium*. It seems likely that spores are occasionally blown into southern Florida, probably from the West Indies, and plants develop as naturalized populations.

2. Polypodium scouleri Hooker & Greville, Icon. Filic. 1: 56. 1829 · Coast polypody

Stems conspicuously whitish pruinose, stout, 3–12 mm diam., bland to slightly sweet-tasting; scales concolored to weakly bicolored, uniformly dark brown or with pale margins and base, lanceolate, symmetric, margins denticulate. **Leaves** to 85 cm. **Petiole** stout, to 3 mm diam. **Blade** ovate-lanceolate, pinnatifid, usually widest just above base, to 27 cm wide, stiff and leathery; rachis sparsely scaly to glabrescent abaxially, glabrous adaxially; scales bicolored, ovate-lanceolate, much more than 6 cells wide. **Segments** oblong to linear, usually more than 12 mm wide; margins entire to crenulate; apex rounded to rarely broadly acute; midrib glabrous adaxially. **Venation** anastomosing, usually forming 1 row of areoles. **Sori** crowded against midrib, usually more than 3 mm diam., circular when immature. **Sporangiasters** absent. **Spores** usually less than 52 μm, rugose, surface projections less than 3 μm tall. $2n = 74, 111$.

Sporulating late fall–spring. Cracks and ledges on cliffs, occasionally epiphytic; on a variety of substrates but preferring volcanic substrates in warmer, drier climates, rarely far from ocean; 0–500 m; B.C.; Calif., Oreg., Wash.; Mexico in Baja California.

The distinctive *Polypodium scouleri* has occasionally been assigned to the genus *Goniophlebium* because of its anastomosing venation and conspicuous areoles. Its venation pattern can be quite variable, however, and cannot be used as the sole feature distinguishing *P. scouleri* from *P. californicum*. Combining venation characteristics with others provided in the key distinguishes it clearly from its congeners in *Polypodium*. Some evidence suggests that *P. scouleri* hybridizes with *P. californicum* (S. A. Whitmore, unpubl.). I. Manton (1951) reported diploid and triploid cytotypes for *P. scouleri*, and variation in spore size suggests that the species may also include tetraploid populations.

3. Polypodium glycyrrhiza D. C. Eaton, Amer. J. Sci. Arts., ser. 2, 22: 138. 1856 · Licorice fern

Polypodium aleuticum A. E. Bobrov; *P. falcatum* Kellogg; *P. occidentale* (Hooker) Maxon; *P. vulgare* Linnaeus subsp. *occidentale* (Hooker) Hultén; *P. vulgare* var. *falcatum* (Kellogg) H. Christ; *P. vulgare* var. *occidentale* Hooker

Stems not whitish pruinose, slender to moderately stout, to 6 mm diam., intensely sweet, licorice-flavored; scales concolored, brown or slightly darker near point of attachment, lanceolate to lanceolate-ovate, symmetric, margins entire. **Leaves** to 75 cm. **Petiole** usually slender, 0.5–2 mm diam. **Blade** lanceolate-ovate to oblong, pinnatifid, widest near middle or just below, to 16 cm wide, herbaceous, rarely slightly leathery; rachis sparsely scaly to glabrescent abaxially, puberulent adaxially; scales linear, usually less than 3 cells wide. **Segments** linear to oblong, less than 12 mm wide; margins serrate; apex acute to attenuate; midrib puberulent adaxially. **Venation** free. **Sori** midway between margin and midrib or slightly closer to midrib, usually less than 3 mm diam., circular to oval when immature. **Sporangiasters** absent. **Spores** less than 58 μm, verrucose, with surface projections less than 3 μm. $2n = 74$.

Sporulating late fall–spring. Cliffs and rocky slopes along coasts, often epiphytic; on a variety of substrates; 0–700 m; B.C., Yukon; Alaska, Calif., Idaho,

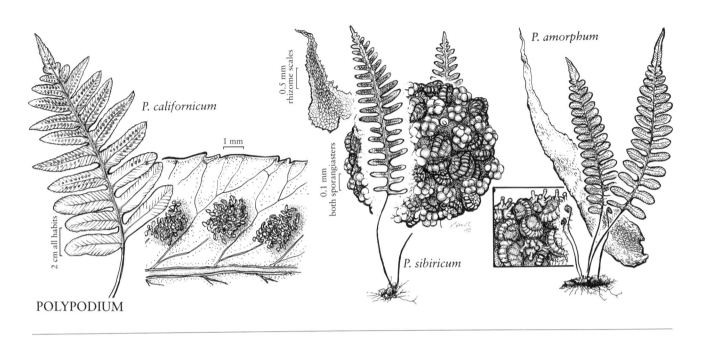

P. californicum

0.5 mm rhizome scales

1 mm

0.1 mm both sporangiasters

P. amorphum

P. sibiricum

2 cm all habits

POLYPODIUM

Oreg., Wash.; Asia in Kamchatka in the former Soviet republics.

Polypodium glycyrrhiza hybridizes with *P. calirhiza* and with *P. hesperium* to produce sterile triploids with misshapen spores. *Polypodium glycyrrhiza* was involved in the origin of both of these allotetraploid species, and some individuals can be difficult to identify. Free versus anastomosing venation distinguishes this species from *P. calirhiza*; the presence of adaxial hairs on the rachis separates it from *P. hesperium*. An additional character for distinguishing these taxa is spore length, which is less than 58 μm in diploid *P. glycyrrhiza* and more than 58 μm in the two tetraploid species. Reports of *P. glycyrrhiza* occurring in Arizona (T. Reeves 1981; D. B. Lellinger 1985) are based on misidentified specimens.

4. **Polypodium californicum** Kaulfuss, Enum. Filic., 102. 1824 · California polypody

Polypodium californicum var. *kaulfussii* D. C. Eaton; *P. vulgare* Linnaeus subsp. *californicum* (Kaulfuss) Hultén; *P. vulgare* var. *kaulfussii* (D. C. Eaton) Fernald

Stems dull or inconspicuously glaucous, moderately stout, to 10 mm diam., acrid or bland-tasting; scales uniformly brown or slightly darker near point of attachment, lanceolate-ovate, symmetric, margins entire to erose. **Leaves** to 70 cm. **Petiole** usually slender, to 3 mm diam. **Blade** deltate to lanceolate-ovate, pinnatifid, usually widest near base, to 20 cm wide, leathery to herbaceous; rachis sparsely scaly to glabrescent abaxially, puberulent adaxially; scales deltate to ovate, usually more than 10 cells wide. **Segments** linear-lanceolate to oblong, usually less than 15 mm wide; margins serrate; apex widely obtuse to rarely attenuate; midrib puberulent adaxially. **Venation** weakly to conspicuously anastomosing, most segments containing several areoles, often forming 1 row. **Sori** midway between margin and midrib or slightly closer to midrib, usually less than 3 mm diam., oval when immature. **Sporangiasters** absent. **Spores** less than 58 μm, verrucose, with surface projections to 3 μm. $2n = 74$.

Sporulating early winter–spring. Cliffs and soil on rocky slopes; on a variety of substrates but usually igneous; 0–1500 m.; Calif.; Mexico in Baja California.

R. M. Lloyd and F. A. Lang (1964) recognized two cytotypes within *Polypodium californicum*. The tetraploid has proved to be an allopolyploid involving *P. californicum* and *P. glycyrrhiza* and is treated here as a separate species, *P. calirhiza*, following S. A. Whitmore and A. R. Smith (1991). *Polypodium californicum* can be confused with *P. calirhiza*, but it usually can be distinguished by blade shape, venation, spore size, and geographic distribution. D. S. Barrington et al. (1986) reported that spores of northern populations of *P. californicum* can be as large as those of *P. calirhiza*, but the former species has veins forming more areoles per segment than does the latter.

5. Polypodium calirhiza S. A. Whitmore & A. R. Smith, Madroño 38: 235. 1991

Polypodium intermedium Hooker & Arnott 1840, not Colla 1836; *P. vulgare* Linnaeus var. *intermedium* (Hooker & Arnott) Fernald

Stems rarely whitish to glaucous, moderately stout to slender, to 8 mm diam., acrid- or slightly sweet-tasting; scales concolored brown or slightly darker near point of attachment, lanceolate-ovate, symmetric, margins entire to erose. **Leaves** to 70 cm. **Petiole** usually slender, to 3 mm diam. **Blade** lanceolate-ovate to oblong, pinnatifid, widest below middle or occasionally at base, to 16 cm wide, leathery to herbaceous; rachis sparsely scaly to glabrescent abaxially, puberulent adaxially; scales lanceolate-ovate, usually more than 3 cells wide. **Segments** linear-lanceolate to oblong, usually less than 15 mm wide; margins conspicuously serrate; apex obtuse to acute; midrib puberulent adaxially. **Venation** weakly to moderately anastomosing, some to many segments lacking areoles. **Sori** midway between margin and midrib or slightly closer to midrib, usually less than 4 mm diam., oval when immature. **Sporangiasters** absent. **Spores** more than 58 μm, verrucose, surface projections less than 3 μm. $2n = 148$.

Sporulating winter–summer. Cliffs and rocky slopes, sometimes epiphytic; on a variety of substrates but usually on granite or other igneous rocks; 0–1500 m; Calif., Oreg.; Mexico.

Although originally considered a cytotype of *Polypodium californicum*, *P. calirhiza* is an allotetraploid involving *P. californicum* and *P. glycyrrhiza* (S. A. Whitmore and A. R. Smith 1991) and therefore should be treated as a distinct species. Some individuals of *P. calirhiza* can be difficult to distinguish from the two parental species (see comments under *P. californicum* and *P. glycyrrhiza*); most collections can be identified based on a combination of blade shape, venation pattern, spore size, and geographic distribution. *Polypodium calirhiza* hybridizes with *P. glycyrrhiza* to produce sterile triploid plants with misshapen spores.

6. Polypodium hesperium Maxon, Proc. Biol. Soc. Wash. 13: 200. 1900 · Western polypody

Polypodium prolongilobum Clute; *P. vulgare* Linnaeus subsp. *columbianum* (Gilbert) Hultén; *P. vulgare* var. *columbianum* Gilbert; *P. vulgare* var. *hesperium* (Maxon) A. Nelson & J. F. Macbride

Stems occasionally whitish pruinose, slender to moderately stout, to 6 mm diam., acrid- to sweet-tasting: scales concolored, brown or slightly mottled, often darker near point of attachment, lanceolate, usually symmetric, margins entire to denticulate. **Leaves** to 35 cm. **Petiole** slender, to 1.5 mm diam. **Blade** oblong to lanceolate-ovate, occasionally deltate, pinnatifid, usually widest at or near middle, to 7 cm wide, herbaceous to somewhat leathery; rachis sparsely scaly to glabrescent abaxially, glabrous adaxially; scales linear-lanceolate, usually less than 6 cells wide. **Segments** oblong to linear-lanceolate, less than 12 mm wide; margins entire to crenulate or obscurely serrate; apex obtuse to acute; midrib glabrous adaxially. **Venation** free. **Sori** midway between margin and midrib, less than 3 mm diam., oval when immature. **Sporangiasters** absent. **Spores** more than 58 μm, rugose to verrucose or tuberculate, surface projections commonly less than 3 μm. $2n = 148$.

Sporulating summer–fall. Cracks and ledges on cliffs; on a variety of noncalcareous substrates, rarely on limestone; 300–3500 m.; B.C.; Ariz., Calif., Colo., Idaho, Mont., Nev., N.Mex., Oreg., Utah, Wash.; Mexico in Chihuahua, Baja California.

Using morphologic and chromosomal data, F. A. Lang (1971) proposed that *Polypodium hesperium* originated through allotetraploidy involving *P. glycyrrhiza* and *P. amorphum*, a hypothesis recently supported by electrophoretic studies (C. H. Haufler, M. D. Windham, and E. W. Rabe, unpubl.). Variations in spore surface morphology and banding patterns observed in isozyme studies indicate that *P. hesperium* may have originated more than once from different individuals of the same species. Some collections of *P. hesperium* can be mistaken for *P. glycyrrhiza,* but the latter species is easily distinguished by its pubescent rachises, linear blade scales, and smaller spores (less than 58 μm). Although *P. amorphum* has sporangiasters and *P. hesperium* lacks them, misshapen sporangia in *P. hesperium* can mimic these distinctive soral structures. Therefore, it is often necessary to use a combination of soral, stem scale, and blade scale features (discussed in the key) to separate *P. hesperium* from *P. amorphum*. Hybridization occurs between *P. hesperium* and each of its progenitor diploids to form triploid individuals with misshapen spores (F. A. Lang 1971). Rare, sterile, tetraploid hybrids with *P. saximontanum* have also been detected (M. D. Windham, unpubl.).

7. Polypodium appalachianum Haufler & Windham, Amer. Fern J. 81: 18. 1991 · Appalachian polypody, polypode des Appalaches

Stems often whitish pruinose, slender, to 6 mm diam., acrid-tasting; scales concolored to weakly bicolored, uniformly golden brown or slightly darker near apex, lanceolate, contorted distally, margins denticulate. **Leaves** to 40 cm. **Petiole** slender, ± 1.5 mm diam. **Blade** elongate-deltate, rarely oblong, pinnatifid, usually widest at or near base, to 9 cm wide, herbaceous to somewhat leathery; rachis sparsely scaly to glabrescent abaxially, glabrous adaxially; scales lanceolate-ovate, usually more than 6 cells wide. **Segments** linear to oblong, less than 8 mm wide, margins entire to crenulate; apex acute to narrowly rounded; midrib glabrous adaxially. **Venation** free. **Sori** midway between margin and midrib to nearly marginal, less than 3 mm diam., circular when immature. **Sporangiasters** present, usually more than 40 per sorus, heads densely covered with glandular hairs. **Spores** less than 52 μm, verrucose, projections less than 3 μm tall. 2n = 74.

Sporulating summer–fall. Cliffs and rocky slopes; on a variety of substrates; 0–1800 m; N.B., Nfld., N.S., Ont., P.E.I., Que.; Ala., Conn., Del., D.C., Ga., Ky., Maine, Md., Mass., N.H., N.J., N.Y., N.C., Ohio, Pa., R.I., S.C., Tenn., Vt., Va., W.Va.

Polypodium appalachianum is a newly recognized species traditionally identified as the diploid cytotype of *P. virginianum* (A. M. Evans 1971; I. Manton and M. G. Shivas 1953). Because the tetraploid cytotype is an allopolyploid (C. H. Haufler and Z. R. Wang 1991), and the type specimen of *P. virginianum* is tetraploid (R. Cranfill and D. M. Britton 1983), the diploid is recognized here as a distinct species, *P. appalachianum*. Some collections of *P. appalachianum* can be difficult to distinguish from *P. virginianum*, but the latter species has spores averaging more than 52 μm, and *P. appalachianum* has spores less than 52 μm. Frequent hybridization between *P. appalachianum* and *P. virginianum* forms morphologically intermediate, triploid individuals with misshapen spores. Particularly confusing is the frequent occurrence of the triploid sympatric with only one parent or with neither parent nearby.

8. Polypodium sibiricum Siplivinsky, Novosti Sist. Vyssh. Rast. 11: 329. 1974 · Polypode de Sibérie

Stems often whitish pruinose, slender, to 6 mm diam., acrid-tasting; scales concolored to weakly bicolored, uniformly dark brown, often lighter near base, lanceolate, contorted distally, margins denticulate. **Leaves** to 25 cm. **Petiole** slender, to 1 mm diam. **Blade** oblong-linear, pinnatifid, usually widest at or near middle, to 4 cm wide, somewhat leathery; rachis sparsely scaly to glabrescent abaxially, glabrous adaxially; scales lanceolate-ovate, usually more than 6 cells wide. **Segments** oblong, less than 7 mm wide; margins entire to crenulate; apex rounded to broadly acute; midrib glabrous adaxially. **Venation** free. **Sori** midway between margin and midrib to nearly marginal, less than 3 mm diam., circular when immature. **Sporangiasters** present, less than 40 per sorus, heads normally without glandular hairs. **Spores** less than 52 μm, tuberculate with tubercles, surface projections more than 3 μm tall. 2n = 74.

Sporulating summer–early fall. Cracks and ledges on rock outcrops; on a variety of substrates including granite and dolomite; 100–1000 m; Greenland; Alta., B.C., Man., N.W.T., Ont., Que., Sask., Yukon; Alaska; n Asia.

This boreal diploid has traditionally been identified as *Polypodium virginianum* (T. M. C. Taylor 1970; F. A. Lang 1971), but recent investigations indicate that it is conspecific with the eastern Eurasian species *P. sibiricum* (C. H. Haufler and M. D. Windham 1991). The sporangiasters of *P. sibiricum* normally lack glands, but some collections have sporangiasters with a few glandular hairs. Although such collections could be misidentified, the spores of *P. sibiricum* are less than 52 μm and clearly distinguish it from *P. virginianum*, *P. amorphum*, and *P. saximontanum*. Hybridization occurs between *P. sibiricum* and *P. virginianum* where these species overlap in Canada, forming triploid individuals with misshapen spores (C. H. Haufler and Z. R. Wang 1991).

9. Polypodium virginianum Linnaeus, Sp. Pl. 2: 1085. 1753 · Rock polypody, tripes-de-roches, polypode de Virginie

Polypodium vinlandicum A. Löve & D. Löve; *P. vulgare* Linnaeus var. *americanum* Hooker; *P. vulgare* var. *virginianum* (Linnaeus) D. C. Eaton

Stems often whitish pruinose, slender, to 6 mm diam., acrid-tasting; scales weakly bicolored, lanceolate, contorted distally, base and margins light brown, sometimes with dark central stripe, margins denticulate. **Leaves** to 40 cm. **Petiole** slender, to 2 mm diam. **Blade** oblong to narrowly lanceolate, pinnatifid, usually widest near middle, occasionally at or near base, to 7 cm wide, somewhat leathery; rachis sparsely scaly to glabrescent abaxially, glabrous adaxially; scales lanceolate-ovate, usually more than 6 cells wide. **Segments** oblong, less than 8 mm wide; margins entire to crenulate; apex rounded to broadly acute; midrib glabrous adaxially. **Venation** free. **Sori** midway between margin and midrib to nearly marginal, less than 3 mm diam., circular when immature. **Sporangiasters** present, usually less than 40 per sorus, heads covered with glandular hairs. **Spores** more than 52 μm, tuberculate, surface projections more than 3 μm tall. $2n = 148$.

Sporulating summer–fall. Cliffs and rocky slopes; on a variety of substrates; 0–1800 m; St. Pierre and Miquelon; Alta., Man., N.B., Nfld., N.W.T., N.S., Ont., P.E.I., Que., Sask.; Ala., Ark., Conn., Del., D.C., Ga., Ill., Ind., Iowa, Ky., Maine, Md., Mass., Mich., Minn., Mo., N.H., N.J., N.Y., N.C., Ohio, Pa., R.I., S.C., S.Dak., Tenn., Vt., Va., W.Va., Wis.

Traditionally, two cytotypes have been recognized within *Polypodium virginianum* (I. Manton and M. G. Shivas 1953). Recent research has demonstrated that the tetraploid cytotype, which properly bears the name *P. virginianum* (R. Cranfill and D. M. Britton 1983), is an allopolyploid produced by hybridization between the diploid cytotype (here called *P. appalachianum*) and *P. sibiricum* (C. H. Haufler and M. D. Windham 1991; C. H. Haufler and Z. R. Wang 1991). Although sometimes similar to its diploid parents in overall leaf morphology, *P. virginianum* has consistently larger spores, typically more than 52 μm (see additional comments under *P. appalachianum* and *P. sibiricum*). Frequent hybridizations between *P. virginianum* and *P. appalachianum* form morphologically intermediate, triploid individuals with misshapen spores. Sterile triploids also result from hybridization between *P. virginianum* and *P. sibiricum*.

10. Polypodium amorphum Suksdorf, Werdenda 1: 16. 1927

Polypodium montense F. A. Lang
Stems often whitish pruinose, slender, to 6 mm diam., acrid-tasting; scales weakly bicolored, lanceolate, contorted distally, bases and margins light brown, sometimes with dark central stripe, margins often coarsely dentate. **Leaves** to 30 cm. **Petiole** slender, to 1.5 mm diam. **Blade** oblong to rarely del-

tate, pinnatifid, usually widest near middle, occasionally at or near base, to 4 cm wide, somewhat leathery; rachis sparsely scaly to glabrescent abaxially, glabrous adaxially; scales lanceolate-ovate, usually more than 6 cells wide. **Segments** oblong, less than 12 mm wide; margins entire to crenulate; apex rounded to broadly acute; midrib glabrous adaxially. **Venation** free. **Sori** midway between margin and midrib to nearly marginal, less than 3 mm diam., circular when immature. **Sporangiasters** present, usually less than 40 per sorus, heads covered with glandular hairs. **Spores** more than 58 μm, rugose to verrucose, surface projections less than 3 μm tall. $2n = 74$.

Sporulating summer–fall. Cliffs and rocky slopes; usually on igneous substrates; 0–1800 m; B.C.; Oreg., Wash.

The diploid *Polypodium amorphum* is one of the progenitors of allotetraploid *P. hesperium*, and these two species are occasionally sympatric. Although *P. amorphum* can be mistaken for *P. hesperium*, consistent differences exist for separating these two species (see comments under *P. hesperium*). Hybridization between *P. amorphum* and *P. hesperium* results in triploid individuals with misshapen spores (F. A. Lang 1971).

11. Polypodium saximontanum Windham, Contr. Univ. Michigan Herb. 19: 47. 1993

Stems often whitish pruinose, slender, to 6 mm diam., acrid-tasting; scales weakly bicolored, lanceolate, contorted distally, bases and margins light brown, sometimes with dark central stripe, margins often coarsely dentate. **Leaves** to 25 cm. **Petiole** slender, to 1.5 mm diam. **Blade** oblong to linear, pinnatifid, usually widest near middle, to 4 cm wide, somewhat leathery; rachis sparsely scaly to glabrescent abaxially, glabrous adaxially; scales lanceolate-ovate, usually more than 6 cells wide. **Segments** oblong, less than 12 mm wide; margins entire to crenulate; apex rounded to broadly acute; midrib glabrous adaxially. **Venation** free. **Sori** midway between margin and midrib to nearly marginal, less than 3 mm diam., circular when immature. **Sporangiasters** present, usually less than 40 per sorus, heads with a few glandular hairs or rarely without glands. **Spores** more than 58 μm, tuberculate, surface projections more than 3 μm tall. $2n = 148$.

Sporulating summer–fall. Cracks and ledges on rocks; apparently confined to granitic substrates; 1800–3000 m; Colo., N.Mex., S.Dak., Wyo.

Polypodium saximontanum is an allotetraploid species whose progenitor diploid species are *P. amorphum* and *P. sibiricum* (M. D. Windham 1993). Prior to its recognition as a distinct species, collections of *P. saximontanum* were variously referred to *P. montense* F. A. Lang (= *P. amorphum*), *P. hesperium*, and/or *P. virginianum*. In addition to its separate geographic range, *P. saximontanum* can be distinguished from *P. virginianum* by having narrower leaves and a reduced fre-

quency of glandular hairs on its sporangiasters. *Polypodium saximontanum* also has a separate range from *P. amorphum* and has spores with large (greater than 3 μm tall) projections. Although *P. saximontanum* overlaps in range with *P. hesperium*, the latter species has no sporangiasters. Tetraploid hybrids of these two species have misshapen spores.

3. PHLEBODIUM (R. Brown) J. Smith, J. Bot. (Hooker) 4: 58. 1841 · Golden spolypodies [Greek *phlebos*, vein, referring to the prominent venation]

Clifton E. Nauman

Polypodium Linnaeus sect. *Phlebodium* R. Brown in J. J. Bennett et al., Pl. Jav. Rar. 1: 4. 1838

Plants epiphytic. **Stems** creeping, branched, not whitish pruinose, 8–30 mm diam.; scales concolored, lanceolate, not clathrate, glabrous, margins dentate. **Leaves** monomorphic, widely spaced, not conspicuously narrowed at tip, to 130 cm. **Petiole** articulate to stem, dark brown, round in cross section, with 2 adaxial grooves. **Blade** ovate to elliptic, not pectinate, pinnatisect with broadly winged rachises and rounded sinuses, with fewer than 20 pairs of segments, glaucous, usually glabrous, scales absent; rachis glabrous. **Segments** linear to lanceolate, margins entire, apex rounded. **Veins** free near margins, well developed near costae, highly reticulate, usually with 1 row of costal areoles extending from vein to vein without included veinlets, and with a series of elongate, polygonal areoles with 1–3 excurrent included veinlets meeting at apices, and similar areoles closer to margin of segment mostly without included veinlets. **Sori** on veins, transverse, 2 in each major areole on included veinlets, circular to oblong; indument absent. **Spores** tuberculate. $x = 37$.

Species 2–4 (1 in the flora): tropical regions, North America, Mexico, West Indies, Central America, South America.

SELECTED REFERENCES Duncan, W. H. 1954. *Polypodium aureum* in Florida and Georgia. Amer. Fern J. 44: 155–158. Lellinger, D. B. 1987. Nomenclatural notes on some ferns of Costa Rica, Panama, and Colombia. III. Amer. Fern J. 77: 101–102. Proctor, G. R. 1985. Ferns of Jamaica. London. Snyder, L. H. Jr. and J. G. Bruce. 1986. Field Guide to the Ferns and Other Pteridophytes of Georgia. Athens, Ga.

1. **Phlebodium aureum** (Linnaeus) J. Smith, J. Bot. (Hooker) 4: 59. 1841 · Goldfoot fern, golden polypody

Polypodium aureum Linnaeus, Sp. Pl. 2: 1087. 1753

Stems creeping, ca. 8–15(–30) mm diam., densely scaly; scales reddish to golden, long-attenuate, 10–20 mm. **Leaves** bright green or glaucous, arching to pendent, scattered, 3–13 dm. **Petiole** 1.5–5 dm, smooth, with a few scales near base. **Blade** pinnately and deeply lobed, 3–8 × 1–5 dm, glabrous, terminal segment conform. **Segments** lanceolate to elliptic, or linear-lanceolate to linear, 6–20 × 1–4 cm, margins entire or sometimes undulate. **Sori** in 1 line on each side of costae, occasionally

2d row present, sori terminal or at junction of free included veinlets. $2n = 148$.

Epiphytic on a variety of trees or on logs, dense piles of humus, but most commonly among old leaf bases of *Sabal palmetto* Loddiges, in various habitats from hammocks to swamps; 0 m; Fla., Ga.; Mexico; West Indies; Central America; South America.

Phlebodium aureum occurs north to Dixie and Nassau counties in Florida, and it is disjunct in Franklin County. It is also found in Georgia (W. H. Duncan 1954; L. H. Snyder Jr. and J. G. Bruce 1986). Two varieties (or subspecies) have been recognized, *Phlebodium aureum* var. *aureum* and *P. aureum* var. *areolatum* (Humboldt & Bonpland ex Willdenow) Farwell. The latter is now often elevated to species rank and given the name *P. pseudoaureum* (Cavanilles) Lellinger.

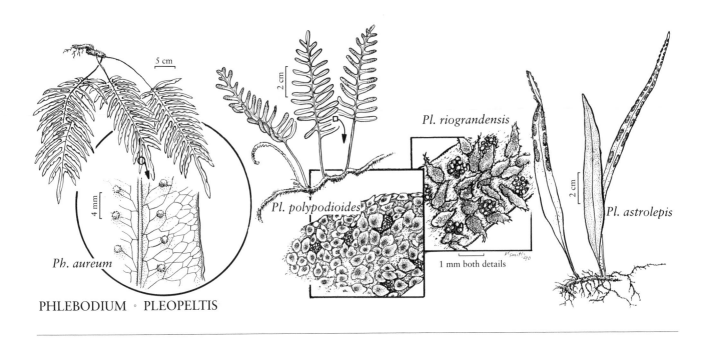

Pl. riograndensis

Pl. polypodioides

Ph. aureum

PHLEBODIUM · PLEOPELTIS

1 mm both details

Pl. astrolepis

Phlebodium pseudoaureum is widespread in Central America and South America (D. B. Lellinger 1987) and has been reported as rare in Florida by G. R. Proctor (1985). I have not seen specimens that could be convincingly referred to *P. pseudoaureum*.

Phlebodium aureum, a tetraploid species, is believed to have arisen through allopolyploidy following hybridization between *P. pseudoaureum* and *P. decumanum* (Willdenow) J. Smith, a widespread species in tropical America.

4. PLEOPELTIS Humboldt & Bonpland ex Willdenow, Sp. Pl. 5(1): 211. 1810 · Shielded-sorus ferns [Greek *pleos*, many, and *pelte*, shield, in reference to the peltate scales covering immature sori]

Elisabeth G. Andrews

Michael D. Windham

Plants epiphytic, on rock, or rarely terrestrial. **Stems** often long-creeping, usually branched, 1–2.5 mm diam., not whitish pruinose; scales bicolored with dark central region, round to ovate-lanceolate or subulate, clathrate toward center, glabrous or pubescent with reddish brown hairs, margins transparent, erose-denticulate to fringed-ciliate. **Leaves** monomorphic, widely spaced, not conspicuously narrowed at tip, to 25 cm. **Petiole** green to dark brown or black, flattened to terete, often grooved proximally and winged distally. **Blade** oblong-linear to deltate, simple and entire or deeply pinnatifid, not pectinate, with fewer than 25 pairs of segments, not glaucous, with conspicuous peltate scales on abaxial surface; rachis sparsely to densely scaly abaxially, glabrous to sparsely scaly adaxially; scales ovate-lanceolate to spheric, peltate, clathrate toward center. **Segments** (when present) linear to oblong; margins entire to crenulate; apex rounded. **Venation** mostly free or complexly anastomosing, with 1–several included veinlets in fertile areoles. **Sori** often confined to distal 1/2 of leaf, discrete or confluent, circular to oval when immature, borne at ends of single veins or at junction of several veinlets, in 1 row on either side of midrib; indument of ephemeral, peltate scales covering immature sori. **Spores** smooth with scattered spheric deposits or slightly papillate or verrucose. *x* = 34, 35, 37.

Species ca. 50 (4 in the flora): tropical regions, North America, Mexico, West Indies, Central America, South America, Africa, Asia in India.

Pleopeltis is a neotropical genus of mostly epiphytic ferns. Its center of diversity is in southern Mexico and Central America. The genus is currently under revision, and preliminary data indicate that certain scaly-leaved species traditionally placed in the genus *Polypodium* are actually more closely related to *Pleopeltis* than they are to *Polypodium*. The two North American species of this scaly *Polypodium* group recently have been transferred to *Pleopeltis* (M. D. Windham 1993) and are similarly treated here.

SELECTED REFERENCES Wendt, T. 1980. Notes on some *Pleopeltis* and *Polypodium* species of the Chihuahuan Desert Region. Amer. Fern J. 70: 5–11. Windham, M. D. 1993. New taxa and nomenclatural changes in the North American fern flora. Contr. Univ. Michigan Herb. 19: 31–61. 1993.

1. Blades simple.
  2. Scales on abaxial blade surface scattered, less than 0.5 mm wide, mostly roundish and deeply fringed (appearing stellate); petioles conspicuously flattened in cross section; sori oval to oblong, often confluent; stem scales mostly round, with numerous reddish brown hairs. . . . . 1. *Pleopeltis astrolepis*
  2. Scales on abaxial blade surface dense and overlapping, 0.5–1 mm wide, broadly ovate-lanceolate with fringed-ciliate margins; petioles round in cross section; sori round to oval, discrete; stem scales lanceolate-attenuate, with only occasional hairs. . . . . . .
    . . . . 2a. *Pleopeltis polylepis* var. *erythrolepis*
1. Blades deeply pinnatifid.
  3. Mature blade scales distinctly bicolored with dark brown centers and broad tan to silvery gray transparent margins, clathrate only near point of attachment; petiole scales often overlapping, margins ± entire; sori deeply embossed, forming bumps on adaxial blade surface; stems usually 1–2 mm diam. . . . . 3a. *Pleopeltis polypodioides* var. *michauxiana*
  3. Mature blade scales not bicolored or occasionally with very narrow transparent margin, dark reddish brown, clathrate throughout with cell luminae large and clear; petiole scales rarely overlapping, margins denticulate to ciliate; sori not embossed, rarely visible on adaxial blade surface; stems usually 2–3 mm diam. . . . . 4. *Pleopeltis riograndensis*

1. **Pleopeltis astrolepis** (Liebmann) E. Fournier, Mexic. Pl. 1: 87. 1872 · Star-scaled fern

    Polypodium astrolepis Liebmann, Kongel. Danske Vidensk. Selsk. Skr., Naturvidensk. Math. Afd., ser. 5, 1: 185. 1849; *Grammitis elongata* Swartz (not *Polypodium elongatum* Aiton); *G. lanceolata* Schkuhr; *G. revoluta* Sprengel ex Willdenow; *Pleopeltis revoluta* (Sprengel ex Willdenow) A. R. Smith

Stems long-creeping, branched, ca. 1 mm diam.; scales round to ovate-lanceolate, centrally clathrate with cell luminae clear, pubescent, hairs reddish brown. **Leaves** to 20 cm, weakly hygroscopic. **Petiole** conspicuously flattened, sparsely scaly; scales rarely overlapping, margins fringed. **Blade** linear-elliptic, simple, to 2 cm wide, margins entire, moderately scaly abaxially, sparsely scaly adaxially; scales bicolored, mostly round, less than 0.5 mm wide, centers brown, clathrate, margins transparent, deeply fringed. **Venation** complexly anastomosing, fertile areoles with several included veinlets. **Sori** oval to oblong, discrete or often confluent, surficial to shallowly embossed, soral scales attached to receptacle. **Spores** shallowly and irregularly papillate to verrucose, ca. 58 μm. 2*n* = 136.

Epiphytic in swamps on pond apple (*Annona glabra* Linnaeus); 0 m; Fla.; Mexico; West Indies; Central America; South America.

Pleopeltis astrolepis is a common species in tropical America, including the West Indies. The only known North American locality, in Broward County, Florida, was discovered in 1977, and (in 1992) the plants are in danger of extirpation from development.

2. Pleopeltis polylepis (Roemer ex Kunze) T. Moore, Index Fil., 348. 1862

Polypodium polylepis Roemer ex Kunze, Linnaea 13: 131. 1839

Varieties 2 (1 in the flora): North America, Mexico, Central America.

Although *Pleopeltis polylepis* is the name most commonly used for this taxon in regional floras, questions remain as to its legitimacy. According to C. Christensen (1937), *Polypodium peltatum* Cavanilles (1802) is conspecific with *Pleopeltis polylepis* (as *Polypodium polylepis* Roemer ex Kunze). There are problems, however, associated with the true origin and identity of Cavanilles's type specimen. If the type originated in Ecuador, as claimed by Christensen, the epithet *peltatum* refers to a different taxon than the one treated in this flora. If the collection locality is not Ecuador, but Mexico (the expedition during which the type was collected visited both countires), the possibility exists that the two taxa are conspecific, and therefore the epithet *peltatum* has priority and must be used.

2a. Pleopeltis polylepis (Roemer ex Kunze) T. Moore var. **erythrolepis** (Weatherby) T. Wendt, Amer. Fern J. 70: 9. 1980

Polypodium erythrolepis Weatherby, Contr. Gray Herb. 65: 11. 1922; *Pleopeltis erythrolepis* (Weatherby) Pichi-Sermolli

Stems long-creeping, branched, 1–2.5 mm diam.; scales lanceolate-attenuate, centrally clathrate with cell luminae clear, glabrescent or with scattered reddish brown hairs. **Leaves** to 10 cm, weakly hygroscopic. **Petiole** round in cross section, sparsely scaly; scales rarely overlapping, margins fringed-ciliate. **Blade** ovate-lanceolate to oblanceolate, simple, 0.5–2 cm wide, margins entire, densely scaly abaxially, sparsely scaly adaxially; scales bicolored, broadly ovate-lanceolate, 0.5–1 mm wide, centers brown, clathrate, margins transparent, irregularly fringed-ciliate. **Venation** complexly anastomosing, fertile areoles with several included veinlets. **Sori** round to oval, discrete, surficial to shallowly embossed, soral scales attached to receptacle. **Spores** smooth with scattered spheric deposits, ca. 59 μm.

Sporulating summer–fall. Forming extensive mats on porphyritic rocks in Upper Limpia Canyon; 2300 m; Texas; n Mexico.

The only known North American population of *Pleopeltis polylepis* occurs in the Davis Mountains of western Texas. It is placed in variety *erythrolepis,* which differs from var. *polylepis,* the variety common in Mexico, in that the scales on the abaxial surface of mature blades are fringed-ciliate rather than entire to erose, and mostly ovate-lanceolate rather than spheric. In addition, the scales of the abaxial blade surface tend to be more abundant than in var. *polylepis.* The two varieties are very similar, however, and a broad region of overlap exists in parts of Mexico where plants of intermediate morphology occur (T. Wendt 1980). In general, var. *erythrolepis* occurs in northern Mexico and Texas, and var. *polylepis* ranges from southern Mexico to Guatemala.

3. Pleopeltis polypodioides (Linnaeus) E. G. Andrews & Windham in Windham, Contr. Univ. Michigan Herb. 19: 46. 1993

Acrostichum polypodioides Linnaeus, Sp. Pl. 2: 1068. 1753; *Marginaria polypodioides* (Linnaeus) Tidestrom; *Polypodium polypodioides* (Linnaeus) Watt

Varieties 6 (1 in the flora): North America, Mexico, West Indies, Central America, South America, Africa.

3a. Pleopeltis polypodioides (Linnaeus) E. G. Andrews & Windham var. **michauxiana** (Weatherby) E. G. Andrews & Windham in Windham, Contr. Univ. Michigan Herb. 19: 46. 1993 · Resurrection fern, Gray's polypody

Polypodium polypodioides (Linnaeus) Watt var. *michauxianum* Weatherby, Contr. Gray Herb. 124: 31. 1939

Stems long-creeping, much branched, 1–2 mm diam.; scales linear to subulate, centrally clathrate with cell luminae occluded, surfaces glabrous, margins denticulate to fringed-ciliate. **Leaves** to 25 cm, strongly hygroscopic. **Petiole** grooved, otherwise round in cross section, densely scaly when young; scales often overlapping, margins mostly entire. **Blade** narrowly triangular to elliptic, deeply pinnatifid, to 5 cm wide, densely scaly abaxially, glabrous adaxially except for a few lanceolate scales along rachis; scales distinctly bicolored, spheric to deltate-ovate, usually less than 0.5 mm wide, centers dark brown, obscurely clathrate, margins broad, transparent, entire to erose. **Venation** mostly free with occasional areoles, never more than 1 included veinlet in fertile areoles. **Sori** round, discrete, deeply embossed, forming conspicuous bumps on adaxial surface, soral scales attached at periphery of receptacle. **Spores** smooth with scattered spheric deposits on surface, 45–52 μm. $2n = 74$.

Sporulating summer–fall. Epiphytic on various species of trees, especially oaks, magnolias, and elms, or on rocks (usually limestone or sandstone), fence posts, buildings, or mossy banks, usually in moist, shady areas; 0–700 m; Ala., Ark., Fla., Ga., Ill., Ind., Kans., Ky., La., Md., Miss., Mo., N.C., Ohio, Okla., S.C., Tenn., Tex., Va., W.Va.; Mexico; Central America in Guatemala.

This species is the most widespread epiphytic fern in the flora, although in some parts of its range (e.g., the north) it often occurs on rock. *Pleopeltis polypodioides* is a common neotropical species, and the North American variety is just one of six that have been recognized. *Pleopeltis polypodioides* var. *michauxiana* differs from the other varieties in having more or less entire blade scales and a glabrous adaxial leaf surface. In the southeastern United States (particularly Florida), some plants grade slightly into var. *polypodioides,* which is common in the West Indies. The latter variety has fringed to denticulate blade scales and scattered scales on the adaxial blade surface. Within its range, *Pleopeltis polypodioides* var. *michauxiana* could only be mistaken for *Polypodium virginianum* Linnaeus, which has similar leaf morphology but lacks scales on the abaxial blade surface.

4. **Pleopeltis riograndensis** (T. Wendt) E. G. Andrews & Windham in Windham, Contr. Univ. Michigan Herb. 19: 46. 1993 · Rio Grande scaly polypody

Polypodium thyssanolepis A. Braun ex Klotzsch var. *riograndense* T. Wendt, Amer. Fern J. 70: 6. 1980

Stems short-creeping, sparingly branched, 2–3 mm diam.; scales subulate to lanceolate-acuminate, centrally clathrate with cell luminae large and clear, surfaces glabrous, margins lacerate-ciliate. **Leaves** to 20 cm, strongly hygroscopic. **Petiole** grooved, otherwise round in cross section, sparsely scaly; scales rarely overlapping, margins denticulate to ciliate. **Blade** triangular-oblong to ovate, deeply pinnatifid, to 5 cm wide, moderately scaly abaxially, glabrous adaxially; scales concolored to obscurely bicolored, usually dark reddish brown throughout, broadly ovate-lanceolate, clathrate, more than 0.5 mm wide, margins fringed-ciliate. **Venation** mostly free with occasional areoles, never more than 1 included veinlet in fertile areoles. **Sori** round, discrete, surficial to shallowly embossed, soral scales attached at periphery of receptacle. **Spores** smooth with scattered spheric deposits on surface, 60–74 μm. $2n = 148$.

Sporulating summer–fall. Growing on rocky slopes and ledges, and in crevices, usually in moist, shaded canyons; 1500–2500 m; Ariz., Texas; n Mexico.

In the past *Pleopeltis riograndensis* has been treated as a variety of *P. thyssanolepis,* but it differs from the latter species in that the petiole and leaf are only sparsely scaly rather than densely so, the blade scales are mostly ovate or ovate-lanceolate rather than nearly spheric, the venation is mostly free rather than mostly areolate, and the basal segments of the blade are alternate rather than opposite to nearly opposite.

5. CAMPYLONEURUM C. Presl, Tent. Pterid., 189. 1836 · Strap ferns [Greek *kampylos,* curved, and *neuron,* nerve, in reference to the venation]

Clifton E. Nauman

Plants epiphytic. **Stems** short- to long-creeping, branched, 2–10 mm diam., sometimes whitish pruinose; scales brown, ovate, often clathrate, entire. **Leaves** monomorphic, tufted, conspicuously narrowed to tip, to 150 cm. **Petiole** absent or present, articulate to stem. **Blade** linear to lanceolate or elliptic, simple, glabrous, sparsely scaly, margins entire; scales clathrate, basifixed. **Primary veins** connected by 1–several secondary veins (cross veins), forming 1–several areoles between each pair of primary veins, areoles with 1–4 excurrent veinlets, sometimes these excurrent veinlets dividing areoles completely. **Sori** in 1–10 rows between midrib and margin, terminal on included veinlets, discrete, round; indument absent. **Spores** verrucose to nearly psilate. $x = 37$.

Species ca. 25–50 (4 in the flora): tropical, North America, Mexico, West Indies, Central America, South America.

SELECTED REFERENCES Eaton, A. A. 1906. Pteridophytes observed during three excursions into southern Florida. Bull. Torrey Bot. Club 33: 455–486. Lellinger, D. B. 1988. Some new species of *Campyloneurum* and a provisional key to the genus. Amer. Fern J. 78: 14–34.

1. Primary veins obscure and slightly to strongly curved; blades generally less than 6 cm wide.
  2. Blades more than 2 cm wide, margins shallowly sinuate to undulate, not revolute; areoles in 4–8 series between costa and margin; petioles always present. . . . 1. *Campyloneurum costatum*
  2. Blades less than 2 cm wide, margins revolute, not shallowly sinuate to undulate; areoles in 1–4 series between costa and margin; petioles present or absent. . . 2. *Campyloneurum angustifolium*
1. Primary veins conspicuous and straight to slightly curved; blades generally more than 6 cm wide.
  3. Petioles absent to ca. 9 cm; blades yellow-green, margins entire and plane to lightly undulate. . . . . 3. *Campyloneurum phyllitidis*
  3. Petioles 5–18 cm; blades dark green, margins undulate. . . . . 4. *Campyloneurum latum*

1. Campyloneurum costatum (Kunze) C. Presl, Tent. Pterid., 190, plate 7. 1836 · Tailed strap fern

Polypodium costatum Kunze, Linnaea 9: 38. 1834

Stems short-creeping, 2–5 mm diam. **Leaves** few, arching to pendent, stiff. **Petiole** 4–15 cm. **Blade** dark green, linear-elliptic or linear-oblanceolate, 20–40 × 2.5–6 cm, leathery; base acuminate; margins shallowly sinuate to undulate; apex abruptly caudate-acuminate. **Veins** obscure, primary veins obscure, slightly to strongly curved, areoles in 4–8 series between costa and margin, with usually 2 free included veinlets per areole. **Sori** in 1–several rows on each side of costa. $2n = 74$.

Epiphytic in swamps, on various rough-barked trees; substrate subacid; 0 m; Fla.; West Indies; Central America; South America in Venezuela, Colombia, Ecuador.

In the flora *Campyloneurum costatum* is currently known only from Collier County, Florida. It had formerly been reported from Dade County, where it now may have been extirpated.

2. Campyloneurum angustifolium (Swartz) Fée, Mém. Foug. 5: 257. 1852 · Narrow strap fern

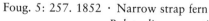

Polypodium angustifolium Swartz, Prodr., 130. 1788

Stems short-creeping, 3–7 mm diam. **Leaves** usually many, arching to pendent. **Petiole** essentially absent to ca. 3–5(–8) cm. **Blade** yellowish to dark green, elongate-linear, ± falcate, 30–60 × 0.5–1.5 cm, papery; base and apex long-attenuate; margins often slightly revolute. **Veins** obscure, primary veins inconspicuous, slightly to strongly curved, areoles in 1–4 series between costa and margin, with free included veinlets. **Sori** in 1–2 rows on each side of costa.

Epiphytic on oaks, pond apples, magnolias, and other rough-barked trees in hammocks and swamps in the everglade keys; substrate subacid; 0 m; Fla.; Mexico; West Indies; Central America; South America.

In the flora *Campyloneurum angustifolium* occurs only in Collier County, Florida. It was reported earlier from Dade and Seminole counties.

3. Campyloneurum phyllitidis (Linnaeus) C. Presl, Tent. Pterid., 190, plate 7, figs. 18–20. 1836 · Long strap fern

Polypodium phyllitidis Linnaeus, Sp. Pl. 2: 1083. 1753

Stems short-creeping, 4–10 mm diam. **Leaves** few to many, erect to arching. **Petiole** essentially absent to ca. 9 cm. **Blade** yellowish green, linear to linear-elliptic, 24–140 × 3–12 cm, leathery; base attenuate; margins entire to slightly undulate; apex acute (sometimes forked). **Veins** obvious, primary veins ± prominent, straight to slightly curved, areoles in 7–16(–18) series between costa and margin, with free included veinlets, a percurrent veinlet often dividing some areoles further. **Sori** in several rows on each side of costa. $2n = 148$.

Epiphytic in hammocks and swamps; sometimes on walls in limestone sinkholes where it is reduced in size; substrate circumneutral to subacid; 0 m; Fla.; Mexico; West Indies; Central America; South America.

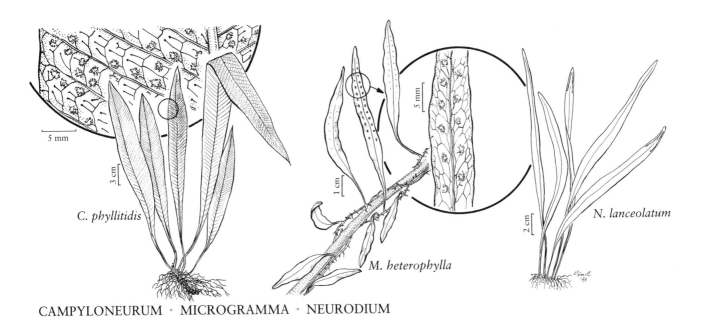

CAMPYLONEURUM · MICROGRAMMA · NEURODIUM

4. Campyloneurum latum T. Moore, Index Fil., 225. 1861 · Wide strap fern

Polypodium latum (T. Moore) Sodiro

Stems short-creeping, to 7 mm diam. **Leaves** few to many, erect to arching. **Petiole** 5–18 cm. **Blade** dark green, oblong, elliptic to oblanceolate, 20–60 cm × 4–9 cm [larger], leathery; base cuneate; apex abruptly acute. **Veins** obvious, primary veins ± prominent, straight to slightly curved, areoles in 8–18 series between costa and margin, with free included veinlets, a percurrent veinlet often dividing some areoles further. **Sori** in several rows on each side of costa.

Epiphytic in tropical hammocks; 0 m; Fla.; Mexico; West Indies; Central America; South America.

The venation of *Campyloneurum latum* is similar to, but more variable than, that of *C. phyllitidis*.

The only record in the flora is based on a collection by J. J. Soar and A. A. Eaton in December 1903 from Dade County, Florida (A. A. Eaton 1906). Extensive searches at the original collection site and many other likely locations in the southern portions of Florida have failed to turn up any plant that could be called *Campyloneurum latum*. The species is apparently extirpated from North America north of Mexico.

6. MICROGRAMMA C. Presl, Tent. Pterid., 185, 213, plate 7, figs. 13–14. 1836 · Vine ferns [Greek *mikros,* small, and *gramme,* line; the sori are elongate in the type species]

Clifton E. Nauman

Plants epiphytic or on rock. **Stems** long-creeping, branched, 0.5–1.5 mm diam., not whitish pruinose; scales bicolored, rhombic, glabrous, margins entire or toothed. **Leaves** monomorphic or dimorphic, well separated, not narrowed toward tip, to 17 cm; fertile leaves often narrower and longer than sterile leaves. **Petiole** articulate to stem, straw-colored to greenish, somewhat flattened. **Blade** simple, linear-elliptic to linear oblanceolate, glabrous, or sparsely scaly on midrib, margins entire to crenate. **Veins** obscure, forming a row of costal areoles and often additional areoles toward blade margins, areoles often with 1–several free included veinlets. **Sori** in 1 row between midrib and margin on included veinlets, discrete, round [elongate]; indument of threadlike scales or absent. **Spores** tuberculate to rugose. $x = 37$.

Species 20–30 (1 in the flora): tropical, North America, Mexico, West Indies, Central America, South America, Africa.

1. **Microgramma heterophylla** (Linnaeus) Wherry, Amer. Fern J. 54: 145. 1964 · Climbing vine fern

Polypodium heterophyllum Linnaeus, Sp. Pl. 2: 1083. 1753 **Stems** appressed to substrate, 0.5–1.2 mm diam.; scales reddish to dark brown, linear-triangular. **Leaves** irregularly spaced. **Petiole** to ca. 2 cm. **Blade** 4–15 × ca. 1 cm, margins entire to undulate and ± shallowly crenulate, gla-

brous. **Veins** forming 1 series of linear polygonal costal areoles without free included veinlets, medial row of larger rectangular to polygonal areoles (usually with 1–2, simple or forked, free included veinlets), and a series of short, casually anastomosing marginal veinlets. **Sori** terminal on included veinlets; indument of few paraphyses longer than sporangia.

Epiphytic on relatively smooth-barked trees, or growing on logs and rock, in tropical hammocks; substrate circumneutral to acidic; 0 m; Fla.; West Indies.

7. **NEURODIUM** Fée, Mém. Soc. Mus. Hist. Nat. Strasbourg 4: 201. 1850 [as "*Nevrodium*"] · Ribbon ferns [Greek *neuron*, nerve, and *-ium*, resemblance; veinlets are embossed]

Clifton E. Nauman

Paltonium C. Presl

Plants epiphytic. **Stems** short-creeping, branched, 2–4 mm diam., not whitish pruinose; scales bicolored, lanceolate, somewhat clathrate, margins dentate. **Leaves** essentially monomorphic, tufted, narrowed toward apex, to 45 cm. **Petiole** articulate to stem, greenish, round in cross section, grooved at base. **Blade** linear to lanceolate-elliptic, simple, not glaucous, glabrous, margins entire. **Veins** obscure, in irregular areoles that are generally longer than wide and parallel to costa, included veinlets 1–2, simple or forked, free. **Sori** in nearly continuous, nearly marginal band (coenosorus) in distal 1/2 of blade, sometimes also with 1–3 separate oblong to circular sori proximally; indument filamentous. **Spores** minutely tuberculate. $x = 37$.

Species 1 (1 in the flora): tropical, North America, Mexico, West Indies, Central America.

1. **Neurodium lanceolatum** (Linnaeus) Fée, Mém. Foug. 3: 28. 1852 · Ribbon fern

Pteris lanceolata Linnaeus, Sp. Pl. 2: 1073. 1753 *Paltonium lanceolatum* (Linnaeus) C. Presl **Roots** brown, in dense mass often covering stems. **Stem** scales black to dark brown, triangular-acuminate to lanceolate-acuminate, clathrate, papillate-ciliolate. **Leaves** erect-arching, 1–4.5

dm. **Petiole** less than 3.5 cm. **Blade** 20 × 1.3 cm, leathery; base narrowly cuneate to attenuate; apex acute to rounded. **Sori** to 6 cm, ending ca. 0.5 cm below apex of blade. $2n = 74$.

Epiphytic in hammocks and mangrove swamps; substrate subacidic to circumneutral; 0 m; Fla.; Mexico; West Indies; Central America.

23. MARSILEACEAE Mirbel
• Water-clover Family

David M. Johnson

Plants aquatic or amphibious, rhizomatous. **Stems** growing on soil surface or subterranean, main stem long-creeping and giving rise to long or short shoots only at nodes; hairs laterally attached, multicellular. **Roots** arising at nodes and also along internodes. **Leaves** distichous, long-petioled, sometimes filiform and lacking expanded blades. **Sori** within hard bean- or pea-shaped bodies (sporocarps) arising on short stalks from near or at base of petioles. **Sporangia** of 2 kinds, borne within the same sorus and sporocarp; megasporangia containing a single megaspore; microsporangia containing 20–64 microspores. **Gametophytes** remaining within spores; microgametophytes of only a few cells; megagametophytes protruding from spores, each bearing 1 simple archegonium.

Genera 3, species ca. 50 (2 genera, 7 species in the flora): nearly worldwide, temperate and tropical regions.

Spore germination in the family occurs after rupture of the sporocarp wall allows the sporocarp contents to be hydrated. A gelatinous structure emerges from the sporocarp, breaking it into valves and carrying the sori into the water. Spore germination (gametophyte growth) and fertilization occur immediately.

Regnellidium diphyllum Lindman was introduced into a wildlife pond in Mahoning County, Ohio, in 1985 and continues to persist (C. F. Chuey, in litt. 1991). *Regnellidium* is similar to *Marsilea* but differs from it in having two leaflets instead of four.

SELECTED REFERENCES Braun, A. 1871. Hr. Braun theilte neuere Untersuchungen über die Gattungen *Marsilea* und *Pilularia*. Monatsber. Königl. Preuss. Akad. Wiss. Berlin 1870: 653–753. Johnson, D. M. 1986. Systematics of the New World species of *Marsilea* (Marsileaceae). Syst. Bot. Monogr. 11: 1–87.

1. Leaves with blades palmately divided into 4 obdeltate or cuneate pinnae. . . . . 1. *Marsilea*, p. 332
1. Leaves filiform, lacking expanded blades. . . . . 2. *Pilularia*, p. 335

1. MARSILEA Linnaeus, Sp. Pl. 2: 1099. 1753; Gen. Pl. ed. 5, 485. 1754, name conserved · Water-clover, pepperwort [for Count Luigi Marsigli (1656–1730), Italian mycologist at Bologna]

Plants aquatic or amphibious, forming diffuse or dense colonies. **Roots** arising at nodes, sometimes also on internodes. **Leaves** deciduous in temperate regions, heteromorphic, floating leaves averaging larger than land leaves. **Petiole** filiform, stiffly erect or procumbent in land leaves, lax in floating leaves. **Blade** palmately divided into 4 pinnae. **Pinnae** cuneate or obdeltate, pulvinate at base, frequently with numerous red or brown streaks abaxially in floating leaves. **Sporocarps** borne on branched or unbranched stalks at or near bases of petioles, aboveground (except in *Marsilea ancylopoda*), attached laterally to stalk apex (attached portion called raphe), tip of stalk often protruding as bump or tooth (proximal tooth), some species also with tooth distal to stalk apex (distal tooth); sporocarps densely to sparsely hairy, less so with age, dehiscing into 2 valves.

Ca. 45 species (6 in the flora with 5 native, 1 introduced): nearly worldwide.

Species identification is virtually impossible without fertile material. The common name water-clover refers to the resemblance of the leaves to those of clover (*Trifolium* spp., Fabaceae); pepperwort refers to the sporocarp, which approximates a peppercorn in size and shape.

SELECTED REFERENCES Johnson, D. M. 1985. New records for longevity of *Marsilea* sporocarps. Amer. Fern J. 75: 30–31. Johnson, D. M. 1988. Proposal to conserve *Marsilea* L. (Pteridophyta: Marsileaceae) with *Marsilea quadrifolia* as *typ. conserv.* Taxon 37: 483–486.

1. Roots both at nodes and sparsely (1–3) along internodes; stalks of sporocarps frequently branched. . . . . 1. *Marsilea quadrifolia*
1. Roots only at nodes; stalks of sporocarps branched or unbranched.
   2. Distal tooth of sporocarps 0.4–1.2 mm, acute. . . . . 6. *Marsilea vestita*
   2. Distal tooth of sporocarps absent or to 0.4 mm and blunt.
      3. Sporocarps 6–9 mm, strongly ascending; stalks usually branched. . . . . 5. *Marsilea macropoda*
      3. Sporocarps 2.4–6 mm, perpendicular to strongly nodding, i.e., stalk curved or bent; stalks unbranched.
         4. Stalks recurved or prostrate, often hooked again at base of raphe; sporocarps borne underground or below stem level. . . . . 2. *Marsilea ancylopoda*
         4. Stalks erect, never hooked at base of raphe; sporocarps borne aboveground or above stem level.
            5. Sporocarps 2–3 mm wide, proximal tooth of sporocarp 0.2 mm or absent. . . . . 4. *Marsilea mollis*
            5. Sporocarps 3.6–4 mm wide, proximal tooth of sporocarp 0.2–0.6 mm and curved away from sporocarp. . . . . 3. *Marsilea oligospora*

1. Marsilea quadrifolia Linnaeus, Sp. Pl. 2: 1099. 1753

Plants forming diffuse clones. **Roots** arising at nodes and 1–3 on internodes. **Petioles** 5.4–16.5 cm, sparsely pubescent to glabrous. **Pinnae** 7–21 × 6–19 mm, sparsely pubescent to glabrous. **Sporocarp stalks** ascending, frequently branched, attached 1–12 mm above base of petiole; unbranched stalks or ultimate branches of stalks 3–16 mm; common trunk of branched stalks 1–4 mm (rarely 2–3 unbranched stalks attached separately to same petiole). **Sporocarps** perpendicular to ascending, 4–5.6 × 3–4 mm, 2.3–2.8 mm thick, rounded, oval, or elliptic in lateral view, pubescent but soon glabrate; raphe 1.4–1.9 mm, proximal tooth usually absent, distal tooth absent or 0.1–0.2 mm. **Sori** 10–17.

Sporocarps produced summer–fall (Jun–Oct). On mud and in shallow water; introduced; Ont.; Conn., Del., Ill., Ind., Iowa, Ky., Maine, Md., Mass., Mich., Mo., N.H., N.J., N.Y., Ohio, Pa., R.I., Vt.; se Europe; Asia.

Marsilea quadrifolia was introduced in Connecticut about 1860. Many of the localities from which it is

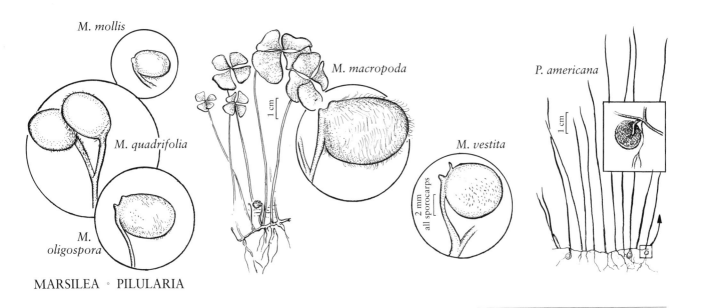

M. mollis

M. quadrifolia

M. macropoda

M. vestita

M. oligospora

P. americana

MARSILEA · PILULARIA

known at present are artificial bodies of water. This may indicate intentional introduction of the plant as a curiosity.

Because its leaves are glabrous to essentially glabrous, *Marsilea quadrifolia* is unlikely to be confused with any other *Marsilea* in the flora. Likewise, the petioles of the land leaves in this species tend to be procumbent rather than stiffly erect as in the others. The branched sporocarp stalks found in *M. quadrifolia* are found elsewhere only in *M. macropoda*; the latter, however, is a hairy plant and has no distal tooth on the very large sporocarp.

Marsilea minuta Linnaeus, a widespread species in the paleotropics, has recently been collected from the Florida Panhandle. It resembles *M. quadrifolia* in having roots both at the nodes and on the internodes and in having relatively glabrous land leaves, but it has sporocarps that are only 1.3–1.7 mm thick, with a distal tooth 0.3–0.6 mm long. *Marsilea minuta* also has a tendency for the terminal margins of the land leaves to be crenate rather than entire.

2. **Marsilea ancylopoda** A. Braun, Monatsber. Königl. Preuss. Akad. Wiss. Berlin 1863: 434. 1864

Plants forming dense clones. **Roots** arising at nodes. **Petioles** 1–18 cm, sparsely pubescent. **Pinnae** 2–17 × 1–16 mm, sparsely pubescent to glabrous. **Sporocarp stalks** recurved or prostrate, unbranched, attached at base of petiole, often hooked near apex, 3–11 mm. **Sporocarps** perpendicular to strongly nodding, underground or below stem level (all other species have sporocarps above stem level), 4–6 × 2.5–5 mm, 2.1–3.2 mm thick, rectangular to round in lateral view, covered with pelt of shaggy hairs but eventually glabrate; raphe 0.8–1.4 mm, proximal tooth 0.2 mm or absent, distal tooth absent or a broad bump 0.1 mm. **Sori** 14–22.

Borders of temporary ponds and ditches; Fla.; Mexico; West Indies; Central America; South America.

Marsilea ancylopoda is not currently known to be extant in the flora; it was known at least at the turn of the century from a single locality in peninsular Florida. Other collections from Florida are *M. vestita*, are of questionable identity or have not been seen by the author. The subterranean sporocarps of *M. ancylopoda* are unique among the New World species of the genus.

3. **Marsilea oligospora** Goodding, Bot. Gaz. 33: 66. 1902

Plants forming dense clones. **Roots** arising at nodes. **Petioles** 3.1–5.7 cm, sparsely pubescent. **Pinnae** 6–15 × 5–14 mm, pilose. **Sporocarp stalks** erect, unbranched, attached at base of petiole, sometimes bent near apex, 5–10 mm. **Sporocarps** nodding, 5–6 × 3.6–4 mm, 1.5 mm thick, ovate in lateral view, covered with long straight hairs when young but eventually glabrate; raphe 0.8–1 mm, proximal tooth 0.2–0.6 mm, curved away from sporocarp, distal tooth up to 0.4 mm, broad, blunt or absent. **Sori** 14–20.

Sporocarps produced summer–fall (Jun–Oct). Around

ponds and marshes, in wet depressions in sagebrush and less commonly on river margins; 700–2400 m; Calif., Idaho, Mont., Nev., Oreg., Utah, Wash., Wyo.

Marsilea oligospora recently has been resegregated from *M. vestita* (D. M. Johnson 1986), from which it differs consistently in its nodding sporocarps that lack a pronounced distal tooth and its pilose leaves and stems. Where their ranges overlap, *M. oligospora* also has longer sporocarp stalks than does *M. vestita*. Plants of this species were recently grown from spores 100 years old (D. M. Johnson 1985).

4. Marsilea mollis B. L. Robinson & Fernald, Proc. Amer. Acad. Arts 30: 123. 1895

Plants forming dense clones. **Roots** arising at nodes. **Petioles** 1–14 cm, sparsely erect-pilose to glabrous. **Pinnae** 2–17 × 1–16 mm, densely pilose abaxially, sparsely pilose adaxially. **Sporocarp stalks** erect, unbranched, attached at base of petiole, or rarely up to 3.8 mm above it, not hooked at apex, 1.7–6.7 mm. **Sporocarps** nodding or perpendicular, 2.4–5 × 2–3 mm, 1.3–1.7 mm thick, ovate in lateral view, covered with needlelike hairs with spreading tips when young but soon glabrate; raphe 0.6–1.4 mm, proximal tooth 0.2 mm, blunt, or absent, distal tooth 0.2 mm or absent. **Sori** 10–14.

Sporocarps produced spring–fall (May–Oct). On mud and in shallow water, in ponds and marshes; 1300–2000 m; Ariz., Tex.; Mexico; South America.

The name *Marsilea mexicana* A. Braun has sometimes been misapplied to this species.

The red or brown streaks on the pinnae, reported as characteristic of *Marsilea mollis* by a number of authors, are found on floating leaves of nearly all species in the genus (D. M. Johnson 1986).

5. Marsilea macropoda Engelmann ex A. Braun in Kunze, Amer. J. Sci. Arts, ser. 2, 6: 88. 1848

Plants forming dense clones. **Roots** arising at nodes. **Petioles** 5–39 cm, hairy. **Pinnae** 9–35 × 8–39 mm, conspicuously white hairy abaxially, hairy adaxially; hairs often exceeding margin to make pinnae appear white margined from above. **Sporocarp stalks** erect, frequently branched, attached 2–4.5 mm above base of petiole; unbranched stalks or ultimate branches of stalks 8–17 mm; common trunk of branched stalks 1–5 mm (rarely 2–3 unbranched stalks attached separately to same petiole).

Sporocarps strongly ascending, 6–9 × 4.5–5.5 mm, 2 mm thick, elliptic or quadrate in lateral view, covered with matted or twisted hairs; raphe 1.6–2.4 mm, proximal tooth 0.4–0.5 mm, blunt, distal tooth absent or present as slightly raised area. **Sori** 19–23.

Sporocarps produced nearly year-round (Feb–Nov). On mud and in shallow water, locally abundant in clay soils of ditches, ponds, marshes; 0–400 m; Ala., La., Tex.; Mexico.

Marsilea macropoda is introduced in Alabama and Louisiana.

Marsilea macropoda, with its tall, hairy leaves and branched sporocarp stalks, is a striking and distinctive species, worthy of cultivation. It is sympatric with *M. vestita* over much of its range, and individuals that appear to be hybrids have been collected in Blanco, Brazos, and Kleberg counties in Texas. These plants combine the large hairy leaves of *M. macropoda* with the shorter-stalked sporocarps bearing a superior tooth found in *M. vestita*.

6. Marsilea vestita Hooker & Greville, Icon. Filic. 2: plate 159. 1830

Marsilea fournieri C. Christensen; *M. mucronata* A. Braun; *M. tenuifolia* Engelmann ex A. Braun; *M. uncinata* A. Braun; *M. vestita* subsp. *tenuifolia* (Engelmann ex A. Braun) D. M. Johnson

Plants forming diffuse or dense clones. **Roots** arising at nodes. **Petioles** 2–20 cm, sparsely pubescent. **Pinnae** 4–19 × 4–16 mm, pubescent to glabrous. **Sporocarp stalks** erect, unbranched, attached at base of petiole (occasionally up to 3 mm above it), not hooked at apex, 0.5–25 mm. **Sporocarps** perpendicular or slightly nodding, 3.6–7.6 × 3–6.5 mm, 1.5–2 mm thick, elliptic to nearly round in lateral view, pubescent but soon glabrate, scars left by fallen trichomes often appearing as purple or brown specks; raphe 1.1–1.7 mm, proximal tooth 0.3–0.6 mm, blunt, distal tooth 0.4–1.2 mm, acute, often hooked at apex. **Sori** 14–22.

Sporocarps produced spring–fall (Apr–Oct). Widespread and variable; in ponds and wet depressions and on river floodplains; 0–2300 m; Alta., B.C., Sask.; Ariz., Ark., Calif., Colo., Idaho, Iowa, Kans., La., Minn., Mont., Nebr., Nev., N.Mex., N.Dak., Okla., Oreg., S.Dak., Tex., Utah, Wash., Wyo.; Mexico; South America in Peru.

A number of segregate species have been named and recognized in regional floras in North America: *Marsilea mucronata* A. Braun (less hairy, found east of Rocky Mountains), *M. uncinata* (glabrous, sporocarp stalks long, distal tooth of sporocarp hooked, south central

United States), *M. tenuifolia* (pinnae very narrow, central Texas), and *M. fournieri* (small plants and pinnae, southwest). The features upon which these species are based intergrade into one another. The species are therefore best treated as conspecific with *M. vestita* (D. M. Johnson 1986).

Putative hybrids between *Marsilea macropoda* and this species are discussed under the former.

2. PILULARIA Linnaeus, Sp. Pl. 2: 1100. 1753; Gen. Pl. ed. 5, 486. 1754 · Pillwort

[Latin *pilula*, a little ball, in reference to the spheric sporocarps]

Plants aquatic, commonly entirely under water and infrequently persisting on bare mud. **Roots** arising at nodes. **Leaves** filiform, terete, lacking expanded blades. **Sporocarps** subterranean, borne on unbranched stalks arising at bases of petioles, attached laterally to stalk apex (attached portion called raphe), lacking teeth; sporocarps densely to sparsely hairy, less so with age, dehiscing into 4 valves.

Species 6 (1 in the flora): North America, South America, Europe, Pacific Islands in New Zealand, Australia.

SELECTED REFERENCES Dennis, W. M. and D. H. Webb. 1981. The distribution of *Pilularia americana* A. Br. (Marsileaceae) in North America, north of Mexico. Sida 9: 19–24. Petrik-Ott, A. J. and F. D. Ott. 1980. *Pilularia americana* new to Tennessee. Amer. Fern J. 70: 29–30.

1. Pilularia americana A. Braun, Monatsber. Königl. Preuss. Akad. Wiss. Berlin 1863: 435. 1864 · American pillwort

Plants forming dense clones. **Leaves** 1.6–10.2 cm, sparsely pubescent. **Sporocarp stalks** attached at base of leaf, 1–3 mm. **Sporocarps** globose, 1.6–2.7 mm diam., covered with matted hairs until mature; raphe minute, teeth lacking. **Sori** 4.

Sporocarps produced spring–fall (Apr–Oct). In shallow water of ponds and temporary pools and on reservoir margins; 50–600 m; Ala., Ark., Calif., Ga., Kans., Mo., Nebr., N.C., Okla., Oreg., S.C., Tenn., Tex.; Mexico in Baja California.

Pilularia americana also has been reported from Alaska. I have seen no vouchers from Alaska, nor have I seen the Oregon vouchers. Because of its grasslike appearance and subterranean sporocarps, *P. americana* is probably overlooked and more common than records indicate. The telltale circinate vernation of the leaves is the best characteristic for distinguishing it from similar plants.

Pilularia caroliniana A. Braun, an invalid name, has been used for this species and may appear on specimens.

24. SALVINIACEAE Reichenbach

• Floating Fern or Water Spangle Family

Clifton E. Nauman

Plants small, floating aquatics. **Stems** creeping, branched, bearing hairs but no true roots. **Leaves** in whorls of 3, with 2 leaves green, sessile or short-petioled, flat, entire, and floating, 1 leaf finely dissected, petiolate, rootlike, and pendent. **Submerged leaves** bearing sori that are surrounded by basifixed membranous indusia (sporocarps); sporocarps of 2 types, bearing either megasporangia that are few in number (ca. 10), each with single megaspore, or many microsporangia, each with 64 microspores. **Spores** of 2 kinds and sizes, both globose, trilete. **Megagametophytes** and **microgametophytes** protruding through sporangium wall; megagametophytes floating on water surface with archegonia directed downward; microgametophytes remaining fixed to sporangium wall.

Genus 1, species 10 (1 species in the flora): mostly tropical, North America, Mexico, West Indies, Central America, South America, Eurasia, Africa including Madagascar.

1. SALVINIA Séguier, Pl. Veron. 3: 52. 1754 · Floating fern, water spangles [for A. W. Salvini (1633–1729), an Italian botanist]

Stems with many multicellular hairs. **Leaves** horizontally spreading. **Blades** of floating leaves green and pubescent abaxially (on side away from water). **Sporocarps** borne on chainlike or cymelike organs or submerged leaves; sporangia indehiscent, dispersed as units when sporocarps decay.

Species ca. 10 (1 in the flora): mostly tropical, North America, Mexico, West Indies, Central America, South America, Eurasia, Africa including Madagascar.

Leaf development in *Salvinia* is unique. The upper side of the floating leaf, which appears to face the stem axis, is morphologically abaxial (J. G. Croxdale 1978, 1979, 1981).

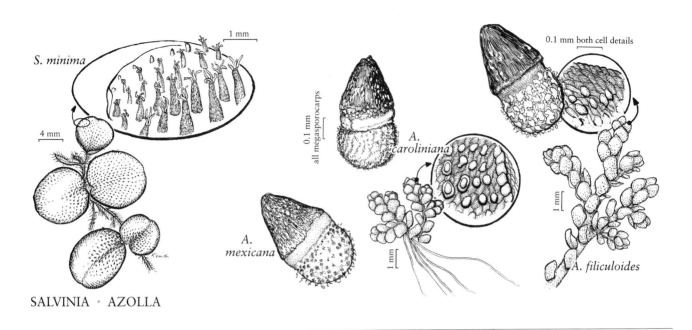

1. Salvinia minima Baker, J. Bot. 24: 98. 1886 · Water spangles, floating fern.

Plants deep green, ± elongate. **Stems** to ca. 6 cm; hairs dark. **Leaves** 1–1.5 cm. **Blades** of floating leaves almost round to elliptic, obtuse or notched at tip, rounded to cordate at base, abaxially (facing upward) with stiff hairs, with 4 separated branches (i.e., not fused at their tips), adaxially (facing into water) brown and pubescent with slender unbranched hairs. **Venation** obscure, areolate, but tips of veins free, ending short of margins. **Sporocarps** in clusters of 4–8, proximal sporocarps with up to 25 megasporangia, distal sporocarps with numerous microsporangia.

Sporulates spring (Apr) and fall (Nov). On still or stagnant waters of ponds, canals, and slow streams; Ala., Fla., Ga., La.; Mexico; West Indies; Central America.

A report of *Salvinia minima* from Minnesota has not been confirmed.

The following names have been variously misapplied to plants in the flora area: *Salvinia auriculata* Aublet; *Salvinia natans* (Linnaeus) Allioni; *Salvinia rotundifolia* Willdenow.

The name *Salvinia auriculata* has been misapplied to species in the United States. *Salvinia auriculata* differs from *S. minima* in the arrangement of the hairs on the abaxial leaf surface. Those in *S. minima* are free, while those of *S. auriculata* (and several other species) are joined at their tips, the hairs resembling an "egg beater." This rather obvious feature is sometimes difficult to assess in herbarium specimens because the hairs are often curled or shriveled into a brownish knot, or have been broken and lost entirely, or never develop. Careful searches must be made to locate intact hairs for identification. Although no previous reports of other species in North America have been verified, these species could be easily overlooked. *Salvinia molesta* D. S. Mitchell, for example, is in cultivation in Columbia County, Florida, and represents a candidate for escape. Species of *Salvinia* are known to escape in various regions (G. R. Proctor 1985).

I have seen only three fertile specimens of *Salvinia minima*, all from Florida, two collected in April and the third collected in November. As with *Azolla*, collectors should make every effort to locate fertile plants.

Material of *Salvinia natans* (Linnaeus) Allioni has been misidentified as *S. rotundifolia* (M. L. Fernald 1950), and as such, might be mistakenly attributed to *S. minima*. In North America, the name *S. natans* appears to have been applied only to an 1886 collection from Perry County, Missouri (J. A. Steyermark 1963). This application may have been correct because the collection is outside the greater part of the range of *S. minima*, at a latitude consistent with *S. natans* from Europe and Asia, and the population was reportedly an escape from cultivation (J. A. Steyermark 1963).

25. AZOLLACEAE Wettstein
• Azolla Family

Thomas A. Lumpkin

Plants aquatic, floating on placid water, occasionally stranded, subsisting on mud; plants heterosporous (producing 2 kinds of spores), leptosporangiate, proliferous by axillary fragmentation. **Roots** translucent to brown, lax, singular [in bundles] without branches, emerging at stem branch points; root hairs to 1 cm, emerging from root cap. **Stems** usually not green, with extensive subdichotomous branching, prostrate and reniform or polyreniform, or nearly erect. **Leaves** sessile, alternate, often imbricate, in 2 ranks along upper side of stem, 0.6–2 mm wide; each leaf with 2 lobes; upper (emersed) lobe greenish or reddish and photosynthetic, with narrow colorless margin, several cells thick, bearing colony of blue-green algae (*Anabaena*) in ovoid cavity at base of lower side; lower lobe often floating or immersed, slightly larger than upper lobe, mostly not green (often colorless and translucent), 1 cell thick except at base, ± cup-shaped. **Sporocarps** in pairs [tetrads] at base of lateral branches, members of pair of same sex or of different sexes. **Megasporocarps** containing 1 megasporangium that produces 1 functional megaspore. **Megaspore** spheric, 0.2–0.6 mm, topped with dark, conic, slightly narrower structure (indusium) covering 3 [9] floats and a blue-green algal colony. **Microsporocarps** globose, apically umbonate, 10–27 μm diam., containing to 130 microsporangia; microsporangia containing 32 or 64 microspores 3 μm diam., aggregated into 3–10 masses covered with arrowlike barbs [glabrous or with needlelike hairs on 1 side].

Genus 1, species ca. 7 (3 species in the flora): worldwide, tropical to temperate regions.

Azollaceae has been included in Salviniaceae, but the relationship is not close.

In this treatment, "upper lobe" refers to the emersed lobe and "lower lobe" refers to the immersed lobe. Developmentally, the emersed lobe is actually abaxial and the immersed lobe is adaxial. To facilitate identification in the field, however, the terms describe the appearance of the lobes to the viewer, not the development of the lobes.

SELECTED REFERENCES Christensen, C. 1938. Azollaceae. In: F. Verdoorn, ed. 1938. Manual of Pteridology. The Hague. P. 550. Reed, C. F. 1954. Index Marsileata et Salviniata. Bol. Soc. Brot., ser. 2a, 28: 1–61.

1. AZOLLA Lamarck in Lamarck et al., Encycl. 1: 343. 1783 · [Greek *azo*, to dry, and *ollyo*, to kill, alluding to death from drought]

Roots 3–5 cm. **Stems** prostrate, 1–3 cm, or nearly erect, 3–5 cm, hairs absent. **Leaves** with 1(–2)-celled hairs on upper surface of upper lobe. **Sporocarps** in pairs. **Megasporocarp** megaspore with 3 floats. **Microsporocarp** masses entirely covered with arrowlike barbs. $x = 22$.

Species ca. 7 (3 in the flora): tropical to temperate regions.

Azolla is divided into sect. *Azolla* and sect. *Rhizosperma* (Meyen) Mettenius, which are sometimes recognized as subgenera. New World species belong to sect. *Azolla* and differ from sect. *Rhizosperma* by having 3 floats per megasporocarp (fig. 25.1), subdichotomous branching, and straight barbs on microsporangial masses (rather than 9 floats per megasporocarp, pinnate branching, and needlelike hairs or hairs absent on microsporangial masses). The genus was more diverse in past geologic ages: 30 species are known from the Cretaceous (J. W. Hall 1974).

The species of *Azolla* are difficult taxonomically because (1) about 80% of the specimens lack sori, which are necessary for identification, and (2) the characteristics needed to identify the species are difficult to observe. A scanning electron microscope is needed to see sculpturing of the megaspores, and a light microscope is needed to see the number of cells per hair on the upper leaf lobe. (These hairs are best seen in profile on mature leaves; at least $40 \times$ magnification is needed.)

Previous workers emphasized the number of septae (internal partitions) in the barbs on the microsporangial masses as a primary differentiating characteristic among species of *Azolla* (e.g., H. K. Svensen 1944). This character is not constant, however, either within a species or within an individual (R. K. Godfrey et al. 1961; L. V. Hill and B. Gopal 1967; K. Seto and T. Nasu 1975). Because nearly all floristic work in North America since the 1940s has been based on Svenson's synoptic treatment, the identity of most specimens is questionable, and therefore ranges are imprecisely known.

The maps for this treatment are tentative. For the most part, I have noted occurrences only in those states or regions from which specimens have been identified using characteristics given in this key. Literature that attributes a particular species of *Azolla* to a particular state or province must be questioned because the specimens were presumably identified using the inconsistent glochidial characteristics given by H. K. Svenson (1944). More work is needed to determine the distribution of *Azolla* species in North America.

Agriculturally, *Azolla* is famous for its symbiosis with the nitrogen-fixing *Anabaena azollae* Strasburger, a cyanobacterium (blue-green alga) found at the stem apices, beneath indusia, and in cavities of the upper leaf lobes. Because the plants fix nitrogen, they are often used as a green fertilizer or mixed with livestock feed as a nutritional supplement. *Azolla pinnata* has been cultivated for many centuries in rice paddies of northern Vietnam and southeastern China, where it acts as a fertilizer after it decomposes.

Azolla is the most frequently studied genus of ferns in the world because of its economic importance. The three North American species are naturalized in Europe and South Africa, and they have been introduced into Hawaii for horticulture and into Asia for agriculture. All species have been studied for agricultural uses in rice-producing areas.

Azolla is usually found in stagnant or slow-moving water of ponds, lakes, marshes, swamps, and streams. Plants turn reddish when under stress, such as from poor nutrition, salinity, or high temperatures. Sporulation needs further investigation.

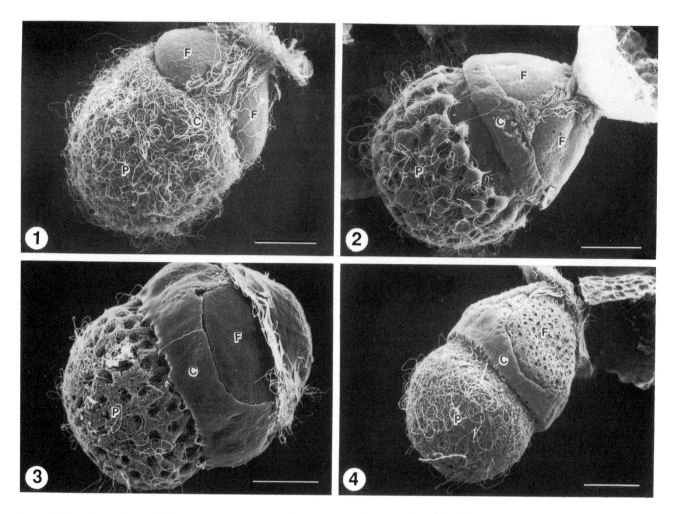

Figure 25.1. Comparison of the megaspore apparatus from four modern species of *Azolla*. Structures that can be seen include floats (F), collar (C), and perispore or perine (P), which covers the megaspore. 1. *Azolla caroliniana*. The collar and perispore are covered by dense filaments. 2. *Azolla filiculoides*. Two of three floats are visible overlying a glabrous collar. Clusters of filaments decorate the perispore. 3. *Azolla mexicana*. Hairlike filaments are visible on surface of perispore underlying a prominent glabrous collar. 4. *Azolla microphylla*. Note glabrous collar and uniform covering of filaments on surface of perispore. Scale bar = 100 μm. (From S. K. Perkins et al. 1985, p. 1724. Reprinted with permission from SEM, Inc.)

SELECTED REFERENCES Hall, J. W. 1974. Cretaceous Salviniaceae. Ann. Missouri Bot. Gard. 61: 354–367. Lumpkin, T. A. and D. L. Plucknett. 1982b. Botany and ecology. In: T. A. Lumpkin and D. L. Plucknett, eds. 1982. *Azolla* as a Green Manure: Use and Management in Crop Production. Boulder. Pp. 15–38. Moore, A. W. 1969. *Azolla*: Biology and agronomic significance. Bot. Rev. (Lancaster) 35: 17–35. Svenson, H. K. 1944. The New World species of *Azolla*. Amer. Fern J. 34: 69–84. Van Hove, C., T. de Waha Baillonville, H. F. Diara, P. Godard, Y. Mai Kodomi, and N. Sanginga. 1987. *Azolla* collection and selection. In: International Rice Research Institute. 1987. *Azolla* Utilization. Proceedings of the Workshop on *Azolla* Use. Fuzhou, Fujian, China, 31 March–5 April 1985. Los Banos, Laguna, Philippines. Pp. 77–87. Zimmerman, W. J., T. A. Lumpkin, and I. Watanabe. 1989. Classification of *Azolla* spp., section *Azolla*. Euphytica 43: 223–232.

1. Largest hairs on upper leaf lobe unicellular; megaspores warty with raised angular bumps. . . . . 3. *Azolla filiculoides*
1. Largest hairs on upper leaf lobe with 2 or more cells; megaspores not covered with raised angular bumps.
  2. Megaspores not pitted, densely covered with tangled filaments (filosum). . . . . . . 1. *Azolla caroliniana*
  2. Megaspores pitted, sparsely covered with a few long filaments (filosum). . . . . 2. *Azolla mexicana*

1. Azolla caroliniana Willdenow, Sp. Pl. 5(1): 541. 1810

Plants dark green or with margins of bright crimson or whole plants dark red, free-floating or forming multilayer mat to 4 cm thick under good conditions; plants infrequently fertile. **Stems** prostrate, 0.5–1 cm. **Largest hairs** on upper leaf lobe near stem with 2 or more cells; broad pedicel cell often 1/2 or more height of hair, apical cell curved, with tip nearly parallel to leaf surface. **Megaspores** without raised angular bumps or pits, densely and uniformly covered with tangled filaments.

Stagnant or slow-moving water in ponds, lakes, marshes, swamps, and streams; Ala., Ark., Conn., Del., Fla., Ga., Ill., Ind., Kans., Ky., La., Md., Mass., Mich., Minn., Miss., Mo., Nebr., N.J., N.Y., N.C., Ohio, Okla., Pa., R.I., S.C., S.Dak., Tenn., Tex., Va., W.Va., Wis.; Mexico; West Indies; South America; Europe; Asia.

The sporophyte of *Azolla caroliniana* commonly survives throughout the year in temperate areas (with hard frosts and prolonged ice cover). It is the best adapted of all species for subsistence on mud. *Azolla caroliniana* is rarely collected with sporocarps.

2. Azolla mexicana C. Presl, Abh. Königl. Böhm. Ges. Wiss., ser. 5, 3: 150. 1845

Plants green or often blue-green to dark red, some red-fringed leaves usually present in nature, free-floating or forming a multilayer mat to 4 cm thick in early summer; plants frequently fertile. **Stems** prostrate, 1–1.5 cm. **Largest hairs** on upper leaf lobe near stem 2(–3)-celled; broad pedicel cell often 1/2 or more height of hair, apical cell curved, with tip nearly parallel to leaf surface. **Megaspores** not covered with raised angular bumps, pitted and sparsely covered with a few long filaments extending over surface.

Stagnant or slow-moving waters; B.C.; Ariz., Ark., Calif., Colo., Idaho, Ill., Iowa, Kans., Minn., Mo., Nebr., Nev., N.Mex., Okla., Oreg., Tex., Utah, Wash., Wis.; Mexico; Central America; South America.

Azolla mexicana is generally less cold tolerant and has a narrower environmental range than *A. caroliniana*. Both species are closely related and are similar vegetatively in culture. In the western United States, *A. mexicana* is often fertile. Distribution in the Great Plains area is tentative and needs further study. In the eastern United States, *A. mexicana* may have been occasionally introduced.

3. Azolla filiculoides Lamarck in Lamarck et al., Encycl. 1: 343. 1783

Plants green to yellowish green or dark red, with 2 growth stages; plants fertile only in mature stage, generally in late spring. **Stems** prostrate when immature, 1–3 cm, internodes elongate to 5 mm, becoming nearly erect to 5 cm or more when mature and crowded. **Hairs** on upper leaf lobes strictly unicellular. **Megaspores** warty with raised angular bumps, each with a tangle of filaments.

Stagnant and slow-moving waters; B.C.; Ariz., Calif., Oreg., Wash.; Mexico; Central America; South America; Europe; ne Asia; s Africa; Pacific Islands in Hawaii.

Azolla filiculoides is cold tolerant, surviving even in fragmented parts under thin ice. It usually reaches a climax population in late spring, becomes fertile, collapses, and is replaced by other more heat-tolerant aquatics such as *Lemna* spp. Hybrids between this species (male) and *A. microphylla* Kaulfuss (female), a species of Central America, South America, and the West Indies, have been reported (Do V. C. et al. 1989). V. M. Bates and E. T. Browne (1981) reported *A. filiculoides* from Georgia, far removed from its main range in western North America. The most likely explanation is that the plants represent escapes from horticulture.

SELECTED REFERENCES Bates, V. M. and E. T. Browne. 1981. *Azolla filiculoides* new to the southeastern United States. Amer. Fern J. 71: 33–34. Duncan, R. E. 1940. The cytology of sporangium development in *Azolla filiculoides*. Bull. Torrey Bot. Club 67: 391–412.

GYMNOSPERMS

Key to Gymnosperm Families

James E. Eckenwalder

John W. Thieret

In the key below, the term "needlelike" as applied to gymnosperms in the flora is descriptive of those leaves that are evergreen (except in *Larix* and *Taxodium*), more or less linear, and either more or less flattened (e.g., *Abies*) or not (e.g., *Pinus*).

1. Leaves pinnately compound, leaflets linear; dioecious; stem below ground, or its leaf-bearing apex exposed; Florida and Georgia. . . . . 1. *Zamiaceae*, p. 347
1. Leaves simple, fan-shaped, needlelike, or scalelike; dioecious or monoecious; stem above-ground; widespread.
    2. Leaves deciduous, fan-shaped, dichotomously veined; ovules 2 at ends of long stalks, often only 1 maturing. . . . . 2. *Ginkgoaceae*, p. 350
    2. Leaves mostly evergreen (deciduous in *Larix* and usually in *Taxodium*, often ephemeral and not photosynthetic in *Ephedra*), needlelike or scalelike, not dichotomously veined; ovules (and seeds) 1–20 on cone scales or 1 on short stalks.
        3. Longest internodes 2–10 cm (shorter at bases of branches); pollen cones compound, each sporangiophore subtended by a pair of bracteoles; seeds inserted on axis of seed cone, subtended by several pairs or whorls of papery, herbaceous, or fleshy bracts; shrubs (sometimes clambering), neither resinous nor fragrant; dry areas in w United States (se Texas to Oregon and California). . . . . 6. *Ephedraceae*, p. 428
        3. Longest internodes 0–1 cm; pollen cones simple, the individual sporophylls not bracteate (usually few sterile bracts at base of cone); seeds inserted on scales of seed cone, scales woody (fleshy and coalescent in *Juniperus*), or the seeds solitary and not enclosed by bracts or scales in Taxaceae; shrubs or trees, usually resinous and fragrant; widespread.
            4. Seed 1, subtended by but not concealed by inconspicuous sterile bracts, with fleshy or juicy arils; leaves needlelike, spreading in 1 plane ("2-ranked") (except on erect shoots), without resin canals. . . . . 5. *Taxaceae*, p. 423
            4. Seeds 1–400 in woody or fleshy cones, enclosed by conspicuous cone-scales, without arils (if cones fleshy then leaves not spreading in 1 plane); leaves needle-like or scalelike, spreading in 1 plane or not, with or without resin canals.

5. Foliage leaves needlelike, alternate or fascicled, individually abcising from branchlets when shed (except that fascicles in *Pinus* are shed as units); cone scales imbricate, seeds 2 per scale. 3. *Pinaceae*, p. 352
5. Foliage leaves needlelike or scalelike, alternate, opposite, or whorled, persistent on branchlets (but most branchlets shed with age); cone scales valvate or imbricate (if imbricate then leaves opposite and scalelike), seeds 1–20 per scale. . . . . 4. *Cupressaceae*, p. 399

1. ZAMIACEAE Horianow

• Sago-palm Family

Garrie P. Landry

Plants superficially palmlike or fernlike, perennial, evergreen, dioecious. **Stems** subterranean with exposed apex or aboveground, fleshy, stout, cylindric, simple or irregularly branched. **Roots** with small secondary roots; coral-like roots developing at base of stem at or below soil surface. **Leaves** pinnately compound, spirally clustered at stem apex, leathery, petiole and rachis unarmed [with stout spines]; leaflets entire or dentate [spinose], venation dichotomous [netted]; resin canals absent. **Cones** axillary, appearing terminal, short-peduncled [sessile], disintegrating at maturity; sporophylls densely crowded, spirally arranged, often covered with indument. **Pollen cones** soon shed, generally smaller and more numerous than seed cones; sporophylls bearing many crowded, small microsporangia (pollen sacs) adaxially; pollen spheric, not winged. **Seed cones** persistent for a year or more, 1(–2) per plant, nearly globose to ovoid, tapering sharply or blunt at apex; sporophylls peltate, thickened and laterally expanded distally, bearing 2(–3) ovules. **Seeds** angular, inner coat hardened, outer coat fleshy, often brightly colored; cotyledons 2.

Genera 9, species ca. 100 (1 genus, 1 species in the flora): primarily tropical to warm temperate regions, North America, Central America, South America, Africa, Australia.

SELECTED REFERENCES Candolle, A. L. P. de. 1868. Cycadaceae. In: A. P. de Candolle and A. L. P. de Candolle, eds. 1823–1873. Prodromus Systematis Naturalis Regni Vegetabilis.... Paris etc. Vol. 16, part 2, pp. 522–547. Johnson, L. A. S. 1959. The families of cycads and the Zamiaceae of Australia. Proc. Linn. Soc. New South Wales 84: 64–117. Schuster, J. 1932. Cycadaceae. In: H. G. A. Engler, ed. 1900–1953. Das Pflanzenreich.... Berlin. Vol. 99[IV,1], pp. l–168.

1. ZAMIA Linnaeus, Sp. Pl. ed. 2, 2: 659. 1762 · [Derivation equivocal, perhaps from misreading of Latin *azania,* a kind of pine cone, or from Latin *zamia,* loss, from the "sterile appearance" of the pollen cones]

Stems often branched, subterranean to aboveground. **Leaves** broadly oblong-elliptic; leaflets entire to coarsely dentate, without midribs, venation dichotomous but appearing parallel. **Cones** distinctly peduncled. **Pollen cones** more slender than seed cones. $x = 8$.

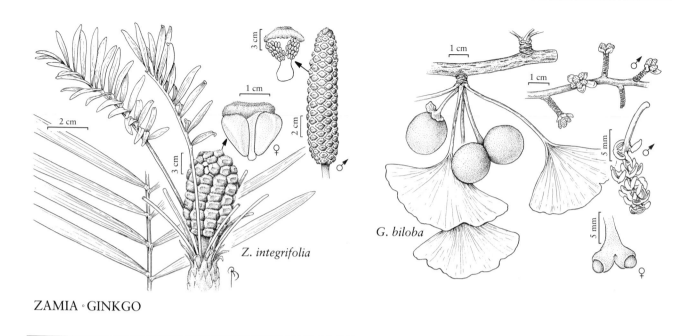

Z. integrifolia

G. biloba

ZAMIA · GINKGO

Species ca. 30 (1 in the flora): subtropics and tropics, North America, Mexico, West Indies, Central America, South America.

SELECTED REFERENCES   Duncan, W. H. 1979. *Zamia* (Cycadaceae) new for Georgia. Sida 8: 115–116. Eckenwalder, J. E. 1980. Dispersal of the West Indian cycad *Zamia pumila* L. Biotropica 12: 79–80. Eckenwalder, J. E. 1980b. Taxonomy of the West Indian cycads. J. Arnold Arbor. 61: 701–722. Hardin, J. W. 1971. Studies of the southeastern United States flora. II. The gymnosperms. J. Elisha Mitchell Sci. Soc. 87: 43–50. Kral, R. 1983b. Cycadaceae: *Zamia integrifolia* Ait. In: R. Kral, ed. 1983. A Report on Some Rare, Threatened, or Endangered Forest-related Vascular Plants of the South. Washington. Pp. 20–23. [U.S.D.A., Techn. Publ. R8-TP 2.] Landry, G. P. 1980. The Ecology and Variation of *Zamia pumila* L. in Florida. M.S. thesis. Louisiana State University. Stevenson, D. W. 1987. Again the West Indian zamias. Fairchild Trop. Gard. Bull. 42(3): 23–27. Stevenson, D. W. 1987b. Comments on character distribution, taxonomy and nomenclature of the genus *Zamia* L. in the West Indies and Mexico. Encephalartos 9: 3–7. Ward, D. B., ed. N.d. Rare and Endangered Biota of Florida. Vol. 5. Plants. Gainesville. Pp. 122–124.

1. Zamia integrifolia Linnaeus f. in Aiton, Hort. Kew. 3: 478. 1789 · Coontie, Florida arrowroot, conti hateka (Seminole)

Zamia floridana A. de Candolle; *Z. silvicola* Small; *Z. umbrosa* Small
Stem subterranean, or leaf-bearing apex exposed. **Leaves** 2–10 dm; petiole unarmed; leaflets 6–17 cm × 2–18 mm, linear, often twisted, very stiff, dark glossy green, 7–23-veined; margins often revolute, entire or with small teeth to slight denticulations near apex. **Pollen cones** generally 2–5 per plant, narrowly cylindric, 5–16 cm, tapering slightly at apex. **Seed cones** cylindric-ellipsoid, 5–19 cm, blunt at apex; ovules 2 per sporophyll. **Seeds** drupelike, oblong to ovoid, somewhat angular, 1.5–2 cm, outer coat bright orange. $2n = 16$.

Period of receptivity and maturation of seeds December–March. Hammocks, pine-oak woodlands, scrub, and shell mounds; 0–30 m; Fla., Ga.; West Indies.

Once common to locally abundant, *Zamia integrifolia* is becoming increasingly uncommon as its habitats are being destroyed. The species is now considered "endangered" in Florida.

The choice of specific epithet to use for our species follows the conclusion reached by D. W. Stevenson (1987).

Controversy has long existed over the classification of *Zamia* in Florida. Recent researchers, however, have concluded that only one species is present in the flora. The several binomials applied to our *Zamia* reflect variability in plant vigor, leaf shape, leaflet width, number of marginal teeth and veins per leaflet, and geographic distribution. Forms with wide leaflets—"*Zamia umbrosa*"—are restricted to coastal hammocks of northeastern Florida and southeastern Georgia and appear to be quite distinct from plants of the remainder of Florida—*Z. integrifolia* and *"Z. floridana."* Especially robust forms have been described as *"Zamia silvicola."* Studies by D. B. Ward (n.d.) indicate that these features have a genetic basis, but formal recognition of these different phases as species does not lead to better understanding of the complex. The variants in Florida may have originated from introductions of divergent forms of *Zamia* from elsewhere. The starchy stems, after treatment to remove a poisonous principle, were a significant part of aboriginal diets, and the plants were presumably dispersed by aborigines.

Zamia angustifolia Jacquin, a species thought to be restricted to the Bahamas and eastern Cuba, was reported in southern Florida by J. K. Small (1933). No voucher specimens were cited or are known to exist. Small also reported *Zamia pumila* Linnaeus from Florida, although erroneously.

2. GINKGOACEAE Engler

· Maidenhair Tree Family

R. David Whetstone

Trees, deciduous, dioecious (possibly rarely monoecious). **Bark** gray, furrowed, with flattened ridges. **Long shoots** and short (spur) shoots present. **Roots** fibrous to woody. **Leaves** simple, alternate to fascicled, stomates abaxial; apices deeply cleft to truncate; petioles equal to or exceeding blades; resin canals absent. **Pollen cones** borne on spurs, catkinlike; sporophylls distributed along axis; bracts absent; pollen spheric, not winged. **Seed cones** absent; ovules 2, pedunculate, subtended by collar believed to comprise 2 modified megasporophylls, borne on spurs. **Seeds** with fleshy outer coat, inner coat hard; cotyledons 2–3.

Genus 1 (6 + known from fossils), species 1: native to China, cultivated in the flora.

SELECTED REFERENCE Page, C. N. 1990. Coniferophytina: Ginkgoaceae. In: K. Kubitzki et al., eds. 1990 +. The Families and Genera of Vascular Plants. 1 + vol. Berlin etc. Vol. 1, pp. 284–289.

1. GINKGO Linnaeus, Mant. Pl. 2: 313. 1771 · Ginkgo [Chinese *yin*, silver, and *hing*, apricot, in reference to appearance of the seed]

Leaf blades broader than long. **Pollen cones** lax, elongate. **Ovules** often abscising before fertilization. **Seeds** 1–2 per peduncle.

Species 1: widely cultivated in temperate regions worldwide.

The genus is known from fossils that date back nearly 200 million years and are nearly identical to present-day trees.

1. Ginkgo biloba Linnaeus, Mant. Pl. 2: 313. 1771

· Maidenhair tree, ginkgo

Trees to 30 m. **Crown** somewhat ovoid to obovoid, tending to be asymmetric, primary branches ascending at ca. 45° from trunk. **Long shoots** faintly striate; spurs thick, knoblike or to 3 cm, gray, covered with bud-scale scars. **Buds** brown, globose, scales imbricate, margins scarious. **Leaves** fan-shaped, glabrous except for tuft of hairs in axils, blades 2–9.5 × 2–12 cm, mostly 1.5 times wider than long, apices cleft to truncate; venation dichotomous, appearing parallel; leaf scars semicircular; petioles channeled on adaxial surface, 2.5–

8.5 cm. **Seeds** obovoid to ellipsoid, yellow to orange, 2.3–2.7 × 1.9–2.3 cm, mostly 1.1–1.2 times longer than broad, glaucous, rugose, with apical scar, maturing in single season, usually 1 per peduncle, occasionally polyembryonic, outer coat foul-smelling; peduncles orange, glaucous, ridged, 3–9.5 cm, collar broadly elliptic, 7.2–8.6 mm broad. $2n = 24$.

Pollination March–April; seeds shed August–November. Cultivated ground, fencerows, and woods; introduced; possible in various provinces and states; worldwide.

Ginkgo is widely planted as an ornamental. The unusual shape of the crown, natural resistance to disease, and yellow leaf color in fall make this a favorite street and park tree. Ovulate trees produce an abundance of seeds, which have a particularly obnoxious odor; the planting of ovulate ginkgoes is often discouraged for this reason. Seeds (canned with fleshy outer coat removed) are sold in ethnic markets as "silver almonds" or "white nuts," the gametophyte and embryo being edible. Oils from the outer coat are known to cause dermatitis in some humans.

In China *Ginkgo biloba* is either extinct in the wild or drastically restricted in range. The species is reported to occur naturally in remote mountain valleys in China's Zhejiang province (C. N. Page 1990). Persistence of trees planted about dwellings, however, when no trace of the dwellings remains, complicates discerning the status of such trees. Most, if not all, ginkgoes exist only in cultivation.

In the flora area seeds of ginkgo, minus the fleshy outer coat, have been found beneath various species of trees up to 150 m from the nearest seed-producing ginkgo. The dispersal agents were almost certainly birds, possibly crows. A cache of ginkgo seeds, in association with scats of raccoons [*Procyon lotor* (Linnaeus), family Procyonidae], was found in a tree crotch about 50 m from the nearest source of the seeds (J. W. Thieret, pers. comm.). Apparent animal dispersal of ginkgo requires further study.

Seedlings or saplings of ginkgo are very rarely found in the vicinity of planted trees and in fencerows and woods (undocumented reports from Kentucky, New Jersey, New York, Ohio, Pennsylvania, and Virginia), hence the inclusion of the species in the flora. Nevertheless, the species is doubtfully naturalized in North America despite about two centuries of cultivation here.

SELECTED REFERENCE Franklin, A. H. 1959. *Ginkgo biloba* L.: Historical summary and bibliography. Virginia J. Sci., n. s. 10: 131–176.

3. PINACEAE Lindley

• Pine Family

John W. Thieret

Trees (occasionally shrubs), evergreen (annually deciduous in *Larix*), resinous and aromatic, monoecious. **Bark** smooth to scaly or furrowed. **Lateral branches** well developed and similar to leading (long) shoots or reduced to well-defined short (spur) shoots (*Larix, Pinus*); twigs terete, sometimes clothed by persistent primary leaves or leaf bases; longest internodes less than 1 cm; buds conspicuous. **Roots** fibrous to woody, unspecialized. **Leaves** (needles) simple, shed singly (except whole fascicles shed in *Pinus*), alternate and spirally arranged but sometimes proximally twisted so as to appear 1- or 2-ranked, or fascicled, linear to needlelike, sessile to short-petiolate; foliage leaves either borne singly (spirally) on long shoots or in tufts (fascicles) on short shoots; juvenile leaves (when present) borne on long shoots, scalelike; resin canals present. **Pollen cones** maturing and shed annually, solitary or clustered, axillary, ovoid to ellipsoid or cylindric; sporophylls overlapping, bearing 2 abaxial microsporangia (pollen sacs); pollen spheric, 2-winged, less commonly with wings reduced to frill (in *Tsuga* sect. *Tsuga*), or not winged (in *Larix* and *Pseudotsuga*). **Seed cones** maturing and shed in 1–3 seasons or long-persistent, sometimes serotinous (not opening upon maturity but much later: *Pinus*), compound, axillary, solitary or grouped; scales overlapping, free from subtending included or exserted bracts for most of length, spirally arranged, strongly flattened, at maturity relatively thin to strongly thickened and woody (in *Pinus*), with 2 inverted, adaxial ovules. **Seeds** 2 per scale, elongate terminal wing partially decurrent on seed body (wing short or absent in some species of *Pinus*); aril lacking; cotyledons 2–12[–18].

Genera 10, species ca. 200 (6 genera, 66 species in the flora with 64 natives and 2 naturalized): almost entirely in the Northern Hemisphere.

The Pinaceae, with a fossil record extending back to the Cretaceous (C. N. Miller Jr. 1988), constitute a clearly defined natural taxon, the basic delimiting features of which are seen in the mature seed cones: bract-scale complexes consisting of well-developed scales that are free for most of their length from the subtending bracts, two inverted ovules on the adaxial face of each scale, and usually an obvious seed wing that develops from the cone scale. The 10 genera, too, are clearly defined.

352

The cones of certain members of the Pinaceae remain on the tree and closed for several to many years until a stimulus (often fire) causes them to open and shed their seeds. This condition, known as serotiny (adjective, serotinous), is seen in various pines (e.g., *Pinus attenuata, P. banksiana, P. contorta*).

This primarily Northern Hemisphere family extends south to the West Indies, Central America, Japan, China, Indonesia, the Himalayas, and North Africa. The family is dominant in the vegetation of large regions including, in the flora area, forests of the boreal and Pacific regions, of the western mountains, and of the southeastern coastal plain. Only one species of the family, *Pinus merkusii*, crosses the equator (in Sumatra).

Members of the Pinaceae are of major economic importance as producers of most of the world's softwood timber. Additionally, they are sources of pulpwood, naval stores (e.g., tar, pitch, turpentine, etc.), essential oils, and other forest products. All members of the family present in the flora, especially pines, are of varying importance to wildlife for food and cover. Many species, including most of the genera, are grown as ornamentals and shelter-belt trees and for revegetation. Most commonly seen in cultivation in the flora area are species of *Abies, Cedrus, Larix, Picea, Pinus, Pseudotsuga,* and *Tsuga,* each of these genera being represented by numerous cultivars. *Keteleeria* and *Pseudolarix* are mainly botanical garden subjects. *Cathaya,* the most recently described genus (1958), is apparently not yet in cultivation in North America.

Among the vegetative features useful for identification of some genera of Pinaceae are the leaf scars. These are best observed on those portions of living branchlets from which leaves have fallen.

SELECTED REFERENCES Burns, R. M. and B. H. Honkala. 1990. Silvics of North America. 1. Conifers. Washington. [Agric. Handb. 654.] Canadian Forestry Service. 1983. Reproduction of conifers. Forest. Techn. Pub. Canad. Forest. Serv. 31. Farjon, A. 1990. A Bibliography of Conifers. Königstein. [Regnum Veg. 122.] Hosie, R. C. 1969. Native Trees of Canada, ed. 7. Ottawa. Pp. 83–95. Little, E. L. Jr. 1979. Checklist of United States Trees (Native and Naturalized). Washington. Pp. 33–36. [Agric. Handb. 541.] Silba, J. 1986. Encyclopaedia Coniferae. Phytologia Mem. 8: 1–127.

1. Leaves on year-old and older branches borne either in clusters (fascicles) of 2–5(–6), each cluster scaly-sheathed at base at least when young, or in clusters of 10–60 on short (spur) shoots, clusters not scaly-sheathed.
    2. Leaves in clusters of (1–)2–5; scales of seed cones with thickened apical portion (apophysis) bearing terminal or central, scarlike to raised umbo and, often, prickle or claw. . 6. *Pinus,* p. 373
    2. Leaves in clusters of 10–60 on short (spur) shoots; scales of seed cones without apophyses, umbos, and prickles. . . . . 4. *Larix,* p. 366
1. Leaves borne singly along branches, not scaly-sheathed at base or, if so when young, then terete.
    3. Leaves terete, scaly-sheathed at base when young; scales of seed cones with thickened apical portion (apophysis) bearing central, scarlike umbo; mature trees mostly round topped, usually less than 15 m; arid areas in w North America. . . . . 6. *Pinus,* p. 373
    3. Leaves neither terete nor scaly-sheathed; scales of seed cones lacking apophyses and umbos; mature trees of various habit, often conic, often more than 15 m; habitat various, but mostly not of arid areas in w North America.
        4. Twigs conspicuously roughened by decurrent, spreading or appressed, peglike projections that persist after leaves fall.
            5. Leaves sharp-pointed or less often bluntly acute, somewhat flattened or ± square in cross section, sessile; leading shoot erect. . . . . 5. *Picea,* p. 369
            5. Leaves rounded or notched, flattened, abruptly narrowed to petiolelike base; leading shoot typically drooping. . . . . 2. *Tsuga,* p. 362

4. Twigs not roughened by decurrent peglike projections, but, if roughened at all, by ± circular to elliptic leaf scars, these flush with twig surface, slightly depressed, slightly raised evenly all around, or slightly raised on proximal side only.

    6. Leaf scars circular to elliptic, flush with twig surface, slightly depressed, or slightly raised evenly all around; seed cones erect, at least before full maturity, not falling whole but scale by scale, each cone axis persisting as an erect "spike" on branch, the fan-shaped scales often littering ground under tree, bracts protruding beyond scales or not. . . . . 1. *Abies,* p. 354

    6. Leaf scars transversely elliptic, slightly raised on proximal side only, thus tilted; seed cones pendent, falling whole, bracts protruding beyond scales. . . . . 3. *Pseudotsuga,* p. 365

1. ABIES Miller, Gard. Dict. Abr. ed. 4, vol. 1. 1754 · Fir [Latin name of a European fir]

Richard S. Hunt

Trees evergreen, crown usually spirelike to conic, sometimes flat to round topped in age. **Bark** initially thin, smooth, bearing resin blisters, in age furrowed and/or flaking in plates. **Branches** whorled, irregular internodal branches occasionally produced by epicormic sprouting (growing from a dormant bud); short (spur) shoots absent; leaf scars prominent, ± circular to broadly elliptic, flush with twig surface, slightly depressed, or slightly raised evenly all around. **Buds** ovate or oblong, resinous or not, apex rounded or pointed. **Leaves** borne singly, persisting 5 or more years, spirally arranged but often proximally twisted so as to appear either 1-ranked (pointing up like toothbrush bristles) or 2-ranked, sessile, typically constricted and often twisted above the somewhat broadened base, sheath absent; leaves on vegetative branches flattened, frequently grooved adaxially, usually notched to rounded at apex; leaves on fertile branches sometimes appearing 4-sided, upright, sharp-pointed to rounded at apex; resin canals 2. **Cones** borne on year-old twigs. **Pollen cones** grouped, ovate or oblong-cylindric, leaving gall-like protuberances after falling, yellow to red, green, blue, or purple. **Seed cones** maturing in 1 season, erect, ovoid to oblong-cylindric or cylindric, not falling whole but scale by scale, cone axis persisting as an erect "spike" on branch; scales shed individually, fan-shaped, lacking apophysis and umbo; bracts included to exserted. **Seeds** winged, the wing-seed juncture bearing resin sac; cotyledons 4–10. $x = 12$.

Species ca. 42 (11 in the flora): widespread in north temperate regions, North America, Mexico, Central America, Eurasia (s to Himalayas, s China, and Taiwan), n Africa.

In *Abies* several traditionally accepted species have closely allied sibling species, e.g., *A. balsamea–A. fraseri, A. bifolia–A. lasiocarpa,* and *A. magnifica–A. procera.* Other species may be more distinct morphologically, but many of these still appear to have evolved in geographic isolation without strong reproductive barriers developing. Thus, when distributions of species overlap, introgression between the taxa is the rule; this may make it difficult to assign certain individuals to a species. In the interests of nomenclatural stability, I have accepted the taxa recognized by the U.S. Forest Service (E. L. Little Jr. 1979). This classification does not recognize varieties based on variations in bract characteristics but recognizes species that perhaps would be treated as varieties in other conifer genera. The only exceptions to this treatment are some necessary changes within *A. concolor* and *A. lasiocarpa.* Cases of introgression are discussed under the taxa involved. Some distinct or possibly distinct geographic populations deserve further study and may warrant future taxonomic recognition.

Most North American firs are major components of vegetation, especially in the boreal, Pacific Coast coniferous, and western montane coniferous forests, where they are important for watershed management. They are cut for pulpwood and lumber and, largely from planta-tions, for Christmas trees. All our species, especially *Abies concolor,* and several exotics are grown—some more than others—as ornamentals. Firs provide cover, and their leaves are im-portant as food, for various birds and mammals. Species of *Abies* frequently have a pleasant odor; their foliage has been used as a stuffing material for pillows. Most commercial products with "pine odors" are in fact scented with essential oils distilled from *Abies* foliage by Russian farmers. A similar oil could be derived from balsam fir in North America.

Character states used in the key are primarily those of the lowermost (i.e., the most acces-sible) branches.

Notes on the following features, made at the time of collection of specimens, are useful in identification.

Size and placement of resin canals in the leaves as seen in cross section with a handlens when a leaf is pulled apart or cut with a sharp knife. In *Abies balsamea, A. bifolia, A. fraseri,* and *A. lasiocarpa* the canals are ± median, placed between the abaxial epidermis and adaxial epidermis (sometimes closer to the abaxial) and in from the leaf margins; they are "large," i.e., up to about one-fourth as wide as the leaf is thick midway between the midvein and margins (each is like a tiny "eye" on each side of the midvein). In our other firs they are placed just above the abaxial margin and are "small," i.e., about one-fifth or less as wide as the leaf is thick.

Stance of the leaves, e.g., whether they are in flat sprays ("2-ranked") or point up like brush bristles ("1-ranked"), and whether some on a twig point in a direction different from others on the same twig.

Differences in color and glaucousness of the abaxial and adaxial leaf surfaces.

Shape of leaf apex as observed with a handlens.

Distribution of stomates—and number of rows of stomates—on the abaxial and adaxial leaf surfaces, particularly midway between base and apex of leaf.

Leaf-scar periderm color. Pull a leaf from a twig and note, with a handlens, the color of the scar's periphery.

Presence or absence of resin on the buds (collect a few extra buds for dissection). If buds are not available (as in the early part of the growing season), collect older branch material bearing old bud scales.

Cone color of both pollen and seed cones (binoculars are handy to note this feature of the seed cones).

SELECTED REFERENCES Liu, T. S. 1971. A Monograph of the Genus *Abies.* Taipei. Matzenko, A. E. 1968. Conspectus generis *Abies* Mill. Novosti Sist. Vyssh. Rast. 5: 9–12.

In the key below, all descriptions are based on leaves from vegetative branches of the lower crown unless otherwise noted.
1. Leaves stiff, apex sharply pointed; stomates absent on adaxial leaf surface; seed cones and twigs glabrous; Santa Lucia Mountains of coastal California. . . . . 1. *Abies bracteata*
1. Leaves flexible, apex acute, rounded, or notched; stomates usually present on adaxial leaf surface; seed cones pubescent, twigs pubescent or glabrous; widespread.
  2. Resin canals of leaves ± median, located well away from the epidermis.
    3. Stomatal rows 0–4 on adaxial surface at midleaf; trans-Canadian (central Alberta to Newfoundland) and e United States.
      4. Stomatal rows (4–)7(–8) in each band on abaxial surface at midleaf; bracts of seed cones included or exserted; trans-Canadian and e United States (s to Iowa, Michigan, West Virginia, and n Virginia). . . . . 4. *Abies balsamea*

4. Stomatal rows (8–)10(–12) in each band on abaxial surface at midleaf; bracts of seed cones exserted; e montane United States (North Carolina, Tennessee, and w Virginia). . . . . 5. *Abies fraseri*
3. Stomatal rows 3–6 on adaxial surface at midleaf; w North America (Alaska and sw Northwest Territories, s to California, Arizona, and New Mexico).
   5. Fresh leaf scars with red periderm; basal bud scales equilaterally triangular, margins crenate or dentate; w Yukon s along coastal mountains to n California. . . 6. *Abies lasiocarpa*
   5. Fresh leaf scars with tan periderm; basal bud scales isosceles triangular, margins entire; e Yukon and sw Northwest Territories s along Rocky Mountains to s Arizona and s New Mexico. 7. *Abies bifolia*
2. Resin canals of leaves marginal, located near the lower epidermis.
  6. Buds not resinous, or slightly so at tips, basal scales densely pubescent; leaves mostly 1-ranked, surfaces similar in color, proximal portion curved, appressed to twig for 2–3 mm (best seen on abaxial surface of twig), distal portion of leaf divergent; California, w Nevada, w Oregon, w Washington.
    7. Basal bud scales pubescent throughout; seed cones 15–20 cm, bracts included or exserted; adaxial surface of leaves usually without longitudinal groove. . . . . 10. *Abies magnifica*
    7. Basal bud scales pubescent centrally, glabrous at margins; seed cones 10–15 cm, bracts exserted; adaxial surface of leaves usually with longitudinal groove. . . . 11. *Abies procera*
  6. Buds resinous, basal scales sparingly pubescent to glabrous (densely pubescent in *Abies amabilis* with leaf surfaces strikingly different in color); leaves mostly 2-ranked, surfaces similar or strikingly different in color, proximal portion ± straight, not appressed to twig; widespread.
    8. Stomatal rows absent on adaxial surface at midleaf; leaves strongly glaucous abaxially, green adaxially; seed cones green, blue, purple, or gray.
      9. Basal bud scales densely pubescent; adaxial surface of twigs (especially in mid to upper crown) ± concealed by forwardly directed, ± appressed leaves, the other, usually longer leaves spreading horizontally; pollen cones at pollination red, becoming reddish yellow; mature seed cones purple. . . . . . . 2. *Abies amabilis*
      9. Basal bud scales slightly pubescent or glabrous; adaxial surface of crown twigs not concealed by leaves, leaves spreading horizontally, 2-ranked, shorter and longer intermixed; pollen cones at pollination ± blue, red, purple, orange, yellow, or green; mature seed cones light green, dark blue, deep purple, or gray. . . . . 3. *Abies grandis*
    8. Stomatal rows 5–18 on adaxial surface at midleaf; leaves glaucous or green abaxially and adaxially; seed cones olive-green.
     10. Adaxial surface at midleaf glaucous, with about (7–)12(–18) rows of stomates; leaves (2–)4–6 cm; leaf apex of lower branches usually rounded; widespread in w US but not in Sierra Nevada. . . . . 8. *Abies concolor*
     10. Adaxial surface at midleaf not glaucous, with about (5–)7(–9) rows of stomates; leaves 2–4(–6) cm; leaf apex of lower branches weakly notched; Sierra Nevada of California and Nevada, north coastal mountains of California. . . 9. *Abies lowiana*

1. **Abies bracteata** (D. Don) Poiteau, Rev. Hort., sér. 2, 4: 7. 1845 · Bristlecone fir

Pinus bracteata D. Don, Trans. Linn. Soc. London 17: 443. 1836; *Abies venusta* (Douglas) K. Koch

Trees to 25 m; trunk to 1 m diam.; crown spirelike, narrow. **Bark** red-brown, thin, smooth, with age slightly fissured and broken into appressed scales. **Branches** diverging from trunk at right angles, the lower often drooping in age; twigs becoming purplish green or brown, glabrous, glaucous when young. **Buds** exposed, brown, ovate to fusiform, extremely large, not resinous, apex pointed; basal scales short, broad, equilaterally triangular, glabrous, not resinous, margins entire, apex sharp-pointed. **Leaves** 2.5–6 cm × 3 mm, 2-ranked to spiraled, stiff; cross section flat, with raised vein abaxially, grooveless to faintly grooved adaxially; odor pungent; abaxial surface with 8–10 stomatal rows on each side of midrib; adaxial surface dark green, lacking stomates; apex sharply

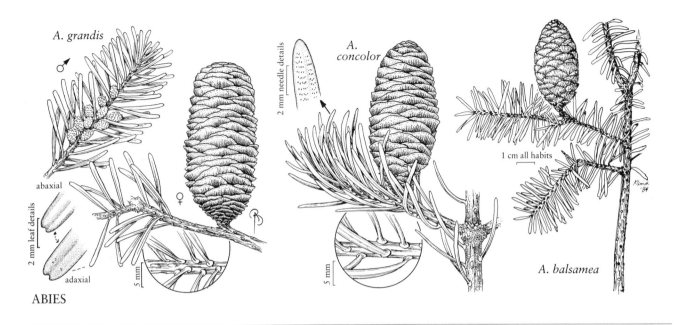

A. grandis
♂
2 mm leaf details
abaxial
adaxial
5 mm
ABIES

2 mm needle details
A. concolor
♀
5 mm
1 cm all habits
A. balsamea
KSmith '84

pointed; resin canals small, near margins and abaxial epidermal layer. **Pollen cones** at pollination yellow to yellow-green. **Seed cones** ovoid, 7–10 × 4–5 cm, pale purplish brown, borne on stout peduncles, apex round; scales ca. 1.5–2 × 2–2.5 cm, glabrous; bracts exserted, not reflexed. **Seeds** 10 × 5 mm, body deep red-brown; wing about as long as body, deep red-brown; cotyledons ca. 7.

Dry, coastal coniferous forests; of conservation concern; 600–900 m; Calif.

Abies bracteata grows in the Santa Lucia Mountains along the coast of California.

2. **Abies amabilis** Douglas ex J. Forbes, Pinet. Woburn., 125, plate 44. 1839 · Pacific silver fir, silver fir, sapin gracieux

Trees to 75 m; trunk to 2.6 m diam.; crown spirelike, with age becoming flat topped, cylindric. **Bark** gray, thin, smooth, with age breaking into scaly plates. **Branches** diverging from trunk at right angles, short, stiff; twigs mostly opposite, darker brown abaxially, light brown adaxially, pubescence tan. **Buds** hidden by leaves or exposed, brown, globose, small, resinous (at least apically), apex rounded; basal scales short, broad, triangular, densely pubescent, usually not resinous, margins entire, apex sharp-pointed. **Leaves** (0.7–)1–2.5 cm × 1–3 mm, mostly 2-ranked, flexible, ± concealing the adaxial surface of the twigs (especially in mid to upper crown),

some leaves forwardly directed, others usually longer and spreading horizontally, proximal portion ± straight; cross section flat, prominently grooved adaxially; odor pungent; abaxial surface with 5–6 stomatal rows on each side of midrib; adaxial surface dark, lustrous green, lacking stomates; apex prominently notched; resin canals small, near margins and abaxial epidermal layer. **Pollen cones** at pollination red, becoming reddish yellow. **Seed cones** cylindric, 8–10(–13) × 3.5–5 cm, purple, sessile, apex round to nipple-shaped; scales ca. 2 × 2 cm, pubescent; bracts included. **Seeds** 10–12 × 4 mm, body tan; wing about as long as body, rose to tan; cotyledons 4–7.

Moist, coastal coniferous forests; 0–2000 m; B.C.; Alaska, Calif., Oreg., Wash.

3. **Abies grandis** (Douglas ex D. Don in Lambert) Lindley, Penny Cycl. 1: 30. 1833 · Grand fir, lowland white fir, sapin grandissime

Pinus grandis Douglas ex D. Don in Lambert, Descr. Pinus [ed. 3] 2: unnumbered page between 144 and 145. 1832

Trees to 75 m; trunk to 1.55 m diam.; crown conic, in age round topped or straggly. **Bark** gray, thin to thick, with age becoming brown, often with reddish periderm visible in furrows bounded by hard flat ridges. **Branches** spreading, drooping; twigs mostly opposite, light brown, pubescent. **Buds** exposed, purple, green, or brown, globose, small to moderately large, resinous, apex round; basal scales short, broad, equilaterally triangular, slightly pubescent or glabrous, resinous, margins entire, apex pointed or slightly rounded. **Leaves** (1–)2–6 cm × 1.5–2.5 mm, 2-ranked, flexible with leaves at center of branch segment longer than those near ends, or with distinct long and short leaves intermixed, proximal portion ± straight, leaves higher in tree spiraled and 1-ranked; cross section flat, grooved adaxially; odor pungent, faintly turpentinelike; abaxial surface with 5–7 stomatal rows on each side of midrib; adaxial surface light to dark lustrous green, lacking stomates or with a few stomates toward leaf apex; apex distinctly notched (rarely rounded); resin canals small, near margins and abaxial epidermal layer. **Pollen cones** at pollination bluish red, purple, orange, yellow, or ± green. **Seed cones** cylindric, (5–)6–7(–12) × 3–3.5 cm, light green, dark blue, deep purple, or gray, sessile, apex rounded; scales ca. 2–2.5 × 2–2.5 cm, densely pubescent; bracts included. **Seeds** 6–8 × 3–4 mm, body tan; wing about 1.5 times as long as body, tan with rosy tinge; cotyledons (4–)5–6(–7). $2n = 24$.

Moist, coastal coniferous forests and mountain slopes; 0–1500 m; B.C.; Calif., Idaho, Mont., Oreg., Wash.

Abies grandis is rather uniform morphologically and chemically. At its southern limit in southern Oregon and northern California, it introgresses with *A. concolor* (J. L. Hamrick and W. J. Libby 1972; E. Zavarin et al. 1975; D. B. Zobel 1973). In the area of introgression, specimens in lower, wetter habitats are best assigned to *A. grandis*; those in higher, drier habitats, to *A. concolor*. Others are best considered to be *A. concolor* × *grandis*.

4. Abies balsamea (Linnaeus) Miller, Gard. Dict. ed. 8, Abies no. 3. 1768 · Balsam fir, sapin baumier

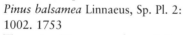

Pinus balsamea Linnaeus, Sp. Pl. 2: 1002. 1753

Trees to 23 m; trunk to 0.6 m diam.; crown spirelike. **Bark** gray, thin, smooth, in age often becoming broken into irregular brownish scales. **Branches** diverging from trunk at right angles, the lower often spreading and drooping; twigs mostly opposite, greenish brown, pubescence sparse. **Buds** hidden by leaves or exposed, brown, conic, small, resinous, apex acute; basal scales short, broad, nearly equilaterally triangular, glabrous, resinous, margins entire, apex sharp-pointed. **Leaves** 1.2–2.5 cm × 1.5–2 mm, 1-ranked (particularly on lower branches) to spiraled, flexible; cross section flat, grooved adaxially; odor pinelike (copious β-pinene); abaxial surface with (4–)6–7(–8) stomatal rows on each side of midrib; adaxial surface dark green, slightly or not

glaucous, with 0–3 stomatal rows at midleaf, these more numerous toward leaf apex; apex slightly notched to rounded; resin canals large, ± median, away from margins, midway between abaxial and adaxial epidermal layers. **Pollen cones** at pollination red, purplish, bluish, greenish, or orange. **Seed cones** cylindric, 4–7 × 1.5–3 cm, gray-purple, turning brown before scale shed, sessile, apex round to obtuse; scales ca. 1–1.5 × 0.7–1.7 cm (relationship reversed in more western collections), pubescent; bracts included or exserted and reflexed over scales. **Seeds** 3–6 × 2–3 mm, body brown; wing about twice as long as body, brown-purple; cotyledons ca. 4. $2n = 24$.

Boreal and northern forests; 0–1700 m; St. Pierre and Miquelon; Alta., Man., N.B., Nfld., N.S., Ont., P.E.I., Que., Sask.; Conn., Iowa, Maine, Mass., Mich., Minn., N.H., N.Y., Pa., Vt., Va., W.Va., Wis.

Balsam fir is frequently segregated into two varieties (e.g., H. J. Scoggan 1978–1979) based on whether the bracts are included (var. *balsamea*) or exserted (var. *phanerolepis* Fernald), the latter considered by Liu T. S. (1971) to be a hybrid between *Abies balsamea* and *A. fraseri*. D. T. Lester (1968) demonstrated, however, that bract length may vary within a cone, annually, and from tree to tree. Nevertheless, a tendency exists for the exserted variety to be found most commonly from Newfoundland south through New England (R. C. Hosie 1969; B. F. Jacobs et al. 1984); it is not found west of Ontario. Western populations lack 3-carene and have other minor chemical differences separating them from eastern balsam fir (E. Zavarin and K. Snajberk 1972; R. S. Hunt and E. von Rudloff 1974). Morphologic variation in balsam fir has been studied mainly east of Ontario; the populations to the west have been ignored for the most part, although they may yield stronger evidence for species subdivision.

In Alberta, populations intermediate between western *Abies balsamea* and *A. bifolia* (E. H. Moss 1953; R. S. Hunt and E. von Rudloff 1974, 1979) may be classified as *A. balsamea* × *bifolia*. In West Virginia and Virginia, populations of balsam fir tend to be more similar to *A. fraseri* than are more northern populations (B. F. Jacobs et al. 1984).

Balsam fir (*Abies balsamea*) is the provincial tree of New Brunswick.

SELECTED REFERENCE   Lester, D. T. 1968. Variation in cone morphology of balsam fir, *Abies balsamea*. Rhodora 70: 83–94.

5. Abies fraseri (Pursh) Poiret in Lamarck et al., Encycl.
5: 35. 1817 · Fraser fir, southern balsam fir

Pinus fraseri Pursh, Fl. Amer. Sept.
2: 639. 1814

Trees to 25 m; trunk to 0.75 m
diam.; crown spirelike. **Bark** gray,
thin, smooth, with age developing appressed reddish scales at
trunk base. **Branches** diverging
from trunk at right angles; twigs
opposite, pale yellow-brown, pu-
bescence reddish. **Buds** exposed, light brown, conic,
small, resinous, apex acute; basal scales short, broad,
equilaterally triangular, glabrous, resinous, margins entire, apex sharp-pointed. **Leaves** 1.2–2.5 cm × 1.5–2
mm, 2-ranked, particularly in lower parts of tree, to
spiraled, flexible; cross section flat, grooved adaxially;
odor turpentinelike, strong; abaxial surface with (8–)
10(–12) stomatal rows on each side of midrib; adaxial
surface dark lustrous green, sometimes slightly glaucous, with 0–3 stomatal rows at midleaf, these more
numerous toward leaf apex; apex slightly notched to
rounded; resin canals large, ± median, away from
margins and midway between abaxial and adaxial epidermal layers. **Pollen cones** at pollination reddish yellow or yellowish green. **Seed cones** cylindric, 3.5–6 ×
2.5–4 cm, dark purple overlaid with yellowish green
bracts, sessile, apex round; scales ca. 0.7–1 × 1–1.3
cm, pubescent; bracts exserted and reflexed over cone
scales. **Seeds** 4–5 × 2–3 mm, body brown; wing about
as long as body, purple; cotyledons ca. 5. $2n = 24$.

Mountain forests; of conservation concern; 1500 m;
N.C., Tenn., Va.

Some (e.g., B. F. Jacobs et al. 1984) have argued that
Fraser fir is at the end of a disjunct cline of balsam fir
and perhaps does not deserve separate specific status.
A. E. Matzenko (1968) took the opposite view, classi-
fying Fraser fir and balsam fir in different taxonomic
series of the genus.

6. Abies lasiocarpa (Hooker) Nuttall, N. Amer. Sylv. 3:
138. 1849 · Subalpine fir, alpine fir, sapin concolore

Pinus lasiocarpa Hooker, Fl. Bor.-
Amer. 2: 163. 1838

Trees to 20 m; trunk to 0.8 m
diam.; crown spirelike. **Bark** gray,
thin, smooth, furrowed in age.
Branches stiff, straight; twigs opposite to whorled, greenish gray
to light brown, bark splitting as
early as 2 years to reveal red-
brown layer, somewhat pubescent; fresh leaf scars with
red periderm. **Buds** hidden by leaves or exposed, tan to
dark brown, nearly globose, small, resinous, apex

rounded; basal scales short, broad, equilaterally triangular, glabrous or with a few trichomes at base, not
resinous, margins crenate to dentate, apex sharp-pointed.
Leaves 1.8–3.1 cm × 1.5–2 mm, spiraled, turned upward, flexible; cross section flat, prominently grooved
adaxially; odor sharp (β-phellandrene); abaxial surface
with 4–5 stomatal rows on each side of midrib; adaxial surface bluish green, very glaucous, with 4–6 stomatal rows at midleaf, rows usually continuous to leaf
base; apex prominently or weakly notched to rounded;
resin canals large, ± median, away from margins and
midway between abaxial and adaxial epidermal layers.
Pollen cones at pollination ± purple to purplish green.
Seed cones cylindric, 6–12 × 2–4 cm, dark purple,
sessile, apex rounded; scales ca. 1.5 × 1.7 cm, densely
pubescent; bracts included (specimens with exserted,
reflexed bracts are insect infested). **Seeds** 6 × 2 mm,
body brown; wing about 1.5 times as long as body,
light brown; cotyledon number 4–5. $2n = 24$.

Coastal, subalpine coniferous forests; 1100–2300 m;
B.C., Yukon; Alaska, Calif., Oreg., Wash.

The only unique populations in this species come from
coastal Alaska (A. S. Harris 1965; C. J. Heusser 1954).
They are found at lower elevations (0–900 m) and appear to be isolated, with no reported introgression between them and the coastal mountain populations. The
population on the Prince of Wales Island has distinct
terpene patterns and needs morphological and developmental studies to see if these patterns contrast with
neighboring populations.

Through central British Columbia and northern
Washington, *Abies lasiocarpa* introgresses with *A. bifolia*. These trees may have morphologic features resembling either species and may have intermediate terpene patterns; they are best classified as interior
subalpine fir (*A. bifolia* × *lasiocarpa*). At the southern
end of its range, *A. lasiocarpa* possibly hybridizes with
A. procera (R. S. Hunt and E. von Rudloff 1979). *Abies
lasiocarpa* shares with *A. procera* a red periderm, crystals in the ray parenchyma (R. W. Kennedy et al. 1968),
and reflexed tips of the bracts, features not shared with
A. bifolia.

Abies lasiocarpa usually exists in small stands at high
elevations and is not often observed. Its differences in
comparison to *A. bifolia* have prompted studies (W. H.
Parker et al. 1979) to see if it is *A. bifolia* introgressed
with the sympatric *A. amabilis*. *Abies lasiocarpa* and
A. amabilis, however, are separated by many morphologic features, and no hybrids have been found (W. H.
Parker et al. 1979).

SELECTED REFERENCES Harris, A. S. 1965. Subalpine fir [*Abies lasiocarpa* (Hook.) Nutt.] on Harris Ridge near Hollis, Prince of Wales
Island, Alaska. NorthW. Sci. 39: 123–128. Heusser, C. J. 1954. Alpine fir at Taku glacier, Alaska, with notes on its post glacial migra-

tion to the territory. Bull. Torrey Bot. Club 81: 83–86. Zavarin, E., K. Snajberk, T. Reichert, and Tsien E. 1970. On the geographic variability of the monoterpenes from the cortical blister oleoresin of *Abies lasiocarpa*. Phytochemistry 9: 377–395.

7. **Abies bifolia** A. Murray bis, Proc. Roy. Hort. Soc. London 3: 320. 1863 · Rocky Mountain alpine fir, Rocky Mountain subalpine fir, corkbark fir

Abies subalpina Engelmann

Trees to 30 m; trunk to 0.45 m diam.; crown spirelike. **Bark** gray, thin, smooth, with age somewhat furrowed and scaly (toward southern end of range bark is corky [corkbark fir]). **Branches** diverging from trunk at right angles, stout, stiff; twigs opposite to whorled, grayish, pubescence sparse, light brown; fresh leaf scars with light brown periderm. **Buds** exposed, brown, globose, small, resinous, apex rounded; basal scales long, narrow, isosceles triangular to spatulate, glabrous, resinous or not resinous, margins entire to rarely crenate, apex sharp-pointed or rounded. **Leaves** 1.1–2.5 cm × 1.25–1.5 mm, spiraled and turned upward, flexible; cross section flat, grooved adaxially, sometimes only slightly so; odor camphorlike; abaxial surface with 3–5 stomatal rows on each side of midrib; adaxial surface light green to bluish green, usually glaucous, with 3–6 stomatal rows at midleaf, rows usually continuous to leaf base, usually more numerous toward leaf apex; apex slightly notched to rounded; resin canals large, ± median, away from margins and midway between abaxial and adaxial epidermal layers. **Pollen cones** at pollination purplish. **Seed cones** cylindric, 5–10 × 3–3.5 cm, dark purple-blue to grayish purple, sessile, apex rounded; scales ca. 1.5 × 2.5 cm, densely pubescent; bracts included. **Seeds** 5–7 × 2–3 mm, body brown; wing about 1.5 times as long as body, grayish brown; cotyledons 3–6.

Continental, subalpine coniferous forests; 600–3600 m; Alta., B.C., N.W.T., Yukon; Ariz., Colo., Idaho, Mont., Nev., N.Mex., Oreg., Utah, Wash., Wyo.

Abies bifolia has been—and by many workers still is—included in synonymy under *A. lasiocarpa* or *A. subalpina* since about 1890, and *A. subalpina* under *A. lasiocarpa* since about the 1920s. *Abies bifolia* is distinct from *A. lasiocarpa*, however, in chemical tests on wood (H. S. Fraser and E. P. Swan 1972), lack of crystals in the ray parenchyma (R. W. Kennedy et al. 1968), lack of lasiocarpenonol (J. F. Manville and A. S. Tracey 1989), and distinct terpene patterns (R. S. Hunt and E. von Rudloff 1979). *Abies bifolia* also tends to have slightly shorter and fewer prominently notched leaves than *A. lasiocarpa*. The two are clearly separated by

the color of their periderm and by the shape of their basal bud scales. These firs may be more distinct than the pairs *A. balsamea*–*A. fraseri* and *A. procera*–*A. magnifica*. A north-south transect, however, from south central Yukon to northern Washington yielded introgressed trees possessing characteristics of both *A. lasiocarpa* and *A. bifolia*, recalling the interior spruce (Canadian Forestry Service 1983), which has characteristics of both *Picea glauca* and *P. engelmannii*. These trees can similarly be called interior subalpine fir, i.e., *A. bifolia* × *lasiocarpa*. Both *A. lasiocarpa* and *A. bifolia* need comparative morphologic studies.

Isolated southern populations of *Abies bifolia* may also have unique characteristics. The taxonomy of corkbark fir, treated by some as *A. lasiocarpa* var. *arizonica* (Merriam) Lemmon, is uncertain. This taxon should probably be a segregate of *A. bifolia*, not *A. lasiocarpa*, a disposition that requires a thorough morphologic and chemical reappraisal, especially since the work of E. Zavarin et al. (1970) suggested that populations south of Wyoming may have unique terpene patterns. In north central Alberta, *A. bifolia* introgresses with *A. balsamea* (R. S. Hunt and E. von Rudloff 1974; E. H. Moss 1953).

8. **Abies concolor** (Gordon & Glendinning) Hildebrand, Verh. Naturhist. Vereines Preuss. Rheinl. Westphalens 18: 261. 1861 · White fir, Rocky Mountain white fir, pino real blanco

Picea concolor Gordon & Glendinning, Pinetum, 155. 1858

Trees to 40 m; trunk to 0.9 m diam.; crown spirelike. **Bark** gray, thin, smooth, with age thickening (to 18 cm) and breaking into deep longitudinal furrows, often revealing yellowish inner periderm, appearing "corky." **Branches** diverging from trunk at right angles, the lower often spreading and drooping in age; twigs mostly opposite, glabrous or with yellowish pubescence. **Buds** exposed, either yellowish and nearly conic (when large) or brownish and nearly globose (when small), resinous, apex rounded to pointed; basal scales equilaterally triangular, glabrous, not resinous, margins entire, apex sharp-pointed. **Leaves** 1.5–6 cm × 2–3 mm, mostly 2-ranked, flexible, proximal portion ± straight; cross section flat, sometimes slightly grooved adaxially; odor pungent, frequently camphorlike; abaxial surface glaucous, with 4–7 stomatal rows on each side of midrib; adaxial surface grayish green, glaucous, with (7–)12 (–18) stomatal rows at midleaf, these usually fewer toward leaf apex; apex usually rounded, sometimes acute or notched; resin canals small, near margins and abaxial epidermal layer. **Pollen cones** at pollination ± red, purple, or ± green. **Seed cones** cylindric, 7–12 × 3–4.5 cm, olive-green, sessile, apex round; scales ca. 2.5–3 × 3–3.5 cm, pubescent; bracts included. **Seeds** 8–12 × 3 mm, body tan; wing about twice as long as body, tan with rosy tinge; cotyledons 5–7. $2n = 24$.

Coniferous forests; 1700–3400 m; Ariz., Calif., Colo., Idaho, Nev., N.Mex., Oreg., Utah; Mexico in Baja California, Sonora.

Abies concolor is a western catchall species for firs with green seed cones and with glaucous adaxial leaf surfaces. Many of these populations have long been isolated geographically and genetically. A geographic cluster of populations in Utah has shorter leaves and slightly different terpene patterns than a similar cluster of populations in Colorado and northern New Mexico (J. W. Wright et al. 1971; E. Zavarin et al. 1975). Another large geographic cluster, in southern New Mexico and Arizona, seems to be strongly linked chemically to Colorado populations (E. Zavarin et al. 1975) and morphologically to southern California populations (J. L. Hamrick and W. J. Libby 1972). Northern California populations with pubescent twigs and notched leaves are unique, as are the Baja California populations with very short, thick leaves and about 18 adaxial stomatal rows. In Los Padres National Forest of coastal southern California and in the Cascades of northern California, apparent introgression with *A. lowiana* (E. Zavarin et al. 1975; J. L. Hamrick and W. J. Libby 1972) has occurred. Many consider *A. lowiana* (given specific rank in this treatment) as a synonym of *A. concolor* or place it in an infraspecific rank under that species.

SELECTED REFERENCES Hamrick, J. L. and W. J. Libby. 1972. Variation and selection in western U.S. montane species. I. White fir. Silvae Genet. 21: 29–35. Wright, J. W., W. A. Lemmien, and J. N. Bright. 1971. Genetic variation in southern Rocky Mountain white fir. Silvae Genet. 20: 148–150. Zavarin, E., K. Snajberk, and J. Fisher. 1975. Geographic variability of monoterpenes from the cortex of *Abies concolor*. Biochem. Syst. & Ecol. 3: 191–203.

9. **Abies lowiana** (Gordon) A. Murray bis, Proc. Roy. Hort. Soc. London 3: 317. 1863 · Sierra white fir, California white fir

Picea lowiana Gordon, Pinetum Suppl., 53. 1862; *Abies concolor* (Gordon & Glendinning) Hildebrand var. *lowiana* (Gordon) Lemmon

Trees to 60 m; trunk to 1.9 m diam.; crown somewhat spirelike. **Bark** gray, thin, smooth, with age thickening and breaking into deep longitudinal furrows ("corky"), which may reveal

inner yellow-caramel periderm. **Branches** diverging from trunk at right angles, or slightly drooping; twigs mostly opposite, yellow-green, pubescent or glabrous. **Buds** exposed, tan, globose, resinous, moderately large, apex rounded; basal scales short, broad, equilaterally triangular, sparsely pubescent, not resinous, margins entire, apex sharp-pointed. **Leaves** 2–6 cm × 2–3 mm, mostly 2-ranked, flexible, proximal portion ± straight; cross section flat, grooved adaxially; odor pungent, pinelike; abaxial surface glaucous or not, with 5–8 stomatal rows on each side of midrib; adaxial surface light green, not glaucous, with ca. (5–)7(–9) stomatal rows at midleaf, these fewer toward leaf apex; apex weakly notched; resin canals small, near margins and abaxial epidermal layer. **Seed cones** cylindric, 8–9 × 4–4.5 cm, olive-green, turning to yellowish brown, then darker brown, sessile, apex round; scales ca. 2.5–3 × 2.8–3.8 cm, pubescent; bracts included. **Seeds** 8–12 × 3 mm, body dull brown; wing about twice as long as body, brown with rosy tinge; cotyledons 5–9.

Mixed coniferous forests; 900–2300 m; Calif., Nev.

10. **Abies magnifica** A. Murray bis, Proc. Roy. Hort. Soc. London 3: 318. 1863 · California red fir, Shasta red fir

Trees to 57 m; trunk to 2.5 m diam.; crown narrowly conic. **Bark** grayish, thin, with age thickening and becoming deeply furrowed with ridges being often 4 times wider than furrows, plates reddish. **Branches** ascending in upper crown, descending in lower crown; twigs opposite to whorled, light yellow to ± tan, reddish pubescent for 1–2 years. **Buds** hidden by leaves or exposed, usually dark brown, ovoid, small, not resinous or with resin drop near tip, apex rounded; basal scales short, broad, equilaterally triangular, densely pubescent, not resinous, margins entire to crenate, apex sharp-pointed. **Leaves** 2–3.7 cm × 2 mm, mostly 1-ranked, flexible, the proximal portion often appressed to twig for 2–3 mm (best seen on abaxial surface of twig), distal portion divergent; cross section flat, with or without weak groove adaxially toward leaf base, or cross section 3–4-sided on fertile branches; odor camphorlike; abaxial surface with 2 glaucous bands, each band with 4–5 stomatal rows; adaxial surface blue-green to silvery blue, with single glaucous band that may divide into 2 toward leaf base, band with (8–)10(–13) stomatal rows at midleaf; apex rounded or, on fertile branches, somewhat pointed; resin canals small, near margins and abaxial epidermal layer. **Pollen cones** at pollination ± purple or reddish brown. **Seed cones** oblong-cylindric, 15–20 × 7–10 cm, purple at first but becoming yellowish brown or greenish brown, sessile, apex round; scales ca. 3 × 4 cm, pubescent; bracts included to exserted and reflexed (Shasta red fir) over scales. **Seeds** 15 × 6 mm, body dark reddish brown; wing about as long as body, rose; cotyledons 7–8. $2n = 24$.

Mixed coniferous forests; 1400–2700 m; Calif., Nev., Oreg.

Abies magnifica often exists in extensive high elevation stands in the Sierra Nevada; its close relative *A. procera* occurs in small mountaintop populations relatively isolated from one another. As expected for isolated populations, *A. procera* produces large interpopulation variation in morphology (J. Maze and W. H. Parker 1983) and chemistry (E. Zavarin et al. 1978). Where the two species meet in southern Oregon and northern California, many populations are intermediate; these have been called *A. magnifica* var. *shastensis* Lemmon. The status of such intermediates is unsettled. They may be accepted as hybrids between *A. magnifica* and *A. procera* (Liu T. S. 1971) or, alternatively, the paleontological record suggests that the two species may have originated from the intermediates (E. Zavarin et al. 1978). Individuals from this region should be assigned to *A. magnifica*, *A. procera*, or *A. magnifica* × *procera* (E. L. Parker 1963), depending on the morphologic criteria selected to differentiate the species, though clearly these individuals are genetically quite different from those near the type localities of the two species.

An extensive study of this variation, as proposed by E. Zavarin et al. (1978), is warranted. Such a study should consider data from the type localities as a basis of comparison. Moreover, to evaluate this situation critically, one should first determine if any genetic exchange occurs between *Abies lasiocarpa* and *A. procera* that may complicate an evaluation.

11. Abies procera Rehder, Rhodora 42: 522. 1940 · Noble fir

Abies nobilis (Douglas ex D. Don) Lindley 1833, not A. Dietrich 1824

Trees to 80 m; trunk to 2.2 m diam.; crown spirelike. **Bark** grayish brown, in age becoming thick and deeply furrowed (furrows and ridges about same width) and reddish brown (especially reddish when plates flake off). **Branches** diverging from trunk at right angles, stiff; twigs reddish brown, finely pubescent for several years. **Buds** hidden by leaves, tan, ovoid, small, not resinous, apex rounded; basal scales short, broad, equilaterally triangular, pubescent centrally, not resinous, margins entire to crenate, apex sharp-pointed. **Leaves** 1–3(–3.5) cm × 1.5–2 mm, 1-ranked, flexible, proximal portion often appressed to twig for 2–3 mm (best seen on abaxial surface of twig), distal portion divergent; cross section flat, with prominent raised midrib abaxially, with or without groove adaxially, or cross section 4-sided on fertile branches; odor pungent, faintly turpentine-like; abaxial surface with 2–4 glaucous bands, each band with (4–)6–7 stomatal rows; adaxial surface bluish green, with 0–2 glaucous bands, each band with 0–7 stomatal rows at midleaf; apex rounded to notched; leaves on fertile branches 4-sided with 4 bands of stomates below; resin canals small, near margins and abaxial epidermal layer. **Pollen cones** at pollination ± purple, ± red, or reddish brown. **Seed cones** oblong-cylindric, 10–15 × 5–6.5 cm, green, red, or purple, overlaid with green bracts, at maturity brown (bracts light-colored and scales dark), sessile, apex rounded; scales ca. 2.5 × 3 cm, pubescent; bracts exserted and reflexed over scales. **Seeds** 12 × 6 mm, body reddish brown; wing slightly longer than body, light brown to straw; cotyledons (4–)5–6(–7). $2n = 24$.

Mixed coniferous forests; 60–2700 m; Calif., Oreg., Wash.

See discussion under *Abies magnifica*.

SELECTED REFERENCE Maze, J. and W. H. Parker. 1983. A study of population differentiation and variation in *Abies procera*. Canad. J. Bot. 61: 1094–1104.

2. TSUGA (Endlicher) Carrière, Traité Gén. Conif., 185. l855 · Hemlock, pruche [Japanese *tsuga*, name for native hemlocks of Japan]

Ronald J. Taylor

Pinus Linnaeus sect. *Tsuga* Endlicher, Syn. Conif., 83. 1847

Trees evergreen; crown conic; leading shoot usually drooping. **Bark** gray to brown, scaly, often deeply furrowed. **Branches** horizontal, often tending to be arranged in flattened "sprays"

and arched downward; short (spur) shoots absent; young twigs and distal portions of stem flexuous and pendent, roughened by peglike projections persisting after leaves fall. **Buds** mostly rounded at apex, not resinous. **Leaves** borne singly, persisting several years, ± 2-ranked or radiating in all directions, flattened to somewhat angular; abruptly narrowed to a petiolelike base, set on peglike projections, these angled, projected forward, sheath absent; apex rounded or notched; resin canals 1. **Cones** borne on year-old twigs. **Pollen cones** solitary, globose, brown. **Seed cones** maturing in 1 year, shedding seeds and falling soon thereafter or persisting for several years, pendent, ovoid, oblong, or oblong-cylindric, sessile or nearly so; scales persistent, shape various, thin, leathery, lacking apophysis and umbo; bracts small, included. **Seeds** winged; cotyledons 4–6. $x = 12$.

Species ca. 10 (4 in the flora): Northern Hemisphere.

Species of *Tsuga* are found naturally in areas of relatively moist climates where water stresses are minimal. Most are conspicuous, if not dominant, members of the communities in which they occur.

Hemlock wood is moderately strong and pliable and lacks resin ducts. With the decline of associated species considered superior in commercial value, hemlocks have become important in the timber industry, especially for pulp. Hemlocks are also widely used for horticultural purposes; numerous cultivars have been developed.

1. Leaves mostly spreading in all directions from twigs, bearing stomates on both surfaces or on abaxial surface only; seed cones ovoid, oblong, or oblong-cylindric, 2.5–6 cm; scales oblong, 8–18 mm.
   2. Leaves bearing stomates on both surfaces, thicker along midline; seed cones 3–6 cm, half as wide; montane tree of w North America. . . . . 2. *Tsuga mertensiana*
   2. Leaves bearing stomates on abaxial surface only, flat, not notably thicker along midline, but slightly revolute; seed cones 2.5–4 cm, more than half as wide; montane tree of se United States. . . . . 1. *Tsuga caroliniana*
1. Leaves mostly appearing 2-ranked, bearing stomates on abaxial surface only; seed cones ovoid, (1–)1.5–2.5(–3) cm; scales ovate to cuneate, 8–15 mm.
   3. Buds 2.5–3.5 mm; crown narrowly conic; tree of w North America. . . . . 4. *Tsuga heterophylla*
   3. Buds 1.5–2.5 mm; crown broadly conic; tree of e North America. . . . . 3. *Tsuga canadensis*

1. Tsuga caroliniana Engelmann, Bot. Gaz. 6: 223. 1881 · Carolina hemlock

Trees to 30 m; trunk to 2 m diam.; crown conic. **Bark** brown, scaly and fissured. **Twigs** light brown, thinly covered with short, dark hairs. **Buds** oblong, 2–3 mm. **Leaves** 10–20 mm, mostly spreading in all directions from twigs, flat but slightly revolute; abaxial surface glaucous, with 2 broad, conspicuous stomatal bands, adaxial surface shiny green; margins entire. **Seed cones** ovoid to oblong, 2.5–4 × 1.5–2.5 cm; scales oblong, 12–18 × 8–12 mm, bases clawed, apex rounded. $2n = 24$.

Rocky montane slopes; 700–1200 m; Ga., N.C., S.C., Tenn., Va.

Tsuga caroliniana is valuable as an attractive ornamental; a number of cultivars have been developed. The wood is of little commercial importance because of the combination of mediocre quality and the relative rarity of the species in nature.

SELECTED REFERENCE James, R. L. 1959. Carolina hemlock—wild and cultivated. Castanea 24: 112–134.

2. Tsuga mertensiana (Bongard) Carrière, Traité Gén. Conif. ed. 2, 250. 1867 · Mountain hemlock, pruche de Patton

Pinus mertensiana Bongard, Mém. Acad. Imp. Sci. St. Pétersbourg, Sér. 6., Sci. Math. 2: 163. 1832; *Abies hookeriana* A. Murray bis; *A. pattoniana* A. Murray bis; *Hesperopeuce mertensiana* (Bongard) Rydberg; *H. pattoniana* (A. Murray bis) Lemmon; *Picea (Tsuga) hookeriana* (A. Murray bis) Bertrand; *Pinus hookeriana* (A. Murray bis) McNab; *Pinus pattoniana* (A.

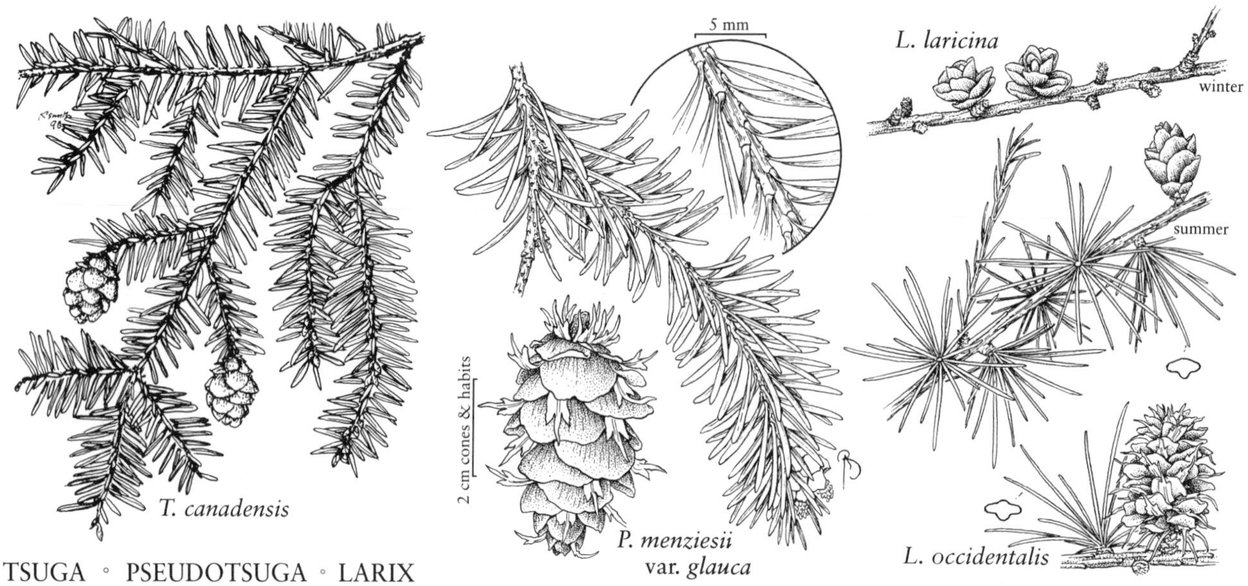
5 mm
L. laricina
winter
summer
2 cm cones & habits
T. canadensis
P. menziesii var. *glauca*
L. occidentalis

TSUGA ◦ PSEUDOTSUGA ◦ LARIX

Murray bis) Parlatore; *Tsuga crassifolia* Flous; *T. hookeriana* (A. Murray bis) Carrière; *T. pattoniana* (A. Murray bis) Engelmann var. *hookeriana* (A. Murray bis) Lemmon; ×*Tsuga-Picea hookeriana* (A. Murray bis) M. Van Campo-Duplan & Gaussen.

Trees to 40 m; trunk to 1.5 m diam.; crown conic. **Bark** charcoal gray to reddish brown, scaly and deeply fissured. **Twigs** yellow-brown, glabrous to densely pubescent. **Buds** oblong, 3–4 mm. **Leaves** 10–25(–30) mm, mostly spreading in all directions from twigs, curved toward twig apex, thickened centrally along midline, somewhat rounded or 4-angled in cross section, both surfaces glaucous, with ± inconspicuous stomatal bands; margins entire. **Seed cones** oblong-cylindric, 3–6 × 1.5–3 cm; scales broadly fan-shaped, 8–15 × 8–15 mm, apex rounded to pointed. $2n = 24$.

Coastal and montane forests to alpine slopes (where it occurs in krummholz form); 0–2400 m; B.C.; Alaska, Calif., Idaho, Mont., Nev., Oreg., Wash.

The wood of *Tsuga mertensiana* is somewhat inferior to that of western hemlock both for building purposes and as pulp. This is a very handsome tree with its branches densely clothed with pale, spreading leaves and is adaptable to a wide variety of climatic conditions.

M. Van Campo-Duplan and H. Gaussen (1948) postulated that this taxon originated by hybridization between *Picea* and *Tsuga*. Although this is unlikely, some characteristics such as leaf arrangement and shape, phenolic chemistry, and pollen grain structure lend some support for this hypothesis.

3. **Tsuga canadensis** (Linnaeus) Carrière, Traité Gén. Conif., 189. 1855 · Eastern hemlock, pruche du Canada

Pinus canadensis Linnaeus, Sp. Pl. ed. 2, 2: 1471. 1763

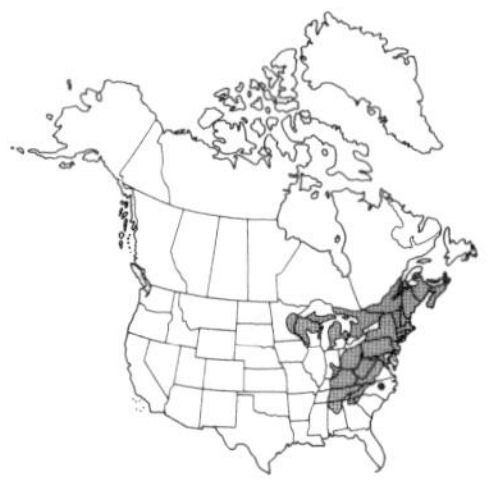

Trees to 30 m; trunk to 1.5 m diam.; crown broadly conic. **Bark** brownish, scaly and fissured. **Twigs** yellow-brown, densely pubescent. **Buds** ovoid, 1.5–2.5 mm. **Leaves** (5–)15–20(–25) mm, mostly appearing 2-ranked, flattened; abaxial surface glaucous, with 2 broad, conspicuous stomatal bands, adaxial surface shiny green (yellow-green); margins minutely dentate, especially toward apex. **Seed cones** ovoid, 1.5–2.5 × 1–1.5 cm; scales ovate to cuneate, 8–12 × 7–10 mm, apex ± round, often projected outward. $2n = 24$.

Moist rocky ridges, ravines, and hillsides; 600–1800 m; N.B., N.S., Ont., P.E.I., Que.; Ala., Conn., Del., Ga., Ind., Ky., Maine, Md., Mass., Mich., Minn., N.H., N.J., N.Y., N.C., Ohio, Pa., R.I., S.C., Tenn., Vt., Va., W.Va., Wis.

Numerous cultivars of *Tsuga canadensis* have been developed, including compact shrubs, dwarfs, and graceful trees. Wood of the species tends to be brittle and inferior to that of the other North American hemlocks.

Eastern hemlock (*Tsuga canadensis*) is the state tree of Pennsylvania.

4. Tsuga heterophylla (Rafinesque) Sargent, Silva 12: 73, plate 605. 1898 · Western hemlock, pruche de l'ouest

Abies heterophylla Rafinesque, Atlantic J. 1: 119. 1832

Trees to 50 m; trunk to 2 m diam.; crown narrowly conic. **Bark** gray-brown, scaly and moderately fissured. **Twigs** yellow-brown, finely pubescent. **Buds** ovoid, gray-brown, 2.5–3.5 mm. **Leaves** (5–)10–20(–30) mm, mostly appearing 2-ranked, flattened; abaxial surface glaucous with 2 broad, conspicuous stomatal bands, adaxial surface shiny green (yellow-green); margins minutely dentate. **Seed cones** ovoid, (1–)1.5–2.5(–3) × 1–2.5 cm; scales ovate, 8–15 × 6–10 mm, apex round to pointed. $2n = 24$.

Coastal to midmontane forests; 0–1500 m; Alta., B.C.; Alaska, Calif., Idaho, Mont., Oreg., Wash.

Tsuga heterophylla is a dominant species over much of its broad distributional range. It has become the most important timber hemlock in North America. The wood is superior to that of other hemlocks for building purposes and it makes excellent pulp for paper production.

Tsuga ×*jeffreyi* (Henry) Henry was described from southwestern British Columbia and western Washington as a hybrid between *T. heterophylla* and *T. mertensiana*. Hybridization is rare, if it occurs at all, and it is therefore of little consequence (R. J. Taylor 1972). At the upper elevational limits of its distribution and under stressful conditions, *T. heterophylla* tends to resemble *T. mertensiana*, e.g., leaves are less strictly 2-ranked and stomatal bands on the abaxial leaf surfaces are less conspicuous than at lower elevations.

Western hemlock *(Tsuga heterophylla)* is the state tree of Washington.

3. PSEUDOTSUGA Carrière, Traité Gén. Conif. ed. 2, 256. 1867 · Douglas-fir [Greek *pseudo*, false, and tsuga, *hemlock*]

Barney Lipscomb

Trees conic, evergreen. **Bark** initially smooth, with resin blisters; in age reddish brown, corky, furrowed. **Branches** often pendulous, irregularly whorled; short (spur) shoots absent; leaf scars transversely elliptic, slightly raised proximally but essentially flush with twig distally. **Buds** elongate, not resinous, apex acute. **Leaves** borne singly, persisting 6–8 years, alternate, short-stalked, flattened; resin canals 2, marginal. **Cones** borne on year-old twigs. **Pollen cones** axillary. **Seed cones** maturing first season, shed whole, deflexed or pendent, ellipsoid, ovoid, or cylindric, nearly sessile; scales persistent, lacking apophysis and umbo; apex rounded; bracts ± exserted, apex 3-lobed, lobes with acute apices, central lobe narrow, longer than lateral lobes. **Seeds** winged; cotyledons 6–12. $x = 12, 13$.

Species 5 (2 in the flora): North America, 3 in e Asia.

SELECTED REFERENCES Flous, F. 1937. Révision du genre *Pseudotsuga*. Bull. Soc. Hist. Nat. Toulouse 71: 33–164. Little, E. L. Jr. 1952. The genus *Pseudotsuga* (Douglas-fir) in North America. Leafl. W. Bot. 6: 181–198.

1. Leaf apex mucronulate; seed cones 9–20 cm; sw California. . . . . .1. *Pseudotsuga macrocarpa*
1. Leaf apex obtuse to acute; seed cones 4–10 cm; widely distributed in w North America.
. . . . 2. *Pseudotsuga menziesii*

1. **Pseudotsuga macrocarpa** (Vasey) Mayr, Wald.
Nordamer., 278, plates 6, 8, 9. 1890 · Bigcone
Douglas-fir, bigcone-spruce

Abies macrocarpa Vasey, Gard.
Monthly & Hort. 18: 21. 1876
(Jan.); *Pseudotsuga californica*
Flous; *P. douglasii* (Lindley) Car-
rière var. *macrocarpa* (Vasey) En-
gelmann

Trees to 44 m; trunk to 2.3 m
diam.; crown broadly conic.
Twigs slender, glabrous or pu-
bescent. **Leaves** (20–)25–45 × 1–1.5 mm, bluish green,
apex mucronulate. **Pollen cones** pale yellow. **Seed cones**
9–20 × 4–7 cm. **Seeds** 9–12 mm. $2n = 24$.

Slopes, cliffs, and canyons, in chaparral and mixed
coniferous forests; 200–2400 m; Calif.

Pseudotsuga macrocarpa, a tree of scattered occurrence and of no concern for timber, is valuable for es-
thetics and watershed protection. The northernmost
stands of the species, in Kern County, are about 35
kilometers east of the closest approach of *P. menziesii*.

SELECTED REFERENCE   Minnich, R. A. 1982. *Pseudotsuga macrocarpa*
in Baja California? Madroño 29: 22–31.

2. **Pseudotsuga menziesii** (Mirbel) Franco, Bol. Soc.
Brot., ser. 2, 24: 74. 1950 · Douglas-fir, Oregon-pine,
sapin de Douglas

Abies menziesii Mirbel, Mém. Mus. Hist. Nat. 13: 63, 70.
1825; *A. mucronata* Rafinesque; *A. taxifolia* Poiret 1805,
not Desfontaines 1804; *Pinus taxifolia* Lambert 1803, not
Salisbury 1796; *Pseudotsuga douglasii* (Lindley) Carrière; *P.
mucronata* (Rafinesque) Sudworth; *P. taxifolia* (Lambert)
Britton

Trees to 90(–100) m; trunk to 4.4 m diam.; crown
narrow to broadly conic, flattened in age. **Twigs** slender, pubescent, becoming glabrous with age. **Leaves** 15–
30(–40) × 1–1.5 mm, yellow-green to dark or bluish
green, apex obtuse to acute. **Pollen cones** yellow-red.
Seed cones 4–10 × 3–3.5 cm. **Seeds** 5–6 mm, wing
longer than seed body.

Varieties 2 (2 in the flora): North America, Mexico.

Pseudotsuga menziesii is a most important timber tree,
valued in both the Old and New worlds. The two intergrading varieties are sympatric in southern British
Columbia and northeastern Washington.

Douglas-fir (*Pseudotsuga menziesii*) is the state tree
of Oregon.

1. Bracts straight, appressed; seed cones 6–10 cm;
   leaves yellowish green; Pacific Coast region. .
   . . . . 2a. *Pseudotsuga menziesii* var. *menziesii*
1. Bracts spreading, often reflexed; seed cones 4–7
   cm; leaves bluish green to dark green or gray-
   green; Rocky Mountain region. . . . .
   . . . . 2b. *Pseudotsuga menziesii* var. *glauca*

2a. **Pseudotsuga menziesii** (Mirbel) Franco var. **menziesii**
· Coast Douglas-fir, Douglas-fir

Trees to 90(–100) m; trunk to 4.4
m diam. **Leaves** yellowish green.
Seed cones 6–10 cm; bracts
straight, appressed. $2n = 26$.

Coniferous or mixed forests; 0–
1800 m; B.C.; Calif., Nev., Oreg.,
Wash.

2b. **Pseudotsuga menziesii** (Mirbel) Franco var. **glauca**
(Mayr) Franco, Bol. Soc. Brot., ser. 2, 24: 77. 1950
· Rocky Mountain Douglas-fir, pino real colorado

Pseudotsuga douglasii (Lindley)
Carrière var. *glauca* Mayr, Wald.
Nordamer., 307, plate 6. 1890
Trees to 40(–50) m; trunk to 1.2
m diam. **Leaves** bluish green to
dark green or gray-green. **Seed
cones** 4–7 cm; bracts spreading,
often reflexed. $2n = 26$.

Coniferous or mixed forests;
600–3000 m; Alta., B.C.; Ariz., Colo., Idaho, Mont.,
Nev., N.Mex., Oreg., Tex., Utah, Wash., Wyo.; Mexico.

4. LARIX   Miller, Gard. Dict. Abr. ed. 4, vol. 2. 1754 · Larch, mélèze [Latin *larix*, name for larch]

William H. Parker

Trees deciduous; crown sparse, open. **Bark** silver-gray to gray-brown on young trees, becoming reddish brown to brown, smooth initially, scaly to thickened and furrowed with age.
Branches whorled; short (spur) shoots prominent on twigs 2 years or more old, each bearing
leaves (needles), and often pollen cone, or seed cone; lateral long shoots (sylleptic branches)

sometimes produced by current-year growth increments; leaf scars many. **Buds** rounded. **Leaves** in tufts of 10–60 on short (spur) shoots or borne singly on 1st-year long shoots, deciduous, ± flattened, with abaxial keel, sessile, base decurrent, sheath absent, apex pointed or rounded; resin canals 2. **Pollen cones** solitary, ovoid-cylindric, yellowish. **Seed cones** maturing in 1 season, persisting several years, erect, globose to ovoid, usually terminal on short shoots and thus appearing stalked, sometimes sessile on 1-year-old long shoots; scales persistent, circular to oblong-obovate, thin, lacking apophysis and umbo; bracts included or exserted. **Seeds** winged; cotyledons 4–6. $x = 12$.

Species 10 (3 in the flora): boreal and cold north temperate areas, North America, Eurasia.

Species of *Larix* are present in most boreal regions; they often form only a minor component of the vegetation. Some are important for their hard, heavy, and decay-resistant wood. Only a few have received any horticultural attention; some cultivars exist for the most commonly cultivated Old World larches, *L. decidua* Miller and *L. kaempferi* (Lambert) Carrière, but almost none for the North American species.

SELECTED REFERENCES Arno, S. F. and J. R. Habeck. 1972. Ecology of alpine larch (*Larix lyallii* Parl.) in the Pacific Northwest. Ecol. Monogr. 42: 417–450. Bakowsky, O. A. 1989. Phenotypic Variation in *Larix lyallii* and Relationships in the Larch Genus. M.Sc.F. thesis. Lakehead University. Carlson, C. 1965. Interspecific Hybridization of *Larix occidentalis* and *Larix lyallii*. M.Sc.F. thesis. University of Montana. Dickinson, T. A., W. H. Parker, and R. E. Strauss. 1987. Another approach to leaf shape comparisons. Taxon 36: 1–20. Knudsen, G. M. 1968. Chemotaxonomic Investigation of Hybridization between *Larix occidentalis* and *Larix lyallii*. M.Sc.F. thesis. University of Montana. Owens, J. N. and S. Simpson. 1986. Pollen from conifers native to British Columbia. Canad. J. Forest Res. 16: 955–967. Parker, W. H. and T. A. Dickinson. 1990. Range-wide morphological and anatomical variation in *Larix laricina*. Canad. J. Bot. 68: 832–840. Powell, G. R. 1987. Syllepsis in *Larix laricina*: Analysis of tree leaders with and without sylleptic long shoots. Canad. J. Forest Res. 17: 490–498.

1. Twigs glabrous; seed cones 1–2 cm, scales 10–30, surpassing bracts at maturity; short-shoot leaves 1–2 cm, thickness 0.3–0.5 mm; pollen 53–65 μm diam. . . . . 1. *Larix laricina*
1. Twigs initially pubescent, becoming glabrous, or strongly tomentose for 2–3 years; seed cones 2–5 cm, scales 45–55, shorter than awn-tipped bracts at maturity; short-shoot leaves 2–5 cm, thickness 0.4–0.6 mm; pollen 71–93 μm diam.
   2. Twigs initially pubescent (but not tomentose), becoming glabrous or very sparsely pubescent during first year; seed cones 2–3 cm, scale margins entire, cone stalk 3.5–5 mm diam.; width-to-thickness ratio of short-shoot leaves 1.3–1.7, adaxial surface with shallow convex midrib; resin canals 20–50 μm from margins, each surrounded by 5–7 epithelial cells. 2. *Larix occidentalis*
   2. Twigs strongly tomentose; seed cones 2.5–4(–5) cm, scale margins erose, cone stalk 2.5–4 mm diam.; width-to-thickness ratio of short-shoot leaves 1.2–1.4, adaxial surface 2-angled; resin canals 40–80 μm from margins, each surrounded by 6–10 epithelial cells. . . . . 3. *Larix lyallii*

1. Larix laricina (Du Roi) K. Koch, Dendrologie 2(2): 263. 1873 · Tamarack, mélèze laricin

Pinus laricina* Du Roi, Diss. Observ. Bot., 49. 1771; *Larix alaskensis* W. Wight; *L. laricina* var. *alaskensis* (W. Wight) Raup

Trees to 20 m; trunk to 0.6 m diam.; crown narrow, branches sparse. **Bark** of young trees gray, smooth, becoming reddish brown and scaly, inner layer red-purple. **Branches** horizontal or slightly ascending; twigs orange-brown, glabrous. **Buds** dark red, subtended by ring of hairlike bracts, glabrous. **Leaves** of short shoots 1–2 cm × 0.5–0.8 mm, 0.3–0.5 mm thick, keeled abaxially, rounded adaxially, pale blue-green; resin canals 10–20 μm from margins. **Seed cones** 1–2 × 0.5–1 cm, usually on curved stalks 2–5 × 2–2.5 mm, sometimes sessile on long shoots; scales 10–30, margins entire, brown-strigose to -tomentose at base; bracts mucronate or tipped by awn to 1 mm, hidden by mature scales, at first dark red to violet, later turning yellow-brown. **Pollen** 53–65 μm diam. **Seeds** with bodies 2–3 mm, wings 4–6 mm. $2n = 24$.

Boreal forests in wet, poorly drained sphagnum bogs and muskegs, also on moist upland mineral soils; 0–1200 m; St. Pierre and Miquelon; Alta., B.C., Man., N.B., Nfld., N.W.T., N.S., Ont., P.E.I., Que., Sask., Yukon; Alaska, Conn., Ill., Ind., Maine, Md., Mass., Mich., Minn., N.H., N.J., N.Y., Ohio, Pa., R.I., Vt., W.Va., Wis.

Disjunct Alaskan populations of *Larix laricina*, originally described as *Larix alaskensis* on the basis of narrower cone scales and bracts, are indistinguishable from other populations of the species.

The wood of tamarack is used for railway ties, pilings, and posts; it formerly was used for boat construction. Slow-growing trees develop wood with high resin content, making it decay resistant but limiting its value for pulpwood. The bark contains a tannin, which has been used for tanning leather. Although tamarack is the most rapidly growing boreal conifer under favorable conditions, it is of little commercial interest because of insect and disease problems and its poor pulping properties. Plants of this species are often stunted in the far north and on mountain slopes.

2. **Larix occidentalis** Nuttall, N. Amer. Sylv. 3: 143, plate 120. 1849 · Western larch, mélèze occidental

Trees to 50 m; trunk to 2 m diam., usually (when forest grown) branch-free over most of height; crown short, conic. **Bark** reddish brown, scaly, with deep furrows between flat, flaky, cinnamon-colored plates. **Branches** horizontal, occasionally drooping in lower crown of open-grown trees; twigs orange-brown, initially pubescent, becoming glabrous or very sparsely pubescent during first year. **Buds** dark brown, generally puberulent, scale margins erose. **Leaves** of short shoots 2–5 cm × 0.65–0.80 mm, 0.4–0.6 mm thick, keeled abaxially, with shallow convex midrib adaxially, pale green; resin canals 20–50 μm from margins, each surrounded by 5–7 epithelial cells. **Seed cones** 2–3 × 1.3–1.6 cm, on curved stalks 2.5–4.5 × 3.5–5 mm; scales 45–55, margins entire, adaxial surface pubescent; bracts tipped by awn to 3 mm, exceeding scales by ca. 4 mm. **Pollen** 71–84 μm diam. **Seeds** reddish brown, body 3 mm, wing 6 mm. $2n = 24$.

Mountain valleys and lower slopes; 500–1600 m; B.C.; Idaho, Mont., Oreg., Wash.

Western larch, when forest grown, is usually branch-free over most of its height. This is one of the most valuable timber-producing species in western North America. Its wood is made into framing, railway ties, pilings, exterior and interior finishing work, and pulp. In some localities it is the preferred firewood.

3. **Larix lyallii** Parlatore, Conif. Nov., 3. 1863 · Subalpine larch, alpine larch, mélèze de Lyall

Trees to 25 m; trunk to 1.2 m diam.; crown sparse, conic. **Bark** furrowed and flaking into red- to purple-brown scales. **Branches** horizontal, occasionally pendulous, persistent on trunk when dead; twigs strongly white- to yellow-tomentose for 2–3 years. **Buds** tomentose, scale margins ciliate. **Leaves** of short shoots 2–3.5 cm × 0.6–0.8 mm, 0.4–0.6 mm thick, keeled abaxially, 2-angled adaxially; resin canals 40–80 μm from margins, each surrounded by 6–10 epithelial cells. **Seed cones** 2.5–4(–5) × 1.1–1.9 cm, on curved stalks 3–7 × 2.5–4 mm; scales 45–55, margins erose, abaxial surface tomentose; bracts tipped by awn 4–5 mm, exceeding mature scales by ca. 6 mm. **Pollen** 78–93 μm diam. **Seeds** yellow to purple, body 3 mm, wing 6 mm.

Subalpine talus slopes; 1800–2400 m; Alta., B.C.; Idaho, Mont., Wash.

Larix lyallii and *L. occidentalis* (*Larix* sect. *Multiseriales*) are similar morphologically and have similar geographic ranges. Just how closely the two species are related has not been determined, but they probably originated from a common ancestor resembling *L. potaninii* Batalin. Although the geographic ranges of the two species overlap considerably, elevational differences of 150 to 300 m usually separate them. Some morphologically intermediate specimens have been collected from Washington and Montana.

Because of its restricted distribution and growth at timberline, alpine larch has no commercial importance; it is often dwarfed and misshapen.

5. PICEA A. Dietrich, Fl. Berlin 2: 794. 1824 · Spruce, épinette [Latin *picis,* pitch, name of a pitchy pine]

Ronald J. Taylor

Trees evergreen; crown broadly conic to spirelike; leading shoot erect. **Bark** gray to reddish brown, thin and scaly (with thin plates), sometimes with resin blisters (especially in *Picea engelmannii* and *P. glauca*), becoming relatively thick and furrowed with age. **Branches** whorled; short (spur) shoots absent; twigs roughened by persistent leaf bases. **Buds** ovoid, apex rounded to acute, sometimes resinous. **Leaves** borne singly, spreading in all directions from twigs, persisting to 10 years, mostly 4-angled and square in cross section (to triangular or ± flattened), mostly rigid, sessile on peglike base; base decurrent, persistent after leaves shed, sheath absent; apex usually sharp-pointed, sometimes bluntly acute; resin canals 1–2. **Cones** borne on year-old twigs. **Pollen cones** grouped, axillary, oblong, yellow to purple. **Seed cones** maturing in 1 season, usually shed at maturity (persisting for several years in *Picea mariana*), borne mostly on upper branches, pendent, ovoid to cylindric, sessile or terminal on leafy branchlets and thus appearing ± stalked; scales persistent, elliptic to fan-shaped, thin, lacking apophysis and umbo; bracts included. **Seeds** winged; cotyledons 5–15. $x = 12$.

Species ca. 35 (8 in the flora with 7 native and 1 naturalized): north temperate regions, North America, Mexico, Eurasia.

SELECTED REFERENCES Roche, L. 1969. A genecological study of the genus *Picea* and seedlings grown in a nursery. New Phytol. 68: 505–554. Taylor, R. J. and T. F. Patterson. 1980. Biosystematics of Mexican spruce species and populations. Taxon 29: 421–469. Wright, J. W. 1955. Species crossability in spruce in relation to distribution and taxonomy. Forest Sci. 1: 319–349.

1. Leaves flattened or broadly triangular in cross section; stomates occurring only on, or more conspicuously on, adaxial leaf surface; Alaska to California.
    2. Leaf apex sharp-pointed; twigs glabrous; seed-cone scales elliptic to narrowly diamond-shaped. . . . . 7. *Picea sitchensis*
    2. Leaf apex blunt, especially on older leaves; twigs finely pubescent; seed-cone scales fan-shaped. . . . . 8. *Picea breweriana*
1. Leaves square in cross section; stomates occurring more or less equally on all leaf surfaces; widespread.
    3. Twigs pubescent; cone scales usually fan-shaped, broadest near apex; seed cones 2.3–4.5(–5) cm; mostly eastern or boreal.
        4. Seed cones 1.5–2.5(–3.5) cm; leaves 0.6–1.5(–2) cm, mostly blunt-tipped, glaucous, blue- to gray-green. . . . . 2. *Picea mariana*
        4. Seed cones 2.3–4.5(–5) cm; leaves 0.8–2.5(–3) cm, mostly acute, sharp-pointed, not glaucous, yellow-green to dark green. . . . . 3. *Picea rubens*
    3. Twigs mostly glabrous; cone scales usually ± diamond-shaped or elliptic, broadest near middle (broadest at apex in *Picea glauca*); seed cones (2.5–)3–16 cm; mostly western or boreal.
        5. Seed cones (10–)12–16 cm; leaves 1–2.5 cm, blunt-tipped; introduced and locally naturalized species. . . . . 1. *Picea abies*
        5. Seed cones 2.5–11(–12) cm; leaves (0.8–)1.2–3(–3.5) cm, mostly sharp-pointed; native species.
            6. Seed-cone scales fan-shaped, margin at apex ± entire, apex extending 0.5–3 mm beyond seed-wing impression; leaves (0.8–)1.5–2(–2.5) cm. . . . . 4. *Picea glauca*
            6. Seed-cone scales diamond-shaped or elliptic, margin at apex irregularly toothed to erose, apex usually extending 3 mm or more beyond seed-wing impression; leaves 1.2–3(–3.5) cm.

7. Seed cones 3–7(–8) cm; cone scales extending 3–8 mm beyond seed-wing impression; twigs finely pubescent. . . . . 5a. *Picea engelmannii* var. *engelmannii*
7. Seed cones (5–)6–11(–12) cm; cone scales extending 8–10 mm beyond seed-wing impression; twigs usually glabrous. . . . . 6. *Picea pungens*

1. **Picea abies** (Linnaeus) H. Karsten, Deut. Fl. 2/3: 324. 1881 · Norway spruce, épinette de Norvège

Pinus abies Linnaeus, Sp. Pl. 2: 1002. 1753

Trees to 30 m; trunk to 2 m diam.; crown conic. **Bark** gray-brown, scaly. **Branches** short and stout, the upper ascending, the lower drooping; twigs stout, reddish brown, usually glabrous. **Buds** reddish brown, 5–7 mm, apex acute. **Leaves** 1–2.5 cm, 4-angled in cross section, rigid, light to dark green, bearing stomates on all surfaces, apex blunt-tipped. **Seed cones** (10–)12–16 cm; scales diamond-shaped, widest near middle, 18–30 × 15–20 mm, thin and flexuous, margin at apex erose to toothed, apex extending 6–10 mm beyond seed-wing impression. 2*n* = 24.

Woods and persisting after cultivation; introduced; Minn., probably elsewhere; Europe.

Norway spruce, native to Europe, has become locally naturalized, at least in north central United States (and adjacent Canada). The species is the most widely cultivated spruce in North America; many cultivars exist, including dwarf shrubs.

2. **Picea mariana** (Miller) Britton, Sterns, & Poggenburg, Prelim. Cat., 71. 1888 · Black spruce, épinette noire

Abies mariana Miller, Gard. Dict. ed. 8, Abies no. 5. 1768; *Picea brevifolia* Peck; *P. mariana* var. *brevifolia* (Peck) Rehder; *P. nigra* (Aiton) Link; *Pinus nigra* Aiton

Trees to 25 m (often shrublike); trunk to 0.25 m diam.; crown narrowly conic to spirelike. **Bark** gray-brown. **Branches** short and drooping, frequently layering; twigs not pendent, rather slender, yellow-brown, pubescent. **Buds** gray-brown, ca. 3 mm, apex acute. **Leaves** 0.6–1.5(–2) cm, 4-angled in cross section, rigid, pale blue-green, glaucous, bearing stomates on all surfaces, apex mostly blunt-tipped. **Seed cones** 1.5–2.5(–3.5) cm; scales fan-shaped, broadest near apex, 8–12 × 8–12 mm, rigid, margin at apex irregularly toothed. 2*n* = 24.

Muskegs, bogs, bottomlands, dry peatlands; 0–1500 m; St. Pierre and Miquelon; Alta., B.C., Man., N.B., Nfld., N.W.T., N.S., Ont., P.E.I., Que., Sask., Yukon; Alaska, Conn., Maine, Mass., Mich., Minn., N.H., N.J., N.Y., Pa., R.I., Vt., Wis.

To a limited extent, *Picea mariana* hybridizes with *P. rubens*, e.g., on disturbed sites in eastern Canada. Natural hybridization with *P. glauca*, though reported, remains unverified (A. G. Gordon 1976).

Because *Picea mariana* is a small tree, it has limited commercial value. Frequently it is harvested with *P. glauca* and used for pulp.

Black spruce (*Picea mariana*) is the provincial tree of Newfoundland.

SELECTED REFERENCES Little, E. L. Jr. and S. S. Pauley. 1958. A natural hybrid between black and white spruce in Minnesota. Amer. Midl. Naturalist 60: 202–211. Morgenstern, E. K. and J. L. Farrar. 1964. Introgressive Hybridization in Red Spruce and Black Spruce. [Toronto.] [Univ. Toronto, Fac. Forest., Techn. Rep. 4.]

3. **Picea rubens** Sargent, Silva 12: 33, plate 597. 1898 · Red spruce, épinette rouge

Picea australis Small; *P. nigra* (Aiton) Link var. *rubra* (Du Roi) Engelmann; *P. rubra* (Du Roi) Link 1831, not A. Dietrich 1824

Trees to 40 m; trunk to 1 m diam.; crown narrowly conic. **Bark** gray-brown to reddish brown. **Branches** horizontally spreading; twigs not pendent, rather stout, yellow-brown, densely pubescent to glabrate. **Buds** reddish brown, 5–8 mm, apex acute. **Leaves** 0.8–2.5(–3) cm, 4-angled in cross section, somewhat flexuous, yellow-green to dark green, not glaucous, bearing stomates on all surfaces, apex mostly acute to sharp-pointed. **Seed cones** 2.3–4.5(–5) cm; scales broadly fan-shaped, broadest near apex, 8–12 × 8–12 mm, stiff, margin at apex entire to irregularly toothed. 2*n* = 24.

Upper montane to subalpine forests; 0–2000 m; St. Pierre and Miquelon; N.B., N.S., Ont., P.E.I., Que.; Conn., Maine, Md., Mass., N.H., N.J., N.Y., N.C., Pa., Tenn., Vt., Va., W.Va.

Throughout the Appalachians, trees of *Picea rubens* are dying, possibly as a consequence of environmental pollution. In eastern Canada this species hybridizes to a limited extent with *P. mariana* (A. G. Gordon 1976).

Red spruce (*Picea rubens*) is the provincial tree of Nova Scotia.

SELECTED REFERENCE Morgenstern, E. K. and J. L. Farrar. 1964. Introgressive Hybridization in Red Spruce and Black Spruce. [Toronto.] [Univ. Toronto, Fac. Forest., Techn. Rep. 4.]

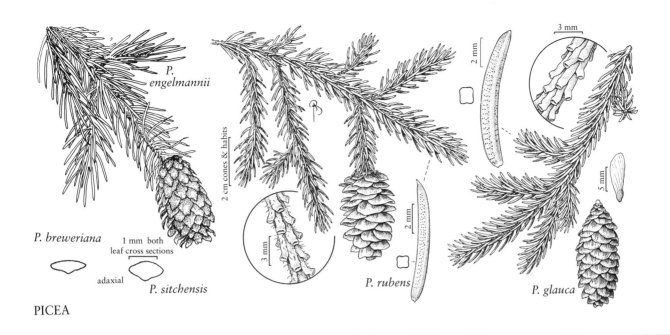

P.
engelmannii

P. breweriana

1 mm both
leaf cross sections

adaxial

P. sitchensis

P. rubens

P. glauca

2 cm cones & habits

3 mm

2 mm

2 mm

3 mm

5 mm

PICEA

4. Picea glauca (Moench) Voss, Mitt. Deutsch. Dendrol. Ges. 16: 93. 1907 [1908] · White spruce, western white spruce, Porsild spruce, Black Hills spruce, épinette blanche

Pinus glauca Moench, Verz. Ausländ. Bäume, 73. 1785; *Abies canadensis* Miller; *Picea alba* (Aiton) Link; *P. alba* var. *albertiana* (S. Brown) Beissner; *P. albertiana* S. Brown; *P. canadensis* (Miller) Britton, Sterns, & Poggenburg; *P. canadensis* var. *glauca* (Moench) Sudworth; *P. glauca* var. *albertiana* (S. Brown) Sargent; *P. glauca* var. *densata* Bailey; *P. glauca* var. *porsildii* Raup; *Pinus alba* Aiton

Trees to 30 m; trunk to 1 m diam.; crown broadly conic to spirelike. **Bark** gray-brown. **Branches** slightly drooping; twigs not pendent, rather slender, pinkish brown, glabrous. **Buds** orange-brown, 3–6 mm, apex rounded. **Leaves** (0.8–)1.5–2(–2.5) cm, 4-angled in cross section, rigid, blue-green, bearing stomates on all surfaces, apex sharp-pointed. **Seed cones** 2.5–6(–8) cm; scales fan-shaped, broadest near rounded apex, 10–16 × 9–13 mm, flexuous, margin at apex ± entire, apex extending 0.5–3 mm beyond seed-wing impression. $2n = 24$.

Muskegs, bogs, and river banks to montane slopes; 0–1000 m; St. Pierre and Miquelon; Alta., B.C., Man., N.B., Nfld., N.W.T., N.S., Ont., P.E.I., Que., Sask.,

Yukon; Alaska, Maine, Mich., Minn., Mont., N.H., N.Y., S.Dak., Vt., Wis., Wyo.

In areas of sympatry *Picea glauca* and *P. engelmannii* regularly hybridize and intergrade completely (R. Daubenmire 1974; E. H. Garman 1957; K. W. Horton 1959; L. Roche 1969; T. M. C. Taylor 1959). This has greatly complicated the taxonomy of *P. glauca*, a dominant tree of interior forests of Canada and Alaska. Three varieties have been recognized. *Picea glauca* var. *albertiana* was described as having unusually prominent leaf bases, cones nearly as broad as long, cone scales acute and broader than long, and an unusually narrow crown. These are common characteristics of hybrids (e.g., R. Daubenmire 1974). *Picea glauca* var. *porsildii* was described as differing from the type variety by having smooth bark with resin blisters, short angular cone scales, an unusually broad crown, and pubescent twigs. These characteristics, also largely intermediate between those of *P. glauca* var. *glauca* and *P. engelmannii*, may reflect hybridization where the species overlap. Although the two varieties noted above are reported from well beyond the range of sympatry, the diagnostic characteristics are not well correlated and occur rather sporadically. Also the most distinctive feature of the varieties, the crown shape, is in part responsive to competitive pressures. Because of the problems of hybridization and sporadic occurrence of key characteristics, *P. glauca* is treated here in the broad sense.

Picea glauca (white spruce) is the provincial tree of

Manitoba and the state tree (as Black Hills spruce) of South Dakota.

SELECTED REFERENCES LaRoi, G. H. and J. R. Dugle. 1968. A systematic and genecological study of *Picea glauca* and *P. engelmannii*, using paper chromatography of needle extracts. Canad. J. Bot. 46: 649–687. Little, E. L. Jr. and S. S. Pauley. 1958. A natural hybrid between black and white spruce in Minnesota. Amer. Midl. Naturalist 60: 202–211.

5. **Picea engelmannii** Parry ex Engelmann, Trans. Acad. Sci. St. Louis 2: 212. 1863 · Engelmann spruce, épinette d'Engelmann, pino real

Varieties 2 (1 in the flora): North America, Mexico.

5a. **Picea engelmannii** Parry ex Engelmann var. **engelmannii**

Picea glauca (Moench) Voss subsp. *engelmannii* (Parry ex Engelmann) T. M. C. Taylor; *P. columbiana* Lemmon; *P. engelmannii* var. *glabra* Goodman

Trees to 60 m; trunk to 2 m diam.; crown narrowly conic. **Bark** gray to reddish brown. **Branches** spreading horizontally to somewhat drooping; twigs not pendent, rather stout, yellow-brown, finely pubescent, occasionally glabrous. **Buds** orange-brown, 3–6 mm, apex rounded. **Leaves** 1.6–3(–3.5) cm, 4-angled in cross section, rigid, blue-green, bearing stomates on all surfaces, apex sharp-pointed. **Seed cones** 3–7(–8) cm; scales diamond-shaped to elliptic, widest above middle, 13–20 × 9–16 mm, flexuous, margin at apex irregularly toothed to erose, apex extending 3–8 mm beyond seed-wing impression. $2n = 24$.

Montane and subalpine forests; 1000–3000 m; Alta., B.C.; Ariz., Calif., Colo., Idaho, Mont., Nev., N.Mex., Oreg., Utah, Wash., Wyo.; Mexico.

Only *Picea engelmannii* var. *engelmannii* occurs in the flora. In Mexico the species is also represented by *P. engelmannii* var. *mexicana* (Martínez) R. J. Taylor & T. F. Patterson.

Although *Picea engelmannii* varies considerably over its broad distributional range, the variation is continuous, militating against recognition of multiple varieties (R. J. Taylor and T. F. Patterson 1980). In the northern part of its range, it hybridizes freely and completely intergrades with *P. glauca*, as noted above.

SELECTED REFERENCE LaRoi, G. H. and J. R. Dugle. 1968. A systematic and genecological study of *Picea glauca* and *P. engelmannii*, using paper chromatography of needle extracts. Canad. J. Bot. 46: 649–687.

6. **Picea pungens** Engelmann, Gard. Chron., n. s. 11: 334. 1879 · Blue spruce, Colorado blue spruce, épinette bleue, pino real

Picea parryana Sargent

Trees to 50 m; trunk to 1.5 m diam.; crown broadly conic. **Bark** gray-brown. **Branches** slightly to strongly drooping; twigs not pendent, stout, yellow-brown, usually glabrous. **Buds** dark orange-brown, 6–12 mm, apex rounded to acute. **Leaves** 1.6–3 cm, 4-angled in cross section, rigid, blue-green, bearing stomates on all surfaces, apex spine-tipped. **Seed cones** (5–)6–11(–12) cm; scales elliptic to diamond-shaped, widest below middle, 15–22 × 10–15 mm, rather stiff, margin at apex erose, apex extending 8–10 mm beyond seed-wing impression. $2n = 24$.

Midmontane forests; 1800–3000 m; Ariz., Colo., Idaho, N.Mex., Utah, Wyo.

Limited hybridization occurs between *Picea pungens* and *P. engelmannii* (R. Daubenmire 1972; R. J. Taylor et al. 1975).

Blue spruce *(Picea pungens)* is the state tree of Colorado (as Colorado blue spruce) and Utah.

7. **Picea sitchensis** (Bongard) Carrière, Traité Gén. Conif., 260. 1855 · Sitka spruce, épinette de Sitka

Pinus sitchensis Bongard, Mém. Acad. Imp. Sci. Saint-Pétersbourg, Sér. 6, Sci. Math. 2: 164. 1832 (Aug.); *Abies falcata* Rafinesque; *A. menziesii* (Douglas ex D. Don) Lindley 1835, not Mirbel 1825; *Picea falcata* (Rafinesque) Suringar; *P. menziesii* (Douglas ex D. Don) Carrière; *Pinus menziesii* Douglas ex D. Don

Trees to 80 m; trunk to 5 m diam.; crown narrowly conic. **Bark** grayish brown to orange-brown. **Branches** somewhat drooping; twigs not pendent, rather stout, pinkish brown, glabrous. **Buds** reddish brown, 5–10 mm, apex rounded. **Leaves** (1.2–)1.5–2.5(–3) cm, flattened or broadly triangular in cross section (abaxial surface rounded or slightly angular), rather rigid, blue-green to light yellow-green, abaxial surface darker green with stomatal bands very narrow or absent, adaxial surface glaucous with conspicuous stomatal bands separated by ridge, apex sharp-pointed. **Seed cones** 5–9(–10) cm; scales variable, elliptic to narrowly diamond-shaped, 15–22 × 12–16 mm, rather rigid, margin at apex erose, apex extending 4–8 mm beyond seed-wing impression. $2n = 24$.

Pacific coastal forests; 0–900 m; B.C.; Alaska, Calif., Oreg., Wash.

Picea sitchensis intergrades extensively with *P. glauca* in the river inlets of north coastal British Columbia and coastal Alaska. The name *P.* × *lutzii* Little is applied to hybrids between the two species (R. Daubenmire 1968).

Sitka spruce *(Picea sitchensis)* is the state tree of Alaska.

SELECTED REFERENCE Daubenmire, R. 1968. Some geographic variations in *Picea sitchensis* and their ecologic interpretation. Canad. J. Bot. 46: 787–798.

8. Picea breweriana S. Watson, Proc. Amer. Acad. Arts 20: 378. 1885 · Brewer spruce, weeping spruce

Trees to 40 m; trunk to 1.5 m diam., typically buttressed; crown conic. **Bark** gray to brown. **Branches** drooping; twigs pendent, elongate, slender, gray-brown, finely pubescent. **Buds** gray-brown, 5–7 mm, apex rounded. **Leaves** 1.5–3 cm, flattened or broadly triangular in cross section (abaxial surface rounded or slightly angular), rather rigid, abaxial surface dark green with stomatal bands absent, adaxial surface glaucous with conspicuous stomatal bands separated by slight ridge or angle, apex blunt (especially on older leaves). **Seed cones** 6.5–12 cm; scales fan-shaped, 15–20 × 15–20 mm, rigid, margin at apex entire to slightly erose.

Montane to subalpine forests of the Siskiyou Mountains; 1000–2300 m; Calif., Oreg.

SELECTED REFERENCE Waring, R. H., W. H. Emmingham, and S. W. Running. 1975. Environmental limits of an endemic spruce, *Picea breweriana*. Canad. J. Bot. 53: 1599–1613.

6. PINUS Linnaeus, Sp. Pl. 2: 1000. 1753; Gen Pl. ed. 5, 434. 1754 · Pine [Latin *pinus*, name for pine]

Robert Kral

Apinus Necker ex Rydberg; *Strobus* (Sweet) Opiz; *Caryopitys* Small

Trees or shrubs aromatic, evergreen; crown usually conic when young, often rounded or flat-topped with age. **Bark** of older stems variously furrowed and plated, plates and/or ridges layered or scaly. **Branches** usually in pseudowhorls; shoots dimorphic with long shoots and short shoots; short shoots borne in close spirals from axils of scaly bracts and bearing fascicles of leaves (needles). **Buds** ovoid to cylindric, apex pointed (blunt), usually resinous. **Leaves** dimorphic, spirally arranged; foliage leaves (needles) (1–)2–5(–6) per fascicle, persisting 2–12 or more years, terete or ± 2–3-angled and rounded on abaxial surface, sessile, sheathed at base by 12–15 overlapping scale leaves, these (at least firmer basal ones) persisting for life of fascicle or shed after first season; resin canals 2 or more. **Pollen cones** in dense, spikelike cluster around base of current year's growth, mostly ovoid to cylindric-conic, tan to yellow, red, blue, or lavender. **Seed cones** maturing in 2(–3) years, shed early or variously persistent, pendent to ± erect, at maturity conic or cylindric, sessile or stalked, shedding seed soon after maturity or variously serotinous (not opening upon maturity but much later); scales persistent, woody or pliable, surface of exposed apical portion of each scale (apophysis) thickened, with umbo (exposed scale surface of young cone) represented by a scar (sometimes apiculate) or extended into a hook, spur, claw, or prickle; bracts included. **Seeds** winged or wingless; cotyledons (3–)6–10(–18). $x = 12$.

Species ca. 100 (38 in the flora with 37 native and 1 widely naturalized): widespread in north temperate and north tropical (mountainous) regions, North America, Mexico, West Indies, Central America, Eurasia, n Africa, Pacific Islands in Sumatra.

In many areas *Pinus* is a forest dominant, either early successional and thus weedy or often longer-lived and part of climax forest. Certain southern pines, especially fire successional species, have a "grass stage," i.e., the stem of the young seedling elongates little during the first several years and bears many long, curved leaves, the plant then reminiscent of a dense clump of grass.

Nomenclature used here, and to a very large degree the taxonomy, follows Elbert L. Little Jr. (1971), former Chief Dendrologist, United States Department of Agriculture, Forest Service. Much work is being done with problematic groups, particularly complexes in *Pinus contorta,* the pinyons, the bristlecone pines, and *P. ponderosa* and related taxa. Considerable chemotaxonomic and genetic data are available on the genus, but coverage is far from comprehensive. Therefore, the conservative approach used in this treatment emphasizes external morphology.

Users of this account of *Pinus* should note the following.

Leaf measurements given herein are based on healthy, fully expanded growth, especially that of cone-bearing branches.

Fascicle-sheath measurements are based on fully developed, unbroken sheaths, not on sheaths as they later break up.

After pollen is shed, pollen cones may lengthen considerably. Measurements given below for pollen cones are those of the cones at the time that pollen is released.

Colors of seed cones are those of mature, closed or newly opened cones, not of old, open, persistent cones or of weathered serotinous cones. Mature, open cones may be hygroscopic, closing partially or completely when wet.

Descriptions of apophyses, too, are based on mature, closed or newly opened cones. Unlike characters of umbos of most species, characters of apophyses are much altered as the cone grows.

The term "twig" is used here to refer to growth of the current season.

Leaf dimorphism is a problem. In some species, for example, low rainfall and beyond-normal stresses in the environment can lead to sets of atypically short leaves. There are also pathologic abnormalities, e.g., "little-leaf" disease in *Pinus echinata.* Such responses are not accounted for.

Two exotic species of *Pinus* have been reported as naturalized in the flora: *P. nigra* Arnott in Illinois and *P. thunbergiana* Franco (*P. thunbergii* Parlatore) in Massachusetts. These are not included in the key below, where they would key to *P. resinosa.* Both are distinguished from that species by their fresh leaves, which bend—rather than break—when bent; by their pale silvery—not reddish brown—winter buds; by their seed-cone scales, some or all of which are minutely armed—rather than unarmed; and by their apophyses, which at the time of seed-shed are cream to light brown or gray—rather than light red-brown. They are distinguished from each other as follows: *P. nigra*—seed cones sessile with base rounded, terminal bud resinous, and leaves sometimes with central resin canals; and *P. thunbergiana*—seed cones stalked with base more or less truncate, terminal bud not resinous, and leaves lacking central resin canals.

Pine *(Pinus)* has been adopted by Arkansas as the state tree. Southern pine *(Pinus* spp.) is the state tree of Alabama.

SELECTED REFERENCES Bailey, D. K. 1970. Phytogeography and taxonomy of *Pinus* subsection *Balfourianae*. Ann. Missouri Bot. Gard. 57: 210–249. Bailey, D. K. 1987. A study of *Pinus* subsection *Cembroides* I: The single-needle pinyons of the Californias and the Great Basin. Notes Roy. Bot. Gard. Edinburgh 44: 275–310. Bailey, D. K. and F. G. Hawksworth. 1979. Pinyons of the Chihuahuan Desert region. Phytologia 44: 129–133. Critchfield, W. B. and E. L. Little Jr. 1966. Geographic Distribution of the Pines of the World. Washington. [U.S.D.A., Misc. Publ. 991.] Duffield, J. W. 1952. Relationships and species hybridization in the genus *Pinus*. Silvae Genet. 1: 93–97. Fowells, H. A. 1965. Silvics of Forest Trees of the United States. Washington. [Agric. Handb. 271.] Kurz, H. and R. K. Godfrey. 1962. Trees of Northern Florida. Gainesville. Little, E. L. Jr. and W. B. Critchfield. 1969. Subdivisions of the genus *Pinus* (pines). Washington. [U.S.D.A., Misc. Publ. 1144.] Mirov, N. T. 1967. The Genus *Pinus*. New York. Peattie, D. C. 1953. A Natural History of Western Trees. Boston. Perry, J. P. Jr. 1991. The Pines of Mexico and Central America. Portland. Preston, R. J. 1976. North American Trees (Exclusive of Mexico and Tropical United States), ed. 3. Ames. Price, R. A. 1989. The genera of Pinaceae in the southeastern United States. J. Arnold Arbor. 70: 247–305. Sargent, C. S. 1922. Manual of the Trees of North America (Exclusive of Mexico), ed. 2. Boston and New York. [Facsimile edition in 2 vols. 1961, reprinted 1965, New York.] Shaw, G. R. 1914. The Genus *Pinus*. Cambridge, Mass. [Publ. Arnold Arbor. 5.] Sudworth, G. B. 1908. Forest Trees of the Pacific Slope. Washington. Sudworth, G. B. 1917. The Pine Trees of the Rocky Mountain Region. Washington. [U.S.D.A. Bull. 460.]

1. Leaves in cross section with 1 fibrovascular bundle; scales of fascicle sheaths shed early, not falling with fascicle, these and bud scales mostly with margins entire, less often finely ciliate or finely fringed; seed cones unarmed (except in *P. balfouriana*, *P. aristata*, *P. longaeva*)(subg. *Strobus*).
    2. Leaves mostly 5 per fascicle, 3-sided, straight or curved; open seed cones (cones of *P. albicaulis* do not open) narrowly ovoid to ± cylindric; seeds winged or wingless.
        3. Fascicles persistent 10 or more years, forming a long brush; leaves curved, connivent, 5 cm or less; apophyses much thickened, umbos central; mature seed cones lanceoloid-cylindric before opening, purple to brown or red-brown, with prickles (prickles much reduced or absent in *P. balfouriana*), open cones lance-ovoid to ovoid or cylindric; seeds winged; alpine or timberline trees.
            4. Abaxial surface of leaves with strong, narrow median groove; prickle slender, long (mostly 6–10 mm). . . . . 12. *Pinus aristata*
            4. Abaxial surface of leaves lacking median groove or, if grooved, with grooves indistinct and more than 1; prickle to 6 mm or very reduced, weak, even absent.
                5. Seed-cone base conic; apophysis rounded, umbo depressed, prickle absent or to 1 mm, weak; resin exudates on seed cone amber. . . . . 11. *Pinus balfouriana*
                5. Seed-cone base rounded; apophysis sharply keeled, umbo raised on low buttress, truncate or umbilicate, abruptly narrowed to slender prickle 1–6 mm; resin exudates on seed cone pale. . . . . 13. *Pinus longaeva*
        3. Fascicles persistent 8 or fewer years, not forming a long brush; leaves straight or curved, of various lengths; apophyses variously thickened, umbos terminal; mature seed cones variously shaped, pale brown to gray or gray-brown (purplish in *P. albicaulis*), without prickles; seeds winged or wingless; trees of various elevations.
            6. Seed wing at least as long as seed body; apophysis mostly not much thicker than subtending part of seed-cone scale; leaves straight, not persisting past 4–5 years; large trees of various elevations.
                7. Stomatal lines evident on all surfaces of leaves (best seen on younger growth); seed cones 25–50 cm; seed body 1–2 cm, wing 2–3 cm. . . . . 4. *Pinus lambertiana*
                7. Stomatal lines evident only on adaxial surface of leaves; seed cones 7–25 cm; seed body 0.5–0.7 cm, wing 1.8–2.5 cm.
                    8. Leaf apices broadly to narrowly acute; apophyses of mature scales creamy brown to yellowish, without purple or gray tints; bark of mature tree distinctly platy; w North America. 3. *Pinus monticola*
                    8. Leaf apices abruptly acute to short-acuminate; apophyses of mature scales with purple or gray tints; bark of mature tree distinctly furrowed; e North America. . . . . 2. *Pinus strobus*

6. Seed wing lacking or shorter than seed body; apophysis mostly distinctly thicker than subtending portion of seed-cone scale; leaves straight or curved, persisting 5 years or more (except *P. strobiformis*); medium-sized to low trees mostly of high elevations.

    9. Mature seed cones broadly ovoid to depressed-ovoid or nearly globose, 4–8 cm, remaining closed, scales readily broken off through animal agency. . . . .   1. *Pinus albicaulis*

    9. Mature seed cones lance-ovoid or lance-cylindric before opening, 7 cm or more, opening at maturity, scales firmly attached.

       10. Apophyses of fertile scales recurved; bark of mature trunk thick, furrowed; stomatal lines not evident on abaxial leaf surface. . . . . . .   6. *Pinus strobiformis*

       10. Apophyses of fertile scales not recurved; bark of mature trunk thin, platy; stomatal lines evident on all leaf surfaces. . . . .   5. *Pinus flexilis*

2. Leaves mostly 1–4 per fascicle, terete or 2–3-sided, stiff and strongly incurved; open seed cones depressed-ovoid to nearly globose; seeds wingless.

    11. Leaves mostly 1 per fascicle, terete (sometimes with a strong groove on each side).

    . . . .   10. *Pinus monophylla*

    11. Leaves mostly 2–4(–5) per fascicle, mostly 2–3-sided.

       12. Leaves 0.6–0.9(–1.0) mm wide, (2–)3(–4) per fascicle. . . . .   7. *Pinus cembroides*

       12. Leaves at least 1 mm wide, (1–)2–4 per fascicle.

          13. Leaves (3–)4(–5) per fascicle; trees in s California and Baja California.

          . . . .   9. *Pinus quadrifolia*

          13. Leaves (1–)2(–3) per fascicle; trees in n Mexico and Rocky Mountains northward, rarely in California Sierra Nevada. . . . .   8. *Pinus edulis*

1. Leaves in cross section with 2 fibrovascular bundles; scales of fascicle sheaths, at least lower ones, persistent and falling with fascicle (except in *P. leiophylla* and *P. torreyana*), these and bud scales mostly with margins long-fringed; seed cones mostly armed (subg. *Pinus*).

14. Leaves 2(–3) per fascicle.

    15. Longer leaves of healthy branches 10–15 cm or more.

       16. Seed cones mostly asymmetric, curved-ovoid when open, mostly in whorls, larger umbos forming claws; fascicle sheath mostly 1.5 cm or less; cones serotinous; California. . . . .   38. *Pinus muricata*

       16. Seed cones symmetric or nearly so, not curved when open, mostly not in whorls, umbos not forming claws; fascicle sheath mostly 1.5 cm or more; cones not serotinous; e North America or widespread in w North America.

          17. Seed cones unarmed, mostly 6 cm or less; fresh leaves brittle, breaking cleanly when bent, ca. 1 mm wide or less; e boreal forest. . . . .   15. *Pinus resinosa*

          17. Seed cones armed, mostly 6 cm or more; fresh leaves pliant, not breaking cleanly when bent, mostly over 1 mm wide; se or w montane North America.

             18. Seed cones stalked, chocolate brown, apophyses lustrous; se North America. . . . .   24. *Pinus elliottii*

             18. Seed cones sessile or nearly sessile, yellow-brown to brown or red-brown, apophyses rarely lustrous; se or w North America.

                19. Healthy twigs to ca. 1 cm thick; terminal bud lance-cylindric, mostly less than 1 cm broad, slightly resinous; open seed cones narrowly ovoid, mostly dull yellow-brown; se North America.

                . . . .   18. *Pinus taeda*

                19. Healthy twigs 1–2 cm thick; terminal bud ovoid, fully 1 cm broad, very resinous; open seed cones broadly ovoid, mostly reddish brown; w North America. . . . .   25. *Pinus ponderosa*

    15. Longer leaves of healthy branches mostly 10–13 cm or less.

       20. Seed cones evidently asymmetric, variably serotinous (except for *P. contorta* var. *murrayana*).

21. Larger umbos extended into stout, curved claws; Appalachian Mountains and associated piedmont, not Alabama or Florida. . . . . 23. *Pinus pungens*
21. Larger umbos not extended into claws; n, w North America, or Alabama and Florida.
    22. Seed cones curved forward on branches, unarmed or with small reflexed apiculi. . . . . 32. *Pinus banksiana*
    22. Seed cones spreading to recurved on branches, mostly armed with prickles.
        23. Twigs aging brownish, rough; w North America. . . . . 33. *Pinus contorta*
        23. Twigs aging gray, smooth or nearly so; Alabama and Florida. . . 35. *Pinus clausa*
20. Seed cones symmetric or nearly so, not serotinous (except *P. clausa*).
    24. Leaves strongly twisted, longer ones mostly (2–)3–8 cm × 1 mm or broader; seed cones persistent or not.
        25. Bark on upper sections of trunk orange, platy; leaves blue- to gray- or yellow-green, stomatal lines conspicuous; twigs at first dull green to orange-brown, not glaucous; seed cones unarmed, not persistent; adaxial surface of seed-cone scales lacking contrasting border distally; introduced Eurasian species. . . . . 16a. *Pinus sylvestris* var. *sylvestris*
        25. Bark on upper sections of trunk reddish, scaly; leaves deep to pale yellow-green, stomatal lines inconspicuous; twigs at first red- or purple-tinged, often glaucous; seed cones armed, persistent; adaxial surface of seed-cone scales with strong purple-red or purple-brown border distally; native. . . . . 34. *Pinus virginiana*
    24. Leaves straight, slightly twisted, fine, longer ones mostly 6–13 cm × ca. 1 mm; seed cones persistent.
        26. Twigs roughened and cracking below leafy portion; bark plates with evident resin pockets. . . . . 19. *Pinus echinata*
        26. Twigs smooth below leafy portion; bark plates lacking resin pockets.
            27. Bark on upper sections of trunk reddish to red-brown, platy; seed cones often long-serotinous, long-persistent, usually armed with prickle; adaxial surface of seed-cone scales with dark red-brown, purple, or purple-gray border distally. . . . . 35. *Pinus clausa*
            27. Bark on upper sections of trunk gray, ± smooth, appearing slick; seed cones not serotinous, semipersistent, unarmed or with weak, short, deciduous prickle; adaxial surface of seed-cone scales lacking contrasting border distally. . . . . 20. *Pinus glabra*
14. Leaves 3(–5) per fascicle.
  28. Pines of e North America (w to Missouri, e Oklahoma, and e Texas).
    29. Longer leaves of healthy growth rarely more than 11 cm, mostly 5–11(–15) cm; seed cones 10 cm or less.
        30. Two-year-old branchlets slender (ca. 5 mm thick or less); twigs at first greenish brown to red-brown, often glaucous; leaves ca. 1 mm wide, not or only slightly twisted; bark plates with evident resin pockets; adaxial surface of seed-cone scales lacking contrasting border distally. . . . . . 19. *Pinus echinata*
        30. Two-year-old branchlets stout (mostly over 5 mm thick); twigs at first orange-brown; leaves 1–1.5(–2 mm) wide, twisted; bark plates without resin pockets; adaxial surface of seed-cone scales with dark red-brown border distally. . . . . 21. *Pinus rigida*
    29. Longer leaves of healthy growth rarely less than 12 cm, mostly 15 cm or more; seed cones variable in length.
        31. Leaves 15–45 cm; pollen cones purplish.
            32. Terminal buds ovoid, silvery white, 3–4 cm; seed cones sessile (rarely short stalked), 15–25 cm, apophyses dull; leaves 20–45 cm. . . . . 17. *Pinus palustris*

32. Terminal buds cylindric, silvery brown, 1.5–2 cm; seed cones stalked, (7–)9–18(–20) cm, apophyses lustrous (as if varnished); leaves 15–20(–23) cm. . . . . 24. *Pinus elliottii*
31. Leaves mostly under 20 cm; pollen cones yellowish to brownish.
    33. Seed cones variably serotinous, long-persistent, broadly ovoid to nearly globose when open; umbos with short, weak prickle or none; adaxial surface of seed-cone scales with dark red-brown border distally; trunks commonly with adventitious shoots. . . . . 22. *Pinus serotina*
    33. Seed cones not serotinous, not persistent, narrowly ovoid when open; umbos with stout-based, sharp prickle; adaxial surface of seed-cone scales lacking dark border distally; trunks not producing adventitious shoots. . . . . 18. *Pinus taeda*
28. Pines of w North America (e to North Dakota, Nebraska, Colorado, and w Texas).
    34. Seed cones massive, heavy; apophyses much extended, particularly toward base of cone, continuous with umbo to produce a curved-tipped pyramid or long claw; seed body 1.5–2.5 cm; leaves 15–32 cm; California and Baja California.
    35. Leaves mostly 5 per fascicle; larger apophyses and umbos forming short, curved-tipped pyramids; seed cones 10–15 cm. . . . . 31. *Pinus torreyana*
    35. Leaves mostly 3 per fascicle; larger apophyses and umbos curved, forming long claws; seed cones 15–35 cm.
    36. Mature seed cones pale yellow-brown; leaves not drooping; larger claws (apophysis plus umbo) 2.5–3 cm; seed wing longer than seed body. . . . . 30. *Pinus coulteri*
    36. Mature seed cones dull brown; leaves drooping; larger claws (apophysis plus umbo) not over 2 cm; seed wing shorter than seed body. . . . . 29. *Pinus sabiniana*
    34. Seed cones, if massive, not markedly heavy; apophyses, if extended, not continuous with umbo, umbo either unarmed or with prickle, outcurved claw, or apiculus; seed body to 1 cm; leaf length various; widespread.
    37. Fascicle sheath to 1.5 cm, completely shed; seed cones slender-stalked, 3.5–5(–9) cm, symmetric, umbo unarmed or producing a short, often deciduous prickle; leaves 6–15 cm × 0.8–1 mm; Arizona, New Mexico. . . . . 14a. *Pinus leiophylla* var. *chihuahuana*
    37. Fascicle sheath mostly over 1.5 cm, its base persistent and falling with fascicle; seed cones sessile or nearly sessile, of various lengths, symmetry, and umbo form; leaves mostly over 1 mm wide and of various lengths; distribution various.
    38. Seed cones strongly asymmetric, persistent, often serotinous; apophyses conspicuously extended, larger toward outside base of cone, there forming large angulate tubercles or mammillae; leaves mostly 9–15 cm.
    39. Seed cones lanceoloid before opening; larger apophyses tubercular-angulate, umbos stout-triangular. . . . . 37. *Pinus attenuata*
    39. Seed cones ovoid before opening; larger apophyses less angular, more rounded, umbos mostly depressed. . . . . 36. *Pinus radiata*
    38. Seed cones symmetric or slightly asymmetric, not persistent and not serotinous; apophyses mostly little extended; leaf length various.
    40. Leaves (20–)25–45 cm, often drooping; fascicle sheaths 3–4 cm; leaf margins harshly serrulate; closed seed cones sometimes curved; larger umbos producing outcurved claws. . . . . . 28. *Pinus engelmannii*
    40. Leaves 7–25(–30) cm, not drooping; fascicle sheaths 1–2.5 (–3) cm; leaf margins finely serrulate; closed seed cones not curved or only slightly so; umbos with prickle or claw not outcurved.

41. Lower scales of seed cones just prior to and after cone fall spreading and reflexed, thus well separated from adjacent scales; seed cones with steep spirals of 5–7 scales per row as viewed from side; buds very resinous; fresh-cut wood smelling of turpentine; umbo broadly pyramidal, narrowing to short, stout prickle or merely acute or depressed; widely distributed. . . . . 25. *Pinus ponderosa*

41. Lower scales of seed cones just prior to and after cone fall not so spreading and reflexed, thus not well separated from adjacent scales; seed cones with low spirals of 8 or more scales per row as viewed from side; buds not resinous; fresh-cut wood with sweet or lemony fragrance; umbo pyramidal but narrowing abruptly at tip to slender, reflexed prickle; California, Nevada, Oregon.

    42. Seed cones (10–)15–30 cm; abaxial surface of seed-cone scales neither darker than nor sharply contrasting in color with exposed adaxial surface; leaves 12–25 cm × ca. 1.5–2 mm. . . . . 27. *Pinus jeffreyi*

    42. Seed cones 7–10 cm; abaxial surface of seed-cone scales darker than and sharply contrasting in color with paler, adaxial surface; leaves 10–15 cm × ca. 1.5 mm. . . 26. *Pinus washoensis*

1. **Pinus albicaulis** Engelmann, Trans. Acad. Sci. St. Louis 2: 209. 1863 · Whitebark pine, pine à blanche écorce

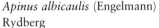

Apinus albicaulis (Engelmann) Rydberg

Trees to 21 m; trunk to 1.5 m diam., straight to twisted and contorted; crown conic, becoming rounded to irregularly spreading. **Bark** pale gray, from distance appearing whitish to light gray and smooth, in age separating into thin plates. **Branches** spreading to ascending, often persistent to trunk base; twigs stout, pale red-brown, with light brown, often glandular puberulence, somewhat roughened by elevated scars, aging gray to pale gray-brown. **Buds** ovoid, light red-brown, 0.8–1 cm; scale margins entire. **Leaves** 5 per fascicle, mostly ascending and upcurved, persisting 5–8 years, 3–7 cm × 1–1.5(–2) mm, mostly connivent, deep yellow-green, abaxial surface less so, adaxial surface conspicuously whitened by stomates, margins rounded, minutely serrulate distally, apex conic-acute; sheath 0.8–1.2 cm, shed early. **Pollen cones** cylindro-ovoid, ca. 10–15 mm, scarlet. **Seed cones** remaining on tree (unless dislodged by animals), not opening naturally but through animal agency, spreading, symmetric, broadly ovoid to depressed-ovoid or nearly globose, 4–8 cm, dull gray to black-purple, sessile to short-stalked; scales thin-based and easily broken off; apophyses much thickened, strongly cross-keeled, tip upcurved, brown; umbo terminal, short, incurved, broadly triangular, tip acute. **Seeds** obovoid; body 7–11 mm, chestnut brown, wingless. $2n = 24$.

Thin, rocky, cold soils at or near timberline, montane forests; 1300–3700 m; Alta., B.C.; Calif., Idaho, Mont., Nev., Oreg., Wash., Wyo.

Although two reliable dendrologists, G. B. Sudworth (1917) and N. T. Mirov (1967), include Utah in the distribution of *Pinus albicaulis,* more recent workers have not found it to occur there.

The fresh-cut wood of *Pinus albicaulis* is sweet-scented. Seeds are dispersed mainly by Clark's nutcracker [*Nucifraga columbiana* (Wilson), family Corvidae].

SELECTED REFERENCE Lanner, R. M. 1982. Adaptations of whitebark pine for seed dispersal by Clark's nutcracker. Canad. J. Forest Res. 12: 391–402.

2. **Pinus strobus** Linnaeus, Sp. Pl. 2: 1001. 1753 · Eastern white pine, northern white pine, pin blanc

Pinus chiapensis (Martínez) Andresen; *P. strobus* var. *chiapensis* Martínez; *Strobus strobus* (Linnaeus) Small

Trees to 67 m; trunk to 1.8 m diam., straight; crown conic, becoming rounded to flattened. **Bark** gray-brown, deeply furrowed, with long, irregularly rectangular, scaly plates. **Branches** whorled, spreading-upswept; twigs slender, pale red-brown, glabrous or pale puberulent, aging gray, ± smooth. **Buds** ovoid-cylindric, light red-brown, 0.4–0.5 cm, slightly resinous. **Leaves** 5 per fascicle, spreading to ascending, persisting 2–3 years, 6–10 cm × 0.7–1 mm, straight, slightly twisted, pliant, deep green to blue-green, pale stomatal lines evident

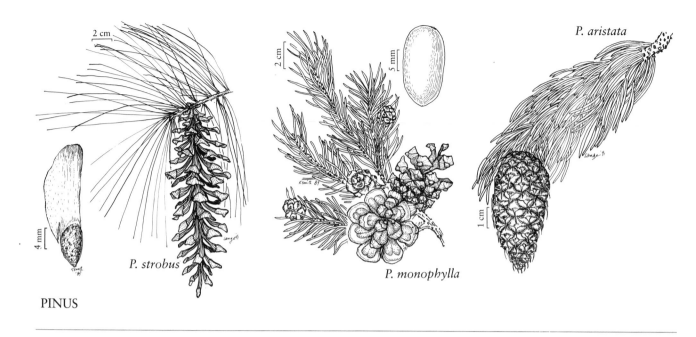

P. aristata

P. strobus

P. monophylla

PINUS

only on adaxial surfaces, margins finely serrulate, apex abruptly acute to short-acuminate; sheath 1–1.5 cm, shed early. **Pollen cones** ellipsoid, 10–15 mm, yellow. **Seed cones** maturing in 2 years, shedding seeds and falling soon thereafter, clustered, pendent, symmetric, cylindric to lance-cylindric or ellipsoid-cylindric before opening, ellipsoid-cylindric to cylindric or lance-cylindric when open, (7–)8–20 cm, gray-brown to pale brown, with purple or gray tints, stalks 2–3 cm; apophyses slightly raised, resinous at tip; umbo terminal, low. **Seeds** compressed, broadly obliquely obovoid; body 5–6 mm, red-brown mottled with black; wing 1.8–2.5 cm, pale brown. $2n = 24$.

Mesic to dry sites; 0–1500 m; St. Pierre and Miquelon; Man., N.B., Nfld., N.S., Ont., P.E.I., Que.; Conn., Del., Ga., Ill., Ind., Iowa, Ky., Maine, Md., Mass., Mich., Minn., N.H., N.J., N.Y., N.C., Pa., Ohio, R.I., S.C., Tenn., Vt., Va., W.Va., Wis.; Mexico; Central America in Guatemala.

Pinus strobus is an important timber tree; because of extensive lumbering, few uncut stands remain. It was once prized as a source for ship masts, and large tracts of it were reserved for the Royal Navy during colonial times.

Pinus strobus var. *chiapensis* appears to be as Martínez saw it: a clinal variant that, compared to the type variety, has finer leaves, different resin canal distribution, and heavier cones when cones of similar sizes are compared.

Eastern white pine (*Pinus strobus*) is the provincial tree of Ontario and the state tree of Maine and Michigan.

3. Pinus monticola Douglas ex D. Don in Lambert, Descr. Pinus [ed. 3] 2: unnumbered page between 144 and 145. 1832 · Western white pine, pin argenté

Strobus monticola (Douglas ex D. Don) Rydberg

Trees to 70 m; trunk to 2.5 m diam., straight; crown narrowly conic, becoming broad and flattened. **Bark** gray, distinctly platy, plates scaly. **Branches** nearly whorled, spreading-ascending; twigs slender, pale red-brown, rusty puberulent and slightly glandular (rarely glabrous), aging purple-brown or gray, smooth. **Buds** ellipsoid or cylindric, rust-colored, 0.4–0.5 cm, slightly resinous. **Leaves** 5 per fascicle, spreading to ascending, persisting 3–4 years, 4–10 cm × 0.7–1 mm, straight, slightly twisted, pliant, blue-green, abaxial surface without evident stomatal lines, adaxial surfaces with evident stomatal lines, margins finely serrulate, apex broadly to narrowly acute; sheath 1–1.5 cm, shed early. **Pollen cones** ellipsoid, 10–15 mm, yellow. **Seed cones** maturing in 2 years, shedding seeds and falling soon thereafter, clustered, pendent, symmetric, lance-cylindric to ellipsoid-cylindric before opening, broadly lanceoloid to ellipsoid-cylindric when open, 10–25 cm, creamy brown to yellowish, without purple or gray tints, resinous, stalks to 2 cm; umbo terminal, depressed. **Seeds** compressed, broadly obovoid-deltoid; body 5–7 mm, red-brown; wing 2–2.5 cm. $2n = 24$.

Montane moist forests, lowland fog forests; 0–3000 m; Alta., B.C.; Calif., Idaho, Mont., Nev., Oreg., Wash.

Pinus monticola is the most important western source for matchwood. Its wood lacks the sugary exudates seen in *P. lambertiana.*

Western white pine *(Pinus monticola)* is the state tree of Idaho.

4. Pinus lambertiana Douglas, Trans. Linn. Soc. London 15: 500. 1827 · Sugar pine

Trees to 75 m; trunk to 3.3 m diam., massive, straight; crown narrowly conic, becoming rounded. **Bark** cinnamon- to gray-brown, deeply furrowed, plates long, scaly. **Branches** spreading, distal branches ascending; twigs gray-green to red-tan, aging gray, mostly puberulent. **Buds** cylindro-ovoid, red-brown, to 0.8 cm, resinous. **Leaves** 5 per fascicle, spreading to ascending, persisting 2–4 years, 5–10 cm × (0.9–)1–1.5(–2) mm, straight, slightly twisted, pliant, blue-green, abaxial surface with only a few lines evident, adaxial surfaces with evident white stomatal lines, margins finely serrulate, apex acuminate; sheath (1–)1.5–2 cm, shed early. **Pollen cones** ellipsoid-cylindric, to 15 mm, yellow. **Seed cones** maturing in 2 years, shedding seeds and falling soon thereafter, often clustered, pendent, symmetric, cylindric before opening, lance-cylindric to ellipsoid-cylindric when open, 25–50 cm, yellow-brown, stalks 6–15 cm; apophyses somewhat thickened; umbo terminal, depressed, resinous, slightly excurved. **Seeds** obovoid, oblique apically; body 1–2 cm, deep brown; wing broad, 2–3 cm. $2n = 24$.

Montane dry to moist forests; 330–3200 m; Calif., Nev., Oreg.; Mexico in n Baja California.

The largest species of the genus, *Pinus lambertiana* also has the longest seed cone in the genus. It is an important timber tree with harvest far exceeding regrowth. It is easily distinguished from *P. monticola* and *P. strobus* by its larger cones and thicker cone scales with larger seeds; it is somewhat less reliably distinguished by its leaves, which are slightly wider and more tapering-tipped and have some stomatal lines evident on the abaxial surfaces (the lines not evident in *P. monticola* and *P. strobus*). A "sugary" resin high in cyclitols exudes from the sweet-scented fresh-cut wood.

5. Pinus flexilis E. James, Account Exped. Pittsburgh 2: 27, 35. 1823 · Limber pine, pin blanc de l'ouest

Apinus flexilis (E. James) Rydberg **Trees** to 26 m; trunk to 2 m diam., straight to contorted; crown conic, becoming rounded. **Bark** gray, nearly smooth, cross-checked in age into scaly plates and ridges. **Branches** spreading to ascending, often persistent to trunk base; twigs pale red-brown, puberulous (rarely glabrous), slightly resinous, aging gray, smooth. **Buds** ovoid, light red-brown, 0.9–1 cm, resinous; lower scales ciliolate along margins. **Leaves** 5 per fascicle, spreading to upcurved and ascending, persisting 5–6 years, 3–7 cm × 1–1.5 mm, pliant, dark green, abaxial surface with less conspicuous stomatal bands than adaxial surfaces, adaxial surfaces with strong, pale stomatal bands, margins finely serrulate, apex conic-acute to acuminate; sheath 1–1.5(–2) cm, shed early. **Pollen cones** broadly ellipsoid-cylindric, ca. 15 mm, pale red or yellow. **Seed cones** maturing in 2 years, shedding seeds and falling soon thereafter, spreading, symmetric, lance-ovoid before opening, cylindro-ovoid when open, 7–15 cm, straw-colored, resinous, sessile to short-stalked, apophyses much thickened, strongly cross-keeled, umbo terminal, depressed. **Seeds** irregularly obovoid; body 10–15 mm, brown, sometimes mottled darker, wingless or nearly so. $2n = 24$.

High montane forests, often at timberline; (1000–)1500–3600 m; Alta., B.C.; Ariz., Calif., Colo., Idaho, Mont., Nebr., Nev., N.Mex., N.Dak., Oreg., S.Dak., Utah, Wyo.

Pinus flexilis, much branched with a strongly tapering trunk, is little utilized because of its form and relative inaccessibility. It reportedly forms intermediates with *P. strobiformis* where the two overlap. The fresh-cut wood has the odor of turpentine.

6. Pinus strobiformis Engelmann in Wislizenus, Mem. Tour N. Mexico, 102. 1848 · Southwestern white pine, Mexican white pine, pino enano

Pinus ayacahuite Ehrenberg var. *brachyptera* G. R. Shaw; *P. ayacahuite* var. *reflexa* (Engelmann) Voss; *P. ayacahuite* var. *strobiformis* (Engelmann) Lemmon; *P. flexilis* E. James var. γ *reflexa* Engelmann; *P. reflexa* (Engelmann) Engelmann **Trees** to 30 m; trunk to 0.9 m diam., slender, straight; crown conic, becoming rounded to irregular. **Bark** gray, aging red-brown, furrowed, with narrow, irregular, scaly ridges. **Branches** spreading-ascending; twigs slender, pale red-brown, puberulous or glabrous, sometimes glaucous, aging gray or gray-brown, smooth. **Buds** ellipsoid, red-brown, ca. 1 cm, resinous. **Leaves** 5 per fascicle, spreading to ascending-upcurved, persisting 3–5 years, 4–9 cm × 0.6–1 mm, straight, slightly twisted, pliant, dark green to blue-green, abaxial surface without evident stomatal lines, adaxial surfaces conspicuously whitened by narrow stomatal lines, margins sharp, razorlike and entire to finely serrulate, apex narrowly acute to short-subulate; sheath 1.5–2 cm, shed early. **Pollen cones** cylindric, ca. 6–10 mm, pale yellow-brown. **Seed cones** maturing in 2 years, shedding seeds and falling soon thereafter, pendent, symmetric, lance-cylindric before opening, broadly lance-cylindric when open, 15–25 cm, creamy brown to light yellow-brown, stalks to 6 cm; apophyses somewhat thickened, strongly cross-keeled, tip reflexed; umbo terminal, low. **Seeds** ovoid; body 10–13 mm, red-brown, essentially wingless. $2n = 24$.

Arid to moist summit elevations, montane forests; 1900–3000 m; Ariz., N.Mex., Tex.; n Mexico.

In the northern part of the range, *Pinus strobiformis* overlaps *P. flexilis* and reportedly hybridizes with it. On average *P. strobiformis* has longer, more slender leaves and thinner, more spreading-tipped apophyses than are found in *P. flexilis,* and stomatal bands are not evident on the abaxial surface of its leaves.

7. Pinus cembroides Zuccarini, Abh. Math.-Phys. Cl. Königl. Bayer. Akad. Wiss. 1: 392. 1832 · Mexican pinyon, piñón, pino piñonero

Pinus cembroides var. *bicolor* Little; *P. cembroides* var. *remota* Little; *P. discolor* D. K. Bailey & Hawksworth; *P. remota* (Little) D. K. Bailey & Hawksworth

Shrubs or trees to 15 m; trunk to 0.3 m diam., strongly tapering, much branched; crown rounded. **Bark** red-brown to dark brown, shallowly and irregularly furrowed, ridges broad, scaly. **Branches** spreading-ascending; twigs red-brown, sometimes finely papillate, aging gray to gray-brown. **Buds** ovoid to short cylindric, pale red-brown, 0.5–1.2 cm, slightly resinous. **Leaves** (2–)3(–4) per fascicle, spreading to upcurved, persisting 3–4 years, 2–6 cm × 0.6–0.9(–1) mm, connivent, 2–3-sided, blue- to gray-green, abaxial surface not conspicuously whitened with stomatal bands or if stomatal bands present, these less conspicuous than on adaxial surfaces, often with 2 subepidermal resin bands evident, adaxial surfaces conspicuously whitened with stomatal lines, margins entire to finely serrulate, apex narrowly conic or subulate; sheath 0.5–0.7 cm, scales soon recurved, forming rosette, shed early. **Pollen cones** ellipsoid, to 10 mm, yellow. **Seed cones** maturing in 2 years, shedding seeds and falling soon thereafter, spreading, symmetric, ovoid before opening, broadly depressed-ovoid to nearly globose when open, 1–3.5 cm, pale yellow- to pale red-brown, resinous, nearly sessile or short-stalked; apophyses thickened, slightly domed, angulate, transversely keeled; umbo subcentral, slightly raised to depressed, truncate or umbilicate. **Seeds** ovoid to obovoid; body (7–)12–15(–20) mm, brown, wingless. $2n = 24$.

Pinyon-juniper woodland, foothills, mesas, tablelands; 700–2300 m; Ariz., N.Mex., Tex.; Mexico.

Pinus cembroides is the common pinyon of Mexican commerce. Populations of the Edwards Plateau, Texas, are disjunct about 150 km east and north of the main area of distribution of the species, and they have been described as a distinct variety, *P. cembroides* var. *remota* Little, on the basis of thin seed shell and a higher frequency of 2-leaved fascicles in contrast to the thicker seed shell and prevalently 3-leaved fascicles in Mexican pinyon populations to the west and south. The strong overlap in nearly all character states between the populations of the Edwards Plateau and other populations makes var. *remota* difficult to maintain.

8. Pinus edulis Engelmann in Wislizenus, Mem. Tour N. Mexico, 88. 1848 · Pinyon, piñón

Caryopitys edulis (Engelmann) Small; *Pinus cembroides* Zuccarini var. *edulis* (Engelmann) Voss

Shrubs or trees to 21 m; trunk to 0.6 m diam., strongly tapering, erect; crown conic, rounded, dense. **Bark** red-brown, shallowly and irregularly furrowed, ridges scaly, rounded. **Branches** persistent to near trunk base; twigs pale red-brown to tan, rarely glaucous, aging gray-brown to gray, glabrous to papillose-puberulent. **Buds** ovoid to ellipsoid, red-brown, 0.5–1 cm, resinous. **Leaves** (1–)2(–3) per fascicle, upcurved, persisting 4–6 years, 2–4 cm × (0.9–)1–1.5 mm, connivent, 2-sided (1-leaved fascicles with leaves 2-grooved, 3-leaved fascicles with leaves 3-sided), blue-green, all surfaces marked with pale stomatal bands, particularly the adaxial, margins entire or finely serrulate, apex narrowly acute to subulate; sheath 0.5–0.7 cm, scales soon recurved, forming rosette, shed early. **Pollen cones** ellipsoid, ca. 7 mm, yellowish to red-brown. **Seed cones** maturing in 2 years, shedding seeds and falling soon thereafter, spreading, symmetric, ovoid before opening, depressed-ovoid to nearly globose when open, ca. (3.5–)4(–5) cm, pale yellow- to pale red-brown, resinous, nearly sessile to short-stalked;

apophyses thickened, raised, angulate; umbo subcentral, slightly raised or depressed, truncate or umbilicate. **Seeds** mostly ellipsoid to obovoid; body 10–15 mm, brown, wingless. $2n = 24$.

Dry mountain slopes, mesas, plateaus, and pinyon-juniper woodland; 1500–2100(–2700) m; Ariz., Calif., Colo., N.Mex., Okla., Tex., Utah, Wyo.; Mexico in Chihuahua.

Pinus edulis var. *fallax* Little (*P. californiarum* subsp. *fallax* (Little) D. K. Bailey) appears to combine features of *P. edulis* and *P. monophylla*. More study is needed.

Seeds of *Pinus edulis,* the commonest southwestern United States pinyon, are much eaten and traded by Native Americans.

Pinyon *(Pinus edulis)* is the state tree of New Mexico.

9. **Pinus quadrifolia** Parlatore ex Sudworth, U.S.D.A. Div. Forest. Bull. 14: 17. 1897 · Parry pinyon, piñón

Pinus cembroides Zuccarini var. *parryana* Voss; *P. juarezensis* Lanner; *P. parryana* Engelmann 1862, not Gordon 1858

Trees to 10 m; trunk to 0.5 m diam., straight, much branched; crown dense, becoming rounded. **Bark** red-brown, irregularly furrowed and cross-checked to irregularly rectangular, plates scaly. **Branches** spreading to ascending, persistent to trunk base; twigs slender, pale orange-brown, puberulent-glandular, aging brown to gray-brown. **Buds** ovoid, light red-brown, ca. 0.4–0.5 cm, slightly resinous. **Leaves** (3–)4(–5) per fascicle, persisting 3–4 years, (2–)3–6 cm × (1–)1.2–1.7 mm, curved, connivent, stiff, green to blue-green, margins entire to minutely scaly-denticulate, finely serrulate, apex subulate, adaxial surfaces mostly strongly whitened with stomatal bands, abaxial surface not so but 2 subepidermal resin bands evident; sheath 0.5–0.6 cm, scales soon recurved, forming rosette, shed early. **Pollen cones** ovoid, ca. 10 mm, yellowish. **Seed cones** maturing in 2 years, shedding seeds and falling soon thereafter, spreading, symmetric, ovoid before opening, broadly ovoid to depressed-globose when open, (3–)4–8(–10) cm, pale yellow-brown, sessile to short-stalked, apophyses thickened, strongly raised, diamond-shaped, transversely keeled, umbo subcentral, low-pyramidal or sunken, blunt. **Seeds** obovoid, body ca. 15 mm, brown, wingless.

Dry rocky sites; 1200–1800 m; Calif.; Mexico in Baja California.

Pinus quadrifolia is the rarest pinyon in the flora. It hybridizes naturally with *P. monophylla.*

10. **Pinus monophylla** Torrey & Frémont in Frémont, Rep. Exped. Rocky Mts. 2: 319, plate 4. 1845 · Singleleaf pinyon, piñón

Caryopitys monophylla (Torrey & Frémont) Rydberg; *Pinus californiarum* D. K. Bailey; *P. cembroides* Zuccarini var. *monophylla* (Torrey & Frémont) Voss

Trees to 14 m; trunk to 0.5 m diam., strongly tapering, much branched; crown usually rounded, dense. **Bark** red-brown, irregularly furrowed or cross-checked, scaly. **Branches** spreading and ascending, persistent to near trunk base; twigs stout, orange-brown, aging brown to gray, sometimes sparsely puberulent. **Buds** ellipsoid, light red-brown, 0.5–0.7 cm, resinous; scale margins fringed. **Leaves** 1(–2) per fascicle, ascending, persisting 4–6(–10) years, 2–6 cm × 1.3–2(–2.5) mm, curved, terete (though often 2-grooved), gray-green, all surfaces with stomatal lines, margins entire, apex subulate; sheath 0.5–1 cm, scales soon recurved, forming rosette, shed early. **Pollen cones** ellipsoid, ca. 10 mm, yellow. **Seed cones** maturing in 2 years, shedding seeds and falling soon thereafter, spreading, symmetric, ovoid before opening, broadly depressed-ovoid to nearly globose when open, 4–6(–8) cm, pale yellow-brown, nearly sessile; apophyses thickened, slightly raised; umbo subcentral, raised or depressed, nearly truncate, apiculate. **Seeds** cylindric-ellipsoid; body 15–20 mm, gray-brown to brown, wingless. $2n = 24$.

Dry low-montane or foothill pinyon-juniper woodland; 1000–2300 m; Ariz., Calif., Idaho, Nev., Utah; Mexico in Baja California.

Pinus monophylla hybridizes with *P. edulis* and *P. quadrifolia.*

Singleleaf pinyon *(Pinus monophylla)* is the state tree of Nevada.

11. **Pinus balfouriana** Greville & Balfour in A. Murray bis, Bot. Exped. Oregon 8: no. 618, plate 3, fig. 1. 1853 · Foxtail pine

Pinus balfouriana var. *austrina* (R. Mastrogiuseppe & J. Mastrogiuseppe) Silba; *P. balfouriana* subsp. *austrina* R. Mastrogiuseppe & J. Mastrogiuseppe

Trees to 22 m; trunk to 2.6 m diam., erect or leaning; crown broadly conic to irregular. **Bark** gray to salmon or cinnamon, platy or irregularly deep-fissured or with irregular blocky plates. **Branches** contorted, ascending to descending; twigs red-brown, aging gray to drab yellow-gray, glabrous or puberulent, young branches resembling long bottlebrushes because of persistent leaves. **Buds** ovoid-acuminate, red-brown, 0.8–1 cm, resinous. **Leaves** 5 per fascicle, upcurved, persisting 10–30 years, 1.5–4 cm × 1–1.4 mm, mostly connivent, deep blue- to deep yellow-green, abaxial surface without median groove but usually with 2 subepidermal but evident resin bands, adaxial surfaces conspicuously whitened by stomates, margins mostly entire to blunt, apex broadly acute to acuminate; sheath 0.5–1 cm, soon forming rosette, shed early. **Pollen cones** ellipsoid, ca. 6–10 mm, red. **Seed cones** maturing in 2 years, shedding seeds and falling soon thereafter, spreading, symmetric, lance-cylindric with conic base before opening, broadly lance-ovoid or ovoid to cylindric or ovoid-cylindric when open, 6–9 (–11) cm, purple, aging red-brown, nearly sessile; apophyses much thickened, rounded, larger toward cone base; umbo central, usually depressed; prickle absent or weak, to 1 mm, resin exudates amber. **Seeds** ellipsoid to narrowly obovoid; body to 10 mm, pale brown, mottled with deep red; wing 10–12 mm. $2n = 24$.

Timberline and alpine meadows; of conservation concern; 1500–3500 m; Calif.

Pinus balfouriana is the true "foxtail pine." In leaf character it is hardly, if at all, distinguishable from *P. longaeva,* but its strongly conic-based cones with distinctly shorter-prickled, sunken-centered umbos at once distinguish it from that species.

Plants shown to be genetically distinct from the type (differences in chemistry, form, foliage, cone orientation, and seeds) have been called *Pinus balfouriana* subsp. *austrina* R. Mastrogiuseppe & J. Mastrogiuseppe. As in several other species or species complexes in *Pinus,* however, there is a problem with a character gradient involving related taxa. The evidence presented by D. K. Bailey (1970) and later by R. J. Mastrogiuseppe and J. D. Mastrogiuseppe (1980) could as well be used to indicate that *P. balfouriana* (with its two infraspecific taxa) and *P. longaeva* represent a single species of three subspecies or three varieties. The more conservative view of Bailey is followed here.

SELECTED REFERENCE Mastrogiussepe, R. J. and J. D. Mastrogiuseppe. 1980. A study of *Pinus balfouriana* Grev. & Balf. (Pinaceae). Syst. Bot. 5: 86–104.

12. **Pinus aristata** Engelmann in Parry & Engelmann, Amer. J. Sci. Arts, ser. 2, 34: 331. 1862 · Colorado bristlecone pine

Pinus balfouriana Greville & Balfour var. *aristata* (Engelmann) Engelmann

Trees to 15 m; trunk to 1 m diam., strongly tapering, twisted; crown rounded, flattened (sheared), or irregular. **Bark** gray to red-brown, shallowly fissured, with long, flat, irregular ridges. **Branches** contorted; twigs pale red-brown, aging gray, puberulent, young branches resembling long bottlebrushes because of persistent leaves. **Buds** ovoid-acuminate, pale red-brown, ca. 1 cm, resinous. **Leaves** 5 per fascicle, upcurved, persisting 10–17 years, (2–)3–4 cm × 0.8–1 mm, mostly connivent, deep blue-green, with drops and scales of resin, abaxial surface with strong, narrow median groove, adaxial surfaces conspicuously whitened by stomates, margins entire or distantly serrulate, apex conic-acute to conic-subulate; sheath 0.5–1.5 cm, scales soon recurving, shed early. **Pollen cones** ellipsoid, ca. 10 mm, bluish to red. **Seed cones** maturing in 2 years, shedding seeds and falling soon thereafter, spreading, symmetric, lance-cylindric before opening, lance-ovoid to ovoid or cylindric when open, 6–11 cm, purple to brown, nearly sessile; apophyses much thickened; umbo central, with triangular base, extended into slender, brittle prickle 4–10 mm. **Seeds** obliquely obovoid; body 5–6 mm, gray-brown to near black; wing ca. 10–13 mm. $2n = 24$.

Subalpine and alpine; 2500–3400 m; Ariz., Colo., N.Mex.

Pinus aristata has leaves usually narrower and sharper than in *P. longaeva* and *P. balfouriana,* and the leaves almost always have a narrow, median groove on the abaxial surface.

13. **Pinus longaeva** D. K. Bailey, Ann. Missouri Bot. Gard. 57: 243. 1970 · Intermountain bristlecone pine

Pinus aristata Engelmann var. *longaeva* (D. K. Bailey) Little

Trees to 16 m; trunk to 2 m diam., strongly tapering; crown rounded, flattened (sheared), or irregular. **Bark** red-brown, shallowly to deeply fissured with thick, scaly, irregular, blocky ridges. **Branches** contorted, pendent; twigs pale red-brown, aging gray to yellow-gray, puberulent, young branches resembling long bottlebrushes because of persistent leaves. **Buds** ovoid-acuminate, pale red-brown, ca. 1 cm, resinous. **Leaves**

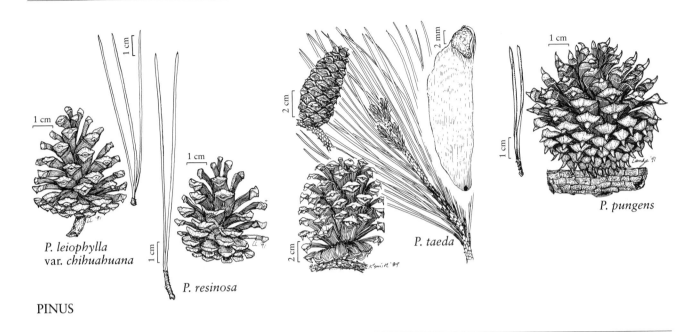

P. leiophylla
var. *chihuahuana*

P. resinosa

P. taeda

P. pungens

PINUS

mostly 5 per fascicle, upcurved, persisting 10–30 years, 1.5–3.5 cm × 0.8–1.2 mm, mostly connivent, deep yellow-green, with few resin splotches but often scurfy with pale scales, abaxial surface without median groove but with 2 subepidermal but evident resin bands, adaxial surfaces conspicuously whitened with stomates, margins entire or remotely and finely serrulate distally, apex bluntly acute to short-acuminate; sheath ca. 1 cm, soon forming rosette, shed early. **Pollen cones** cylindro-ellipsoid, 7–10 mm, purple-red. **Seed cones** maturing in 2 years, shedding seeds and falling soon thereafter, spreading, symmetric, lance-cylindric with rounded base before opening, lance-cylindric to narrowly ovoid when open, 6–9.5 cm, purple, aging red-brown, nearly sessile; apophyses much thickened, sharply keeled; umbo central, raised on low buttress, truncate to umbilicate, abruptly narrowed to slender but stiff, variable prickle 1–6 mm, resin exudate pale. **Seeds** ellipsoid-obovoid; body 5–8 mm, pale brown, mottled with dark red; wing 10–12 mm.

Subalpine and alpine; 1700–3400 m; Calif, Nev., Utah.

Pinus longaeva is considered by dendrochronologists to be the longest-lived tree. One tree was estimated to be 5000 years old.

14. Pinus leiophylla Schiede & Deppe in Schlechtendal & Chamisso, Linnaea 6: 354. 1831 · Pino chino, ocote chino

Varieties 2 (1 in the flora): North America, Mexico.

14a. Pinus leiophylla Schiede & Deppe var. **chihuahuana** (Engelmann) G. R. Shaw, Pines Mexico 14, plate 7, figs. 10, 11. 1909 · Chihuahua pine, pino real, pino prieto

Pinus chihuahuana Engelmann in Wislizenus, Mem. Tour N. Mexico, 103. 1848

Trees to 25 m; trunk to 0.9 m diam., slender; crown conic, becoming rounded. **Bark** brown to red-brown, narrowly furrowed, cross-checked into long, irregularly and narrowly rectangular, flat, scaly ridges. **Branches** ascending; twigs slender, orange-brown or glaucous and violet, aging red-brown, ± smooth or cracking. **Buds** ovoid, light red-brown, ca. 0.6–0.7(–1) cm, slightly resinous. **Leaves** (2–)3(–4) per fascicle, spreading-ascending, persisting 2 years, 6–15 cm × 0.8–1 mm, straight to slightly twisted, dull gray-green, all surfaces with fine stomatal lines, margins finely serrulate, apex acute to acuminate; sheath to 1.5 cm, shed early and completely. **Pollen cones** broadly ellipsoid, ca. 10–15 mm, brown or yellow. **Seed cones** maturing in 3 years, shedding seeds soon thereafter but long-persistent, paired or solitary, symmetric, lateral, narrowly ovoid before opening, broadly ovoid to nearly globose when open, 3.5–5(–9) cm, chestnut brown or greenish brown, aging gray to gray-brown, stalks to 1.5 cm; apophyses slightly thickened and raised, not keeled; umbo central, slightly raised or depressed, with short, often deciduous prickle or unarmed. **Seeds** obovoid; body ca. 2 mm, gray, mottled darker; wing ca. 10 mm, dark-lined. $2n = 24$.

Dry slopes and plateaus; 1500–2500 m; Ariz., N.Mex.; Mexico.

Pinus leiophylla var. *chihuahuana* is one of the few pines that produce sprouts from stumps. It differs from the type variety in its dark, less roughened bark, its shorter range of leaf length, and its slightly broader leaves that occur more consistently in threes. (The narrower, often longer leaves of the type variety are in fours and fives.)

The type variety, *Pinus leiophylla* var. *leiophylla*, is exclusively Mexican.

15. Pinus resinosa Aiton, Hort. Kew. 3: 367. 1789 · Red pine, Norway pine, pin rouge

Trees to 37 m; trunk to 1.5 m diam., straight; crown narrowly rounded. **Bark** light red-brown, furrowed and cross-checked into irregularly rectangular, scaly plates. **Branches** spreading-ascending; twigs moderately slender (to 1 cm thick), orange- to red-brown, aging darker brown, rough. **Buds** ovoid-acuminate, red-brown, to ca. 2 cm, resinous; scale margins fringed. **Leaves** 2 per fascicle, straight or slightly twisted, brittle, breaking cleanly when bent, deep yellow-green, all surfaces with narrow stomatal bands, margins serrulate, apex short-conic, acute; sheath 1–2.5 cm, base persistent. **Pollen cones** ellipsoid, ca. 15 mm, dark purple. **Seed cones** maturing and opening in 2 years, spreading, symmetric, ovoid before opening, broadly ovoid to nearly globose when open, 3.5–6 cm, light red-brown, nearly sessile; apophyses slightly thickened, slightly raised, transversely low-keeled; umbo central, centrally depressed, unarmed. **Seeds** ovoid; body 3–5 mm, brown; wing to 20 mm. $2n = 24$.

Sandy soils, eastern boreal forests; 200–800(–1300) m; Man., N.B., Nfld., N.S., Ont., P.E.I., Que.; Conn., Ill., Maine, Mass., Mich., Minn., N.H., N.J., N.Y., Pa., Vt., W.Va., Wis.

Pinus resinosa was once the most important timber pine in the Great Lakes region.

Norway pine *(Pinus resinosa)* is the state tree of Minnesota.

16. Pinus sylvestris Linnaeus, Sp. Pl. 2: 1000. 1753 · Scotch pine, pin d'Écosse

Varieties ca. 20 (1 introduced in the flora): North America, Eurasia.

16a. Pinus sylvestris Linnaeus var. sylvestris

Trees to 40 m (usually much shorter in North America); trunk to 0.6 m diam. (usually less in North America), straight or contorted, erect or leaning; crown broad-conic to irregular or flattened. **Bark** ashy gray to brown, furrowed, ridges irregularly rectangular, scaly, orange on upper sections of trunk, platy. **Branches** spreading to ascending, poorly self-pruning; twigs slender, dull green- to orange-brown, not glaucous, aging gray-brown, rough. **Buds** conic-ovoid, red-brown, 0.6–1.1 cm, resinous. **Leaves** 2 per fascicle, spreading to ascending-upcurved, persisting 2–4 years, 4–6(–8) cm × 2 mm, strongly twisted, somewhat flattened, blue- to gray- or yellow-green, all surfaces with evident stomatal lines, margins entire to finely serrulate, apex acute to abruptly acuminate; sheath 0.3–0.6 cm, base semipersistent. **Pollen cones** ovoid, to 10 mm, yellow or pale pink. **Seed cones** maturing in 2 years, shedding seeds and mostly falling soon thereafter, nearly symmetric, narrowly ovoid or lanceoloid before opening, broadly ovoid to depressed-globose when open, 3–6 cm, dull gray-brown to tan or greenish gray, nearly sessile or on stalks to 1 cm, scales lacking contrasting border on adaxial surface distally; apophyses slightly raised, isodiametric, 4-keeled, more elongate abaxially toward cone base; umbo central, raised to depressed, truncate, mostly umbilicate, unarmed. **Seeds** asymmetrically obovoid; body 3–5 mm, gray to near black; wing to 15 mm.

Cultivated ground, abandoned fields, fencerows, and woods; introduced; possible in various provinces and states; Europe.

Pinus sylvestris is a Eurasian species widely planted and escaping from cultivation over much of northern United States and southern Canada. Representatives of the species in the flora originated from mostly poor seed stock of var. *sylvestris*. They are not much used for timber (as the tree is in Europe and Asia, where several varieties are known) but are used for pulpwood and Christmas trees.

17. Pinus palustris Miller, Gard. Dict. ed. 8, Pinus no. 14. 1768 · Longleaf pine

Pinus australis F. Michaux.

Trees to 47 m; trunk to 1.2 m diam., straight; crown rounded. **Bark** orange-brown, with coarse, rectangular, scaly plates. **Branches** spreading-descending, upcurved at tips; twigs stout (to 2 cm thick), orange-brown, aging darker brown, rough. **Buds** ovoid, silvery white, 3–4 cm; scales narrow, margins fringed. **Leaves** (2)–3 per fascicle, spreading-recurved, persisting 2 years, 20–45 cm × ca. 1.5 mm, slightly twisted, lustrous yellow-green, all surfaces with fine stomatal lines, margins finely serrulate, apex abruptly acute to acuminate; sheath 2–2.5(–3) cm, base persistent. **Pollen cones** cylindric, 30–80 mm, purplish. **Seed cones** maturing in 2 years, quickly shedding seeds and falling, solitary or paired toward branchlet tips, symmetric, lanceoloid before opening, ovoid-cylindric when open, 15–25 cm, dull brown, sessile (rarely short-stalked); apophyses dull, slightly thickened, slightly raised, nearly rhombic, strongly cross-keeled; umbo central, broadly triangular, with short, stiff, reflexed prickle. **Seeds** truncate-obovoid; body ca. 10 mm, pale brown, mottled darker; wing 30–40 mm. $2n = 24$.

Dry sandy uplands, sandhills, and flatwoods; 0–700 m; Ala., Fla., Ga., La., Miss., N.C., S.C., Tex., Va.

Pinus palustris is fire successional, with a deep taproot and a definite grass stage. It is a valued species for lumber and pulpwood and was once important for naval stores (e.g., turpentine, pine oil, tar, pitch). It is fast disappearing over much of its natural range, partly through overharvesting but especially because of difficulties in adapting it to current plantation and management techniques.

Longleaf pine *(Pinus palustris)* is the state tree of North Carolina.

18. Pinus taeda Linnaeus, Sp. Pl. 2: 1000. 1753

· Loblolly pine

Trees to 46 m; trunk to 1.6 m diam., usually straight, without adventitious shoots; crown broadly conic to rounded. **Bark** red-brown, forming square or irregularly rectangular, scaly plates, resin pockets absent. **Branches** spreading-ascending; twigs moderately slender (to ca. 1 cm thick), orangish to yellow-brown, aging darker brown, rough. **Buds** lance-cylindric, pale red-brown, 1–1.2(–2) cm, mostly less than 1 cm broad, slightly resinous; scale margins white-fringed, apex acuminate. **Leaves** 2–3 per fascicle, ascending to spreading, persisting 3 years, (10–)12–18(–23) cm × 1–2 mm, straight, slightly twisted, pliant, deep yellow-green, all surfaces with narrow stomatal lines, margins finely serrulate, apex acute to abruptly conic-subulate; sheath 1–2.5 cm, base persistent. **Pollen cones** cylindric, 20–40 mm, yellow to yellow-brown. **Seed cones** maturing in 2 years, shedding seeds soon thereafter, not persistent, solitary or in small clusters, nearly terminal, symmetric, lanceoloid before opening, narrowly ovoid when open, 6–12 cm, mostly dull yellow-brown, sessile to nearly sessile, scales without dark border on adaxial surface distally; apophyses dull, slightly thickened, variously raised (more

so toward cone base), rhombic, strongly transversely keeled; umbo central, recurved, stoutly pyramidal, tapering to stout-based, sharp prickle. **Seeds** obdeltoid; body 5–6 mm, red-brown; wing to 20 mm. $2n = 24$.

Mesic lowlands and swamp borders to dry uplands; 0–700 m; Ala., Ark., Del., Fla, Ga., Ky., La., Md., Miss., N.J., N.C., Okla., S.C., Tenn., Tex., Va.

Originally most races of *Pinus taeda* were in the lowlands. Following disturbance of the natural vegetation after settlement by Europeans, the species spread to fine-textured, fallow, upland soils, where it now occurs intermixed with *P. echinata* and *P. virginiana*. In the Southeast *P. taeda* is commonly used in plantation forestry, along with *P. elliottii* and *P. echinata*. *Pinus taeda* frequently forms hybrids with *P. echinata* and *P. palustris* (*P.* ×*sondereggeri* H. H. Chapman). Commercially, it is a valuable pulpwood and timber species.

19. Pinus echinata Miller, Gard. Dict. ed. 8, Pinus no. 12. 1768 · Shortleaf pine

Trees to 40 m; trunk to 1.2 m diam., straight; crown rounded to conic. **Bark** red-brown, scaly-plated, plates with evident resin pockets. **Branches** spreading-ascending; 2-year-old branchlets slender (ca. 5 mm or less), greenish brown to red-brown, often glaucous, aging red-brown to gray, roughened and cracking below leafy portion. **Buds** ovoid to cylindric, red-brown, 0.5–0.7(–1) cm, resinous. **Leaves** 2(–3) per fascicle, spreading-ascending, persistent 3–5 years, (5–)7–11(–13) cm × ca. 1 mm, straight, slightly twisted, gray- to yellow-green, all surfaces with fine stomatal lines, margins finely serrulate, apex abruptly acute; sheath 0.5–1(–1.5) cm, base persistent. **Pollen cones** cylindric, 15–20 mm, yellow- to pale purple-green. **Seed cones** maturing in 2 years, semipersistent, solitary or clustered, spreading, symmetric, lanceoloid or narrowly ovoid before opening, ovoid-conic when open, 4–6(–7) cm, red-brown, aging gray, nearly sessile or on stalks to 1 cm, scales lacking contrasting dark border on adaxial surfaces distally; umbo central, with elongate to short, stout, sharp prickle. **Seeds** ellipsoid; body ca. 6 mm, gray to nearly black; wing 12–16 mm. $2n = 24$.

Uplands, dry forests; 200–610 m; Ala., Ark., Del., Fla., Ga., Ill., Ky., La., Md., Miss., Mo., N.J., N.C., Ohio, Okla., Pa., S.C., Tenn., Tex., Va., W.Va.

Although *Pinus echinata* is highly valued for timber and pulpwood, it is afflicted by root rot. It hybridizes with *P. taeda*, the pine most commonly associated with it.

20. Pinus glabra Walter, Fl. Carol., 237. 1788

· Spruce pine

Trees to 30 m; trunk to 1 m diam., straight; crown conic to rounded. Bark gray, fissured and cross-checked into elongate, irregular, scaly plates, resin pockets absent, on upper sections of trunk ± smooth, gray, looking slick. Branches whorled, spreading to ascending; twigs slender, purple-red to red-brown, occasionally glaucous, aging gray, smooth. Buds ovoid to ovoid-cylindric, red-brown, ca. 0.5–1 cm, slightly resinous; scale margins finely fringed. Leaves 2 per fascicle, spreading to ascending, persisting 2–3 years, 4–8(–10) cm × 0.7–1.2 mm, straight, slightly twisted, dark green, all surfaces with fine stomatal lines, margins finely serrulate, apex sharply conic; sheath 0.5–1 cm, base persistent. Pollen cones lance-cylindric, 10–15 mm, purple-brown. Seed cones maturing in 2 years, shedding seeds soon thereafter, semipersistent, spreading to recurved, nearly symmetric, lance-ovoid before opening, ovoid-cylindric when open, 3.5–7 cm, red-brown, aging gray, nearly sessile or on stalks to 1 cm, scales lacking contrasting border on adaxial surfaces (as in *P. echinata*); apophyses but slightly thickened and raised; umbo central, depressed, unarmed or with small, curved, weak, deciduous, short-incurved prickle. Seeds deltoid-obovoid; body ca. 6 mm, brown, mottled darker; wing to ca. 12 mm. $2n = 24$.

Sandy alluvium and mesic woodland; 0–150 m; Ala., Fla., Ga., La., Miss., S.C.

Pinus glabra is more shade tolerant than most yellow pines. Although the trees grow large, the wood is not much valued. The species is similar in tree form to *P. strobus*. It resembles *P. echinata* in shoot and leaf but has less prickly cones and deeper green leaves.

21. Pinus rigida Miller, Gard. Dict. ed. 8, Pinus no. 10.

1768 · Pitch pine, pin rigide

Trees to 31 m; trunk to 0.9 m diam., straight or crooked, commonly with adventitious sprouts; crown rounded or irregular. Bark red-brown, deeply and irregularly furrowed, with long, irregularly rectangular, flat, scaly ridges, resin pockets absent. Branches arching-spreading to ascending, poorly self-pruning; 2-year-old branchlets stout (mostly over 5 mm thick), orange-brown, aging darker brown, rough. Buds ovoid to ovoid-cylindric, red-brown, ca. 1–1.5 cm, resinous; scale margins fringed, apex cuspidate. Leaves 3(–5) per fascicle, spreading to

ascending, persisting 2–3 years, 5–10(–15) cm × 1–1.5(–2) mm, straight, twisted, deep to pale yellow-green, all surfaces with fine stomatal lines, margins serrulate, apex abruptly subulate-acuminate; sheath 0.9–1.2 cm, base persistent. Pollen cones cylindric, ca. 20 mm, yellow. Seed cones maturing in 2 years, shedding seeds soon thereafter or variously serotinous and long-persistent, often clustered, symmetric, conic to ovoid before opening, broadly ovoid with flat or slightly convex base when open, 3–9 cm, creamy brown to light red-brown, sessile to short-stalked, base truncate, scales firm, with dark red-brown border on adaxial surface distally; apophyses slightly raised, rhombic, with strong transverse keels; umbo central, low-triangular, with slender, downcurved prickle. Seeds broadly obliquely obovoid-deltoid; body 4–5(–6) mm, dark brown, mottled darker, or near black; wing 15–20 mm. $2n = 24$.

Upland or lowland, sterile, dry to boggy soils; 0–1400 m; Ont., Que.; Conn., Del., Ga., Ky., Maine, Md., Mass., N.H., N.J., N.Y., N.C., Ohio, Pa., R.I., S.C., Tenn., Vt., Va., W.Va.

Pinus rigida often has poor form and is not valued highly as saw timber. It is fire successional, sprouts adventitiously, and is frequently shrubby in the northern part of its range. It is known to hybridize naturally with *P. echinata*.

22. Pinus serotina Michaux, Fl. Bor.-Amer. 2: 205. 1803

· Pond pine

Pinus rigida Miller subsp. *serotina* (Michaux) R. T. Clausen; *P. rigida* var. *serotina* (Michaux) Hoopes

Trees to 21 m; trunk to 0.6 m diam., straight or more often crooked, commonly with adventitious sprouts; crown becoming ragged, thin, often broadly rounded or flat. Bark red-brown, irregularly furrowed and cross-checked into rectangular, flat, scaly plates. Branches spreading to ascending; twigs stout, orange- to yellow-orange, frequently glaucous, aging darker. Buds ovoid to narrowly ovoid, red-brown, 1–1.5(–2) cm, resinous. Leaves 3 per fascicle (to 5 in adventitious or disturbed growth), spreading to ascending, persisting 2–3 years, (12–)15–20(–21) cm × 1.3–1.5(–2) mm, slightly twisted, tufted at twig tips, straight, yellow-green, all surfaces with fine stomatal lines, margins serrulate, apex acuminate; sheath 1–2 cm, base persistent. Pollen cones cylindric, to 30 mm, yellow-brown. Seed cones maturing in 2 years, in some populations beginning to shed seeds then but more often variably serotinous, long-persistent, often whorled, symmetric, ovoid to lanceloid before opening, broadly ovoid to nearly globose when open, 5–8 cm, pale red-brown to creamy brown, sessile or on stalks to 1 cm; scales with dark red-brown border on adaxial surface distally; apophyses slightly thickened, low, rhombic, low cross-keeled; umbo central, low-conic, with short, weak prickle, sometimes unarmed. **Seeds** ellipsoid, oblique at tip, somewhat compressed; body 5–6 mm, pale brown, mottled darker or nearly black; wing to 20 mm. $2n = 24$.

Flatwoods, flatwoods bogs, savannas, and barrens; 0–200 m; Ala., Del., Fla., Ga., Md., N.J., N.C., S.C., Va.

Pinus serotina is fire successional and sprouts adventitiously after crown fires. It is part of a distinct forest type including *Taxodium distichum* (Linnaeus) Richard, *Nyssa biflora* Walter, *Magnolia virginiana* Linnaeus, *Persea* sp., and *Ilex* sp. Of good form when protected from fire, *P. serotina* then much resembles *P. taeda*, with which it hybridizes naturally. It is of increasing importance as pulpwood.

23. Pinus pungens Lambert, Ann. Bot. (London) 2: 198. 1805 · Table mountain pine, mountain pine

Trees to 12 m; trunk to 0.6 m diam., straight to crooked, erect to leaning, poorly self-pruning; crown irregularly rounded or flattened. **Bark** red- to gray-brown, irregularly checked into scaly plates. **Branches** horizontally spreading; twigs slender, orange- to yellow-brown, aging darker brown, rough. **Buds** ovoid to cylindric, red-brown, 0.6–0.9 cm, resinous. **Leaves** 2(–3) per fascicle, spreading or ascending, persisting 3 years, 3–6(–8) cm × 1–1.5 mm, twisted, deep yellow-green, all surfaces with fine stomatal lines, margins harshly serrulate, apex acute to short-acuminate; sheath 0.5–1 cm, base persistent. **Pollen cones** ellipsoid, ca. 15 mm, yellow. **Seed cones** maturing in 2 years, variably serotinous, mostly whorled, downcurved, asymmetric, ovoid before opening, broadly ovoid when open, (4–)6–10 cm, gray- to pale red-brown, nearly sessile or on stalks to 1 cm; apophyses thickened, diamond-shaped, strongly keeled, elongate, mammillate at cone base abaxially; umbo central, a stout, curved, sharp claw. **Seeds** deltoid-obovoid, oblique; body ca. 6 mm, deep purple-brown to black; wing 10–20(–30) mm. $2n = 24$.

Dry, mostly sandy or shaly uplands; Appalachians and associated Piedmont; 500–1350 m; Del., Ga., Md., N.J., N.C., Pa., S.C., Tenn., Va., W.Va.

Pinus pungens is a scrub pine and is too small and knotty to be much utilized except for pulpwood and firewood. Its common name refers to a general type of landform, not to a specific, named mountain.

24. Pinus elliottii Engelmann, Trans. Acad. Sci. St. Louis 4: 186, plates 1–3. 1880 · Slash pine

Pinus heterophylla (Elliott) Sudworth, 1893, not K. Koch, 1849; *P. taeda* Linnaeus var. *heterophylla* Elliott

Trees to 30 m; trunk to 0.8 m diam., straight to contorted; crown conic, becoming rounded or flattened. **Bark** orange- to purple-brown, irregularly furrowed and cross-checked into large, irregularly rectangular, papery-scaly plates. **Branches** spreading to ascending; twigs stout (to ca. 1 cm thick), orange-brown, aging darker brown, rough-scaly. **Buds** cylindric, silvery brown, 1.5–2 cm; scale margins fringed. **Leaves** 2 or 3 per fascicle, spreading or ascending, persisting ca. 2 years, 15–20 (–23) cm × 1.2–1.5 mm, straight, slightly twisted, pliant, yellow- to blue-green, all surfaces with stomatal lines, margins finely serrulate, apex abruptly acute to acuminate; sheath 1–2 cm, base persistent. **Pollen cones** cylindric, 30–40 mm, purplish. **Seed cones** maturing in 2 years, falling the year after seed-shed, single or in pairs, symmetric, lance-ovoid before opening, ovoid or ovoid-cylindric when open, (7–)9–18(–20) cm, light chocolate brown, on stalks to 3 cm; apophyses lustrous (as if varnished), slightly raised, strongly cross-keeled; umbo central, depressed-pyramidal, with short, stout prickle. **Seeds** ellipsoid, oblique-tipped; body 6–7 mm, dark brown; wing to 20 mm.

Varieties 2 (native only in the flora): introduced in subtropical and warm temperate areas worldwide.

1. Seedlings essentially without grass stage, buds thus scattered on the stem; leaves mostly in 3s, sometimes in 2s on same shoot; resin canals 3–5 per leaf; base of open cone ± truncate. 24a. *Pinus elliottii* var. *elliottii*
1. Seedlings tending toward a grass stage, buds thus crowded on contracted stems; leaves mostly in 2s, sometimes in 3s on same shoot; resin canals 3–9 per leaf; base of open cone rounded. . . . . 24b. *Pinus elliottii* var. *densa*

24a. Pinus elliottii Engelmann var. elliottii · Slash pine

Seedlings essentially without grass stage, height growth uniform after seed germination, buds scattered upstem. **Leaves** mostly in 3s, sometimes in 2s on same shoot, resin canals per leaf 3–5, hypodermis 2–3 cell-layers thick. **Seed-cone base** ± truncate when open. $2n = 24$.

Lowland to upland forests, old fields, and fine white sands, mostly long-hydroperiod soils; 0–150 m; Ala., Fla., Ga., La., Miss., S.C.

Pinus elliotti var. *elliottii* is the fastest growing of the southern yellow pines, much planted in the United States

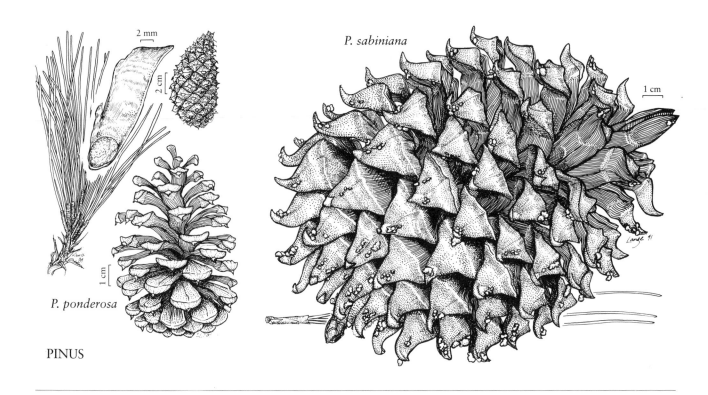

P. sabiniana

P. ponderosa

PINUS

outside its range. It is very susceptible, however, to ice damage and fusiform gall inland. This is a naval stores pine, but it is considered increasingly important in plantations as a lumber and pulpwood pine. It is much planted in subtropical and warm temperate climates worldwide, particularly in Brazil.

24b. Pinus elliottii Engelmann var. **densa** Little & Dorman, J. Forest. (Washington) 50: 921, fig. 1, 2. 1952 · South Florida slash pine

Pinus densa (Little & Dorman) de Laubenfels & Silba

Seedlings with vertical growth interrupted by grass stage, stem then more thickened, fascicles much more numerous and crowded around bud, and other buds more approximate on stem. **Leaves** mostly in 2s, sometimes in 3s on same shoot, resin canals per leaf 3–9, hypodermis (2–)3–4(–5) cell-layers thick. **Seed-cone base** mostly rounded when open.

Flatwoods, mostly over limestone; 0–10 m; Fla.

The name *Pinus caribaea* Morelet has been applied in error to *P. elliottii* var. *densa*. This variety is best distinguished by its wood, which is heavier and harder than that of typical slash pine, and by its having a grass stage comparable to that of *P. palustris*. This variety is not used for naval stores as is the type variety; neither is it commercially planted.

25. Pinus ponderosa Douglas ex Lawson & C. Lawson, Agric. Man., 354. 1836 · Ponderosa pine, western yellow pine, pin à bois lourd, pino real, pinabete

Trees to 72 m; trunk to 2.5 m diam., straight; crown broadly conic to rounded. **Bark** yellow- to red-brown, deeply irregularly furrowed, cross-checked into broadly rectangular, scaly plates. **Branches** descending to spreading-ascending; twigs stout (to 2 cm thick), orange-brown, aging darker orange-brown, rough. **Buds** ovoid, to 2 cm, fully 1 cm broad, red-brown, very resinous; scale margins white-fringed. **Leaves** 2–5 per fascicle, spreading to erect, persisting (2–)4–6(–7) years, 7–25 (–30) cm × (1–)1.2–2 mm, slightly twisted, tufted at twig tips, pliant, deep yellow-green, all surfaces with evident stomatal lines, margins serrulate, apex abruptly to narrowly acute or acuminate; sheath 1.5–3 cm, base persistent. **Pollen cones** ellipsoid-cylindric, 1.5–3.5 cm, yellow or red. **Seed cones** maturing in 2 years, shedding seeds soon thereafter, leaving rosettes of scales on branchlets, solitary or rarely in pairs, spreading to reflexed, symmetric to slightly asymmetric, conic-ovoid before opening, broadly ovoid when open, 5–15 cm, mostly reddish brown, sessile to nearly sessile, scales in steep spirals (as compared to *Pinus jeffreyi*) of 5–7 per row as viewed from side, those of cones just prior to and after cone fall spreading and reflexed, thus well separate from adjacent scales; apophyses dull to lustrous, thickened and variously raised and transversely keeled; umbo central, usually pyramidal to truncated, rarely depressed, merely acute, or with a very short apiculus, or with a stout-based spur or prickle. **Seeds** ellipsoid-obovoid; body (3–)4–9 mm, brown to yellow-brown, often mottled darker; wing 15–25 mm.

Varieties 3 (3 in the flora): North America, Mexico.

Pinus ponderosa is the most economically important western yellow pine. Its wood is more similar in character to the white pines, and it is often referred to as white pine. The taxonomy of this complex is far from resolved.

Ponderosa pine (*Pinus ponderosa*) is the state tree of Montana.

1. Leaves mainly 2–3 per fascicle, (7–)10–17 cm.
. . . . 25c. *Pinus ponderosa* var. *scopulorum*
1. Leaves mainly 3–5 per fascicle, (7–)12–25(–30) cm.
  2. Buds very resinous; leaves commonly 3 per fascicle, 12–25(–30) cm × (1.2–)1.5–2 mm; pollen cones mostly red; seed cones 8–15 cm; twigs commonly not glaucous. . . . .
. . . . 25a. *Pinus ponderosa* var. *ponderosa*
  2. Buds slightly resinous; leaves commonly 4–5 per fascicle, 7–17 cm × 1–1.2(–1.5) mm; pollen cones mostly yellow; seed cones 5–8 cm; twigs usually glaucous. . . . .
. . . . 25b. *Pinus ponderosa* var. *arizonica*

25a. Pinus ponderosa Douglas ex Lawson & C. Lawson var. **ponderosa** · Ponderosa pine

Trees to 72 m; trunk to 2.5 m diam. **Twigs** commonly red-brown, not glaucous. **Buds** very resinous. **Leaves** mainly 3 per fascicle, 12–25(–30) cm × (1.2–)1.5–2 mm. **Pollen cones** mostly red. **Seed cones** 8–15 cm, symmetric; apophyses of fertile scales moderately raised; umbo low-pyramidal, tapering acuminately to short broad-based prickle. **Seed** body 6–9 mm; wing 15–25 mm.

Montane, dry, open forests; 0–2300 m; B.C.; Calif., Nev., Oreg., Wash.

Pinus ponderosa var. *ponderosa* is sympatric with and hybridizes with *P. jeffreyi,* but in areas of sympatry it usually occurs at lower elevations than *P. jeffreyi.* Loggers often distinguish the two by the strong turpentine odor of fresh-cut wood of *P. ponderosa,* which contrasts with the sweeter odor of *P. jeffreyi.* Harvest of var. *ponderosa* far exceeds regrowth because of high timber value and multiple uses of the wood. *Pinus pon-*

derosa var. *ponderosa* is the largest and stateliest yellow pine in the flora.

25b. Pinus ponderosa Douglas ex Lawson & C. Lawson var. **arizonica** (Engelmann) Shaw, Pines Mexico, 24, plates 4, 17, fig. 4. 1909 · Arizona pine

Pinus arizonica Engelmann in Rothrock, U.S. Geogr. Surv., Rep. Wheeler, 260. 1878[1879]

Trees to 30 m; trunk to 1.2 m diam. **Twigs** mostly purple or red-brown, usually glaucous. **Buds** slightly resinous. **Leaves** mainly 4–5 per fascicle, 7–17 cm × 1–1.2(–1.5) mm. **Pollen cones** mostly yellow. **Seed cones** often asymmetric, 5–8 cm; apophyses at abaxial base often strongly raised, frequently mammillate; umbo low, broadly pryamidal, and merely acute or with very short apiculus. **Seed** body 3–4 mm; wing to 15 mm. $2n = 24.$

Slopes, canyons and rims, and tablelands; 2100–2500 m; Ariz., N.Mex.; Mexico.

The least common, least accessible ponderosa pine in the flora is *Pinus ponderosa* var. *arizonica.* It has leaves ranging widely in number per fascicle and in length, perhaps an expression of intergradation with *P. ponderosa* var. *scopulorum,* with which it is sympatric over broad areas.

Texas (Chisos Mountains) pines that are referred by some workers to *P. ponderosa* var. *arizonica,* belong to *P. ponderosa* var. *scopulorum.*

25c. Pinus ponderosa Douglas ex Lawson & C. Lawson var. **scopulorum** Engelmann in S. Watson, Bot. California 2: 126. 1880 · Rocky Mountain ponderosa pine

Pinus brachyptera Engelmann; *P. scopulorum* (Engelmann) Lemmon

Trees to 24 m; trunk to 1.5 m diam. **Twigs** mostly red-brown, rarely glaucous. **Leaves** mainly 2–3 per fascicle, (7–)10–17 cm × (1.2–)1.4–2 mm. **Pollen cones** yellow. **Seed cones** mostly symmetric, 5–10 cm; apophyses of fertile scales moderately raised; umbo low pyramidal, narrowing acuminately to a stout-based prickle or short sharp spur. **Seed** body 3–4 mm; wing to 15 mm.

Tablelands, canyon slopes and rims, and foothills, western Great Plains, Rocky Mountains; 1000–3000 m; B.C.; Ariz., Colo., Idaho, Mont., Nebr., Nev., N.Mex., N.Dak., Okla., Oreg., S.Dak., Tex., Utah, Wash., Wyo.; Mexico.

The most important timber pine of the Rocky Mountains is *Pinus ponderosa* var. *scopulorum*. It intergrades with *P. ponderosa* var. *ponderosa* in Idaho, Montana, and Washington, and with *P. ponderosa* var. *arizonica* in Arizona, New Mexico, and Mexico.

26. Pinus washoensis H. Mason & Stockwell, Madroño 8: 62. 1945 · Washoe pine

Trees to 60 m; trunk to 1 m diam., straight; crown pyramidal. Bark yellow-brown to reddish, fissured, plates scaly. Branches spreading-ascending; twigs stout, orangish, aging gray, rough. Buds ovoid, red-brown, 1.5–2 cm, not resinous; scale margins fringed. Leaves (2–)3 per fascicle, spreading-ascending, persisting (2–)4–6(–7) years, 10–15 cm × ca. 1.5 mm, slightly twisted, gray-green, all surfaces with stomatal lines, margins finely serrulate, apex acuminate; sheath 1–2 cm, base persistent. Pollen cones cylindric, 10–20 mm, red-purple. Seed cones maturing in 2 years, shedding seeds soon thereafter, not persistent, spreading, slightly asymmetric, ovoid-conic before opening, broadly ovoid when open, 7–10 cm, tan or pale red-brown, sessile, abaxial surface of scales darker and sharply contrasting in color with adaxial surface; apophyses slightly raised, low pyramidal; umbo central, narrowly pyramidal, tapering into short, reflexed, fine prickle. Seeds ellipsoid; body ca. 0.8 cm, gray-brown; wing to 16 mm. $2n = 24$.

Dry montane forests; of conservation concern; 2100–2500 m; Calif., Nev.

Pinus washoensis often occurs in large stands and resembles *P. jeffreyi*. The number and posture of seed-cone scales fall within the ranges given for *P. jeffreyi*. The abaxial surface of these scales has a significantly darker pigmentation, however; such a color contrast is not apparent in *P. jeffreyi*. Forest geneticists have developed hybrids between *P. washoensis* and related yellow pines, but no natural hybrids have been observed. Some workers regard *P. washoensis* as closely related to—or even conspecific with—*P. ponderosa*.

27. Pinus jeffreyi Greville & Balfour in A. Murray bis, Bot. Exped. Oregon 8: 2 plates. 1853 · Jeffrey pine

Pinus deflexa Torrey; *P. jeffreyi* var. *deflexa* (Torrey) Lemmon; *P. ponderosa* Douglas ex Lawson & C. Lawson var. *jeffreyi* Balfour ex Vasey

Trees to 61 m; trunk to 2.5 m diam., usually straight; crown conic to rounded. Bark yellow-brown to cinnamon, deeply fur-

rowed and cross-checked, forming large irregular scaly plates. Branches spreading-ascending; twigs stout (to 2 cm thick), purple-brown, often glaucous, aging rough. Buds ovoid, tan to pale red-brown, 2–3 cm, not resinous; scale margins conspicuously fringed. Leaves 3 per fascicle, spreading-ascending, persisting (2–)4–6(–7) years, 12–22(–25) cm × ca. 1.5–2 mm, slightly twisted, gray- to yellow-green, all surfaces with fine stomatal lines, margins finely serrulate, apex acute to acuminate; sheath (1–)1.5–2.5(–3) cm, base persistent. Pollen cones lance-cylindric, 20–35 mm, yellow to yellow- or purple-brown. Seed cones maturing in 2 years, shedding seeds and falling soon thereafter, nearly terminal, spreading, slightly asymmetric at base, ovoid-conic before opening, cylindro-ovoid when open, (10–)15–30 cm, light red-brown, nearly sessile or on stalks to 0.5 cm, abaxial surface of scales not darker than or sharply contrasting in color with adaxial surface, scales in low spirals (as compared to *Pinus ponderosa*) of 8 or more per row as viewed from side, those of cones just prior to and after cone fall not so spreading and deflexed, thus not so much separated from adjacent scales; apophyses slightly thickened and raised, not keeled; umbo central, slightly raised, with short, slender, reflexed prickle. Seeds ellipsoid-obovoid; body ca. 1 cm, brown or gray-brown, mottled darker; wing to 2.5 cm. $2n = 24$.

High, dry montane forests mostly above the *Pinus ponderosa* zone; 2000–2500 m; Calif., Nev., Oreg.; Mexico in Baja California.

Pinus jeffreyi has a form very similar to that of *P. ponderosa*, but it is a smaller species when compared with sympatric populations of the latter. It is cut and sold under the same name as *P. ponderosa*, but the sweetish odor of the fresh-cut wood contrasts sharply with the turpentine odor of ponderosa pine. The resin chemistry of the two species is significantly different.

28. Pinus engelmannii Carrière, Rev. Hort., sér. 4, 3: 227. 1854 · Apache pine

Pinus macrophylla Engelmann in Wislizenus 1848, not Lindley 1839; *P. apacheca* Lemmon; *P. latifolia* Sargent

Trees to 35 m; trunk to 0.6 m diam., straight; crown irregularly rounded, rather thin. Bark dark brown, at maturity deeply furrowed, ridges becoming yellowish, of narrow, elongate, scaly plates. Branches straight to ascending; twigs stout (1–2 cm thick), pale gray-brown, aging darker brown, rough. Buds ovoid-conic, to 2 cm, resinous; scale margins pale fringed. Leaves 3(–5) per fascicle, spreading-ascending, often drooping, forming a brush at twig tips, persisting 2 years, (20–)25–45 cm × 2 mm, dull green, all surfaces with fine stomatal lines, margins coarsely serrulate, apex conic-subulate; sheath 3–4 cm, base persistent. **Pollen cones** cylindric, ca. 25 mm, yellow to yellow-brown. **Seed cones** maturing in 2 years and shedding seeds soon thereafter, not persistent, terminal, sometimes curved, often asymmetric, lance-ovoid before opening, ovoid when open, 11–14 cm, light dull brown, nearly sessile or short-stalked; apophyses rhombic, somewhat to quite elongate, strongly raised toward outer cone base, sometimes curved, strongly cross-keeled, narrowed to thick, curved, broadly triangular-based umbo, this often producing outcurved claw. **Seeds** obovoid; body ca. 8–9 mm, dark brown; wing to 20 mm. $2n = 24$.

High and dry mountain ranges, valleys, and plateaus; 1500–2500 m; Ariz., N.Mex.; Mexico.

In general appearance *Pinus engelmannii* much resembles *P. palustris* with its short-persistent, long leaves (but in this species drooping) and in its tendency to form a grass stage. It has a deep taproot as do *P. palustris* and *P. ponderosa*.

29. Pinus sabiniana Douglas ex D. Don in Lambert, Descr. Pinus [ed. 3] 2: unnumbered page between 144 and 145, plate 80. 1832 · Digger pine, gray pine

Trees to 25 m; trunk to 1.2 m diam., straight to crooked, often forked; crown conic to raggedly lobed, sparse. **Bark** dark brown to near black, irregularly and deeply furrowed, ridges irregularly rectangular or blocky, scaly, often breaking away, bases of furrows and underbark orangish. **Branches** often ascending; cone-bearing branchlets stout, twigs comparatively slender, both pale purple-brown and glaucous, aging gray, rough. **Buds** ovoid, red-brown, ca. 1 cm, resinous; scale margins white-fringed. **Leaves** mostly 3 per fascicle, drooping, persisting 3–4 years, 15–32 cm × 1.5 mm, slightly twisted, dull blue-green, all surfaces with pale, narrow stomatal lines, margins serrulate, apex short-acuminate; sheath to 2.4 cm, base persistent. **Pollen cones** ellipsoid, 10–15 mm, yellow. **Seed cones** maturing in 2 years, shedding seeds soon thereafter, persisting to 7 years, pendent, massive, heavy, nearly symmetric, ovoid before opening, broadly to narrowly ovoid or ovoid-cylindric when open, 15–25 cm, dull brown, resinous, stalks to 5 cm; apophyses elongate, curved, continuous with umbos to form long, upcurved claws to 2 cm. **Seeds** narrowly obovoid; body ca. 20 mm, dark brown; wing broad, short, ca. 10 mm. $2n = 24$.

Dry foothills on the west slope of the Sierra Nevada, and in the coast ranges, nearly ringing the Central Valley of California; 30–1900 m; Calif.

Seeds of *Pinus sabiniana* were an important food source for many Indian groups in California, sometimes collectively referred to as "Digger Indians." Because the name "Digger" has been used as a derogatory ethnic term, many people prefer to avoid using the vernacular name Digger pine.

SELECTED REFERENCE   Griffin, J. R. 1964. Cone morphology in *Pinus sabiniana*. J. Arnold Arbor. 45: 260–273.

30. Pinus coulteri D. Don, Trans. Linn. Soc. London 17: 440. 1836 · Coulter pine

Trees to 24 m; trunk to 1 m diam., straight to contorted; crown broad, thin, irregular. **Bark** dark gray-brown to near black, deeply furrowed, with long, scaly, irregularly anastomosing, rounded ridges. **Branches** often ascending; twigs stout to moderately slender, violet-brown, often glaucous, aging gray-brown, rough. **Buds** ovoid, deep red-brown, 1.5(–3) cm, resinous; scale margins white-fringed, apex cuspidate. **Leaves** 3 per fascicle, slightly spreading, not drooping, mostly ascending in a brush, persisting 3–4 years, 15–30 cm × ca. 2 mm, slightly curved or straight, twisted, dusty gray-green, all surfaces with pale, fine stomatal lines, margins serrulate, apex abruptly subulate; sheath 2–4 cm, base persistent. **Pollen cones** ovoid to cylindric, to 25 mm, light purple-brown, aging orange-brown. **Seed cones** maturing in 2 years, gradually shedding seeds thereafter and moderately persistent, massive, heavy, drooping, asymmetric at base, narrowly ovoid before opening, ovoid-cylindric when open, 20–35 cm, pale yellow-brown, resinous, stalks to 3 cm; apophyses transverse-rhombic, strongly and sharply cross-keeled, elongate, curved, continuous with umbos to form long, upcurved claws 2.5–3 cm. **Seeds** obovoid; body 15–22 mm, dark brown; wing to 25 mm. $2n = 24$.

Dry rocky slopes, flats, ridges, and chaparral, transitional to oak-pine woodland; 300–2100 m; Calif.; Mexico in Baja California.

Pinus coulteri is the heaviest-coned pine; one who seeks its shade should wear a hardhat.

31. Pinus torreyana Parry ex Carrière, Traité Gén. Conif., 326. 1855 · Torrey pine

Trees to 15(–23) m; trunk to 1 m diam., in nature mostly crooked and leaning; crown rounded to flattened or irregular. **Bark** red-brown to purple-red, deeply furrowed with irregular, elongate, flat, scaly ridges. **Branches** irregular, spreading-ascending, candelabra-like; twigs stout (1–2 cm thick), greenish, aging deep gray-brown to near black, rough. **Buds** conic-ovoid, pale brown, to 2.5 cm; scale margins white-fringed. **Leaves** mostly 5 per fascicle, ascending or spreading, persisting

3–4 years, 15–30 cm × ca. 2 mm, straight or curved, slightly twisted, dull gray-green, all surfaces with fine stomatal lines, margins serrulate, apex abruptly acute; sheath to 2 cm, shed early, base persistent. **Pollen cones** ovoid, 20–30 mm, yellow. **Seed cones** maturing in 3 years, shedding seeds soon thereafter, persisting to 5 years, lateral, massive, heavy, symmetric, ovoid before opening, broadly ovoid when open, 10–15 cm, yellow- to red-brown, lustrous, stalks to 4 cm; apophyses thick, angulately dome-shaped, with 5 low convergent keels; umbo central, forming short, curved-tipped pyramid. **Seeds** narrowly obovoid; body 16–24 mm, brown, apically dark brown; wing broad, oblique-tipped, to 15 mm.

Subspecies 2: only in the flora.

Pinus torreyana is a rare and local Tertiary relic species whose present range is reduced to two small areas of southern California: near Del Mar (San Diego County) and on the northeastern shore of Santa Rosa Island (Santa Barbara County). Its distribution in Oligocene and Miocene (or at least that of its near ancestor) extended north to Oregon. Its harsh natural habitat elicits an unusually contorted and often sparse form, quite unlike the cleaner and taller form the species takes in cultivation.

In terms of numbers of individuals in the wild, as well as the small area occupied by natural populations, *Pinus torreyana* is without a doubt the rarest North American pine. As such it is under protection. Artificial crosses between it and another, more widespread Tertiary relic, *P. sabiniana*, have been successful.

SELECTED REFERENCE Haller, J. R. 1986. Taxonomy and relationships of the mainland and island populations of *Pinus torreyana* (Pinaceae). Syst. Bot. 11: 39–50.

1. Crown of sheltered tree narrower than height of mature tree, open with well-spaced branches in natural sites; seed cones less than 13.5 cm broad, as broad as long or longer; only tip of umbo curved outward; maximum seed width ca. 12 mm. . . . . 31a. *Pinus torreyana* subsp. *torreyana*
1. Crown of sheltered tree broader than height of mature tree, compact in natural sites; seed cones mostly more than 13.5 cm broad, broader than long; entire umbo curved outward; maximum seed width ca. 14 mm. 31b. *Pinus torreyana* subsp. *insularis*

31a. Pinus torreyana Parry ex Carrière subsp. **torreyana** · Mainland Torrey pine

Trees to 15 m in native stands (to 23 m in sheltered sites); mature crown of sheltered trees narrower than tree height, fairly open in natural sites. **Seed cones** mostly broadly ovoid, as broad as long or longer, mostly less than 13.5 cm; umbos mostly shorter than 6 mm, terminal portion curved outward. **Seeds** to 12 mm wide, averaging less than 11 mm, light to medium brown with dark mottling. $2n = 24$.

Dry fogbelt zone on eroding, mostly dry slopes; of conservation concern; 0–125 m; Calif.

Pinus torreyana subsp. *torreyana* occurs naturally only in a relict stand near Del Mar, California.

31b. Pinus torreyana Parry ex Carrière subsp. **insularis** J. R. Haller, Syst. Bot. 11: 45. 1986 · Island Torrey pine

Trees 10(–15) m in native stands; mature crown of sheltered trees broader than tree height, compact in natural sites. **Seed cones** broader than long, mostly over 13.5 cm; umbos mostly more than 6 mm, whole umbo curved outward. **Seeds** to 14 mm wide, averaging over 11 mm, brown to dark brown with darker mottling.

Dry fogbelt zone, ravines and low ridges; of conservation concern; 130–180 m; Calif.

Pinus torreyana subsp. *insularis* occurs naturally only on Santa Rosa Island, California.

32. Pinus banksiana Lambert, Descr. Pinus 1: 7, plate 3. 1803 · Jack pine, pin gris

Pinus divaricata (Aiton) Sudworth; *P. sylvestris* Linnaeus [var.] δ *divaricata* Aiton

Trees to 27 m; trunk to 0.6 m diam., straight to crooked; crown becoming irregularly rounded or spreading and flattened. **Bark** orange- to red-brown, scaly. **Branches** descending to spreading-ascending, poorly self-pruning; twigs slender, orange-red to red-brown, aging gray-brown, rough. **Buds** ovoid, red-brown, 0.5–1 cm, resinous; scale margins nearly entire. **Leaves** 2 per fascicle, spreading or ascending, persisting 2–3 years, 2–5 cm × 1–1.5(–2) mm, twisted, yellow-green, all surfaces with fine stomatal lines, mar-

P. *attenuata*

P. *clausa*

P. *banksiana*

P. *muricata*

PINUS

gins finely serrulate, apex acute to short-subulate; sheath 0.3–0.6 cm, semipersistent. **Pollen cones** cylindric, 10–15 mm, yellow to orange-brown. **Seed cones** maturing in 2 years, shedding seeds soon thereafter or often long-serotinous and shedding seeds only through age or fire, upcurved, asymmetric, lanceoloid before opening, ovoid when open, 3–5.5 cm, tan to light brown or greenish yellow, slick, nearly sessile or short-stalked, most apophyses depressed but increasingly mammillate toward outer cone base; umbo central, depressed, small, sunken centrally, unarmed or with a small, reflexed apiculus. **Seeds** compressed-obovoid, oblique; body 4–5 mm, brown to near black; wing 10–12 mm. $2n = 24$.

Fire successional in boreal forests, tundra transition, dry flats, and hills, sandy soils; 0–800 m; Alta., B.C., Man., N.B., N.W.T., N.S., Ont., P.E.I., Que., Sask.; Ill., Ind., Maine, Mich., Minn., N.H., N.Y., Pa., Vt., Wis.

Pinus banksiana reaches its largest size and best form in Canada. In western Alberta and in northeastern British Columbia, it is sympatric with *P. contorta* and forms hybrid swarms with that species.

Jack pine *(Pinus banksiana)* is the territorial tree of the Northwest Territories.

33. **Pinus contorta** Douglas ex Loudon, Arbor. Frutic.
Brit. 4: 2292, figs. 2210, 2211. 1838 · Lodgepole pine

Shrubs or trees to 50 m; trunk to 0.9 m diam., straight to contorted; crown various according to genetic race. **Bark** brown to gray- or red-brown, platy to furrowed.

Lower branches often descending, the upper spreading or ascending. **Twigs** slender, orange to red-brown, aging darker brown, rough. **Buds** narrowly to broadly ovoid, dark red-brown, to 1.2 cm, slightly resinous. **Leaves** 2 per fascicle, spreading or ascending, persisting 3–8 years, 2–8 cm × 0.7–2(–3) mm, twisted, yellow-green to dark green, all surfaces with fine stomatal lines, margins finely serrulate, apex blunt to acute or narrowly acuminate; sheath 0.3–0.6(–1) cm, persistent. **Pollen cones** ellipsoid to cylindric, 5–15 mm, orange-red. **Seed cones** maturing in 2 years or variably serotinous, variably persistent, spreading to reflexed, often curved, nearly symmetric or variably asymmetric, lanceoloid to ovoid before opening, broadly ovoid to nearly globose when open, 2–6 cm, tan to pale red-brown, lustrous, nearly sessile or on stalks to 1 cm; apophyses nearly rhombic, variously elongate, cross-keeled, often mammillate toward outer cone base and on inside above middle; umbo central, depressed-triangular, prickle barely elongate to stubby or slender and to 6 mm. **Seeds** compressed, obovoid; body ca. 5 mm, red-brown, mottled with black, or all black; wing 10–14 mm. $2n = 24$ (variety not indicated).

Varieties 3: only in the flora.

Pinus contorta is fire successional over most of its range and is characterized by prolific seeding and high seed viability in disturbed habitats, often resulting in extremely slow-growing, overly dense stands. Some authors consider it to consist of 4 races; these have been given various infraspecific ranks, but perhaps they are more conventionally treated as 3 varieties.

1. Leaves 2–7 cm × 0.7–0.9(–1.1) mm, dark green; mature trunk with bark evidently furrowed; seed cones strongly asymmetric, strongly recurved, persistent or variously serotinous. . . . . 33a. *Pinus contorta* var. *contorta*
1. Leaves (4–)5–8 cm × (0.7–)1–2(–3) mm, yellow-green; mature trunk with bark not evidently furrowed; seed cones asymmetric to nearly symmetric, recurved to spreading, variously serotinous or soon shed.
   2. Seed cones asymmetric, recurved, variously serotinous, long-persistent; mid and lower apophyses mostly much domed; main branches mostly horizontally spreading, not ascending at tip. . . . . . . 33b. *Pinus contorta* var. *latifolia*
   2. Seed cones nearly symmetric, mostly spreading, not serotinous, not persistent; mid and lower apophyses mostly shallowly domed; main branches ascending at tips. . . . . 33c. *Pinus contorta* var. *murrayana*

33a. Pinus contorta Douglas ex Loudon var. **contorta** · Shore pine

Pinus bolanderi Parlatore; *P. contorta* subsp. *bolanderi* (Parlatore) Critchfield; *P. contorta* var. *bolanderi* Lemmon

Trees to 10 m; trunk to 0.5 m diam., straight, leaning landward in protected stands, or contorted and bent when not, or reduced to shrub form by wind shear and salt spray; mature crown irregularly rounded or flat. **Bark** evidently and irregularly furrowed, cross-checked into small, square or rectangular, orange-brown to purple-brown scaly plates. **Branches** spreading, often contorted. **Leaves** 2–7 cm × 0.7–0.9(–1.1) mm, dark green, apex acute to broadly acute or blunt. **Seed cones** maturing and shedding seeds in 2 years, persistent or variously serotinous, strongly asymmetric, strongly recurved, often in whorls, proximal outer apophyses strongly domed.

Maritime fog forests, bogs, and dry foothills; 0–600 m; B.C.; Alaska, Calif., Oreg., Wash.

Pinus contorta var. *contorta* is fire successional. It is mostly a low-quality tree for timber use.

33b. Pinus contorta Douglas ex Loudon var. **latifolia** Engelmann in S. Watson, Botany (Fortieth Parallel), 331. 1871 · Lodgepole pine

Pinus contorta subsp. *latifolia* (Engelmann) Critchfield

Trees to 46 m; trunk to 0.8 m diam., mostly straight and evenly tapering, or at or above timberline reduced to shrub form by wind shear; crown usually conic at maturity. **Bark** gray- to red-brown, not evidently furrowed, separating into loose plates. **Branches** mostly horizontally spreading, not ascending at tip. **Leaves** (4–)5–8 cm × 1–2(–3) mm, yellow-green, apex narrowly acute to short-acuminate. **Seed cones** maturing in 2 years, then shedding seeds or variously serotinous, long-persistent, strongly asymmetric, mostly recurved, seldom whorled, mostly in 2s or solitary, mid and lower apophyses mostly much domed.

Low to high montane forests, often to timberline; 0–3500 m; Alta., B.C., N.W.T., Sask., Yukon; Alaska, Colo., Idaho, Mont., Oreg., S.Dak., Utah, Wash., Wyo.

Pinus contorta var. *latifolia* is fire successional. It is the most wide-ranging and commercially utilized variety. Its poor self-pruning character makes it less desirable for lumber but adequate for mine timbers, fences, and pulpwood.

Lodgepole pine (*Pinus contorta* var. *latifolia*) is the provincial tree of Alberta.

33c. Pinus contorta Douglas ex Loudon var. **murrayana** (Greville & Balfour) Engelmann in S. Watson, Bot. California 2: 126. 1880 · Sierra lodgepole pine

Pinus murrayana Greville & Balfour in A. Murray bis, Bot. Exped. Oregon 8: no. 740, plate 1853; *P. contorta* subsp. *murrayana* (Greville & Balfour) Critchfield

Trees to 50 m; trunk to 0.9 m diam., straight, little tapering; crown mostly conic at maturity. **Bark** scaly, not evidently furrowed, orange- to purple-brown. **Branches** spreading, ascending at tips. **Leaves** 5–8 cm × 1–2 mm, yellow-green, apex acute. **Seed cones** maturing in 2 years, shedding seeds and falling soon thereafter, nearly symmetric, mostly spreading, rarely in whorls, more often paired or solitary, mid and lower apophyses mostly shallowly domed.

Montane forests; 400–3500 m; Calif., Nev., Oreg., Wash.; Mexico in Baja California.

Pinus contorta var. *murrayana* is the tallest and best-formed variety of this species.

34. Pinus virginiana Miller, Gard. Dict. ed. 8, Pinus no. 9. 1768 · Virginia pine

Trees to 18 m; trunk to 0.5 m diam., straight or contorted to erect or leaning; crown irregularly rounded or flattened. **Bark** gray-brown with irregular, scaly-plated ridges, on upper sections of trunk reddish, scaly. **Branches** spreading-ascending to spreading-descending; twigs slender, red- or purple-tinged, often glaucous, aging red-brown to gray, rough. **Buds** ovoid to cylindric, red-brown, 0.6–1 cm, resinous or not resinous; scale margins white-fringed. **Leaves** 2 per fascicle, spreading or ascending, persisting 3–4 years, 2–8 cm × 1–1.5 mm, strongly twisted, deep to pale yellow-green, all surfaces with inconspicuous stomatal lines, margins serrulate, apex narrowly acute; sheath 0.4–1 cm, base persistent. **Pollen cones** ellipsoid-cylindric, 10–20 mm, red-brown or yellow. **Seed cones** maturing in 2 years, shedding seeds soon thereafter, persisting to 5 years, symmetric, lance-ovoid or lanceoloid before opening, ovoid when open, 3–7(–8) cm, dull red-brown, nearly sessile or on stalks to 1 cm, scales rigid, with strong purple-red or purple-brown border on adaxial surface distally; apophyses slightly thickened, slightly elongate; umbo central, low-pyramidal, with slender, stiff prickle. **Seeds** compressed-obovoid, oblique apically; body 4–7 mm, pale brown, mottled darker; wing narrow, to 20 mm. $2n = 24$.

Dry uplands, sterile sandy or shaly barrens, old fields, and lower mountains; 0–900 m; Ala., Del., Ga., Ind., Ky., Md., Miss., N.J., N.Y., N.C., Ohio, Pa., S.C., Tenn., Va., W.Va.

Pinus virginiana is weedy and fire successional and often forms large stands. It is mostly too small and too profusely branched to be valued except as pulpwood.

35. Pinus clausa (Chapman ex Engelmann) Sargent, Rep. For. N. America, 199. 1884 · Sand pine

Pinus inops Aiton var. *clausa* Chapman ex Engelmann, Bot. Gaz. 2: 125. 1877; *P. clausa* var. *immuginata* D. B. Ward

Trees to 21 m; trunk to 0.5 m diam., straight and erect to leaning and crooked, much branched; crown mostly rounded or irregular. **Bark** gray to gray-brown, furrowed, with narrow, flat, irregular ridges, resin pockets absent, on upper sections of the trunk reddish to red-brown, platy becoming smooth distally. **Branches** spreading to ascending, poorly self-pruning; twigs slender, violet- to red-brown, rarely glaucous, aging gray, smooth. **Buds** cylindric, purple-brown, to 1 cm; scale margins white-fringed. **Leaves** 2 per fascicle, spreading-ascending, persisting 2–3 years, (3–)6–9(–10) cm × ca. 1 mm, straight, slightly twisted, dark green, all surfaces with fine, inconspicuous stomatal lines, margins finely serrulate, apex short-conic; sheath 0.3–0.5(–0.7) cm, base persistent. **Pollen cones** ellipsoid, ca. 10 mm, brownish yellow. **Seed cones** maturing in 2 years, shedding seeds soon thereafter or often long-serotinous, long-persistent, solitary or whorled, spreading, symmetric (rarely slightly asymmetric, reflexed), lanceoloid before opening, ovoid to broadly ovoid when open, 3–8 cm, red-brown, sessile or on stalks to 1 cm, scales with dark red-brown, purple, or purple-gray border distally on adaxial surface; apophyses thickened, shallowly and angulately raised, transversely rhombic, cross-keeled; umbo central, low-pyramidal, tapering to sharp tip or weak, often deciduous prickle. **Seeds** obovoid-oblique; body ca. 4 mm, dark brown to nearly black; wing to 17 mm. $2n = 24$.

Fire successional in sand dunes and white sandhills; 0–60 m; Ala., Fla.

Although *Pinus clausa* is too profusely branched to be important for saw timber, it is managed to produce a high volume of pulpwood in northern peninsular Florida.

36. Pinus radiata D. Don, Trans. Linn. Soc. London 17: 442. 1836 · Monterey pine

Pinus insignis Douglas ex Loudon

Trees to 30 m; trunk to 0.9 m diam., contorted to straight; crown broadly conic, becoming rounded to flattened. **Bark** gray, deeply V-furrowed, furrow bases red, ridges irregularly elongate-rectangular, their flattened surfaces scaly. **Branches** level to downcurved or ascending, poorly self-pruning; twigs slender, red-brown, sometimes glaucous, aging gray, rough. **Buds** ovoid to ovoid-cylindric, red-brown, ca. 1.5 cm, resinous. **Leaves** (2–)3 in a fascicle, spreading-ascending, persisting 3–4 years, (8–)9–15(–20) cm × 1.3–1.8(–2) mm, straight, slightly twisted, deep yellow-green, all surfaces with fine stomatal lines, margins serrulate, apex conic-subulate; sheath (1–)1.5–2 cm, base persistent. **Pollen cones** ellipsoid-cylindric, 10–15 mm, orange-brown. **Seed cones** maturing in 2 years, shedding seeds soon thereafter, but often serotinous and persistent 6–20 years, solitary to whorled, spreading to recurved, curved, very asymmetric, ovoid before opening, broadly ovoid when open, 7–14 cm, pale red-brown and lustrous, scales rigid, stalks to 1 cm; apophyses toward outer cone base increasingly mammillate, those on inward cone side and middle and apex of cone more level; umbo central, mostly depressed, with small central boss or occasionally with slender, deciduous prickle. **Seeds** compressed-ellipsoid; body ca. 6 mm, dark brown; wing 20–30 mm. $2n = 24$.

Coastal fog belt; of conservation concern; 30–400 m; Calif.; Mexico in Baja California [600–1200 m].

Pinus radiata has an extremely narrow natural range: three coastal areas in California (one in San Mateo and Santa Cruz counties, one in Monterey County, and one in San Luis Obispo County) and off the coast of Baja California, Mexico (Guadalupe Island and debatably also on Cedros Island). Some natural populations of the species are under protection. Along the California coast it has escaped from cultivation, and from there into southern coastal Oregon it shows signs of naturalizing.

Pinus radiata is a much better-formed tree and of greater silvicultural value within its introduced range (Africa, Australia, Europe, and New Zealand, where it is a principal timber tree) than in its native range. It hybridizes naturally with *P. attenuata* (*P.* × *attenuiradiata* Stockwell & Righter).

37. Pinus attenuata Lemmon, Mining Sci. Press 64: 45. 1892 · Knobcone pine

Pinus tuberculata Gordon 1849, not D. Don 1836

Shrubs or trees to 24 m; trunk to 0.8 m diam., usually straight; crown mostly narrowly to broadly conic. **Bark** purple-brown to dark brown, shallowly and narrowly fissured, with irregular, flat, loose-scaly plates, on upper sections of trunk nearly smooth. **Branches** ascending; twigs slender, red-brown. **Buds** ovoid to ovoid-cylindric, dark red-brown, aging darker, ca. 1.5 cm, resinous; scale margins fringed, apex attenuate. **Leaves** 3 per fascicle, spreading or ascending, persisting 4–5 years, (8–)9–15(–20) cm × (1–)1.3–1.8 mm, straight or slightly curved, twisted, yellow-green, all surfaces with fine stomatal lines, margins serrulate, apex abruptly conic-subulate; sheath (1–)1.5–2 cm, base persistent. **Pollen cones** ellipsoid-cylindric, 10–15 mm, orange-brown. **Seed cones** maturing in 2 years, serotinous, long-persistent, remaining closed for 20 years or more, or opening on burning, in whorls, hard and heavy, very asymmetric, lanceoloid before opening, ovoid-cylindric when open, 8–15 cm, yellow- or pale red-brown, stalks to 1 cm; apophyses toward outside base increasingly elongate, mammillate or raised-angled-conic, downcurved near

base, scarcely raised on branchlet side, rhombic; umbo central, low-pyramidal, sharp, upcurved. **Seeds** compressed-oblique-obovoid; body ca. 6–7 mm, nearly black; wing narrow, to 20 mm. $2n = 24$.

Fire successional on dry slopes and foothills of Sierra Nevada and the Cascade and Coast ranges; 300–1200 m; Calif., Oreg.; Mexico in Baja California.

Pinus attenuata, mostly a chaparral species, bears cones at an early age. Its seed crops are heavy, and a hot fire permits the seeds to be released. It forms hybrids with *P. muricata* and *P. radiata*.

38. Pinus muricata D. Don, Trans. Linn. Soc. London 17: 441. 1836 · Bishop pine

Pinus muricata var. *borealis* Axelrod; *P. muricata* var. *cedrosensis* J. T. Howell; *P. muricata* var. *stantonii* Axelrod; *P. radiata* var. *binata* (Engelmann) Brewer & S. Watson; *P. remorata* H. Mason

Trees to 24 m; trunk to 0.9 m diam., straight to contorted; crown becoming rounded, flattened, or irregular. **Bark** dark gray, deeply furrowed, ridges long, scaly-plated. **Branches** spreading-ascending, often contorted; twigs stout to slender, orange-brown, aging darker brown, rough. **Buds** ovoid-cylindric, dark brown, 1–2.5 cm, resinous. **Leaves** 2 per fascicle, spreading to upcurved, persisting 2–3 years, 8–15 cm × (1.2–)1.5(–2) mm, slightly twisted, dark yellow-green, all surfaces with stomatal lines, margins strongly serrulate, apex abruptly conic-acute; sheath to 1.5 cm, base persistent. **Pollen cones** ellipsoid, to 5 mm, orange. **Seed cones** maturing in 3 years, serotinous, long-persistent, mostly in whorls, mostly asymmetric, lanceoloid-ovoid before opening, curved-ovoid when open, 4–9 cm, glossy bright to pale red-brown, sessile or on stalks to 1 cm, mostly downcurved, scales with deep red-brown border distally on adaxial surface; apophyses much thickened, the abaxial ones progressively more angulately dome-shaped toward base of cone; umbo central, a stout-based, curved claw. **Seeds** obliquely ellipsoid; body 6–7 mm, dark brown to near black; wing 15–20 mm. $2n = 24$.

Dry ridges to coastal, windshorn forests, often in or around bogs; of conservation concern; 0–300 m; Calif.; Mexico in Baja California.

The several varieties described for *Pinus muricata* reflect the high variability in leaf characters and in degree of elaboration of apophysis and umbo in this species. The extremes can sometimes occur together.

4. CUPRESSACEAE Bartlett
• Redwood or Cypress Family

Frank D. Watson

James E. Eckenwalder

Trees or shrubs evergreen (usually deciduous in *Taxodium*), generally resinous and aromatic, monoecious (usually dioecious in *Juniperus*). **Bark** fibrous and furrowed (smooth or exfoliating in plates in some *Cupressus* and *Juniperus* species). **Lateral branches** well developed, similar to leading shoots, twigs terete, angled, or flattened dorsiventrally (with structurally distinct lower and upper surfaces; *Thuja, Calocedrus*), densely clothed by scalelike leaves or by decurrent leaf bases; longest internodes to 1 cm; buds undifferentiated and inconspicuous (except in *Sequoia*). **Roots** fibrous to woody (bearing aboveground "knees" in *Taxodium*). **Leaves** simple, usually persisting 3–5 years and shed with lateral shoots (cladoptosic) (shed annually in *Taxodium*), alternate and spirally arranged but sometimes twisted so as to appear 2-ranked, or opposite in 4 ranks, or whorled, deltate-scalelike to linear, decurrent, sessile or petioled; adult leaves appressed or spreading, often differing between lateral and leading shoots (twigs heterophyllous), sometimes strongly dimorphic on each twig (*Thuja, Calocedrus*) with lateral scale-leaf pairs conspicuously keeled; juvenile leaves linear, flattened, spreading; often with solitary abaxial resin gland; resin canal present. **Pollen cones** maturing and shed annually, solitary, terminal (rarely in clusters of 2–5, axillary in *Juniperus communis*; usually in terminal panicles in *Taxodium*), simple, spheric to oblong; sporophylls overlapping, bearing 2–10 abaxial microsporangia (pollen sacs); pollen spheric, not winged. **Seed cones** maturing in 1–2 seasons, shed with short shoots or persisting indefinitely on long-lived axes (shattering at maturity in *Taxodium*), compound, solitary, terminal (rarely in clusters of 2–5, axillary in *Juniperus communis*); scales overlapping or abutting, fused to subtending bracts with only bract apex sometimes free; each scale-bract complex peltate, oblong or cuneate, at maturity woody or fleshy, with 1–20 erect (inverted with age in *Sequoia* and *Sequoiadendron*), adaxial ovules. **Seeds** 1–20 per scale, not winged or with 2–3 symmetric or asymmetric wings; aril lacking; cotyledons 2–9.

Genera 25–30, species 110–130 (9 genera, 30 species in the flora): widespread in temperate regions.

Pollination usually occurs in late winter or spring but may occur anytime from late summer to early winter for some species of *Juniperus*. Seed maturation occurs in late summer or autumn. Species of *Cupressus* have serotinous cones that remain closed for many years, some opening only after exposure to fire.

The Cupressaceae, with a known fossil record extending back to the Jurassic (C. N. Miller Jr. 1988), constitute a diverse family often divided between Cupressaceae in the strict sense (for genera with leaves opposite in four ranks or whorled) and Taxodiaceae (leaves mostly alternate), but they are best kept together (J. E. Eckenwalder 1976; R. A. Price 1989). The unity of the family is best shown in the structure of the mature seed cones: the bract-scale complexes are intimately fused for most of their common length, the 1–20 ovules are erect at first but may invert with maturity, and the paired seed wings, if present, are derived from the seed coat. A majority of genera are monotypic and most others display disjunct or relictual distributions, even though individual species may be widely distributed. Only bird-dispersed *Juniperus* is species rich, with a wide, nearly continuous Northern Hemisphere distribution. Because of their uniformity, seedlings and juvenile specimens may not be determinable to genus. Foliage of cultivars may deviate greatly from forms found in wild plants.

Although no members of the family attain dominance over immense geographic spans as do some species of the Pinaceae in the boreal forests, they can achieve considerable local and regional prominence. Examples include redwood *(Sequoia sempervirens)* along the coast of northern California, several species of *Juniperus* (together with pinyons) at moderate elevations in the southwestern United States and Mexico, and baldcypress *(Taxodium distichum)* in deep swamps of the southeastern United States. Their ranges and regions of dominance were considerably greater during the early Tertiary.

The heartwood of many species of Cupressaceae is resistant to termite damage and fungal decay, and therefore it is widely used in contact with soil. Most prominent in the flora are redwood and baldcypress; the premier coffin wood of China, *Cunninghamia lanceolata*, is another member of the family. Other genera, usually called cedars, may have aromatic woods with a variety of specialty uses. Wooden pencils are made from incense-cedar *(Calocedrus decurrens)* and eastern redcedar *(Juniperus virginiana)*, which is also used for lining cedar chests. Wood from species of *Thuja* is still used for cedar roofing shingles.

In addition to the taxa treated below (including one naturalized species), several additional species and genera are cultivated to a greater or lesser extent and may persist without spreading after abandonment of cultivation. Some of the more important cultivated species of genera not treated in the flora include: *Cryptomeria japonica* (Linnaeus f.) D. Don (Japanese-cedar), differing from *Sequoiadendron* in its smaller, globose cones with bract/scale complexes bearing five to eight teeth; *Cunninghamia lanceolata* (Lambert) Hooker (China-fir), unlike all North American native taxa in its pointed, flat, lanceolate, drooping leaves to 7 cm; *Metasequoia glyptostroboides* Hu & W. C. Cheng (dawn-redwood), differing from *Sequoia* in its opposite leaves and deciduous branchlets and from *Taxodium* in its opposite leaves and persistent seed-cone scales; *Microbiota decussata* Komarov (microbiota), differing from spreading junipers in its minute, opening, one-seeded cone; *Platycladus orientalis* (Linnaeus) Franco [*Thuja orientalis* Linnaeus; *Biota orientalis* (Linnaeus) Endlicher; oriental arborvitae], which may escape locally from cultivation, differing from *Thuja* in its vertical sprays of branchlets, thicker, fleshier bract/scale complexes with prominent hornlike umbos, and unwinged seeds; and *Thujopsis dolabrata* (Linnaeus f.) Siebold & Zuccarini (hiba arborvitae), differing from *Thuja* in its much broader lateral leaf pairs and its sprays of branchlets with prominent white waxy markings beneath.

SELECTED REFERENCES Burns, R. M. and B. H. Honkala. 1990. Silvics of North America. 1. Conifers. Washington. [Agric. Handb. 654.] Canadian Forestry Service. 1983. Reproduction of Conifers. Forest. Techn. Pub. Canad. Forest. Serv. 31. Ecken-walder, J. E. 1976. Re-evaluation of Cupressaceae and Taxodiaceae: A proposed merger. Madroño 23: 237–256. Farjon, A. 1990. A Bibliography of Conifers. Königstein. [Regnum Veg. 122.] Hosie, R. C. 1969. Native Trees of Canada, ed. 7. Ottawa. Pp. 83–95. Krajina, V. J., K. Klinka, and J. Worrall. 1982. Distribution and Ecological Characteristics of Trees and Shrubs of British Columbia. Vancouver. Little, E. L. Jr. 1979. Checklist of United States Trees (Native and Naturalized). Washington. Pp. 33–36. [Agric. Handb. 541.] Rehder, A. J. 1949. Bibliography of Cultivated Trees and Shrubs Hardy in the Cooler Temperate Regions of the Northern Hemisphere. Jamaica Plain. Silba, J. 1986. Encyclopaedia Coniferae. Phytologia Mem. 8: 1–127.

1. Leaves alternate; leaves of adults usually with expanded needlelike or linear or linear-lanceolate blade.
  2. Leafy branchlets falling annually or seasonally; pollen cones mostly in pendent axillary panicles; seed cones nearly globose, shattering at maturity, scales each bearing (1–)2 irregularly 3-angled, wingless seeds; se United States. . . . . 3. *Taxodium*, p. 403
  2. Leafy branchlets persisting for several years; pollen cones mostly terminal and solitary; seed cones oblong or globose, opening but scales persistent at maturity, scales each bearing 2–9 lenticular, winged seeds; w United States.
    3. Branchlets with leaves mostly in 2 ranks, with obvious annual growth constrictions; leaves linear or linear-lanceolate to deltate, flattened, free portion to ca. 30 mm; mature seed cones 1.3–3.5 cm. . . . . 1. *Sequoia*, p. 401
    3. Branchlets with radiating leaves, without obvious annual growth constrictions; leaves mostly needlelike, triangular in cross section, free portion to ca. 15 mm; mature seed cones 4–9 cm. . . . . 2. *Sequoiadendron*, p. 403
1. Leaves opposite in 4 ranks or whorled; leaves of adults usually ± scalelike to subulate.
  4. Seed cones berrylike, remaining closed, seeds retained; scales generally fleshy or fibrous; monoecious or dioecious. . . . . 8. *Juniperus*, p. 412
  4. Seed cones opening, seeds shed; scales woody; monoecious.
    5. Leaves in whorls of 3. . . . . 9. *Callitris*, p. 421
    5. Leaves opposite in 4 ranks or seemingly in whorls of 4.
      6. Branchlets in radial arrays (partially comblike in *C. macnabiana*); seed cones globose or oblong with peltate scales, at least 10 mm diam. . . . . 4. *Cupressus*, p. 405
      6. Branchlets in flattened sprays; seed cones ellipsoid with oblong, basifixed scales or globose with rounded, peltate or basifixed scales, less than 12 mm diam.
        7. Branchlets terete or rhombic in cross section, facial and lateral leaves similar; seed cones globose, their scales usually peltate (basifixed in *C. nootkatensis*). . . . . 5. *Chamaecyparis*, p. 408
        7. Branchlets flattened, facial and lateral leaves clearly differentiated; seed cones ellipsoid, their scales basifixed.
          8. Scales of seed cones 4–6 pairs; seed wings equal; leaves clearly opposite in 4 ranks. . . . . 6. *Thuja*, p. 410
          8. Scales of seed cones 2–3 pairs; seed wings markedly unequal; leaves seemingly in whorls of 4. . . . . 7. *Calocedrus*, p. 412

1. SEQUOIA Endlicher, Syn. Conif., 197. 1847 · [For Sequoyah, also known as George Guess, inventor and publisher of the Cherokee alphabet]

Frank D. Watson

Trees giant, evergreen. **Branchlets** terete, with obvious annual growth constrictions. **Leaves** alternate, mostly in 2 ranks. **Adult leaves** linear or linear-lanceolate to deltate, generally flattened, divergent to strongly appressed; abaxial glands absent. **Pollen cones** with 6–12 sporophylls, each sporophyll with 2–6 pollen sacs. **Seed cones** maturing and opening in 1 season,

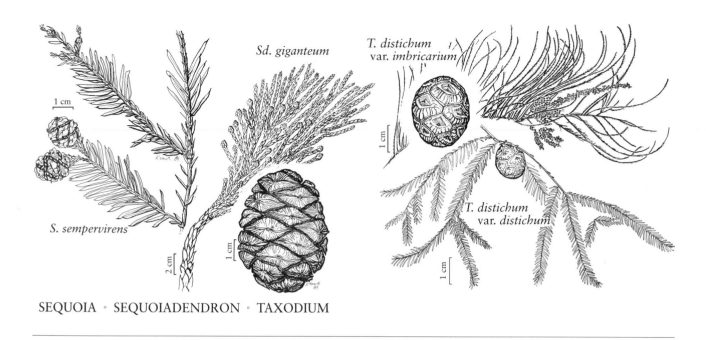

Sd. giganteum

T. distichum var. *imbricarium*

S. sempervirens

T. distichum var. *distichum*

SEQUOIA · SEQUOIADENDRON · TAXODIUM

oblong to globose; scales persistent, 15–30, valvate, ± peltate, thick and woody. **Seeds** 2–7 per scale, lenticular, narrowly 2-winged; cotyledons 2(–4). $x = 11$.

Species 1: only in the flora.

SELECTED REFERENCE   Schwarz, O. and H. Weide. 1962. Systematische Revision der Gattung *Sequoia* Endl. Feddes Repert. Spec. Nov. Regni Veg. 66: 159–192.

1. **Sequoia sempervirens** (D. Don) Endlicher, Syn. Conif., 198. 1847 · Redwood

Taxodium sempervirens D. Don in Lambert, Descr. Pinus 2: 24. 1824
Trees to ca. 110 m; trunk to 9 m diam.; crown conic and monopodial when young, narrowed conic in age. **Bark** reddish brown, to ca. 35 cm thick, fibrous, ridged and furrowed. **Branches** downward sweeping to slightly ascending. **Leaves** 1–30 mm, generally with stomates on both surfaces, the free portion to 30 mm, those on leaders, ascending branchlets, and fertile shoots divergent to strongly appressed, short-lanceolate to deltate, those on horizontally spreading to drooping branchlets mostly linear to linear-lanceolate, divergent and in 2 ranks, with 2 prominent, white abaxial stomatal bands. **Pollen cones** nearly globose to ovoid, 2–5 mm, borne singly on short terminal or axillary stalks. **Seed cones** 1.3–3.5 cm. **Seeds** flattened, 3–6 mm, leathery. $2n = 66$.

Coastal redwood forests; generally below 300 m, occasionally to 1000 m; Calif., Oreg.

Redwood is the only naturally occurring hexaploid conifer. It is one of only a few vegetatively reproducing conifers (from stump sprouts) and possibly the tallest tree species known. Winter buds, though small, are evident.

Redwood, including *Sequoia sempervirens* and *Sequoiadendron giganteum*, is the state tree of California.

2. SEQUOIADENDRON J. Buchholz, Amer. J. Bot. 26: 536. 1939 · [*Sequoia*, generic name of coast redwood, and Greek *dendros*, tree]

Frank D. Watson

Trees giant, evergreen. **Branchlets** terete. **Leaves** alternate, radiating. **Adult leaves** mostly needlelike, triangular in cross section, somewhat divergent to strongly appressed; abaxial glands absent. **Pollen cones** with 12–20 sporophylls, each sporophyll with 2–5 pollen sacs. **Seed cones** maturing and opening in 2 years, persistent to 20 years, oblong; scales persistent, 25–45, valvate, ± peltate, thick and woody. **Seeds** 3–9 per scale, lenticular, subequally 2-winged; cotyledons (3–)4(–6). $x = 11$.

Species 1: only in the flora.

1. Sequoiadendron giganteum (Lindley) J. Buchholz, Amer. J. Bot. 26: 536. 1939 · Giant sequoia, bigtree, Sierra-redwood

Wellingtonia gigantea Lindley, Gard. Chron. 10: 823. 1853; *Sequoia gigantea* (Lindley) Decaisne 1854, not Endlicher 1847

Trees to 90 m; trunk to 11 m diam.; crown conic and monopodial when young, narrowed and somewhat rounded in age. **Bark** reddish brown, to ca. 60 cm thick, fibrous, ridged and furrowed. **Branches** generally horizontal to downward-sweeping with upturned ends. **Leaves** generally with stomates on both surfaces, the free portion to ca. 15 mm. **Pollen cones** nearly globose to ovoid, 4–8 mm. **Seed cones** 4–9 cm. **Seeds** 3–6 mm. $2n = 22$.

Mixed montane coniferous forests, in isolated groves on the w slopes of the Sierra Nevada; 900–2700 m; Calif.

Mature individuals of this species are the most voluminous living organisms and among the most long-lived trees. *Sequoiadendron giganteum* was formerly included in *Sequoia*, under the later homonym *Sequoia gigantea* (Lindley) Decaisne, a conservative placement that still has merit (J. Doyle 1945; O. Schwarz and H. Weide 1962).

Redwood, including *Sequoiadendron giganteum* and *Sequoia sempervirens*, is the state tree of California.

3. TAXODIUM Richard, Ann. Mus. Natl. Hist. Nat. 16: 298. 1810 · [*Taxus*, generic name of yew, and Greek *-oides*, like]

Frank D. Watson

Trees deciduous or evergreen. **Branchlets** terete. **Lateral roots** commonly producing erect, irregularly conic to rounded "knees" in periodically flooded habitats. **Leaves** alternate, in 2 ranks or not. **Adult leaves** divergent to strongly appressed, linear or linear-lanceolate to deltate, generally flattened, free portion to ca. 17 mm; abaxial glands absent. **Pollen cones** with 10–20 sporophylls, each sporophyll with 2–10 pollen sacs. **Seed cones** maturing and shattering in 1 season, nearly globose; scales falling early, 5–10, valvate, ± peltate, thin and woody. **Seeds** (1–)2 per scale, irregularly 3-angled, wingless; cotyledons 4–9. $x = 11$.

Species 1(–3) (1 in the flora): North America, Mexico, Central America in Guatemala.

The pollen cones are usually borne at the base of alternate leaves, forming pendent axillary panicles; they occur less commonly singly or in racemes.

Taxodium is variously treated as one to three species but treated here as one polymorphic species with two varieties in the flora. A third taxon tentatively accorded varietal rank—*Taxodium distichum* (Linnaeus) Richard var. *mexicanum* Gordon (= *Taxodium mucronatum* Tenore)—and recognized by some authors as occurring within our range, appears to differ only in minor phenological characters. Whether or not populations from farther south in

Mexico and Guatemala differ sufficiently for formal taxonomic recognition at any rank has yet to be determined.

SELECTED REFERENCE   Watson, F. D. 1985. The nomenclature of pondcypress and baldcypress. Taxon 34: 506–509.

1. **Taxodium distichum** (Linnaeus) Richard, Ann. Mus. Natl. Hist. Nat. 16: 298. 1810

Cupressus disticha Linnaeus, Sp. Pl. 2: 1003. 1753

Trees seasonally cladoptosic; trunk enlarged basally and often conspicuously buttressed; crown monopodial and conic when young, often becoming irregularly flat-topped or deliquescent (branched and so divided that the main axis cannot be determined) with age. **Shoot system** conspicuously dimorphic, long shoots indeterminate, bearing individually abscising, linear to lanceolate leaves, short shoots determinate, abscising in autumn with their leaves, variable, intergrading, at one extreme pendent to horizontally spreading, bearing decurrent, narrowly linear and laterally divergent leaves in 2 rows, at other extreme strictly ascending to occasionally pendent, bearing short-lanceolate to deltate and tightly appressed leaves. **Pollen cones** in pendent panicles to ca. 25 cm, 2–3 mm, conspicuous in winter prior to pollination. **Seed cones** 1.5–4 cm.

Varieties 3 (2 in the flora): North America, Mexico, Central America in Guatemala.

The two varieties recognized in the flora are indistinguishable in reproductive characteristics and continuously intergrading in morphologic and phenologic characteristics, although pure populations of the extremes appear morphologically and ecologically distinct. Unlike the varieties in the flora, var. *mexicanum* is annually cladoptosic, with determinate short shoots abscising concomitantly with expansion of shoots of the following year. Specimens from juvenile individuals, stump sprouts, fertile branchlets, terminal vegetative branchlets, or late-season growth may not be determinable to variety.

Taxodium distichum (baldcypress) is the state tree of Louisiana.

1. Determinate short shoots with leaves mostly in 2 ranks, pendent to horizontally spreading; leaves mostly narrowly linear, ca. 5–17 mm, laterally divergent, free portion contracted and twisted basally. . . . . 1a. *Taxodium distichum* var. *distichum*
1. Determinate short shoots with leaves not in 2 ranks, mostly ascending vertically; leaves mostly narrowly lanceolate, ca. 3–10 mm, appressed and overlapping, free portion not contracted and twisted basally. . . . . 1b. *Taxodium distichum* var. *imbricarium*

1a. **Taxodium distichum** (Linnaeus) Richard var. **distichum** · Bald-cypress, southern-cypress

Taxodium distichum var. *nutans* (Aiton) Sweet

Trees to 50 m; trunk to 4 m diam. **Bark** usually dark reddish brown to light brown with shallow furrows. **Branchlets** mostly with leaves in 2 ranks, pendent to horizontally spreading. **Leaves** mostly narrowly linear, ca. 5–17 mm, laterally divergent, free portion contracted and twisted basally. $2n = 22$.

Brownwater rivers, lake margins, and swamps, occasionally in slightly brackish water; 0–160 m (to 500 m in Tex.); Ala., Ark., Del., Fla., Ga., Ill., Ind., Ky., La., Md., Miss., Mo., N.C., Okla., S.C., Tenn., Tex., Va.

This variety and var. *mexicanum* exhibit continuous morphologic integradation and are distinguished on the basis of phenology and distribution.

1b. **Taxodium distichum** (Linnaeus) Richard var. **imbricarium** (Nuttall) Croom, Cat. Pl. New Bern, 30. 1837 · Pondcypress

Cupressus disticha Linnaeus var. *imbricaria* Nuttall, Gen. N. Amer. Pl. 2: 224. 1818; *Taxodium ascendens* Brongniart

Trees to ca. 30 m; trunk to 2 m diam. **Bark** brown to light gray, typically somewhat thicker and more deeply furrowed than that of other varieties. **Branchlets** with leaves not in 2 ranks, mostly ascending vertically. **Leaves** ca. 3–10 mm, appressed and overlapping, mostly narrowly lanceolate, free portion not contracted or twisted basally. $2n = 22$.

Blackwater rivers, lake margins, swamps, Carolina Bay lakes, pocosins, and wet, poorly drained, pine flatwoods; 0–100 m; Ala., Fla., Ga., La., Miss., N.C., S.C.

The name *Taxodium distichum* (Linnaeus) Richard var. *nutans* (Aiton) Sweet has been misapplied to this taxon; the type of this name belongs to var. *distichum* (F. D. Watson 1985).

4. CUPRESSUS Linnaeus, Sp. Pl. 2: 1002. 1753; Gen. Pl. ed. 5, 435. 1754 · Cypress

[Latin name of *C. sempervirens*]

James E. Eckenwalder

Trees or large shrubs evergreen. **Branchlets** terete or quadrangular, in decussate arrays (or partially comblike in *Cupressus macnabiana*). **Leaves** opposite in 4 ranks. **Adult leaves** appressed to divergent, scalelike, rhomboid, free portion of long-shoot leaves to 4 mm; abaxial gland present or absent. **Pollen cones** with 4–10 pairs of sporophylls, each sporophyll with 3–10 pollen sacs. **Seed cones** maturing in 1–2 years, generally persisting closed many years or until opened by fire, globose or oblong, 1–4 cm; scales persistent, 3–6 pairs, valvate, peltate, thick and woody. **Seeds** 5–20 per scale, lenticular or faceted, narrowly 2-winged; cotyledons 2–5. $x = 11$.

Species 10–26 (7 in the flora): warm north temperate regions.

The genus *Cupressus* in North America consists mainly of small, disjunct, relictual populations, many differing from related populations in color and size of leaves and seeds, activity of leaf glands, glaucousness of various parts, form of growth, and characteristics of bark. Disagreements on the number and rank of taxa reflect these variations. This treatment, with seven taxa, approaches the more conservative end of a spectrum; anywhere from 6 to 15 taxa—species, subspecies, and varieties—might be accepted in the flora. The taxonomy of the genus would benefit from detailed studies of variation in and among populations (cf. J. F. Goggans and C. E. Posey 1968).

The Mediterranean *Cupressus sempervirens* Linnaeus, usually with a fastigiate habit, is commonly cultivated in California, often away from dwellings, but it does not appear to have become naturalized. Other introduced Eurasian and Mexican species are clearly associated with cultivated landscapes.

SELECTED REFERENCES Goggans, J. F. and C. E. Posey. 1968. Variation in seeds and ovulate cones of some species and varieties of *Cupressus*. Circ. Agric. Exp. Sta., Alabama 160: 1–23. Little, E. L. Jr. 1966. Varietal transfers in *Cupressus* and *Chamaecyparis*. Madroño 18: 161–167. Little, E. L. Jr. 1970. Names of New World cypresses (*Cupressus*). Phytologia 20: 429–445. Silba, J. 1981. Revision of *Cupressus* L. (Cupressaceae). Phytologia 49: 390–399. Wolf, C. B. 1948. Taxonomic and distributional studies of the New World cypresses. Aliso 1: 1–250.

1. Most leaves with conspicuous, pitlike, abaxial gland.
  2. Branchlets comblike. . . . . 3. *Cupressus macnabiana*
  2. Branchlets decussate.
    3. Branchlets less than 1.3 mm diam.; seed cones mostly 1–2 cm, with scales covered with resin blisters; seeds 3–4 mm. . . . . 2. *Cupressus bakeri*
    3. Branchlets 1.3 mm diam. or more; seed cones mostly 2–3 cm, with scales smooth or with scattered resin blisters; seeds 4–6 mm.
      4. Leaves rarely resin-dotted; c coastal California. . . . . 4. *Cupressus sargentii*
      4. Leaves resin-dotted; interior s California to Texas. . . . . 1. *Cupressus arizonica*
1. Most leaves without conspicuous, pitlike, abaxial gland, some with inconspicuous shallow or embedded gland.
  5. Seed cones 1–2.5(–3) cm.
    6. Leaves often glaucous; branchlets (1.5–)2–2.5 mm diam.; seeds 4–6 mm. . . . . . 4. *Cupressus sargentii*
    6. Leaves not glaucous; branchlets 1–1.5 mm diam.; seeds 3–4(–5) mm. . . . . . 5. *Cupressus goveniana*
  5. Seed cones (2–)2.5–4 cm.
    7. Seed cones oblong; bark fibrous, branchlets 1.5–2 mm diam. . . . . 6. *Cupressus macrocarpa*
    7. Seed cones globose; bark smooth, branchlets 1–1.5 mm diam. . . . .
    . . . . 7a. *Cupressus guadalupensis* var. *forbesii*

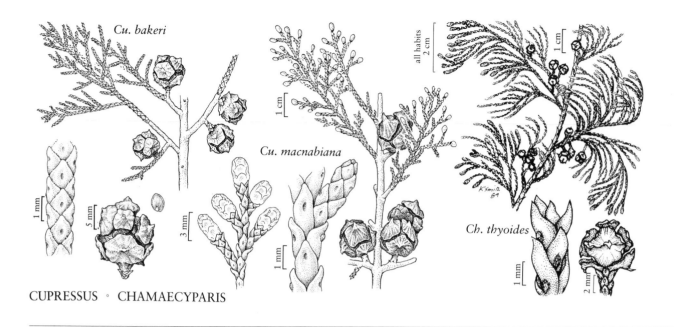

Cu. bakeri

Cu. macnabiana

Ch. thyoides

CUPRESSUS · CHAMAECYPARIS

1. **Cupressus arizonica** Greene, Bull. Torrey Bot. Club 9: 64. 1882 · Arizona cypress, Arizona smooth cypress, Cuyamaca cypress, Piute cypress, cedro, cedro blanco, ciprés de Arizona

Cupressus arizonica var. *glabra* (Sudworth) Little; *C. arizonica* var. *nevadensis* (Abrams) Little; *C. arizonica* var. *stephensonii* (C. B. Wolf) Little; *C. glabra* Sudworth; *C. nevadensis* Abrams; *C. stephensonii* C. B. Wolf

Trees to 23 m, shrubby where subject to fires; crown conic at first, broadly columnar with age, dense. **Bark** smooth at first, remaining so or becoming rough, furrowed, fibrous. **Branchlets** decussate, 1.3–2.3 mm diam. **Leaves** usually with conspicuous, pitlike, abaxial gland that produces drop of resin, often highly glaucous. **Pollen cones** 2–5 × 2 mm; pollen sacs mostly 4–6. **Seed cones** globose or oblong, mostly 2–3 cm, gray or brown, often glaucous at first; scales mostly 3–4 pairs, smooth or with scattered resin blisters, sometimes with erect conic umbos to 4 mm, especially on apical scales. **Seeds** mostly 4–6 mm, light tan to dark brown, not glaucous to heavily glaucous. $2n = 22$.

Canyon bottoms, pinyon-juniper woodland, and chaparral; 750–2000 m; Ariz., Calif., N.Mex., Tex.; Mexico.

Bark texture and foliage features have been used to distinguish geographic varieties or segregate species. Although bark texture may be consistent within populations, over the species as a whole there is complete intergradation between smooth and fibrous barks. Various forms are commonly cultivated and sometimes persistent in the southern United States.

2. **Cupressus bakeri** Jepson, Fl. Calif. 1: 61. 1909 · Baker cypress, Modoc cypress

Cupressus bakeri Jepson subsp. *matthewsii* C. B. Wolf; *C. macnabiana* A. Murray bis var. *bakeri* (Jepson) Jepson

Trees to 30 m; crown broadly columnar, sparse. **Bark** smooth at first, later building up in layers. **Branchlets** decussate, 0.5–1.3 mm diam. **Leaves** with conspicuous, pitlike, abaxial gland that produces drop of resin, slightly glaucous. **Pollen cones** 2–3 × 2–2.5 mm; pollen sacs 3–5. **Seed cones** globose, mostly 1–2 cm, silvery, not glaucous; scales 3–4 pairs, usually covered with resin blisters, umbos often prominent, those of distal scales erect, to 4 mm. **Seeds** mostly 3–4 mm, light tan to medium brown, not glaucous to slightly glaucous.

Mixed evergreen forests; of conservation concern; 1100–2000 m; Calif., Oreg.

3. Cupressus macnabiana A. Murray bis, Edinburgh New Philos. J., ser. 2, 1: 293, plate 11. 1855 · MacNab cypress

Shrubby trees to 12 m; crown broadly conical, dense. **Bark** rough, furrowed, fibrous. **Branchlets** comblike, 0.5–1 mm diam. **Leaves** with conspicuous, pitlike, abaxial gland that produces drop of resin, sometimes glaucous. **Pollen cones** 2–3 × 2 mm; pollen sacs 3–5. **Seed cones** globose, mostly 1.5–2.5 cm, brown or gray, not glaucous; scales 3–4 pairs, smooth except for erect conic umbos, 2–4 mm. **Seeds** 2–5 mm, light to medium brown, sometimes slightly glaucous.

Chaparral and foothill woodland, often on serpentine; 300–850 m; Calif.

In the inner north Coast Ranges *Cupressus macnabiana* and *C. sargentii* produce the only known natural hybrids in *Cupressus* (L. Lawrence et al. 1975).

4. Cupressus sargentii Jepson, Fl. Calif. 1: 61. 1909 · Sargent cypress

Cupressus sargentii Jepson var. *duttonii* Jepson

Trees to 25 m, but often shrubby and less than 10 m; crown broader than tall or columnar, dense or open. **Bark** rough, furrowed, fibrous. **Branchlets** decussate, (1.5–)2–2.5 mm diam. **Leaves** usually with inconspicuous, shallow, pitlike, abaxial gland that usually does not produce drop of resin, often glaucous. **Pollen cones** mostly 3–4 × 2 mm; pollen sacs 3–4. **Seed cones** usually globose, mostly 2–2.5 cm, brown or gray, not glaucous; scales mostly 3–4 pairs, with scattered resin blisters, umbos inconspicuous or to 4 mm. **Seeds** 4–6 mm, dark brown, faintly to prominently glaucous.

Chaparral, foothill woodland, and lower montane forests, on serpentine; 200–1100 m; Calif.

5. Cupressus goveniana Gordon, J. Hort. Soc. London 4: 295. 1849 · Gowen cypress, Mendocino cypress, Santa Cruz cypress

Cupressus abramsiana C. B. Wolf; *C. goveniana* var. *abramsiana* (C. B. Wolf) Little; *C. goveniana* var. *pigmaea* Lemmon; *C. pigmaea* (Lemmon) Sargent

Shrubs or small trees usually to 10 m, but to 50 m under favorable conditions, or bearing cones at as little as 2 dm on shallow hardpan soils; crown globose to columnar, dense or sparse. **Bark** smooth or rough, fibrous. **Branchlets** decussate, 1–1.5 mm diam. **Leaves** without abaxial gland or sometimes with embedded abaxial gland that does not produce drop of resin, not glaucous. **Pollen cones** 3–4 × 1.5–2 mm; pollen sacs 3–6. **Seed cones** globose, 1–2.5(–3) cm, grayish brown, not glaucous; scales 3–5 pairs, smooth, umbo nearly flat at maturity. **Seeds** 3–4(–5) mm, dark brown to jet black, sometimes slightly glaucous.

Coastal closed-cone pine forests, especially on sterile soils; 60–800 m; Calif.

Populations from the three regions of *Cupressus goveniana*—north coast, Santa Cruz Mountains, and Monterey Peninsula—differ in foliage and seed characters and have been treated as varieties or species; additional interpopulational variation occurs within these regions. Trees from Santa Cruz Mountain populations may have originated through hybridization with *C. sargentii* (E. Zavarin et al. 1971). The pygmy forests of this species and *Pinus contorta* Douglas ex Loudon on the shallow hardpan soils of coastal terraces of the Mendocino white plains are a remarkable example of phenotypic plasticity.

6. Cupressus macrocarpa Hartweg, J. Hort. Soc. London 2: 187. 1847 · Monterey cypress, ciprés Monterrey

Trees to 25 m; crown generally broadly spreading, especially on exposed headlands, fairly sparse, often composed of few major limbs from near ground, more upright in sheltered locations. **Bark** rough, fibrous. **Branchlets** decussate, 1.5–2 mm diam. **Leaves** without gland or sometimes with inconspicuous, shallow, pitlike, abaxial gland that does not produce drop of resin, not glaucous. **Pollen cones** 4–6 × 2.5–3 mm; pollen sacs 6–10. **Seed cones** oblong, 2.5–4 cm, grayish brown, not glaucous; scales 4–6 pairs, smooth, umbo nearly flat at maturity.

Seeds mostly 5–6 mm, dark brown, not glaucous. $2n$ = 22.

Coastal bluffs; of conservation concern; 5–35 m; Calif.

The geographically most restricted taxon recognized here, *Cupressus macrocarpa* is confined today to two picturesque groves near Monterey, but it is also known from fossils to have been in other regions. It is much planted and commonly naturalized near the coast from central California north to Washington and in warm temperate and subtropical regions worldwide.

7. Cupressus guadalupensis S. Watson, Proc. Amer. Acad. Arts 14: 300. 1879

Varieties 2 (1 in the flora): North America, Mexico.

7a. Cupressus guadalupensis S. Watson var. **forbesii** (Jepson) Little, Phytologia 20: 435. 1970 · Tecate cypress

Cupressus forbesii Jepson, Madroño 1: 75. 1922

Trees to 10 m, often shrubby; crown globose, sparse. **Bark** smooth. **Branchlets** decussate, mostly 1–1.5 mm diam. **Leaves** often with inconspicuous, embedded, abaxial gland that does not produce drop of resin, not glaucous. **Pollen cones** 3–4 × 2 mm; pollen sacs 3–5. **Seed cones** globose, mostly (2–)2.5–3.5 cm, brown, not glaucous; scales mostly 4–5 pairs, smooth, umbos flat or to 5 mm. **Seeds** mostly 5–6 mm, dark brown, not glaucous.

Chaparral; of conservation concern; 450–1000 m; Calif.; Mexico in Baja California.

The typical variety is endemic to Guadalupe Island, Baja California, Mexico.

5. CHAMAECYPARIS Spach, Hist. Nat. Vég. 11: 329. 1841 · White-cedar, false-cypress, faux-cypres [Greek *chamai,* on the ground, or dwarf, and *cyparissos,* cypress]

David C. Michener

Trees (rarely shrubs). **Branchlets** terete or rhombic in cross section, in fan-shaped or pinnately flattened sprays. **Leaves** opposite in 4 ranks. **Adult leaves** usually appressed, lateral and facial pairs similar, closely overlapping, scalelike, free portion of long-shoot leaves to ca. 7 mm; abaxial glands present or absent, circular to linear. **Pollen cones** with 2–3 pairs of sporophylls, each sporophyll with 2–4 pollen sacs. **Seed cones** maturing and opening in 1–2 years, nearly globose, glaucous, 4–12 mm; scales persistent, 2–5(–6) pairs, valvate, peltate or basifixed, thick and woody, terminal pair often fused. **Seeds** 1–4 per cone scale, lenticular, equally 2-winged; cotyledons 2–3. $x = 11$.

Species 6–7 (3 in the flora): North America, e Asia.

Two Japanese species are widely cultivated and may become established locally. *Chamaecyparis obtusa* (Siebold & Zuccarini) Endlicher (hinoki-cypress) has obtuse, glandless leaves and seed cones ca. 10–12 mm broad; *C. pisifera* (Siebold & Zuccarini) Endlicher (sawara-cypress) has acuminate, obscurely glandular leaves and seed cones 6–8 mm broad. Cultivated juvenile forms of several species have been referred to by the superfluous *Retinospora* Siebold & Zuccarini. Some authors include species of *Chamaecyparis* in the genus *Cupressus.*

1. Seed cones 4–9 mm broad; leaves usually with circular abaxial glands; Atlantic and Gulf coasts. . . . . 1. *Chamaecyparis thyoides*
1. Seed cones 8–12 mm broad; leaves with linear to circular abaxial glands or glands absent; Pacific Coast.
   2. Seed cones with 5–9 scales, opening at end of first year, not notably resinous; leaves usually with linear glands; facial leaves of branchlets frequently separated from each other by paired bases of lateral leaves. . . . . 2. *Chamaecyparis lawsoniana*

2. Seed cones with 4–6 scales, sometimes remaining closed at end of first year, becoming resinous; leaves usually without glands, these circular when present; apices of facial leaves of branchlets often overlapping the base of next facial leaf. . . . . 3. *Chamaecyparis nootakatensis*

1. **Chamaecyparis thyoides** (Linnaeus) Britton, Sterns, & Poggenburg, Prelim. Cat., 71. 1888 · Atlantic white-cedar, southern white-cedar

Cupressus thyoides Linnaeus, Sp. Pl. 2: 1003. 1753; *Chamaecyparis henryae* H. L. Li; *C. thyoides* subsp. *henryae* (H. L. Li) E. Murray; *C. thyoides* var. *henryae* (H. L. Li) Little

Trees to 20(–28) m; trunk to 0.8(–1.5) m diam. **Bark** dark brownish red, less than 3 cm thick, irregularly furrowed and ridged. **Branchlet** sprays fan-shaped. **Leaves** of branchlets to 2 mm, apex acute to acuminate, bases of facial leaves often overlapped by apices of subtending facial leaves; glands usually present, circular. **Pollen cones** 2–4 mm, dark brown; pollen sacs yellow. **Seed cones** maturing and opening the first year, 4–9 mm broad, glaucous, bluish purple to reddish brown, not notably resinous; scales 5–7. **Seeds** 1–2 per scale, 2–3 mm, wing narrower than body.

Bogs and swamps of the Atlantic and Gulf coasts (primarily Coastal Plain); 0–500 m; Ala., Conn., Del., Fla., Ga., Maine, Md., Mass., Miss., N.H., N.J., N.Y., N.C., Pa., R.I., S.C., Va.

H.-L. Li (1962) segregated some populations at the extreme southwestern limit of the species in Alabama, Florida, and Mississippi as *Chamaecyparis henryae* based on smoother bark, less flattened branchlets, lighter yellowish green foliage, steeper angle of leaf appression to the stem, more prominently keeled but less glandular leaves, and slightly larger cones, seeds, and seed wings. These features were contrasted with phenotypes found in the "northern and mid-Atlantic" populations, and Li proposed a relationship to *C. nootkatensis* rather than to *C. thyoides*. Preliminary comparison of herbarium material from the Southeast (including populations in Georgia and Florida) leads to retention of *C. thyoides* as a subtly variable complex with the imperfectly differentiated *C. henryae* at one end of the range.

A. J. Rehder (1949) listed, with bibliographic citations, 30 published varieties and forms best considered as cultivars.

2. **Chamaecyparis lawsoniana** (A. Murray bis) Parlatore, preprinted from Ann. Mus. Imp. Fis. Firenze, n. s. 1: 181. [preprint p. 29] 1864 · Port-Orford-cedar, ginger-pine

Cupressus lawsoniana A. Murray bis, Edinburgh New Philos. J., ser. 2, 1: 299, plate 10. 1855

Trees to 50 m; trunk to 3 m diam. **Bark** reddish brown, 10–20(–25) cm thick, divided into broad, rounded ridges. **Branchlet sprays** predominantly pinnate. **Leaves** of branchlets mostly 2–3 mm, apex acute to acuminate, facial leaves frequently separated by paired bases of lateral leaves; glands usually present, linear. **Pollen cones** 2–4 mm, dark brown; pollen sacs red. **Seed cones** maturing and opening first year, 8–12 mm broad, glaucous, purplish to reddish brown, not notably resinous; scales 5–9. **Seeds** 2–4 per scale, 2–5 mm, wing equal to or broader than body. $2n = 22$.

Forests of the Coast Ranges with isolated inland populations at higher elevations in the Siskiyou Mountains and on Mt. Shasta; 0–1500 m; Calif., Oreg.

A. J. Rehder (1949) listed, with bibliographic citations, 66 published varieties and forms best considered as cultivars.

3. **Chamaecyparis nootkatensis** (D. Don) Sudworth, U.S.D.A. Div. Forest. Bull. 14: 79. 1897 · Alaska-cedar, yellow-cypress

Cupressus nootkatensis D. Don in Lambert, Descr. Pinus 2: 113. 1824

Trees to 40 m or dwarfed at high elevations; trunk to 2 m diam. **Bark** grayish brown, 1–2 cm thick, irregularly fissured. **Branchlet** sprays pinnate. **Leaves** of branchlets mostly 1.5–2.5 mm, stout, occasionally glandular on keel, apex rounded to acute or acuminate, bases of facial leaves often overlapped by apices of subtending facial leaves; glands usually absent (circular when present). **Pollen cones** 2–5 mm, grayish brown; pollen sacs yellow. **Seed cones** maturing and opening the first year, in some populations the second year (J. N. Owens and M. Molder 1975), 8–12 mm broad, glaucous, dark reddish brown, becoming resinous; scales 4–6. **Seeds** 2–4 per scale, 2–5 mm, wing equal to or broader than body.

Coastal mountain ranges; 0–1500 m; B.C.; Alaska, Calif., Oreg., Wash.

Disjunct inland populations of *Chamaecyparis nootkatensis* occur in British Columbia and Oregon (V. J. Krajina et al. 1982).

In addition to variation in habit within the species, occasional plants have divergent forms of foliage. One collection (Canada, British Columbia, dry woods near Victoria, *S. Flowers s.n.*, 1 Aug 1950, UC, WIU) has older foliage typical of the species, with all newer foliage strongly flattened, with facial and lateral leaves of strongly unequal size, and with smaller cones. In light of the foliar and habit phenotypes recognized in the horticultural literature (for example, A. J. Rehder [1949] listed, with full bibliographic citations, 22 published varieties and forms best considered as cultivars), no taxonomic significance is attached to this variation here.

SELECTED REFERENCE Owens, J. N. and M. Molder. 1975. Pollination, female gametophyte, and embryo and seed development in yellow cedar (*Chamaecyparis nootkatensis*). Canad. J. Bot. 53: 186–199.

6. THUJA Linnaeus, Sp. Pl. 2: 1002. 1753; Gen. Pl. ed. 5, 435. 1754 ["Thuya"]
· Arborvitae, thuya, cèdre [Greek name for some evergreen, resinous trees]

Kenton L. Chambers

Trees evergreen, small to large. **Branchlets** flattened, in fan-shaped, flattened, frondlike sprays. **Leaves** opposite in 4 ranks. **Adult leaves** heteromorphic; those on larger branchlets with sharp, erect, free apices to ca. 2 mm; those on flattened lateral branchlets crowded, appressed, scale-like, lateral pairs keeled, facial pairs flat; abaxial glands present or absent. **Pollen cones** with 2–6 pairs of sporophylls, each sporophyll with 2–4 pollen sacs. **Seed cones** maturing and opening first year, ellipsoid, (6–)9–14(–18) mm; scales persistent, 4–6 pairs, overlapping, oblong and basifixed, thin and woody, 2–3 central pairs fertile, uniformly thin or with slightly enlarged apex, remaining scales sterile. **Seeds** 1–3 per scale, lenticular, equally 2-winged; cotyledons 2. $x = 11$.

Species 5 (2 in the flora): North America, e Asia.

Two Asiatic species, *Thuja koraiensis* Nakai (Korean arborvitae) and *T. standishii* (Gordon) Carrière (Japanese arborvitae), are occasional in cultivation. They differ from the native species in having more closely spaced long-shoot leaves with shorter, spreading points.

1. Foliage dull yellowish green abaxially and adaxially; seed-cone scales minutely mucronate; e United States and e Canada. . . . . 1. *Thuja occidentalis*
1. Foliage white-striped abaxially when fresh, glossy green adaxially; seed-cone scales with evident, nearly terminal, deltate projection; w United States and w Canada. . . . . 2. *Thuja plicata*

1. Thuja occidentalis Linnaeus, Sp. Pl. 2: 1002. 1753
· Northern white-cedar, thuier cèdre

Trees to 15(–38) m, stunted or prostrate in harsh environments; trunk to 0.9(–1.8) m diam., sometimes divided into 2–3 secondary stems, often reproducing by layering or forming erect, rooted branches from fallen trunks; crown conical. **Bark** reddish brown or grayish brown, 6–9 mm thick, fibrous, fissured. **Leaves** of branchlets (1.5–)3–5 mm, acute, dull yellowish green on both surfaces of branchlets. **Pollen cones** 1–2 mm, reddish. **Seed cones** ellipsoid, (6–)9–14 mm, brown; fertile scales usually 2 pairs, each minutely mucronate. **Seeds** ca. 8 per cone, 4–7 mm (including wings), reddish brown. $2n = 22$.

On mostly calcareous substrates, neutral to basic swamps, shores of lakes and rivers, uplands, cliffs, and talus; 0–900 m; Man., N.B., N.S., Ont., P.E.I., Que.; Conn., Ill., Ind., Ky., Maine, Md., Mass., Mich., Minn., N.H., N.Y., N.C., Ohio, Pa., Tenn., Vt., Va., W.Va., Wis.

Isolated stands of *Thuja occidentalis* occur north and east of its general range in Canada (to 51° 31′ N latitude in Ontario, 50° N in Quebec). In the United States south of the Great Lakes and in southern New Eng-

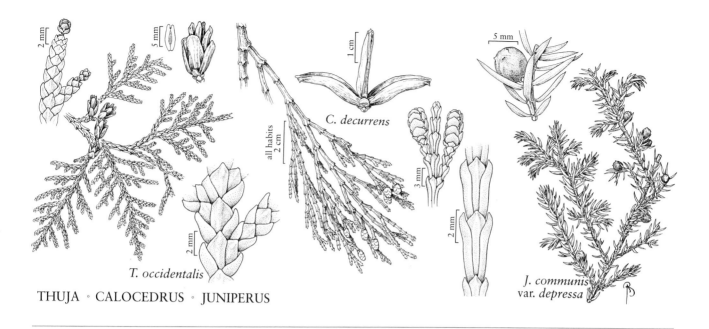

T. occidentalis

C. decurrens

J. communis
var. *depressa*

THUJA · CALOCEDRUS · JUNIPERUS

land, it occurs locally in scattered stands and is rare or extirpated at numerous former sites. In some areas, heavy winter browsing by deer greatly reduces reproductive success through elimination of seedlings or saplings.

Thuja occidentalis is widely utilized in ornamental silviculture and has more than 120 named cultivars. It was probably the first North American tree introduced into Europe (ca. 1566). It is an important timber tree; the wood is used for applications requiring decay resistance.

2. Thuja plicata Donn ex D. Don in Lambert, Descr.
Pinus 2: [19]. 1824 · Western redcedar

Trees to 50(–75) m, sometimes stunted in harsh environments; trunk to 2(–5) m diam., often buttressed at base; crown conical. **Bark** reddish brown or grayish brown, 10–25 mm thick, fibrous, fissured. **Branches** arching, branchlets pendent. **Leaves** of branchlets (1–)3–6 mm (sprays sometimes bearing only very small leaves), apex acute, with white markings on abaxial surface when fresh, glossy green on adaxial surface of branchlets. **Pollen cones** 1–3 mm, reddish. **Seed cones** ellipsoid, 10–14 mm, brown; fertile scales 2–3 pairs, each with evident, nearly terminal, deltate projection. **Seeds** 8–14 per cone, 4–7.5 mm (including wings), reddish brown. $2n = 22$.

On various substrates, commonly in moist sites, mixed coniferous forests, usually not in pure stands; 0–1500 (–2000) m; Alta., B.C.; Alaska, Calif., Idaho, Mont., Oreg., Wash.

The range of *Thuja plicata* consists of a Coast Range–Cascade Range segment from southeastern Alaska to northwestern California (between 56° 30′ and 40° 30′ N latitude) and a Rocky Mountains segment from British Columbia to Idaho and Montana (between 54° 30′ and 45° 50′ N latitude).

Thuja plicata is an important timber tree. Its soft but extremely durable wood is valued for home construction, production of shakes and shingles, and many other uses. Native Americans of the Northwest Coast used it to build lodges, totem poles, and seagoing canoes. Many cultivars are grown for ornament, and the species is managed for timber in Europe and New Zealand.

Western redcedar *(Thuja plicata)* is the provincial tree of British Columbia.

7. CALOCEDRUS   Kurz, J. Bot. 11: 196. 1873 · [Greek *callos*, beautiful, and *kedros*, cedar]

John W. Thieret

Trees evergreen, large. **Branchlets** flattened, in fan-shaped flattened sprays. **Leaves** opposite in 4 ranks (although apparently in whorls of 4). **Adult leaves** dimorphic, appressed, overlapping, scalelike, lateral leaves overlapping facial leaves, free portion of long-shoot leaves to ca. 3 mm; abaxial glands present. **Pollen cones** with 6–8 pairs of sporophylls, each sporophyll with 4 pollen sacs. **Seed cones** maturing and opening first year, ellipsoid, 17–30 mm; scales persistent, (2–)3 pairs, oblong and basifixed, thin and woody; proximal pair reduced, sterile, often reflexed or lacking; median pair fertile; distal pair connate, sterile. **Seeds** 2 per scale, lenticular, unequally 2-winged; cotyledons 2. *x* = 11.

Species 3 (1 in the flora): North America, Asia with 1 in Taiwan and 1 in s China and Burma.

1. **Calocedrus decurrens** (Torrey) Florin, Taxon 5: 192. 1956 · Incense-cedar, cedro incienso

Libocedrus decurrens Torrey, Smithsonian Contr. Knowl. 5(1) [6(2)]: 7, plate 3. 1853

Trees to 57 m; trunk to 3.6 m diam. **Bark** cinnamon brown, fibrous, furrowed and ridged. **Branchlet** segments mostly 2 or more times longer than wide, broadening distally. **Leaves** 3–14 mm, including long-decurrent base, rounded abaxially, apex acute (often abruptly), usually mucronate. **Pollen cones** red-brown to light brown. **Seed cones** oblong-ovate when closed, red-brown to golden brown, prox-imal scales often reflexed at cone maturity, median scales then widely spreading to recurved, distal scales erect. **Seeds** 4 or fewer in cone, 14–25 mm (including wings), light brown. 2*n* = 22.

Montane forests; 300–2800 m; Calif., Nev., Oreg.; Mexico in Baja California.

Incense-cedar is an important commercial softwood species. Its wood, exceptionally resistant to decay and highly durable when exposed to weather, is manufactured into many products, including lumber, pencil stock (for which it is the major United States source), fence posts, shakes, and landscape timbers, which are attractive because of punky spots resulting from fungus. The tree is widely grown as a handsome ornamental.

8. JUNIPERUS   Linnaeus, Sp. Pl. 2: 1038. 1753; Gen. Pl. ed. 5, 461. 1754 · Juniper, cedar, redcedar, cedro, sabino [Latin *juniperus*, name for juniper]

Robert P. Adams

Shrubs or trees evergreen. **Branchlets** terete, 3–6 angled, variously oriented, but not in flattened sprays. **Leaves** opposite in 4 ranks or in whorls of 3. **Adult leaves** closely appressed to divergent, scalelike to subulate, free portion to ca. 10 mm (to ca. 15 mm in *Juniperus communis*); abaxial gland visible or not, elongate to hemispheric (*J. ashei*), sometimes exuding white crystalline deposit. **Pollen cones** with 3–7 pairs or trios of sporophylls, each sporophyll with 2–8 pollen sacs. **Seed cones** maturing in 1 or 2 years, globose to ovoid and berrylike, 3–20 mm, remaining closed, usually glaucous; scales persistent, 1–3 pairs, peltate, tightly coalesced, thick and fleshy or fibrous to obscurely woody. **Seeds** 1–3 per scale, round to faceted, wingless; cotyledons 2–6. *x* = 11.

Species ca. 60 (13 in the flora): primarily Northern Hemisphere, 1 in e Africa.

Juniperus is the only dioecious (sometimes monoecious) genus of Cupressaceae in the flora. Cones, generally terminal, are axillary in *J. communis*.

Numerous cultivars of *Juniperus* species are widely used for landscaping. Mutants, or "sports," affecting plant habit and foliage are present in all species and are likely related to single-gene mutations. Many have been given formal names or incorrectly ascribed to hybridization. Gymnocarpy (bare seeds protruding from the cone), caused by insect larvae (T. A. Zanoni 1978), is occasionally found in most junipers, particularly in the southwestern United States. Specimens with such aberrations may be almost impossible to identify without chemical data.

SELECTED REFERENCES Adams, R. P. 1969. Chemosystematic and Numerical Studies in Natural Populations of *Juniperus*. Ph.D. thesis. University of Texas. Adams, R. P., E. von Rudloff, and L. Hogge. 1983. Chemosystematic studies of the western North American junipers based on their volatile oils. Biochem. Syst. & Ecol. 11: 85–89. Adams, R. P. and T. A. Zanoni. 1979. The distribution, synonymy, and taxonomy of three junipers of the southwest United States and northern Mexico. SouthW. Naturalist 24: 323–330. Fassett, N. C. 1945. *Juniperus virginiana, J. horizontalis,* and *J. scopulorum*. V. Taxonomic treatment. Bull. Torrey Bot. Club 72: 480–482. Hall, M. T. 1952. Variation and hybridization in *Juniperus*. Ann. Missouri Bot. Gard. 39: 1–64. Van Haverbeke, D. F. 1968. A population study of *Juniperus* in the Missouri River basin. Univ. Nebraska Stud., n. s. 38: 1–82. Vasek, F. C. 1966. The distribution and taxonomy of three western junipers. Brittonia 18: 350–372. Zanoni, T. A. 1978. The American junipers of the section *Sabina* (*Juniperus*, Cupressaceae)—A century later. Phytologia 38: 433–454. Zanoni, T. A. and R. P. Adams. 1979. The genus *Juniperus* (Cupressaceae) in Mexico and Guatemala: Synonymy, key, and distributions of the taxa. Bol. Soc. Bot. México 38: 83–131.

1. Leaves of 1 kind, subulate, with basal abscission zone, spreading; cones axillary (*Juniperus* sect. *Juniperus*). . . . . 1. *Juniperus communis*
1. Leaves of 2 kinds, whip (subulate, without basal abscission zone) and scalelike; cones terminal (*Juniperus* sect. *Sabina* Spach).
  2. Margins of leaves entire (at 20×) or, if with irregular teeth (at 40×), then scalelike leaves acuminate at apex.
    3. Margins of leaves with irregular teeth (at 40×), scalelike leaves acuminate at apex; seed cones with 4–13 seeds, tan-brown to brownish purple when mature; branches as well as branchlets flaccid. 2a. *Juniperus flaccida* var. *flaccida*
    3. Margins of leaves entire (at 20× and 40×), scalelike leaves obtuse to acute or apiculate at apex; seed cones with 1–3 seeds, blue-black to brownish blue when mature; branches not drooping, but branchlets often flaccid.
      4. Prostrate to decumbent shrubs; scalelike leaves apiculate at apex; peduncles generally curved. . . . . 3. *Juniperus horizontalis*
      4. Upright trees; scalelike leaves obtuse to acute at apex; peduncles generally straight.
        5. Scalelike leaves not overlapping, or, if so, by not more than 1/5 their length, apex obtuse to acute; bark of larger branchlets exfoliating in plates; seed cones maturing in 2 years, of 2 distinct sizes. . . . . 4. *Juniperus scopulorum*
        5. Scalelike leaves overlapping by more than 1/4 their length, apex acute (to bluntly obtuse in var. *silicicola*, se United States); bark of larger branchlets usually exfoliating in strips; seed cones maturing in 1 year, of 1 size. . . 5. *Juniperus virginiana*
  2. Margins of leaves denticulate (at 20×), scalelike leaves usually obtuse to acute at apex (sometimes mucronate in *Juniperus deppeana* or acuminate in *J. monosperma*).
    6. Seed cones with 3–6 seeds, fibrous to obscurely woody; bark exfoliating in rectangular plates, branchlets erect (rarely bark exfoliating in thin strips, branchlets then flaccid). . . . . 6a. *Juniperus deppeana* var. *deppeana*
    6. Seed cones with 1–3 seeds, fleshy or resinous to fibrous; bark exfoliating in thin strips, branchlets erect.
      7. Leaves with raised hemispheric glands (particularly obvious on whip leaves). . . 7. *Juniperus ashei*
      7. Leaves with ovate, elliptic, elongate, or inconspicuous glands (or glands round on scalelike leaves only).
        8. Seed cones blue to blue-black, with 2(–3) seeds, maturing in 2 years; bark of branchlets greater than 10 mm diam. exfoliating in scales or flakes; single-stemmed trees to 20(–30) m. . . . . 8. *Juniperus occidentalis*

8. Seed cones variously colored, but with brownish or reddish hue, even when some blue is present, with 1(–2) seeds, usually maturing in 1 year; bark of branchlets greater than 10 mm diam. usually exfoliating in strips or smooth; usually multistemmed shrubs or trees to 12 m.

    9. Abaxial glands inconspicuous because embedded in leaf; monoecious. . . . . 9. *Juniperus osteosperma*

    9. Abaxial glands conspicuous; dioecious (very rarely monoecious).

        10. Seed cones mostly 9–10 mm diam., bluish brown, glaucous; branchlets about as wide as length of scalelike leaves; scalelike leaves closely appressed, abaxial surface generally flattened; branchlets terete. . . . . 10. *Juniperus californica*

        10. Seed cones mostly 6–8 mm diam., reddish blue to brownish blue, rose to pinkish, or copper to copper-red, glaucous or not; branchlets ca. 2/3 as wide as length of scalelike leaves; scalelike leaves with apex spreading, abaxial surface raised; branchlets generally 3–4(–6)-sided.

            11. Seed cones reddish blue to brownish blue; fewer than 1/5 of whip-leaf glands with evident white exudate. . . . . 11. *Juniperus monosperma*

            11. Seed cones rose to pinkish or copper to copper-red; 1/4 or more of whip-leaf glands with evident white exudate.

                12. Seed cones rose to pinkish, glaucous; adaxial surface of leaves glaucous. . . . . 12. *Juniperus coahuilensis*

                12. Seed cones copper to copper-red, not glaucous; adaxial surface of leaves not glaucous. . . . . 13. *Juniperus pinchotii*

8a. JUNIPERUS Linnaeus sect. JUNIPERUS

Juniperus Linnaeus sect. *Oxycedrus* Spach

Shrubs or small trees, if shrubs, decumbent or rarely upright. **Adult leaves** in whorls of 3, of 1 kind, subulate, spreading, with basal abscission zone. **Cones** axillary.

1. **Juniperus communis** Linnaeus, Sp. Pl. 2: 1040. 1753

· Common juniper, genévrier comun

Shrubs or small trees dioecious, to 4 m (if trees, to 10 m), multistemmed, decumbent or rarely upright; crown generally depressed. **Bark** brown, fibrous, exfoliating in thin strips, that of small branchlets (5–10 mm diam.) smooth, that of larger branchlets exfoliating in strips and plates. **Branches** spreading or ascending; branchlets erect, terete. **Leaves** green but sometimes appearing silver when glaucous, spreading, abaxial glands very elongate; adaxial surface with glaucous stomatal band; apex acute to obtuse, mucronate. **Seed cones** maturing in 2 years, of 2 distinct sizes, with straight peduncles, globose to ovoid, 6–13 mm, bluish black, glaucous, resinous to obscurely woody, with 2–3 seeds. **Seeds** 4–5 mm. $2n = 22$.

Varieties 5 (3 in the flora): North America, Eurasia.

Juniperus communis is the most widespread juniper species, and many subspecies and varieties have been described. A major study, including chemical charac-

ters, is needed to clarify the taxonomy. J. D. A. Franco (1962) recognized four subspecies (here considered varieties); two of these—var. *communis* and var. *hemisphaerica* (J. Presl & C. Presl) Parlatore—do not occur in the flora and a fifth, recognized here, was not treated by Franco.

The seed cones of *Juniperus communis* are used to flavor gin.

1. Seed cones 9–13 mm, longer than leaves. . . . . 1a. *Juniperus communis* var. *megistocarpa*

1. Seed cones 6–9 mm, shorter than leaves.

    2. Glaucous stomatal band on adaxial leaf surface 2 or more times as wide as each green marginal band; spreading to matlike shrubs; leaves linear-lanceolate, to 2 mm wide, apex acute to obtuse and mucronate. . . . . 1b. *Juniperus communis* var. *montana*

    2. Glaucous stomatal band on adaxial leaf surface about as wide as each green marginal band; prostrate, low shrubs with ascending

branchlet tips (occasionally spreading shrubs, rarely small trees); leaves linear, to 1.6 mm wide, apex acute and mucronate to acuminate. . . . . . 1c. *Juniperus communis* var. *depressa*

1a. Juniperus communis Linnaeus var. **megistocarpa** Fernald & H. St. John, Proc. Boston Soc. Nat. Hist. 36: 58. 1921

Shrubs prostrate. **Leaves** upturned, lanceolate, apex acute and mucronate, to 12 × 1.6 mm, glaucous stomatal band about 1.5 times width of each green marginal band. **Seed cones** 9–13 mm, longer than leaves.

Sand dunes, serpentine, and limestone barrens; 0–500 m; Nfld., N.S., Que.

In Nova Scotia this variety is known from Sable Island. Magdalen Island, Quebec, is the type locality.

Although this taxon appears to be distinct, it might be more appropriately treated at the rank of *forma*.

1b. Juniperus communis Linnaeus var. **montana** Aiton, Hort. Kew. 3: 414. 1789

Juniperus communis subsp. *alpina* (Smith) Čelakovsky; *J. communis* subsp. *nana* (Willdenow) Syme; *J. communis* var. *jackii* Rehder; *J. communis* var. *saxatilis* Pallas; *J. sibirica* Burgsdorff

Shrubs spreading to matlike, 0.5–1 m. **Leaves** upturned or upcurled, to 15 × 2 mm, linear-lanceolate, sometimes almost overlapping, glaucous

stomatal band on adaxial leaf surface 2 or more times width of each green marginal band, apex acute to obtuse and mucronate. **Seed cones** 6–9 mm, shorter than leaves. $2n = 22$.

Dry rocky soil and rock crevices on slopes and summits; 0–2500 m; Greenland; B.C.; Calif., Oreg., Wash.

Juniperus communis var. *montana* is widespread throughout the Northern Hemisphere. Although the proposed var. *jackii* is quite distinct in the field (prostrate shrub with sparsely branched, whiplike, trailing branches), transplants indicate that the unusual growth form is environmentally induced (Steve Edwards, pers. comm.).

1c. Juniperus communis Linnaeus var. **depressa** Pursh, Fl. Amer. Sept. 2: 646. 1814

Juniperus communis subsp. *depressa* (Pursh) Franco

Shrubs prostrate or low with ascending branchlet tips (occasionally spreading shrubs to 3 m, rarely small trees to 10 m). **Leaves** upturned, to 15 × 1.6 mm, rarely spreading, linear, glaucous stomatal band about as wide as each green marginal band, apex acute and mucronate to acuminate. **Seed cones** 6–9 mm, shorter than leaves. $2n = 22$.

Rocky soil, slopes, and summits; 0–2800 m; Alta., B.C., Man., N.B., Nfld., N.W.T., N.S., Ont., P.E.I., Que., Sask., Yukon; Alaska, Ariz., Calif., Colo., Conn., Ga., Idaho, Ill., Ind., Maine, Mass., Mich., Minn., Mont., Nev., N.H., N.Mex., N.Y., N.C., N.Dak., Ohio, Oreg., Pa., R.I., S.C., S.Dak., Utah, Vt., Va., Wash., Wis., Wyo.

In the flora, larger individuals of this variety (to 10 m) have been misidentified as var. *communis*.

8b. JUNIPERUS Linnaeus sect. **SABINA** Spach, Ann. Sci. Nat., Bot., sér. 2, 16: 291. 1841.

Trees or shrubs, if shrubs, prostrate to decumbent or upright. **Adult leaves** opposite in 4 ranks or in whorls of 3, of 2 kinds on same branchlet, whip leaves (subulate, spreading) and scalelike (appressed or with spreading apex), both without basal abscission zone. **Cones** terminal.

2. Juniperus flaccida Schlechtendal, Linnaea 12: 495. 1838 · Drooping juniper, tascate

Varieties 3 (1 in the flora): North America, Mexico.

2a. Juniperus flaccida Schlechtendal var. **flaccida**

Trees dioecious, to 12 m, single-stemmed to 1–2 m; crown globose. **Bark** cinnamon to reddish brown or gray to reddish brown, exfoliating in broad interlaced fibrous strips, that of small branchlets (5–10 mm diam.) smooth, that of larger branchlets exfoliating in wide strips or plates. **Branches** drooping; branchlets flaccid, 3–4-sided in cross section, ca. 2/3 or less as wide as length of scalelike leaves. **Leaves** green, abaxial gland variable, elongate, conspicuous, exudate absent, margins appearing entire at 20× but with irregular teeth at 40×; whip leaves 4–6 mm, not glaucous adaxially; scalelike leaves 1.5–2 mm, overlapping by 1/4–1/5 their length, apex rounded to acuminate, spreading. **Seed cones** maturing in 1 year, of 1 size, with straight to curved peduncles, globose, 9–20 mm, tan-brown to brownish purple when mature, glaucous, obscurely woody, with (4–)6–10 (–13) seeds. **Seeds** 5–6 mm.

Rocky soils and slopes, 900–2900 m; Tex.; Mexico.

This variety is found in Big Bend National Park, Texas. It is abundant in Mexico, where two additional varieties occur: var. *poblana* Martínez and var. *martinezii* (Pérez de la Rosa) Silba.

3. Juniperus horizontalis Moench, Methodus, 699. 1794 · Creeping juniper, savinier

Juniperus horizontalis var. *douglasii* hort.; *J. horizontalis* var. *variegata* Beissner

Shrubs dioecious, prostrate to decumbent; crown depressed. **Bark** brown, exfoliating in thin strips, that of small branchlets (5–10 mm diam.) smooth, that of larger branchlets exfoliating in wide strips or plates. **Branches** creeping; branchlets erect, 3–4-sided in cross section, ca. 2/3 or less as wide as length of scalelike leaves. **Leaves** green but turning reddish purple in winter, abaxial gland elliptic, conspicuous, exudate absent, margins entire (at 20× and 40×); whip leaves 4–8 mm, not glaucous adaxially; scalelike leaves 1.5–2 mm, mostly overlapping to 1/3 their length, apex rounded or obtuse to acute and apiculate, spreading. **Seed cones** mostly maturing in 2 years, of 2 distinct sizes, generally with curved peduncles, globose to ovoid, 5–7 mm, blue-black to brownish blue when mature, lightly glaucous, soft and resinous, with 1–2(–3) seeds. **Seeds** 4–5 mm. 2*n* = 22.

Sand dunes, sandy and gravelly soils, prairies, slopes, rock outcrops, and stream banks; 0–1000 m; St. Pierre and Miquelon; Alta., B.C., Man., N.B., Nfld., N.W.T.,

N.S., Ont., P.E.I., Que., Sask., Yukon; Alaska, Ill., Iowa, Maine, Mass., Mich., Minn., Mont., Nebr., N.H., N.Y., N.Dak., S.Dak., Vt., Wis., Wyo.

Juniperus horizontalis, a prostrate species, hybridizes with the trees *J. virginiana* and *J. scopulorum* (R. P. Adams 1983; N. C. Fassett 1945; M. Palma-Otal et al. 1983) and is closely related to both. The hybrid between *J. horizontalis* and *J. scopulorum* has been named *J.* × *fassettii* Boivin.

4. Juniperus scopulorum Sargent, Gard. & Forest 10: 420, fig. 54. 1897 · Rocky Mountain juniper, Rocky Mountain redcedar

Sabina scopulorum (Sargent) Rydberg

Trees dioecious, to 20 m, single-stemmed (rarely multistemmed); crown conic to occasionally rounded. **Bark** brown, exfoliating in thin strips, that of small branchlets (5–10 mm diam.) smooth, that of larger branchlets exfoliating in plates. **Branches** spreading to ascending; branchlets erect to flaccid, 3–4-sided in cross section, ca. 2/3 or less as wide as length of scalelike leaves. **Leaves** light to dark green but often glaucous blue or blue-gray, abaxial gland elliptic, conspicuous, exudate absent, margins entire (at 20× and 40×); whip leaves 3–6 mm, not glaucous adaxially; scalelike leaves 1–3 mm, not overlapping to overlapping by not more than 1/5 their length, keeled to rounded, apex obtuse to acute, appressed or spreading. **Seed cones** maturing in 2 years, of 2 distinct sizes, generally with straight peduncles, globose to 2-lobed, 6–9 mm, appearing light blue when heavily glaucous, but dark blue-black beneath glaucous coating when mature (or tan beneath glaucous coating when immature), resinous to fibrous, with (1–)2(–3) seeds. **Seeds** 4–5 mm. 2*n* = 22.

Rocky soils, slopes, and eroded hillsides; 1200–2700 m (0 m at Vancouver Island and Puget Sound); Alta., B.C.; Ariz., Colo., Idaho, Mont., Nebr., Nev., N.Mex., N.Dak., Oreg., S.Dak., Utah, Wash., Wyo.; n Mexico.

Juniperus scopulorum hybridizes with its eastern relative *J. virginiana* in zones of contact in the Missouri River basin (C. W. Comer et al. 1982) and with *J. horizontalis* (*J.* × *fassettii* Boivin; N. C. Fassett 1945). Relictual hybridization with *J. virginiana* is known in the Texas panhandle (R. P. Adams 1983).

5. Juniperus virginiana Linnaeus, Sp. Pl. 2: 1039. 1753 · Eastern redcedar

Trees dioecious, to 30 m, single-stemmed; crown narrowly erect to conical, round, or flattened. **Bark** brown, exfoliating in thin strips, that of small branchlets (5–

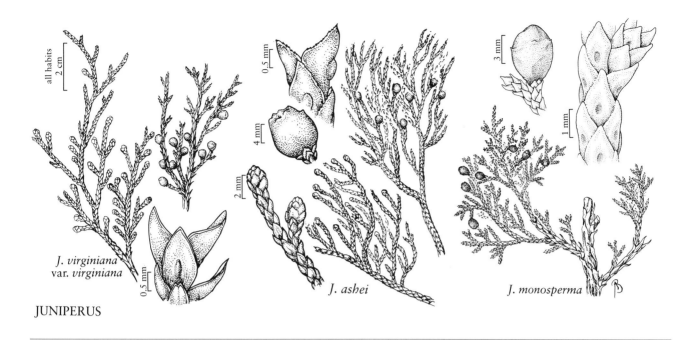

J. virginiana
var. virginiana

JUNIPERUS

J. ashei

J. monosperma

10 mm diam.) smooth, that of larger branchlets usually not exfoliating in plates. **Branches** pendulous to ascending; branchlets generally erect, sometimes lax to flaccid, 3–4-sided in cross section, ca. 2/3 or less as wide as length of scalelike leaves. **Leaves** green but sometimes turning reddish brown in winter, abaxial gland elliptic or elongate, conspicuous, exudate absent, margins entire (at 20× and 40×); whip leaves 3–6 mm, not glaucous adaxially; scalelike leaves 1–3 mm, overlapping by more than 1/4 their length, keeled, apex obtuse to acute, spreading. **Seed cones** maturing in 1 year, of 1 size, generally with straight peduncles, globose to ovoid, 3–6(–7) mm, blue-black to brownish blue when mature, glaucous, soft and resinous, with 1–2(–3) seeds. **Seeds** 1.5–4 mm.

Varieties 2: only in the flora.

1. Seed cones 4–6(–7) mm; crown narrowly erect to conic or round; bark reddish brown; scalelike leaves acute at apex; pollen cones 3–4 mm. . . . . 5a. *Juniperus virginiana* var. *virginiana*
1. Seed cones 3–4 mm; crown flattened; bark cinnamon reddish; scalelike leaves bluntly obtuse to acute at apex; pollen cones 4–5 mm. . . . . 5b. *Juniperus virginiana* var. *silicicola*

5a. Juniperus virginiana Linnaeus var. **virginiana**

· Eastern redcedar, cèdre rouge

Juniperus virginiana var. *crebra* Fernald & Griscom; *Sabina virginiana* (Linnaeus) Antoine

Trees to 30 m; crown narrowly erect (in young, fast-growing trees) to conic or occasionally round. **Bark** reddish brown. **Branches** erect, spreading, or pendulous. **Scalelike leaves** acute at apex. **Pollen cones** 3–4 mm. **Seed cones** globose to ovoid, 4–6(–7) mm. **Seeds** 2–4 mm. $2n = 22, 33$.

Upland to low woods, old fields, glades, fencerows, and river swamps; 0–1400 m; Ont., Que.; Ala., Ark., Conn., Del., D.C., Fla., Ga., Ill., Ind., Iowa, Kans., Ky., La., Maine, Md., Mass., Mich., Minn., Miss., Mo., Nebr., N.H., N.J., N.Y., N.C., Ohio, Okla., Pa., R.I., S.C., S.Dak., Tenn., Tex., Vt., Va., W.Va., Wis.

Eastern redcedar hybridizes with the related species *Juniperus horizontalis* (M. Palma-Otal et al. 1983) and *J. scopulorum* (C. W. Comer et al. 1982). Reported hybridization with *J. ashei* has been refuted in subsequent studies (R. P. Adams 1977).

The wood of *Juniperus virginiana* is used for production of eastern redcedarwood oil, fenceposts, and cedar chests.

5b. Juniperus virginiana Linnaeus var. **silicicola** (Small) E. Murray, Kalmia 13: 8. 1983 · Southern red-cedar, coastal redcedar

Sabina silicicola Small, Mem. New York Bot. Gard. 24: 5. 1923; *Juniperus silicicola* (Small) L. H. Bailey

Trees to 10 m; crown flattened or conic (when young and protected or crowded). **Bark** cinnamon reddish. **Branches** spreading to pendulous. **Scalelike leaves** bluntly obtuse to acute at apex. **Pollen cones** 4–5 mm. **Seed cones** 3–4 mm. **Seeds** 1.5–3 mm.

Coastal foredunes and coastal river sandbanks; 0–15 m; Fla., Ga., N.C., S.C.

This southern variety of *Juniperus virginiana* appears to be restricted to coastal foredunes but differs little in morphology or leaf terpenoids from upland *J. virginiana* and appears to intergrade with that variety in Georgia (R. P. Adams 1986). These taxa are distinct from the Caribbean junipers *J. lucayana* Britton of the Bahamas, Jamaica, and Cuba, and *J. bermudiana* Linnaeus of Bermuda (R. P. Adams et al. 1987). Reports of *J. virginiana* var. *silicicola* from west of Florida are questionable.

6. Juniperus deppeana Steudel, Nomencl. Bot. ed. 2, 1: 835. 1841 · Alligator juniper, cedro chino

Varieties 5 (1 in the flora): North America, Mexico.

6a. Juniperus deppeana Steudel var. **deppeana**

Trees dioecious, to 10–15(–30) m, single-stemmed; crown rounded. **Bark** brown, exfoliating in rectangular plates (rarely in thin strips in f. *sperryi*, but then branchlets flaccid), that of small branchlets (5–10 mm diam.) smooth, that of larger branchlets exfoliating in plates. **Branches** spreading to ascending; branchlets erect, rarely flaccid, 3–4-sided in cross section, ca. 2/3 or less as wide as length of scalelike leaves. **Leaves** green, but sometimes appearing silvery when glaucous, abaxial gland ovate to elliptic, conspicuous, exudate absent, margins denticulate (at 20×); whip leaves 3–6 mm, not glaucous adaxially; scalelike leaves 1–2 mm, not overlapping, keeled, apex acute to mucronate, appressed. **Seed cones** maturing in 2 years, of 2 distinct sizes, with straight to curved peduncle, globose, 8–15 mm, reddish tan to dark reddish brown, glaucous, fibrous to obscurely woody, with (3–)4–5(–6) seeds. **Seeds** 6–9 mm.

Rocky soils, slopes, and mountains; 2000–2900 m; Ariz., N.Mex., Tex.; Mexico.

Although four additional varieties are found in Mexico, the relationships among the *J. deppeana* taxa are poorly understood and need additional study (R. P. Adams et al. 1984). The very rare *J. deppeana* Steudel var. *deppeana* f. *sperryi* (Correll) R. M. Adams (= *J. deppeana* Steudel var. *sperryi* Correll) is endemic to the Davis Mountains, Texas, where only two or three individuals are known to exist. This form is characterized by bark that exfoliates in thin strips and by flaccid branchlets.

7. Juniperus ashei J. Buchholz, Bot. Gaz. 90: 329. 1930 · Ashe juniper, mountain-cedar

Trees dioecious, to 15 m, single-stemmed to 1–3 m, occasionally branching at base; crown rounded to irregular and open. **Bark** brown, exfoliating in thin strips, that of small branchlets (5–10 mm diam.) smooth, that of larger branchlets exfoliating in strips. **Branches** spreading to ascending; branchlets erect, 3–4-sided in cross section, ca. 2/3 or less as wide as length of scalelike leaves. **Leaves** dark green, abaxial glands hemispheric, raised (particularly obvious on whip leaves), exudate absent, margins denticulate (at 20×); whip leaves 3–6 mm, not glaucous adaxially; scalelike leaves 1–2 mm, not overlapping or overlapping to 1/4 their length, keeled, apex acute to obtuse, spreading. **Seed cones** maturing in 1 year, of 1 size, with straight peduncles, ovoid to nearly globose, 6–9 mm, dark blue, glaucous, fleshy and resinous, with 1(–3) seeds. **Seeds** 4–6 mm. $2n = 22$.

Limestone glades and bluffs; 150–600 m; Ark., Mo., Okla., Tex.; Mexico.

The name *Juniperus mexicana* Sprengel has been misapplied to this species. Reports of hybridization with *J. virginiana* and *J. pinchotii* have been refuted using numerous chemical and morphologic characters (R. P. Adams 1977).

Ashe juniper is a source of Texas-cedarwood oil and fence posts.

8. Juniperus occidentalis Hooker, Fl. Bor.-Amer. 2: 166. 1838 · Western juniper

Trees monoecious or dioecious, to 20(–30) m, single-stemmed; crown rounded to conical. **Bark** red-brown to brown, exfoliating in thin strips, that of small branchlets (5–10 mm diam.) smooth, that of larger branchlets exfoliating in scales or flakes. **Branches** spreading to ascending; branchlets erect, 3–4-sided in cross section, ca. 2/3 or less as wide as length of scalelike leaves. **Leaves** green, abaxial glands ovate to elliptic, conspicuous, with yellow or white exudate, margins denticulate (at 20×); whip leaves 3–6 mm, not glaucous adaxially; scalelike leaves 1–3 mm, not overlapping, rounded, apex acute to obtuse, appressed. **Seed cones** maturing in 2 years, of 2 distinct sizes, with straight peduncle, ovoid, 5–10 mm, blue to blue-black, glaucous, fleshy and resinous, with 2(–3) seeds. **Seeds** 2–4 mm.

Varieties 2: only in the flora.

1. Bark red-brown; seed cones (5–)7.5(–9) mm; plants often (50%) monoecious. . . . .
. . . . . . 8a. *Juniperus occidentalis* var. *occidentalis*
1. Bark brown; seed cones (7–)8.5(–10) mm; plants mostly (90%) dioecious. . . . .
. . . . 8b. *Juniperus occidentalis* var. *australis*

8a. Juniperus occidentalis Hooker var. occidentalis

Sabina occidentalis (Hooker) Antoine

Plants often (50%) monoecious. **Bark** red-brown. **Seed cones** (5–)7.5(–9) mm.

Dry rocky foothill and mountain slopes; (0–)1500–3000 m; Calif., Idaho, Oreg., Nev., Wash.

8b. Juniperus occidentalis Hooker var. australis (Vasek) A. H. Holmgren & N. H. Holmgren in Cronquist et al., Intermount. Fl. 1: 239. 1972

Juniperus occidentalis Hooker subsp. *australis* Vasek, Brittonia 18: 352. 1966

Plants mostly (90%) dioecious. **Bark** brown. **Seed cones** (7–)8.5(–10) mm.

Dry rocky slopes; 1000–3000 m; Calif.

F. C. Vasek (1966) reported hybridization of this variety with *Juniperus osteosperma* in northwestern Nevada.

9. Juniperus osteosperma (Torrey) Little, Leafl. W. Bot. 5: 125. 1948 · Utah juniper, sabina morena

Juniperus tetragona Schlechtendal var. *osteosperma* Torrey, Pacif. Railr. Rep. 4(5): 141. 1857; *J. californica* Carrière var. *utahensis* Engelmann; *Sabina osteosperma* (Torrey) Antoine; *S. utahensis* (Engelmann) Rydberg

Shrubs or trees monoecious, to 6(–12) m, multi- or single-stemmed; crown rounded. **Bark** exfoliating in thin gray-brown strips, that of smaller and larger branchlets smooth. **Branches** spreading to ascending; branchlets erect, 3–4-sided in cross section, about as wide as length of scalelike leaves. **Leaves** light yellow-green, abaxial glands inconspicuous and embedded, exudate absent, margins denticulate (at 20×); whip leaves 3–5 mm, glaucous adaxially; scalelike leaves 1–2 mm, not overlapping, or, if so, by less than 1/10 their length, keeled, apex rounded, acute or occasionally obtuse, appressed. **Seed cones** maturing in 1–2 years, of 1–2 sizes, with straight peduncles, globose, (6–)8–9(–12) mm, bluish brown, often almost tan beneath glaucous coating, fibrous, with 1(–2) seeds. **Seeds** 4–5 mm.

Dry, rocky soil and slopes; 1300–2600 m; Ariz., Calif., Colo., Idaho, Mont., Nev., N.Mex., Utah, Wyo.

Juniperus osteosperma is the dominant juniper of Utah. It is reported to hybridize with *J. occidentalis* in northwestern Nevada (F. C. Vasek 1966).

10. Juniperus californica Carrière, Rev. Hort., sér. 4, 3: 352. 1854 · California juniper, huata, cedro

Sabina californica (Carrière) Antoine

Shrubs or trees dioecious (rarely monoecious), to 8 m, multistemmed (seldom single-stemmed); crown rounded. **Bark** gray, exfoliating in thin strips, that of smaller and larger branchlets smooth. **Branches** spreading to ascending; branchlets erect, terete, about as wide as length of scalelike leaves. **Leaves** light green, abaxial glands elliptic to ovate, conspicuous, exudate absent, margins denticulate (at 20×); whip leaves 3–5 mm, not glaucous adaxially; scalelike leaves 1–2 mm, not overlapping, or rarely overlapping by ca. 1/5 their length, generally flattened, apex acute to obtuse, closely appressed. **Seed cones** maturing in 1 year, of 1 size, with straight peduncles, globose, (7–)9–10(–13) mm, bluish brown, glaucous, fibrous, with 1(–2) seeds. **Seeds** 5–7 mm.

Dry, rocky slopes and flats; 750–1600 m; Ariz., Calif., Nev.; Mexico in Baja California.

Although two races, differing in volatile leaf oils, were described by F. C. Vasek and R. W. Scora (1967) and confirmed by R. P. Adams et al. (1983), no differences were found in volatile wood oils (R. P. Adams 1987). To date, no morphological characters appear to be correlated with the chemical races. No other Western Hemisphere species of *Juniperus* has been found to have leaf-oil races.

11. Juniperus monosperma (Engelmann) Sargent, Silva 10: 89. 1896 · One-seed juniper, sabina

Juniperus occidentalis Hooker var. *monosperma* Engelmann, Trans. Acad. Sci. St. Louis 3: 590. 1878

Shrubs or trees dioecious, to 7 (–12) m, usually branching near base; crown rounded to flattened-globose. **Bark** gray to brown, exfoliating in thin strips, that of small branchlets (5–10 mm diam.) smooth, that of larger branchlets exfoliating in either flakes or strips. **Branches** ascending to erect; branchlets erect, 4–6-sided, ca. 2/3 as wide as length of scalelike leaves. **Leaves** green to dark green, abaxial glands elongate, fewer than 1/5 of glands (on whip leaves) with an evident white crystalline exudate, margins denticulate (at 20×); whip leaves 4–6 mm, glaucous adaxially; scalelike leaves 1–3 mm, not overlapping, or if so, by less than 1/4 their length, keeled, apex acute to acuminate, spreading. **Seed cones** maturing in 1 year, of 1 size, with straight peduncles, globose to ovoid, 6–8 mm, reddish blue to brownish blue, glaucous, fleshy and resinous, with 1(–3) seeds. **Seeds** 4–5 mm.

Dry, rocky soils and slopes; 1000–2300 m; Ariz., Colo., N.Mex., Okla., Tex.

Reports of hybridization with *J. pinchotii* have been refuted by use of numerous chemical and morphologic characters (R. P. Adams 1975); the two species have nonoverlapping pollination seasons.

12. Juniperus coahuilensis (Martinez) Gaussen ex R. P. Adams, Phytologia 74:450. 1993 · Roseberry

Juniperus erythrocarpa Cory var. *coahuilensis* Martinez, Anales Inst. Biol. Univ. Nac. México 17: 115–116. 1946.

Shrubs or trees dioecious, to 8 m, single-stemmed to 1 m or branched at base; crown flattened-globose to round or irregular. **Bark** gray to brown, exfoliating in long ragged strips, that of small branchlets (5–10 mm diam.) smooth, that of larger branchlets exfoliating in strips or occasionally in flakes. **Branches** spreading to ascending; branchlets erect, 3–4-sided in cross section, ca. 2/3 as wide as length of scalelike leaves. **Leaves** green to light green, abaxial glands elliptic to ovate, at least 1/4 of glands (on whip leaves) with an evident white crystalline exudate, margins denticulate (at 20×); whip leaves 4–6 mm, glaucous adaxially; scalelike leaves 1–3 mm, not overlapping or if so, by less than 1/4 their length, keeled, apex acute, spreading. **Seed cones** maturing in 1 year, of 1 size, with straight peduncles, globose to ovoid, 6–7 mm, rose to pinkish but yellow-orange, orange, or dark red beneath glaucous coating, fleshy and somewhat sweet, with 1(–2) seeds. **Seeds** 4–5 mm.

Bouteloua grasslands and adjacent rocky slopes; 980–1600(–2200) m; Ariz., N.Mex., Tex.; Mexico.

Roseberry juniper is unusual in that it sprouts from the stump after burning or cutting. Hybridization with *Juniperus pinchotii* occurs in Big Bend National Park, Texas (R. P. Adams and J. R. Kistler 1991), and possibly near Saltillo, Mexico. Reports of hybridization with *J. ashei* have been refuted (R. P. Adams 1975).

13. Juniperus pinchotii Sudworth, Forest. Irrig. 10: 204. 1905 · Pinchot juniper, redberry juniper

Juniperus erythrocarpa Cory

Shrubs or shrubby trees dioecious, to 6 m, usually multistemmed; crown flattened-globose to irregular. **Bark** ashy gray to brown, exfoliating in long strips, that of small branchlets (5–10 mm diam.) smooth, that of larger branchlets exfoliating in strips or sometimes in flakes. **Branches** spreading to ascending; branchlets erect, 3–4-sided in cross section, ca. 2/3 as wide as length of scalelike leaves. **Leaves** yellow-green, abaxial glands elliptic to elongate, many with an evident white crystalline exudate, margins denticulate (at 20×); whip leaves 4–6 mm, not glaucous adaxially; scalelike leaves 1–2 mm, not overlapping or overlapping by not more than 1/5 their length, keeled, apex acute, spreading. **Seed cones** maturing in 1 year, of 1 size, with straight peduncles, globose to ovoid, 6–8 (–10) mm, copper to copper-red, not glaucous, fleshy and sweet, not resinous, with 1(–2) seeds. **Seeds** 4–5 mm.

Gravelly soils on rolling hills and in ravines, limestone, gypsum; 300–1000(–1700) m; N.Mex., Okla., Tex.; Mexico.

Pinchot juniper hybridizes with *Juniperus coahuilensis* (R. P. Adams and J. R. Kistler 1991) but not with *J. ashei* (R. P. Adams 1977) or *J. monosperma* (R. P. Adams 1975). The type specimen of *J. erythrocarpa* is merely an individual with brighter red seed cones.

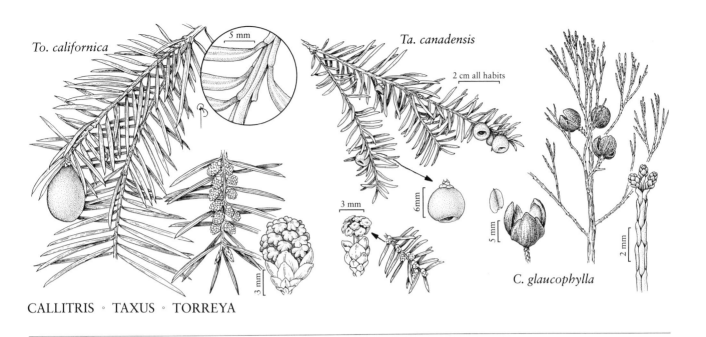

CALLITRIS · TAXUS · TORREYA

9. CALLITRIS Ventenat, Dec. Gen. Nov., 10. 1808 · Cypress-pine [Greek *callos*, beautiful, and *treis*, three, referring to the beauty of the plants and the three-whorled leaves and cone scales]

Richard P. Wunderlin

Shrubs or trees evergreen. **Branchlets** angled or furrowed-cylindrical, variously oriented. **Leaves** in whorls of 3–5. **Adult leaves** scalelike, appressed, abaxial surface keeled or rounded, free portion to 1 mm, abaxial glands absent. **Pollen cones** solitary or in small clusters, with 4–15 whorls of sporophylls, each with 2–4 pollen sacs. **Seed cones** maturing in 1–2 years, of 1–2 sizes, often remaining unopened for many years, ovoid or globose, 1–3.5 cm; scales persistent, in 2 equally inserted whorls of 3(–4), valvate, rhombic-deltate, basifixed, thick and woody. **Seeds** 2–9 per scale, round or 3-angled, broadly 1–3-winged; cotyledons 2.

Species 16 (1 naturalized in the flora): North America, Australia and New Caledonia.

SELECTED REFERENCES Blake, S. T. 1959. New or noteworthy plants, chiefly from Queensland, l. Proc. Roy. Soc. Queensland. 70(6): 33–46. Garden, J. 1957. A revision of the genus *Callitris* Vent. Contr. New South Wales Natl. Herb. 2(5): 363–392. Thompson, J. 1961. Cupressaceae. Contr. New South Wales Natl. Herb., Fl. Ser. 1/18: 46–55. Thompson, J. and L. A. S. Johnson. 1986. *Callitris glaucophylla*, Australia's 'white cypress pine'—A new name for an old species. Telopea 2: 731–736.

1. **Callitris glaucophylla** J. Thompson & L. P. Johnson, Telopea 2: 731. 1986 · White cypress-pine

Callitris columellaris F. Mueller var. *campestris* Silba

Shrubs or trees to 30 m. **Bark** brown, furrowed. **Leaves** in whorls of 3 (sometimes 4 or 5 when juvenile), usually glaucous, juvenile leaves 7–8 mm, mature leaves 1–3 mm with apex broadly acute. **Pollen cones** cylindric-oblong, 5–10 × 2–5 mm. **Seed cones** depressed-globose to ovoid, 1.2–2 cm, dark brown; peduncle 7–8 mm; scales thin, indistinctly dentate along margin, separat-ing almost to base when mature, alternate ones short and narrow, larger ones angled, often sharply, toward apex. **Seeds** 4–5 mm, chestnut brown.

Sand pine scrub and thickets, often near plantings of the species; introduced; 0–10 m; Fla.; Australia.

Callitris glaucophylla is naturalized in Brevard, Indian River, Orange, and Seminole counties in Florida.

This inland Australian species is sometimes united with the eastern coastal *Callitris columellaris* F. Mueller under that name or distinguished at varietal rank (var. *campestris* Silba). The names *C. glauca* R. Brown ex R. T. Baker & H. G. Smith and *C. hugelii* (Carrière) Franco have been applied erroneously to it.

5. TAXACEAE Gray

• Yew Family

Matthew H. Hils

Trees or shrubs evergreen, usually neither resinous nor aromatic (sharp- or foul-odored in *Torreya*), dioecious or monoecious. **Bark** scaly or fissured. **Lateral branches** well developed, similar to leading shoots; twigs terete, not densely clothed by leaves but ± ridged by decurrent leaf bases; longest internodes less than 1 cm; buds ± inconspicuous. **Roots** fibrous to woody. **Leaves** (needles) simple, persisting several years, shed singly, alternate [opposite], spirally arranged but often twisted so as to appear 2-ranked, linear to linear-lanceolate, decurrent; resin canals present or absent. **Pollen cones** maturing and shed annually, solitary or clustered, axillary on year-old branches, globose to ovoid, sporophylls bearing 2–16 microsporangia (pollen sacs); pollen ± spheric, not winged. **Seed cones** reduced to 1–2 ovules subtended by inconspicuous, decussate bracts, maturing in 1–2 seasons, axillary on year-old branches. **Seeds** 1 per "cone," erect, not winged, hard seed coat partially or wholly surrounded by a juicy, fleshy or leathery aril; cotyledons 2.

Genera 5, species 17–20 (2 genera, 5 species in the flora): mainly Northern Hemisphere.

SELECTED REFERENCES Burns, R. M. and B. H. Honkala. 1990. Silvics of North America. 1. Conifers. Washington. [Agric. Handb. 654.] Canadian Forestry Service. 1983. Reproduction of conifers. Forest. Techn. Pub. Canad. Forest. Serv. 31. Farjon, A. 1990. A Bibliography of Conifers. Königstein. [Regnum Veg. 122.] Florin, R. 1948. On the morphology and relationship of the Taxaceae. Bot. Gaz. 110: 31–39. Hosie, R. C. 1969. Native Trees of Canada, ed. 7. Ottawa. Pp. 110–111. Krüssmann, G. 1972. Handbuch der Nadelgehölze. Berlin. Little, E. L. Jr. 1979. Checklist of United States Trees (Native and Naturalized). Washington. Pp. 283, 287–288. [Agric. Handb. 541.] Pilger, R. K. F. 1903. Taxaceae. In: H. G. A. Engler, ed. 1900–1953. Das Pflanzenreich . . . Berlin. Vol. 18[IV,5], pp. 1–124. Pilger, R. K. F. 1916. Die Taxales. Mitt. Deutsch. Dendrol. Ges. 25: 1–28. Pilger, R. K. F. 1926. Coniferae: Taxaceae. In: H. G. A. Engler et al., eds. 1924+. Die natürlichen Pflanzenfamilien . . . , ed. 2. Leipzig and Berlin. Vol. 13, pp. 199–211. Price, R. A. 1990. The genera of Taxaceae in the southeastern United States. J. Arnold Arbor. 71: 69–91. Silba, J. 1986. Encyclopaedia Coniferae. Phytologia Mem. 8: 1–127.

1. Leaves flexible, without resin canal, apex mucronate, soft-pointed, not sharp to touch; aril scarlet to orange-scarlet, soft, mucilaginous, thick, cup-shaped, open at apex, exposing hard seed coat. . . . . 1. *Taxus*, p. 424
1. Leaves rigid, stiff, with central resin canal, apex acute, spine-tipped, sharp to touch; aril green or green with purple streaks, leathery, resinous, thin, completely enclosing hard seed coat. . . . . 2. *Torreya*, p. 426

1. TAXUS Linnaeus, Sp. Pl. 2: 1040. 1753; Gen. Pl. ed. 5, 462. 1754 · Yew [Latin name for yew]

Trees or shrubs dioecious or monoecious. **Bark** reddish brown, scaly. **Branches** ascending to drooping; twigs irregularly alternate, green or yellow-green when young, reddish brown in age. **Leaves** often appearing 2-ranked, flexible; stomates abaxial, in 2 broad, pale bands; apex soft-pointed, mucronate, not sharp to touch; resin canal absent. **Pollen cones** globose, yellowish, with 4–16 peltate sporophylls, each bearing 2–9 sporangia. **Ovule** 1. **Seed** maturing in 1 season, brown; aril scarlet to orange-scarlet, soft, mucilaginous, thick, cup-shaped, open at apex, exposing hard seed coat; albumen uniform. $x = 12$.

Species 6–10 (3 in the flora): mainly north temperate regions.

The species of *Taxus*, discouragingly similar, are more geographically than morphologically separable; they were all treated by R. K. F. Pilger (1903) as subspecies of *T. baccata* Linnaeus. Detailed study of the genus (not neglecting the cultivated representatives), including extensive fieldwork, is much needed and long overdue.

The foliage, bark, and seeds—but not the fleshy red aril—of most *Taxus* species are toxic due to the presence of taxine (M. R. Cooper and A. W. Johnson 1984; J. M. Kingsbury 1964); this alkaloid, however, was not found in *T. brevifolia* (I. Jones and E. V. Lynn 1933). Two Eurasian species, *T. baccata* Linnaeus (English yew) and *T. cuspidata* Siebold & Zuccarini (Japanese yew), are best known and documented for toxicity. Cattle have been poisoned by *T. canadensis* planted in British Columbia, but toxicity of *T. brevifolia* has not been conclusively recorded (J. M. Kingsbury 1964). Although horses, cattle, and humans have been poisoned by ingesting yew leaves and seeds, the fresh foliage of *T. canadensis* is browsed by deer, and that of *T. brevifolia* by moose.

The only other yew native to the New World is *Taxus globosa* Schlechtendahl of Mexico and Honduras. The Old World *T. baccata* Linnaeus and *T. cuspidata* Siebold & Zuccarini—and *T. ×media* Rehder, the alleged hybrid between these two—are common in cultivation in the flora area.

Although no extralimital species of *Taxus* is naturalized in North America, spontaneous, immature (sapling) exotic yews have been noted in a very few localities in the northeastern United States within the range of *Taxus canadensis*. Apparently originating from seeds dispersed (probably by birds) from cultivated yews, these plants differ from *T. canadensis* in having typically erect (rather than sprawling) stems. Immature volunteer yews are, with the use of macromorphological characters and with our present knowledge, probably unidentifiable to species.

Although species of *Taxus* are much cultivated in the Pacific Northwest, spontaneous yews have not been recorded there away from cultivated individuals, in the vicinity of which (or under which) they may reseed. Should such volunteers be found, the shape of their leaf epidermal cells as viewed in cross section—wider than tall (rather than taller than wide) or ± isodiametric—may be used to distinguish them from *T. brevifolia*.

Anatomical features of the leaves of yews, helpful in identification of the species, have been contributed by R. W. Spjut (unpublished data).

SELECTED REFERENCES Chadwick, L. C. and R. A. Keen. 1976. A study of the genus *Taxus*. Ohio Agric. Exp. Sta. Bull. 1086. Cooper, M. R. and A. W. Johnson. 1984. Poisonous Plants in Britain and Their Effects on Animals and Man. London. [Minist. Agric., Fisheries & Food Ref. Book 161.] Jones, I. and E. V. Lynn. 1933. Differences in species of *Taxus*. J. Amer. Pharm. Assoc. 22: 528–531. Keen, R. A. 1956. A Study of the Genus *Taxus*. Ph.D. thesis. Ohio State University. Keen, R. A. and L. C. Chadwick. 1955. Sex reversal in *Taxus*. Amer. Nurseryman 100(6): 13–14. Kingsbury, J. M. 1964. Poisonous Plants of the United States and Canada. Englewood Cliffs.

1. Shrubs to 2 m, usually monoecious, low, sprawling, typically without a central stem; abaxial leaf surface mostly without cuticular papillae along stomatal bands; e North America (Manitoba to Newfoundland, south to Missouri, Kentucky, and Virginia). . . . . 2. *Taxus canadensis*
1. Shrubs or trees to 15(–25) m, dioecious, usually upright, typically with a central stem; abaxial leaf surface with cuticular papillae along stomatal bands; w North America or Florida.
   2. Shrubs or small trees to 15(–25) m, trunk diam. to 6(–12) dm; leaves yellow-green adaxially, epidermal cells as viewed in cross section of leaf mostly taller than wide; western North America (Alaska south to Montana and California). . . . . 1. *Taxus brevifolia*
   2. Shrubs or small trees to 6(–10) m, trunk diam. to 3.8 dm; leaves dark green adaxially, epidermal cells as viewed in cross section of leaf wider than tall or ± isodiametric; northwestern Florida. . . . . 3. *Taxus floridana*

1. Taxus brevifolia Nuttall, N. Amer. Sylv. 3: 86, plate 108. 1849 · Pacific yew

Taxus baccata Linnaeus subsp. *brevifolia* (Nuttall) Pilger; *T. baccata* var. *brevifolia* (Nuttall) Koehne; *T. baccata* var. *canadensis* Bentham; *T. bourcieri* Carrière; *T. lindleyana* A. Murray bis

Shrubs or small trees to 15 (–25) m, dioecious; trunk to 6 (–12) dm diam., straight to contorted, fluted; crown open-conical. **Bark** scaly, outer scales purplish to purplish brown, inner ones reddish to reddish purple. **Branches** horizontal to drooping. **Leaves** 1–2.9 cm × 1–3 mm, pale green abaxially, cuticular papillae present along stomatal bands, shiny yellow-green adaxially, epidermal cells as viewed in cross section of leaf mostly taller than wide. **Seed** ovoid, 2–4-angled, 5–6.5 mm.

Seeds maturing late summer–fall. Open to dense forests, along streams, moist flats, slopes, deep ravines, and coves; 0–2200 m; Alta., B.C.; Alaska, Calif., Idaho, Mont., Oreg., Wash.

The name *Taxus baccata* Hooker has been misapplied to this species.

The leaves of *Taxus brevifolia* are usually somewhat falcate.

The wood of *Taxus brevifolia* is hard and durable, yet easily worked, making it popular for construction of novelty items by local woodworkers. Because of this, large trees are unscrupulously poached; in some areas the species has been nearly extirpated. The bark of the tree is a promising natural source of taxol, a drug for treating various cancers; exploitation of the species for medicinal purposes is further threatening it.

SELECTED REFERENCE Taylor, R. L. and S. Taylor. 1981. *Taxus brevifolia* in British Columbia. Davidsonia 12(4): 89–94.

2. Taxus canadensis Marshall, Arbust. Amer., 151. 1785 · Canada yew, American yew, ground-hemlock, if du Canada, sapin trainard

Taxus baccata Linnaeus subsp. *canadensis* (Marshall) Pilger; *T. baccata* var. *minor* Michaux; *T. minor* (Michaux) Britton; *T. procumbens* Loddiges

Shrubs to 2 m, usually monoecious, low, diffusely branched, straggling, spreading to prostrate. **Bark** reddish, very thin. **Branches** spreading and ascending. **Leaves** 1–2.5 cm × 1–2.4 mm, pale green abaxially, mostly without cuticular papillae along stomatal bands, dark green to yellow-green adaxially, epidermal cells as viewed in cross section of leaf wider than tall or ± isodiametric. **Seed** somewhat flattened, 4–5 mm. 2n = 24.

Seeds maturing late summer–early fall. Understory shrub in rich forests (deciduous, mixed, or coniferous), bogs, swamps, gorges, ravine slopes, and rocky banks; 0–1500 m; St. Pierre and Miquelon; Man., N.B., Nfld., N.S., Ont., P.E.I., Que.; Conn., Ill., Ind., Iowa, Ky., Maine, Mass., Mich., Minn., N.H., N.Y., Ohio, Pa., R.I., Tenn., Vt., Va., W.Va., Wis.

3. Taxus floridana Nuttall ex Chapman, Fl. South. U.S., 436. 1860 · Florida yew

Taxus baccata Linnaeus var. *floridana* (Nuttall ex Chapman) Pilger

Shrubs or small trees to 6(–10) m, dioecious, trunk to 3.8 dm diam. **Bark** purplish brown, thin, scaly. **Branches** stout, spreading. **Leaves** 1–2.6(–2.9) cm × 1–2 (–2.2) mm, mostly slightly falcate, light green with 2 grayish bands abaxially, with cuticular papillae along stomatal bands, dark green adaxially, epidermal cells as viewed in cross section of leaf wider than tall or ± isodiametric. **Seed** ellipsoid, 5–6 mm.

Seeds maturing in early fall. Moist, shaded ravines in hardwood forests; of conservation concern; 15–30 m; Fla.

Taxus floridana is a rare endemic along the Appalachicola River in Florida.

2. TORREYA Arnott, Ann. Nat. Hist. 1: 130. 1838, name conserved · Torreya, stinking-cedar [After John Torrey (1796–1873), distinguished U.S. botanist]

Tumion Rafinesque

Trees dioecious. **Bark** brown to grayish brown, tinged with orange, fissured. **Branches** spreading to drooping; twigs nearly opposite. **Leaves** mostly appearing 2-ranked, rigid; stomates abaxial, in 2 narrow, glaucous, whitish or brownish bands; apex sharp-pointed, spine-tipped, sharp to touch; resin canal central. **Pollen cones** ovoid or oblong, with 6–8 whorls of 4 sporophylls, each bearing 4 sporangia. **Ovules** 2, only 1 of each pair maturing. **Seed** maturing in 2 years; aril green or green with purple streaks, resinous, leathery, thin, completely enclosing woody seed coat, splitting into 2 parts at maturity; albumen ruminate. $x = 11$.

Species 4(–6) (2 in the flora): North America, Asia in China and Japan.

Two Asian species are planted as ornamentals in North America: *Torreya nucifera* Siebold & Zuccarini (kaya-nut, Japanese torreya), which yields edible seeds and cooking oil, and *T. grandis* Fortune (Chinese torreya).

SELECTED REFERENCES Buchholz, J. T. 1940. The embryogeny of *Torreya*, with a note on *Austrotaxus*. Bull. Torrey Bot. Club 67: 731–754. Burke, J. G. 1975. Human use of the California nutmeg tree, *Torreya californica*, and of other members of the genus. Econ. Bot. 29: 127–139.

1. Two-year-old branches reddish brown; leaves 3–8 cm, flattened on adaxial side, with 2 deeply impressed, glaucous bands of stomates abaxially, emitting pungent odor when crushed; aril light green streaked with purple; California. . . . . 1. *Torreya californica*
1. Two-year-old branches yellowish green, yellowish brown, or gray; leaves 1.5–3.8 cm, rounded on adaxial side, with 2 scarcely impressed, grayish bands of stomates abaxially, emitting fetid odor when crushed; aril dark green streaked with purple; Florida, Georgia.
 · 2. *Torreya taxifolia*

1. Torreya californica Torrey, New York J. Pharm. (1852–54) 3: 49. 1854 · California torreya, California-nutmeg

Torreya myristica Hooker; *Tumion californicum* (Torrey) Greene
Trees to 20(–25) m; trunk to 9 (–12) dm diam.; crown conic or, in age, round-topped. **Branches** spreading to slightly drooping; 2-year-old branches reddish brown. **Leaves** 3–8 cm, abaxial side with 2 deeply impressed, glaucous bands of stomates, flattened on adaxial side, emitting pungent odor when crushed. **Pollen cones** whitish. **Seed** (including aril) 2.5–3.5 cm; aril light green streaked with purple. $2n = 16$.

Rare and local along mountain streams, protected slopes, creek bottoms, and moist canyons of the Coastal Range and Sierra Nevada; 0–2000 m; Calif.

2. Torreya taxifolia Arnott, Ann. Nat. Hist. 1: 130. 1838 · Florida torreya, stinking-cedar, gopherwood

Tumion taxifolium (Arnott) Greene
Trees to 13(–18) m; trunk to 8 dm diam.; crown rather open-conical. **Branches** spreading to slightly drooping; 2-year-old branches yellowish green, yellowish brown, or gray. **Leaves** 1.5–3.8 cm, abaxial side with 2 scarcely impressed, grayish bands of stomates, rounded on adaxial side, emitting fetid odor when crushed. **Pollen cones** pale yellow. **Seed** (includ-

ing aril) 2.5–3.5 cm; aril glaucous, dark green, streaked with purple.

River bluffs, slopes, and moist ravines; of conservation concern; 15–30 m; Fla., Ga.

Torreya taxifolia is a rare endemic mainly along the Apalachicola River.

Populations of *Torreya taxifolia* were thriving until the 1950s, but since then they have been decimated by fungal disease (R. L. Godfrey and H. Kurz 1962). Only nonreproductive stump sprouts remain in the wild. The Florida torreya was listed as federally endangered in 1984 under the U.S. Endangered Species Act, and efforts are underway to reestablish this once thriving species in its native habitat (L. R. McMahan 1989).

SELECTED REFERENCES Coulter, J. M. and W. J. G. Land. 1905. Gametophytes and embryo of *Torreya taxifolia*. Bot. Gaz. 39: 161–178. Fish and Wildlife Service [U.S.D.I.]. 1986. Florida Torreya (*Torreya taxifolia*) Recovery Plan. Atlanta. Godfrey, R. L. and H. Kurz. 1962. The Florida torreya destined for extinction. Science 136: 900, 902. Kurz, H. 1939. Torreya west of the Apalachicola River. Proc. Florida Acad. Sci. 3: 66–77. McMahan, L. R. 1989. Conservationists join forces to save Florida torreya. Center Pl. Conservation 4(1): 1, 8. Savage, T. 1983. A Georgia station for *Torreya taxifolia* Arn. survives. Florida Sci. 46: 62–64. Stalter, R. and S. Dial. 1984. Environmental status of the stinking cedar. Bartonia 50: 40–42.

6. EPHEDRACEAE Dumortier
• Mormon-tea or Joint-fir Family

Dennis Wm. Stevenson

Shrubs or vines, erect or clambering, *Equisetum*-like, dioecious (very rarely monoecious). **Bark** gray to reddish brown, cracked and fissured. **Branches** generally many, terete, whorled to fascicled, finely longitudinally grooved, internodes 1–10 cm. **Roots** generally fibrous. **Leaves** simple, opposite and decussate or whorled, scalelike, connate at base to form sheath, generally ephemeral, mostly not photosynthetic; resin canals absent. **Pollen cones** compound, 1–10 in whorls at nodes; each compound cone composed of 2–8 sets of opposite or whorled membranous bracts, proximal bracts empty, each distal bract subtending a small cone composed of 2 basally fused bracteoles subtending a sporangiophore bearing 2–10(–15) sessile to long-stalked, bilocular, apically dehiscent, pollen-producing microsporangia. **Pollen** prolate, with 6–12 longitudinal furrows, not winged. **Seed cones** compound, 1–10 in whorls at nodes of twigs; each compound cone sessile or on short to long peduncle, composed of 2–10 sets of overlapping, opposite or whorled, membranous or papery to fleshy bracts, proximal bracts empty, most distal bracts subtending 1 axillary cone composed of a pair of fused bracteoles enclosing a single-integumented ovule with integument projecting as tube from bracteole-envelope, envelope forming a leathery "seed coat" that is shed with seed. **Seeds** 1–2(–3) per compound seed cone, yellow to dark brown, smooth or furrowed; cotyledons 2.

Genus 1, species ca. 60 (12 species in the flora): temperate and warm regions worldwide except Australia.

In addition to the characters given in the key to families, wood anatomy can be used to distinguish *Ephedra* from the other gymnosperms in the flora. Only *Ephedra* has small cones, ring porous wood, wide multiseriate rays, and vessels in older stems.

Since antiquity, several species of *Ephedra* have been used medicinally worldwide. Such uses include cough medicines, an antipyretic, an antisyphilitic, a stimulant for poor circulation, and an antihistamine. These uses are based on the presence of tannins and alkaloids, particularly ephedrines.

SELECTED REFERENCES Markgraf, F. 1926. Ephedraceae. In: H. G. A. Engler et al., eds. 1924+. Die natürlichen Pflanzenfamilien . . ., ed. 2. Leipzig and Berlin. Vol. 13, pp. 409–419. Meyer, C. A. von. 1846. Versuch einer Monographie der Gattung *Ephedra*. Mém. Acad. Imp. Sci. Saint-Pétersbourg, Sér. 6, Sci. Math., Seconde Pt. Sci. Nat. 5(2): 225–298. Kubitzki, K. 1990. Ephedraceae. In: K. Kubitzki et al., eds. 1990+. The Families and Genera of Vascular Plants. Berlin etc. Vol. 1, pp. 379–382. Parlatore, F. 1868. Gnetaceae. In: A. P. and A. L. P. de Candolle, eds. 1823–1873. Prodromus Systematis Naturalis Regni Vegetabilis. . . . Paris etc. Vol. 16, part 2, pp. 352–359. Stapf, O. 1889. Die Arten der Gattung *Ephedra*. Denkschr. Kaiserl. Akad. Wiss., Wien. Math.-Naturwiss. Kl. 56(2): 1–112.

1. EPHEDRA Linnaeus, Sp. Pl. 2: 1040. 1753; Gen. Pl. ed. 5, 462. 1754 · Mormon-tea, joint-fir, cañatilla, popotillo, tepopote [Greek *ep-*, upon, and *hédra,* seat or sitting upon a place; from the ancient name used by Pliny for *Equisetum*; the stems resemble the jointed stems of *Equisetum,* the segments of which appear to sit one upon the other]

Shrubs or occasionally clambering vines. **Branches** jointed, yellowish green to olive-green when young. **Leaves** opposite or in whorls of 3, apex obtuse to setaceous from an adaxial-median thickening. **Pollen cones** lanceoloid or ellipsoid to ovoid or obovoid. **Seed cones** ellipsoid to ovoid, obovoid, or nearly globose. **Seeds** ellipsoid to globose, yellow to dark brown, smooth to scabrous or furrowed. $x = 7$.

Species ca. 60 (12 in the flora): generally dry areas in temperate, tropical North America and Mediterranean regions, Mexico, South America (Ecuador to Patagonia and lowland Argentina), s Europe, Asia, n Africa (including Canary Islands).

The species of *Ephedra* are presented here in alphabetical order for three reasons. First, no modern monographic treatment has been written for all species of the genus since that of O. Stapf (1889). Second, it appears that the species occurring in North America belong to at least three wholly different groups within the genus, but this is not yet supported by thorough systematic studies. Third, interspecific relationships within any putative infrageneric group occurring in North America are at best vague and ill defined.

The North American species of *Ephedra* are well defined based on combinations of vegetative and reproductive characters. Putative hybrids reported and described by H. C. Cutler (1939) appear to be products of singular events; these hybrids are discussed under the parental species. Infraspecific taxa are not recognized in this treatment because there appear to be no consistent defining characters and no geographic correlations; previous recognition of infraspecific taxa (H. C. Cutler 1939) appears to be based on random variability.

SELECTED REFERENCES Benson, L. D. 1943. Revisions of status of southwestern trees and shrubs. Amer. J. Bot. 30: 230–240. Cutler, H. C. 1939. Monograph of the North American species of the genus *Ephedra*. Ann. Missouri Bot. Gard. 26: 373–427. Mussayev, I. 1978. On geography and phylogeny of some representatives of the genus *Ephedra*. Bot. Žurn. (Moscow & Leningrad) 63: 523–543. [In Russian.]

1. Leaves and bracts mostly in whorls of 3.
   2. Cones always sessile; seeds scabrous. . . . . 10. *Ephedra torreyana*
   2. Cones usually with short, scaly peduncles (rarely sessile); seeds usually smooth (sometimes scabrous in *E. funerea*).
      3. Terminal buds spinelike; leaf bases shredding with age; cone bracts reddish brown.
      . . . . 11. *Ephedra trifurca*
      3. Terminal buds acute at apex; leaf bases persistent or completely deciduous; cone bracts yellow, green-yellow, or orange-yellow.

4. Leaf bases deciduous; twigs yellow-green; cone bracts as broad as long; seeds nearly globose. . . . . 3. *Ephedra californica*

4. Leaf bases persistent, forming a black, thickened collar; twigs gray-green; cone bracts longer than broad; seeds ellipsoid. . . . . 7. *Ephedra funerea*

1. Leaves and bracts mostly opposite.

  5. Branches lax, vinelike, trailing or clambering; microsporangial stalks 1–2 mm. . . . .
     . . . . 9. *Ephedra pedunculata*

  5. Branches rigid; microsporangial stalks less than 1 mm.

    6. Twigs viscid. . . . . 5. *Ephedra cutleri*

    6. Twigs not viscid.

      7. Leaf bases persistent, forming a black, thickened collar; nodes obviously swollen; seeds 2.

        8. Twigs with smooth ridges; seed cones sessile or on short, scaly peduncles, inner bracts membranous, with yellow center and base. . . . . 12. *Ephedra viridis*

        8. Twigs with slightly scabrous ridges; seed cones usually on long, smooth peduncles, inner bracts fleshy (at least in center) and orange. . . . . 4. *Ephedra coryi*

      7. Leaf bases completely deciduous or becoming gray and shredded with age; nodes not or only inconspicuously swollen; seeds 1–2.

        9. Leaf bases completely deciduous, brown when shed; seeds 1–2.

          10. Bracts of pollen cones yellow to light brown; inner bracts of seed cones herbaceous; seeds (1–)2. . . . . 8. *Ephedra nevadensis*

          10. Bracts of pollen cones pale green to red; inner bracts of seed cones fleshy and red; seeds 1(–2). . . . . 1. *Ephedra antisyphilitica*

        9. Leaf bases persistent and shredding, brown, becoming gray with age; seed 1.

          11. Twigs usually scabrous; bracts of pollen cones yellow to red-brown; seeds smooth to slightly scabrous. . . . . 2. *Ephedra aspera*

          11. Twigs smooth or very slightly scabrous; bracts of pollen cones light yellow; seeds furrowed. 6. *Ephedra fasciculata*

1. Ephedra antisyphilitica Berlandier ex C. A. Meyer, Mém. Acad. Imp. Sci. Saint-Pétersbourg, Sér. 6, Sci. Math., Seconde Pt. Sci. Nat. 5(2): 291. 1846 · Clap-weed, popote, tepopote

Ephedra antisyphilitica var. *brachy-carpa* Cory; *E. occidentalis* Torrey ex Parlatore; *E. texana* E. L. Reed **Shrubs** erect or spreading, 0.25–1 m. **Bark** gray, slightly cracked and irregularly fissured. **Branches** alternate or whorled, rigid, angle of divergence about 45°. **Twigs** green, becoming yellow-green with age, glaucous, not viscid, with numerous longitudinal grooves; internodes 2–5 cm. **Terminal buds** conic, 1–3 mm, apex obtuse. **Leaves** opposite, 1–3 mm, connate to 2/3–7/8 their length; bases thickened, brown, completely deciduous; apex obtuse. **Pollen cones** 1–2 at node, lance-ellipsoid, 5–8 mm, sessile to nearly sessile; bracts opposite, 5–8 pairs, pale green to red, ob-ovate, 2–4 × 2–3 mm, membranous, connate at base; bracteoles slightly exceeding bracts; sporangiophores 4–5 mm, 1/2 exserted, with 4–6 sessile to very short-stalked (less than 1 mm) microsporangia. **Seed cones** 1–2 at node, ellipsoid, 6–12 mm, sessile to nearly sessile; bracts opposite, 4–6 pairs, ovate, 5–7 × 5–10 mm, connate to 1/8–7/8 their length, inner pairs becoming fleshy and red. **Seeds** 1(–2), ellipsoid, 6–9 × 2–4 mm, light brown to chestnut, smooth.

Coning late winter–early spring (Mar–Apr). Arid soils and rocky slopes; 100–1200 m; Okla., Tex.; Mexico in Nuevo León and San Luis Potosí.

Mexican populations of *Ephedra antisyphilitica* are disjunct from those in Texas.

2. Ephedra aspera Engelmann ex S. Watson, Proc. Amer. Acad. Arts 18: 157. 1883 · Boundary ephedra, pitamoreal

Ephedra nevadensis S. Watson var. *aspera* (Engelmann ex S. Watson) L. D. Benson; *E. peninsularis* I. M. Johnston; *E. reedii* Cory **Shrubs** erect, 0.5–1.5 m. **Bark** gray, cracked and fissured. **Branches** opposite or whorled, rigid, angle of divergence about 30°. **Twigs** pale to dark green, becoming yellow with age, not viscid, slightly to strongly scabrous, with numerous longitudinal grooves; internodes 1–6 cm. **Terminal buds** conic, 1–2 mm, apex

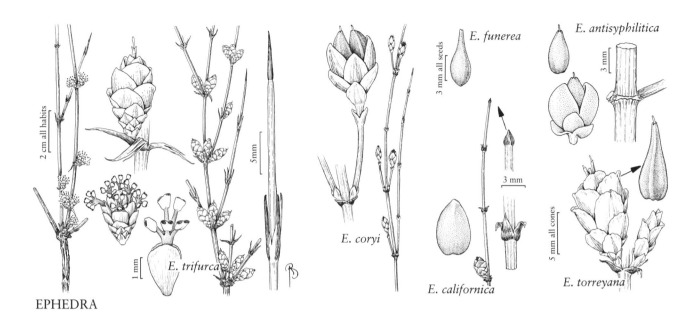

E. funerea

3 mm all seeds

E. antisyphilitica

3 mm

E. coryi

E. californica

3 mm

5 mm all cones

E. torreyana

E. trifurca

1 mm

2 cm all habits

5mm

EPHEDRA

obtuse. **Leaves** opposite (rarely in whorls of 3), 1–3(–5) mm, connate to 1/2–7/8 their length; bases thickened, brown, shredding with age, ± persistent; apex obtuse. **Pollen cones** 2 (rarely 1 or whorled) at node, obovoid, 4–7 mm, sessile or rarely on short peduncles; bracts opposite, 6–10 pairs, yellow to red-brown, obovate, 3–4 × 2–3 mm, membranous; bracteoles slightly exceeding bracts; sporangiophores 4–5 mm, 1/2 exserted, with 4–6 sessile to short-stalked (less than 1 mm) microsporangia. **Seed cones** usually 2 at node, ovoid, 6–10 mm, sessile or on short, scaly peduncles; bracts opposite, 5–7 pairs, circular, 4–7 × 2–4 mm, membranous, with red-brown thickened center and base, margins entire. **Seed** 1, ellipsoid, 5–8 × 2–4 mm, light brown to brown, smooth to slightly scabrous.

Coning late winter–early spring. Dry rocky slopes, ravines, and fans; 500–1800 m; Ariz., Calif., N.Mex., Tex.; n Mexico.

apex acute. **Leaves** in whorls of 3, 2–6 mm, connate to 1/2–3/4 their length; bases at first membranous, then becoming thickened, completely deciduous; apex acute. **Pollen cones** 1–several at node, ovoid, 6–8 mm, on short, scaly peduncles; bracts in 8–12 whorls of 3, light orange-yellow, ovate, 2–3 × 2–3 mm, membranous, slightly united at base; bracteoles equaling or slightly exceeding bracts; sporangiophores 3–5 mm, exserted to 1/3 their length, with 3–7 sessile to short-stalked microsporangia. **Seed cones** 1–several at node, ovoid, 8–10 mm, on very short, scaly peduncles; bracts in 4–6 whorls of 3, circular, 5–7 × 5–10 mm, papery, with orange- to green-yellow center and base, slightly clawed, margins entire. **Seeds** 1(–2), nearly globose, 7–10 mm diam., brown, smooth.

Coning late winter–early spring. Dry slopes and fans to valley grasslands; 50–1000 m; Calif.; Mexico in Baja California.

3. **Ephedra californica** S. Watson, Proc. Amer. Acad. Arts 14: 300. 1879 · California ephedra, cañatillo

Shrubs erect, 0.25–1 m. **Bark** gray-brown, cracked and irregularly fissured. **Branches** alternate or whorled, semiflexible to rigid, angle of divergence about 45°. **Twigs** yellow-green, becoming yellow, then yellow-brown with age, glaucous, with numerous very fine longitudinal grooves; internodes 3–10 cm. **Terminal buds** conic, 2–3 mm,

4. **Ephedra coryi** E. L. Reed, Bull. Torrey Bot. Club 63: 351. 1936 · Cory's ephedra

Shrubs rhizomatous, forming clumps, erect, 0.25–1.5 m, 3–5 m diam. **Bark** red-brown, cracked and irregularly fissured. **Branches** alternate or whorled, becoming rigid, angle of divergence about 25°. **Twigs** bright green, becoming yellow-green with age, not viscid, ridges between longitudinal grooves papillate, slightly scabrous; internodes 2–

5 cm. **Terminal buds** conic, 1–3 mm. **Leaves** opposite, 2–5 mm, connate to 1/2–3/4 their length; bases thickened, brown, persistent, becoming hard, enlarged, and black; apex acute. **Pollen cones** 2–several at node, obovoid, 4–6 mm, on very short, scaly peduncles (rarely sessile); bracts opposite, 5–9 pairs, light yellow, ovate, 2–4 × 2–3 mm, membranous, slightly connate at base; bracteoles slightly exceeding bracts; sporangiophores 2–4 mm, 1/4 exserted, with 5–7 sessile to short-stalked (less than 1 mm) microsporangia. **Seed cones** 2–several at node, obovoid to nearly globose, 7–15 mm, on smooth peduncles, 5–25 mm, with 1 pair of basal and 1 pair of nearly terminal bracts, at least in early cones; bracts opposite, 3–4 pairs, ovate to circular, 5–8 × 5–12 mm, inner pairs becoming fleshy (at least centrally) and orange at maturity. **Seeds** 2, ellipsoid, 5–8 × 2–4 mm, brown to chestnut, smooth.

Coning late winter–early spring. Sandy, semiarid areas; 500–2300 m; N.Mex., Tex.

In New Mexico *Ephedra coryi* occurs only in an isolated population in the San Andreas Mountains and represents the shorter extreme (5–10 mm) in the range of peduncle length.

5. Ephedra cutleri Peebles, J. Wash. Acad. Sci. 30: 473. 1940 · Navajo ephedra

Ephedra coryi E. L. Reed var. *viscida* Cutler; *E. viridis* Coville var. *viscida* (Cutler) L. D. Benson

Shrubs rhizomatous, forming clumps, erect, 0.25–1.5 m, 3–5 m diam. **Bark** reddish brown, cracked and irregularly fissured. **Branches** alternate or whorled, rigid, angle of divergence about 25°. **Twigs** bright green, becoming yellow-green with age, viscid, ridges between longitudinal grooves papillate; internodes 1–5 cm. **Terminal buds** conic, 1–3 mm. **Leaves** opposite, 2–5 mm, connate to 1/4–1/2 their length; bases thickened, brown, persistent; apex setaceous. **Pollen cones** 2–several at node, obovoid, 4–6 mm, on very short, scaly peduncles (rarely sessile); bracts opposite, 5–9 pairs, light yellow, ovate, 2–4 × 2–3 mm, membranous, slightly connate at base; bracteoles slightly exceeding bracts; sporangiophores 2–4 mm, barely exserted, with 5–7 sessile to short-stalked (less than 1 mm) microsporangia. **Seed cones** 2–several at node, obovoid to nearly globose, 7–15 mm, peduncles 5–25 mm, with 1 pair of basal and 1 pair of nearly terminal bracts; bracts opposite, 3 or 4 pairs, ovate, 3–6 × 2–5 mm, membranous, with yellow center and base, margins entire. **Seeds** 2, ellipsoid, 5–8 × 2–4 mm, brown to chestnut, smooth.

Coning late winter–midspring (Mar–May). Dry, flat,

sandy areas, occasionally on rocky slopes; 1400–2300 m; Ariz., Colo., N.Mex., Utah.

The hybrid *Ephedra* ×*arenicola* is discussed under *E. torreyana*.

6. Ephedra fasciculata A. Nelson, Amer. J. Bot. 21: 573. 1934 · Fasciculate ephedra

Ephedra clokeyi Cutler; *E. fasciculata* var. *clokeyi* (Cutler) Clokey

Shrubs erect or prostrate, 0.5–1 m. **Bark** gray, cracked and fissured. **Branches** opposite or whorled, rigid, angle of divergence about 30°. **Twigs** pale green, becoming yellow with age, not viscid, usually smooth or very slightly scabrous, with numerous longitudinal grooves; internodes 1–5 cm. **Terminal buds** conic, 1–3 mm, apex obtuse. **Leaves** opposite, 1–3 mm, connate to 1/2–3/4 their length; bases membranous, brown, shredding and becoming gray with age, ± persistent; apex obtuse. **Pollen cones** 2–several at node, ellipsoid to obovoid, 4–8 mm, sessile; bracts opposite, 4–8 pairs, light yellow, obovate, 2–3 × 2 mm, membranous, slightly connate at base; bracteoles exceeding bracts; sporangiophores 3–9 mm, 1/4–3/4 exserted, with 6–10 sessile to short-stalked (less than 1 mm) microsporangia. **Seed cones** 2–several at node, obovoid to ellipsoid, 6–13 mm, sessile or on short peduncles; bracts opposite, 4–7 pairs, elliptic, 3–7 × 2–4 mm, membranous with light brown to green, thickened center and base, slightly connate at base, margins entire. **Seeds** 1(–2), ellipsoid, 5–12 × 3–5 mm, light brown, longitudinally furrowed.

Coning late winter–early spring. Dry rocky slopes, washes, and sandy areas; 300–1200 m; Ariz., Calif., Nev., Utah.

7. Ephedra funerea Coville & C. V. Morton, J. Wash. Acad. Sci. 25: 307. 1935 · Death Valley ephedra

Ephedra californica S. Watson var. *funerea* (Coville & C. V. Morton) L. D. Benson

Shrubs erect, 0.25–1.5 m. **Bark** gray, slightly cracked and irregularly fissured. **Branches** alternate or whorled, rigid, angle of divergence about 60°. **Twigs** gray-green, becoming gray with age, glaucous, slightly scabrous, with numerous very fine longitudinal grooves; internodes 2–6 cm. **Terminal buds** conic, 1–4 mm, apex acute. **Leaves** in whorls of 3, 2–6 mm, connate to 2/3–3/4 their length; bases splitting at margins, persistent, forming black, thickened collar;

apex acute. **Pollen cones** 1–3 at node, narrowly ellipsoid, 5–8 mm, on very short, scaly peduncles (rarely sessile); bracts in 6–9 whorls of 3, light yellow, ovate, 3–4 × 2–3 mm, membranous, base short-clawed; bracteoles equaling bracts; sporangiophores 3–5 mm, exserted to 1/3 their length, with 3–7 sessile to short-stalked microsporangia. **Seed cones** 1–3 at node, lance-obovoid, 8–15 mm, on short, scaly peduncles (rarely sessile); bracts in 6–9 whorls of 3, obovate, 4–8 × 3–5 mm, papery, yellow-translucent with green-yellow center and base, base broadly clawed, margins slightly dentate. **Seeds** 1(–3), ellipsoid, 6–10 × 2–4 mm, pale green to light brown, smooth to scabrous.

Coning late winter–early spring. Sandy, dry soil and rocky scrub areas; of conservation concern; 500–1500 m; Calif., Nev.

8. **Ephedra nevadensis** S. Watson, Proc. Amer. Acad. Arts 14: 298. 1879 · Nevada ephedra

Ephedra antisyphilitica S. Watson 1871, not Berlandier ex C. A. Meyer 1846; *E. antisyphilitica* var. *pedunculata* S. Watson; *E. nevadensis* subvar. *paucibracteata* Stapf

Shrubs erect, 0.25–1.5 m. **Bark** gray, fissured. **Branches** alternate or whorled, rigid, angle of divergence about 45°. **Twigs** pale green, becoming yellow with age, not viscid, glaucous, with numerous longitudinal grooves; internodes 1–6 cm. **Terminal buds** conic, 1–3 mm, apex obtuse. **Leaves** opposite (rarely in whorls of 3), 2–4(–8) mm, connate to 1/2–3/4 their length; bases thickened, brown, completely deciduous; apex obtuse. **Pollen cones** 1–several at node, ellipsoid, 4–8 mm, sessile or on short peduncles with 2 pairs of basal bracts; bracts opposite, 5–9 pairs, yellow to light brown, obovate, 3–4 × 2–3 mm, membranous; bracteoles slightly exceeding bracts; sporangiophores 3–5 mm, exserted to 1/4–1/2 their length, with 6–9 sessile to short-stalked (less than 1 mm) microsporangia. **Seed cones** 1–several at node, nearly globose, 5–11 mm, on long peduncles, with 1–2 pairs of basal bracts; bracts opposite, 3–5 pairs, nearly circular, 4–8 × 3–6 mm, herbaceous, with light brown to yellow-green center, occasionally pinkish tinged, margins entire. **Seeds** (1–)2, ellipsoid, 6–9 × 2–4 mm, brown, smooth. $2n = 14, 28$.

Coning late winter–midspring. Dry, rocky slopes and hills, rarely in sandy flat areas; 700–1900 m; Ariz., Calif., Nev., Oreg., Utah.

9. **Ephedra pedunculata** Engelmann ex S. Watson, Proc. Amer. Acad. Arts 18: 157. 1883 · Clap-weed

Shrubs vinelike, trailing or clambering, 6–7 m. **Bark** gray, slightly cracked and fissured. **Branches** alternate (rarely whorled), lax, angle of divergence about 55°. **Twigs** gray-green, becoming green, then yellow-green with age, glaucous, with several moderately deep longitudinal grooves; internodes 1–8 cm. **Terminal buds** attenuate, 1–3 mm. **Leaves** opposite, 1–3 mm, connate to 2/3–7/8 their length; bases splitting at margins, persistent; apex obtuse. **Pollen cones** 1–2 at node, lanceoloid, 4–8 mm, sessile or on short to long, smooth peduncles; bracts opposite, 6–12 pairs, light yellow to reddish, obovate, 2–4 × 2–3 mm, membranous, free or slightly connate at base; bracteoles barely exceeding bracts; sporangiophores 3–5 mm, exserted to 1/2 their length, with 4–6 long-stalked (1–2 mm) microsporangia. **Seed cones** 1–2 at node, ovoid, 6–10 mm, on short to long, smooth peduncles; bracts opposite, 3–6 pairs, 3–4 × 2–3 mm, ovate, connate to 1/8–7/8 their length, inner pairs becoming succulent and red when ripe. **Seeds** 2, ellipsoid, 4–10 × 2–4 mm, brown, smooth.

Coning midwinter–early spring. Dry, sandy to rocky areas and slopes; of conservation concern; 100–1000 m; Tex.; n Mexico.

10. **Ephedra torreyana** S. Watson, Proc. Amer. Acad. Arts 14: 299. 1879 · Mormon-tea

Shrubs erect, 0.25–1 m. **Bark** gray, cracked and irregularly fissured. **Branches** alternate or whorled, rigid, angle of divergence about 45°. **Twigs** blue-green, becoming gray with age, glaucous, with numerous very fine longitudinal grooves; internodes 2–5 cm. **Terminal buds** conic, less than 4 mm. **Leaves** in whorls of 3, 2–5 mm, connate to 2/3 their length; bases becoming gray and shredded with age; apex acute. **Pollen cones** 1–4 at node, ovoid, 6–8 mm, sessile; bracts in 6–9 whorls of 3, cream to pale yellow, ovate, slightly clawed, 2–4 × 2–4 mm, membranous; bracteoles slightly exceeding bracts; sporangiophores 2–4 mm, exserted to 1/2 their length, with 5–8 sessile to short-stalked microsporangia. **Seed cones** 1–several at node, ovoid, 9–15 mm, sessile; bracts in 5 or 6 whorls of 3, obovate, 6–9 × 6–10 mm, papery, translucent with orange-yellow to greenish yellow center and base, base clawed, margins minutely dentate, undulate. **Seeds** 1–2(–3), ellipsoid, 7–10 × 1.5–3 mm, light brown to yellowish green, scabrous.

Coning spring. Dry, rocky to sandy areas; 500–2000 m; Ariz., Colo., Nev., N.Mex., Tex., Utah; Mexico in Chihuahua.

Ephedra torreyana is known to form hybrids with two other species of *Ephedra* as reported and described by H. C. Cutler (1939). The first of these is *E.* × *intermixta* Cutler, the hybrid between *E. torreyana* and *E. trifurca*. This hybrid occurs in a small area of southwestern New Mexico (near Engle, Sierra County) within the zone of sympatry of the two parental species; it may be fertile (mature seeds are formed). It is intermediate in most characters but can be identified by its combination of the spinelike terminal buds of *E. trifurca* and the scabrous, light yellow seeds of *E. torreyana*.

The second hybrid is *Ephedra* × *arenicola* Cutler, the hybrid between *E. torreyana* and *E. cutleri*. This hybrid is known only from the type locality in extreme northeastern Arizona (near Dennehotso, Apache County) in an area of sympatry of the parental species. This hybrid is intermediate in most characters, but it can be distinguished by its combination of the setaceous leaves, viscid stems, and long-pedunculate seed cones of *E. cutleri* with the persistent, whorled leaves of *E. torreyana*.

11. Ephedra trifurca Torrey ex S. Watson, Botany (Fortieth Parallel): 329. 1871 · Mexican-tea

Shrubs erect, 0.5–5 m. **Bark** gray, cracked and irregularly fissured. **Branches** alternate or whorled, rigid, angle of divergence about 30°. **Twigs** pale green, becoming yellow, then gray with age, glaucous, with numerous very fine longitudinal grooves; internodes 3–10 cm. **Terminal buds** spinelike, to 10 mm. **Leaves** in whorls of 3, 5–15 mm, connate to 1/2–3/4 their length; bases becoming gray and shredded with age; apex spinose. **Pollen cones** 1–several at node, obovoid, 6–10 mm, on short, scaly peduncles; bracts in 8–12 whorls of 3, reddish brown, obovate, slightly clawed, 3–4 × 2–3 mm, membranous; bracteoles nearly equaling bracts; sporangio-

phores 4–5 mm, exserted to 1/4 their length, with 4–5 short-stalked microsporangia. **Seed cones** 1–several at node, obovoid, 10–15 mm, on short, scaly peduncles (rarely sessile); bracts in 6–9 whorls of 3, circular, 8–12 × 8–12 mm, papery, translucent with reddish brown center and base, base clawed, margins entire. **Seeds** 1(–3), ellipsoid, 8–15 × 1.5–3 mm, light brown, smooth.

Coning late winter–early spring. Dry rocky slopes to flat sandy areas; 500–2000 m; Ariz., Calif., N.Mex., Tex.; Mexico in Baja California, Chihuahua, Coahuila, Sonora.

The hybrid *Ephedra* × *intermixta* is discussed under *E. torreyana*.

12. Ephedra viridis Coville, Contr. U.S. Natl. Herb. 4: 220. 1893 · Green ephedra

Ephedra nevadensis S. Watson subvar. *pluribracteata* Palmer ex Stapf; *E. nevadensis* var. *viridis* (Coville) M. E. Jones

Shrubs erect, 0.5–1 m. **Bark** gray, cracked and irregularly fissured. **Branches** alternate or whorled, rigid, angle of divergence about 30°. **Twigs** bright green to yellow-green, becoming yellow with age, not viscid, ridges between longitudinal grooves barely papillate, smooth; internodes 1–5 cm. **Terminal buds** conic, 1–2 mm, apex obtuse. **Leaves** opposite, 2–5 mm, connate to 1/2–3/4 their length, not photosynthetic; bases thickened, brown, persistent, becoming hard, enlarged, and black; apex setaceous. **Pollen cones** 2–several at node, obovoid, 5–7 mm, sessile; bracts opposite, 6–10 pairs, light yellow and slightly reddened, ovate, 2–4 × 2–3 mm, membranous, slightly connate at base; bracteoles slightly exceeding bracts; sporangiophores 2–4 mm, 1/4–1/2 exserted, with 5–8 sessile to nearly sessile microsporangia. **Seed cones** 2–several at node, obovoid, 6–10 mm, sessile or on short, scaly peduncles; bracts opposite, 4–8 pairs, ovate, 4–7 × 2–4 mm, membranous or papery, with yellow center and base, margins entire. **Seeds** 2, ellipsoid, 5–8 × 2–4 mm, light brown to brown, smooth. $2n = 28$.

Coning spring. Dry rocky slopes and canyon walls; 800–2500 m; Ariz., Calif., Colo., Nev., N.Mex., Oreg., Utah, Wyo.

Literature Cited

Robert W. Kiger, Editor

This is a consolidated list of all works cited in volume 2, whether as selected references, in text, or in nomenclatural contexts. In citations of articles, both here and in the taxonomic treatments, and also in nomenclatural citations, the titles of serials are rendered in the abbreviated forms recommended in G. D. R. Bridson and E. R. Smith (1991). Cross references to the corresponding full serial titles are interpolated here alphabetically by abbreviated form. In nomenclatural citations (only), book titles are rendered in the abbreviated forms recommended in F. A. Stafleu and R. S. Cowan (1976–1988) and F. A. Stafleu and E. A. Mennega (1992), which are indicated parenthetically following the full citations of those works here. Cross references to the full citations are also interpolated in the list alphabetically by abbreviated form. Two or more works published in the same year by the same author or group of coauthors will be distinguished uniquely and consistently throughout all volumes of *Flora of North America* by lowercase letters (b, c, d, . . .) suffixed to the date for the second and subsequent works in the set. The suffixes are assigned in order of editorial encounter, and do not reflect actual chronological sequence of publication. The first work by any particular author or group from any given year carries the implicit date suffix "a"; thus, the sequence of explicit suffixes begins with "b". In three cases, this list includes citations with dates suffixed "b" that are not preceded by citations of "a" works (i.e., ones with no date suffix) for the same year. This does not reflect omissions here but rather that there are corresponding "a" works cited (and encountered first from) elsewhere in the *Flora* that are not pertinent here.

Abh. Königl. Böhm. Ges. Wiss. = Abhandlungen der Königlichen böhmischen Gesellschaft der Wissenschaften.

Abh. Math.-Phys. Cl. Königl. Bayer. Akad. Wiss. = Abhandlungen der Mathematisch-physikalischen Classe der Königlich bayerischen Akademie der Wissenschaften.

Abrams, L. and R. S. Ferris. 1923–1960. Illustrated Flora of the Pacific States: Washington, Oregon, and California. 4 vols. Stanford. (Ill. Fl. Pacific States)

Account Exped. Pittsburgh — See: E. James 1823.

Acta Bot. Neerl. = Acta Botanica Neerlandica.

Acta Bot. Yunnan. = Acta Botanica Yunnanica. [Yunnan Zhiwu Yanjiu.]

Acta Phytotax. Geobot. = Acta Phytotaxonomica et Geobotanica. [Shokubutsu Bunrui Chiri.]

Acta Phytotax. Sin. = Acta Phytotaxonomica Sinica. [Chih Wu Fen Lei Hsüeh Pao.]

Acta Phytotax. Sin., Addit. = Acta Phytotaxonomica Sinica. Additamentum. [Zhiwu Fenlei Xuebao.]

Acta Univ. Lund. = Acta Universitatis Lundensis. Nova Series. Sectio 2, Medica, Mathematica, Scientiae Rerum Naturalium. [Lunds Universitets Årsskrift N.F., Avd. 2.]

Actes Soc. Hist. Nat. Paris = Actes de la Société d'Histoire Naturelle de Paris.

Adams, D. C. and P. B. Tomlinson. 1979. *Acrostichum* in Florida. Amer. Fern J. 69: 42–46.

Adams, R. P. 1969. Chemosystematic and Numerical Studies in Natural Populations of *Juniperus*. Ph.D. thesis. University of Texas.

Adams, R. P. 1975. Numerical-chemosystematic studies of infraspecific variation in *Juniperus pinchotii* Sudw. Biochem. Syst. & Ecol. 3: 71–74.

Adams, R. P. 1977. Chemosystematics: Analysis of populational differentiation and variability of ancestral and modern *Juniperus ashei*. Ann. Missouri Bot. Gard. 64: 184–209.

Adams, R. P. 1983. Infraspecific terpenoid variation in *Juniperus scopulorum*: Evidence for Pleistocene refugia and recolonization in western North America. Taxon 32: 30–46.

Adams, R. P. 1986. Geographic variation in *Juniperus silicicola* and *J. virginiana* of the southeastern United States: Multivariate analyses of morphology and terpenoids. Taxon 35: 61–75.

Adams, R. P. 1987. Investigation of *Juniperus* species of the United States for new sources of cedar wood oil. Econ. Bot. 41: 48–54.

Adams, R. P., A. L. Almirall, and L. Hogge. 1987. The volatile leaf oils of the junipers of Cuba: *Juniperus lucayana* Britton and *Juniperus saxicola* Britton and Wilson. Flav. Fragr. J. 2: 33–36.

Adams, R. P. and J. R. Kistler. 1991. Hybridization between *Juniperus erythrocarpa* Cory and *Juniperus pinchotii* Sudworth in the Chisos Mountains, Texas. SouthW. Naturalist 36: 295–301.

Adams, R. P., E. von Rudloff, and L. Hogge. 1983. Chemosystematic studies of the western North American junipers based on their volatile oils. Biochem. Syst. & Ecol. 11: 85–89.

Adams, R. P. and T. A. Zanoni. 1979. The distribution, synonymy, and taxonomy of three junipers of the southwest United States and northern Mexico. SouthW. Naturalist 24: 313–330.

Adams, R. P., T. A. Zanoni, and L. Hogge. 1984. Analysis of the volatile oils of *Juniperus deppeana* and its infraspecific taxa: Chemosystematic implications. Biochem. Syst. & Ecol. 12: 23–28.

Adanson, M. 1763[–1764]. Familles des Plantes. 2 vols. Paris. (Fam. Pl.)

Adansonia = Adansonia; Recueil Périodique d'Observations Botaniques.

Agardh, J. G. 1839. Recensio Specierum Generis Pteridis. Lund and Leipzig. (Recens. Spec. Pter.)

Agric. Man. — See: P. Lawson and C. Lawson 1836.

Aiton, W. 1789. Hortus Kewensis; or, a Catalogue of the Plants Cultivated in the Royal Botanic Garden at Kew. 3 vols. London. (Hort. Kew.)

Alt, K. S. and V. Grant. 1960. Cytotaxonomic observations on the goldback fern. Brittonia 12: 153–170.

Alverson, E. R. 1989. A new species of parsley fern, *Cryptogramma* (Adiantaceae), from western North America. Amer. Fern J. 79: 95–102.

Amer. Fern J. = American Fern Journal; a Quarterly Devoted to Ferns.

Amer. J. Bot. = American Journal of Botany.

Amer. J. Sci. Arts = American Journal of Science, and Arts.

Amer. Midl. Naturalist = American Midland Naturalist; Devoted to Natural History, Primarily That of the Prairie States.

Amer. Naturalist = American Naturalist....

Amer. Nurseryman = American Nurseryman.

Analecta Pteridogr. — See: G. Kunze 1837

Anales Hist. Nat. = Anales de Historia Natural.

Anales Inst. Biol. Univ. Nac. México = Anales del Instituto de Biológia de la Universidad Nacional de México.

Anales Jard. Bot. Madrid = Anales del Jardin Botánico de Madrid.

Anleit. Kenntn. Gew. — See: K. Sprengel 1802–1804.

Ann. Bot. Fenn. = Annales Botanici Fennici.

Ann. Bot. (London) = Annals of Botany. (London.)

Ann. Carnegie Mus. = Annals of the Carnegie Museum.

Ann. Missouri Bot. Gard. = Annals of the Missouri Botanical Garden.

Ann. Mus. Imp. Fis. Firenze = Annali del Musèo Imperiale di Fisica e Storia Naturale di Firenze. [N. s. vol. 1, 1865, under title: Annali del R. Museo di Fisica e Storia Naturale di Firenze.]

Ann. Mus. Natl. Hist. Nat. = Annales du Muséum National d'Histoire Naturelle. ["National" dropped with vol. 5.]

Ann. Nat. Hist. = Annals of Natural History; or, Magazine of Zoology, Botany and Geology.

Ann. Sci. Nat., Bot. = Annales des Sciences Naturelles. Botanique.

Arbor. Frutic. Brit. — See: J. C. Loudon [1835–]1838.

Arbust. Amer. — See: H. Marshall 1785.

Arch. Bot. (Leipzig) = Archiv für die Botanik.

Arctic Alpine Res. = Arctic and Alpine Research.

Arno, S. F. and J. R. Habeck. 1972. Ecology of alpine larch (*Larix lyallii* Parl.) in the Pacific Northwest. Ecol. Monogr. 42: 417–450.

Arq. Mus. Nac. Rio de Janeiro = Arquivos do Museu Nacional do Rio de Janeiro.

Arreguín-Sánchez, M. L. and R. Aguirre-Claverán. 1986. Una nueva localidad para Norteamérica de *Phyllitis scolopendrium* (L.) Newman var. *americana* Fernald. Phytologia 60: 339–403.

Atlantic J. = Atlantic Journal, and Friend of Knowledge.

Atti Soc. Ital. Sci. Nat. = Atti della Società Italiana di Scienze Naturali.

Austral. Syst. Bot. = Australian Systematic Botany.

Bailey, D. K. 1970. Phytogeography and taxonomy of *Pinus* subsection *Balfourianae*. Ann. Missouri Bot. Gard. 57: 210–249.

Bailey, D. K. 1987. A study of *Pinus* subsection *Cembroides* I: The single-needle pinyons of the Californias and the Great Basin. Notes Roy. Bot. Gard. Edinburgh 44: 275–310.

Bailey, D. K. and F. G. Hawksworth. 1979. Pinyons of the Chihuahuan Desert region. Phytologia 44: 129–133.

Baillon, H. E. 1876–1892. Dictionnaire de Botanique. 4 vols. in 34 fasc. Paris. (Dict. Bot.)

Baker, J. G. 1883. A synopsis of the genus *Selaginella*, pt. 1. J. Bot. 21: 1–5.

Baker, J. G. 1887. Handbook of the Fern-allies: A Synopsis of the Genera and Species of the Natural Orders Equisetaceae, Lycopodiaceae, Selaginellaceae, Rhizocarpeae. London. (Handb. Fern-allies)

Bakowsky, O. A. 1989. Phenotypic Variation in *Larix lyallii* and Relationships in the Larch Genus. M.Sc.F. thesis. Lakehead University.

Barrington, D. S., C. H. Haufler, and C. R. Werth. 1989. Hybridization, reticulation, and species concepts in the ferns. Amer. Fern J. 79: 55–64.

Barrington, D. S., C. A. Paris, and T. A. Ranker. 1986. Systematic inferences from spore and stomate size in the ferns. Amer. Fern J. 76: 149–159.

Bartonia = Bartonia; a Botanical Annual.

Bates, V. M. and E. T. Browne. 1981. *Azolla filiculoides* new to the southeastern United States. Amer. Fern J. 71: 33–34.

Beck, C. B., ed. 1988. Origin and Evolution of Gymnosperms. New York.

Beckner, J. 1968. *Lygodium microphyllum,* another fern escaped in Florida. Amer. Fern J. 58: 93–94.

Beddome, R. 1863[–1864]. The Ferns of Southern India. Being Descriptions and Plates of the Ferns of the Madras Presidency. 16 fasc. Madras. (Ferns S. India)

Beih. Nova Hedwigia = Beihefte zur Nova Hedwigia.

Beitel, J. M. 1979. The clubmosses *Lycopodium sitchense* and *L. sabinaefolium* in the upper Great Lakes area. Michigan Bot. 18: 3–13.

Beitel, J. M., W. H. Wagner Jr., and K. S. Walter. 1981. Unusual frond development in sensitive fern *Onoclea sensibilis* L. Amer. Midl. Naturalist 105(2): 396–400.

Beitr. Pflanzenk. Russ. Reiches = Beiträge zur Pflanzenkunde des russischen Reiches.

Bemerk. Botrychium — See: F. Ruprecht 1859

Benedict, R. C. 1909. The genus *Ceratopteris*: A preliminary revision. Bull. Torrey Bot. Club 36: 463–476.

Benedict, R. C. 1911. The genera of the fern tribe Vittarieae, their external morphology, venation, and relationships. Bull. Torrey Bot. Club 38: 153–190.

Benedict, R. C. 1914. A revision of the genus *Vittaria* J. E. Smith. I. The species of the subgenus *Radiovittaria*. Bull. Torrey Bot. Club 41: 391–410.

Benham, D. M. 1982. Biology of *Cheilanthes leucopoda* and *Cheilanthes kaulfussii* (Filicales). M.S. thesis. Angelo State University.

Benham, D. M. 1989. A Biosystematic Revision of the Fern Genus *Astrolepis* (Adiantaceae). Ph.D. dissertation. Northern Arizona University.

Benham, D. M. 1992. Additional combinations in *Astrolepis*. Amer. Fern J. 82: 59–62.

Benham, D. M. and M. D. Windham. 1992. Generic affinities of the star-scaled cloak ferns. Amer. Fern J. 82: 47–58.

Bennett, J. J., R. Brown, and T. Horsfield. 1838–1852. Plantae Javanicae Rariores, Descriptae Iconibus Illustratae, Quas in Insula Java, Annis 1802–1818.... 4 parts. London. (Pl. Jav. Rar.)

Benson, L. D. 1943. Revisions of status of southwestern trees and shrubs. Amer. J. Bot. 30: 230–240.

Ber. Deutsch. Bot. Ges. = Berichte der Deutschen botanischen Gesellschaft.

Biblioth. Bot. — See: K. Domin 1914–1930.

Bierhorst, D. W. 1971. Morphology of Vascular Plants. New York.

Bi-Monthly Res. Notes Forest. Serv. Canada = Bi-monthly Research Notes, Forestry Service, Canada.

Biochem. Syst. & Ecol. = Biochemical Systematics and Ecology.

Biol. Arbejder Tilegn. Eug. Warming — See: L. K. Rosenvinge 1911.

Biol. Skr. = Biologiske Skrifter.

Bishop, L. E. 1978. Revision of the genus *Cochlidium* (Grammitidaceae). Amer. Fern J. 68: 76–94.

Blake, S. T. 1959. New or noteworthy plants, chiefly from Queensland, 1. Proc. Roy. Soc. Queensland 70(6): 33–46.

Blasdell, R. F. 1963. A monographic study of the fern genus *Cystopteris*. Mem. Torrey Bot. Club 21: 1–102.

Blue Jay = Blue Jay; Bulletin of the Yorktown Natural History Society.

Blume, C. L. 1827–1828. Enumeratio Plantarum Javae et Insularum Adjacentium.... 2 fasc. Leiden. (Enum. Pl. Javae)

Blumea = Blumea; Tidjschrift voor die Systematiek en die Geografie der Planten (A Journal of Plant Taxonomy and Plant Geography).

Bobrov, A. E. 1967. The family Osmundaceae (R. Br.) Kaulf. Its taxonomy and geography. Bot. Zhurn. (Moscow & Leningrad) 52: 1600–1610.

Bol. Soc. Bot. México = Boletín de la Sociedad Botánica de México.

Bol. Soc. Brot. = Boletim da Sociedade Broteriana.

Bolton, J. [1785–]1790. Filices Britannicae: An History of the British Proper Ferns. 2 parts. Leeds and Huddersfield. [Parts paged consecutively.] (Fil. Brit.)

Boom, B. M. 1982. Synopsis of *Isoëtes* in the southeastern United States. Castanea 47: 38–59.

Bory de Saint-Vincent, J. B. et al. 1822–1831. Dictionnaire Classique d'Histoire Naturelle. 17 vols. Paris. (Dict. Class. Hist. Nat.)

Bot. California — See: S. Watson 1876–1880.

Bot. Exped. Oregon — See: A. Murray 1849–1859.

Bot. Gaz. = Botanical Gazette; Paper of Botanical Notes.

Bot. Helv. = Botanica Helvetica.

Bot. J. Linn. Soc. = Botanical Journal of the Linnean Society.

Bot. Jahrb. Syst. = Botanische Jahrbücher für Systematik, Pflanzengeschichte und Pflanzengeographie.

Bot. Mag. = Botanical Magazine; or, Flower-garden Displayed.... [Edited by Wm. Curtis]. [With vol. 15, 1801, title became Curtis's Botanical Magazine; or.... Vols. 1–184, 1787–1983. Vols. 54–70, 1827–44, also as n. s. vols. 1–17; vols. 71–130, 1845–1904, also as ser. 3, vols. 1–60; vols. 131–146, 1905–20, also as ser. 4, vols. 1–16; continuous volumation only, without series identity, vols. 147–184. Plates in vols. 1–164 numbered consecutively 1–9688; plates in vols. 165–184, 1948–83, numbered 1–882.]

Bot. Mag. (Tokyo) = Botanical Magazine. [Shokubutsu-gaku Zasshi.]

Bot. Misc. = Botanical Miscellany.

Bot. Not. = Botaniska Notiser.

Bot. Rev. (Lancaster) = Botanical Review, Interpreting Botanical Progress.

Bot. Taschenb. — See: D. H. Hoppe 1790–1849.

Bot. Taschenbuch — See: F. Weber and D. M. H. Mohr 1807.

Bot. Zeitung (Berlin) = Botanische Zeitung.

Bot. Zhurn. (Moscow & Leningrad) = Botanicheskii Zhurnal.

Botany (Fortieth Parallel) — See: S. Watson 1871.

Bouchard, A. and S. G. Hay. 1976. *Thelypteris limbosperma* in eastern North America. Rhodora 78: 552–553.

Bower, F. O. 1908. The Origin of a Land Flora, a Theory Based upon the Facts of Alternation. London.

Bradea = Bradea; Boletim do Herbarium Bradeanum.

Brass, L. J. 1955. Report of the southern Florida field trip. Amer. Fern J. 45: 48–54.

Braun, A. 1871. Hr. Braun theilte neuere Untersuchungen über die Gattungen *Marsilea* und *Pilularia.* Monatsber. Königl. Preuss. Akad. Wiss. Berlin 1870: 653–753.

Breedlove, D. E., ed. 1981+. Flora of Chiapas. 2+ parts. San Francisco.

Bridson, G. D. R. and E. R. Smith. 1991. B-P-H/S. Botanico-Periodicum-Huntianum/Supplementum. Pittsburgh.

Brit. Fern Gaz. = British Fern Gazette.

Britton, D. M. 1953. Chromosome studies in ferns. Amer. J. Bot. 40: 575–583.

Britton, E. C. and A. Taylor. 1902. The life history of *Vittaria lineata.* Mem. Torrey Bot. Club 8: 185–220.

Britton, N. L., E. E. Sterns, J. F. Poggenburg, A. Brown, T. C. Porter, and C. A. Hollick. 1888. Preliminary Catalogue of Anthophyta and Pteridophyta Reported As Growing Spontaneously Within One Hundred Miles of New York City. New York. [Authorship often attributed as B.S.P. in nomenclatural contexts.] (Prelim. Cat.)

Britton, N. L., L. M. Underwood, W. A. Murrill, J. H. Barnhart, and H. W. Rickett, eds. 1905–1972. North American Flora.... 42 vols. New York. [Vols. 1–34, 1905–1957; ser. 2, vols. 1–8, 1954–1972.]

Brittonia = Brittonia; a Journal of Systematic Botany....

Broun, M. 1938. Index to North American Ferns. Lancaster, Pa. (Index N. Amer. Ferns)

Brown, D. F. M. 1964. A monographic study of the fern genus *Woodsia.* Nova Hedwigia 16: 1–154.

Brown, R. 1810. Prodromus Florae Novae Hollandiae et Insulae van-Diemen.... London. (Prodr.)

Brownsey, P. J. 1983. Polyploidy and aneuploidy in *Hypolepis* and the evolution of the Dennstaedtiales. Amer. Fern J. 73: 97–108.

Brownsey, P. J. 1989. The taxonomy of bracken (*Pteridium:* Dennstaedtiaceae) in Australia. Austral. Syst. Bot. 2: 113–128.

Brownsey, P. J. and J. D. Lovis. 1987. Chromosome numbers for the New Zealand species of *Psilotum* and *Tmesipteris* and the phylogenetic relationships of the Psilotales. New Zealand J. Bot. 25: 439–454.

Bruce, J. G. 1975. Systematics and Morphology of Subgenus *Lepidotus* of the Genus *Lycopodium* (Lycopodiaceae). Ph.D. thesis. University of Michigan.

Bruce, J. G. 1976. Comparative studies of *Lycopodium carolinianum.* Amer. Fern J. 66: 125–137.

Bruce, J. G., W. H. Wagner Jr., and J. M. Beitel. 1991. Two new species of bog clubmoss, *Lycopodiella* (Lycopodiaceae) from southwestern Michigan. Michigan Bot. 30: 3–10.

Brummitt, R. K. and C. E. Powell, eds. 1992. Authors of Plant Names. A List of Authors of Scientific Names of Plants, with Recommended Standard Forms of Their Names, Including Abbreviations. Kew.

Buchholz, J. T. 1940. The embryogeny of *Torreya,* with a note on *Austrotaxus.* Bull. Torrey Bot. Club 67: 731–754.

Buck, W. R. 1977. A new species of *Selaginella* in the *Selaginella apoda* complex. Canad. J. Bot. 55: 366–371.

Buck, W. R. 1978. The taxonomic status of *Selaginella eatonii.* Amer. Fern J. 68: 33–36.

Buck, W. R. and T. W. Lucansky. 1976. An anatomical and morphological comparison of *Selaginella apoda* and *Selaginella ludoviciana.* Bull. Torrey Bot. Club 103: 9–16.

Bull. Acad. Int. Géogr. Bot. = Bulletin de l'Académie Internationale de Géographie Botanique.

Bull. Acad. Roy. Sci. Bruxelles = Bulletins de l'Académie Royale des Sciences et Belles-lettres de Bruxelles.

Bull. Bot. Dept., Jamaica = Bulletin of the Botanical Department. [Kingston, Jamaica.]

Bull. Brit. Mus. (Nat. Hist.), Bot. = Bulletin of the British Museum (Natural History). Botany.

Bull. Chin. Bot. Soc. = Bulletin of the Chinese Botanical Society. [Chung Kuo Chih-wu Hsüeh Hui Hui-pao.]

Bull. Fan Mem. Inst. Biol. = Bulletin of the Fan Memorial Institute of Biology. [Ts'ing Cheng Cheng Wou Tiao Tch'a So Houei Pao.]

Bull. Herb. Boissier = Bulletin de l'Herbier Boissier.

Bull. Illinois Nat. Hist. Surv. = Bulletin of the Illinois Natural History Survey.

Bull. Inst. Jamaica, Sci. Ser. = Bulletin of the Institute of Jamaica. Science Series.

Bull. Misc. Inform. Kew = Bulletin of Miscellaneous Information, Royal Gardens, Kew.

Bull. Osaka Mus. Nat. Hist. = Bulletin of the Osaka Museum of Natural History. [Osaka-shiritsu Shizenkagaku Hakubutsukan Kenkyu Hokoku.]

Bull. Sci. Soc. Philom. Paris = Bulletin des Sciences, par la Société Philomatique.

Bull. Soc. Bot. France = Bulletin de la Société Botanique de France.

Bull. Soc. Hist. Nat. Toulouse = Bulletin de la Société d'Histoire Naturelle de Toulouse.

Bull. Torrey Bot. Club = Bulletin of the Torrey Botanical Club.

Burke, J. G. 1975. Human use of the California nutmeg tree, *T. californica,* and of other members of the genus. Econ. Bot. 29: 127–139.

Burkhalter, J. R. 1985. A new station for *Dicranopteris flexuosa* in Bay County, Florida. Amer. Fern J. 75: 79.

Burns, R. M. and B. H. Honkala. 1990. Silvics of North America. 1. Conifers. Washington. [Agric. Handb. 654.]

Butters, F. K. 1917. Taxonomic and geographic studies in North American ferns. Rhodora 19: 169–216.

Calcutta J. Nat. Hist. = Calcutta Journal of Natural History, and Miscellany of the Arts and Sciences in India.

Callan, A. D. 1972. A Cytotaxonomic Study of a Triploid *Polystichum munitum × californicum* Hybrid from Douglas County, Oregon, with Some Observations on the Varieties of *Polystichum munitum*. M.S. thesis. Southern Oregon State College.

Canad. Field-Naturalist = Canadian Field-Naturalist.

Canad. J. Bot. = Canadian Journal of Botany.

Canad. J. Forest Res. = Canadian Journal of Forest Research.

Canad. Naturalist & Quart. J. Sci. = Canadian Naturalist and Quarterly Journal of Science, with Proceedings of the Natural History Society of Montreal.

Canad. Naturalist Geol. = Canadian Naturalist and Geologist.

Canadian Forestry Service. 1983. Reproduction of conifers. Forest. Techn. Pub. Canad. Forest. Serv. 31.

Candolle, A. L. P. de. 1868. Cycadaceae. In: A. P. de Candolle and A. L. P. de Candolle, eds. 1823–1873. Prodromus Systematis Naturalis Regni Vegetabilis.... 17 vols. Paris etc. Vol. 16, part 2, pp. 522–547.

Candolle, A. P. de and A. L. P. de Candolle, eds. 1823–1873. Prodromus Systematis Naturalis Regni Vegetabilis.... 17 vols. Paris etc. [Vols. 1–7 edited by A. P. de Candolle, vols. 8–17 by A. L. P. de Candolle.] (Prodr.)

Carlson, C. 1965. Interspecific Hybridization of *Larix occidentalis* and *Larix lyallii*. M.Sc.F. thesis. University of Montana.

Carlson, T. J. and W. H. Wagner Jr. 1982. The North American distribution of the genus *Dryopteris*. Contr. Univ. Michigan Herb. 15: 141–162.

Carrière, E. A. 1855. Traité Général des Conifères.... Paris. (Traité Gén. Conif.)

Carrière, E. A. 1867. Traité Général des Conifères..., ed. 2. 1 part only. Paris. (Traité Gén. Conif. ed. 2)

Caryologia = Caryologia; Giornale di Citologia, Citosistematica e Citogenetica.

Castanea = Castanea; Journal of the Southern Appalachian Botanical Club.

Cat. N. Amer. Pl. ed. 2 — See: A. A. Heller 1900.

Cat. N. Amer. Pl. ed. 3 — See: A. A. Heller 1909–1914.

Cat. Pl. New Bern — See: H. B. Croom 1837.

Cavanilles, A. J. 1791–1801. Icones et Descriptiones Plantarum, Quae aut Sponte in Hispania Crescunt, aut in Hortis Hospitantur. 6 vols. Madrid. (Icon.)

Cavanilles, A. J. [1801–]1802. Descripción de las Plantas.... Madrid. (Descr. Pl.)

Center Pl. Conservation = The Center for Plant Conservation [Newsletter].

Chadwick, L. C. and R. A. Keen. 1976. A study of the genus *Taxus*. Ohio Agric. Exp. Sta. Bull. 1086.

Chapman, A. W. 1860. Flora of the Southern United States ..... New York. (Fl. South. U.S.)

Ching, R. C. 1934. A revision of the compound leaved Poly-

sticha and other related species in the continental Asia including Japan and Formosa. Sinensia 5: 23–91.

Ching, R. C. 1936. On the genus *Cyrtomium* Presl. Bull. Chin. Bot. Soc. 2: 85–106.

Christensen, C. [1905–]1906. Index Filicum sive Enumeratio Omnium Generum Specierumque Filicum et Hydropteridum ab Anno 1753 ad Finem Anni 1905.... Copenhagen.

Christensen, C. 1913. A monograph of the genus *Dryopteris*. Part I. The tropical American pinnatifid-bipinnatifid species. Kongel. Danske Vidensk. Selsk. Skr., Naturvidensk. Math. Afd., ser. 7, 10: 55–282.

Christensen, C. 1916. *Maxonia*, a new genus of tropical American ferns. Smithsonian Misc. Collect. 66(9): 1–4.

Christensen, C. 1930. The genus *Cyrtomium*. Amer. Fern J. 20: 41–52.

Christensen, C. 1937. Taxonomic fern studies. IV. Revision of the Bornean and New Guinean ferns, etc. Dansk Bot. Ark. 9(3): 33–52.

Christensen, C. 1938. Azollaceae. In: F. Verdoorn, ed. 1938. Manual of Pteridology. The Hague. P. 550.

Circ. Agric. Exp. Sta., Alabama = Circular, Agricultural Experiment Station, Alabama.

Clairville, J. P. de. 1811. Manuel d'Herborisation en Suisse et en Valais.... Winterthur. (Man. Herbor. Suisse)

Clausen, R. T. 1938. A monograph of the Ophioglossaceae. Mem. Torrey Bot. Club 19(2): 1–177.

Clausen, R. T. 1946. *Selaginella* subgenus *Euselaginella* in the southeastern United States. Amer. Fern J. 36: 65–81.

Cody, W. J. and D. M. Britton. 1989. Ferns and Fern Allies of Canada. Ottawa.

COHMAP Members. 1988. Climatic changes of the last 18,000 years: Observations and model simulations. Science 241: 1043–1052.

Comer, C. W., R. P. Adams, and D. F. Van Haverbeke. 1982. Intra- and inter-specific variation of *Juniperus virginiana* L. and *J. scopulorum* Sarg. seedlings based on volatile oil composition. Biochem. Syst. & Ecol. 10: 297–306.

Conif. Nov. — See: F. Parlatore 1863.

Contr. Gray Herb. = Contributions from the Gray Herbarium of Harvard University. [Some numbers reprinted from (or in?) other periodicals, e.g. Rhodora.]

Contr. New South Wales Natl. Herb. = Contributions from the New South Wales National Herbarium.

Contr. New South Wales Natl. Herb., Fl. Ser. = Contributions from the New South Wales National Herbarium, Flora Series.

Contr. Univ. Michigan Herb. = Contributions from the University of Michigan Herbarium.

Contr. U.S. Natl. Herb. = Contributions from the United States National Herbarium.

Cooper, M. R. and A. W. Johnson. 1984. Poisonous Plants in Britain and Their Effects on Animals and Man. London. [Minist. Agric., Fisheries and Food Ref. Book 161.]

Cooper-Driver, G. 1977. Chemical evidence for separating the Psilotaceae from the Filicales. Science 198: 1260–1262.

Cooperrider, T. S. 1968. Exclusion of the New York fern and hayscented fern from the flora of Iowa. Amer. Fern J. 58: 176–178.

Copeland, E. B. 1947. Genera Filicum, the Genera of Ferns. Waltham, Mass. (Gen. Fil.)

Correll, D. S. 1956. Ferns and Fern Allies of Texas. Renner, Tex.

Coulter, J. M. and W. J. G. Land. 1905. Gametophytes and embryo of *Torreya taxifolia*. Bot. Gaz. 39: 161–178.

Cranfill, R. 1980. Ferns and Fern Allies of Kentucky. Frankfort, Ky.

Cranfill, R. 1983. The distribution of *Woodwardia areolata*. Amer. Fern J. 73: 46–52.

Cranfill, R. and D. M. Britton. 1983. Typification within the *Polypodium virginianum* complex (Polypodiaceae). Taxon 32: 557–560.

Critchfield, W. B. and E. L. Little Jr. 1966. Geographic Distribution of the Pines of the World. Washington. [U.S.D.A., Misc. Publ. 991.]

Cronquist, A., A. H. Holmgren, N. H. Holmgren, J. L. Reveal, P. K. Holmgren, and R. C. Barneby. 1972+. Intermountain Flora. Vascular Plants of the Intermountain West, U.S.A. 4+ vols. New York and London. [Vol. 1, 1972; vol. 3, part B, 1989; vol. 4, 1984; vol. 6, 1977.] (Intermount. Fl.)

Croom, H. B. 1837. A Catalogue of Plants, Native or Naturalized, in the Vicinity of New Bern, North Carolina; with Remarks and Synonyms. New York. (Cat. Pl. New Bern)

Croxdale, J. G. 1978. *Salvinia* leaves. I. Origin and early morphogenesis of floating and submerged leaves. Canad. J. Bot. 56: 1982–1991.

Croxdale, J. G. 1979. *Salvinia* leaves. II. Morphogenesis of the floating leaf. Canad. J. Bot. 57: 1951–1959.

Croxdale, J. G. 1981. *Salvinia* leaves. III. Morphogenesis of the submerged leaf. Canad. J. Bot. 59: 2065–2072.

Cusick, A. W. 1987. A binomial for a common hybrid *Lycopodium*. Amer. Fern J. 77: 100–101.

Cutler, H. C. Monograph of the North American species of the genus *Ephedra*. Ann. Missouri Bot. Gard. 26: 373–427.

Cycl. — See: A. Rees [1802–]1819–1820.

Dansk Bot. Ark. = Dansk Botanisk Arkiv Udgivet af Dansk Botanisk Forening.

Darling, T. Jr. 1961. Florida rarities. Amer. Fern J. 51: 1–15.

Darling, T. Jr. 1982. The deletion of *Nephrolepis pectinata* from the flora of Florida. Amer. Fern J. 72: 63.

Daubenmire, R. 1968. Some geographic variations in *Picea sitchensis* and their ecologic interpretation. Canad. J. Bot. 46: 787–798.

Daubenmire, R. 1972. On the relation between *Picea pungens* and *Picea engelmannii* in the Rocky Mountains. Canad. J. Bot. 50: 733–742.

Daubenmire, R. 1974. Taxonomic and ecologic relationships between *Picea glauca* and *Picea engelmannii*. Canad. J. Bot. 52: 1545–1560.

Dec. Gen. Nov. — See: E. P. Ventenat 1808.

Decken, C. C. von der. 1869–1879. Reisen in Ost-Afrika in 1851–1861. 3 vols. in parts. Leipzig and Heidelberg. [Vol. 3(3), 1879, is Botanik von Ost-Afrika by P. Ascherson, O. Böckeler, F. W. Klatt, M. Kuhn, P. G. Lorentz, and W. Sonder, also reprinted separately under that title in the same year.] (Reisen Ost-Afrika)

Delchamps, C. E. 1966. *Trichomanes holopterum* — A filmy fern new to the United States. Amer. Fern J. 56: 138–139.

Dendrologie — See: K. H. E. Koch 1873.

Denkschr. Kaiserl. Akad. Wiss., Wien. Math.-Naturwiss. Kl. = Denkschriften der Kaiserlichen Akademie der Wissenschaften, Wien. Mathematisch-naturwissenschaftliche Klasse.

Dennis, W. M. and D. H. Webb. 1981. The distribution of *Pilularia americana* A. Br. (Marsileaceae) in North America, north of Mexico. Sida 9: 19–24.

Descr. Pinus — See: A. B. Lambert 1803–1824.

Descr. Pinus [ed. 3] — See: A. B. Lambert 1832.

Descr. Pl. — See: A. J. Cavanilles [1801–]1802.

Desfontaines, R. L. [1798–1799.] Flora Atlantica sive Historia Plantarum, Quae in Atlante, Agro Tunetano et Algeriensi Crescunt. 2 vols. in 9 parts. Paris. (Fl. Atlant.)

Deut. Fl. — See: H. Karsten 1880–1883.

Deutschl. Fl. — See: G. F. Hoffmann [1791–1804].

DeVol, C. E. 1957. The geographical distribution of *Ceratopteris pteridoides*. Amer. Fern J. 47: 67–72.

Dickinson, T. A., W. H. Parker, and R. E. Strauss. 1987. Another approach to leaf shape comparisons. Taxon 36: 1–20.

Dict. Bot. — See: H. E. Baillon 1876–1892.

Dict. Class. Hist. Nat. — See: J. B. Bory de Saint-Vincent et al. 1822–1831.

Dietrich, A. G. 1824. Flora der Gegend um Berlin.... 2 parts. Berlin. (Fl. Berlin)

Diss. Observ. Bot. — See: J. P. Du Roi 1771.

Distr. Crypt. Vasc. Ross. — See: F. Ruprecht 1845.

Do V. C., I. Watanabe, W. J. Zimmerman, T. A. Lumpkin, and T. de Waha Baillonville. 1989. Sexual hybridization among *Azolla* species. Canad. J. Bot. 67: 3482–3485.

Döll, J. C. 1843. Rheinische Flora. Frankfurt am Main. (Rhein. Fl.)

Domin, K. 1914–1930. Beiträge zur Flora und Pflanzengeographie Australiens, Bibliotheca Botanica. I. Teil. Systematische Bearbeitung des eigenen sowie auch fremden, besonders des von Frau Amalie Dietrich in Queensland (1868–1873) und von Dr. Clement in Nordwest-Australien gesammelten Materiales mit teilweiser Berücksichtigung der gesammten Flora Australiens. 2 vols. in fascicles. Stuttgart. [Volumes comprise (only): Band 20, Heft 85, Abteilungen 1–2, Lieferungen 1–4, and Band 22, Heft 89, Abteilung 3, Lieferungen 1–8, with consecutive pagination throughout.] (Biblioth. Bot.)

Doyle, J. 1945. Naming of the redwoods. Nature 155: 254–257.

Du Roi, J. P. 1771. Dissertatio Inauguralis Observationes Botanicas Sistens. Helmstad. (Diss. Observ. Bot.)

Duffield, J. W. 1952. Relationships and species hybridization in the genus *Pinus*. Silvae Genet. 1: 93–97.

Duncan, R. E. 1940. The cytology of sporangium development in *Azolla filiculoides*. Bull. Torrey Bot. Club 67: 391–412.

Duncan, W. H. 1954. *Polypodium aureum* in Florida and Georgia. Amer. Fern J. 44: 155–158.

Duncan, W. H. 1979. *Zamia* (Cycadaceae) new for Georgia. Sida 8: 115–116.

Eaton, A. A. 1900. The genus *Isoëtes* in New England. Fernwort Pap. 2: 1–16.

Eaton, A. A. 1906. Pteridophytes observed during three excursions into southern Florida. Bull. Torrey Bot. Club 33: 455–486.

Eaton, D. C. [1877–]1879–1880. The Ferns of North America. Coloured Figures and Descriptions, with Synonymy and Geographical Distribution.... 2 vols. in 28 fasc. Salem, Mass. and Boston. (Ferns N. Amer.)

Eckenwalder, J. E. 1976. Re-evaluation of Cupressaceae and Taxodiaceae: A proposed merger. Madroño 23: 237–256.

Eckenwalder, J. E. 1980. Dispersal of the West Indian cycad *Zamia pumila* L. Biotropica 12: 79–80.

Eckenwalder, J. E. 1980b. Taxonomy of the West Indian cycads. J. Arnold Arbor. 61: 701–722.

Ecol. Monogr. = Ecological Monographs.

Econ. Bot. = Economic Botany; Devoted to Applied Botany and Plant Utilization.

Edinburgh New Philos. J. = Edinburgh New Philosophical Journal.

Edinburgh Philos. J. = Edinburgh Philosophical Journal.

Emory, W. H. 1857–1859. Report on the United States and Mexican Boundary Survey, Made under the Direction of the Secretary of the Interior. 2 vols. in parts. Washington. (Rep. U.S. Mex. Bound.)

Encephalartos = Encephalartos; Journal of the Cycad Society of Southern Africa.

Encycl. — See: J. Lamarck et al. 1783–1817.

Endlicher, S. L. 1847. Synopsis Coniferarum. Sankt-Gallen. (Syn. Conif.)

Engelmann, G. 1882. The genus *Isoëtes* in North America. Trans. Acad. Sci. St. Louis 4: 358–390.

Engler, H. G. A., ed. 1900–1953. Das Pflanzenreich.... 107 vols. Berlin. [Sequence of volume (Heft) numbers (order of publication) is independent of the sequence of series and family (Roman and Arabic) numbers (taxonomic order).]

Engler, H. G. A., H. Harms, J. Mattfeld, H. Melchior, and E. Werdermann, eds. 1924+. Die natürlichen Pflanzenfamilien..., ed. 2. 26+ vols. Leipzig and Berlin.

Engler, H. G. A. and K. Prantl, eds. 1887–1915. Die natürlichen Pflanzenfamilien.... 254 fasc. Leipzig. [In this work's complex and inconsistently applied numbering scheme, the sequence of fascicle (Lieferung) numbers (order of publication) is independent of the sequence of division (Teil) and subdivision (Abteilung) numbers (taxonomic order).] (Nat. Pflanzenfam.)

Enum. Filic. — See: G. F. Kaulfuss 1824.

Enum. Pl. Javae — See: C. L. Blume 1827–1828.

Enum. Stirp. Vindob. — See: N. J. Jacquin 1762.

Euphytica = Euphytica. Netherlands Journal of Plant Breeding.

Evans, A. M. 1969. Interspecific relationships in the *Polypodium pectinatum–plumula* complex. Ann. Missouri Bot. Gard. 55(3): 193–293.

Evans, A. M., ed. 1971. A Review of Systematic Studies of the Pteridophytes of the Southern Appalachians. Blacksburg.

Evans, A. M. 1971b. *Polypodium*. In: A. M. Evans, ed. 1971. A Review of Systematic Studies of the Pteridophytes of the Southern Appalachians. Blacksburg. Pp. 130–140.

Evolution = Evolution, International Journal of Organic Evolution.

Ewan, J. 1944. Annotations on west American ferns III. Amer. Fern J. 34: 107–120.

Exot. Fl. — See: W. J. Hooker [1822–]1823–1827.

Fairchild Trop. Gard. Bull. = Fairchild Tropical Garden Bulletin.

Fam. Pl. — See: M. Adanson 1763[–1764].

Farjon, A. 1990. A Bibliography of Conifers. Königstein. [Regnum Veg. 122.]

Farrar, D. R. 1967. Gametophytes of four tropical fern genera reproducing independently of their sporophytes in the southern Appalachians. Science 155: 1266–1267.

Farrar, D. R. 1974. Gemmiferous fern gametophytes—Vittariaceae. Amer. J. Bot. 61: 146–155.

Farrar, D. R. 1978. Problems in the identity and origin of the Appalachian *Vittaria* gametophyte, a sporophyteless fern of the eastern United States. Amer. J. Bot. 65: 1–12.

Farrar, D. R. 1985. Independent fern gametophytes in the wild. Proc. Roy. Soc. Edinburgh, B 86: 361–369.

Farrar, D. R. 1990. Species and evolution in asexually reproducing independent fern gametophytes. Syst. Bot. 15: 98–111.

Farrar, D. R. 1992. *Trichomanes intricatum*: The independent *Trichomanes* gametophyte in the eastern United States. Amer. Fern J. 82: 68–74.

Farrar, D. R. and G. P. Landry. 1987. *Vittaria graminifolia* in the United States, again. [Abstract.] Amer. J. Bot. 74: 709–710.

Farrar, D. R., J. C. Parks, and B. W. McAlpin. 1982. The fern genera *Vittaria* and *Trichomanes* in the northeastern United States. Rhodora 85: 83–92.

Farrar, D. R. and W. H. Wagner Jr. 1968. The gametophyte of *Trichomanes holopterum* Kunze. Bot. Gaz. 129: 210–219.

Farrnkräuter — See: G. Kunze 1840–1851.

Fassett, N. C. 1945. *Juniperus virginiana, J. horizontalis,* and *J. scopulorum*. V. Taxonomic treatment. Bull. Torrey Bot. Club 72: 480–482.

Feddes Repert. = Feddes Repertorium.

Feddes Repert. Spec. Nov. Regni Veg. = Feddes Repertorium Specierum Novarum Regni Vegetabilis.

Fée, A. L. A. 1844–1866. Mémoires sur les Familles des Fougères. 11 parts. Strasbourg and Paris. [Each part also with individual title. Some parts issued also or only in journals.] (Mém. Foug.)

Fern Bull. = Fern Bulletin; a Quarterly Devoted to Ferns.

Fern Gaz. = Fern Gazette; Journal of the British Pteridological Society.

Fernald, M. L. 1928. The American representatives of *Asplenium ruta-muraria*. Rhodora 30: 37–43.

Fernald, M. L. 1935. Critical plants of the upper Great Lakes region of Ontario and Michigan. Rhodora 37: 197–222.

Fernald, M. L. 1950. Gray's Manual of Botany, ed. 8. New York.

Fernald, M. L. 1950b. *Adiantum capillus-veneris* in the United States. Rhodora 52: 201–208.

Ferns Brit. For. — See: J. Smith 1866.

Ferns N. Amer. — See: D. C. Eaton [1877–]1879–1880.

Ferns Pacif. Coast — See: J. G. Lemmon 1882.

Ferns S. India — See: R. Beddome 1863[–1864].

Ferns S.E. States — See: J. K. Small 1938.

Ferns Trop. Florida — See: J. K. Small 1918.

Fernwort Pap. = Fernwort Papers.

Fiddlehead Forum = Fiddlehead Forum; Bulletin of the American Fern Society.

Fieldiana, Bot. = Fieldiana: Botany.

Fil. Brit. — See: J. Bolton [1785–]1790.

Fil. Eur. — See: C. Milde 1867.

Fil. Spec. — See: J. H. F. Link 1841.

Fish and Wildlife Service [U.S.D.I.]. 1986. Florida Torreya *(Torreya taxifolia)* Recovery Plan. Atlanta.

Fl. Aegypt.-Arab. — See: P. Forsskål 1775.

Fl. Amer. Sept. — See: F. Pursh 1814.

Fl. Atlant. — See: R. L. Desfontaines [1798–1799].

Fl. Berlin — See: A. G. Dietrich 1824.

Fl. Bor.-Amer. — See: W. J. Hooker [1829–]1833–1840; A. Michaux 1803.

Fl. Bras. — See: C. F. P. von Martius et al. 1840–1906.

Fl. Calif. — See: W. L. Jepson 1909–1943.

Fl. Carniol. — See: J. A. Scopoli 1760.

Fl. Carol. — See: T. Walter 1788.

Fl. Franç. — See: J. Lamarck 1778[1779].

Fl. Friburg. — See: F. Spenner 1825–1829.

Fl. Murmansk. Obl. = Flora Murmanskoi Oblasti.

Fl. South. U.S. — See: A. W. Chapman 1860.

Flav. Fragr. J. = Flavour and Fragrance Journal.

Flexner, S. B. and L. C. Hauck, eds. 1987. The Random House Dictionary of the English Language, ed. 2 unabridged. New York.

Flora = Flora; oder (allgemeine) botanische Zeitung. [Vols. 1–16, 1818–33, include "Beilage" and "Ergänzungsblätter"; vols. 17–25, 1834–42, include "Beiblatt" and "Intelligenzblatt."]

Florida Sci. = Florida Scientist.

Florin, R. 1948. On the morphology and relationship of the Taxaceae. Bot. Gaz. 110: 31–39.

Florin, R. 1956. Nomenclatural notes on genera of living gymnosperms. Taxon 5: 188–192.

Flous, F. 1937. Revision du genre *Pseudotsuga*. Bull. Soc. Hist. Nat. Toulouse 71: 33–164.

Folia Geobot. Phytotax. = Folia Geobotanica et Phytotaxonomica.

Forbes, J. 1839. Pinetum Woburnense; or, a Catalogue of Coniferous Plants, in the Collection of the Duke of Bedford, at Woburn Abbey; Systematically Arranged. [London.] (Pinet. Woburn.)

Forest Sci. = Forest Science.

Forest. Irrig. = Forestry and Irrigation.

Forest. Techn. Pub. Canad. Forest. Serv. = Forestry Technical Publications, Canadian Forestry Service.

Forsskål, P. 1775. Flora Aegyptiaco-Arabica. Copenhagen. (Fl. Aegypt.-Arab.)

Fournier, E. 1872–1886. Mexicanas Plantas.... 2 vols. Paris. (Mexic. Pl.)

Fowells, H. A. 1965. Sylvics of Forest Trees of the United States. Washington. [Agric. Handb. 271.]

Franco, J. D. A. 1962. Taxonomy of the common juniper. Bol. Soc. Brot., ser. 2a, 36: 107–120.

Franklin, A. H. 1959. *Ginkgo biloba* L.: Historical summary and bibliography. Virginia J. Sci., n. s. 10: 131–176.

Franklin, J., J. Richardson, R. Brown, W. J. Hooker, C. F. Schwägrichen, et al. 1823. Narrative of a Journey to the Shores of the Polar Sea, in the Years 1819, 20, 21 and 22. London. [Richardson: Appendix VII. Botanical appendix, pp. [729]–778, incl. bryophytes by Schwägrichen, algae and lichens by Hooker. Brown: Addenda [to Appendix VII], pp. 779–783.] (Narr. Journey Polar Sea)

Fraser, H. S. and E. P. Swan. 1972. A chemical test to differentiate *Abies amabilis* from *A. lasiocarpa* wood. Bi-Monthly Res. Notes Forest. Serv. Canada 28: 32.

Fraser-Jenkins, C. R. 1989. A classification of the genus *Dryopteris* (Pteridophyta: Dryopteridaceae). Bull. Brit. Mus. (Nat. Hist.), Bot. 18: 323–477.

Frémont, J. C., J. Torrey, and J. Hall. 1843–1845. Report of the Exploring Expedition to the Rocky Mountains in the Year 1842, and to Oregon and North California in the Year 1843–44. 2 parts. Washington. [Parts paged consecutively.] (Rep. Exped. Rocky Mts.)

García de López, I. 1978. Revisión del género *Acrostichum* en la República Dominicana. Moscosoa 1: 64–70.

Gard. & Forest = Garden and Forest; a Journal of Horticulture, Landscape Art and Forestry.

Gard. Bull. Singapore = Gardens' Bulletin. Singapore.

Gard. Chron. = Gardener's Chronicle.

Gard. Dict. ed. 8 — See: P. Miller 1768.

Gard. Dict. Abr. ed. 4 — See: P. Miller 1754.

Gard. Monthly & Hort. = Gardener's Monthly and Horticulturist.

Gard. Monthly & Hort. Advertiser = Gardener's Monthly and Horticultural Advertiser.

Garden, J. 1957. A revision of the genus *Callitris* Vent. Contr. New South Wales Natl. Herb. 2(5): 363–392.

Garman, E. H. 1957. The Occurrence of Spruce in the Interior of British Columbia. Victoria. [B.C. Forest Serv., Techn. Publ. T 49.]

Gastony, G. J. 1977. Chromosomes of the independently reproducing Appalachian gametophyte: A new source of taxonomic evidence. Syst. Bot. 2: 43–48.

Gastony, G. J. 1980. The deletion of *Vittaria graminifolia* from the flora of Florida. Amer. Fern J. 70: 12–14.

Gastony, G. J. 1986. Electrophoretic evidence for the origin of a fern species by unreduced spores. Amer. J. Bot. 73: 1563–1569.

Gastony, G. J. 1988. The *Pellaea glabella* complex: Electrophoretic evidence for the derivations of the agamosporous taxa and a revised synonymy. Amer. Fern J. 78: 44–67.

Gastony, G. J. and L. D. Gottlieb. 1985. Genetic variation in the homosporous fern *Pellaea andromedifolia*. Amer. J. Bot. 72: 257–267.

Gastony, G. J. and C. H. Haufler. 1976. Chromosome numbers and apomixis in the fern genus *Bommeria*. Biotropica 8: 1–11.

Gastony, G. J. and M. D. Windham. 1989. Species concepts in pteridophytes: The treatment and definition of agamosporous species. Amer. Fern J. 79: 65–77.

Gaudichaud-Beaupré, C. 1826[–1830]. Voyage Autour du Monde... Exécuté sur les Corvettes de S.M. l'Uranie et la Physicienne... Publié... par M. Louis Freycinet. Botanique par M. Charles Gaudichaud.... 12 parts, atlas. Paris. [Botanical portion of larger work by H. L. C. Freycinet.] (Voy. Uranie)

Gen. Fil. — See: E. B. Copeland 1947; W. J. Hooker [1838–]1842; H. W. Schott 1834[–1836].

Gen. N. Amer. Pl. — See: T. Nuttall 1818.

Gen. Pl. — See: J. C. Schreber 1789–1791.

Gen. Pl. ed. 5 — See: C. Linnaeus 1754.

Ges. Naturf. Freunde Berlin Mag. Neuesten Entdeck. Gesammten Naturk. = Der Gesellschaft naturforschender Freunde zu Berlin Magazin für die neuesten Entdeckungen in der gesammten Naturkunde.

Giorn. Sci. Nat. Econ. Palermo = Giornale de Scienze Naturali ed Economiche di Palermo.

Gmelin, J. F. 1791[–1792]. Caroli à Linné... Systema Naturae per Regna Tria Naturae.... Tomus II. Editio Decima Tertia, Aucta, Reformata. 2 parts. Leipzig. (Syst. Nat.)

Godfrey, R. K., G. W. Reinert, and R. D. Houk. 1961. Observations on microsporocarpic material of *Azolla caroliniana*. Amer. Fern J. 51: 89–92.

Godfrey, R. K. and H. Kurz. 1962. The Florida torreya destined for extinction. Science 136: 900, 902.

Goggans, J. F. and C. E. Posey. 1968. Variation in seeds and ovulate cones of some species and varieties of *Cupressus*. Circ. Agric. Exp. Sta., Alabama 160: 1–23.

Gordon, A. G. 1976. The taxonomy and genetics of *Picea rubens* and its relationship to *Picea mariana*. Canad. J. Bot. 54: 781–813.

Gordon, G. 1862. A Supplement to Gordon's Pinetum.... London. (Pinetum Suppl.)

Gordon, G. and R. Glendenning. 1858. The Pinetum: Being a Synopsis of All the Coniferous Plants at Present Known.... London. (Pinetum)

Gordon, J. E. 1981. *Arachniodes simplicior* new to South Carolina and the United States. Amer. Fern J. 71: 65–68.

Gött. Gel. Anz. = Göttingische gelehrte Anzeigen (unter der Aufsicht der Königl. Gesellschaft der Wissenschaften).

Gray, A. 1848. A Manual of the Botany of the Northern United States.... Boston, Cambridge, and London. (Manual)

Gray, A. 1856. A Manual of the Botany of the Northern United States..., ed. 2. New York. (Manual ed. 2)

Gray, A. 1867. A Manual of the Botany of the Northern United States..., ed. 5. New York and Chicago. [Pteridophytes by D. C. Eaton.] (Manual ed. 5)

Gray, S. F. 1821. A Natural Arrangement of British Plants .... 2 vols. London. (Nat. Arr. Brit. Pl.)

Griffin, J. R. 1964. Cone morphology in *Pinus sabiniana*. J. Arnold Arbor. 45: 260–273.

Hall, J. W. 1974. Cretaceous Salviniaceae. Ann. Missouri Bot. Gard. 61: 354–367.

Hall, M. T. 1952. Variation and hybridization in *Juniperus*. Ann. Missouri Bot. Gard. 39: 1–64.

Haller, J. R. 1986. Taxonomy and relationships of the mainland and island populations of *Pinus torreyana* (Pinaceae). Syst. Bot. 11: 39–50.

Hamrick, J. L. and W. J. Libby. 1972. Variation and selection in western U. S. montane species. I. White fir. Silvae Genet. 21: 29–35.

Handb. Fern-allies — See: J. G. Baker 1887.

Handb. Skand. Fl. ed. 7 — See: C. J. Hartman 1858.

Handbuch — See: J. H. F. Link 1829–1833.

Hannover. Mag. = Hannoverisches Magazin worin kleine Abhandlungen...gesamlet (gesammelt) und aufbewahret sind.

Hanson, M. G. and R. P. Hanson. 1979. Northernmost station for *Asplenium pinnatifidum*. Trans. Wisconsin Acad. Sci. 67: 165–170.

Hardin, J. W. 1971. Studies of the southeastern United States flora II. The gymnosperms. J. Elisha Mitchell Sci. Soc. 87: 43–50.

Harriman Alaska Exped. — See: C. H. Merriam, ed. 1901–1914.

Harris, A. S. 1965. Subalpine fir [*Abies lasiocarpa* (Shook.) Nutt.] on Harris Ridge near Hollis, Prince of Wales Island, Alaska. NorthW. Sci. 39: 123–128.

Hartman, C. J. 1858. Handbok i Skandinaviens Flora, Innefettande Sveriges och Norriges Vexter, till och med Mossorna. Stockholm. (Handb. Skand. Fl. ed. 7)

Haufler, C. H. 1979. A biosystematic revision of *Bommeria*. J. Arnold Arbor. 60: 445–476.

Haufler, C. H. 1985. Pteridophyte evolutionary biology: The electrophoretic approach. Proc. Roy. Soc. Edinburgh, B 86: 315–323.

Haufler, C. H. and Wang Z. R. 1991. Chromosomal analyses and the origin of allopolyploid *Polypodium virginianum*. Amer. J. Bot. 78: 624–629.

Haufler, C. H. and M. D. Windham. 1991. New species of North American *Cystopteris* and *Polypodium*, with comments on their reticulate relationships. Amer. Fern J. 81: 7–23.

Haufler, C. H., M. D. Windham, D. M. Britton, and S. J. Robinson. 1985. Triploidy and its evolutionary significance in *Cystopteris protrusa*. Canad. J. Bot. 63: 1855–1863.

Haufler, C. H., M. D. Windham, and T. A. Ranker. 1990. Biosystematic analysis of the *Cystopteris tennesseensis* (Dryopteridaceae) complex. Ann. Missouri Bot. Gard. 77: 314–329.

Hauke, R. L. 1963. A taxonomic monograph of *Equisetum* subgenus *Hippochaete*. Beih. Nova Hedwigia 8: 1–123.

Hauke, R. L. 1966. A systematic study of *Equisetum arvense*. Nova Hedwigia 13: 81–109.

Hauke, R. L. 1978. A taxonomic monograph of *Equisetum* subgenus *Equisetum*. Nova Hedwigia 30: 385–455.

Hauke, R. L. 1979. *Equisetum ramosissimum* in North America. Amer. Fern J. 69: 1–5.

Heller, A. A. 1900. Catalogue of North American Plants North of Mexico, Exclusive of the Lower Cryptogams, ed. 2. [Lancaster, Pa.] (Cat. N. Amer. Pl. ed. 2)

Heller, A. A. 1909–1914. Catalogue of North American Plants North of Mexico, Exclusive of the Lower Cryptogams, ed. 3. [Lancaster, Pa.] (Cat. N. Amer. Pl. ed. 3)

Heusser, C. J. 1954. Alpine fir at Taku glacier, Alaska, with notes on its post glacial migration to the territory. Bull. Torrey Bot. Club 81: 83–86.

Hevly, R. H. 1965. Studies of the sinuous cloak fern (*Notholaena sinuata*) complex. J. Arizona Acad. Sci. 3: 205–208.

Hewitson, W. 1962. Comparative morphology of the Osmundaceae. Ann. Missouri Bot. Gard. 49: 57–93.

Heywood, V. H., ed. 1968. Modern Methods in Plant Taxonomy. London and New York.

Hickey, R. J. 1977. The *Lycopodium obscurum* complex in North America. Amer. Fern J. 67: 45–49.

Hickey, R. J. 1986. *Isoëtes* megaspore surface morphology: Nomenclature, variation, and systematic importance. Amer. Fern J. 76: 1–16.

Hickey, R. J., W. C. Taylor, and N. T. Luebke. 1989. The species concept in pteridophyta with special reference to *Isoëtes*. Amer. Fern J. 79: 78–89.

Hickok, L. G. 1977. Cytological relationships between three diploid species of the fern genus *Ceratopteris* Brongn. Canad. J. Bot. 55: 1660–1667.

Hickok, L. G. and E. J. Klekowski Jr. 1974. Inchoate speciation in *Ceratopteris*: An analysis of the synthesized hybrid *C. richardii* × *C. pteridoides*. Evolution 28: 439–446.

Hill, L. V. and B. Gopal. 1967. *Azolla primaeva* and its phylogenetic significance. Canad. J. Bot. 45: 1179–1191.

Hill, R. W. and W. H. Wagner Jr. 1974. Seasonality and spore type of pteridophytes in Michigan. Michigan Bot. 13: 40–44.

Hist. Fil. — See: J. Smith 1875.

Hist. Nat. Vég. — See: E. Spach 1834–1838.

Hist. Pl. Dauphiné — See: J. Lamarck and C. de Mirbel 1803[1802?]; D. Villars 1786–1789.

Hitchcock, C. L., A. Cronquist, M. Ownbey, and J. W. Thompson. 1955–1969. Vascular Plants of the Pacific Northwest. 5 vols. Seattle.

Hoffmann, G. F. [1791–1804.] Deutschland Flora oder botanisches Taschenbuch.... 4 vols. Erlangen. (Deutschl. Fl.)

Höh. Sporenpfl. Deutschl. — See: C. Milde 1865.

Holt, P. C., ed. 1971. The Distributional History of the Biota of the Southern Appalachians. Part 2. Flora. Blacksburg, Va. [Virginia Polytechnic Inst. and State Univ., Res. Div. Monogr. 2.]

Holttum, R. E. 1940. New species of *Lomariopsis*. Bull. Misc. Inform. Kew 1939: 613–628.

Holttum, R. E. 1947. A revised classification of the leptosporangiate ferns. J. Linn. Soc., Bot. 51: 123–158.

Holttum, R. E. 1957. Morphology, growth habit, and classification in the family Gleicheniaceae. Phytomorphology 7: 168–184.

Holttum, R. E. 1969. Studies in the family Thelypteridaceae. The genera *Phegopteris*, *Pseudophegopteris*, and *Macrothelypteris*. Blumea 17: 5–32.

Holttum, R. E. 1971. Studies in the family Thelypteridaceae III. A new system of genera in the Old World. Blumea 19: 17–52.

Holttum, R. E. 1976. Some new names in Thelypteridaceae, with comments on cytological reports relating to this family. Webbia 30: 191–195.

Holttum, R. E. 1981. The genus *Oreopteris* (Thelypteridaceae). Kew Bull. 36: 223–226.

Holttum, R. E. 1982. Thelypteridaceae. In: C. G. G. J. van Steenis and R. E. Holttum, eds. 1959–1982. Flora Malesiana, Being an Illustrated Systematic Account of the Malesian Flora. Series II. Pteridophyta. The Hague, Boston, and London. Vol. 1, part 5.

Holub, J. 1975. *Diphasiastrum*, a new genus in Lycopodiaceae. Preslia 14: 97–100.

Hooker, W. J. [1822–]1823–1827. Exotic Flora, Containing Figures and Descriptions of New, Rare, or Otherwise Interesting Exotic Plants.... 3 vols. in 38 fasc. Edinburgh. (Exot. Fl.)

Hooker, W. J. [1829–]1833–1840. Flora Boreali-Americana; or, the Botany of the Northern Parts of British America.... 2 vols. in 12 parts. London, Paris, and Strasbourg. (Fl. Bor.-Amer.)

Hooker, W. J. [1838–]1842. Genera Filicum; or Illustrations of the Ferns, and Other Allied Genera.... 12 parts. London. [120 plates with accompanying letterpress.] (Gen. Fil.)

Hooker, W. J. [1844–]1846–1864. Species Filicum.... 5 vols. in 20 parts. London. [Parts numbered consecutively.] (Sp. Fil.)

Hooker, W. J. and J. G. Baker. [1865–]1868. Synopsis Filicum; or, a Synopsis of All Known Ferns.... 10 parts. London. [Parts paged consecutively.] (Syn. Fil.)

Hooker, W. J. and J. G. Baker. 1874. Synopsis Filicum; or, a

Synopsis of All Known Ferns..., ed. 2. London. (Syn. Fil. ed. 2)

Hooker, W. J. and R. K. Greville. [1827–]1831[–1832]. Icones Filicum.... Figures and Descriptions of Ferns.... 2 vols. in 12 parts. London and Strasbourg. (Icon. Filic.)

Hooker, W. J., J. D. Hooker, D. Oliver, W. T. Thiselton-Dyer, D. Prain, A. W. Hill, E. J. Salisbury, G. Taylor, and J. P. M. Brenan, eds. [1836–]1837–1975 +. Icones Plantarum; or Figures with Brief Descriptive Characters and Remarks, of New or Rare Plants.... 38 + vols. in 5 series. London, Oxford, and Kew. (Icon. Pl.)

Hoppe, D. H. 1790–1849. Botanisches Taschenbuch.... 23 vols. Regensburg, Nuremburg, and Altdorf. [Volumes numbered by year, 1790–1811, and 1849, the latter edited by A. E. Fürnrohr.] (Bot. Taschenb.)

Horner, H. T. Jr. and H. J. Arnott. 1963. Sporangial arrangement in North American species of *Selaginella*. Bot. Gaz. 124: 371–383.

Hort. Berol. — See: J. H. F. Link 1827–1833.

Hort. Kew. — See: W. Aiton 1789.

Hort. Reg. Monac. — See: C. F. P. von Martius and F. Schrank 1829.

Horton, K. W. 1959. Characteristics of Subalpine Spruce in Alberta. Ottawa. [Forest Res. Div., Techn. Note 76.]

Hosie, R. C. 1969. Native Trees of Canada, ed. 7. Ottawa.

Hunt, R. S. and E. von Rudloff. 1974. Chemosystematic studies in the genus *Abies*. I. Leaf and twig oil analysis of alpine and balsam firs. Canad. J. Bot. 52: 477–487.

Hunt, R. S. and E. von Rudloff. 1979. Chemosystematic studies in the genus *Abies*. IV. Introgression in *Abies lasiocarpa* and *Abies bifolia*. Taxon 28: 297–305.

Icon. — See: A. J. Cavanilles 1791–1801.

Icon. Filic. — See: W. J. Hooker and R. K. Greville [1827–]1831[–1832].

Icon. Pl. — See: W. J. Hooker et al. [1836–]1837–1975 +.

Icon. Pl. ed. Keller — See: C. C. Schmidel 1762–1771.

Ill. Fl. Pacific States — See: L. Abrams and R. S. Ferris 1923–1960.

Index Fil. — See: T. Moore 1857[–1862].

Index N. Amer. Ferns — See: M. Broun 1938.

Index Pl. Jap. — See: J. Matsumura 1904–1912.

Index Seminum (Berlin), App. = Index Seminum (Berlin), Appendix.

Intermount. Fl. — See: A. Cronquist et al. 1972 +.

International Rice Research Institute. 1987. *Azolla* Utilization. Proceedings of the Workshop on *Azolla* Use. Fuzhou, Fujian, China, 31 March–5 April 1985. Los Banos, Laguna, Philippines.

Iowa State Coll. J. Sci. = Iowa State College Journal of Science.

Iwatsuki, K. 1964. An American species of *Stegnogramma*. Amer. Fern J. 54: 141–153.

J. Amer. Pharm. Assoc. = Journal of the American Pharmaceutical Association.

J. Arizona Acad. Sci. = Journal of the Arizona Academy of Science.

J. Arnold Arbor. = Journal of the Arnold Arboretum.

J. Bot. = Journal of Botany, British and Foreign.

J. Bot. (Hooker) = Journal of Botany, (Being a Second Series of the Botanical Miscellany), Containing Figures and Descriptions....

J. Bot. (Schrader) = Journal für die Botanik. [Edited by H. A. Schrader.] [Volumation indicated by nominal year date and volume number for that year (1 or 2); e.g. 1800(2).]

J. Elisha Mitchell Sci. Soc. = Journal of the Elisha Mitchell Scientific Society.

J. Fac. Sci. Univ. Tokyo, Sect. 3, Bot. = Journal of the Faculty of Science, University of Tokyo. Section III. Botany.

J. Forest. (Washington) = Journal of Forestry.

J. Hort. Soc. London = Journal of the Horticultural Society of London.

J. Jap. Bot. = Journal of Japanese Botany.

J. Linn. Soc., Bot. = Journal of the Linnean Society. Botany.

J. Tennessee Acad. Sci. = Journal of the Tennessee Academy of Science.

J. Wash. Acad. Sci. = Journal of the Washington Academy of Sciences.

Jacobs, B. F., C. R. Werth, and S. I. Guttman. 1984. Genetic relationships in *Abies* (fir) of eastern United States: An electrophoretic study. Canad. J. Bot. 62: 609–616.

Jacquin, N. J. 1762. Enumeratio Stirpium Plerarumque, Quae Sponte Crescunt in Agro Vindobonensi, Montibusque Confinibus. Vienna. (Enum. Stirp. Vindob.)

Jahrb. Hamburg. Wiss. Anst. Beih. = Jahrbuch der Hamburgischen wissenschaftlichen Anstalten. Beihefte.

Jahrb. Königl. Bot. Gart. Berlin = Jahrbuch des Königlichen botanischen Gartens und des botanischen Museums zu Berlin.

James, E. 1823. Account of an Expedition from Pittsburgh to the Rocky Mountains, Performed in the Years 1819 and '20... under the Command of Major Stephen H. Long. 2 vols. + atlas. Philadelphia. (Account Exped. Pittsburgh)

James, R. L. 1959. Carolina hemlock—wild and cultivated. Castanea 24: 112–134.

Jap. J. Bot. = Japanese Journal of Botany.

Jepson, W. L. 1909–1943. A Flora of California.... 3 vols. in 12 parts. San Francisco etc. [Pagination and part numbering sequential within each vol.; vol. 1 page sequence independent of part number sequence (chronological); part 8 of vol. 1 (pp. 1–32, 579–index) never published.] (Fl. Calif.)

Jermy, A. C. 1986. Subgeneric names in *Selaginella*. Fern Gaz. 13: 117–118.

Jermy, A. C. 1990. Isoëtaceae. In: K. Kubitzki et al., eds. 1990 +. The Families and Genera of Vascular Plants. 1 + vol. Berlin etc. Vol. 1, pp. 26–31.

Jermy, A. C. 1990b. Selaginellaceae. In: K. Kubitzki et al., eds. 1990 +. The Families and Genera of Vascular Plants. 1 + vol. Berlin etc. Vol. 1, pp. 39–45.

Jermy, A. C., J. A. Crabbe, and B. A. Thomas, eds. 1973.

The Phylogeny and Classification of the Ferns. London. [Bot. J. Linn. Soc., Suppl. 1.]

Jermy, A. C. and L. Harper. 1971. Spore morphology of the *Cystopteris fragilis* complex. Fern Gaz. 10: 211–213.

Johnson, D. M. 1985. New records for longevity of *Marsilea* sporocarps. Amer. Fern J. 75: 30–31.

Johnson, D. M. 1986. Systematics of the New World species of *Marsilea* (Marsileaceae). Syst. Bot. Monogr. 11: 1–87.

Johnson, D. M. 1986b. Trophopods in North American species of *Athyrium* (Aspleniaceae). Syst. Bot. 11: 26–31.

Johnson, D. M. 1988. Proposal to conserve *Marsilea* L. (Pteridophyta: Marsileaceae) with *Marsilea quadrifolia* as *typ. conserv.* Taxon 37: 483–486.

Johnson, L. A. S. 1959. The families of cycads and the Zamiaceae of Australia. Proc. Linn. Soc. New South Wales 84: 64–117.

Jones, I. and E. V. Lynn. 1933. Differences in species of *Taxus*. J. Amer. Pharm. Assoc. 22: 528–531.

Kalmia = Kalmia; Botanic Journal.

Karrfalt, E. E. 1981. The comparative and developmental morphology of the root system of *Selaginella selaginoides* (L.) Link. Amer. J. Bot. 68: 224–253.

Karsten, H. 1880–1883. Deutsche Flora. Pharmaceutisch-medicinische Botanik.... 13 Lieferungen. Berlin. (Deut. Fl.)

Kato, M. 1977. Classification of *Athyrium* and allied genera of Japan. Bot. Mag. (Tokyo) 90: 23–40.

Kato, M. 1984. A taxonomic study of the athyrioid fern genus *Deparia* with main reference to the Pacific species. J. Fac. Sci. Univ. Tokyo, Sect. 3, Bot. 13: 375–430.

Kato, M. and D. Darnaedi. 1988. Taxonomic and phytogeographic relationships of *Diplazium flavoviride, D. pycnocarpon,* and *Diplaziopsis*. Amer. Fern J. 78: 77–85.

Kaulfuss, G. F. 1824. Enumeratio Filicum Quas in Itinere Circa Terram Legit Cl. Adalbertus de Chamisso.... Leipzig. (Enum. Filic.)

Keen, R. A. 1956. A Study of the Genus *Taxus*. Ph.D. thesis. Ohio State University.

Keen, R. A. and L. C. Chadwick. 1955. Sex reversal in *Taxus*. Amer. Nurseryman 100(6): 13–14.

Kennedy, R. W., C. B. R. Sastry, G. M. Barton, and E. L. Ellis. 1968. Crystals in the wood of the genus *Abies* indigenous to Canada and the United States. Canad. J. Bot. 46: 1221–1228.

Kew Bull. = Kew Bulletin.

Khoshoo, T. N. 1961. Chromosome numbers in gymnosperms. Silvae Genet. 10: 1–9.

Kingsbury, J. M. 1964. Poisonous Plants of the United States and Canada. Englewood Cliffs.

Knobloch, I. W. 1967. Chromosome numbers in *Cheilanthes, Notholaena, Llavea,* and *Polypodium*. Amer. J. Bot. 54: 461–464.

Knobloch, I. W., W. Tai, and T. A. Ninan. 1973. The cytology of some species of the genus *Notholaena*. Amer. J. Bot. 60: 92–95.

Knudsen, G. M. 1968. Chemotaxonomic Investigation of Hybridization between *Larix occidentalis* and *Larix lyallii*. M.Sc.F. thesis. University of Montana.

Koch, K. H. E. 1869–1873. Dendrologie. Bäume, Sträucher und Halbsträucher, welche in Mittel- und Nord-Europa im Freien kultivirt werden. 2 vols. in 3. Erlangen. (Dendrologie)

Koch, W. D. J. 1846–1847. Synopsis der deutschen und schweizer Flora..., ed. 2. 7 parts. Leipzig. (Syn. Deut. Schweiz. Fl. ed. 2)

Koller, A. L. and S. E. Scheckler. 1986. Variation in microsporangia and micospore dispersal in *Selaginella*. Amer. J. Bot. 73: 1274–1288.

Kongel. Danske Vidensk. Selsk. Skr., Naturvidensk. Math. Afd. = Kongelige Danske Videnskabernes Selskabs Skrifter. Naturvidenskabelig og Mathematisk Afdeling.

Kott, L. S. and D. M. Britton. 1983. Spore morphology and taxonomy of *Isoëtes* in northeastern North America. Canad. J. Bot. 61: 3140–3163.

Krajina, V. J., K. Klinka, and J. Worrall. 1982. Distribution and Ecological Characteristics of Trees and Shrubs of British Columbia. Vancouver.

Kral, R., ed. 1983. A Report on Some Rare, Threatened, or Endangered Forest-related Vascular Plants of the South. Washington. [U.S.D.A., Techn. Publ. R8-TP 2.]

Kral, R. 1983b. Cycadaceae: *Zamia integrifolia* Ait. In: R. Kral, ed. 1983. A Report on Some Rare, Threatened, or Endangered Forest-related Vascular Plants of the South. Washington. Pp. 20–23.

Kramer, K. U. 1957. A revision of the genus *Lindsaea* in the New World with notes on allied genera. Acta Bot. Neerl. 6: 97–290.

Kramer, K. U. 1990. *Pteris*. In: K. Kubitzki et al., eds. 1990 +. The Families and Genera of Vascular Plants. 1 + vol. Berlin etc. Vol. 1, pp. 250–252.

Kramer, K. U. and R. Viane. 1990. Aspleniaceae. In: K. Kubitzki et al., eds. 1990 +. The Families and Genera of Vascular Plants. 1 + vol. Berlin etc. Vol. 1, pp. 52–56.

Krüssmann, G. 1972. Handbuch der Nadelgehölze. Berlin.

Kubitzki, K. 1990. Ephedraceae. In: K. Kubitzki et al., eds. 1990 +. The Families and Genera of Vascular Plants. 1 + vol. Berlin etc. Vol. 1, pp. 379–382.

Kubitzki, K., K. U. Kramer, and P. S. Green, eds. 1990 +. The Families and Genera of Vascular Plants. 1 + vol. Berlin etc.

Kuntze, O. 1891–1898. Revisio Generum Plantarum Vascularium Omnium atque Cellularium Multarum.... 3 vols. Leipzig etc.

Kunze, G. 1837. Analecta Pteridographica seu Descriptio et Illustratio Filicum aut Novarum, aut Minus Cognitarum. Leipzig. (Analecta Pteridogr.)

Kunze, G. 1840–1851. Die Farnkräuter in kolorirten Abbildungen naturgetreu erläutert und beschrieben.... Schkuhr's Farnkräuter, Supplement. 2 vols. in 14 fasc. Leipzig. (Farrnkräuter)

Kunze, R. 1834. [Description.] Linnaea 9: 92.

Kurz, H. 1939. *Torreya* west of the Appalachicola River. Proc. Florida Acad. Sci. 3: 66–77.

Kurz, H. and R. K. Godfrey. 1962. Trees of Northern Florida. Gainesville.

Labouriau, L. G. 1958. Studies on the initiation of sporangia in ferns. Arq. Mus. Nac. Rio de Janeiro 46: 119–202.

Lakela, O. and R. W. Long. 1976. Ferns of Florida. Miami.

Lamarck, J. 1778[1779]. Flore Françoise ou Description Succincte de Toutes les Plantes Qui Croissent Naturellement en France.... 3 vols. Paris. (Fl. Franç.)

Lamarck, J. and C. de Mirbel. 1803[1802?]. Histoire Naturelle des Végétaux, Classés par Familles.... 15 vols. Paris. (Hist. Nat. Vég.)

Lamarck, J., J. Poiret, A. P. de Candolle, L. Desrousseaux, and M. Savigny. 1783–1817. Encyclopédie Méthodique. Botanique.... 13 vols. Paris and Liège. [Vols. 1–8, suppls. 1–5.] (Encycl.)

Lambert, A. B. 1803–1824. A Description of the Genus Pinus.... 2 vols. London. (Descr. Pinus)

Lambert, A. B. 1832. A Description of the Genus Pinus.... 2 vols. London. (Descr. Pinus [ed. 3])

Landry, G. P. 1980. The Ecology and Variation of Zamia pumila L. in Florida. M.S. thesis. Louisiana State University.

Landry, G. P. and W. D. Reese. 1991. Ctenitis submarginalis (Langsd. & Fisch.) Copel. new to Louisiana: First record in the U.S. outside of Florida. Amer. Fern J. 81: 105–106.

Lang, F. A. 1971. The Polypodium vulgare complex in the Pacific Northwest. Madroño 21: 235–254.

Langsdorff, G. H. and F. E. L. von Fischer. 1810[–1818]. Plantes Recueillies Pendant le Voyage des Russes Autour du Monde. Première Partie. Icones Filicum. 2 parts. Tübingen. (Pl. Voy. Russes Monde)

Lanner, R. M. 1982. Adaptations of whitebark pine for seed dispersal by Clark's nutcracker. Canad. J. Forest Res. 12: 391–402.

LaRoi, G. H. and J. R. Dugle. 1968. A systematic and gen-ecological study of Picea glauca and P. engelmannii, using paper chromatography of needle extracts. Canad. J. Bot. 46: 649–687.

Lawrence, L., R. Bartschot, E. Zavarin, and J. R. Griffin. 1975. Natural hybridization of Cupressus sargentii and C. macnabiana and the composition of the derived essential oils. Biochem. Syst. & Ecol. 2: 113–119.

Lawrence, G. H. M. 1951. Taxonomy of Vascular Plants. New York.

Lawson, P. and C. Lawson. 1836. The Agriculturist's Manual; Being a Familiar Description of the Agricultural Plants Cultivated in Europe... and Forming a Report of Lawson's Agricultural Museum in Edinburgh. Edinburgh, London, and Dublin. (Agric. Man.)

Leafl. W. Bot. = Leaflets of Western Botany.

Lee, D. W. 1977. On iridescent plants. Gard. Bull. Singapore 30: 21–29.

Lellinger, D. B. 1968. A note on Aspidotis. Amer. Fern J. 58: 140–141.

Lellinger, D. B. 1981. Notes on North American ferns. Amer. Fern J. 71: 90–94.

Lellinger, D. B. 1985. A Field Manual of the Ferns & Fern-allies of the United States & Canada. Washington.

Lellinger, D. B. 1987. Nomenclatural notes on some ferns of Costa Rica, Panama, and Columbia. III. Amer. Fern J. 77: 101–102.

Lellinger, D. B. 1988. Some new species of Campyloneurum and a provisional key to the genus. Amer. Fern J. 78: 14–34.

Lemmon, J. G. 1882. Ferns of the Pacific Coast, Including Arizona. San Francisco. (Ferns Pacif. Coast)

Leonard, S. W. 1972. The distribution of Thelypteris torresiana in the southeastern United States. Amer. Fern J. 62: 97–99.

Lester, D. T. 1968. Variation in cone morphology of balsam fir, Abies balsamea. Rhodora 70: 83–94.

Li, H. L. 1962. A new species of Chamaecyparis. Morris Arbor. Bull. 13: 43–46.

Liew, F. S. 1972. Numerical taxonomic studies on North American lady ferns and their allies. Taiwania 17: 190–221.

Link, J. H. F. 1827–1833. Hortus Regius Botanicus Berolinensis.... 2 vols. Berlin. (Hort. Berol.)

Link, J. H. F. 1829–1833. Handbuch zur Erkennung der nutzbarsten und am häufigsten vorkommenden Gewächse.... 3 vols. Berlin. (Handbuch)

Link, J. H. F. 1841. Filicum Species in Horto Regio Botanico Berolinensi Cultae.... Berlin. (Fil. Spec.)

Linnaea = Linnaea. Ein Journal für die Botanik in ihrem ganzen Umfange.

Linnaeus, C. 1753. Species Plantarum.... 2 vols. Stockholm. (Sp. Pl.)

Linnaeus, C. 1754. Genera Plantarum, ed. 5. Stockholm. (Gen. Pl. ed. 5)

Linnaeus, C. 1758[–1759]. Systema Naturae per Regna Tria Naturae..., ed. 10. 2 vols. Stockholm. (Syst. Nat. ed. 10)

Linnaeus, C. 1762–1763. Species Plantarum..., ed. 2. 2 vols. Stockholm. (Sp. Pl. ed. 2)

Linnaeus, C. 1767[–1771]. Mantissa Plantarum. 2 parts. Stockholm. [Mantissa [1] and Mantissa [2] Altera paged consecutively.] (Mant. Pl.)

Linnaeus, C. f. 1781[1782]. Supplementum Plantarum Systematis Vegetabilium Editionis Decimae Tertiae, Generum Plantarum Editionis Sextae, et Specierum Plantarum Editionis Secundae. Braunschweig. (Suppl. Pl.)

Little, E. L. Jr. 1952. The genus Pseudotsuga (Douglas-fir) in North America. Leafl. W. Bot. 6: 181–198.

Little, E. L. Jr. 1953. A natural hybrid in spruce. J. Forest. (Washington) 51: 745–747.

Little, E. L. Jr. 1966. Varietal transfers in Cupressus and Chamaecyparis. Madroño 18: 161–167.

Little, E. L. Jr. 1970. Names of New World cypresses (Cupressus). Phytologia 20: 429–445.

Little, E. L. Jr. 1971. Atlas of United States Trees. I. Conifers and Important Hardwoods. Washington. [U.S.D.A., Misc. Publ. 1146.]

Little, E. L. Jr. 1979. Checklist of United States Trees (Native and Naturalized). Washington. [Agric. Handb. 541.]

Little, E. L. Jr. and W. B. Critchfield. 1969. Subdivisions of the genus *Pinus* (pines). Washington. [U.S.D.A., Misc. Publ. 1144.]

Little, E. L. Jr. and S. S. Pauley. 1958. A natural hybrid between black and white spruce in Minnesota. Amer. Midl. Naturalist 60: 202–211.

Liu, T. S. 1971. A Monograph of the Genus *Abies*. Taipei.

Lloyd, R. M. 1971. Systematics of the onocleoid ferns. Univ. Calif. Publ. Bot. 61: 1–86, 5 plates.

Lloyd, R. M. 1974. Systematics of the genus *Ceratopteris* Brongn. (Parkeriaceae) II. Taxonomy. Brittonia 26(2): 139–160.

Lloyd, R. M. 1980. Reproductive biology and gametophyte morphology of New World populations of *Acrostichum aureum*. Amer. Fern J. 70: 99–110.

Lloyd, R. M., C. H. Haufler, and F. A. Lang. 1992. Clarifying the history of *Polypodium australe* Fée reported from San Clemente Island, California. Amer. Fern J. 82: 39–40.

Lloyd, R. M. and J. E. Hohn. 1969. Occurrence of the European *Polypodium australe* Fée on San Clemente Island, California. Amer. Fern J. 59: 56–60.

Lloyd, R. M. and E. J. Klekowski Jr. 1970. Spore germination and viability in Pteridophyta: Evolutionary significance of chlorophyllous spores. Biotropica 2: 129–137.

Lloyd, R. M. and F. A. Lang. 1964. The *Polypodium vulgare* complex in North America. Brit. Fern Gaz. 9: 168–177.

Long, G., ed. 1832–1858. The Penny Cyclopaedia of The Society for the Diffusion of Useful Knowledge. 27 vols. + 3 vols. suppl. London. (Penny Cycl.)

Loudon, J. C. [1835–]1838. Arboretum et Fruticetum Britannicum; or, the Trees and Shrubs of Britain, Native and Foreign.... 8 vols. London. (Arbor. Frutic. Brit.)

Löve, A., D. Löve, and B. M. Kapoor. 1971. Cytotaxonomy of a century of Rocky Mountain orophytes. Arctic Alpine Res. 3: 139–165.

Löve, A., D. Löve, and R. E. G. Pichi-Sermolli. 1977. Cytotaxonomical Atlas of the Pteridophyta. Vaduz.

Lumpkin, T. A. and D. L. Plucknett, eds. 1982. *Azolla* As a Green Manure: Use and Management in Crop Production. Boulder.

Lumpkin, T. A. and D. L. Plucknett. 1982b. Botany and ecology. In: T. A. Lumpkin and D. L. Plucknett, eds. 1982. *Azolla* As a Green Manure: Use and Management in Crop Production. Boulder. Pp. 15–38.

Madroño = Madroño; Journal of the California Botanical Society [from vol. 3: a West American Journal of Botany.]

Magill, R. E., ed. 1990. Glossarium polyglottum bryologiae, a multilingual glossary for bryology. Monogr. Syst. Bot. Missouri Bot. Gard. 33.

Man. Herbor. Suisse — See: J. P. de Clairville 1811.

Man. Pteridol. — See: F. Verdoorn 1938.

Mant. Pl. — See: C. Linnaeus 1767[–1771].

Manton, I. 1950. Problems of Cytology and Evolution in the Pteridophyta. London.

Manton, I. 1951. Cytology of *Polypodium* in America. Nature 167: 37.

Manton, I. and M. G. Shivas. 1953. Two cytological forms of *Polypodium virginianum* in eastern North America. Nature 172: 410.

Manual — See: A. Gray 1848.

Manual ed. 2 — See: A. Gray 1856.

Manual ed. 5 — See: A. Gray 1867.

Manville, J. F. and A. S. Tracey. 1989. Chemical differences between alpine firs of British Columbia. Phytochemistry 28(10): 2681–2686.

Markgraf, F. 1926. Ephedraceae. In: H. G. A. Engler et al., eds. 1924+. Die natürlichen Pflanzenfamilien..., ed. 2. 26+ vols. Leipzig and Berlin. Vol. 13, pp. 409–419.

Marshall, H. 1785. Arbustrum Americanum: The American Grove.... Philadelphia. (Arbust. Amer.)

Martius, C. F. P. von, A. W. Eichler, and I. Urban, eds. 1840–1906. Flora Brasiliensis. 15 vols. in 40 parts, 130 fasc. Munich, Vienna, and Leipzig. [Volumes and parts numbered in systematic sequence, fascicles numbered independently in chronological sequence.] (Fl. Bras.)

Martius, C. F. P. von and F. Schrank. 1829. Hortus Regius Monacensis. Leipzig. (Hort. Reg. Monac.)

Mastrogiuseppe, R. J. and J. D. Mastrogiuseppe. 1980. A study of *Pinus balfouriana* Grev. & Balf. (Pinaceae). Syst. Bot. 5: 86–104.

Mathews, F. P. 1945. A comparison of the toxicity of *Notholaena sinuata* and *N. sinuata* var. *cochisensis*. Rhodora 47: 393–395.

Matsumura, J. 1904–1912. Index Plantarum Japonicarum sive Enumeratio Plantarum Omnium ex Insulis Kurile, Yezo, Nippon, Sikoku, Kiusiu, Liukiu et Formosa.... 2 vols. in 3. Tokyo. (Index Pl. Jap.)

Matzenko, A. E. 1968. Conspectus generis *Abies* Mill. Novosti Sist. Vyssh. Rast. 5: 9–12.

Maxon, W. R. 1909. Gleicheniaceae. In: N. L. Britton et al., eds. 1905–1972. North American Flora.... 42 vols. New York. Vol. 16, pp. 53–63.

Maxon, W. R. 1912. Notes on the North American species of *Phanerophlebia*. Bull. Torrey Bot. Club 39: 23–28.

Mayr, H. 1890[1889]. Die Waldungen von Nordamerika.... Munich. (Wald. Nordamer.)

Maze, J. and W. H. Parker. 1983. A study of population differentiation and variation in *Abies procera*. Canad. J. Bot. 61: 1094–1104.

McAlpin, B. W. 1971. *Ophioglossum* leaf sheaths: Development and morphological nature. Bull. Torrey Bot. Club 98: 194–199.

McMahan, L. R. 1989. Conservationists join forces to save Florida torreya. Center Pl. Conservation 4(1): 1, 8.

McNeill, J. and K. M. Pryer. 1985. The status and typification of *Phegopteris* and *Gymnocarpium*. Taxon 34: 136–143.

Mém. Acad. Imp. Sci. Saint-Pétersbourg, Sér. 6, Sci. Math., Seconde Pt. Sci. Nat. = Mémoires de l'Académie Impériale des Sciences de Saint-Pétersbourg. Sixième Série. Sciences Mathématiques, Physiques et Naturelles. Seconde Partie: Sciences Naturelles.

Mém. Acad. Imp. Sci. St.-Pétersbourg, Sér. 6, Sci. Math. = Mémoires de l'Académie Impériale des Sciences de St.-Pétersbourg. Sixième Série. Sciences Mathématiques, Physiques et Naturelles.

Mém. Acad. Roy. Sci. Belgique = Mémoires de l'Académie Royale des Sciences, Lettres et Beaux Arts de Belgique.

Mém. Acad. Roy. Sci. (Turin) = Mémoires de l'Académie Royale des Sciences.

Mém. Foug. — See: A. L. A. Fée 1844–1866.

Mém. Mus. Hist. Nat. = Mémoires du Muséum d'Histoire Naturelle.

Mem. New York Bot. Gard. = Memoirs of the New York Botanical Garden.

Mém. Soc. Linn. Paris = Mémoires de la Société Linnéenne de Paris, Précédés de Son Histoire.

Mém. Soc. Mus. Hist. Nat. Strasbourg = Mémoires de la Société du Muséum d'Histoire Naturelle de Strasbourg.

Mem. Torrey Bot. Club = Memoirs of the Torrey Botanical Club.

Mem. Tour N. Mexico — See: F. A. Wislizenus 1848.

Merriam, C. H., ed. 1901–1914. Harriman Alaska Expedition with Cooperation of Washington Academy of Sciences. 14 vols. New York. (Harriman Alaska Exped.)

Merriam-Webster. 1988. Webster's New Geographical Dictionary. Springfield, Mass.

Mesler, M. R. 1973. Sexual reproduction in *Ophioglossum crotalophoroides*. Amer. Fern J. 63: 28–33.

Mesler, M. R., R. D. Thomas, and J. G. Bruce. 1975. Mature gametophytes and young sporophytes of *Ophioglossum nudicaule*. Phytomorphology 25: 156–166.

Methodus — See: C. Moench 1794.

Mexic. Pl. — See: E. Fournier 1872–1886.

Meyer, C. A. von. 1846. Versuch einer Monographie der Gattung *Ephedra*. Mém. Acad. Imp. Sci. Saint-Pétersbourg, Sér. 6, Sci. Math., Seconde Pt. Sci. Nat. 5(2): 225–298.

Michaux, A. 1803. Flora Boreali-Americana.... 2 vols. Paris and Strasbourg. (Fl. Bor.-Amer.)

Michigan Bot. = Michigan Botanist.

Mickel, J. T. 1962. Monographic study of the fern genus *Anemia* subgenus *Coptophyllum*. Iowa State Coll. J. Sci. 36: 349–482.

Mickel, J. T. 1973. Position of and classification within the Dennstaedtiaceae. In: A. C. Jermy et al., eds. 1973. The Phylogeny and Classification of the Ferns. London. Pp. 135–144.

Mickel, J. T. 1976. Vegetative propagation in *Asplenium exiguum*. Amer. Fern J. 66: 81–82.

Mickel, J. T. 1979. How to Know the Ferns and Fern Allies. Dubuque.

Mickel, J. T. 1979b. The fern genus *Cheilanthes* in the continental United States. Phytologia 41: 431–437.

Mickel, J. T. 1981. The fern genus *Anemia* (Schizaeaceae) subgenus *Anemiorrhiza*. Brittonia 33: 413–429.

Mickel, J. T. 1982. The genus *Anemia* (Schizaeaceae) in Mexico. Brittonia 34: 388–413.

Mickel, J. T. and J. M. Beitel. 1988. Pteridophyte flora of Oaxaca, Mexico. Mem. New York Bot. Gard. 46: 1–568.

Milde, C. 1865. Die höheren Sporenpflanzen Deutschland's und der Schweiz.... Leipzig. (Höh. Sporenpfl. Deutschl.)

Milde, C. 1867. Filices Europae et Atlantidis, Asiae Minoris et Sibiriae.... Leipzig. (Fil. Eur.)

Miller, C. N. Jr. 1988. The origin of modern conifer families. In: C. B. Beck, ed. 1988. Origin and Evolution of Gymnosperms. New York. Pp. 448–486.

Miller, P. 1754. The Gardeners Dictionary.... Abridged..., ed. 4. 3 vols. London. (Gard. Dict. Abr. ed. 4)

Miller, P. 1768. The Gardeners Dictionary..., ed. 8. London. (Gard. Dict. ed. 8)

Mining Sci. Press = Mining and Scientific Press.

Minnich, R. A. 1982. *Pseudotsuga macrocarpa* in Baja California? Madroño 29: 22–31.

Mirov, N. T. 1967. The Genus *Pinus*. New York.

Mitt. Deutsch. Dendrol. Ges. = Mitteilungen der Deutschen dendrologischen Gesellschaft.

Moench, C. 1785. Verzeichniss ausländischer Bäume und Stauden des Lustschlosses Weissenstein bey Cassel. Frankfurt and Leipzig. (Verz. Ausländ. Bäume)

Moench, C. 1794. Methodus Plantas Horti Botanici et Agri Marburgensis.... Marburg. (Methodus)

Mokry, F., S. Rasbach, and T. Reichstein. 1986. *Asplenium adulterinum* Milde subsp. *presolensis* subsp. nova. Bot. Helv. 96: 7–18.

Monatsber. Königl. Preuss. Akad. Wiss. Berlin = Monatsberichte der Königlich preussischen Akademie der Wissenschaften zu Berlin.

Montgomery, J. D. 1982. *Dryopteris* in North America. Part II. The hybrids. Fiddlehead Forum 9(4): 23–30.

Montgomery, J. D. and E. M. Paulton. 1981. *Dryopteris* in North America. Fiddlehead Forum 8(4): 25–31.

Moore, A. W. 1969. *Azolla*: Biology and agronomic significance. Bot. Rev. (Lancaster) 35: 17–35.

Moore, T. 1857[–1862]. Index Filicum: A Synopsis, with Characters, of the Genera, and an Enumeration of the Species of Ferns.... 20 parts. London. (Index Fil.)

Moran, R. C. 1981. ×*Asplenosorus shawneensis*, a new natural fern hybrid between *Asplenium trichomanes* and *Camptosorus rhizophyllus*. Amer. Fern J. 71: 85–89.

Moran, R. C. 1982. The *Asplenium trichomanes* complex in the United States and adjacent Canada. Amer. Fern J. 72: 5–11.

Moran, R. C. 1982b. *Cystopteris* ×*illinoensis*: A new natural hybrid fern. Amer. Fern J. 72: 41–44.

Moran, R. C. 1983. *Cystopteris* ×*wagneri*: A new naturally occurring hybrid between *C. tennesseensis* and *C. tenuis*. Castanea 48: 224–229.

Moran, R. C. 1983b. *Cystopteris tenuis* (Michx.) Desv.: A poorly understood species. Castanea 48: 218–223.

Moran, R. C. 1987. Monograph of the neotropical fern genus *Polybotria* (Dryopteridaceae). Bull. Illinois Nat. Hist. Surv. 34: 1–138.

Morgenstern, E. K. and J. L. Farrar. 1964. Introgressive Hybridization in Red Spruce and Black Spruce. [Toronto.] [Univ. Toronto, Fac. Forest., Techn. Rep. 4.]

Morris Arbor. Bull. = Morris Arboretum Bulletin.

Morton, C. V. 1951. A fern new to the United States. Amer. Fern J. 41: 86–87.

Morton, C. V. 1957. Observations on cultivated ferns, I. Amer. Fern J. 47: 7–14.

Morton, C. V. 1966. The Mexican species of *Tectaria*. Amer. Fern J. 56: 120–137.

Morton, C. V. 1967. The fern herbarium of André Michaux. Amer. Fern J. 57: 166–182.

Morton, C. V. 1968. The genera, subgenera, and sections of the Hymenophyllaceae. Contr. U.S. Natl. Herb. 38: 153–214.

Morton, C. V. and R. K. Godfrey. 1958. *Diplazium japonicum* naturalized in Florida. Amer. Fern J. 48: 28–30.

Morzenti, V. M. 1966. Morphological and cytological data on southeastern United States species of the *Asplenium heterochroum–resiliens* complex. Amer. Fern J. 56: 167–177.

Morzenti, V. M. 1967. *Asplenium plenum*, a fern which suggests an unusual method of species formation. Amer. J. Bot. 54: 1061–1068.

Moscosoa = Moscosoa; Contribuciones Científicas del Jardin Botánico Nacional "Dr. Raphael M. Moscosa".

Moss, E. H. 1953. Forest communities in northwestern Alberta. Canad. J. Bot. 31: 212–252.

Moyroud, R. and C. E. Nauman. 1989. A new station for *Dicranopteris flexuosa* in Florida. Amer. Fern J. 79: 155.

Muhlenbergia = Muhlenbergia; a Journal of Botany.

Mulligan, G. A. and W. J. Cody. 1979. Chromosome numbers in Canadian *Phegopteris*. Canad. J. Bot. 57: 1815–1819.

Murray, A. 1849–1859. Botanical Expedition to Oregon.... 11 pamphlets. Edinburgh. (Bot. Exped. Oregon)

Murray, J. A. 1784. Caroli à Linné Equitis Systema Vegetabilium.... Editio Decima Quarta.... Göttingen. (Syst. Veg. ed. 14)

Mussayev, I. 1978. On geography and phylogeny of some representatives of the genus *Ephedra*. Bot. Zhurn. (Moscow & Leningrad) 63: 523–543. [In Russian.]

N. Amer. Sylv. — See: T. Nuttall 1842–1849.

Nakai, T. and M. Honda, eds. 1938–1951. Nova Flora Japonica.... 10 parts. Tokyo and Osaka. (Nov. Fl. Jap.)

Narr. Journey Polar Sea — See: J. Franklin 1823.

Nat. Arr. Brit. Pl. — See: S. F. Gray 1821.

Nat. Pflanzenfam. — See: H. G. A. Engler and K. Prantl 1887–1915.

Native Ferns ed. 6 — See: L. M. Underwood 1900.

Naturaleza (Mexico City) = Naturaleza. [Mexico, D.F.]

Nature = Nature; a Weekly Illustrated Journal of Science.

Nauman, C. E. 1981. The genus *Nephrolepis* in Florida. Amer. Fern J. 71: 35–40.

Nauman, C. E. 1985. A Systematic Revision of the Neotropical Species of *Nephrolepis* Schott. Ph.D. dissertation. University of Tennessee.

Nauman, C. E. 1987. Schizaeaceae in Florida. Sida 12: 69–74.

Nauman, C. E. and D. F. Austin. 1978. Spread of the exotic fern *Lygodium microphyllum* in Florida. Amer. Fern J. 68: 65–66.

Ned. Kruidk. Arch. = Nederlandsch Kruidkundig Archief. Verslagen en Mededelingen der Nederlandsche Botanische Vereeniging.

Neues J. Bot. = Neues Journal für die Botanik.

New Phytol. = New Phytologist; a British Botanical Journal.

New York J. Pharm. (1852–54) = New York Journal of Pharmacy.

New York State. 1840. Communication from the Governor, Transmitting Several Reports Relative to the Geological Survey of the State. Albany. [Assembly Doc. 50. Catalogue of plants by J. Torrey, pp. 111–197.] (Rep. Geol. Surv.)

New Zealand J. Bot. = New Zealand Journal of Botany.

Nomencl. Bot. ed. 2 — See: E. G. Steudel 1840–1841.

NorthW. Sci. = Northwest Science.

Notes Roy. Bot. Gard. Edinburgh = Notes from the Royal Botanic Garden, Edinburgh.

Nouv. Ann. Mus. Hist. Nat. = Nouvelles Annales du Muséum d'Histoire Naturelle.

Nouv. Mém. Acad. Roy. Sci. Bruxelles = Nouveaux Mémoires de l'Académie Royale des Sciences et Belles-lettres de Bruxelles.

Nov. Fl. Jap. — See: T. Nakai and M. Honda 1938–1951.

Nova Hedwigia = Nova Hedwigia. Zeitschrift für Kryptogamenkunde.

Novi Comment. Acad. Sci. Imp. Petrop. = Novi Commentarii Academiae Scientiarum Imperalis Petropolitanae.

Novon = Novon; a Journal for Botanical Nomenclature.

Novosti Sist. Vyssh. Rast. = Novosti Sistematiki Vysshikh Rastenii.

Numer. List — See: N. Wallich 1828[–1849].

Nuttall, T. 1818. The Genera of North American Plants, and Catalogue of the Species, to the Year 1817.... 2 vols. Philadelphia. (Gen. N. Amer. Pl.)

Nuttall, T. 1842–1849. The North American Sylva.... 3 vols. Philadelphia. (N. Amer. Sylv.)

Observ. Bot. — See: A. J. Retzius [1779]–1791.

Ohio Agric. Exp. Sta. Bull. = Ohio Agricultural Experiment Station Bulletin.

Øllgaard, B. 1987. A revised classification of the Lycopodiaceae s. lat. Opera Bot. 92: 153–178.

Øllgaard, B. 1989. Index of the Lycopodiaceae. Biol. Skr. 34: 1–135.

Øllgaard, B. 1990. Lycopodiaceae. In: K. Kubitzki et al., eds. 1990+. The Families and Genera of Vascular Plants. 1+ vol. Berlin etc. Vol. 1, pp. 31–39.

Opera Bot. = Opera Botanica a Societate Botanice Lundensi.

Owens, J. N. and M. Molder. 1975. Pollination, female gametophyte, and embryo and seed development in yellow cedar (*Chamaecyparis nootkatensis*). Canad. J. Bot. 53: 186–199.

Owens, J. N. and S. Simpson. 1986. Pollen from conifers native to British Columbia. Canad. J. Forest Res. 16: 955–967.

Pacif. Railr. Rep. 4(5) — See: J. Torrey 1857.

Page, C. N. 1976. The taxonomy and phytogeography of bracken—A review. Bot. J. Linn. Soc. 73: 1–34.

Page, C. N. 1989. Compression and slingshot megaspore ejection in *Selaginella selaginoides*—A new phenomenon in pteridophytes. Fern Gaz. 13: 267–275.

Page, C. N. 1990. Coniferophytina. In: K. Kubitzki et al., eds. 1990+. The Families and Genera of Vascular Plants. 1+ vol. Berlin etc. Vol. 1, pp. 282–361.

Palisot de Beauvois, A. 1805. Prodrome des Cinquième et Sixième Familles de l'Aethéogamie. Les Mousses. Les Lycopodes. Paris. (Prodr. Aethéogam.)

Palma-Otal, M., W. S. Moore, R. P. Adams, and G. R. Joswiak. 1983. Genetic and biographical analyses of natural hybridization between *Juniperus virginiana* L. and *J. horizontalis* Moench. Canad. J. Bot. 61: 2733–2746.

Pap. Michigan Acad. Sci. = Papers of the Michigan Academy of Sciences, Arts and Letters.

Paris, C. A. 1991. *Adiantum viridimontanum,* a new maidenhair fern in eastern North America. Rhodora 93: 105–122.

Paris, C. A. and M. D. Windham. 1988. A biosystematic investigation of the *Adiantum pedatum* complex in eastern North America. Syst. Bot. 13: 240–255.

Parker, E. L. 1963. The geographic overlap of noble fir and red fir. Forest Sci. 9: 207–216.

Parker, W. H., G. E. Bradfield, J. Maze, and Lin S. C. 1979. Analysis of variation in leaf and twig characters of *Abies lasiocarpa* and *A. amabilis* from north-coastal British Columbia. Canad. J. Bot. 57: 1354–1366.

Parker, W. H. and T. A. Dickinson. 1990. Range-wide morphological and anatomical variation in *Larix laricina.* Canad. J. Bot. 68: 832–840.

Parlatore, F. 1863. Coniferas Novas Nullas Descripsit.... Florence. (Conif. Nov.)

Parlatore, F. 1868. Gnetaceae. In: A. P. de Candolle and A. L. P. de Candolle, eds. 1823–1873. Prodromus Systematis Naturalis Regni Vegetabilis.... 17 vols. Paris etc. Vol. 16, part 2, pp. 352–359.

Peattie, D. C. 1953. A Natural History of Western Trees. Boston.

Peirce, J. S. 1937. Systematic anatomy of the woods of Cupressaceae. Trop. Woods 49: 5–21.

Penny Cycl. — See: G. Long 1832–1858.

Perkins, S. K., G. A. Peters, T. A. Lumpkin, and H. E. Calvert. 1985. Scanning electron microscopy of perine architecture as a taxonomic tool in the genus *Azolla* Lamarck. Scan. Electron Microscop. 4: 1719–1734.

Perring, F. H. and B. G. Gardener, eds. 1976. The biology of bracken. [Symposium.] Bot. J. Linn. Soc. 73.

Perry, J. P. Jr. 1991. The Pines of Mexico and Central America. Portland.

Petersen, R. L. and D. E. Fairbrothers. 1983. Flavonols of the fern genus *Dryopteris*: Systematic and morphological implications. Bot. Gaz. 144: 104–109.

Petrik-Ott, A. J. 1979. The Pteridophytes of Kansas, Nebraska, South Dakota, and North Dakota, U.S.A. Vaduz.

Petrik-Ott, A. J. and F. D. Ott. 1980. *Pilularia americana* new to Tennessee. Amer. Fern J. 70: 29–30.

Pfeiffer, N. E. 1922. Monograph of the Isoetaceae. Ann. Missouri Bot. Gard. 9: 79–233.

Phillips, T. L. and G. A. Leisman. 1966. *Paurodendron,* a rhizomorphic lycopod. Amer. J. Bot. 53: 1086–1100.

Phytographia — See: C. L. Willdenow 1794.

Phytologia = Phytologia; Designed to Expedite Botanical Publication.

Phytologia Mem. = Phytologia Memoirs.

Phytomorphology = Phytomorphology; an International Journal of Plant Morphology.

Pichi-Sermolli, R. E. G. 1971. Names and types of the genera of fern-allies. Webbia 26: 157–174.

Pichi-Sermolli, R. E. G. 1983. Fragmenta pteridologiae—VII. Webbia 37: 111–140.

Pichi-Sermolli, R. E. G. 1989. Again on the typification of the generic name *Notholaena* R. Brown. Webbia 43(2): 301–310.

Pilger, R. K. F. 1903. Taxaceae. In: H. G. A. Engler, ed. 1900–1953. Das Pflanzenreich.... 107 vols. Berlin. Vol. 18[IV,5], pp. 1–124.

Pilger, R. K. F. 1916. Die Taxales. Mitt. Deutsch. Dendrol. Ges. 25: 1–28.

Pilger, R. K. F. 1926. Coniferae. In: H. G. A. Engler et al., eds. 1924+. Die natürlichen Pflanzenfamilien . . . , ed. 2. 26+ vols. Leipzig and Berlin. Vol. 13, pp. 121–407.

Pines Mexico — See: G. R. Shaw 1909.

Pinet. Woburn. — See: J. Forbes 1839.

Pinetum — See: G. Gordon and R. Glendenning 1858.

Pinetum Suppl. — See: G. Gordon 1862.

Pl. Jav. Rar. — See: J. J. Bennett et al. 1838–1852.

Pl. Syst. Evol. = Plant Systematics and Evolution.

Pl. Veron. — See: J. F. Séguier 1745–1754.

Pl. Voy. Russes Monde — See: G. H. Langsdorff and F. E. L. von Fischer 1810[–1818].

Porter, D. M., R. W. Kiger, and J. E. Monahan. 1973. A Guide for Contributors to Flora North America. Part II. An Outline and Glossary of Terms for Morphological and Habitat Description. (Provisional Edition.) Washington.

Powell, G. R. 1987. Syllepsis in *Larix laricina*: Analysis of tree leaders with and without sylleptic long shoots. Canad. J. Forest Res. 17: 490–498.

Précis Découv. Somiol. — See: C. S. Rafinesque 1814.

Prelim. Cat. — See: N. L. Britton et al. 1888.

Presl, C. B. 1825–1835. Reliquiae Haenkeanae seu Descriptiones et Icones Plantarum, Quas in America Meridionali et Boreali, in Insulis Philippinis et Marianis Collegit Thaddeus Haenke.... 2 vols. in 7 parts. Prague. (Reliq. Haenk.)

Presl, C. B. 1836. Tentamen Pteridographiae, seu Genera Filicacearum Praesertim Juxta Venarum Decursum et Distributionem Exposita.... Prague. (Tent. Pterid.)

Presl, C. B. 1845. Supplementum Tentaminis Pteridographiae.... Prague. (Suppl. Tent. Pterid.)

Preslia = Preslia. Věstník (Časopis) Československé Botanické Společnosti.

Preston, R. J. 1976. North American Trees (Exclusive of Mexico and Tropical United States), ed. 3. Ames.

Price, M. G. 1983. *Pecluma,* a new tropical American fern genus. Amer. Fern J. 73: 109–116.

Price, R. A. 1989. The genera of Pinaceae in the southeastern United States. J. Arnold Arbor. 70: 247–305.

Price, R. A. 1990. The genera of Taxaceae in the southeastern United States. J. Arnold Arbor. 71: 69–91.

Proc. & Trans. Roy. Soc. Canada = Proceedings and Transactions of the Royal Society of Canada.

Proc. Amer. Acad. Arts = Proceedings of the American Academy of Arts and Sciences.

Proc. Biol. Soc. Wash. = Proceedings of the Biological Society of Washington.

Proc. Boston Soc. Nat. Hist. = Proceedings of the Boston Society of Natural History.

Proc. Florida Acad. Sci. = Proceedings of the Florida Academy of Sciences.

Proc. Linn. Soc. New South Wales = Proceedings of the Linnean Society of New South Wales.

Proc. Natl. Acad. Sci. U.S.A. = Proceedings of the National Academy of Sciences of the United States of America.

Proc. Roy. Hort. Soc. London = Proceedings of the Royal Horticultural Society of London.

Proc. Roy. Soc. Edinburgh, B = Proceedings of the Royal Society of Edinburgh. Series B, Biology [later: Biological Sciences].

Proc. Roy. Soc. Queensland = Proceedings of the Royal Society of Queensland.

Proc. Staten Island Assoc. Arts = Proceedings of the Staten Island Association of Arts and Sciences.

Proctor, G. R. 1985. Ferns of Jamaica. London.

Prodr. — See: R. Brown 1810; A. P. de Candolle and A. L. P. de Candolle 1823–1873; O. P. Swartz 1788.

Prodr. Aethéogam. — See: A. Palisot de Beauvois 1805.

Pryer, K. M. 1990. The limestone oak fern: New to Manitoba. Blue Jay 48(4): 192–195.

Pryer, K. M. 1992. The status of Gymnocarpium heterosporum and G. robertianum in Pennsylvania. Amer. Fern J. 82: 34–39.

Pryer, K. M. and D. M. Britton. 1983. Spore studies in the genus Gymnocarpium. Canad. J. Bot. 61: 377–388.

Pryer, K. M., D. M. Britton, and J. McNeill. 1983. A numerical analysis of chromatographic profiles in North American taxa of the fern genus Gymnocarpium. Canad. J. Bot. 61: 2592–2602.

Pryer, K. M. and C. H. Haufler. 1993. Isozymic and chromosomal evidence for the allotetraploid origin of Gymnocarpium dryopteris (Dryopteridaceae). Syst. Bot. 18: 150–172.

Pursh, F. 1814. Flora Americae Septentrionalis; or, a Systematic Arrangement and Description of the Plants of North America. 2 vols. London. (Fl. Amer. Sept.)

Rafinesque, C. S. 1814. Précis des Découvertes et Travaux Somiologiques.... Palermo. (Précis Découv. Somiol.)

Rafinesque, C. S. 1836[–1838b.] New Flora and Botany of North America.... 4 parts. Philadelphia.

Raine, C. A., D. R. Farrar, and E. S. Sheffield. 1991. A new Hymenophyllum species in the Appalachians represented by independent gametophyte colonies. Amer. Fern J. 81: 109–118.

Ranker, T. A. and C. H. Haufler. 1990. A new combination in Bommeria (Adiantaceae). Amer. Fern J. 80: 1–3.

Recens. Spec. Pter. — See: J. G. Agardh 1839.

Reed, C. F. 1953. Index Isoetales. Bol. Soc. Brot., ser. 2a, 27: 5–72.

Reed, C. F. 1954. Index Marsileata et Salviniata. Bol. Soc. Brot., ser. 2a, 28: 1–61.

Reed, C. F. 1965. Isoëtes in southeastern United States. Phytologia 12: 369–400.

Rees, A. [1802–]1819–1820. The Cyclopaedia; or, Universal Dictionary of Arts, Sciences, and Literature.... 39 vols. in 79 parts. London. [Pages unnumbered.] (Cycl.)

Reeves, T. 1979. A Monograph of the Fern Genus Cheilanthes subgenus Physapteris (Adiantaceae). Ph.D. dissertation. Arizona State University.

Reeves, T. 1981. Notes on North American lower vascular plants—II. Amer. Fern J. 71: 62–64.

Rehder, A. J. 1949. Bibliography of Cultivated Trees and Shrubs Hardy in the Cooler Temperate Regions of the Northern Hemisphere. Jamaica Plain.

Reichstein, T. 1981. Hybrids in European Aspleniaceae (Pteridophyta). Bot. Helv. 91: 89–139.

Reisen Ost-Afrika — See: C. C. von der Decken 1869–1879.

Reliq. Haenk. — See: C. B. Presl 1825–1835.

Rep. Exped. Rocky Mts. — See: J. C. Frémont 1843–1845.

Rep. For. N. America — See: C. S. Sargent 1884.

Rep. Geol. Surv. — See: New York State 1840.

Rep. U.S. Geogr. Surv., Wheeler — See: J. T. Rothrock 1878[1879].

Rep. U.S. Mex. Bound. — See: W. H. Emory 1857–1859.

Retzius, A. J. [1779]–1791. Observationes Botanicae.... 6 vols. Leipzig. (Observ. Bot.)

Rev. Hort. = Revue Horticole; Journal d'Horticulture Pratique.

Revista Latinoamer. Quím. = Revista Latinoamericana de Química.

Rhein. Fl. — See: J. C. Döll 1843.

Rhodora = Rhodora; Journal of the New England Botanical Club.

Roche, L. 1969. A genecological study of the genus Picea and seedlings grown in a nursery. New Phytol. 68: 505–554.

Rosenvinge, L. K., ed. 1911. Biologiske Arbejder Tilegnede Eug. Warming paa Hans 70 Aars Fødseldag den 3. November 1911. Copenhagen. (Biol. Arbejder Tilegn. Eug. Warming)

Roth, A. W. 1788–1800. Tentamen Florae Germanicae.... 3 vols. in 5 parts. Leipzig. (Tent. Fl. Germ.)

Rothmaler, W. 1944. Pteridophyten-Studien I. Feddes Repert. 54: 55–82.

Rothrock, J. T. 1878[1879]. Report upon United States Geographical Surveys West of the One Hundredth Meridian, in Charge of First Lieut. Geo. M. Wheeler.... Vol. 6 — Botany. Washington. (Rep. U.S. Geogr. Surv., Wheeler)

Rouffa, A. S. 1971. An appendageless Psilotum. Introduction to aerial shoot morphology. Amer. Fern J. 61: 75–86.

Rumsey, F. J., E. S. Sheffield, and D. R. Farrar. 1990. British filmy fern gametophytes. Pteridologist 2: 40–42.

Ruprecht, F. 1845. Distributio Cryptogamarum Vascularium in Imperio Rossico. St. Petersburg. [Alternate title: Beiträge zur Pflanzenkunde des Russischen Reiches.... Dritte Lieferung.] (Distr. Crypt. Vasc. Ross.)

Ruprecht, F. 1859. Bemerkungen über einige Arten der Gattung *Botrychium*. St. Petersburg. [Alternate title: Beiträge zur Pflanzenkunde des Russischen Reiches.... Eilfte und letzte Lieferung.] (Bemerk. Botrychium)

Sargent, C. S. 1884. Department of the Interior, Census Office.... Report on the Forests of North America (Exclusive of Mexico).... Washington. [47 Congr., 2 Sess., House Misc. Doc. 42(9).] (Rep. For. N. America)

Sargent, C. S. 1890–1902. The Silva of North America.... 14 vols. Boston and New York. (Silva)

Sargent, C. S. 1922. Manual of the Trees of North America (Exclusive of Mexico), ed. 2. Boston and New York. [Facsimile edition in 2 vols. 1961, reprinted 1965, New York.] (Man. Trees ed. 2)

Sarvela, J. 1978. A synopsis of the fern genus *Gymnocarpium*. Ann. Bot. Fenn. 15: 101–106.

Sarvela, J. 1980. *Gymnocarpium* hybrids from Canada and Alaska. Ann. Bot. Fenn. 17: 292–295.

Sarvela, J., D. M. Britton, and K. M. Pryer. 1981. Studies on the *Gymnocarpium robertianum* complex in North America. Rhodora 83: 421–431.

Savage, T. 1983. A Georgia station for *Torreya taxifolia* Arn. survives. Florida Sci. 46: 62–64.

Scan. Electron Microscop. = Scanning Electron Microscopy; International Journal of Scanning Electron Microscopy, Related Techniques, and Applications.

Schmidel, C. C. 1762–1771. Icones Plantarum et Analyses Partium Aeri.... [Nuremberg]. (Icon. Pl. ed. Keller)

Schneller, J. J. 1976. The position of the megaprothallus of *Salvinia natans*. Fern Gaz. 11: 217–219.

Schott, H. W. 1834[–1836]. Genera Filicum. 4 fasc. Vienna. (Gen. Fil.)

Schreber, J. C. 1789–1791. Caroli a Linné ... Genera Plantarum.... 2 vols. Frankfurt am Main. (Gen. Pl.)

Schuster, J. 1932. Cycadaceae. In: H. G. A. Engler, ed. 1900–1953. Das Pflanzenreich.... 107 vols. Berlin. Vol. 99[IV,1], pp. 1–168.

Schwarz, O. and H. Weide. 1962. Systematische Revision der Gattung *Sequoia* Endl. Feddes Repert. Spec. Nov. Regni Veg. 66: 159–192.

Science = Science; an Illustrated Journal [later: a Weekly Journal Devoted to the Advancement of Science]. [American Association for the Advancement of Science.]

Scoggan, H. J. 1978–1979. The Flora of Canada. 4 parts. Ottawa. [Natl. Mus. Nat. Sci. Publ. Bot. 7.]

Scopoli, J. A. 1760. Flora Carniolica.... Vienna. (Fl. Carniol.)

Séguier, J. F. 1745–1754. Plantae Veronenses, seu Stirpium Quae in Agro Veronensi Reperiuntur Methodica Synopsis.... 3 vols. Verona. (Pl. Veron.)

Seto, K. and T. Nasu. 1975. Discovery of fossil *Azolla* massulae from Japan and some notes on recent Japanese species. Bull. Osaka Mus. Nat. Hist. 29: 51–60.

Shaver, J. M. 1954. Ferns of Tennessee, with the Fern Allies Excluded. Nashville. [Reprint 1970, New York under the title: Ferns of the Eastern Central States, with Special Reference to Tennessee.]

Shaw, G. R. 1909. The Pines of Mexico.... Boston. (Pines Mexico)

Shaw, G. R. 1914. The Genus *Pinus*.... Cambridge, Mass. [Publ. Arnold Arbor. 5.]

Sheffield, E. S. and D. R. Farrar. 1988. Cryo SEM examination of gemma formation in *Vittaria graminifolia*. Amer. J. Bot. 75: 894–899.

Shing, K. H. 1965. A taxonomical study of the genus *Cyrtomium* Presl. Acta Phytotax. Sin., Addit. 1: 1–48.

Shivas, M. G. 1969. A cytotaxonomic study of the *Asplenium adiantum-nigrum* complex. Brit. Fern Gaz. 10: 68–80.

Sida = Sida; Contributions to Botany.

Siegler, D. S. and E. Wollenweber. 1983. Chemical variation in *Notholaena standleyi*. Amer. J. Bot. 70: 790–798.

Silba, J. 1981. Revision of *Cupressus* L. (Cupressaceae). Phytologia 49: 390–399.

Silba, J. 1986. Encyclopaedia Coniferae. Phytologia Mem. 8: 1–127.

Silva — See: C. S. Sargent 1890–1902.

Silvae Genet. = Silvae Genetica.

Sleep, A. and T. Reichstein. 1967. Der Farnbastard *Polystichum* ×*Meyeri* hybr. nov. = *Polystichum braunii* (Spenner) Fée × *P. lonchitis* (L.) Roth und seine Cytologie. Bauhinia 3: 299–309, 363–374.

Slosson, M. 1902. The origin of *Asplenium ebenoides*. Bull. Torrey Bot. Club 29: 487–495.

Small, J. K. 1918. Ferns of Tropical Florida, Being Descriptions of and Notes on the Ferns and Fern-allies Growing Naturally on the Everglade Keys and Florida Keys.... New York. (Ferns Trop. Florida)

Small, J. K. 1933. Manual of the Southeastern Flora, Being Descriptions of the Seed Plants Growing Naturally in Florida, Alabama, Mississippi, Eastern Louisiana, Tennessee, North Carolina, South Carolina and Georgia. New York.

Small, J. K. 1938. Ferns of the Southeastern States. Lancaster, Pa. [Facsimile edition 1964, New York and London.] (Ferns S.E. States)

Smith, A. R. 1971. Systematics of the Neotropical species of *Thelypteris* section *Cyclosorus*. Univ. Calif. Publ. Bot. 59: 1–143.

Smith, A. R. 1974. Taxonomic and cytological notes on ferns from California and Arizona. Madroño 22: 376–378.

Smith, A. R. 1975. The California species of *Aspidotis*. Madroño 23: 15–24.

Smith, A. R. 1980. New taxa and combinations of pteridophytes from Chiapas, Mexico. Amer. Fern J. 70: 15–27.

Smith, A. R. 1981. Pteridophytes. In: D. E. Breedlove, ed. 1981 +. Flora of Chiapas. 2 + parts. San Francisco. Part 2.

Smith, D. M. 1956. A collection of *Asplenium montanum* in Indiana. Amer. Fern J. 46: 94–95.

Smith, D. M. 1980. Flavonoid analysis of the *Pityrogramma*

triangularis complex. Bull. Torrey Bot. Club 107: 134–145.

Smith, D. M. and D. A. Levin. 1963. A chromatographic study of reticulate evolution in the Appalachian aspleniums. Amer. J. Bot. 50: 952–958.

Smith, J. 1866. Ferns: British & Foreign. London. (Ferns Brit. For.)

Smith, J. 1875. Historia Filicum.... London. (Hist. Fil.)

Smithsonian Contr. Knowl. = Smithsonian Contributions to Knowledge.

Smithsonian Misc. Collect. = Smithsonian Miscellaneous Collections.

Snyder, L. H. Jr. and J. G. Bruce. 1986. Field Guide to the Ferns and Other Pteridophytes of Georgia. Athens, Ga.

Soltis, P. S. and D. E. Soltis. 1987. Population structure and estimates of gene flow in the homosporous fern *Polystichum munitum*. Evolution 41: 620–629.

Soltis, P. S., D. E. Soltis, and E. R. Alverson. 1987. Electrophoretic and morphological confirmation of interspecific hybridization between *Polystichum kruckebergii* and *P. munitum*. Amer. Fern J. 77: 42–49.

Soltis, P. S., D. E. Soltis, P. G. Wolf, and J. M. Riley. 1989. Electrophoretic evidence for interspecific hybridization in *Polystichum*. Amer. Fern J. 79: 7–13.

Somers, P. 1982. A unique type of microsporangium in *Selaginella* series *Articulatae*. Amer. Fern J. 72: 88–92.

Somers, P. and W. R. Buck. 1975. *Selaginella ludoviciana, S. apoda* and their hybrids in the southeastern United States. Amer. Fern J. 65: 76–82.

Soper, J. H. and S. Rao. 1958. *Isoëtes* in eastern Canada. Amer. Fern J. 48: 97–102.

SouthW. Naturalist = Southwestern Naturalist.

Sp. Fil. — See: W. J. Hooker [1844–]1846–1864.

Sp. Pl. — See: C. Linnaeus 1753; C. L. Willdenow 1797–1830.

Sp. Pl. ed. 2 — See: C. Linnaeus 1762–1763.

Spach, E. 1834–1838. Histoire Naturelle des Végétaux. Phanérogames.... 14 vols., atlas. Paris. (Hist. Nat. Vég.)

Spenner, F. 1825–1829. Flora Friburgensis et Regionum Proxime Adjacentium.... 3 vols. Friburg. (Fl. Friburg.)

Sprengel, K. 1802–1804. Anleitung zur Kenntniss der Gewächse.... 3 parts. Halle. (Anleit. Kenntn. Gew.)

Sprengel, K. [1824–]1825–1828. Caroli Linnaei... Systema Vegetabilium. Editio Decima Sexta.... 4 vols., suppl. Göttingen. [Vol. 4 in 2 parts, each paged separately.] (Syst. Veg.)

Spring, A. F. 1850. Monographie famille des Lycopodiacées, seconde partie. Mém. Acad. Roy. Sci. Belgique 24: 1–358.

Stafleu, F. A. and R. S. Cowan. 1976–1988. Taxonomic Literature: A Selective Guide to Botanical Publications and Collections with Dates, Commentaries and Types, ed. 2. 7 vols. Utrecht, Antwerp, The Hague, and Boston.

Stafleu, F. A. and E. A. Mennega. 1992. Taxonomic Literature: A Selective Guide to Botanical Publications and Collections with Dates, Commentaries and Types. Supplement 1. Königstein.

Stalter, R. and S. Dial. 1984. Environmental status of the stinking cedar. Bartonia 50: 40–42.

Stapf, O. 1889. Die Arten der Gattung *Ephedra*. Denkschr. Kaiserl. Akad. Wiss., Wien. Math.-Naturwiss. Kl. 56(2): 1–112.

Steenis, C. G. G. J. van and R. E. Holttum, eds. 1959–1982. Flora Malesiana, Being an Illustrated Systematic Account of the Malesian Flora. Series II. Pteridophyta. 1 vol. in 5 parts. The Hague, Boston, and London.

Steudel, E. G. 1840–1841. Nomenclator Botanicus Enumerans Ordine Alphabetico Nomina atque Synonyma tum Generica tum Specifica.... 2 vols. Stuttgart and Tubingen. (Nomencl. Bot. ed. 2)

Stevenson, D. W. 1976. Observations on phyllotaxis, stelar morphology, the shoot apex and gemmae of *Lycopodium lucidulum* Michaux (Lycopodiaceae). Bot. J. Linn. Soc. 72: 81–100.

Stevenson, D. W. 1987. Again the West Indian zamias. Fairchild Trop. Gard. Bull. 42(3): 23–27.

Stevenson, D. W. 1987b. Comments on character distribution, taxonomy and nomenclature of the genus *Zamia* L. in the West Indies and Mexico. Encephalartos 9: 3–7.

Steyermark, J. A. 1963. Flora of Missouri. Ames.

Stokey, A. G. 1940. Spore germination and vegetative stages of the gametophytes of *Hymenophyllum* and *Trichomanes*. Bot. Gaz. 101: 759–790.

Stolze, R. G. 1981. Ferns and fern allies of Guatemala. Part 2. Polypodiaceae. Fieldiana, Bot., n. s. 6: 1–522.

Stolze, R. G. 1987. *Schizaea pusilla* discovered in Peru. Amer. Fern J. 77: 64–65.

Sudworth, G. B. 1908. Forest Trees of the Pacific Slope. Washington.

Sudworth, G. B. 1917. The Pine Trees of the Rocky Mountain Region. Washington. [U.S.D.A. Bull. 460.]

Sunyatsenia = Sunyatsenia. Journal of the Botanical Institute; College of Agriculture, Sun Yatsen University.

Suppl. Pl. — See: C. Linnaeus f. 1781[1782].

Suppl. Tent. Pterid. — See: C. B. Presl 1845.

Svensk Bot. Tidskr. = Svensk Botanisk Tidskrift Utgifven af Svenska Botaniska Föreningen.

Svenson, H. K. 1944. The New World species of *Azolla*. Amer. Fern J. 34: 69–84.

Swartz, O. P. 1788. Nova Genera & Species Plantarum seu Prodromus.... Stockholm, Uppsala, and Åbo. (Prodr.)

Swartz, O. P. 1806. Synopsis Filicum.... Kiel. (Syn. Fil.)

Syn. Conif. — See: S. L. Endlicher 1847.

Syn. Deut. Schweiz. Fl. ed. 2 — See: W. D. J. Koch 1846–1847.

Syn. Fil. — See: W. J. Hooker and J. G. Baker [1865–]1868; O. P. Swartz 1806.

Syn. Fil. ed. 2 — See: W. J. Hooker and J. G. Baker 1874.

Syst. Bot. = Systematic Botany; Quarterly Journal of the American Society of Plant Taxonomists.

Syst. Bot. Monogr. = Systematic Botany Monographs; Monographic Series of the American Society of Plant Taxonomists.

Syst. Nat. — See: J. F. Gmelin 1791[–1792].

Syst. Nat. ed. 10 — See: C. Linnaeus 1758[–1759].

Syst. Veg. — See: K. Sprengel [1824–]1825–1828.

Syst. Veg. ed. 14 — See: J. A. Murray 1784.

Taxon = Taxon; Journal of the International Association for Plant Taxonomy.

Taylor, M. S. 1938. Filmy-ferns in South Carolina. J. Elisha Mitchell Sci. Soc. 54: 345–348.

Taylor, R. J. 1972. The relationship and origin of *Tsuga heterophylla* and *Tsuga mertensiana* based on phytochemical and morphological interpretations. Amer. J. Bot. 59: 149–157.

Taylor, R. J. and T. F. Patterson. 1980. Biosystematics of Mexican spruce species and populations. Taxon 29: 421–469.

Taylor, R. L. and S. Taylor. 1981. *Taxus brevifolia* in British Columbia. Davidsonia 12(4): 89–94.

Taylor, R. J., S. Williams, and R. Daubenmire. 1975. Interspecific relationships and the question of introgression between *Picea engelmannii* and *Picea pungens*. Canad. J. Bot. 53: 2547–2555.

Taylor, T. M. C. 1947. New species and combinations in *Woodsia* section *Perrinia*. Amer. Fern J. 37: 84–88.

Taylor, T. M. C. 1959. The taxonomic relationship between *Picea glauca* (Moench) Voss and *P. engelmannii* Parry. Madroño 15: 111–115.

Taylor, T. M. C. 1967. *Mecodium wrightii* in British Columbia and Alaska. Amer. Fern J. 57: 1–6.

Taylor, T. M. C. 1970. Pacific Northwest Ferns and Their Allies. Toronto.

Taylor, W. C. and R. J. Hickey. 1992. Habitat, evolution, and speciation of *Isoëtes*. Ann. Missouri Bot. Gard. 79: 613–622.

Taylor, W. C., N. T. Luebke, and M. B. Smith. 1985. Speciation and hybridization in North American quillworts. Proc. Roy. Soc. Edinburgh, B 86: 259–263.

Tent. Fl. Germ. — See: A. W. Roth 1788–1800.

Tent. Pterid. — See: C. B. Presl 1836.

Thieret, J. W. 1980. Louisiana Ferns and Fern Allies. Lafayette.

Thompson, J. 1961. Cupressaceae. Contr. New South Wales Natl. Herb., Fl. Ser. 1/18: 46–55.

Thompson, J. and L. A. S. Johnson. 1986. *Callitris glaucophylla*, Australia's 'white cypress pine'—A new name for an old species. Telopea 2: 731–736.

Tindale, M. D. 1960. Pteridophyta of south eastern Australia. Contr. New South Wales Natl. Herb., Fl. Ser. 211: 47–78.

Torrey, J. 1857. Explorations and Surveys for a Railroad Route from the Mississippi River to the Pacific Ocean. War Department. Route Near the Thirty-fifth Parallel, Explored by Lieutenant A. W. Whipple, Topographical Engineers, in 1853 and 1854. Report on the Botany of the Expedition. Washington. [Reprinted from Pacif. Railr. Rep. 4(5)[no. 4]: [59]–182, plates 1–25. 1857.] (Pacif. Railr. Rep. 4(5))

Torreya = Torreya; a Monthly Journal of Botanical Notes and News.

Traité Gén. Conif. — See: E. A. Carrière 1855.

Traité Gén. Conif. ed. 2 — See: E. A. Carrière 1867.

Trans. Acad. Sci. St. Louis = Transactions of the Academy of Science of St. Louis.

Trans. Linn. Soc. London = Transactions of the Linnean Society of London.

Trans. Wisconsin Acad. Sci. = Transactions of the Wisconsin Academy of Sciences, Arts and Letters.

Trav. Lab. Forest. Toulouse = Travaux du Laboratoire Forestiere de Toulouse.

Trop. Woods = Tropical Woods....

Tryon, A. F. 1949. Spores of the genus *Selaginella* in North America, north of Mexico. Ann. Missouri Bot. Gard. 36: 413–431.

Tryon, A. F. 1957. A revision of the fern genus *Pellaea* section *Pellaea*. Ann. Missouri Bot. Gard. 44: 125–193.

Tryon, A. F. 1968. Comparisons of sexual and apogamous races in the fern genus *Pellaea*. Rhodora 70: 1–24.

Tryon, A. F. and D. M. Britton. 1958. Cytotaxonomic studies on the fern genus *Pellaea*. Evolution 12: 137–145.

Tryon, A. F., R. M. Tryon, and F. Badré. 1980. Classification, spores, and nomenclature of the marsh fern. Rhodora 82: 461–474.

Tryon, R. M. 1941. A revision of the genus *Pteridium*. Rhodora 43: 1–31, 37–67.

Tryon, R. M. 1955. *Selaginella rupestris* and its allies. Ann. Missouri Bot. Gard. 42: 1–99, plates 1–6.

Tryon, R. M. 1956. A revision of the American species of *Notholaena*. Contr. Gray Herb. 179: 1–106.

Tryon, R. M. 1960. A review of the genus *Dennstaedtia* in America. Contr. Gray Herb. 187: 23–52.

Tryon, R. M. 1962. Taxonomic fern notes. II. *Pityrogramma* (including *Trismeria*) and *Anogramma*. Contr. Gray Herb. 189: 52–76.

Tryon, R. M. 1964. Taxonomic fern notes. IV. Some American vittarioid ferns. Rhodora 66: 110–117.

Tryon, R. M. 1971. The process of evolutionary migration in species of *Selaginella*. Brittonia 23: 89–100.

Tryon, R. M. and A. F. Tryon. 1982. Ferns and Allied Plants, with Special Reference to Tropical America. New York, Heidelberg, and Berlin.

Ulrike, R. 1987. Growth patterns of gemmlings of *Lycopodium lucidulum*. Amer. Fern J. 77: 50–57.

Underwood, L. M. 1899. North American ferns—II. The genus *Phanerophlebia*. Bull. Torrey Bot. Club 26: 205–216.

Underwood, L. M. 1900. Our Native Ferns and Their Allies with Synoptical Descriptions of the American Pteridophyta North of Mexico, ed. 6. New York. (Native Ferns ed. 6)

Underwood, L. M. 1906. American ferns VII A. The American species of *Stenochlaena*. Bull. Torrey Bot. Club 33: 591–603.

Underwood, L. M. 1907. American ferns—VIII. A preliminary review of the North American Gleicheniaceae. Bull. Torrey Bot. Club 34: 243–262.

Univ. Calif. Publ. Bot. = University of California Publications in Botany.

Univ. Nebraska Stud. = University of Nebraska Studies.

University of Chicago Press. 1982. The Chicago Manual of

Style, for Authors, Editors, and Copywriters, ed. 13. Chicago.

U.S. Expl. Exped. — See: C. Wilkes 1854–1876.

U.S.D.A. Div. Forest. Bull. = U. S. Department of Agriculture, Division of Forestry Bulletin.

Van Campo-Duplan, M. and H. Gaussen. 1948. Sur quatre hybrides de genres chez les Abietinées. Trav. Lab. Forest. Toulouse 1(4, 24): 1–15.

Van Eseltine, G. P. 1918. The allies of *Selaginella rupestris* in the southeastern United States. Contr. U.S. Natl. Herb. 20(5): 159–172.

Van Haverbeke, D. F. 1968. A population study of *Juniperus* in the Missouri River basin. Univ. Nebraska Stud., n. s. 38: 1–82.

Van Hove, C., T. de Waha Baillonville, H. F. Diara, P. Goddard, Y. Mai Godomi, and N. Sanginga. 1987. *Azolla* collection and selection. In: International Rice Research Institute. 1987. *Azolla* Utilization. Proceedings of the Workshop on *Azolla* Use. Fuzhou, Fujian, China, 31 March–5 April 1985. Los Banos, Laguna, Philippines. Pp. 77–87.

Vasek, F. C. 1966. The distribution and taxonomy of three western junipers. Brittonia 18: 350–372.

Vasek, F. C. and R. W. Scora. 1967. Analysis of the oils of western North American junipers by gas-liquid chromatography. Amer. J. Bot. 54: 781–789.

Ventenat, E. P. 1808. Decas Generum Novorum.... Paris. (Dec. Gen. Nov.)

Verdoorn, F., ed. 1938. Manual of Pteridology. The Hague. [Collaborators: A. H. G. Alston, I. Andersson-Kottö, L. R. Atkinson, H. Burgeff, H. G. du Buy, C. Christensen, W. Döpp, W. M. Doctors van Leeuwen, H. Gams, M. J. F. Gregor, M. Hirmer, R. E. Holttum, R. Kräusel, E. L. Neuernbergk, J. C. Schoute, J. Walton, K. Wetzel, S. Williams, H. Winkler, and W. Zimmermann.] (Man. Pteridol.)

Verh. K. K. Zool.-Bot. Ges. Wien = Verhandlungen der Kaiserlich-königlichen zoologisch-botanischen Gesellschaft in Wien.

Verh. Naturhist. Vereines Preuss. Rheinl. Westphalens = Verhandlungen des Naturhistorischen Vereines der preussischen Rheinlande und Westphalens.

Verh. Zool.-Bot. Ges. Wien = Verhandlungen der Zoologisch-botanischen Gesellschaft in Wien.

Verz. Ausländ. Bäume — See: C. Moench 1785.

Viane, R. L. 1986. Taxonomical significance of the leaf indument in *Dryopteris* (Pteridophyta): I. Some North American, Macronesian and European taxa. Pl. Syst. Evol. 153: 77–105.

Vida, G. 1970. The nature of polyploidy in *Asplenium ruta-muraria* L. and *A. lepidum* C. Presl. Caryologia 23: 525–547.

Villars, D. 1786–1789. Histoire des Plantes de Dauphiné. 3 vols. Grenoble, Lyon, and Paris. (Hist. Pl. Dauphiné)

Virginia J. Sci. = Virginia Journal of Science.

Von Aderkas, P. 1984. Economic history of ostrich fern, *Matteuccia struthiopteris*, the edible fiddlehead. Econ. Bot. 38: 14–23.

Voy. Uranie — See: C. Gaudichaud-Beaupré 1826[–1830].

Wagner, D. H. 1979. Systematics of *Polystichum* in western North America north of Mexico. Pteridologia 1: 1–64.

Wagner, F. S. 1987. Evidence for the origin of the hybrid cliff fern, *Woodsia × abbeae* (Aspleniaceae: Athyrioideae). Syst. Bot. 12: 116–124.

Wagner, F. S. 1992. Cytological problems in *Lycopodium* s.l. Ann. Missouri Bot. Gard. 79: 718–729.

Wagner, F. S. 1993. Chromosomes of North American grapeferns and moonworts (Ophioglossaceae: *Botrychium*). Contr. Univ. Michigan Herb. 19: 83–92.

Wagner, W. H. Jr. 1954. Reticulate evolution in the Appalachian aspleniums. Evolution 8: 103–118.

Wagner, W. H. Jr. 1955b. Should the American hart's tongue be interpreted as a distinct species? Amer. Fern J. 45: 127–128.

Wagner, W. H. Jr. 1956. A natural hybrid, × *Adiantum tracyi* C. C. Hall. Madroño 13: 195–205.

Wagner, W. H. Jr. 1961. Nomenclature and typification of two botrychiums of the southeastern United States. Taxon 10: 165–169.

Wagner, W. H. Jr. 1961b. Roots and the taxonomic differences between *Botrychium oneidense* and *B. dissectum*. Rhodora 63: 164–175.

Wagner, W. H. Jr. 1962. Cytological observations on *Adiantum × tracyi* C. C. Hall. Madroño 16: 158–161.

Wagner, W. H. Jr. 1963. A biosystematic survey of United States ferns—A preliminary abstract. Amer. Fern J. 53: 1–16.

Wagner, W. H. Jr. 1965. *Pellaea wrightiana* in North Carolina and the question of its origin. J. Elisha Mitchell Sci. Soc. 81: 95–103.

Wagner, W. H. Jr. 1966b. New data on North American oak ferns, *Gymnocarpium*. Rhodora 68: 121–138.

Wagner, W. H. Jr. 1968. Hybridization, taxonomy, and evolution. In: V. H. Heywood, ed. 1968. Modern Methods in Plant Taxonomy. London and New York. Pp. 113–138.

Wagner, W. H. Jr. 1971. Evolution of *Dryopteris* in relation to the Appalachians. In: P. C. Holt, ed. 1971. The Distributional History of the Biota of the Southern Appalachians. Part 2. Flora. Blacksburg, Va. Pp. 147–192.

Wagner, W. H. Jr. 1971b. The southeastern adder's-tongue, *Ophioglossum vulgatum* var. *pycnostichum* found for the first time in Michigan. Michigan Bot. 10: 67–74.

Wagner, W. H. Jr. 1973. Reticulation of holly ferns (*Polystichum*) in the western United States and adjacent Canada. Amer. Fern J. 63: 99–115.

Wagner, W. H. Jr., C. M. Allen, and G. P. Landry. 1984. *Ophioglossum ellipticum* Hook. & Grev. in Louisiana and the taxonomy of *O. nudicaule* L.f. Castanea 49: 99–110.

Wagner, W. H. Jr. and J. M. Beitel. 1992. Generic classification of modern North American Lycopodiaceae. Ann. Missouri Bot. Gard. 79: 676–686.

Wagner, W. H. Jr., J. M. Beitel, and R. C. Moran. 1989. *Lycopodium hickeyi*: A new species of North American clubmoss. Amer. Fern J. 79: 119–121.

Wagner, W. H. Jr., D. R. Farrar, and B. W. McAlpin. 1970.

Pteridology of the Highlands Biological Station area, southern Appalachians. J. Elisha Mitchell Sci. Soc. 86: 1–27.

Wagner, W. H. Jr. and D. M. Johnson. 1981. Natural history of the ebony spleenwort, Asplenium platyneuron (Aspleniaceae), in the Great Lakes area. Canad. Field-Naturalist 95: 156–166.

Wagner, W. H. Jr. and D. M. Johnson. 1983. Trophopod, a commonly overlooked storage structure of potential systematic value in ferns. Taxon 32: 268–269.

Wagner, W. H. Jr. and C. E. Nauman. 1982. Pteris × delchampsii, a spontaneous fern hybrid from southern Florida. Amer. Fern J. 72: 97–102.

Wagner, W. H. Jr. and V. Quevedo. 1985. Polymorphism in Actinostachys pennula (Swartz) Hooker and the taxonomic status of A. germanii (Fée) Prantl. [Abstract.] Amer. J. Bot. 72: 927–928.

Wagner, W. H. Jr. and A. J. Sharp. 1963. A remarkably reduced vascular plant in the United States. Science 142: 1483–1484, cover.

Wagner, W. H. Jr., A. R. Smith, and T. R. Pray. 1983. A cliff brake hybrid, Pellaea bridgesii × mucronata, and its systematic significance. Madroño 30: 69–83.

Wagner, W. H. Jr. and F. S. Wagner. 1966. Pteridophytes of the Mountain Lake area, Giles Co., Virginia. Biosystematic studies 1964–1965. Castanea 31: 121–140.

Wagner, W. H. Jr. and F. S. Wagner. 1983. Genus communities as a systematic tool in the study of New World Botrychium (Ophioglossaceae). Taxon 32: 51–63.

Wagner, W. H. Jr. and F. S. Wagner. 1990. Moonworts (Botrychium subg. Botrychium) of the Upper Great Lakes region. Contr. Univ. Michigan Herb. 17: 313–325.

Wagner, W. H. Jr., F. S. Wagner, C. H. Haufler, and J. L. Emerson. 1984. A new nothospecies of moonwort (Ophioglossaceae, Botrychium). Canad. J. Bot. 62: 629–634.

Wagner, W. H. Jr., F. S. Wagner, J. A. Lankalis, and J. F. Matthews. 1973. Asplenium montanum × Platyneuron. A new primary member of the Appalachian spleenwort complex from Crowder's Mountain, N.C. J. Elisha Mitchell Sci. Soc. 89: 218–223.

Wagner, W. H. Jr., F. S. Wagner, A. A. Reznicek, and C. R. Werth. 1992. × Dryostichum singulare (Dryopteridaceae), a new fern nothogenus from Ontario. Canad. J. Bot. 70: 245–253.

Wagner, W. H. Jr. and R. S. Whitmire. 1957. Spontaneous production of a morphologically distinct, fertile allopolyploid by a fertile diploid of Asplenium ebenoides. Bull. Torrey Bot. Club 84: 79–89.

Wald. Nordamer. — See: H. Mayr 1890[1889].

Walker, T. G. 1962. The Anemia adiantifolia complex in Jamaica. New Phytol. 61: 291–298.

Walker, T. G. 1972. The anatomy of Maxonia apiifolia: A climbing fern. Brit. Fern Gaz. 10: 241–250.

Wallich, N. 1828[–1849]. A Numerical List of Dried Specimens of Plants, in the East India Company's Museum Collected Under the Superintendence of Dr. Wallich of the Company's Botanic Garden at Calcutta.... London. (Numer. List)

Walter, K. S., W. H. Wagner Jr., and F. S. Wagner. 1982.

Ecological, biosystematic, and nomenclatural notes on Scott's spleenwort, × Asplenosorus ebenoides. Amer. Fern J. 72: 65–75.

Walter, T. 1788. Flora Caroliniana, secundum Systema Vegetabilium Perillustris Linnaei Digesta.... London. (Fl. Carol.)

Ward, D. B., ed. N.d. Rare and Endangered Biota of Florida. Vol. 5. Plants. Gainesville.

Waring, R. H., W. H. Emmingham, and S. W. Running. 1975. Environmental limits of an endemic spruce, Picea breweriana. Canad. J. Bot. 53: 1599–1613.

Waterway, M. J. 1986. A reevaluation of Lycopodium porophilum and its relationship to L. lucidulum (Lycopodiaceae). Syst. Bot. 11: 263–276.

Watson, F. D. 1985. The nomenclature of pondcypress and baldcypress. Taxon 34: 506–509.

Watson, S., W. H. Brewer, and A. Gray. 1876–1880. Geological Survey of California.... Botany.... 2 vols. Cambridge, Mass. (Bot. California)

Watson, S., D. C. Eaton, et al. 1871. United States Geological Exploration [sic] of the Fortieth Parallel. Clarence King, Geologist-in-charge. [Vol. 5] Botany. By Sereno Watson.... Washington. [Botanical portion of a larger work by C. King.] (Botany (Fortieth Parallel))

Weatherby, C. A. 1920. Varieties of Pityrogramma triangularis. Rhodora 22: 113–120.

Weatherby, C. A. 1943. The group of Selaginella parishii. Amer. Fern J. 33: 113–119.

Webbia = Webbia; Raccolta di Scritti Botanici.

Weber, F. and D. M. H. Mohr. 1807. Botanisches Taschenbuch auf das Jahr 1807. Deutschland's kryptogamische Gewächse. Erste Abteilung. Filices, Musci Frondosi et Hepatici.... Kiel. (Bot. Taschenbuch)

Webster, T. R. 1990. Selaginella apoda × ludoviciana, a synthesized hybrid spikemoss. Amer. J. Bot. 77(6, suppl.): 108.

Welsh, S. L. 1974. Anderson's Flora of Alaska and Adjacent Parts of Canada. Provo.

Wendt, T. 1980. Notes on some Pleopeltis and Polypodium species of the Chihuahuan Desert Region. Amer. Fern J. 70: 5–11.

Werdenda = Werdenda. Beiträge zur Pflanzenkunde.

Werth, C. R. 1991. Isozyme studies on the Dryopteris "spinulosa" complex. I: The origin of the log fern Dryopteris celsa. Syst. Bot. 16(3): 446–461.

Werth, C. R., S. I. Guttman, and W. H. Eshbaugh. 1985. Electrophoretic evidence of reticulate evolution in the Appalachian Asplenium complex. Syst. Bot. 10: 184–192.

Werth, C. R., S. I. Guttman, and W. H. Eshbaugh. 1985b. Recurring origins of allopolyploid species in Asplenium. Science 228: 731–733.

Werth, C. R. and M. D. Windham. 1991. A model for divergent allopatric speciation of polyploid pteridophytes resulting from silencing of duplicate gene expression. Amer. Naturalist 137: 515–526.

Wessels Boer, J. G. 1962. The New World species of Trichomanes sect. Didymoglossum and Microgonium. Acta Bot. Neerl. 11: 277–330.

Wherry, E. T. 1964. The Southern Fern Guide. Garden City, N.Y.

White, R. A., D. W. Bierhorst, P. G. Gensel, D. R. Kaplan, and W. H. Wagner Jr. 1977. Taxonomic and morphological relationships of the Psilotaceae: A symposium. Brittonia 29(1): 1–68.

Whitmore, S. A. and A. R. Smith. 1991. Recognition of the tetraploid, *Polypodium calirhiza* sp. nov. (Polypodiaceae), in western North America. Madroño 38: 233–248.

Wilce, J. H. 1965. Section *Complanata* of the genus *Lycopodium*. Beih. Nova Hedwigia 19: i–ix, 1–233, plate 40.

Wilkes, C., A. Gray, W. D. Brackenridge, J. Torrey, C. Pickering, et al. 1854–1876. United States Exploring Expedition. During the years 1838, 1839, 1840, 1841, 1842. Under the Command of Charles Wilkes, U.S.N. ... 18 vols. (1–17, 19). Philadelphia. [Vol. 15: Botany, Phanerogamia (Gray), 1854; Atlas, 1856. Vol. 16: Botany, Cryptogamia, Filices (Brackenridge), 1854; Atlas, 1855. Vol. 17: incl. Phanerogamia of Pacific North America (Torrey), 1874. Vol. 19 (2 parts): Geographical Distribution of Animals and Plants (Pickering), 1854. Vol. 18: Botany, Phanerogamia, part 2 (Gray) not published.] (U.S. Expl. Exped.)

Willdenow, C. L. 1794. Phytographia seu Descriptio Rariorum Minus Cognitarum Plantarum. 1 fasc. only. Erlangen. (Phytographia)

Willdenow, C. L., C. F. Schwägrichen, and J. H. F. Link. 1797–1830. Caroli a Linné Species Plantarum.... Editio Quarta.... 6 vols. Berlin. [Vols. 1–5(1), 1797–1810, by Willdenow; vol. 5(2), 1830, by Schwägrichen; vol. 6, 1824–1825, by Link.] (Sp. Pl.)

Windham, M. D. 1983. The ferns of Elden Mountain, Arizona. Amer. Fern J. 73: 85–93.

Windham, M. D. 1986. Reassessment of the phylogenetic relationships of *Notholaena*. Amer. J. Bot. 73: 742.

Windham, M. D. 1987. *Argyrochosma*, a new genus of cheilanthoid ferns. Amer. Fern J. 77: 37–41.

Windham, M. D. 1987b. Chromosomal and electrophoretic studies of the genus *Woodsia* in North America. [Abstract.] Amer. J. Bot. 74: 715.

Windham, M. D. 1988. The Origin and Genetic Diversification of Polyploid Taxa in the *Pellaea wrightiana* complex (Adiantaceae). Ph.D. dissertation. University of Kansas.

Windham, M. D. 1993. New taxa and nomenclatural changes in the North American fern flora. Contr. Univ. Michigan Herb. 19: 31–61.

Wislizenus, F. A. 1848. Memoir of a Tour to Northern Mexico, connected with Col. Doniphan's Expedition, in 1846 and 1847.... Washington. (Mem. Tour N. Mexico)

Wolf, C. B. 1948. Taxonomic and distributional studies of the New World cypresses. Aliso 1: 1–250.

Wollenweber, E. 1984. Exudate flavonoids of Mexican ferns as chemotaxonomic markers. Revista Latinoamer. Quím. 15: 3–11.

Wright, J. W. 1955. Species crossability in spruce in relation to distribution and taxonomy. Forest Sci. 1: 319–349.

Wright, J. W., W. A. Lemmien, and J. N. Bright. 1971. Genetic variation in southern Rocky Mountain white fir. Silvae Genet. 20: 148–150.

Wunderlin, R. P. 1982. Guide to the Vascular Plants of Central Florida. Tampa.

Yatskievych, G., D. B. Stein, and G. J. Gastony. 1988. Chloroplast DNA evolution and systematics of *Phanerophlebia* (Dryopteridaceae) and related fern genera. Proc. Natl. Acad. Sci. U.S.A. 85: 2589–2593.

Yatskievych, G., M. D. Windham, and E. Wollenweber. 1990. A reconsideration of the genus *Pityrogramma* (Adiantaceae) in western North America. Amer. Fern J. 80: 9–17.

Zanoni, T. A. 1978. The American junipers of the section *Sabina* (*Juniperus*, Cupressaceae)—A century later. Phytologia 38: 433–454.

Zanoni, T. A. and R. P. Adams. 1979. The genus *Juniperus* (Cupressaceae) in Mexico and Guatemala: Synonymy, key, and distributions of the taxa. Bol. Soc. Bot. México 38: 83–131.

Zavarin, E., W. B. Critchfield, and K. Snajberk. 1978. Geographic differentiation of monoterpenes from *Abies procera* and *Abies magnifica*. Biochem. Syst. & Ecol. 6: 267–278.

Zavarin, E., L. Lawrence, and M. C. Thomas. 1971. Compositional variations of leaf monoterpenes in *Cupressus macrocarpa*, *C. pygmaea*, *C. goveniana*, *C. abramsiana* and *C. sargentii*. Phytochemistry 10: 379–393.

Zavarin, E. and K. Snajberk. 1972. Geographic variability of monoterpenes from *Abies balsamea* and *A. fraseri*. Phytochemistry 11: 1407–1421.

Zavarin, E., K. Snajberk, and J. Fisher. 1975. Geographic variability of monoterpenes from the cortex of *Abies concolor*. Biochem. Syst. & Ecol. 3: 191–203.

Zavarin, E., K. Snajberk, T. Reichert, and Tsien E. 1970. On the geographic variability of the monoterpenes from the cortical blister oleoresin of *Abies lasiocarpa*. Phytochemistry 9: 377–395.

Zimmerman, W. J., T. A. Lumpkin, and I. Watanabe. 1989. Classification of *Azolla* spp., Section *Azolla*. Euphytica 43: 223–232.

Zobel, D. B. 1973. Local variation in intergrading *Abies grandis*–*Abies concolor* populations in the central Oregon Cascades: Needle morphology and periderm color. Bot. Gaz. 134: 209–220.

A Note from the Flora of North America Editorial Committee

We thank Francis Boudreau, Coordinator of the Quebec CDC and chief of the Division of Biological Diversity, for providing the Flora of North America project with the Quebec Government's recently adopted list of French names for thirty fern species. He also sent two lists of French names for ferns, fern allies, and gymnosperms prepared by Fleurbec, a Quebec publisher specializing in botany. We also thank Gildo Lavoie, botanist, Division of Biological Diversity, for responding to specific questions we had in regard to several taxa.

FRENCH NAMES ACCEPTED BY THE MINISTRY OF THE ENVIRONMENT OF QUEBEC

Latin name	French name	Latin name	French name
Adiantum aleuticum	adiante des aléoutiennes	Gymnocarpium robertianum	gymnocarpe de robert
Aspidotis densa	aspidote touffue	Isoëtes hieroglyphica	isoète hiéroglyphique
Asplenium platyneuron	doradille ébène	Isoëtes tuckermanii	isoète de tuckerman
Asplenium rhizophyllum	doradille ambulante	Pellaea atropurpurea	pelléade à stipe pourpre
Asplenium ruta-muraria	doradille des murailles	Pellaea glabella	pelléade glabre
Athyrium alpestre subsp. americanum	athyrie alpestre sous-espèce américaine	Phegopteris hexagonoptera	phégoptère à hexagones
Botrychium campestre	botryche champêtre	Polystichum lonchitis	polystic faux-lonchitis
Botrychium oneidense	botryche d'oneida	Polystichum scopulinum	polystic des rochers
Botrychium pallidum	botryche pâle	Selaginella apoda	sélaginelle apode
Botrychium rugulosum	botryche à limbe rugueux	Thelypteris simulata	thélyptère simulatrice
Botrychium spathulatum	botryche à segments spatulés	Woodsia alpina	woodsie alpine
Diplazium pycnocarpon	diplazie à sores denses	Woodsia obtusa	woodsie à lobes arrondis
Dryopteris clintoniana	dryoptère de clinton	Woodsia oregana	woodsie de l'oregon
Dryopteris filix-mas	dryoptère fougère-mâle	Woodsia scopulina	woodsie des rochers
Gymnocarpium jessoense subsp. parvulum	gymnocarpe du japon sous-espèce frêle	Woodwardia virginica	woodwardie de virginie

FRENCH NAMES ADOPTED BY FLEURBEC—PTERIDOPHYTES

Latin name	French name	Latin name	French name
Adiantum pedatum	adiante du canada	Botrychium lunaria	botryche lunaire
Asplenium trichomanes	doradille chevelue	Botrychium matricariifolium	botryche à feuille de matricaire
Asplenium viride	doralille verte	Botrychium multifidum	botryche à feuille couchée
Athyrium filix-femina	athyrie fougère-femelle	Botrychium simplex	botryche simple
Athyrium thelypteridioides	athyrie fausse-thélyptère	Botrychium virginianum	botryche de virginie
Botrychium dissectum	botryche découpé	Cryptogramma stelleri	cryptogramme de steller
Botrychium lanceolatum	botryche élancé	Cystopteris bulbifera	cystoptère bulbifère

Latin name	French name	Latin name	French name
Cystopteris fragilis	cystoptère fragile	Lycopodium clavatum	lycopode à massue
Dennstaedtia punctilobula	dennstaedtie à lobules ponctués	Lycopodium complanatum	lycopode aplati
Dryopteris cristata	dryoptère à crêtes	Lycopodium obscurum	lycopode obscur
Dryopteris fragrans	dryoptère fragrante	Lycopodium sabinifolium	lycopode à feuilles de genévrier
Dryopteris goldiana	dryoptère de goldie	Lycopodium tristachyum	lycopode à trois épis
Dryopteris marginalis	dryoptère à sores marginaux	Matteuccia struthiopteris	matteuccie fougère-à-l'autruche
Dryopteris spinulosa	dryoptère spinuleuse		variété de pensylvanie
Equisetum arvense	prêle des champs	Onoclea sensibilis	onoclée sensible
Equisetum fluviatile	prêle fluviatile	Osmunda cinnamomea	osmunde cannelle
Equisetum hyemale	prêle d'hiver	Osmunda claytoniana	osmunde de clayton
Equisetum palustre	prêle des marais	Osmunda regalis	osmunde royale
Equisetum pratense	prêle des prés	Polypodium virginianum	polypode de virginie
Equisetum scirpoides	prêle faux-scirpe	Polystichum acrostichoides	polystic faux-acrostic
Equisetum sylvaticum	prêle des bois	Polystichum braunii	polystic de braun
Equisetum variegatum	prêle panachée	Pteridium aquilinum	fougère-aigle commune
Gymnocarpium dryopteris	gymnocarpe fougère-du-chêne	Selaginella rupestris	sélaginelle des rochers
Huperzia lucidula	huperzie brillant	Selaginella selaginoides	sélaginelle fausse-sélagine
Huperzia selago	huperzie sélagine	Thelypteris noveboracensis	thélyptère de new york
Isoëtes	isoète	Thelypteris palustris	thélyptère des marais
Lycopodiella inundata	lycopodielle inondé	Thelypteris phegopteris	thélyptère fougère-du-hêtre
Lycopodium alpinum	lycopode alpin	Woodsia glabella	woodsie glabre
Lycopodium annotinum	lycopode interrompu	Woodsia livensis	woodsie de l'île d'elbe

FRENCH NAMES ADOPTED BY FLEURBEC–GYMNOSPERMS

Taxus canadensis	if du canada
Abies balsamea	sapin baumier
Larix laricina	mélèze laricin
Picea glauca	épinette blanche
Picea mariana	épinette noire
Picea × rubens	épinette rouge
Pinus banksiana	pin gris
Pinus resinosa	pin rouge
Pinus rigida	pin rigide
Pinus strobus	pin blanc
Tsuga canadensis	pruche du canada
Juniperus communis	genévrier commun
Juniperus horizontalis	genévrier horizontal
Juniperus virginiana	genévrier rouge
Thuja occidentalis	cèdre-thuya occidental

Index

Names in *italics* are synonyms, casually mentioned hybrids, or plants not established in the flora. Page numbers in **boldface** indicate the primary entry for a taxon. Page numbers in *italics* indicate an illustration. Roman type is used for all other entries, including author names, vernacular names, and accepted scientific names for plants treated as established members of the flora.

Political Map of North America North of Mexico

Canadian Provinces

Alta.	Alberta	N.S.	Nova Scotia
B.C.	British Columbia	Ont.	Ontario
Man.	Manitoba	P.E.I.	Prince Edward Island
N.B.	New Brunswick	Que.	Quebec
Nfld.	Newfoundland (incl. Labrador)	Sask.	Saskatchewan
N.W.T.	Northwest Territories	Yukon	

United States

Ala.	Alabama	Mont.	Montana
Alaska		Nebr.	Nebraska
Ariz.	Arizona	Nev.	Nevada
Ark.	Arkansas	N.H.	New Hampshire
Calif.	California	N.J.	New Jersey
Colo.	Colorado	N. Mex.	New Mexico
Conn.	Connecticut	N.Y.	New York
Del.	Delaware	N.C.	North Carolina
D.C.	District of Columbia	N. Dak.	North Dakota
Fla.	Florida	Ohio	
Ga.	Georgia	Okla.	Oklahoma
Idaho		Oreg.	Oregon
Ill.	Illinois	Pa.	Pennsylvania
Ind.	Indiana	R.I.	Rhode Island
Iowa		S.C.	South Carolina
Kans.	Kansas	S. Dak.	South Dakota
Ky.	Kentucky	Tenn.	Tennessee
La.	Louisiana	Tex.	Texas
Maine		Utah	
Md.	Maryland	Vt.	Vermont
Mass.	Massachusetts	Va.	Virginia
Mich.	Michigan	Wash	Washington
Minn.	Minnesota	W. Va.	West Virginia
Miss.	Mississippi	Wis.	Wisconsin
Mo.	Missouri	Wyo.	Wyoming